Oxford Graduate Texts in Mathematics

Series Editors
R. Cohen   S. K. Donaldson   S. Hildebrandt
T. J. Lyons   M. J. Taylor

T0177654

OXFORD GRADUATE TEXTS IN MATHEMATICS

Books in the series

Matroid Theory

Second Edition

James Oxley
Department of Mathematics
Louisiana State University

OXFORD
UNIVERSITY PRESS

OXFORD

UNIVERSITY PRESS

Great Clarendon Street, Oxford OX2 6DP

Oxford University Press is a department of the University of Oxford.
It furthers the University's objective of excellence in research, scholarship,
and education by publishing worldwide in

Oxford New York

Auckland   Cape Town Dar es Salaam   Hong Kong   Karachi
Kuala Lumpur Madrid Melbourne Mexico City   Nairobi
New Delhi Shanghai Taipei Toronto
With offices in
Argentina Austria Brazil Chile Czech Republic France Greece
Guatemala Hungary Italy Japan South Korea Poland Portugal
Singapore Switzerland Thailand Turkey Ukraine Vietnam

Oxford is a registered trade mark of Oxford University Press
in the UK and in certain other countries

Published in the United States
by Oxford University Press Inc., New York

© James Oxley 2011

ISBN 978-0-19-960339-8

Printed and bound by CPI Group (UK) Ltd, Croydon, CR0 4YY

PREFACE

I first heard mention of matroids in 1973 from a friend at the University of Tasmania who had just read Wilson's (1973) enticing survey paper. Since then, several people have taught me about the many fascinating properties of these structures: Don Row in Tasmania; Dominic Welsh and Aubrey Ingleton in Oxford; and Tom Brylawski in Chapel Hill. While I learnt much from all of my teachers, it was my doctoral supervisor, Dominic Welsh, who was the strongest single influence on this book. His own book, *Matroid theory* (1976), appeared while I was a student and it has been my constant companion ever since. When I considered writing this book, the first question I had to answer was how should it differ from Dominic's book. It is a particular pleasure to see that, after being out of print for many years, Dominic's book was reprinted by Dover in 2010.

This book attempts to blend Welsh's very graph-theoretic approach to matroids with the geometric approach of Rota's school, which I learnt from Brylawski. Unfortunately, I cannot emulate Welsh's feat of providing, in a single volume, a complete survey of the current state of knowledge in matroid theory; the subject has grown too much. Therefore I have had to be selective. While the basic topics virtually select themselves, the more advanced topics covered here reflect my own research interests in the areas of matroid structure and connectivity, and matroid representability.

Chapters 1–6 are intended to provide a basic overview of matroid theory. By omitting certain sections and skimming through others, one can cover this material in a one-semester American graduate course. The later chapters are intended for reference and for those who have the opportunity to teach a full year's course on matroids. These chapters treat more specialized topics or present difficult proofs delayed from earlier chapters. In particular, proofs of all of the major theorems of the subject from the twentieth century are given in reasonably full detail. I have never enjoyed reading proofs in which numerous intermediate steps are left to the reader, so I have tried to avoid writing such proofs.

Exercises have been included at the end of each section except in the last chapter. The first few exercises in each set tend to be straightforward; the last few are often quite difficult. I have used an asterisk (*) to denote harder problems and a degree symbol (°) to denote unsolved problems. Some of the harder problems present theorems that have not been included in the text and these may have quite intricate proofs. In such instances, I have tried to provide an appropriate reference. In many places, proofs have been omitted from the text. Except where otherwise indicated, these proofs are relatively straightforward and they provide a good additional source of exercises.

In order to distinguish the relative importance of the many results in the book, I have adopted the convention of reserving the term 'theorem' for the main results; those results of lesser importance are called 'propositions'. Another convention that will apply throughout is that, except where otherwise stated, M will denote an arbitrary matroid.

The last chapter of the book gathers together numerous unsolved problems, some of which have appeared earlier in the book. The beginning of that chapter also contains a brief review of some of the other books on matroid theory. The references appear after the last chapter, each accompanied by a list of the sections in which it is cited. Following the references, there is an appendix that summarizes the properties of certain interesting matroids that have appeared in earlier sections. The book concludes with a list of the most frequently used notation and with an index of terms. Notation that appears in just one section has usually been omitted from this list. I have adopted certain conventions concerning the index and have included numerous redundancies in an attempt to maximize its usefulness. Generally the listings associated with a particular entry have been limited to the most important ones, with an overall cap of about ten listings per entry. The reader should note, however, that certain terms appear very frequently in particular sections. In such cases, I have usually given only one listing per section. Items such as '2-sum' and '3-connected' appear in the alphabetical list as if they were written 'two-sum' and 'three-connected'.

One of my colleagues commented that it is somewhat intimidating to begin a book with a chapter entitled 'Basic definitions and examples' that is 57 pages long. With this in mind, the following remarks on Chapters 1–6 are made to guide the reader who is seeking a shorter introduction to matroid theory than these chapters provide. For completeness, the proofs of most of the results in the early chapters have been included. Many of these can be omitted on a first reading or by someone who is mainly interested in getting a feel for the subject. A brief overview of matroid theory appears in my paper 'What is a matroid?' (Oxley 2003), an updated version of which appears on my website.

The main ideas in Chapter 1 are that matroids can be axiomatized in many equivalent ways and that the two fundamental classes of examples of matroids arise from graphs and matrices. Most of the proofs of the equivalence of the various axiom systems could be omitted as indeed could most of Sections 1.6–1.8. The first of these three sections deals with a class of matroids that arises in the consideration of scheduling problems. The reader omitting this section should then omit Section 2.4 and 3.2.8–3.2.12. All of Section 1.7 could be omitted beyond the first two pages. If this is done, then 3.3.8–3.3.10 along with the material on lattices in Sections 6.1 and 6.2, should also be left out. Section 1.8 indicates why matroids are of basic importance in combinatorial optimization, but it stands alone from what follows it.

Chapter 2 is much shorter than Chapter 1. The reader seeking to shorten it further could omit 2.1.13–2.1.23 and 2.1.25–2.1.29 in Section 2.1, everything after 2.2.10 in Section 2.2 except for 2.2.26, and all of Section 2.4. In Chapter 3, the reader could omit all of Sections 3.2 and 3.3 after 3.2.7 and 3.3.2, respectively. In Chapter 4, Section 4.3 contains more specialized results than the other two sections of that chapter and it too could be omitted except for the first two results. For a similar reason, the last two sections of Chapter 5 could be omitted although the reader should note at least the statement of Theorem 5.3.1.

Chapter 6, which deals with matroid representations over fields, is the longest in the book but the reader should concentrate on Sections 6.1, 6.3–6.6, and 6.10. Even within these sections, material can be omitted. In Section 6.1, everything after 6.1.3 could be left out. If one restricts attention to representations over prime fields or $\mathbb{R}$, then one can omit all of Section 6.3 after 6.3.9 except for 6.3.12. In Section 6.4, the main result is Theorem 6.4.7 and, after that, the rest of the section consists of examples illustrating the use of that result. The major theorems and unsolved problems in matroid representability are surveyed in Sections 6.5 and 6.6. As such, these sections deserve the attention of even the casual reader, although everything in Section 6.5 after 6.5.12, except 6.5.17, could be omitted. These sections are not difficult to read since the proofs of the main results from these sections are postponed until later in the book. Section 6.10, whose key ideas are contained in 6.10.1–6.10.10, considers a very important generalization of the class of graphic matroids.

This book is a major revision of the first edition, which appeared in 1992. For the reader familiar with that edition, I shall now discuss the changes. Chapters 1–7 are close to their original forms but have many minor revisions and some interesting additions. In particular, Chapter 6 contains a new section (6.10) on Dowling matroids, while Chapter 7 includes a new section (7.4) on a non-commutative way to stick two matroids together. Chapter 8, which deals with higher connectivity, has been significantly revised and lengthened with the inclusion of Tutte's Linking Theorem (Section 8.5), which is a matroid analogue of Menger's Theorem. Also added to that chapter is a proof of Cunningham and Edmonds's tree decomposition of matroids that are 2-connected but not 3-connected. Chapter 9, on binary matroids, includes an expanded discussion of the 3-sum operation, which had previously been in a later chapter. Chapter 10 essentially combines and revises Chapters 10 and 13 of the first edition. It now includes proofs of the excluded-minor theorems for regular, ternary, and graphic matroids. The second and third of these proofs have been completely altered from the first edition. Changes to Chapters 11 and 12 necessitated reversing their order from the first edition, and an extra section on generalized delta-wye exchange has been added to the new Chapter 11. Except for Section 13.1, most of Chapters 13 and 14 is new. The first of these chapters proves Seymour's Decomposition Theorem for regular matroids; the second surveys recent research in representability and structure, and most proofs are omitted. Chapter 13 is the most difficult in the book, but the theorem it proves is probably the most significant result in the subject to have ever been published. The final chapter (15) on unsolved problems includes a number of problems that did not appear in the corresponding chapter (14) of the first edition. A number of the problems from the latter have now been solved and some updates on these appear in Chapters 14 and 15 of this edition. Some other updates appear on the author's website.

While I was writing the first edition of this book, my work was partially supported by a grant from the Louisiana Education Quality Support Fund through the Board of Regents. When I was preparing the second edition, I received support from the National Security Agency and from Merton College, Oxford.

Many people have assisted in the preparation of this book. I have lectured from most of it at Louisiana State University, and the students in those classes have helped to eliminate many mistakes and to clarify the exposition. Several people have also read substantial parts of the manuscript of the first or second edition and have offered many valuable suggestions. In particular, I thank Safwan Akkari, Joe Bonin, Bob Lax, Jon Lee, Manoel Lemos, Dillon Mayhew, James Reid, Dave Wagner, Geoff Whittle, Haidong Wu, and Tom Zaslavsky for their help in this regard. Nell Castleberry spent many painstaking hours carefully preparing the LaTeX version of the first edition of the book. I am also very grateful to Mike Newman, who first taught me about how to write mathematics, and to Bogdan Oporowski, Don Row, and Geoff Whittle for their sustained encouragement and sage counsel.

I offer particular thanks to two people who have given me an extraordinary amount of help with this book, Dominic Welsh and Jim Geelen. Dominic was instrumental in getting me to start the first edition, and he provided much valuable advice and encouragement throughout the preparation of both editions. Jim provided constant support during the preparation of this new edition and very generously offered me much unpublished material of his own for use in the later part of it. In particular, the proof of Seymour's Decomposition Theorem in Chapter 13 follows his outline, and the reader will notice many other attributions to private communications from him.

Dedicated to Judith and to the memory of Margaret and David

CONTENTS

PRELIMINARIES

Matroid theory draws heavily on both graph theory and linear algebra for its motivation, its basic examples, and its notation. The ideal background for a student using this book would include undergraduate courses in linear algebra, graph theory, and abstract algebra. However, I have used the book with students having no previous exposure to graphs and very little abstract algebra; most of the ideas needed from these areas, particularly in the earlier chapters, are relatively straightforward. The rest of this section introduces some notation and outlines some basic concepts. After reading this, the reader who is familiar with vector spaces should have enough background to handle at least the first half of the book.

Except where otherwise indicated, all sets considered in this book will be *finite*. For a set E, its collection of subsets and its cardinality will be denoted by 2^E and $|E|$, respectively. Frequently, when given a set $\mathcal{A}$ of subsets of E, we shall be interested in the *maximal* or *minimal members* of $\mathcal{A}$. The former are those members of $\mathcal{A}$ that are not properly contained in any member of $\mathcal{A}$; the latter are the members of $\mathcal{A}$ that do not properly contain any member of $\mathcal{A}$. The sets of positive integers, integers, rational numbers, real numbers, and complex numbers will be denoted by $\mathbb{Z}^+$, $\mathbb{Z}$, $\mathbb{Q}$, $\mathbb{R}$, and $\mathbb{C}$, respectively. If X and Y are sets, then $X - Y$ denotes the set $\{x \in X : x \notin Y\}$, while $X \triangle Y$ denotes $(X - Y) \cup (Y - X)$, the *symmetric difference* of X and Y. For sets $X_1, X_2, \ldots, X_n$, the notation $X_1 \dot\cup X_2 \dot\cup \cdots \dot\cup X_n$ will refer to the set $X_1 \cup X_2 \cup \cdots \cup X_n$ and will also imply that $X_1, X_2, \ldots, X_n$ are pairwise disjoint. Often, we shall want to add a single element e to a set X or to remove e from X. In such cases, we shall usually abbreviate $X \cup \{e\}$ and $X - \{e\}$ to $X \cup e$ and $X - e$, respectively.

The concept of a multiset is useful when working with collections of objects in which repetitions can occur. Formally, if S is a set, a *multiset* chosen from S is a function m from S into the set of non-negative integers. For example, if $S = \{a, b, c, d, e\}$ and $(m(a), m(b), m(c), m(d), m(e)) = (2, 3, 0, 1, 0)$, then we denote this multiset by $\{a, a, b, b, b, d\}$. A *permutation* of a set S is a bijection from S into itself. A permutation of a finite set will often be written using cycle notation. So, for example, if $S = \{1, 2, \ldots, 9\}$, then the permutation $(1, 5, 2, 9)(7, 8)$ maps 1 to 5, 5 to 2, 2 to 9, 9 to 1, 7 to 8, 8 to 7, 3 to 3, 4 to 4, and 6 to 6.

Let f and g be functions that are defined for all sufficiently large positive integers n. We write $g(n) = O(f(n))$ to mean that $g(n)/f(n)$ remains bounded as $n \to \infty$. The function g is *asymptotic* to f, written $g \sim f$, if $\lim_{n \to \infty} \frac{g(n)}{f(n)} = 1$.

Fields will appear quite often in the book, even in the earlier chapters. We now note some basic facts about these structures. Proofs of these facts can be found in most graduate algebra texts. If $\mathbb{F}$ is a finite field, then $\mathbb{F}$ has exactly p^k elements for some prime number p and some positive integer k. Indeed, for all such p and k, there is a unique field $GF(p^k)$ having exactly p^k elements. This field is called the *Galois field of order p^k*. The field $GF(p)$ coincides with $\mathbb{Z}_p$, the ring of integers modulo p. Thus $GF(p)$ has as its elements $0, 1, \ldots, p-1$, and addition and multiplication of these elements is performed modulo p. When $k > 1$, the field $GF(p^k)$ can be constructed as follows. Let $h(\omega)$ be a polynomial of degree k with coefficients in $GF(p)$ and suppose that this polynomial is irreducible, that is, $h(\omega)$ is not the product of two lower-degree polynomials over $GF(p)$. Such irreducible polynomials are known to exist. Consider the set S of all polynomials in ω that have degree at most $k - 1$ and have coefficients in $GF(p)$. There are exactly p choices for each of the k coefficients of a member of S. Hence $|S| = p^k$. Moreover, under addition and multiplication, both of which are performed modulo $h(\omega)$, the set S forms a field, namely $GF(p^k)$.

The finite fields that will appear most often in this book are $GF(2)$, $GF(3)$, $GF(5)$, and $GF(4)$. The first three of these are $\mathbb{Z}_2$, $\mathbb{Z}_3$, and $\mathbb{Z}_5$, respectively. The elements of $\mathbb{Z}_3$ will usually be written as 0, 1, and -1 rather than 0, 1, and 2. We shall now use the above construction to obtain $GF(4)$, which the reader will easily see is not isomorphic to $\mathbb{Z}_4$. Let $h(\omega) = \omega^2 + \omega + 1$. It is straightforward to show that this polynomial is irreducible over $GF(2)$, and that the addition and multiplication tables for $GF(4)$ are as follows:

+	0	1	ω	$\omega+1$
0	0	1	ω	$\omega+1$
1	1	0	$\omega+1$	ω
ω	ω	$\omega+1$	0	1
$\omega+1$	$\omega+1$	ω	1	0

$\times$	0	1	ω	$\omega+1$
0	0	0	0	0
1	0	1	ω	$\omega+1$
ω	0	ω	$\omega+1$	1
$\omega+1$	0	$\omega+1$	1	ω

Let $\mathbb{F}$ be a field and consider the sequence $1, 1+1, 1+1+1, \ldots$ in $\mathbb{F}$. If these elements are all distinct, then $\mathbb{F}$ is said to have *characteristic* 0; otherwise the *characteristic* of $\mathbb{F}$ is the smallest positive integer p for which the sum of p ones is zero. In the latter case, p is clearly a prime; the subset $\{0, 1, \ldots, p-1\}$ of $\mathbb{F}$ is a subfield of $\mathbb{F}$ isomorphic to $\mathbb{Z}_p$; and $p\alpha = 0$ for all α in $\mathbb{F}$. The intersection of all subfields of a field $\mathbb{F}$ is itself a field and is called the *prime subfield* of $\mathbb{F}$. When $\mathbb{F}$ has characteristic 0, its prime subfield is $\mathbb{Q}$, the field of rational numbers. When $\mathbb{F}$ has characteristic p for $p \neq 0$, its prime subfield is $\mathbb{Z}_p$.

Throughout the book, an arbitrary finite field will be denoted by $GF(q)$, it being implicit that q is a power of a prime. For an arbitrary field $\mathbb{F}$, the multiplicative group of non-zero elements of $\mathbb{F}$ will be denoted by $\mathbb{F}^\times$. It is well known that $GF(q)^\times$ is a cyclic group.

If $\mathbb{F}$ is a field and r is a natural number, then $V(r, \mathbb{F})$ will denote the r-dimensional vector space over $\mathbb{F}$. This vector space will also be written as $V(r, q)$ when $\mathbb{F} = GF(q)$. The r vectors in $V(r, \mathbb{F})$ that have one coordinate equal to one and the remaining coordinates equal to zero will be called *standard basis vectors*. A *non-singular linear transformation* of $V(r, \mathbb{F})$ is a permutation η of $V(r, \mathbb{F})$ such that $\eta(c\underline{x} + d\underline{y}) = c\eta(\underline{x}) + d\eta(\underline{y})$ for all $\underline{x}$ and $\underline{y}$ in $V(r, \mathbb{F})$ and all c and d in $\mathbb{F}$.

The matrix notation used here is relatively standard. The matrices I_r, J_r, and 0_r are, respectively, the $r \times r$ identity matrix, the $r \times r$ matrix of all ones, and the $r \times r$ matrix of all zeros. The transpose of a matrix A and of a vector $\underline{v}$ will be denoted by A^T and $\underline{v}^T$. Some use will be made here of determinants and a quick summary of their properties can be found in Section 2.2. If $\{\underline{v}_1, \underline{v}_2, \ldots, \underline{v}_n\}$ is a set X of vectors in a vector space V over a field $\mathbb{F}$, then $\langle \underline{v}_1, \underline{v}_2, \ldots, \underline{v}_n \rangle$ denotes the subspace of V *spanned* by X, that is, the set of all linear combinations of $\underline{v}_1, \underline{v}_2, \ldots, \underline{v}_n$. The *dimension* of V is denoted by $\dim V$. If W is also a vector space over $\mathbb{F}$, then $V + W$ denotes the vector space over $\mathbb{F}$ that is spanned by $\{\underline{v} + \underline{w} : \underline{v} \in V, \underline{w} \in W\}$.

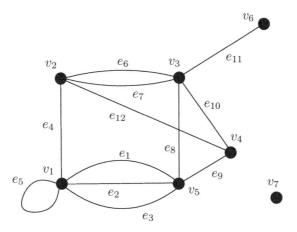

FIG. 0.1. An example of a graph.

Figure 0.1 is a pictorial representation of a particular graph. The vertex set and edge set of this graph are $\{v_1, v_2, \ldots, v_7\}$ and $\{e_1, e_2, \ldots, e_{12}\}$, respectively. An edge, such as e_5, that joins a vertex to itself is called a *loop*. Edges such as e_1, e_2, and e_3 that join the same pair of distinct vertices are called *parallel edges*. The vertex v_7, which does not meet any edges, is an *isolated vertex*. The *ends* of the edge e_9 are v_4 and v_5. Formally, a *graph* G consists of a set $V(G)$ of *vertices* and a multiset $E(G)$ of *edges* each of which consists of an unordered pair of (possibly identical) vertices. If $e \in E(G)$ and $e = \{u, v\}$, where u and v are in $V(G)$, then we say that u and v are *neighbours* or are *adjacent*, and that e is *incident* with u and v. For example, in Figure 0.1, e_8 is the unique edge incident with both v_3 and v_5, so we could denote it unambiguously by $v_3 v_5$, or equivalently, by $v_5 v_3$. We call $V(G)$ and $E(G)$ the *vertex set* and *edge set*, respectively, of the graph G. A graph is *simple* if it has no loops and no parallel edges. If $V(G)$ is empty, then so is $E(G)$, and G is called the *null* or *empty graph*. We shall adopt a common convention of assuming, unless forced by other requirements, that all graphs considered are non-null.

Sometimes we will specify a graph by its *mod-2 vertex–edge incidence matrix*. For the graph in Figure 0.1, this matrix is the following:

$$
\begin{array}{c@{\quad}ccccccccccccc}
 & e_1 & e_2 & e_3 & e_4 & e_5 & e_6 & e_7 & e_8 & e_9 & e_{10} & e_{11} & e_{12} \\
v_1 & 1 & 1 & 1 & 1 & 0 & 0 & 0 & 0 & 0 & 0 & 0 & 0 \\
v_2 & 0 & 0 & 0 & 1 & 0 & 1 & 1 & 0 & 0 & 0 & 0 & 1 \\
v_3 & 0 & 0 & 0 & 0 & 0 & 1 & 1 & 1 & 0 & 1 & 1 & 0 \\
v_4 & 0 & 0 & 0 & 0 & 0 & 0 & 0 & 0 & 1 & 1 & 0 & 1 \\
v_5 & 1 & 1 & 1 & 0 & 0 & 0 & 0 & 1 & 1 & 0 & 0 & 0 \\
v_6 & 0 & 0 & 0 & 0 & 0 & 0 & 0 & 0 & 0 & 0 & 1 & 0 \\
v_7 & 0 & 0 & 0 & 0 & 0 & 0 & 0 & 0 & 0 & 0 & 0 & 0 \\
\end{array}
$$

In general, the rows of such a matrix $[a_{ij}]$ are indexed by the vertices of the graph, the columns are indexed by the edges, and a_{ij} is the number of times (mod 2) that the jth edge, e_j, is incident with the ith vertex, v_i. Hence a_{ij} is 0 or 1 unless e_j is a loop incident with v_i, in which case, $a_{ij} = 0$. The *degree*, $d(v_i)$, of the vertex v_i is the number of edges incident with v_i where each loop counts as two edges.

In the matrix above, each row can be viewed as corresponding to a subset of $\{e_1, e_2, \ldots, e_{12}\}$. For example, the second row is the *incidence vector* of the set $\{e_4, e_6, e_7, e_{12}\}$ since it takes the value 1 in the columns indexed by e_4, e_6, e_7, and e_{12} while taking the value 0 in all other columns. Likewise, the first column of the matrix is the incidence vector of the subset $\{v_1, v_5\}$ of $\{v_1, v_2, \ldots, v_7\}$.

A graph H is a *subgraph* of a graph G if $V(H)$ and $E(H)$ are subsets of $V(G)$ and $E(G)$, respectively. If V' is a subset of $V(G)$, then $G[V']$ denotes the subgraph of G whose vertex set is V' and whose edge set consists of those edges of G that have both endpoints in V'. We say that $G[V']$ is the subgraph of G *induced* by V'. Similarly, if E' is a subset of $E(G)$, then $G[E']$, the subgraph of G *induced* by E', has E' as its edge set and the set of endpoints of edges in E' as its vertex set.

If G_1 and G_2 are graphs, their *union* $G_1 \cup G_2$ is the graph with vertex set $V(G_1) \cup V(G_2)$ and edge set $E(G_1) \cup E(G_2)$. If $V(G_1)$ and $V(G_2)$ are disjoint, then so are $E(G_1)$ and $E(G_2)$, and G_1 and G_2 are called *disjoint graphs*.

Graphs G and H are *isomorphic*, written $G \cong H$, if there are bijections $\psi : V(G) \to V(H)$ and $\theta : E(G) \to E(H)$ such that a vertex v of G is incident with an edge e of G if and only if $\psi(v)$ is incident with $\theta(e)$. Two graphs that play an important role in matroid theory, and indeed throughout graph theory, are K_5 and $K_{3,3}$. These graphs are shown in Figure 0.2. In general, if n is a positive integer, there is, up to isomorphism, a unique graph on n vertices in which each pair of distinct vertices is joined by a single edge. This graph, K_n, is called the *complete graph* on n vertices. The graph $K_{3,3}$ is a special type of *bipartite graph*, the latter being a graph whose vertex set can be partitioned into two *classes* X and Y such that each edge has one endpoint in X and the other in Y.

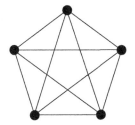 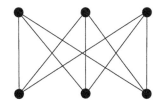

FIG. 0.2. K_5 and $K_{3,3}$.

In Chapter 2, we shall consider the role played by K_5 and $K_{3,3}$ when one considers embedding graphs in the Euclidean plane. In this discussion, the topology of the plane is important. We shall not go into this in detail here although we shall require some basic facts. A *simple curve* is the image of a continuous function from $[0,1]$ into $\mathbb{R}^2$ that does not intersect itself except that its endpoints may coincide. In the exceptional case, the simple curve is *closed* and is also called a *Jordan curve*. If all the points of a Jordan curve are deleted from the plane, the remaining points are partitioned into two open sets. The *Jordan Curve Theorem* asserts that a line joining a point in one of these open sets to a point in the other open set must intersect the Jordan curve.

A *walk* in a graph is a sequence $v_0 e_1 v_1 e_2 \cdots v_{k-1} e_k v_k$ such that $v_0, v_1, \ldots, v_k$ are vertices, $e_1, e_2, \ldots, e_k$ are edges, and each vertex or edge in the sequence, except v_k, is incident with its successor in the sequence. Now suppose that the vertices $v_0, v_1, \ldots, v_k$ are distinct. Then $e_1, e_2, \ldots, e_k$ are also distinct and the walk is a *path*. The *end-vertices* or *ends* of this path are v_0 and v_k, and the path is said to be a (v_0, v_k)-*path* or to *join* v_0 and v_k. The vertices $v_1, v_2, \ldots, v_{k-1}$ are the *internal vertices* of the path. The *length* of a path is the number of edges that it contains. In a simple graph, a path is uniquely determined by its vertex set and so will usually be specified by listing the appropriate sequence of vertices.

A graph is *connected* if each pair of distinct vertices is joined by a path. A graph that is not connected is *disconnected*. In any graph G, the maximal connected subgraphs are called (*connected*) *components*. Evidently the vertex sets of the components of G partition $V(G)$. The number of these components will be denoted by $\omega(G)$.

If P is a (u, v)-path in a graph G and e is an edge of G that joins u to v but is not in P, then the subgraph of G whose vertex set is $V(P)$ and whose edge set in $E(P) \cup e$ is called a *cycle*. The *length* of a cycle is the number of edges it contains. A graph having no cycles is a *forest*, while a connected forest is a *tree*. Clearly a graph is a forest if and only if each of its components is a tree. A *spanning tree* of a connected graph G is a subgraph T of G such that T is a tree and $V(T) = V(G)$. Trees have many attractive properties. In particular, for all trees T,

$$|E(T)| = |V(T)| - 1.$$

Hence if T is a spanning tree of a graph G, then $|E(T)| = |V(G)| - 1$.

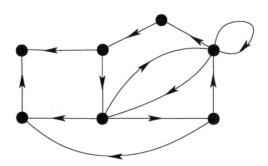

FIG. 0.3. An example of a digraph.

A directed graph is a graph in which every edge is directed from one endpoint to the other as, for example, in Figure 0.3. More formally, a *directed graph* or *digraph* D consists of a non-empty set $V(D)$ of *vertices* and a multiset $A(D)$ of *arcs*, each of which consists of an ordered pair of (possibly identical) vertices. If $a \in A(D)$ and $a = (u, v)$, then a *joins* u *to* v and we say that u is the *tail* of a and v is its *head*. If $u = v$, then a is a *loop* of D. In general, the terminology for directed graphs mimics that for graphs. For instance, a *directed path* is a sequence $(v_0, a_1, v_1, a_2, \ldots, v_{k-1}, a_k, v_k)$ where $v_0, v_1, \ldots, v_k$ are vertices, $a_1, a_2, \ldots, a_k$ are arcs, and, for all i in $\{1, 2, \ldots, k\}$, the arc a_i joins v_{i-1} to v_i. A digraph D' is a *subdigraph* of D if $V(D') \subseteq V(D)$ and $A(D') \subseteq A(D)$.

A digraph that is obtained from a graph G by specifying an order on the endpoints of each edge is called an *orientation* of G. On the other hand, if D is a digraph and G is the graph that is obtained from D by replacing each arc by an edge with the same endpoints, then G is the *underlying graph* of D.

Numerous other graph-theoretic concepts that appear in the book will be defined as needed. For any remaining undefined terminology involving graphs, we refer the reader to Bondy and Murty's (2008) book, which also contains a chapter on the complexity of algorithms. That chapter will be a useful reference for the reader unfamiliar with the basics of computational complexity, although our discussions of complexity issues will be relatively brief and will not begin until Chapter 6.

1

BASIC DEFINITIONS AND EXAMPLES

The study of matroids is an analysis of an abstract theory of dependence. Indeed, Whitney's (1935) founding paper in the subject was entitled 'On the abstract properties of linear dependence'. In defining a matroid, Whitney tried to capture the fundamental properties of dependence that are common to graphs and matrices. He was not alone in such attempts to extract the essence of dependence. The second edition of Van der Waerden's *Moderne Algebra* (1937) distinguished three fundamental properties that are common to algebraic and linear dependence. Other pioneering work was done by Nakasawa (1935, 1936a,b), Birkhoff (1935), and Mac Lane (1936, 1938). We make no attempt here to survey these important early contributions but instead refer the reader to the books of Kung (1986a) and Schrijver (2003, Section 39.10), each of which provides an interesting historical survey of the origins of matroid theory that includes reprinting parts, or, in the first case, all, of the text of selected early papers.

A characteristic of matroids is that they can be defined in many different but equivalent ways. One important task of this opening chapter is to introduce these different axiom systems and to prove their equivalence. The other main purpose of the chapter is to present various fundamental examples of matroids and to illustrate the basic concepts in the context of these examples.

1.1 Independent sets and circuits

The two fundamental classes of matroids that appear in Whitney's paper (1935) arise from matrices and from graphs. In this section, we introduce two equivalent axiom systems for matroids and indicate how each is naturally associated with one of these classes.

A *matroid* M is an ordered pair $(E, \mathcal{I})$ consisting of a finite set E and a collection $\mathcal{I}$ of subsets of E having the following three properties:

(I1) $\emptyset \in \mathcal{I}$.

(I2) *If $I \in \mathcal{I}$ and $I' \subseteq I$, then $I' \in \mathcal{I}$.*

(I3) *If I_1 and I_2 are in $\mathcal{I}$ and $|I_1| < |I_2|$, then there is an element e of $I_2 - I_1$ such that $I_1 \cup e \in \mathcal{I}$.*

We shall call (I2) and (I3) the *hereditary* and *independence augmentation properties*. If M is the matroid $(E, \mathcal{I})$, then M is called a matroid *on* E. The members of $\mathcal{I}$ are the *independent sets* of M, and E is the *ground set* of M. We shall often write $\mathcal{I}(M)$ for $\mathcal{I}$ and $E(M)$ for E, particularly when several matroids are being considered. A subset of E that is not in $\mathcal{I}$ is called *dependent*.

The name 'matroid' was coined by Whitney (1935) because a class of fundamental examples of such objects arises from matrices in the following way.

Proposition 1.1.1 *Let E be the set of column labels of an $m \times n$ matrix A over
a field $\mathbb{F}$, and let $\mathcal{I}$ be the set of subsets X of E for which the multiset of columns
labelled by X is a set and is linearly independent in the vector space $V(m, \mathbb{F})$.
Then $(E, \mathcal{I})$ is a matroid.*

Proof Clearly $\mathcal{I}$ satisfies (I1) and (I2). To prove that (I3) holds, let I_1 and I_2
be linearly independent subsets of E with $|I_1| < |I_2|$. Let W be the subspace of
$V(m, \mathbb{F})$ spanned by $I_1 \cup I_2$. Then $\dim W$, the dimension of W, is at least $|I_2|$.
Now suppose $I_1 \cup e$ is linearly dependent for all e in $I_2 - I_1$. Then W is contained
in the span of I_1. Thus $|I_2| \leq \dim W \leq |I_1| < |I_2|$; a contradiction. Hence $I_2 - I_1$
contains an element e such that $I_1 \cup e \in \mathcal{I}$, that is, (I3) holds.   □

The matroid just obtained from the matrix A will be denoted by $M[A]$. This
matroid is called the *vector matroid* of A. As a particular example of such a
matroid, consider the following.

Example 1.1.2 Let A be the following matrix over the field $\mathbb{R}$ of real numbers.

$$\begin{array}{ccccc} 1 & 2 & 3 & 4 & 5 \end{array}$$
$$\begin{bmatrix} 1 & 0 & 0 & 1 & 1 \\ 0 & 1 & 0 & 0 & 1 \end{bmatrix}$$

Then $E = \{1, 2, 3, 4, 5\}$ and $\mathcal{I} = \{\emptyset, \{1\}, \{2\}, \{4\}, \{5\}, \{1, 2\}, \{1, 5\}, \{2, 4\}, \{2, 5\},
\{4, 5\}\}$. Thus the set of dependent sets of this matroid is

$$\{\{3\}, \{1, 3\}, \{1, 4\}, \{2, 3\}, \{3, 4\}, \{3, 5\}\} \cup \{X \subseteq E : |X| \geq 3\}.$$

The set of *minimal dependent sets*, that is, dependent sets all of whose proper
subsets are independent, is $\{\{3\}, \{1, 4\}, \{1, 2, 5\}, \{2, 4, 5\}\}$.   □

A minimal dependent set in an arbitrary matroid M will be called a *circuit*
of M and we shall denote the set of circuits of M by $\mathcal{C}$ or $\mathcal{C}(M)$. A circuit of
M having n elements will also be called an *n-circuit*. Clearly, once $\mathcal{I}(M)$ has
been specified, $\mathcal{C}(M)$ can be determined. Similarly, $\mathcal{I}(M)$ can be determined
from $\mathcal{C}(M)$: the members of $\mathcal{I}(M)$ are those subsets of $E(M)$ that contain no
member of $\mathcal{C}(M)$. Thus a matroid is uniquely determined by its set $\mathcal{C}$ of circuits.
We now examine some properties of $\mathcal{C}$ with a view to characterizing those
subsets of 2^E that can occur as the set of circuits of a matroid on E. Clearly

(C1) $\emptyset \notin \mathcal{C}$; and
(C2) *if C_1 and C_2 are members of $\mathcal{C}$ and $C_1 \subseteq C_2$, then $C_1 = C_2$.*

Furthermore:

Lemma 1.1.3 *The set $\mathcal{C}$ of circuits of a matroid has the following property:*

(C3) *If C_1 and C_2 are distinct members of $\mathcal{C}$ and $e \in C_1 \cap C_2$, then there is a member C_3 of $\mathcal{C}$ such that $C_3 \subseteq (C_1 \cup C_2) - e$.*

Proof Assume that $(C_1 \cup C_2) - e$ does not contain a circuit. Then $(C_1 \cup C_2) - e$ is in $\mathcal{I}$. By (C2), the set $C_2 - C_1$ is non-empty, so we can choose an element f from this set. As C_2 is a minimal dependent set, $C_2 - f \in \mathcal{I}$. Now choose a subset I of $C_1 \cup C_2$ which is maximal with the properties that it contains $C_2 - f$ and is independent. Evidently $f \notin I$. Moreover, as C_1 is a circuit, some element g of C_1 is not in I. As $f \in C_2 - C_1$, the elements f and g are distinct. Hence

$$|I| \leq |(C_1 \cup C_2) - \{f, g\}| = |C_1 \cup C_2| - 2 < |(C_1 \cup C_2) - e|.$$

Now apply (I3), taking I_1 and I_2 to be I and $(C_1 \cup C_2) - e$, respectively. The resulting independent set contradicts the maximality of I. $\qquad\square$

Condition (C3) is called the *circuit elimination axiom* or, sometimes, the *weak circuit elimination axiom* to contrast it with another circuit elimination axiom which will be introduced in Section 1.4 (see also Exercise 14). We show next that (C1)–(C3) characterize those collections of sets that can occur as the set of circuits of a matroid. In this and many other matroid arguments, drawing a Venn diagram makes the argument easier to follow.

Theorem 1.1.4 *Let E be a set and $\mathcal{C}$ be a collection of subsets of E satisfying* (C1)–(C3). *Let $\mathcal{I}$ be the collection of subsets of E that contain no member of $\mathcal{C}$. Then $(E, \mathcal{I})$ is a matroid having $\mathcal{C}$ as its collection of circuits.*

Proof We shall first show that $\mathcal{I}$ satisfies (I1)–(I3). By (C1), $\emptyset$ does not contain a member of $\mathcal{C}$, so $\emptyset \in \mathcal{I}$ and (I1) holds. If I contains no member of $\mathcal{C}$ and I' is contained in I, then I' contains no member of $\mathcal{C}$. Thus (I2) holds.

To prove (I3), suppose that I_1 and I_2 are members of $\mathcal{I}$ and $|I_1| < |I_2|$. Assume that (I3) fails for the pair (I_1, I_2). Now $\mathcal{I}$ has a member that is a subset of $I_1 \cup I_2$ and has more elements than I_1. Choose such a subset I_3 for which $|I_1 - I_3|$ is minimal. As (I3) fails, $I_1 - I_3$ is non-empty, so we can choose an element e from this set. Now, for each element f of $I_3 - I_1$, let T_f be $(I_3 \cup e) - f$. Then $T_f \subseteq I_1 \cup I_2$ and $|I_1 - T_f| < |I_1 - I_3|$. Therefore $T_f \notin \mathcal{I}$, so T_f contains a member C_f of $\mathcal{C}$. Hence $C_f \subseteq (I_3 \cup e) - f$, so $f \notin C_f$. Moreover, $e \in C_f$, otherwise $C_f \subseteq I_3$ contradicting the fact that $I_3 \in \mathcal{I}$.

Suppose $g \in I_3 - I_1$. If $C_g \cap (I_3 - I_1) = \emptyset$, then $C_g \subseteq ((I_1 \cap I_3) \cup e) - g \subseteq I_1$; a contradiction. Thus there is an element h in $C_g \cap (I_3 - I_1)$, and $C_h \neq C_g$ since $h \notin C_h$. Now $e \in C_g \cap C_h$, so (C3) implies that $\mathcal{C}$ contains a member C such that $C \subseteq (C_g \cup C_h) - e$. But, both C_g and C_h are subsets of $I_3 \cup e$, so $C \subseteq I_3$; a contradiction. We conclude that (I3) holds. Thus $(E, \mathcal{I})$ is a matroid M.

To prove that $\mathcal{C}$ is the set $\mathcal{C}(M)$ of circuits of M, we note that the following statements are equivalent:

(i) C is a circuit of M;
(ii) $C \notin \mathcal{I}(M)$ and $C - x \in \mathcal{I}(M)$ for all x in C;

(iii) C has a member C' of $\mathcal{C}$ as a subset, but C' is not a proper subset of C;
(iv) $C \in \mathcal{C}$. $\square$

On combining Theorem 1.1.4 with Lemma 1.1.3 and the remarks preceding it, we obtain the following result.

Corollary 1.1.5 *Let $\mathcal{C}$ be a set of subsets of a set E. Then $\mathcal{C}$ is the collection of circuits of a matroid on E if and only if $\mathcal{C}$ has the following properties:*

(C1) $\emptyset \notin \mathcal{C}$.
(C2) *If C_1 and C_2 are members of $\mathcal{C}$ and $C_1 \subseteq C_2$, then $C_1 = C_2$.*
(C3) *If C_1 and C_2 are distinct members of $\mathcal{C}$ and $e \in C_1 \cap C_2$, then there is a member C_3 of $\mathcal{C}$ such that $C_3 \subseteq (C_1 \cup C_2) - e$.* $\square$

Next we note an elementary but frequently used property of matroids.

Proposition 1.1.6 *Suppose that I is an independent set in a matroid M and e is an element of M such that $I \cup e$ is dependent. Then M has a unique circuit contained in $I \cup e$, and this circuit contains e.*

Proof Clearly $I \cup e$ contains a circuit, and all such circuits must contain e. Let C and C' be distinct such circuits. Then (C3) implies that $(C \cup C') - e$ contains a circuit. As $(C \cup C') - e \subseteq I$, this is a contradiction; so C is unique. $\square$

We noted earlier that the class of vector matroids is one of two fundamental classes of matroids that were considered in Whitney's (1935) seminal paper. The second such class consists of matroids derived from graphs in a way that will be described in a moment. One of the more obvious indications of the pervasive influence of these two classes of examples throughout matroid theory is in the terminology of the subject, which borrows heavily from both linear algebra and graph theory. Recall that a cycle of a graph is a connected subgraph all of whose vertices have degree two.

Proposition 1.1.7 *Let E be the set of edges of a graph G and $\mathcal{C}$ be the set of edge sets of cycles of G. Then $\mathcal{C}$ is the set of circuits of a matroid on E.*

Proof Clearly $\mathcal{C}$ satisfies (C1) and (C2). To prove that it satisfies (C3), let C_1 and C_2 be the edge sets of two distinct cycles of G that have e as a common edge. Let u and v be the endpoints of e. We now construct a cycle of G whose edge set is contained in $(C_1 \cup C_2) - e$. For each i in $\{1,2\}$, let P_i be the path from u to v in G whose edge set is $C_i - e$. Beginning at u, traverse P_1 towards v letting w be the first vertex at which the next edge of P_1 is not in P_2. Continue traversing P_1 from w towards v until the first time a vertex x is reached that is distinct from w but is also in P_2. Since P_1 and P_2 both end at v, such a vertex must exist. Now adjoin the section of P_1 from w to x to the section of P_2 from x to w. The result is a cycle (see Figure 1.1), the edge set of which is contained in $(C_1 \cup C_2) - e$. Hence $\mathcal{C}$ satisfies (C3). $\square$

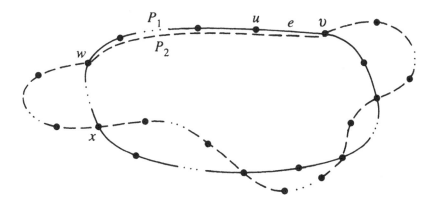

FIG. 1.1. The cycles in a graph satisfy the circuit elimination axiom.

The matroid derived above from the graph G is called the *cycle matroid* or *polygon matroid* of G. It is denoted by $M(G)$. Clearly a set X of edges is independent in $M(G)$ if and only if X does not contain the edge set of a cycle or, equivalently, $G[X]$, the subgraph induced by X, is a forest. Since the ground set of $M(G)$ is the edge set of G, we shall often refer to certain subgraphs of G, such as cycles, when we mean just their edge sets. This commonplace practice should not cause confusion.

Example 1.1.8 Let G be the graph shown in Figure 1.2 and let $M = M(G)$. Then $E(M) = \{e_1, e_2, e_3, e_4, e_5\}$ and $\mathcal{C}(M) = \{\{e_3\}, \{e_1, e_4\}, \{e_1, e_2, e_5\}, \{e_2, e_4, e_5\}\}$. Comparing M with the matroid $M[A]$ in Example 1.1.2, we see that, under the bijection ψ from $\{1, 2, 3, 4, 5\}$ to $\{e_1, e_2, e_3, e_4, e_5\}$ defined by $\psi(i) = e_i$, a set X is a circuit in $M[A]$ if and only if $\psi(X)$ is a circuit in M. Equivalently, a set Y is independent in $M[A]$ if and only if $\psi(Y)$ is independent in M. Thus the matroids $M[A]$ and M have the same structure or are isomorphic. Formally, two matroids M_1 and M_2 are *isomorphic*, written $M_1 \cong M_2$, if there is a bijection ψ from $E(M_1)$ to $E(M_2)$ such that, for all $X \subseteq E(M_1)$, the set $\psi(X)$ is independent in M_2 if and only if X is independent in M_1. We call such a bijection ψ an *isomorphism* from M_1 to M_2 □

A matroid that is isomorphic to the cycle matroid of a graph is called *graphic*,

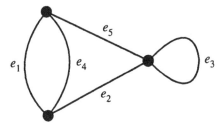

FIG. 1.2. A graph G for which $M(G)$ is isomorphic to $M[A]$ from 1.1.2.

so the matroid $M[A]$ in Example 1.1.2 is graphic. If M is isomorphic to the vector matroid of a matrix D over a field $\mathbb{F}$, then M is *representable over* $\mathbb{F}$ or $\mathbb{F}$-*representable*; and D is a *representation* for M *over* $\mathbb{F}$ or an $\mathbb{F}$-*representation* for M. A matroid that is representable over some field is called *representable* or, sometimes, *coordinatizable* or *linear* or *matric*. Thus the matroid $M(G)$ in Example 1.1.8 is representable, the matrix A being both a $GF(2)$-representation and an $\mathbb{R}$-representation for it. Indeed, as we shall see in Chapter 5, every graphic matroid is representable over every field.

Confronted with just these two fundamental classes of matroids, one naturally asks: *Is every matroid graphic or representable?* In the next section, we shall see an example of a smallest non-graphic matroid, while in Section 2.2, we shall prove that a certain 8-element matroid is non-representable.

When looking for examples of matroids with certain properties, we tend not to look for collections of sets satisfying (I1)–(I3). Instead, we look for a relatively succinct way of presenting the information needed to determine the matroid. Not every matroid is graphic or representable, but, as we shall see later, several other types of matroids have similarly compact presentations. Rarely do we specify a matroid by explicitly listing every independent set or every circuit.

All matroids on three or fewer elements are graphic. In Table 1.1 on the next page, we list all these matroids by presenting a corresponding graph for each. The latter need not be unique. The reader is urged to check the completeness of this table. Doubtless, the reader is now convinced that there are exactly 2^n non-isomorphic matroids on an n-element set! In fact, however, the four values of n shown in the table are the *only* four values for which there are exactly 2^n non-isomorphic matroids. Later in this chapter, we shall present a technique that helps one to show that there are exactly 17 non-isomorphic matroids on a 4-element set. Indeed, if $f(n)$ denotes the number of non-isomorphic matroids on an n-element set, then $f(n)$ is much closer to 2^{2^n} than it is to 2^n. More precisely, on combining bounds of Knuth (1974) and Piff (1973), we get

$$n - (3/2)\log_2 n + O(\log_2 \log_2 n) \leq \log_2 \log_2 f(n)$$
$$\leq n - \log_2 n + O(\log_2 \log_2 n). \quad (1.1)$$

In Example 1.1.8, the loop e_3 and the pair $\{e_1, e_4\}$ of parallel edges give rise to circuits in $M(G)$ of sizes one and two, respectively. Borrowing this graph-theoretic terminology, we call an element e a *loop* of an arbitrary matroid M if $\{e\}$ is a circuit of M. Moreover, if f and g are elements of M such that $\{f, g\}$ is a circuit, then f and g are *parallel* in M. A *parallel class* of M is a maximal subset X of $E(M)$ such that any two distinct members of X are parallel and no member of X is a loop. A parallel class is *trivial* if it contains just one element. If M has no loops and no non-trivial parallel classes, it is called a *simple matroid* or sometimes a *combinatorial geometry* or just a *geometry*.

We have already seen that a matroid M on a fixed ground set E can be specified by a list of its independent sets or by a list of its circuits. Evidently M is also uniquely determined by its collection of maximal independent sets

TABLE 1.1. Matroids with three or fewer elements.

Number n of elements	A corresponding graph	Number of non-isomorphic n-element matroids
0		1
1		
1		2
2		
2		
2		4
2		
3		
3		
3		
3		8
3		
3		
3		
3		

and, in the next section, we shall use this collection to give a third character-ization of matroids. In subsequent sections, we shall look at yet other ways to characterize matroids. It is one of the great beauties of the subject of matroid theory that there are so many equivalent descriptions of matroids. These many different approaches, motivated as they are by numerous different examples of matroids, enable one to bring to the subject intuition developed in other areas of mathematics including graph theory, linear algebra, and geometry.

Like matroids, topological spaces have several equivalent descriptions. For example, they can be defined in terms of their open sets or their closed sets or via a (topological) closure operator. Just as each of these approaches is indispensable in topology, each of the many descriptions of matroids that we shall introduce provides an invaluable tool for the study of these objects.

Exercises

1. Prove that $(E, \mathcal{I})$ is a matroid if and only if $\mathcal{I}$ satisfies (I2) together with
   (I1)$'$ $\mathcal{I} \neq \emptyset$; and
   (I3)$'$ *if I_1 and I_2 are in $\mathcal{I}$ and $|I_2| = |I_1| + 1$, then there is an element e of $I_2 - I_1$ such that $I_1 \cup e \in \mathcal{I}$.*

2. Let A be the matrix

$$
\begin{array}{cccccc}
1 & 2 & 3 & 4 & 5 & 6 \\
\end{array}
$$
$$
\begin{bmatrix}
1 & 0 & 0 & 1 & 1 & 0 \\
0 & 1 & 0 & 1 & 0 & 1 \\
0 & 0 & 1 & 0 & 1 & 1
\end{bmatrix}.
$$

   For q in $\{2, 3\}$, let $M_q[A]$ be the vector matroid of A when A is viewed over $GF(q)$, the field of q elements. Show that:
   (a) The sets of circuits of $M_2[A]$ and $M_3[A]$ are different.
   (b) $M_2[A]$ is graphic but $M_3[A]$ is not.
   (c) $M_2[A]$ is $GF(3)$-representable, but $M_3[A]$ is not $GF(2)$-representable.

3. Prove that $(E, \mathcal{I})$ is a matroid if and only if $\mathcal{I}$ satisfies (I1), (I2), and
   (I3)$''$ *if $X \subseteq E$ and I_1 and I_2 are maximal members of $\{I : I \in \mathcal{I}$ and $I \subseteq X\}$, then $|I_1| = |I_2|$.*

4. Let A be a matrix over a field $\mathbb{F}$. In terms of A, specify precisely when an element of $M[A]$ is a loop, and when two elements of $M[A]$ are parallel.

5. Let C_1 and C_2 be circuits of a matroid M such that $C_1 \cup C_2 = E(M)$ and $C_1 - C_2 = \{e\}$. Prove that if C_3 is a circuit of M, then either $C_3 = C_1$ or $C_3 \supseteq C_2 - C_1$.

6. For each of the matroids M in Table 1.1, find the following:
   (a) all graphs G for which $M(G) \cong M$; and
   (b) a matrix A so that $M[A] \cong M$ when A is viewed over any field.

7. Let M_1 and M_2 be matroids on disjoint sets E_1 and E_2. Let $E = E_1 \cup E_2$ and $\mathcal{I} = \{I_1 \cup I_2 : I_1 \in \mathcal{I}(M_1), I_2 \in \mathcal{I}(M_2)\}$. Prove that $(E, \mathcal{I})$ is a matroid. This matroid, the *direct sum* $M_1 \oplus M_2$ of M_1 and M_2, will be looked at more closely in Chapter 4.

8. Let $\mathcal{D}$ be a set of subsets of a set E. Characterize when $\mathcal{D}$ is the set of dependent sets of a matroid on E.

9. Let M_1 and M_2 be matroids on a set E. Give an example to show that $(E, \mathcal{I}(M_1) \cap \mathcal{I}(M_2))$ need not be a matroid.

10. Show that a set $\mathcal{I}$ of subsets of a set E satisfying (I1), (I2), and the following condition need not be the set of independent sets of a matroid on E.

   (I3)$^-$ *If I_1 and I_2 are in $\mathcal{I}$ and $|I_1| < |I_2|$, then there is an element e of I_2 such that $I_1 \cup e \in \mathcal{I}$.*

11. Let G be the graph obtained from K_5 by deleting two non-adjacent edges. Find representations for $M(G)$ over $GF(2)$ and $GF(3)$.

12. (Nakasawa 1935) Prove that $(E, \mathcal{I})$ is a matroid if and only if $\mathcal{I}$ satisfies (I1), (I2), and the following condition:

   (I3)$'''$ *If I_1 and I_2 are members of $\mathcal{I}$ with $|I_1 - I_2| = 1$ and $|I_2 - I_1| = 2$, then there is an element e of $I_2 - I_1$ such that $I_1 \cup e \in \mathcal{I}$.*

13. (a) Prove that the number of non-isomorphic n-element graphic matroids is at most $(\binom{n+1}{2} + 1)^n$ and the number of non-isomorphic n-element $GF(2)$-representable matroids is at most 2^{n^2}.

   (b) Deduce that, as $n \to \infty$, the proportions of non-isomorphic n-element matroids that are graphic or are $GF(2)$-representable both tend to 0.

14. *Prove that the circuits of a matroid M satisfy the following strong elimination axiom: if C_1 and C_2 are circuits such that $e \in C_1 \cap C_2$ and $f \in C_1 - C_2$, then M has a circuit C_3 such that $f \in C_3 \subseteq (C_1 \cup C_2) - e$. (Hint: Let C_1 and C_2 be a pair of circuits for which the proposition fails so that, among such pairs, $|C_1 \cup C_2|$ is minimal. Apply (C3) several times.)

1.2 Bases

A list of the maximal independent sets in a matroid M is clearly a more efficient way to specify M than a list of all the independent sets. We call a maximal independent set in M a *basis* or a *base* of M. In this section, we study the bases of a matroid, showing that these sets have much in common with the bases in a vector space. For instance:

Lemma 1.2.1 *If B_1 and B_2 are bases of a matroid M, then $|B_1| = |B_2|$.*

Proof Suppose that $|B_1| < |B_2|$. Then, by (I3), as both B_1 and B_2 are independent, there is an element e of $B_2 - B_1$ such that $B_1 \cup e \in \mathcal{I}$. This contradicts the maximality of B_1. Hence $|B_1| \geq |B_2|$ and, similarly, $|B_2| \geq |B_1|$. $\square$

If M is a matroid and $\mathcal{B}$ is its collection of bases, then, by (I1),

(B1) $\mathcal{B}$ *is non-empty.*

The next result contains a far less obvious property of $\mathcal{B}$.

Lemma 1.2.2 *The set $\mathcal{B}$ of bases of a matroid has the following property:*

(B2) *If B_1 and B_2 are members of $\mathcal{B}$ and $x \in B_1 - B_2$, then there is an element y of $B_2 - B_1$ such that $(B_1 - x) \cup y \in \mathcal{B}$.*

Proof Both $B_1 - x$ and B_2 are independent sets. Moreover, $|B_1 - x| < |B_2|$ since, by the preceding lemma, $|B_1| = |B_2|$. Therefore, by (I3), there is an element y of $B_2 - (B_1 - x)$ such that $(B_1 - x) \cup y \in \mathcal{I}$. Evidently $y \in B_2 - B_1$. Furthermore, as $(B_1 - x) \cup y$ is independent, it is contained in a maximal independent set B_1'. By Lemma 1.2.1 again, $|B_1'| = |B_1|$. Moreover, $|B_1| = |(B_1 - x) \cup y|$. Hence $(B_1 - x) \cup y = B_1'$, so $(B_1 - x) \cup y$ is a basis of M, and $\mathcal{B}$ satisfies (B2). $\square$

Condition (B2) is one of several *basis exchange axioms* obeyed by matroids. We shall see further examples of such conditions in Chapter 2 and in the exercises. We now prove that (B1) and (B2) characterize the bases of a matroid.

Theorem 1.2.3 *Let E be a set and $\mathcal{B}$ be a collection of subsets of E satisfying (B1) and (B2). Let $\mathcal{I}$ be the collection of subsets of E that are contained in some member of $\mathcal{B}$. Then $(E, \mathcal{I})$ is a matroid having $\mathcal{B}$ as its collection of bases.*

Proof Since $\mathcal{B}$ satisfies (B1), the collection $\mathcal{I}$ satisfies (I1). Moreover, if $I \in \mathcal{I}$, then $I \subseteq B$ for some set B in $\mathcal{B}$. Thus if $I' \subseteq I$, then $I' \subseteq B$, so $I' \in \mathcal{I}$. Hence $\mathcal{I}$ satisfies (I2). The proof that $\mathcal{I}$ satisfies (I3) uses the following.

Lemma 1.2.4 *All the members of $\mathcal{B}$ have the same cardinality.*

Proof Suppose that B_1 and B_2 are distinct members of $\mathcal{B}$ for which $|B_1| > |B_2|$ so that, among all such pairs, $|B_1 - B_2|$ is minimal. Clearly $B_1 - B_2 \neq \emptyset$. Thus, choosing x in $B_1 - B_2$, we can find an element y of $B_2 - B_1$ so that $(B_1 - x) \cup y \in \mathcal{B}$. Evidently $|(B_1 - x) \cup y| = |B_1| > |B_2|$ and $|((B_1 - x) \cup y) - B_2| < |B_1 - B_2|$. Thus the choice of B_1 and B_2 is contradicted and the lemma is proved. $\square$

Returning to the proof of Theorem 1.2.3, suppose that (I3) fails for $\mathcal{I}$. Then $\mathcal{I}$ has members I_1 and I_2 with $|I_1| < |I_2|$ such that, for all e in $I_2 - I_1$, the set $I_1 \cup e \notin \mathcal{I}$. By definition, $\mathcal{B}$ contains members B_1 and B_2 such that $I_1 \subseteq B_1$ and $I_2 \subseteq B_2$. Assume that such a set B_2 is chosen so that $|B_2 - (I_2 \cup B_1)|$ is minimal. By the choice of I_1 and I_2,

$$I_2 - B_1 = I_2 - I_1. \tag{1.2}$$

Now suppose that $B_2 - (I_2 \cup B_1)$ is non-empty. Then we can choose an element x from this set. By (B2), there is an element y of $B_1 - B_2$ such that $(B_2 - x) \cup y \in \mathcal{B}$. But then $|((B_2 - x) \cup y) - (I_2 \cup B_1)| < |B_2 - (I_2 \cup B_1)|$ and the choice of B_2 is contradicted. Hence $B_2 - (I_2 \cup B_1)$ is empty and so $B_2 - B_1 = I_2 - B_1$. Thus, by (1.2),

$$B_2 - B_1 = I_2 - I_1. \tag{1.3}$$

Next we show that $B_1 - (I_1 \cup B_2)$ is empty. If not, then there is an element x in this set and an element y in $B_2 - B_1$ so that $(B_1 - x) \cup y \in \mathcal{B}$. Now $I_1 \cup y \subseteq (B_1 - x) \cup y$ so $I_1 \cup y \in \mathcal{I}$. Since $y \in B_2 - B_1$, it follows by (1.3) that $y \in I_2 - I_1$ and so we have a contradiction to our assumption that (I3) fails. We conclude that $B_1 - (I_1 \cup B_2)$ is empty. Hence $B_1 - B_2 = I_1 - B_2$. Since the last set is contained in $I_1 - I_2$, it follows that

$$B_1 - B_2 \subseteq I_1 - I_2. \tag{1.4}$$

By Lemma 1.2.4, $|B_1| = |B_2|$, so $|B_1 - B_2| = |B_2 - B_1|$. Therefore, by (1.3) and (1.4), $|I_1 - I_2| \geq |I_2 - I_1|$, so $|I_1| \geq |I_2|$. This contradiction completes the proof that $(E, \mathcal{I})$ is a matroid. Since $\mathcal{B}$ is clearly the set of bases of this matroid, the theorem is proved. □

On combining Theorem 1.2.3 with Lemma 1.2.2 and the remarks preceding it, we get the following.

Corollary 1.2.5 *Let $\mathcal{B}$ be a set of subsets of a set E. Then $\mathcal{B}$ is the collection of bases of a matroid on E if and only if it has the following properties:*

(B1) *$\mathcal{B}$ is non-empty.*

(B2) *If B_1 and B_2 are in $\mathcal{B}$ and $x \in B_1 - B_2$, then there is an element y of $B_2 - B_1$ such that $(B_1 - x) \cup y \in \mathcal{B}$.* □

It was noted in Proposition 1.1.6 that if I is an independent set in a matroid M, and $I \cup e$ is dependent in M, then M has a unique circuit contained in $I \cup e$, and this circuit contains e. The next result is an immediate consequence of this.

Corollary 1.2.6 *Let B be a basis of a matroid M. If $e \in E(M) - B$, then $B \cup e$ contains a unique circuit, $C(e, B)$. Moreover, $e \in C(e, B)$.* □

We call $C(e, B)$ the *fundamental circuit of e with respect to B* and we leave it to the reader to show that every circuit in a matroid is the fundamental circuit of some element with respect to some basis.

Example 1.2.7 Let m and n be non-negative integers with $m \leq n$. Let E be an n-element set and $\mathcal{B}$ be the collection of m-element subsets of E. Then it is easy to check that $\mathcal{B}$ is the set of bases of a matroid on E. We denote this matroid by $U_{m,n}$ and call it the *uniform matroid* of rank m on an n-element set. Clearly

$$\mathcal{I}(U_{m,n}) = \{X \subseteq E : |X| \leq m\}$$

and

$$\mathcal{C}(U_{m,n}) = \begin{cases} \emptyset, & \text{if } m = n, \\ \{X \subseteq E : |X| = m+1\}, & \text{if } m < n. \end{cases}$$

If m is 0, then $U_{m,n}$ consists of n loops, while if m is 1, then $U_{m,n}$ consists of a single parallel class of n elements. If $m \geq 2$, then $U_{m,n}$ is simple. The matroids of the form $U_{n,n}$ are precisely the matroids having no dependent sets. We call such matroids *free*. One of these matroids, $U_{0,0}$, is the unique matroid on the empty set. It is called the *empty matroid*. □

Example 1.2.8 Let $(E_1, E_2, \ldots, E_r)$ be a partition π of a set E into non-empty sets, and let $\mathcal{B}$ consist of those subsets of E that contain exactly one element from each E_i. It is straightforward to check that $\mathcal{B}$ is the set of bases of a matroid M_π on E. We call M_π the *partition matroid* associated with π. Clearly

$$\mathcal{I}(M_\pi) = \{X \subseteq E : |X \cap E_i| \leq 1 \text{ for all } i \text{ in } \{1, 2, \ldots, r\}\}$$

and

$$\mathcal{C}(M_\pi) = \{\{a, b\} : a \neq b \text{ and } \{a, b\} \subseteq E_i \text{ for some } i \text{ in } \{1, 2, \ldots, r\}\}.$$

Thus if $|E_i| = 1$ for all i, then $M_\pi \cong U_{r,r}$. In general, M_π is graphic as it is isomorphic to the cycle matroid of the vertex-disjoint union of graphs $G_1, G_2, \ldots, G_r$, where G_i consists of two vertices joined by $|E_i|$ distinct edges.   □

In a matroid M, we have already specified the relationship between its collections, $\mathcal{B}(M)$ and $\mathcal{I}(M)$, of bases and independent sets, and between $\mathcal{I}(M)$ and $\mathcal{C}(M)$. To relate $\mathcal{B}(M)$ and $\mathcal{C}(M)$ directly, we note that $\mathcal{B}(M)$ is the collection of maximal subsets of $E(M)$ that contain no member of $\mathcal{C}(M)$, while $\mathcal{C}(M)$ is the collection of minimal sets that are contained in no member of $\mathcal{B}(M)$.

The fact that all bases of a matroid have the same cardinality enables us to define a matroid generalization of the dimension function in vector spaces. In the next section, we shall examine this function in detail. Before doing this, however, let us distinguish the bases in graphic and representable matroids. Let G be a graph. We observed earlier that a subset X of $E(G)$ is independent in $M(G)$ exactly when $G[X]$, the subgraph of G induced by X, is a forest. Thus X is a basis of $M(G)$ precisely when $G[X]$ is a forest such that, for all $e \notin X$, the graph $G[X \cup e]$ contains a cycle. It follows that, when G is connected, X is a basis of $M(G)$ if and only if $G[X]$ is a spanning tree of G. In general, X is a basis of $M(G)$ if and only if, for each component H of G having at least one non-loop edge, $H[X \cap E(H)]$ is a spanning tree of H.

As an example, consider the graphs G_1 and G_2 shown in Figure 1.3. Each has edge set $\{1, 2, 3, 4, 5\}$. Moreover, each of $M(G_1)$ and $M(G_2)$ has $\{2, 3, 4, 5\}$ as its unique basis. Thus $M(G_1) = M(G_2)$, although clearly $G_1 \not\cong G_2$. In Chapter 5, we shall solve the problem of determining precisely when two non-isomorphic graphs have isomorphic cycle matroids. Before leaving this problem for the moment, we make the following elementary but useful observation.

Proposition 1.2.9 *Let M be a graphic matroid. Then $M \cong M(G)$ for some connected graph G.*

Proof As M is graphic, $M \cong M(H)$ for some graph H. If H is connected, the result is proved. If not, suppose $H_1, H_2, \ldots, H_n$ are the connected components of H. For each i in $\{1, 2, \ldots, n\}$, choose a vertex v_i in H_i. Form a new graph G by identifying $v_1, v_2, \ldots, v_n$ as a single vertex. Clearly $E(H) = E(G)$ and G is connected. Moreover, if $X \subseteq E(H)$, then X is the set of edges of a cycle in H if and only if X is the set of edges of a cycle in G. Thus $M(G) \cong M(H)$ and the proposition is proved.   □

We remark that, in the preceding proof, the key point about the way in which $H_1, H_2, \ldots, H_n$ were joined is that no new cycles were formed. Indeed, provided this condition is satisfied when the components of H are stuck together, the cycle matroid of the resulting graph is isomorphic to $M(H)$.

We now consider representable matroids. Let A be an $m \times n$ matrix over a field $\mathbb{F}$. The columns of A span a subspace W of $V(m, \mathbb{F})$ of dimension r, say. A subset B of the set of column labels of A is a basis of $M[A]$ if and only if $|B| = r$ and the columns labelled by B form a basis for W. For example, let A be

$$
\begin{array}{cccc}
1 & 2 & 3 & 4
\end{array}
$$
$$
\begin{bmatrix}
1 & 0 & 1 & 1 \\
0 & 1 & 1 & -1
\end{bmatrix}
$$

over $GF(3)$, the field of three elements. In this case, $\dim W$ is 2 and the bases of $M[A]$ are all of the 2-element subsets of $\{1, 2, 3, 4\}$. Thus $M[A] \cong U_{2,4}$. Hence $U_{2,4}$ is *ternary*, that is, $U_{2,4}$ is representable over $GF(3)$.

Ternary matroids are one of the most commonly studied classes of representable matroids. Two other important such classes are binary and regular matroids. A *binary matroid* is one that is representable over $GF(2)$. A *regular matroid* is one that can be represented by a *totally unimodular matrix*, the latter being a matrix over $\mathbb{R}$ for which every square submatrix has determinant in $\{0, 1, -1\}$. Some authors, for example White (1987a), refer to regular matroids as *unimodular matroids*.

We leave it to the reader (Exercise 4) to check that $U_{2,4}$ is not graphic, not regular, and not binary. Indeed, as we shall see in Chapter 6, $U_{2,4}$ is the unique obstruction to a matroid being binary where, of course, we shall need to be precise about what we mean here by the term 'obstruction'. In Chapters 5 and 6, we shall prove that every graphic matroid is regular, and that a matroid is regular if and only if it is representable over every field.

Exercises

1. Prove that a set $\mathcal{B}$ of subsets of a set E is the collection of bases of a matroid on E if and only if $\mathcal{B}$ satisfies (B1) and the following two conditions:

    (B2) *If $B_1, B_2 \in \mathcal{B}$ and $e \in B_1$, then there is an element f of B_2 such that $(B_1 - e) \cup f \in \mathcal{B}$.*

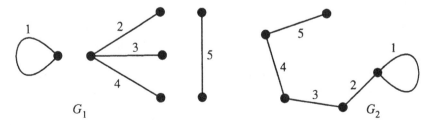

FIG. 1.3. Two non-isomorphic graphs with equal cycle matroids.

(B3) *If $B_1, B_2 \in \mathcal{B}$ and $B_1 \subseteq B_2$, then $B_1 = B_2$.*

2. Prove that the following are equivalent for an element e of a matroid:

   (a) e is a loop.
   (b) e is in no bases.
   (c) e is in no independent sets.

3. Prove that if C is a circuit of a matroid M and $e \in C$, then M has a basis B such that $C = C(e, B)$.

4. (a) Prove that $U_{2,4}$ is representable over a field $\mathbb{F}$ if and only if $|\mathbb{F}| \geq 3$.
   (b) For $n \geq 2$, generalize (a) by finding necessary and sufficient conditions for $U_{2,n}$ to be $\mathbb{F}$-representable.
   (c) Prove that $U_{2,4}$ is neither graphic nor regular.
   (d) Determine all values of m and n for which $U_{m,n}$ is graphic.

5. In a matroid M, suppose B is a basis, $f \in E(M)$, and $e \in E(M) - B$. Prove that $(B \cup e) - f$ is a basis of M if and only if $f \in C(e, B)$.

6. Let M be a matroid and X and Y be subsets of $E(M)$, each of which contains a basis. Prove that if $|X| > |Y|$, then, for some element x of $X - Y$, there is a basis contained in $X - x$.

7. Prove that a set $\mathcal{B}$ of subsets of a set E is the collection of bases of a matroid on E if and only if $\mathcal{B}$ satisfies (B1), (B3) from Exercise 1, and the following condition:

   (B2)′ For $X \subseteq Y \subseteq E$, if $\mathcal{B}$ has members B_1 and B_2 such that $X \subseteq B_1$ and $B_2 \subseteq Y$, then $\mathcal{B}$ has a member B_3 such that $X \subseteq B_3 \subseteq Y$.

8. (Bixby 1981) Show that a set $\mathcal{B}$ of subsets of a set E satisfying (B1), (B3), and the next condition need not be the set of bases of a matroid on E:

   (B2)⁻ *If $B_1 \in \mathcal{B}$ and $e \in E - B_1$, then there is an element f such that $(B_1 - f) \cup e \in \mathcal{B}$.*

9. (a) Give an example of two matroids M_1 and M_2 on a set E such that a subset X of E is a circuit of M_1 if and only if it is a basis of M_2.
   (b) Characterize all pairs (M_1, M_2) of matroids for which the condition in (a) holds.

10. *In a matroid M, let $\mathcal{D}$ be the set of minimal subsets of $E(M)$ that meet every circuit of M. Prove that $\mathcal{D}$ is the set of bases of a matroid on $E(M)$.

1.3 Rank

Two very useful notions in linear algebra are those of dimension of a vector space and span of a set of vectors. In this section and the next, we introduce the matroid generalizations of these ideas.

We begin by defining a fundamental and very natural matroid construction. Let M be the matroid $(E, \mathcal{I})$ and suppose that $X \subseteq E$. Let $\mathcal{I}|X$ be $\{I \subseteq X : I \in \mathcal{I}\}$. Then it is easy to see that the pair $(X, \mathcal{I}|X)$ is a matroid. We call this matroid the *restriction of M to X* or the *deletion of $E - X$ from M*. It is denoted by $M|X$ or $M \backslash (E-X)$. One easily checks that $\mathcal{C}(M|X) = \{C \subseteq X : C \in \mathcal{C}(M)\}$.

As $M|X$ is a matroid, Lemma 1.2.1 implies that all its bases are equicardinal. We define the *rank* $r(X)$ of X to be the cardinality of a basis B of $M|X$ and call such a set B a *basis of* X. Clearly the function r, the *rank function* of M, maps 2^E into the set of non-negative integers. We often write r as r_M. In addition, we usually write $r(M)$ for $r(E(M))$. It is clear that r has the following properties:

(R1) *If $X \subseteq E$, then $0 \le r(X) \le |X|$.*

(R2) *If $X \subseteq Y \subseteq E$, then $r(X) \le r(Y)$.*

We refer to the latter as the *increasing property* of r. Next we note that:

Lemma 1.3.1 *The rank function r of a matroid M on a set E has the following property:*

(R3) *If X and Y are subsets of E, then*
$$r(X \cup Y) + r(X \cap Y) \le r(X) + r(Y).$$

Before proving this lemma, we remark that (R3) is reminiscent of the identity

$$\dim(U + W) + \dim(U \cap W) = \dim U + \dim W, \tag{1.5}$$

which holds for subspaces U and W of a finite-dimensional vector space V, where $U + W$ consists of all vectors of the form $u + w$ with u in U and w in W. The inequality in (R3) is referred to as the *submodular* or *semimodular inequality*. It is straightforward to find examples where this inequality is strict.

Proof of Lemma 1.3.1 Let $B_{X \cap Y}$ be a basis of $X \cap Y$. Then $B_{X \cap Y}$ is an independent set in $M|(X \cup Y)$. It is therefore contained in a basis $B_{X \cup Y}$ of this matroid. Now $B_{X \cup Y} \cap X$ and $B_{X \cup Y} \cap Y$ are independent in $M|X$ and $M|Y$, respectively. Therefore $r(X) \ge |B_{X \cup Y} \cap X|$ and $r(Y) \ge |B_{X \cup Y} \cap Y|$, so

$$
\begin{aligned}
r(X) + r(Y) &\ge |B_{X \cup Y} \cap X| + |B_{X \cup Y} \cap Y| \\
&= |(B_{X \cup Y} \cap X) \cup (B_{X \cup Y} \cap Y)| \\
&\quad + |(B_{X \cup Y} \cap X) \cap (B_{X \cup Y} \cap Y)| \\
&= |B_{X \cup Y} \cap (X \cup Y)| + |B_{X \cup Y} \cap X \cap Y|.
\end{aligned}
$$

But $B_{X \cup Y} \cap (X \cup Y) = B_{X \cup Y}$ and $B_{X \cup Y} \cap X \cap Y = B_{X \cap Y}$. Thus

$$
\begin{aligned}
r(X) + r(Y) &\ge |B_{X \cup Y}| + |B_{X \cap Y}| \\
&= r(X \cup Y) + r(X \cap Y), \quad \text{as required.} \qquad \square
\end{aligned}
$$

Following the pattern of the earlier sections, we now establish that conditions (R1)–(R3) characterize the rank function of a matroid.

Theorem 1.3.2 *Let E be a set and r be a function that maps 2^E into the set of non-negative integers and satisfies (R1)–(R3). Let $\mathcal{I}$ be the collection of subsets X of E for which $r(X) = |X|$. Then $(E, \mathcal{I})$ is a matroid having rank function r.*

The proof of this theorem will use the following.

Lemma 1.3.3 *Let E be a set and r be a function on 2^E satisfying (R2) and (R3). If X and Y are subsets of E such that $r(X \cup y) = r(X)$ for all y in $Y - X$, then $r(X \cup Y) = r(X)$.*

Proof Let $Y - X = \{y_1, y_2, \ldots, y_k\}$. We shall argue by induction on k. If $k = 1$, the result is immediate. Assume it is true for $k = n$ and let $k = n+1$. Then, by the induction assumption and (R3),

$$
\begin{aligned}
r(X) + r(X) &= r(X \cup \{y_1, y_2, \ldots, y_n\}) + r(X \cup y_{n+1}) \\
&\geq r((X \cup \{y_1, y_2, \ldots, y_n\}) \cup (X \cup y_{n+1})) \\
&\quad + r((X \cup \{y_1, y_2, \ldots, y_n\}) \cap (X \cup y_{n+1})) \\
&= r(X \cup \{y_1, y_2, \ldots, y_{n+1}\}) + r(X) \\
&\geq r(X) + r(X),
\end{aligned}
$$

where the last step follows by (R2). Since the first and last lines of the above are equal, equality must hold throughout and, therefore,

$$
r(X \cup \{y_1, y_2, \ldots, y_{n+1}\}) = r(X).
$$

Thus, by induction, the lemma holds.   □

Proof of Theorem 1.3.2 By (R1), $0 \leq r(\emptyset) \leq |\emptyset| = 0$, so $r(\emptyset) = |\emptyset|$ and $\emptyset \in \mathcal{I}$. Hence $\mathcal{I}$ satisfies (I1). Now suppose that $I \in \mathcal{I}$ and $I' \subseteq I$. Then $r(I) = |I|$. By (R3), $r(I' \cup (I - I')) + r(I' \cap (I - I')) \leq r(I') + r(I - I')$, that is,

$$
r(I) + r(\emptyset) \leq r(I') + r(I - I'). \tag{1.6}
$$

But $r(I) = |I|$ and $r(\emptyset) = 0$. Moreover, by (R2), $r(I') \leq |I'|$ and $r(I - I') \leq |I - I'|$. Therefore, by (1.6),

$$
|I| \leq r(I') + r(I - I') \leq |I'| + |I - I'| = |I|.
$$

Since the first and last quantities here are equal, equality must hold throughout. Hence $r(I') = |I'|$, that is, $I' \in \mathcal{I}$.

To prove that $\mathcal{I}$ satisfies (I3), we assume the contrary letting I_1 and I_2 be members of $\mathcal{I}$ with $|I_1| < |I_2|$ such that, for all e in $I_2 - I_1$, the set $I_1 \cup e$ is not in $\mathcal{I}$, so $r(I_1 \cup e) \neq |I_1 \cup e|$. Then, by (R1), (R2), and the fact that $I_1 \in \mathcal{I}$, we get, for all such e, that $|I_1| + 1 > r(I_1 \cup e) \geq r(I_1) = |I_1|$, so $r(I_1 \cup e) = |I_1|$.

Now applying Lemma 1.3.3 with $X = I_1$ and $Y = I_2$, we immediately get that $r(I_1) = r(I_1 \cup I_2)$. But $|I_1| = r(I_1)$ and $r(I_1 \cup I_2) \geq r(I_2) = |I_2|$, so $|I_1| \geq |I_2|$; a contradiction. We conclude that $\mathcal{I}$ satisfies (I3), so $(E, \mathcal{I})$ is a matroid.

To complete the proof of the theorem, we need to show that $r(X) = r_M(X)$ for all $X \subseteq E$. Suppose first that $X \in \mathcal{I}$. Then, by definition, $r(X) = |X|$, while $r_M(X) = |X|$ since X is a basis of $M|X$. Now suppose that $X \notin \mathcal{I}$ and let B be a basis for $M|X$. Then $r_M(X) = |B|$. Moreover, $B \cup x \notin \mathcal{I}$ for all x in $X - B$. Hence $|B| = r(B) \leq r(B \cup x) < |B \cup x|$, so $r(B \cup x) = r(B)$. By Lemma 1.3.3, it follows that $r(B \cup X) = r(B)$, that is, $r(X) = r(B) = |B| = r_M(X)$. We conclude that $r = r_M$. □

The next corollary is obtained by combining the last theorem with Lemma 1.3.1 and the remarks preceding it.

Corollary 1.3.4 *Let E be a set. A function $r : 2^E \to \mathbb{Z}^+ \cup \{0\}$ is the rank function of a matroid on E if and only if r has the following properties:*

(R1) *If $X \subseteq E$, then $0 \leq r(X) \leq |X|$.*
(R2) *If $X \subseteq Y \subseteq E$, then $r(X) \leq r(Y)$.*
(R3) *If X and Y are subsets of E, then*

$$r(X \cup Y) + r(X \cap Y) \leq r(X) + r(Y).$$ □

Independent sets, bases, and circuits are easily characterized in terms of the rank function. We leave it to the reader to prove the following.

Proposition 1.3.5 *Let M be a matroid with rank function r and suppose that $X \subseteq E(M)$. Then*

(i) *X is independent if and only if $|X| = r(X)$;*
(ii) *X is a basis if and only if $|X| = r(X) = r(M)$; and*
(iii) *X is a circuit if and only if X is non-empty and, for all x in X, $r(X - x) = |X| - 1 = r(X)$.* □

We now specify the rank function explicitly for the classes of uniform, representable, and graphic matroids. In each case, we shall suppose that $X \subseteq E(M)$. If $M = U_{m,n}$, then clearly

$$r(X) = \begin{cases} |X|, & \text{if } |X| < m, \\ m, & \text{if } |X| \geq m. \end{cases}$$

If $M = M[A]$ where A is an $m \times n$ matrix over a field $\mathbb{F}$, then, clearly, $r(X)$ is the rank of A_X, the $m \times |X|$ submatrix of A consisting of those columns of A that are labelled by members of X. Equivalently, $r(X)$ equals the dimension of the subspace of $V(m, \mathbb{F})$ that is spanned by the columns of A_X.

Now let $M = M(G)$ where G is a graph. First we determine $r(M)$. Assume initially that G is connected. Then a basis of $M(G)$ is the set of edges of a spanning tree of G. It is well known and can easily be proved by induction that, for a tree T,

$$|V(T)| = |E(T)| + 1.$$

Hence if G is connected, then

1.3.6 $r(M) = |V(G)| - 1$.

Extending this, it is clear that, when G has $\omega(G)$ connected components,

1.3.7 $r(M) = |V(G)| - \omega(G)$.

It follows that the rank in $M(G)$ of a subset X of $E(G)$ is given by

1.3.8 $r(X) = |V(G[X])| - \omega(G[X])$.

Example 1.3.9 Let $M = M(G)$ where G is the graph in Figure 1.4(a). Then, as G is connected, $r(M) = |V(G)| - 1 = 4$. If $X = \{4, 5, 6, 7, 8\}$, then a basis for $M|X$ is $\{4, 5, 6\}$, so $r(\{4, 5, 6, 7, 8\}) = 3$. Equivalently, we can see from Figure 1.4(b) that $|V(G[X])| = 4$ and $\omega(G[X]) = 1$, so, by 1.3.8, $r(X) = 3$. If $Y = \{1, 3, 7\}$, then $|V(G[Y])| = 4$ and $\omega(G[Y]) = 2$, so $r(Y) = 2$. Finally, if $Z = \{1, 2\}$, then $r(Z) = 0$ because $\emptyset$ is a basis for $M|Z$.   □

To close this section, we introduce another important class of matroids. Let M be a matroid. Clearly every circuit of M has size at most $r(M) + 1$. It is straightforward to show that M is uniform if and only if it has no circuits of size less than $r(M) + 1$. We call M *paving* if it has no circuits of size less than $r(M)$. Thus the class of uniform matroids is contained in the class of paving matroids. There are numerous examples, such as $M(K_4)$, of paving matroids that are not uniform. Indeed, as we shall see in Section 15.5, it has been conjectured that most matroids are paving. The next result characterizes paving matroids in terms of their collections of circuits. The proof is left to the reader.

Proposition 1.3.10 *Let $\mathcal{D}$ be a collection of non-empty subsets of a set E. Then $\mathcal{D}$ is the set of circuits of a paving matroid on E if and only if there are a positive integer k with $k \leq |E|$ and a subset $\mathcal{D}'$ of $\mathcal{D}$ such that*

(i)   *every member of $\mathcal{D}'$ has k elements, and if two distinct members D_1 and D_2 of $\mathcal{D}'$ have $k - 1$ common elements, then every k-element subset of $D_1 \cup D_2$ is in $\mathcal{D}'$; and*

(ii) *$\mathcal{D} - \mathcal{D}'$ consists of all of the $(k + 1)$-element subsets of E that contain no member of $\mathcal{D}'$.*   □

Exercises

1. In terms of the rank function, characterize when an element is a loop of a matroid and when two elements are parallel.

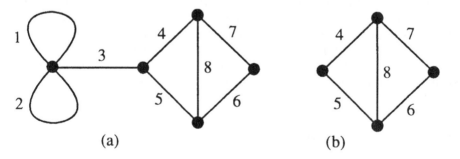

(a)   (b)

FIG. 1.4. (a) A graph G. (b) An edge-induced subgraph $G[X]$ of G.

2. (a) Find a matroid in which equality does not always hold in (R3).
   (b) Find all matroids for which equality does always hold in (R3).

3. Let M be a matroid on a set E and k be a non-negative integer not exceeding $r(M)$. Define $r_{(k)} : 2^E \to \mathbb{Z}^+ \cup \{0\}$ by $r_{(k)}(X) = \min\{k, r(X)\}$.
   (a) Prove that $r_{(k)}$ is the rank function of a matroid on E of rank k. This matroid is obtained from M by a sequence of truncations. The operation of *truncation* will be discussed in Section 7.3.
   (b) Find the collection of independent sets of the matroid in (a).

4. Prove that a matroid M is uniform if and only if it has no circuits of size less than $r(M) + 1$.

5. Let B_X be a basis of a set X in a matroid M. Suppose that $I \subseteq E(M) - X$ and $I \cup B_X \in \mathcal{I}(M)$. Prove that if X' is an independent subset of X, then $I \cup X'$ is an independent set.

6. (a) Characterize paving matroids in terms of their collections of independent sets and in terms of their collections of bases.
   (b) Characterize uniform matroids in terms of their collections of circuits.
   (c) Modify the statement of Proposition 1.3.10 so that it characterizes the set of circuits of a non-uniform rank-r paving matroid.

7. Let M be a matroid on a set E and k be an integer with $r(M) \leq k \leq |E|$. Let $\mathcal{S}_k$ denote the set of k-element subsets of E having rank $r(M)$. Prove that $\mathcal{S}_k$ is the set of bases of a matroid on E. This matroid is called the *elongation of M to rank k* (Welsh 1976) or, when $k = r(M) + 1$, the *Higgs lift* of M (Brylawski 1986a).

8. *(Edmonds and Rota 1966) Let f be a function from 2^E into $\mathbb{Z}^+ \cup \{0\}$ such that
   (a) $f(\emptyset) = 0$;
   (b) if $X \subseteq Y \subseteq E$, then $f(X) \leq f(Y)$; and
   (c) if $X, Y \subseteq E$, then $f(X \cup Y) + f(X \cap Y) \leq f(X) + f(Y)$.
   Let $\mathcal{I}(f) = \{I \subseteq E : f(X) \geq |X| \text{ for all } X \subseteq I\}$. Prove that $(E, \mathcal{I}(f))$ is a matroid, and find its rank function. Matroids of this type will be looked at more closely in Chapter 11.

1.4 Closure

In a vector space V, a vector $\underline{v}$ is in the span of $\{\underline{v}_1, \underline{v}_2, \ldots, \underline{v}_m\}$ if the subspaces spanned by $\{\underline{v}_1, \underline{v}_2, \ldots, \underline{v}_m\}$ and $\{\underline{v}_1, \underline{v}_2, \ldots, \underline{v}_m, \underline{v}\}$ have the same dimension. Now let M be an arbitrary matroid having ground set E and rank function r. Let cl be the function from 2^E into 2^E defined, for all $X \subseteq E$, by

1.4.1 $\text{cl}(X) = \{x \in E : r(X \cup x) = r(X)\}$.

This function is called the *closure operator* of M, and we call $\text{cl}(X)$ the *closure* or *span* of X in M, sometimes writing it as $\text{cl}_M(X)$ Thus, in Example 1.3.9, $\text{cl}(\emptyset) = \{1, 2\}$, $\text{cl}(\{1, 3, 5\}) = \{1, 2, 3, 5\}$, and $\text{cl}(\{4, 5, 6\}) = \{1, 2, 4, 5, 6, 7, 8\}$.

We show next that a set and its closure have the same rank.

Lemma 1.4.2 *For every subset X of the ground set of a matroid M,*

$$r(X) = r(\mathrm{cl}(X)).$$

Proof Let B_X be a basis of X. For each x in $\mathrm{cl}(X) - X$, we have

$$r(B_X \cup x) \le r(X \cup x) = r(X) = |B_X| = r(B_X) \le r(B_X \cup x),$$

so $r(B_X \cup x) = r(B_X) = |B_X| < |B_X \cup x|$. Hence $B_X \cup x$ is a dependent set. It follows that B_X is a basis of $\mathrm{cl}(X)$. Thus $r(X) = r(\mathrm{cl}(X))$.   □

One of the main results of this section will characterize matroids in terms of their closure operators. The following is an important preliminary to that result.

Lemma 1.4.3 *The closure operator of a matroid on a set E has the following properties:*
(CL1) *If $X \subseteq E$, then $X \subseteq \mathrm{cl}(X)$.*
(CL2) *If $X \subseteq Y \subseteq E$, then $\mathrm{cl}(X) \subseteq \mathrm{cl}(Y)$.*
(CL3) *If $X \subseteq E$, then $\mathrm{cl}(\mathrm{cl}(X)) = \mathrm{cl}(X)$.*
(CL4) *If $X \subseteq E$ and $x \in E$, and $y \in \mathrm{cl}(X \cup x) - \mathrm{cl}(X)$, then $x \in \mathrm{cl}(X \cup y)$.*

The proof of (CL4) will use the following straightforward result.

Lemma 1.4.4 *For a matroid M on a set E, if $X \subseteq E$ and $x \in E$, then*

$$r(X) \le r(X \cup x) \le r(X) + 1.$$

Proof The first inequality follows by (R2), the second by (R3).   □

Proof of Lemma 1.4.3 The fact that the closure operator satisfies (CL1) is immediate from the definition. To show that (CL2) holds, suppose that $X \subseteq Y$ and $x \in \mathrm{cl}(X) - X$. Then $r(X \cup x) = r(X)$. Thus, if B_X is a basis of X, then B_X is a basis of $X \cup x$. Therefore $Y \cup x$ has a basis $B_{Y \cup x}$ that contains B_X but does not contain x. Since $B_{Y \cup x}$ must also be a basis of Y, it follows that $r(Y \cup x) = |B_{Y \cup x}| = r(Y)$ and so $x \in \mathrm{cl}(Y)$. We conclude that $\mathrm{cl}(X) \subseteq \mathrm{cl}(Y)$, that is, (CL2) holds.

To prove (CL3), first note that, by (CL1), $\mathrm{cl}(X) \subseteq \mathrm{cl}(\mathrm{cl}(X))$. To establish the reverse inclusion, choose x in $\mathrm{cl}(\mathrm{cl}(X))$. Then $r(\mathrm{cl}(X) \cup x) = r(\mathrm{cl}(X))$. Thus, by Lemma 1.4.2, $r(\mathrm{cl}(X) \cup x) = r(X)$. But, by (R2), $r(\mathrm{cl}(X) \cup x) \ge r(X \cup x) \ge r(X)$. Hence equality holds throughout the last statement, so $x \in \mathrm{cl}(X)$. Thus $\mathrm{cl}(\mathrm{cl}(X)) \subseteq \mathrm{cl}(X)$ and (CL3) follows.

Now suppose $y \in \mathrm{cl}(X \cup x) - \mathrm{cl}(X)$. Then $r(X \cup x \cup y) = r(X \cup x)$ and $r(X \cup y) \ne r(X)$. From the last inequality and Lemma 1.4.4, we deduce that $r(X \cup y) = r(X) + 1$. Thus

$$r(X) + 1 = r(X \cup y) \le r(X \cup y \cup x) = r(X \cup x) \le r(X) + 1.$$

Hence $r(X \cup y \cup x) = r(X \cup y)$, so $x \in \mathrm{cl}(X \cup y)$ as required.   □

Conditions (CL1)–(CL4) characterize those functions that can be closure operators of matroids. Sometimes (CL4) is referred to as the *Mac Lane–Steinitz exchange property*.

Theorem 1.4.5 *Let E be a set and* cl *be a function from 2^E into 2^E satisfying* (CL1)–(CL4). *Let*

$$\mathcal{I} = \{X \subseteq E : x \notin \mathrm{cl}(X - x) \text{ for all } x \text{ in } X\}.$$

Then $(E, \mathcal{I})$ is a matroid having closure operator cl.

Proof Evidently $\emptyset \in \mathcal{I}$, so $\mathcal{I}$ satisfies (I1). Now suppose $I \in \mathcal{I}$ and $I' \subseteq I$. If $x \in I'$, then $x \in I$ so $x \notin \mathrm{cl}(I - x)$. By (CL2), $\mathrm{cl}(I - x)$ contains $\mathrm{cl}(I' - x)$, so $x \notin \mathrm{cl}(I' - x)$ and therefore $I' \in \mathcal{I}$. Thus $\mathcal{I}$ satisfies (I2).

The rest of the proof will make frequent use of the following result.

Lemma 1.4.6 *Suppose $X \subseteq E$ and $x \in E$. If X is in $\mathcal{I}$, but $X \cup x$ is not, then $x \in \mathrm{cl}(X)$.*

Proof As $X \cup x \notin \mathcal{I}$, some element y of $X \cup x$ is in $\mathrm{cl}((X \cup x) - y)$. If $y = x$, then the lemma holds. If $y \neq x$, then $(X \cup x) - y = (X - y) \cup x$, and y is in $\mathrm{cl}((X - y) \cup x) - \mathrm{cl}(X - y)$. Thus, by (CL4), $x \in \mathrm{cl}((X - y) \cup y) = \mathrm{cl}(X)$. $\square$

We now prove that $\mathcal{I}$ satisfies (I3). Suppose that I_1 and I_2 are members of $\mathcal{I}$ such that $|I_1| < |I_2|$ and (I3) fails for the pair (I_1, I_2). Assume, moreover, that among all such pairs, $|I_1 \cap I_2|$ is maximal. Choose y in $I_2 - I_1$ and consider $I_2 - y$. Assume that $I_1 \subseteq \mathrm{cl}(I_2 - y)$. Then, by (CL2) and (CL3), $\mathrm{cl}(I_1) \subseteq \mathrm{cl}(I_2 - y)$. Hence $y \notin \mathrm{cl}(I_1)$ as $y \notin \mathrm{cl}(I_2 - y)$. Therefore, by Lemma 1.4.6, $I_1 \cup y \in \mathcal{I}$ and so (I3) holds for the pair (I_1, I_2); a contradiction. We conclude that $I_1 \not\subseteq \mathrm{cl}(I_2 - y)$ and so some element t of I_1 is not in $\mathrm{cl}(I_2 - y)$. Evidently $t \in I_1 - I_2$. Moreover, by Lemma 1.4.6 again, $(I_2 - y) \cup t \in \mathcal{I}$. Since $|I_1 \cap ((I_2 - y) \cup t)| > |I_1 \cap I_2|$, (I3) holds for the pair $(I_1, (I_2 - y) \cup t)$, that is, for some x in $((I_2 - y) \cup t) - I_1$, the set $I_1 \cup x \in \mathcal{I}$. But $x \in I_2 - I_1$, so (I3) holds for (I_1, I_2). This contradiction completes the proof that $(E, \mathcal{I})$ is a matroid M.

We must now check that cl and the closure operator cl_M of M coincide. Suppose first that $x \in \mathrm{cl}_M(X) - X$. Then $r_M(X \cup x) = r_M(X)$. Let B be a basis of X. Then $B \in \mathcal{I}$ and $B \cup x \notin \mathcal{I}$, so, by Lemma 1.4.6, $x \in \mathrm{cl}(B)$. But, by (CL2), $\mathrm{cl}(B) \subseteq \mathrm{cl}(X)$. Therefore $x \in \mathrm{cl}(X)$ and so $\mathrm{cl}_M(X) \subseteq \mathrm{cl}(X)$. To prove the reverse inequality, suppose that x is in $\mathrm{cl}(X) - X$. Let B be a basis of X. Then $B \cup y \notin \mathcal{I}$ for all y in $X - B$, so, by Lemma 1.4.6, $X \subseteq \mathrm{cl}(B)$. Hence $\mathrm{cl}(X) \subseteq \mathrm{cl}(B)$. Since $x \in \mathrm{cl}(X)$, it follows that $x \in \mathrm{cl}(B)$. Thus $B \cup x \notin \mathcal{I}$, so B is a basis for $X \cup x$ and $r_M(X \cup x) = |B| = r_M(X)$. Hence $x \in \mathrm{cl}_M(X)$ and we conclude that $\mathrm{cl}(X) \subseteq \mathrm{cl}_M(X)$ and therefore that $\mathrm{cl}(X) = \mathrm{cl}_M(X)$. $\square$

Corollary 1.4.7 *Let E be a set. A function* cl $: 2^E \to 2^E$ *is the closure operator of a matroid on E if and only if it satisfies the following conditions:*

(CL1) *If $X \subseteq E$, then $X \subseteq \mathrm{cl}(X)$.*
(CL2) *If $X \subseteq Y \subseteq E$, then $\mathrm{cl}(X) \subseteq \mathrm{cl}(Y)$.*
(CL3) *If $X \subseteq E$, then $\mathrm{cl}(\mathrm{cl}(X)) = \mathrm{cl}(X)$.*
(CL4) *If $X \subseteq E$ and $x \in E$, and $y \in \mathrm{cl}(X \cup x) - \mathrm{cl}(X)$, then $x \in \mathrm{cl}(X \cup y)$.* $\square$

In a matroid M, a subset X of $E(M)$ for which $\mathrm{cl}(X) = X$ is called a *flat* or a *closed set* of M. A *hyperplane* of M is a flat of rank $r(M) - 1$. A subset X of $E(M)$ is a *spanning set* of M if $\mathrm{cl}(X) = E(M)$. We also say that X *spans* a subset Y of $E(M)$ if $Y \subseteq \mathrm{cl}(X)$.

Example 1.4.8 Let $M = M(K_5)$ where the edges of K_5 are as labelled in Figure 1.5. Then M has a unique flat of rank 0, namely $\emptyset$; M has ten rank-1 flats, the ten edges of K_5. The rank-2 flats of M are of two types: edge sets of K_3-subgraphs, of which there are ten; and pairs of non-adjacent edges, of which there are fifteen. The rank-3 flats or hyperplanes are also of two types: edge sets of K_4-subgraphs, of which there are five; and edge sets of subgraphs isomorphic to the disjoint union of a K_2 and a K_3, of which there are ten. Finally, the only rank-4 flat is $E(K_5)$.   □

The next proposition shows that the bases and hyperplanes of a matroid are special spanning and non-spanning sets, respectively. To prove this, we shall use the following lemma whose straightforward proof is left as an exercise.

Lemma 1.4.9 *Suppose that M is a matroid and $X \subseteq E(M)$. If $x \in \mathrm{cl}(X)$, then $\mathrm{cl}(X \cup x) = \mathrm{cl}(X)$.*   □

Proposition 1.4.10 *Let M be a matroid and X be a subset of $E(M)$. Then*

(i) *X is a spanning set if and only if $r(X) = r(M)$;*

(ii) *X is a basis if and only if it is both spanning and independent;*

(iii) *X is a basis if and only if it is a minimal spanning set; and*

(iv) *X is a hyperplane if and only if it is a maximal non-spanning set.*

Proof For (i), let X be a spanning set of M. Then $\mathrm{cl}(X) = E$. By Lemma 1.4.2, $r(X) = r(\mathrm{cl}(X))$, so $r(X) = r(E) = r(M)$. Conversely, if $r(X) = r(M)$, then, by (R2), $r(X) = r(X \cup x)$ for all x in E. Hence $\mathrm{cl}(X) = E$, and (i) holds. Part (ii) is immediate from Proposition 1.3.5(ii). For (iii), let X be a minimal spanning set of M. Then, by (i), $r(X) = r(M)$. Moreover, if $x \in X$, then $X - x$ is not spanning, so, by Lemma 1.4.9, $x \notin \mathrm{cl}(X - x)$. Thus $X \in \mathcal{I}(M)$, so, by (ii), X is a basis. The proofs of the converse of (iii) and of (iv) are left as exercises.   □

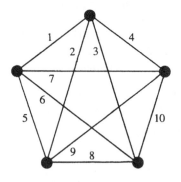

FIG. 1.5. K_5.

By the last result, a set X of edges in a connected graph G is spanning in $M(G)$ if and only if X contains the edge set of some spanning tree of G.

We have already related the closure operator to independent sets, bases, spanning sets, and hyperplanes. Next we relate it to circuits.

Proposition 1.4.11 *Let M be a matroid and X be a subset of $E(M)$. Then*

(i)   *X is a circuit if and only if X is a minimal non-empty set such that $x \in \mathrm{cl}(X - x)$ for all x in X.*

(ii) $\mathrm{cl}(X) = X \cup \{x : M \text{ has a circuit } C \text{ such that } x \in C \subseteq X \cup x\}.$

Proof   Part (i) just restates the fact that a circuit is a minimal dependent set. To prove (ii), suppose that $x \in \mathrm{cl}(X) - X$. Then $r(X \cup x) = r(X)$. Hence if B is a basis of X, then $B \cup x$ is dependent. Thus, by Corollary 1.2.6, there is a circuit C such that $x \in C \subseteq B \cup x$. Hence $x \in C \subseteq X \cup x$. On the other hand, if such a circuit exists, then, by (i) and (CL2), $x \in \mathrm{cl}(C - x) \subseteq \mathrm{cl}(X)$, so (ii) holds. $\square$

We now use the second part of the last proposition to obtain a strengthening of (C3) and thereby a different characterization of matroids in terms of circuits.

Proposition 1.4.12 *The set $\mathcal{C}$ of circuits of a matroid satisfies the following:*

(C3)' *If C_1 and C_2 are members of $\mathcal{C}$ with $e \in C_1 \cap C_2$ and $f \in C_1 - C_2$, then there is a member C_3 of $\mathcal{C}$ such that $f \in C_3 \subseteq (C_1 \cup C_2) - e$.*

Proof   As $e \in \mathrm{cl}(C_2 - e)$ and $C_2 - e \subseteq (C_1 \cup C_2) - \{e, f\}$, the element e is in $\mathrm{cl}((C_1 \cup C_2) - \{e, f\})$. Therefore, by Lemma 1.4.9, $\mathrm{cl}((C_1 \cup C_2) - \{e, f\}) = \mathrm{cl}((C_1 \cup C_2) - f)$. But $f \in \mathrm{cl}(C_1 - f) \subseteq \mathrm{cl}((C_1 \cup C_2) - f)$, so $f \in \mathrm{cl}((C_1 \cup C_2) - \{e, f\})$. It follows, by Proposition 1.4.11(ii), that M has a circuit C such that $f \in C \subseteq (C_1 \cup C_2) - e$, that is, $\mathcal{C}$ satisfies (C3)'. $\square$

Condition (C3)' is known as the *strong circuit elimination axiom*. Notice that the choice of C_3 may depend on each of e and f, so we cannot guarantee either that $C_3 \subseteq (C_1 \cup C_2) - (C_1 \cap C_2)$ or that C_3 contains $C_1 - C_2$. As we shall see in Chapter 9, the condition that $(C_1 \cup C_2) - (C_1 \cap C_2)$ contains a circuit whenever C_1 and C_2 are distinct circuits is a characterization of binary matroids. An alternative proof of (C3)' is sketched in Exercise 14 of Section 1.1. On combining Proposition 1.4.12 with Corollary 1.1.5, we deduce the following.

Corollary 1.4.13 *Let $\mathcal{C}$ be a collection of subsets of a set E. Then $\mathcal{C}$ is the set of circuits of a matroid on E if and only if $\mathcal{C}$ satisfies (C1), (C2), and (C3)'.* $\square$

The next result establishes that a matroid of fixed rank is uniquely determined by its set of non-spanning circuits, that is, by its circuits that are non-spanning sets. It is important here that the rank of the matroid be specified. For example, for all $n \geq 1$, both $U_{n-1,n}$ and $U_{n,n}$ have no non-spanning circuits.

Proposition 1.4.14 *Let M be a rank-r matroid having $\mathcal{C}'$ as its set of non-spanning circuits. Then*

$$\mathcal{C}(M) = \mathcal{C}' \cup \{X \subseteq E(M) : |X| = r + 1 \text{ and } X \text{ contains no member of } \mathcal{C}'\}.$$

Proof Let X be an $(r+1)$-element subset of $E(M)$. Clearly X is dependent in M. Thus X is a circuit of M if and only if it contains no member of $\mathcal{C}'$.   □

How can we recognize flats in general and hyperplanes in particular in graphic, representable, and uniform matroids? In $U_{m,n}$, the flats consist of all sets with fewer than m elements along with $E(U_{m,n})$ itself. The hyperplanes are thus the sets with exactly $m-1$ elements. If $M = M[A]$ where A is an $m \times n$ matrix over a field $\mathbb{F}$, then X is a flat of M if and only if there is a subspace W of $V(m, \mathbb{F})$ such that X is the set of labels for the multiset consisting of those columns of A that are members of W. Moreover, although the subspace W can be chosen to have dimension $r(X)$, it is clear that, in general, the intersection of a subspace of $V(m, \mathbb{F})$ with the multiset of columns of A need not span the subspace.

If $M = M(G)$ for some graph G, then, by Proposition 1.4.11(ii), a subset X of $E(G)$ is a flat in M if and only if G has no cycles C such that X contains all but exactly one edge of C. The latter holds if and only if, for some positive integer n, there is a partition $\{V_1, V_2, \ldots, V_n\}$ of $V(G)$ such that X is the union of the edge sets of the induced graphs $G[V_1], G[V_2], \ldots, G[V_n]$. We now characterize the hyperplanes of $M(G)$ in terms of their complements in $E(G)$. As we shall see in Chapter 2, hyperplane complements play an important role in matroid theory.

Proposition 1.4.15 *Let G be a graph. Then H is a hyperplane in $M(G)$ if and only if $E(G) - H$ is a minimal set of edges whose removal from G increases the number of connected components.*

Proof Let $E(G) - H$ be a minimal set of edges whose removal from G increases the number of connected components. Then an edge e of G that is not in H joins two distinct components G_1 and G_2 of $G[H]$. Moreover, every other edge of $E(G) - H$ must also join G_1 and G_2. Hence $H \cup e$ spans $M(G)$. Thus H is a maximal non-spanning set of $M(G)$, so, by Proposition 1.4.10(iii), H is a hyperplane. The proof of the converse is similar and is left to the reader.   □

The final result of this section gives an alternative characterization of the rank function r of a matroid in terms of conditions that are more local than (R1)–(R3). We have already noted in Corollary 1.3.4 and Lemmas 1.3.3 and 1.4.4 that r obeys the three conditions below. We leave the proof of the converse to the reader (Exercise 9).

Theorem 1.4.16 *Let E be a set. A function r from 2^E into the set of non-negative integers is the rank function of a matroid on E if and only if it has the following properties:*

(R1)′ $r(\emptyset) = 0$.

(R2)′ *If $X \subseteq E$ and $x \in E$, then $r(X) \leq r(X \cup x) \leq r(X) + 1$.*

(R3)′ *If $X \subseteq E$ and x and y are elements of E such that $r(X \cup x) = r(X \cup y) = r(X)$, then $r(X \cup x \cup y) = r(X)$.*   □

Exercises

1. Let M be a matroid, and r and cl be its rank function and its closure operator. Prove the following:

   (a) If $X \subseteq \mathrm{cl}(Y)$ and $\mathrm{cl}(Y) \subseteq \mathrm{cl}(X)$, then $\mathrm{cl}(X) = \mathrm{cl}(Y)$.

   (b) If $Y \subseteq \mathrm{cl}(X)$, then $\mathrm{cl}(X \cup Y) = \mathrm{cl}(X)$.

   (c) The intersection of all of the flats containing X equals $\mathrm{cl}(X)$.

   (d) $r(X \cup Y) = r(X \cup \mathrm{cl}(Y)) = r(\mathrm{cl}(X) \cup \mathrm{cl}(Y)) = r(\mathrm{cl}(X \cup Y))$.

   (e) If $X \subseteq Y$ and $r(X) = r(Y)$, then $\mathrm{cl}(X) = \mathrm{cl}(Y)$.

2. Show that a subset X of a matroid is a hyperplane if and only if X is a maximal non-spanning set.

3. (a) Find all matroids having a unique basis.

   (b) Find all matroids having a unique spanning set.

4. Find all simple connected graphs G such that, whenever a simple connected graph G_1 has G as a subgraph, $E(G)$ is a flat of $M(G_1)$.

5. The *nullity* of a set X in a matroid M is $|X| - r_M(X)$. Find two sets of axioms that characterize the nullity function of a matroid, one corresponding to Corollary 1.3.4 and the other to Theorem 1.4.16.

6. Prove that (a)–(g) are equivalent for an element e of a matroid M:

   (a) e is in every basis.

   (b) e is in no circuits.

   (c) If $X \subseteq E(M)$ and $e \in \mathrm{cl}(X)$, then $e \in X$.

   (d) $r(E(M) - e) = r(E(M)) - 1$.

   (e) $E(M) - e$ is a flat.

   (f) $E(M) - e$ is a hyperplane.

   (g) If I is an independent set, then so is $I \cup e$.

7. Let X be a subset of a matroid M. Prove that X is a hyperplane if and only if $E(M) - X$ is a minimal set intersecting every basis.

8. Let X and Y be flats of a matroid M such that $Y \subseteq X$ and $r(Y) = r(X) - 1$. Prove that M has a hyperplane H such that $Y = H \cap X$.

9. Prove Theorem 1.4.16.

10. Prove that a collection $\mathcal{S}$ of subsets of a set E is the set of spanning sets of a matroid on E if and only if the following conditions hold:

    (S1) $\mathcal{S}$ *is non-empty.*

    (S2) *If $S_1 \in \mathcal{S}$ and $S_2 \supseteq S_1$, then $S_2 \in \mathcal{S}$.*

    (S3) *If $S_1, S_2 \in \mathcal{S}$ and $|S_1| > |S_2|$, then there is an element e in $S_1 - S_2$ such that $S_1 - e \in \mathcal{S}$.*

11. Let $\mathcal{F}$ be a collection of subsets of a set E. Prove that $\mathcal{F}$ is the set of flats of a matroid on E if and only if the following conditions hold:

    (F1) $E \in \mathcal{F}$.

    (F2) *If $F_1, F_2 \in \mathcal{F}$, then $F_1 \cap F_2 \in \mathcal{F}$.*

(F3) *If $F \in \mathcal{F}$ and $\{F_1, F_2, \ldots, F_k\}$ is the set of minimal members of $\mathcal{F}$ that properly contain F, then the sets $F_1 - F$, $F_2 - F, \ldots, F_k - F$ partition $E - F$.*

12. (Asche 1966) Let $C_1, C_2, \ldots, C_n$ be distinct circuits of a matroid M such that $C_j \nsubseteq \bigcup_{i \neq j} C_i$ for all j in $\{1, 2, \ldots, n\}$. Prove that if $D \subseteq E(M)$ and $|D| < n$, then M has a circuit C such that $C \subseteq (\bigcup_{i=1}^{n} C_i) - D$.

13. (Seymour 1980a) Let $C_1, C_2, \ldots, C_k$ be pairwise disjoint circuits of a matroid M and suppose that $x_i \in C_i$ for all i in $\{1, 2, \ldots, k\}$. Assume also that M has at least one circuit that is not in $\{C_1, C_2, \ldots, C_k\}$.
    (a) Prove that M has a circuit C such that $C \cap \{x_1, x_2, ..., x_k\} = \emptyset$.
    (b) *Prove that M has a circuit D that is not in $\{C_1, C_2, \ldots, C_k\}$ such that, for all i in $\{1, 2, ..., k\}$, either $x_i \in D$, or $D \cap C_i = \emptyset$.

14. *(Brylawski 1973, Greene 1973, Woodall 1974) Let B and B' be bases of a matroid M and $(B_1, B_2, \ldots, B_k)$ be a partition of B. Prove that there is a partition $(B_1', B_2', \ldots, B_k')$ of B' such that $(B - B_i) \cup B_i'$ is a basis for all i in $\{1, 2, ..., k\}$.

1.5 Geometric representations of matroids of small rank

One attractive feature of graphic matroids is that we can determine many of their properties from the pictures of the graphs. In this section, we show that all matroids of small rank have a geometric representation that is similarly useful.

We begin our discussion by introducing another class of matroids. A multiset $\{\underline{v}_1, \underline{v}_2, \ldots, \underline{v}_k\}$, the members of which are in $V(m, \mathbb{F})$, is *affinely dependent* if $k \geq 1$ and there are elements $a_1, a_2, \ldots, a_k$ of $\mathbb{F}$ that are not all zero such that $\sum_{i=1}^{k} a_i \underline{v}_i = \underline{0}$ and $\sum_{i=1}^{k} a_i = 0$. Equivalently, $\{\underline{v}_1, \underline{v}_2, \ldots, \underline{v}_k\}$ is affinely dependent if the multiset $\{(1, \underline{v}_1), (1, \underline{v}_2), \ldots, (1, \underline{v}_k)\}$ is linearly dependent in $V(m+1, \mathbb{F})$, where $(1, \underline{v}_i)$ is the $(m+1)$-tuple of elements of $\mathbb{F}$ whose first entry is 1 and whose remaining entries are the entries of $\underline{v}_i$. A multiset of elements from $V(m, \mathbb{F})$ is *affinely independent* if it is not affinely dependent. Clearly an affinely independent multiset must be a set.

Proposition 1.5.1 *Suppose that E is a set that labels a multiset of elements from $V(m, \mathbb{F})$. Let $\mathcal{I}$ be the collection of subsets X of E such that X labels an affinely independent subset of $V(m, \mathbb{F})$. Then $(E, \mathcal{I})$ is a matroid.*

Proof Suppose that E labels the multiset $\{\underline{v}_1, \underline{v}_2, \ldots, \underline{v}_n\}$. Then, from the second definition of affine dependence, we deduce that $(E, \mathcal{I}) = M[A]$ where A is the $(m+1) \times n$ matrix over $\mathbb{F}$, the ith column of which is $(1, \underline{v}_i)^T$. Alternatively, one can prove this result directly by using the first definition of affine dependence, and we leave this to the reader. □

The matroid $(E, \mathcal{I})$ in the last proposition is called the *affine matroid* on E, and if M is isomorphic to such a matroid, we say that M is affine *over* $\mathbb{F}$.

Example 1.5.2 Let E be the subset $\{(0,0), (1,0), (2,0), (0,1), (0,2), (1,1)\}$ of $V(2, \mathbb{R})$ and consider the affine matroid M on E. The six elements of E can

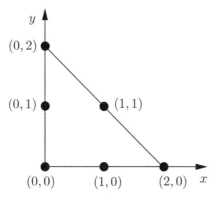

FIG. 1.6. A rank-3 affine matroid.

be represented as points in the Euclidean plane $\mathbb{R}^2$ as in Figure 1.6. It is not difficult to check that the dependent sets of M consist of all subsets of E with four or more elements together with all 3-element subsets of E such that the corresponding three points in Figure 1.6 are collinear.   □

In general, if M is an affine matroid over $\mathbb{R}$ of rank $m+1$ where $m \leq 3$, then a subset X of $E(M)$ is dependent in M if, in the representation of X by points in $\mathbb{R}^m$, there are two identical points, or three collinear points, or four coplanar points, or five points in space. Hence the flats of M of ranks one, two, and three are represented geometrically by points, lines, and planes, respectively. The next example gives a typical geometric representation of such an affine matroid.

Example 1.5.3 Consider the affine matroid M on the subset E of $V(3, \mathbb{R})$ where $E = \{(0,0,0), (1,0,0), (0,1,0), (0,0,1), (1,1,0), (0,1,1)\}$. This matroid has the representation shown in Figure 1.7. From that diagram, we see that the only dependent subsets of E with fewer than five elements are the three 4-point planes $\{(0,0,0),\ (1,0,0),\ (0,1,0),\ (1,1,0)\}$, $\{(0,0,0),\ (0,1,0),\ (0,0,1),\ (0,1,1)\}$, and $\{(1,0,0),\ (1,1,0),\ (0,0,1),\ (0,1,1)\}$.   □

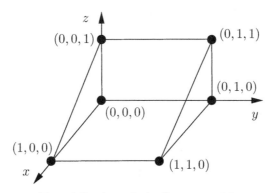

FIG. 1.7. A rank-4 affine matroid.

FIG. 1.8. A geometric representation of a rank-2 matroid.

We now have a geometric way to represent real affine matroids of rank at most four. Next we show how to extend the use of this type of diagram to represent arbitrary matroids of rank at most four.

Example 1.5.4 The matroid M in Example 1.1.2 can be represented by the diagram in Figure 1.8. In such a diagram, we represent a 2-element circuit by two touching points, and we represent a 3-element circuit by a line through the corresponding points. Loops, which cannot occur in an affine matroid, are represented in an inset as shown.   □

In general, such diagrams are governed by the following rules. All loops are marked in a single inset. Parallel elements are represented by touching points, or sometimes by a single point labelled by all the elements in the parallel class. Corresponding to each element that is not a loop and is not in a non-trivial parallel class, there is a distinct point in the diagram that touches no other points. If three elements form a circuit, the corresponding points are collinear. Likewise, if four elements form a circuit, the corresponding points are coplanar. In such a diagram, the lines need not be straight and the planes may be twisted. Moreover, sometimes, to simplify the diagram, certain lines and planes will be listed rather than drawn. At other times, certain lines with fewer than three points on them will be marked as part of the indication of a plane, or as construction lines. We call such a diagram a *geometric representation* for the matroid. The reader is warned that such a representation is not to be confused with the diagram of a graph. Where ambiguity could arise in what follows, we shall always indicate how a particular diagram is to be interpreted.

One needs to be careful not to assume that an arbitrary diagram involving points, lines, and planes is actually a geometric representation for some matroid.

Example 1.5.5 'The Escher matroid' (Brylawski and Kelly 1980). Consider the diagram shown in Figure 1.9 with dependent lines $\{1, 2, 3\}$ and $\{1, 4, 5\}$, and

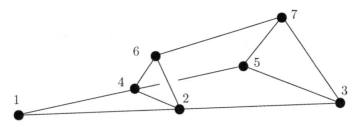

FIG. 1.9. The Escher matroid.

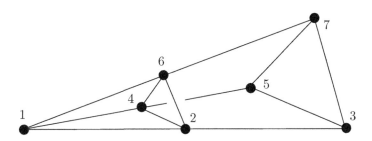

FIG. 1.10. The Escher matroid should have an extra line.

dependent planes $\{1, 2, 3, 4, 5\}$, $\{1, 2, 3, 6, 7\}$, and $\{1, 4, 5, 6, 7\}$. With the rules that govern diagrams being as specified above, this diagram does not represent a matroid on $E = \{1, 2, \ldots, 7\}$. To see this, assume the contrary and let $X = \{1, 2, 3, 6, 7\}$ and $Y = \{1, 4, 5, 6, 7\}$. Then $r(X) = 3 = r(Y)$ and $r(X \cup Y) = 4$. Thus, by (R3), $r(\{1, 6, 7\}) = r(X \cap Y) \leq 2$. But 1, 6, and 7 are distinct non-collinear points, so $r(\{1, 6, 7\}) = 3$; a contradiction. Thus the Escher matroid is not a matroid. However, if we make 1, 6, and 7 collinear as in Figure 1.10, the reader can check that the resulting diagram does represent a rank-4 matroid. □

The next result implies that we get a simple matroid of rank at most three from any set of points in the plane and any set of (possibly curved) lines provided that any two distinct such lines have at most one common point. The straightforward proof is left as an exercise.

Proposition 1.5.6 *Let E be a set and Λ be a collection of subsets of E each having at least three elements such that every two distinct members of Λ meet in at most one element. Let $\mathcal{I}$ be the set of subsets X of E having at most three elements such that no member of Λ contains three elements of X. Then $(E, \mathcal{I})$ is a simple matroid of rank at most three whose rank-one flats are the one-element subsets of E and whose rank-two flats are the members of Λ together with all two-element subsets Y of E for which no member of Λ contains Y. Moreover, every simple matroid of rank at most three arises in this way.* □

Example 1.5.7 The pictures shown in Figure 1.11 are representations of the 7-point and 13-point projective planes, $PG(2, 2)$ and $PG(2, 3)$. Interpreting these pictures as diagrams subject to the above rules, it follows from the last proposition that each represents a rank-3 matroid. The 7-point projective plane is called the *Fano plane*. The corresponding matroid, the *Fano matroid*, will be denoted by F_7 or $PG(2, 2)$. This matroid is of fundamental importance and will occur frequently throughout this book. Indeed, all projective geometries play an important role in matroid theory; we shall discuss this in detail in Chapter 6.

The diagram in Figure 1.12(a) represents the matroid that is obtained from F_7 by deleting the element 7. Note that no line has been drawn through 3 and 6

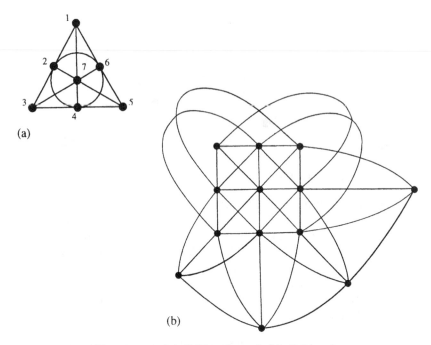

(a)

(b)

FIG. 1.11. (a) $PG(2,2)$ and (b) $PG(2,3)$.

even though $\{3,6\}$ is a rank-2 flat of the matroid. Such 2-point lines are usually omitted from these diagrams, as are 3-point planes.

It is not difficult to check that $F_7\backslash 7 \cong M(K_4)$, where the edges of the graph K_4 are labelled as in Figure 1.12(b). Similarly, $F_7\backslash 2$, for which a geometric representation is shown in Figure 1.13, is also isomorphic to $M(K_4)$. Indeed, as the reader can easily check, $F_7\backslash x \cong M(K_4)$ for all x in $E(F_7)$. This is one of the many attractive features of F_7.   □

Example 1.5.8 Consider the *affine* matroid on $V(3,2)$, the 3-dimensional vector space over $GF(2)$. We denote this affine matroid by $AG(3,2)$. It has eight

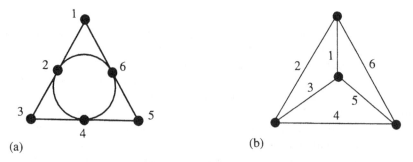

(a)   (b)

FIG. 1.12. (a) $F_7\backslash 7$. (b) A graph whose cycle matroid is $F_7\backslash 7$.

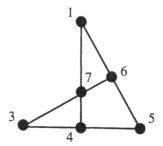

FIG. 1.13. $F_7 \backslash 2$.

elements, corresponding to the eight points in Figure 1.14(a). In addition, it has fourteen 4-point planes, not all of which are marked in Figure 1.14(a). These planes consist of the six faces of the cube, the six diagonal planes such as $\{(0,0,0),(1,0,0),(1,1,1),(0,1,1)\}$, and the two twisted planes, $\{(0,0,0),(1,0,1),(1,1,0),(0,1,1)\}$ and its complement. We call $AG(3,2)$ the *binary affine cube*. Note that if Figure 1.14(a) is viewed as an affine matroid over $\mathbb{R}$ instead of over $GF(2)$, then the resulting matroid R_8 has twelve rather than fourteen planes with the two twisted planes disappearing. We call R_8 the *real affine cube*.

An alternative representation for $AG(3,2)$ can be obtained from the 11-point matroid shown in Figure 1.14(b), which is obtained by 'sticking together' two copies of F_7 along a line. The restriction of this matroid to the set $\{1,2,\ldots,8\}$ is isomorphic to the binary affine cube in Figure 1.14(a). We leave it to the reader to check the details of this. To see the fourteen 4-point planes in the second representation for $AG(3,2)$, first note that $\{1,2,3,4\}$ and $\{5,6,7,8\}$ are two of them. The other twelve break into three groups of four according to whether the corresponding planes in the original 11-point matroid contain a, b, or c. For example, the four such planes containing a arise by taking the union of two lines containing a, one from each copy of F_7, and neither equal to $\{a,b,c\}$. □

We saw in Example 1.5.5 that a diagram involving points, lines, and planes need not correspond to a matroid. Next we state necessary and sufficient conditions for such a diagram to actually be a geometric representation of a simple matroid of rank at most four in which the rank-1, rank-2, and rank-3 flats correspond to the points, lines, and planes in the diagram. These rules are stated just for simple matroids because we already know how to recognize loops and parallel elements in such a diagram. The rules, all of which are very natural from our current geometric perspective, include the following straightforward non-degeneracy conditions: the sets of points, lines, and planes are disjoint; there are no sets of touching points; every line contains at least two points; any two distinct points lie on a line; every plane contains at least three non-collinear points; and any three distinct non-collinear points lie on a plane. For a diagram having at most one plane, as noted in Proposition 1.5.6, there is only one other condition:

1.5.9 *Any two distinct lines meet in at most one point.*

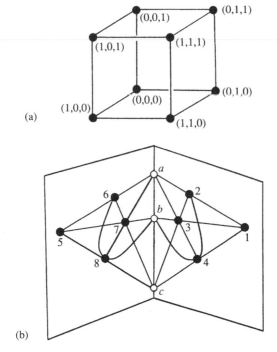

FIG. 1.14. Two geometric representations of $AG(3,2)$.

For a diagram having two or more planes, there are three rules in addition to the non-degeneracy conditions (Mason 1971):

1.5.10 *Any two distinct planes meeting in more than two points do so in a line.*

1.5.11 *Any two distinct lines meeting in a point do so in at most one point and lie on a common plane.*

1.5.12 *Any line not lying on a plane intersects it in at most one point.*

We leave the proofs of these results as exercises. Using geometric representations, the reader should be able to check that there are 17 non-isomorphic matroids on a 4-set, and 35 non-isomorphic matroids on a 5-set.

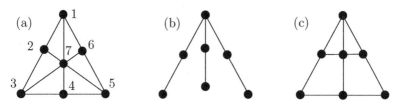

FIG. 1.15. (a) F_7^-. (b) The rank-3 free spike. (c) P_7.

Example 1.5.13 The diagram in Figure 1.15(a) obeys 1.5.9 and so is a geometric representation for a matroid, which we denote by F_7^-. Comparing Figure 1.15(a) with the geometric representation for F_7 in Figure 1.11(a), we see that $\{2, 4, 6\}$ is both a circuit and a hyperplane in F_7, whereas, in F_7^-, this set is a basis. We say that F_7^- has been obtained from F_7 by *relaxing* the circuit–hyperplane $\{2, 4, 6\}$. This operation can be performed more generally. □

Proposition 1.5.14 *Let M be a matroid having a subset X that is both a circuit and a hyperplane. Let $\mathcal{B}' = \mathcal{B}(M) \cup \{X\}$. Then $\mathcal{B}'$ is the set of bases of a matroid M' on $E(M)$. Moreover,*

$$\mathcal{C}(M') = (\mathcal{C}(M) - \{X\}) \cup \{X \cup e : e \in E(M) - X\}.$$

Proof We leave this as an exercise. □

The matroid M' in the last result is called a *relaxation* of M. The matroid F_7^- in Figure 1.15(a), which is a relaxation of the Fano matroid F_7, is called the *non-Fano matroid*. As the reader can check, each matroid that is obtained by relaxing one of the seven circuit–hyperplanes of F_7 is isomorphic to F_7^-.

Example 1.5.15 The diagrams in Figure 1.16 obey 1.5.9 and are therefore geometric representations for rank-3 matroids. We call these matroids the *Pappus* and *non-Pappus matroids*, respectively, because of their relationship to the well-known Pappus configuration in projective geometry. As we shall see in Proposition 6.1.10, the non-Pappus matroid is not representable over any field. A smallest matroid with this property can be obtained from $AG(3, 2)$ by relaxing a circuit–hyperplane. The proof that this matroid is non-representable will be delayed until Example 6.4.11. □

Example 1.5.16 Figure 1.17 contains another familiar object from projective geometry, the 3-dimensional Desargues configuration. One can check that the points, lines, and planes of this diagram obey 1.5.10–1.5.12, so this diagram is indeed a geometric representation for a 10-element rank-4 matroid. Alternatively, one can show that this diagram is the geometric representation for $M(K_5)$ with the edges of K_5 being labelled as in Figure 1.5. □

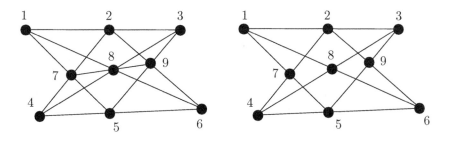

FIG. 1.16. The Pappus and non-Pappus matroids.

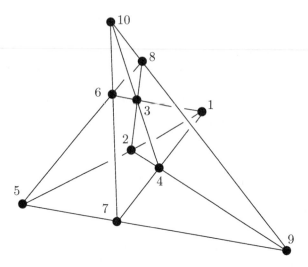

FIG. 1.17. $M(K_5)$.

In an arbitrary matroid, the terms *point, line,* and *plane* are often used to refer to flats of ranks one, two, and three, respectively. A line is *non-trivial* if it contains at least three points.

All three matroids shown in Figure 1.15 and the Fano matroid, F_7, are rank-3 members of an important family of matroids called spikes. These matroids will appear often throughout the book and we now define them. Let $E = \{t, x_1, y_1, x_2, y_2, \ldots, x_r, y_r\}$ for some $r \geq 3$. Let $\mathcal{C}_1 = \{\{t, x_i, y_i\} : 1 \leq i \leq r\}$ and $\mathcal{C}_2 = \{\{x_i, y_i, x_j, y_j\} : 1 \leq i < j \leq r\}$. The set of circuits of every spike on E includes $\mathcal{C}_1 \cup \mathcal{C}_2$. Let $\mathcal{C}_3$ be a, possibly empty, subset of $\{\{z_1, z_2, \ldots, z_r\} : z_i \text{ is in } \{x_i, y_i\} \text{ for all } i\}$ such that no two members of $\mathcal{C}_3$ have more than $r - 2$ common elements. Finally, let $\mathcal{C}_4$ be the collection of all $(r + 1)$-element subsets of E that contain no member of $\mathcal{C}_1 \cup \mathcal{C}_2 \cup \mathcal{C}_3$.

Proposition 1.5.17 *There is a rank-r matroid M on E whose collection $\mathcal{C}$ of circuits is $\mathcal{C}_1 \cup \mathcal{C}_2 \cup \mathcal{C}_3 \cup \mathcal{C}_4$.*

Proof Evidently $\emptyset \notin \mathcal{C}$, and no member of $\mathcal{C}$ is a proper subset of another. If $\mathcal{C}$ is indeed the collection of circuits of a matroid M on E, then the definition of $\mathcal{C}_4$ ensures that $r(M) \leq r$. Moreover, by the definition of $\mathcal{C}_3$, at most one of $\{y_1, y_2, \ldots, y_{r-1}, y_r\}$ and $\{y_1, y_2, \ldots, y_{r-1}, x_r\}$ is in $\mathcal{C}$, so at least one of these sets is independent. Hence $r(M) \geq r$ and therefore $r(M) = r$. It remains to check that $\mathcal{C}$ satisfies (C3).

Suppose that C_1 and C_2 are distinct members of $\mathcal{C}$ and $e \in C_1 \cap C_2$. It is straightforward to check that (C3) holds when both C_1 and C_2 are in $\mathcal{C}_1 \cup \mathcal{C}_2$. Hence we may suppose that $|C_1| \leq |C_2|$ and $C_2 \in \mathcal{C}_3 \cup \mathcal{C}_4$. Then

$$|(C_1 \cup C_2) - e| = |C_1 \cup C_2| - 1$$
$$= |C_1| + |C_2| - |C_1 \cap C_2| - 1$$
$$= |C_1 - C_2| + (|C_2| - 1).$$

Since $|C_2|$ is r or $r+1$ and $C_1 \not\subseteq C_2$, we deduce that either $|(C_1 \cup C_2) - e| \geq r+1$, or $C_2 \in \mathcal{C}_3$ and $|C_1 - C_2| = 1$. In the first case, $(C_1 \cup C_2) - e$ must contain a member of $\mathcal{C}$. In the second case, the definitions of $\mathcal{C}_1, \mathcal{C}_2$, and $\mathcal{C}_3$ ensure that $C_1 \in \mathcal{C}_4$. But then $|C_1| > |C_2|$; a contradiction. $\square$

The matroid M whose existence we have just proved is called a *rank-r spike* with *tip* t and *legs* $L_1, L_2, \ldots, L_r$ where $L_i = \{t, x_i, y_i\}$ for all i. We shall also call M an *r-spike*, a *tipped spike*, or a *tipped r-spike*. The name 'spike' derives from the fact that these matroids can be characterized geometrically as in the next result, the proof of which is left as an exercise.

Proposition 1.5.18 *Let r be an integer exceeding two. A matroid M is a rank-r spike with tip t if and only if M has the following properties:*

(i) *$E(M)$ is the union of r lines $L_1, L_2, \ldots, L_r$ each of which is a 3-element circuit containing the point t;*

(ii) *for all k in $\{1, 2, \ldots, r-1\}$, the union of any k of $L_1, L_2, \ldots, L_r$ has rank $k+1$; and*

(iii) *$r(L_1 \cup L_2 \cup \ldots \cup L_r) = r$.* $\square$

In the construction of a spike, if $\mathcal{C}_3$ is empty, the corresponding spike is called the rank-r *free spike with tip*. In an arbitrary spike M, each circuit in $\mathcal{C}_3$ is also a hyperplane of M. Evidently, when such a circuit–hyperplane is relaxed, we obtain another spike. Repeating this procedure until all of the circuit–hyperplanes in $\mathcal{C}_3$ have been relaxed will produce the free spike.

In the free spike, we can interchange the labels on the elements x_i and y_i without changing the matroid. We call two such elements *clones*. Formally, distinct elements e and f of a matroid M are *clones* if the map that interchanges e and f but fixes every other element of $E(M)$ is an isomorphism from M to M. Thus, for example, two parallel elements in a matroid are clones, as are two loops. Hence, in the matroid shown in Figure 1.8, the elements 1 and 4 are clones; so too are 2 and 5. However, despite the symmetry of each matroid, none of $M(K_4)$, $AG(3, 2)$, or $M(K_5)$ has any clones.

As we can see from Figures 1.11(a) and 1.15, both the Fano matroid, F_7, and the non-Fano matroid, F_7^-, are rank-3 spikes with tip 1. In fact, the reader can easily check that any one of the seven elements of F_7 can be taken as the tip, while any one of $1, 3, 5$, and 7 can be taken as the tip of F_7^-. In Figure 1.14(b), a geometric representation of $AG(3, 2)$ is shown together with three extra points, a, b, and c. If we adjoin any one of a, b, and c to $AG(3, 2)$, the resulting matroid is a 4-spike whose tip is the added point.

Parts (b) and (c) of Figure 1.15 show the rank-3 free spike and another rank-3 spike, P_7. Unlike F_7 and F_7^-, both of these spikes, along with all spikes of rank at least four, have a unique element that can be viewed as the tip.

We often consider spikes from which the tip has been deleted. Such a matroid, which has $2r$ elements and rank r, is called a *tipless spike*. Thus, both $AG(3,2)$ and $M(K_4)$ are tipless spikes. Moreover, $U_{3,6}$ is an example of *tipless free spike* since it can be obtained by deleting the tip from a free spike. We noted above that, in Figure 1.14(b), any one of a, b, and c could be added to $AG(3,2)$ to become the tip of a 4-spike. This ambiguity arises because the circuits in $\mathcal{C}_2$ and $\mathcal{C}_3$ have the same cardinality in a 4-spike. For a tipless spike M of rank at least five, there is a unique spike N with tip t such that $N\backslash t = M$. We remark here that, when the term 'spike' is used in the literature, it sometimes means a tipped spike and sometimes a tipless spike. In this book, we adopt the convention that a 'spike' is a tipped spike.

Exercises

1. Show that a non-empty multiset $\{\underline{v}_1, \underline{v}_2, \ldots, \underline{v}_k\}$ of vectors from $V(m, \mathbb{F})$ is affinely dependent if and only if $\{\underline{v}_1 - \underline{v}_k, \underline{v}_2 - \underline{v}_k, \ldots, \underline{v}_{k-1} - \underline{v}_k\}$ is linearly dependent.
2. Show that if e and f are clones in a matroid, and f and g are clones, then e and g are clones.
3. Determine which of conditions 1.5.10–1.5.12 is violated in Figure 1.9.
4. Prove Proposition 1.5.1 by using the first definition of affine dependence.
5. Prove Proposition 1.5.6.
6. Let D be a diagram involving points, lines, and planes and satisfying 1.5.10–1.5.12 as well as the non-degeneracy conditions. Prove that there is a simple matroid of rank at most four on the set of points of D that has as its rank-1, rank-2, and rank-3 flats the points, lines, and planes, respectively, of D.
7. Prove that a matroid is uniform if and only if every two elements are clones.
8. Show that, in a simple matroid, an element that is on at least two non-trivial lines cannot have a clone.
9. Show that no tipped 3-spike is graphic and that there is exactly one tipless 3-spike that is graphic.
10. Prove that an affine matroid over $GF(2)$ has no circuits with an odd number of elements.
11. Prove that every relaxation of F_7 is isomorphic to F_7^-.
12. Let M be a matroid and X be a circuit–hyperplane of M. Let M' be the matroid obtained from M by relaxing X. Find, in terms of M, the independent sets, the rank function, the hyperplanes, and the flats of M'.
13. (a) Find the number of non-isomorphic 3-spikes.
    (b) Determine how many of these spikes can be derived from the Fano matroid by a sequence of circuit–hyperplane relaxations.
    (c) Find the number of non-isomorphic rank-3 tipless spikes.
14. Prove that the following statements are equivalent for a rank-r matroid M.
    (a) M is a relaxation of some matroid.

(b) M has a basis B such that $C(e, B) = B \cup e$ for every e in $E(M) - B$ and neither B nor $E(M) - B$ is empty.

(c) M has a non-empty basis B such that $B \neq E(M)$ and every $(r-1)$-element subset of B is a flat.

15. Geometric representations for graphic matroids M_1 and M_2 are shown in Figure 1.18. For each i, find a graph G_i such that $M(G_i) \cong M_i$ and verify the answer.

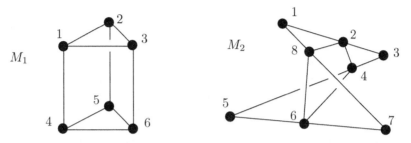

FIG. 1.18. Geometric representations of two graphic matroids.

16. Find geometric representations for the vector matroids of the following matrices where I_r is the $r \times r$ identity matrix, and each matrix is over $GF(3)$ except the first which is over $GF(2)$.

$$A_1 = \begin{bmatrix} I_4 & \begin{array}{ccccc} 1 & 0 & 0 & 1 & 1 \\ 1 & 1 & 0 & 0 & 1 \\ 0 & 1 & 1 & 0 & 1 \\ 0 & 0 & 1 & 1 & 1 \end{array} \end{bmatrix} \qquad A_2 = \begin{bmatrix} I_4 & \begin{array}{cccc} 1 & 0 & 0 & 1 \\ 1 & 1 & 0 & 0 \\ 0 & 1 & 1 & 0 \\ 0 & 0 & 1 & 1 \end{array} \end{bmatrix}$$

$$A_3 = \begin{bmatrix} I_4 & \begin{array}{cccc} 1 & 0 & 0 & 1 \\ 1 & 1 & 0 & 0 \\ 0 & 1 & 1 & 0 \\ 0 & 0 & 1 & -1 \end{array} \end{bmatrix} \qquad A_4 = \begin{bmatrix} I_3 & \begin{array}{cccc} 0 & 1 & 1 & 1 \\ 1 & 0 & 1 & -1 \\ 1 & 1 & 0 & -1 \end{array} \end{bmatrix}$$

17. Let J_r be the $r \times r$ matrix with every entry equal to 1, and let $\mathbf{1}$ be the column vector of r ones.

(a) Show that, for all $r \geq 3$ and all fields $\mathbb{F}$, the matrix $[I_r | J_r - I_r | \mathbf{1}]$ is an $\mathbb{F}$-representation of a spike $M_r(\mathbb{F})$.

(b) Show that $M_4(GF(2))$, $M_4(GF(3))$, and $M_4(\mathbb{R})$ are all non-isomorphic but that $M_3(GF(3)) = M_3(\mathbb{R}) \cong F_7^-$.

1.6   Transversal matroids

So far, we have considered three important classes of matroids: graphic, representable, and uniform matroids. In this section, we consider a fourth such class, that of transversal matroids. This class includes not only all uniform matroids but also all partition matroids.

For a finite set S, a *family* of subsets of S is a finite sequence $(A_1, A_2, \ldots, A_m)$ such that $A_j \subseteq S$ for all j in $\{1, 2, \ldots, m\}$. Note that the terms of this sequence, the *members* of the family, need not be distinct. If $J = \{1, 2, \ldots, m\}$, we shall frequently abbreviate $(A_1, A_2, \ldots, A_m)$ as $(A_j : j \in J)$. A *transversal* or *system of distinct representatives* of $(A_1, A_2, \ldots, A_m)$ is a subset $\{e_1, e_2, \ldots, e_m\}$ of S such that $e_j \in A_j$ for all j in J, and $e_1, e_2, \ldots, e_m$ are distinct. Equivalently, T is a transversal of $(A_j : j \in J)$ if there is a bijection $\psi : J \to T$ such that $\psi(j) \in A_j$ for all j in J. If $X \subseteq S$, then X is a *partial transversal* of $(A_j : j \in J)$ if X is a transversal of $(A_j : j \in K)$ for some subset K of J.

In the special case that $(A_1, A_2, \ldots, A_m)$ is a partition π of S, the set of partial transversals of $\mathcal{A}$ coincides with the set of independent sets of the partition matroid M_π. The main result of this section is that, for all families $\mathcal{A}$ of subsets of S, the set of partial transversals of $\mathcal{A}$ is the set of independent sets of a matroid on S.

Another way to view partial transversals uses the idea of a matching in a bipartite graph. If $\mathcal{A}$ is a family $(A_1, A_2, \ldots, A_m)$ of subsets of a set S and $J = \{1, 2, \ldots, m\}$, then the *bipartite graph* $\Delta[\mathcal{A}]$ *associated with* $\mathcal{A}$ has vertex set $S \cup J$; its edge set is $\{xj : x \in S, \ j \in J, \text{ and } x \in A_j\}$. A *matching* in a graph is a set of edges in the graph no two of which have a common endpoint. It is not difficult to see that a subset X of S is a partial transversal of $\mathcal{A}$ if and only if there is a matching in $\Delta[\mathcal{A}]$ in which every edge has one endpoint in X. When such a matching exists, we say that X is matched *into* J. Moreover, if J_X consists of those vertices of J that meet an edge of the matching, we say that X is matched *to* or *onto* J_X.

Example 1.6.1 Let $S = \{x_1, x_2, \ldots, x_6\}$ and $\mathcal{A} = (A_1, A_2, A_3, A_4)$ where $A_1 = \{x_1, x_2, x_6\}$, $A_2 = \{x_3, x_4, x_5, x_6\}$, $A_3 = \{x_2, x_3\}$, and $A_4 = \{x_2, x_4, x_6\}$. Then the bipartite graph $\Delta[\mathcal{A}]$ is as shown in Figure 1.19.

The set $\{x_1, x_2, x_3, x_4\}$ is a transversal of $\mathcal{A}$. To check this, one needs only check that $\{x_1 1, x_4 2, x_3 3, x_2 4\}$ is a matching in $\Delta[\mathcal{A}]$. Similarly, as $\{x_6 1, x_2 3, x_4 2\}$ is a matching in $\Delta[\mathcal{A}]$, the set $\{x_6, x_2, x_4\}$ is a partial transversal of $\mathcal{A}$. Clearly $\mathcal{A}$ has many other partial transversals. $\square$

In solving scheduling or timetabling problems, one is led naturally to the consideration of partial transversals of families of sets. As we shall see in Section 1.8, the next result of Edmonds and Fulkerson (1965) provides a useful tool in the study of such problems.

Theorem 1.6.2 *Let $\mathcal{A}$ be a family $(A_1, A_2, \ldots, A_m)$ of subsets of a set S. Let $\mathcal{I}$ be the set of partial transversals of $\mathcal{A}$. Then $\mathcal{I}$ is the collection of independent sets of a matroid on S.*

Proof The empty set is a transversal of the empty subfamily of $\mathcal{A}$, so (I1) holds. Moreover, if I is a partial transversal of $\mathcal{A}$ and $I' \subseteq I$, then I' is also a partial transversal of $\mathcal{A}$. Thus (I2) holds.

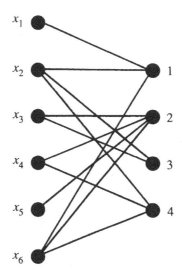

FIG. 1.19. The bipartite graph associated with a family of sets.

Now, to prove that $\mathcal{I}$ satisfies (I3), suppose that I_1 and I_2 are partial transversals of $\mathcal{A}$ such that $|I_1| < |I_2|$. Then, in $\Delta[\mathcal{A}]$, there are matchings W_1 and W_2 that match I_1 and I_2, respectively, into J. Colour the edges of $W_1 - W_2$, $W_2 - W_1$, and $W_1 \cap W_2$ red, blue, and purple, respectively, and let W be the subgraph of $\Delta[\mathcal{A}]$ induced by those edges that are red or blue. Since $|I_1| = |W_1|$ and $|I_2| = |W_2|$, there are more blue edges than red in W.

As W_1 and W_2 are both matchings, every vertex in W has degree one or two. It is routine to check (see Exercise 2) that every connected component of W is a cycle or a path. Moreover, as W is bipartite, every cycle is even. As no two like-coloured edges meet at a vertex, there are equal numbers of red and blue edges in every cycle and in every even path of W. Since W has more blue edges than red, it must have as a component an odd path, P, whose first and last edges are blue. Let the vertices of P, in order, be $v_1, v_2, \ldots, v_{2k}$. Clearly one of v_1 and v_{2k} is in S and the other is in J. Assume, without loss of generality, that $v_1 \in S$. Then, since v_1 meets a blue but no red edge, $v_1 \in I_2 - I_1$. Moreover, $\{v_2, v_4, \ldots, v_{2k}\} \subseteq J$ and $\{v_3, v_5, \ldots, v_{2k-1}\} \subseteq I_1 \cap I_2$. Now interchange the colours on the red and blue edges of P leaving the rest of $\Delta[\mathcal{A}]$ unchanged. In the recoloured graph, there is now one more red edge than before. Indeed, every vertex in $I_1 \cup v_1$ is the endpoint of a red or a purple edge. Moreover, this set of red and purple edges forms a matching. We conclude that $I_1 \cup v_1$ is a partial transversal of $\mathcal{A}$. Hence (I3) holds and $\mathcal{I}$ is indeed the set of independent sets of a matroid. □

The matroid obtained above from the set of partial transversals of $\mathcal{A}$ will be denoted by $M[\mathcal{A}]$. Its ground set was denoted above by S to avoid the potential confusion of using E for a set of vertices in a graph. If M is an arbitrary matroid and $M \cong M[\mathcal{A}]$ for some family $\mathcal{A}$ of sets, then we shall call M a *transversal matroid* and $\mathcal{A}$ a *presentation* of M. It is not difficult to check that all uniform matroids are transversal. Moreover, as we shall see in Chapter 11, every transversal matroid is representable over all sufficiently large fields. In the next example, we exhibit a graphic transversal matroid and a graphic non-transversal matroid. A characterization of precisely which graphic matroids are transversal will be presented in Chapter 10.

Example 1.6.3 Let G_1 and G_2 be the graphs shown in Figure 1.20. Let $A_1 = \{1, 2, 7\}$, $A_2 = \{3, 4, 7\}$, and $A_3 = \{5, 6, 7\}$. Then, for $\mathcal{A} = (A_1, A_2, A_3)$ and $S = \{1, 2, \ldots, 7\}$, it is routine to check that $M[\mathcal{A}] = M(G_1)$.

In contrast, $M(G_2)$ is not transversal. To show this, assume that $M(G_2) = M[\mathcal{A}']$ for some family $\mathcal{A}'$ of subsets of $\{1, 2, \ldots, 6\}$. As $\{1\}$ and $\{2\}$ are independent but $\{1, 2\}$ is dependent, there is a unique member, say A_1', of $\mathcal{A}'$ meeting $\{1, 2\}$. Moreover, A_1' contains both 1 and 2. Similarly, $\mathcal{A}'$ has a unique member A_2' meeting $\{3, 4\}$ and a unique member A_3' meeting $\{5, 6\}$, and these members contain $\{3, 4\}$ and $\{5, 6\}$, respectively. As $\{1, 3\}$, $\{1, 5\}$, and $\{3, 5\}$ must be partial transversals of $\mathcal{A}'$, the sets A_1', A_2', and A_3' are distinct. This implies that $\{1, 3, 5\}$ is a partial transversal of $\mathcal{A}'$; a contradiction. We conclude that $M(G_2)$ is indeed non-transversal.   □

It will sometimes be convenient to view transversal matroids as coming directly from bipartite graphs. It follows from our earlier discussion that if Δ is a bipartite graph with vertex classes S and J, then the set of subsets X of S that are matched into J is precisely the set of independent sets of a transversal matroid. Indeed, this matroid is $M[\mathcal{A}]$ where $\mathcal{A}$ is *the family $(A_j : j \in J)$ of sets associated with* Δ, that is, $A_j = \{x \in S : xj \in E(\Delta)\}$ for all j in J.

Exercises

1. Show the following:
   (a) All uniform matroids are transversal.
   (b) A transversal matroid need not be paving.

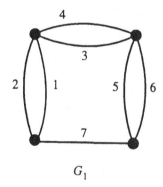
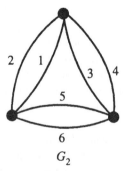

FIG. 1.20. $M(G_1)$ is transversal; $M(G_2)$ is not.

(c) A paving matroid need not be transversal.

2. Prove the following for a graph G.

    (a) If every vertex of G has degree at most two, then G is a disjoint union of paths and cycles.

    (b) If every vertex of G has degree at least two, then G has a cycle.

    (c) If every vertex of G has degree exactly two, then G is a union of vertex-disjoint cycles.

3. Let $S = \{1, 2, \ldots, 6\}$ and $\mathcal{A} = (A_1, A_2, A_3)$ where $A_1 = \{1, 2, 3\}$, $A_2 = \{2, 3, 4\}$, and $A_3 = \{4, 5, 6\}$.

    (a) Find $\Delta[\mathcal{A}]$.

    (b) Give a geometric representation for $M[\mathcal{A}]$.

4. Characterize the circuits of $M[\mathcal{A}]$ in terms of the bipartite graph $\Delta[\mathcal{A}]$.

5. Find a restriction of F_7 that is non-transversal and is minimal with this property.

6. Give a family $\mathcal{A}$ of subsets of $\{1, 2, \ldots, 7\}$ such that Figure 1.21 is a geometric representation for $M[\mathcal{A}]$.

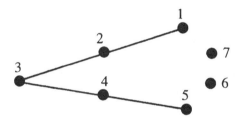

FIG. 1.21. A geometric representation of a transversal matroid.

7. Let M be a rank-r spike for some $r \geq 3$. By trying to construct a presentation for M, prove that it is not transversal.

8. For each integer n exceeding one, give an example of two different families $\mathcal{A}_1$ and $\mathcal{A}_2$ of distinct non-empty subsets of $\{1, 2, \ldots, n\}$ such that $M[\mathcal{A}_1] = M[\mathcal{A}_2]$, and $\mathcal{A}_2$ cannot be obtained by permuting the members of $\mathcal{A}_1$.

9. Let $\mathcal{A} = (A_j : j \in J)$ be a family of subsets of a set S. Prove that $S - A_j$ is a flat of $M[\mathcal{A}]$ for all j in J.

10. Let Δ be a bipartite graph with vertex classes S and J. Show that:

    (a) A matching in Δ need not be contained in a maximum-sized matching.

    (b) The set of edge sets of matchings in Δ need not be the set of independent sets of a matroid on $E(\Delta)$.

    (c) When $\mathcal{A}$ is the family of sets associated with Δ, every partial transversal of $\mathcal{A}$ is contained in a maximum-sized partial transversal of $\mathcal{A}$.

11. Show that both the matroids for which geometric representations are shown in Figure 1.22 are ternary, non-binary, non-graphic, and non-transversal.

12. (Hoffman and Kuhn 1956) Let $\mathcal{A}$ be a family of subsets of a set S and suppose $X \subseteq S$. Prove that $\mathcal{A}$ has a transversal containing X if and only if $\mathcal{A}$ has some transversal, and X is a partial transversal of $\mathcal{A}$.

13. *(Welsh 1969a) A presentation $(A_1, A_2, \ldots, A_m)$ of a transversal matroid is *nested* if $A_1 \subseteq A_2 \subseteq \ldots \subseteq A_m$. Prove that, for a set S, there are exactly $2^{|S|}$ non-isomorphic transversal matroids on S having nested presentations.

1.7 The lattice of flats

The set of flats of a matroid, ordered by inclusion, has a special structure which will be characterized in this section. This approach to matroids has proved attractive to a number of authors (see, for example, Birkhoff 1935, Dilworth 1944, and Crapo and Rota 1970). It will not be our primary focus here, although we shall find it indispensable in certain contexts. After seeing the notion of the simple matroid associated with a given matroid, the reader unfamiliar with partially ordered sets may wish to skim the rest of this section, delaying detailed consideration of it until its techniques are first used in Chapter 6. This material has been included in this chapter both for reference and because of its historical importance in the development of matroid theory.

One feature that will emerge from our study of the order structure of flats is that this structure is unaffected by the presence of loops and parallel elements. We begin by looking at an elementary construction that removes loops and parallel elements from a matroid. This generalizes a graph-theoretic construction. The *simple graph $\widetilde{G}$ associated with a* given *graph G* is obtained from G by deleting all loops and, for each parallel class of edges, identifying all the edges of the class as a single edge. An example of this construction is given in Figure 1.23. Before giving a precise description of the corresponding matroid construction, we make the following elementary observations for an arbitrary matroid M. For the first of these, see Exercise 2 of Section 1.2.

1.7.1 *An element e is a loop of M if and only if e is in no bases of M.*

1.7.2 *If e_1 is not a loop of M, then e_2 is parallel with e_1 if and only if, for every basis B containing e_1, the set $(B - e_1) \cup e_2$ is also a basis.*

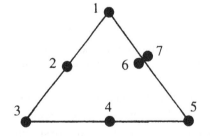 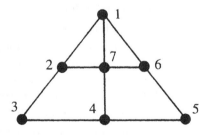

FIG. 1.22. Geometric representations of two non-transversal matroids.

We deduce that no basis intersects the set of loops of M. Now let X be a non-trivial parallel class of M and, for each x in X, let $\mathcal{B}_x$ denote the set of bases of M containing x. If x and y are distinct members of X, then $\mathcal{B}_x \cap \mathcal{B}_y = \emptyset$, but $\{B - x : B \in \mathcal{B}_x\} = \{B - y : B \in \mathcal{B}_y\}$. Thus, if we delete all the loops from M and then, in each non-trivial parallel class X, we distinguish one element and delete all the other elements of X, the matroid we obtain is uniquely determined up to a renaming of the distinguished elements. We denote this matroid by $\mathrm{si}(M)$ and call it the *simple matroid associated with M* or the *simplification of M*. Formally, the ground set of $\mathrm{si}(M)$ is the set of all parallel classes of M, while a subset $\{X_1, X_2, \ldots, X_k\}$ of these parallel classes is independent in $\mathrm{si}(M)$ if and only if $r_M(X_1 \cup X_2 \cup \ldots \cup X_k) = k$. Evidently, for a graph G whose associated simple graph is $\widetilde{G}$, we have $\mathrm{si}(M(G)) = M(\widetilde{G})$.

An equivalent way to view $\mathrm{si}(M)$ is to take its ground set to be the set of rank-1 flats of M with a set $\{Y_1, Y_2, \ldots, Y_k\}$ of these flats being independent in $\mathrm{si}(M)$ if and only if $r_M(Y_1 \cup Y_2 \cup \ldots \cup Y_k) = k$. We remark that $\mathrm{si}(M)$ has been denoted in a variety of ways elsewhere, including by $\widetilde{M}$ in the first edition of this book. The new notation used here reflects current usage in the literature.

We now examine more closely the structure of the set of flats of a matroid. A *partially ordered set* or *poset* is a (possibly infinite) set P together with a *partial order*, a binary relation, $\leq$, that is reflexive, antisymmetric, and transitive. In other words, for all x, y, and z in P, the following hold:

(P1) $x \leq x$.

(P2) *If $x \leq y$ and $y \leq x$, then $x = y$.*

(P3) *If $x \leq y$ and $y \leq z$, then $x \leq z$.*

Posets abound in mathematics. As examples, we may take the real numbers with the usual less-than-or-equal-to relation; the subsets of a set with the relation of set inclusion; and the set D_n of positive integral divisors of a positive integer n under the relation $x \leq y$ if x divides y.

In a partially ordered set P, if $x \leq y$, we shall sometimes write $y \geq x$. If $x \leq y$ but $x \neq y$, we write $x < y$ or $y > x$. If $x < y$ but there is no element z of P such that $x < z < y$, then we say that y *covers* x in P. We can represent P by a *Hasse diagram*. Such a diagram is a simple graph, the vertices of which correspond to elements of P. In this graph, if $x > y$, then the vertex corresponding to x is

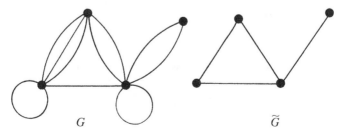

G $\qquad\qquad\qquad\qquad\qquad$ $\widetilde{G}$

FIG. 1.23. A graph and its associated simple graph.

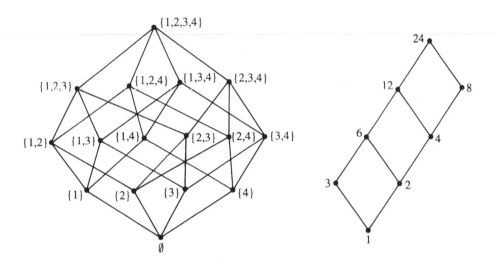

FIG. 1.24. Two Hasse diagrams.

placed higher on the page than that corresponding to y. Two vertices x and y are joined by an edge, a straight line segment, whenever x covers y. Figure 1.24 shows two Hasse diagrams, one corresponding to the set of subsets of $\{1, 2, 3, 4\}$ ordered by inclusion; and the other corresponding to D_{24}, the set of positive integral divisors of 24 under the divisibility order described earlier.

The Hasse diagram in Figure 1.24 represents both the partially ordered set D_{30} of positive integral divisors of 30, and the set of subsets of $\{1, 2, 3\}$ ordered by inclusion, these posets being isomorphic. In general, posets P_1 and P_2 are *isomorphic* if there is a bijection $\psi : P_1 \to P_2$ such that $x \le y$ if and only if $\psi(x) \le \psi(y)$.

The posets represented in Figures 1.24 and 1.25 are all examples of lattices. A *lattice* is a poset $\mathcal{L}$ such that, for every pair of elements, the least upper bound

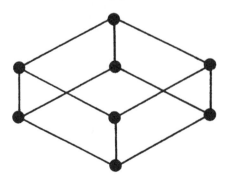

FIG. 1.25. D_{30} and the lattice of subsets of $\{1, 2, 3\}$.

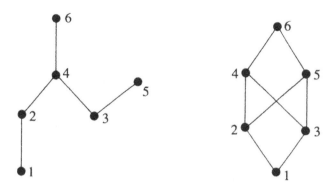

FIG. 1.26. Two posets that are not lattices.

and greatest lower bound of the pair exists. Formally, if x and y are arbitrary elements of $\mathcal{L}$, then $\mathcal{L}$ contains elements $x \vee y$ and $x \wedge y$, the *join* and *meet* of x and y, such that

(L1) $x \vee y \geq x$ and $x \vee y \geq y$; and if $z \geq x$ and $z \geq y$, then $z \geq x \vee y$; and
(L2) $x \wedge y \leq x$ and $x \wedge y \leq y$; and if $z \leq x$ and $z \leq y$, then $z \leq x \wedge y$.

It is routine to check that both of the operations $\vee$ and $\wedge$ are commutative and associative.

The Hasse diagrams shown in Figure 1.26 represent posets that are not lattices. In the first, $2 \wedge 3$ does not exist; in the second, $2 \vee 3$ does not exist.

For a matroid M, we denote by $\mathcal{L}(M)$ the set of flats of M ordered by inclusion. Evidently $\mathcal{L}(M)$ is a partially ordered set.

Lemma 1.7.3 *For all matroids M, the poset $\mathcal{L}(M)$ is a lattice and, for all flats X and Y of M,*

$$X \wedge Y = X \cap Y \quad and \quad X \vee Y = \mathrm{cl}(X \cup Y).$$

Proof Clearly, if X and Y are flats of M, so is $X \cap Y$. To see this, assume to the contrary that x is in $\mathrm{cl}(X \cap Y) - (X \cap Y)$. Then, by Proposition 1.4.11(ii), there is a circuit C such that $x \in C \subseteq (X \cap Y) \cup x$. Then $x \in \mathrm{cl}(X) \cap \mathrm{cl}(Y) = X \cap Y$; a contradiction. We conclude that the meet of X and Y is $X \cap Y$. It is easy to see that the join of X and Y is $\mathrm{cl}(X \cup Y)$. Although $X \cup Y$ need not be closed, $\mathrm{cl}(X \cup Y)$ certainly must be. $\square$

We shall show next that $\mathcal{L}(M)$ is, in fact, a rather special type of lattice. All of the Hasse diagrams shown in Figure 1.27 represent lattices, but none is isomorphic to the lattice of flats of a matroid. To characterize matroid lattices, we shall require some more terminology. Let P be a finite partially ordered set. A *chain* in P *from x_0 to x_n* is a subset $\{x_0, x_1, \ldots, x_n\}$ of P such that $x_0 < x_1 < \ldots < x_n$. The *length* of such a chain is n, and the chain is *maximal* if x_i covers x_{i-1} for all i in $\{1, 2, \ldots, n\}$. If, for every pair $\{a, b\}$ of elements of P with $a < b$, all maximal chains from a to b have the same length, then P is said to satisfy the *Jordan–Dedekind chain condition*. Thus, of the three lattices represented in Figure 1.27, only (b) fails to satisfy this condition.

If the poset P has an element z such that $z \leq x$ for all x in P, then we call z the *zero* of P and denote it by 0. Similarly, if P has an element w such that $w \geq x$ for all x in P, then w is called the *one* of P. The first poset in Figure 1.26 has neither a zero nor a one; the second has both.

Now suppose that P is a partially ordered set having a zero. An element x is called an *atom* of P if x covers 0. The *height* $h(y)$ of an element y of P is the maximum length of a chain from 0 to y. Thus, in particular, the atoms of P are precisely the elements of height one.

It is not difficult to check that every finite lattice has a zero and a one. In particular, for a matroid M, the zero of $\mathcal{L}(M)$ is $\mathrm{cl}(\emptyset)$, while the one is $E(M)$. A finite lattice $\mathcal{L}$ is called *semimodular* if it satisfies the Jordan–Dedekind chain condition and, for every pair x and y of elements of $\mathcal{L}$,

1.7.4 $h(x) + h(y) \geq h(x \vee y) + h(x \wedge y)$.

A *geometric lattice* is a finite semimodular lattice in which every element is a join of atoms.

The following theorem, the main result of this section, motivates the lattice-theoretic approach to matroids.

Theorem 1.7.5 *A lattice $\mathcal{L}$ is geometric if and only if it is the lattice of flats of a matroid.*

We show first that the lattice of flats of a matroid satisfies the Jordan–Dedekind chain condition.

Lemma 1.7.6 *If X and Y are flats of a matroid M and $X \subseteq Y$, then every maximal chain of flats from X to Y has length $r(Y) - r(X)$.*

Proof Suppose that Y covers X. The lemma will follow if we can show, in this case, that $r(Y) = r(X) + 1$. As $X \subsetneq Y$, we can choose an element x from $Y - X$. Clearly $X \subsetneq \mathrm{cl}(X \cup x) \subseteq Y$. Therefore, since Y covers X, we must have that $\mathrm{cl}(X \cup x) = Y$. But, by (R2)', $r(\mathrm{cl}(X \cup x)) = r(X \cup x) \leq r(X) + 1$. Hence $r(Y) = r(X) + 1$ as required. $\square$

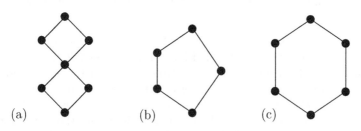

(a)   (b)   (c)

FIG. 1.27. Three posets that are lattices.

Proof of Theorem 1.7.5 Let M be a matroid. We begin by showing that $\mathcal{L}(M)$ is geometric. By the last lemma, $\mathcal{L}(M)$ satisfies the Jordan–Dedekind chain condition. Moreover, the height function h in $\mathcal{L}(M)$ is given by $h(X) = r(X)$ for all flats X. Thus, for flats X and Y,

$$h(X) + h(Y) = r(X) + r(Y) \geq r(X \cup Y) + r(X \cap Y), \quad \text{by (R3)};$$
$$= r(\mathrm{cl}(X \cup Y)) + r(X \cap Y)$$
$$= h(\mathrm{cl}(X \cup Y)) + h(X \cap Y)$$
$$= h(X \vee Y) + h(X \wedge Y).$$

Hence $\mathcal{L}(M)$ is a semimodular lattice. If X is a flat and $\{b_1, b_2, \ldots, b_k\}$ is a basis for X, then

$$X = \mathrm{cl}(\{b_1, b_2, \ldots, b_k\}) = \mathrm{cl}(\{b_1\}) \vee \mathrm{cl}(\{b_2\}) \vee \cdots \vee \mathrm{cl}(\{b_k\}).$$

Since $\mathrm{cl}(\{b_i\})$ is an atom, we conclude that every flat of M is a join of atoms, with $\mathrm{cl}(\emptyset)$ being the join of the empty set of atoms. Thus $\mathcal{L}(M)$ is indeed a geometric lattice.

Conversely, suppose that $\mathcal{L}$ is an arbitrary geometric lattice. If the zero and one of $\mathcal{L}$ coincide, then $\mathcal{L} \cong \mathcal{L}(U_{0,0})$. Now suppose that the zero and one of $\mathcal{L}$ are distinct. Then $\mathcal{L}$ has a non-empty set E of atoms. If $X \subseteq E$, define $r(X) = h(\bigvee_{x \in X} x)$. We shall show that r is the rank function of a matroid M on E whose lattice of flats is isomorphic to $\mathcal{L}$. First we note that $r(\emptyset) = 0$, that is, r satisfies (R1)′. Now suppose $X \subseteq E$ and $e \in E$. Then

$$r(X) = h(\bigvee_{x \in X} x) \leq h(\bigvee_{x \in X \cup e} x) = r(X \cup e).$$

Moreover,

$$r(X \cup e) \leq h(\bigvee_{x \in X} x) + h(e) - h((\bigvee_{x \in X} x) \wedge e) \leq r(X) + 1.$$

Thus r satisfies (R2)′. Finally, if $X \subseteq E$ and e, f are elements of E such that $r(X \cup e) = r(X \cup f) = r(X)$, then

$$r(X \cup e \cup f) = h(\bigvee_{x \in X \cup e \cup f} x)$$
$$= h((\bigvee_{x \in X \cup e} x) \vee (\bigvee_{x \in X \cup f} x))$$
$$\leq h(\bigvee_{x \in X \cup e} x) + h(\bigvee_{x \in X \cup f} x) - h((\bigvee_{x \in X \cup e} x) \wedge (\bigvee_{x \in X \cup f} x))$$
$$\leq r(X) + r(X) - h(\bigvee_{x \in X} x) = r(X).$$

Hence r satisfies (R3)′ and so, by Theorem 1.4.16, r is the rank function of a matroid M on E.

To complete the proof, we need to show that $\mathcal{L}(M)$ and $\mathcal{L}$ are isomorphic posets. To do this, let $\psi : \mathcal{L} \to \mathcal{L}(M)$ be defined by $\psi(X) = \{x \in E : x \leq X\}$. For ψ to be well-defined, we must check that $\{x \in E : x \leq X\}$ is a flat of M.

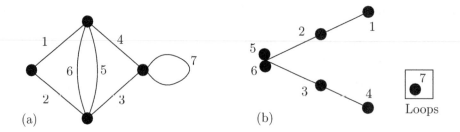

FIG. 1.28. (a) A graph G. (b) A geometric representation for $M(G)$.

Suppose not. Then there is an element y of E such that $y \not\leq X$ and $r(\{x \in E : x \leq X\} \cup y) = r(\{x \in E : x \leq X\})$. But then $h((\bigvee_{x \leq X} x) \vee y) = h(\bigvee_{x \leq X} x)$. Since $(\bigvee_{x \leq X} x) \vee y \geq \bigvee_{x \leq X} x$, it follows that equality holds here, so $y \leq \bigvee_{x \leq X} x = X$; a contradiction. Thus ψ is well-defined. It clearly preserves order and is one-to-one. To show it is onto, let Y be a flat of $\mathcal{L}(M)$. We shall show that $Y = \psi(Y')$ where $Y' = \bigvee_{y \in Y} y$. Since $\psi(Y') = \{x \in E : x \leq Y'\}$, it is clear that $\psi(Y') \supseteq Y$. On the other hand, if $z \in \psi(Y')$, then $z \leq \bigvee_{y \in Y} y$. Therefore $h((\bigvee_{y \in Y} y) \vee z) = h(\bigvee_{y \in Y} y)$, that is, $r(Y \cup z) = r(Y)$, so $z \in \mathrm{cl}(Y) = Y$. Hence $Y \supseteq \psi(Y')$. We conclude that ψ is onto. Hence, $\mathcal{L} \cong \mathcal{L}(M)$ and the theorem is proved. □

Example 1.7.7 Let $M = M(G)$ where G is the graph shown in Figure 1.28(a). A geometric representation for this matroid is shown in Figure 1.28(b), and $\mathcal{L}(M)$ is as shown in Figure 1.29. It is clear from that diagram that the Hasse diagram for $\mathcal{L}(M)$ can become quite cluttered even when M has a relatively small number of elements. □

An important example of a geometric lattice is Π_n, the lattice of flats of the matroid $M(K_n)$. This lattice is often referred to as a *partition lattice* since it can also be derived as follows. Let V be the vertex set of K_n. If F is a flat of $M(K_n)$, we denote by π_F the partition of V in which i and j are in the same class if and only if the edge ij is in F. This determines a map from the set of flats of $M(K_n)$ into the set of partitions of the n-set V. Moreover, this map is easily shown to be a bijection. Indeed, the map is an isomorphism from $\mathcal{L}(M(K_n))$ to the set of partitions of V where, for partitions α and β, we define $\alpha \leq \beta$ if α is a refinement of β, that is, if every class of α is contained in a class of β.

The lattice shown in Figure 1.30(b) is both the lattice of flats of $M(K_4)$, with the edges labelled as in Figure 1.30(a), as well as the lattice Π_4 of partitions of the 4-set $\{a, b, c, d\}$. Every element of the lattice has been labelled by the corresponding flat and some representative elements have also been labelled by the corresponding partitions.

The lattice of flats of a matroid M does not uniquely determine M for clearly $\mathcal{L}(\mathrm{si}(M)) \cong \mathcal{L}(M)$. However, a consequence of the proof of Theorem 1.7.5 is that if M_1 and M_2 are simple matroids for which $\mathcal{L}(M_1) \cong \mathcal{L}(M_2)$, then $M_1 \cong M_2$.

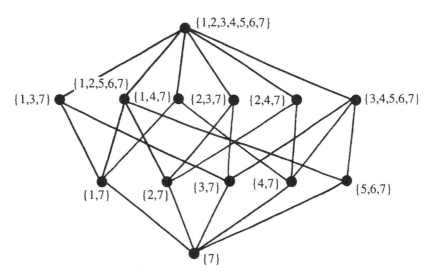

FIG. 1.29. $\mathcal{L}(M(G))$ where G is as shown in Figure 1.28(a).

Every element x in a geometric lattice $\mathcal{L}$ is a join of atoms. Indeed, if $h(x) = k$, there is a set of k atoms whose join is x. The next result asserts that the same result holds in the lattice obtained from $\mathcal{L}$ by reversing all order relations. We state it without explicit reference to the lattice.

Proposition 1.7.8 *Let X be a flat in a matroid M and suppose that $r(X) = r(M) - k$ where $k \geq 1$. Then M has a set $\{H_1, H_2, \ldots, H_k\}$ of hyperplanes such that $X = \cap_{i=1}^{k} H_i$.*

Proof We argue by induction on k. The result is immediate for $k = 1$. Assume it true for $k = n$ and let $k = n + 1$. Evidently M has an element y that is not in X. As X is a flat, $\mathrm{cl}(X \cup y)$ is a flat of M covering X. Thus $r(\mathrm{cl}(X \cup y)) = r(M) - n$, so, by the induction assumption, M has hyperplanes $H_1, H_2, \ldots, H_n$ such that $\cap_{i=1}^{n} H_i = \mathrm{cl}(X \cup y)$. As X is non-spanning, M has a maximal non-spanning set H_{n+1} that contains X and is contained in $E(M) - y$. Then H_{n+1} is a hyperplane. Since $y \in \cap_{i=1}^{n} H_i$, but $y \notin H_{n+1}$, we have $\cap_{i=1}^{n+1} H_i \neq \cap_{i=1}^{n} H_i$. Therefore $\mathrm{cl}(X \cup y) = \cap_{i=1}^{n} H_i \supsetneq \cap_{i=1}^{n+1} H_i \supseteq X$. But $\mathrm{cl}(X \cup y)$ covers X, so $X = \cap_{i=1}^{n+1} H_i$. The proposition now follows by induction. $\square$

While the intersection of two distinct hyperplanes in a rank-r matroid M has rank at most $r - 2$, it may have rank as low as 0. For example, in $U_{k+1,2k}$, the hyperplanes are precisely the k-element subsets of the ground set and, for any j in $\{0, 1, 2, \ldots, k - 2\}$, one can easily find two hyperplanes whose intersection has rank j. One may suspect that intersecting more hyperplanes will lower the rank. But, in $U_{4,n}$, one can easily find as many as $n - 2$ distinct hyperplanes, the intersection of all of which has rank $r(U_{4,n}) - 2$. We defer to the exercises consideration of some of the many other interesting properties of geometric lattices.

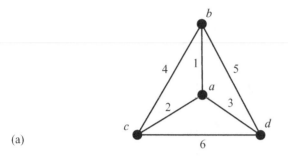

(a)

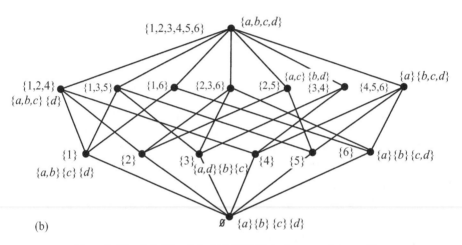

(b)

FIG. 1.30. (a) K_4. (b) $\mathcal{L}(M(K_4))$ is isomorphic to Π_4.

Exercises

1. Suppose that $A_1 = \{1, 2, 3\}$, $A_2 = \{3, 4, 5\}$, $A_3 = \{3, 6, 7\}$, and $A_4 = \{8\}$. Let $S = \{1, 2, \ldots, 10\}$ and $M[\mathcal{A}]$ be the transversal matroid on S associated with $\mathcal{A} = (A_1, A_2, A_3, A_4)$. Find the simple matroid associated with M, keeping the lowest-labelled element in every non-trivial parallel class.

2. (a) Draw the Hasse diagram for D_{54}, the set of positive integral divisors of 54 under the divisibility order.
   (b) Show that $D_{54} \cong D_{24}$ and generalize this to determine all n for which $D_n \cong D_{54}$.

3. Show that a matroid in which every two distinct hyperplanes are disjoint has rank at most two.

4. Find all positive integers n for which D_n is a geometric lattice.

5. Prove that a finite lattice $\mathcal{L}$ is semimodular if and only if, for all x, y in $\mathcal{L}$, the element $x \vee y$ covers both x and y whenever both x and y cover $x \wedge y$.

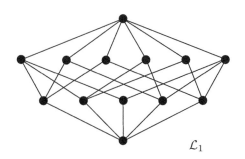

$\mathcal{L}_1$

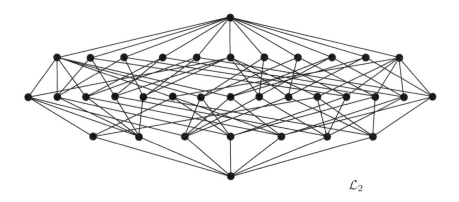

$\mathcal{L}_2$

FIG. 1.31. Two geometric lattices.

6. Draw Hasse diagrams for the lattices of flats of $U_{3,5}$ and the matroids M_1 and M_2 from Exercise 15 of Section 1.5.

7. A lattice $\mathcal{L}$ is *complemented* if, for every element x of $\mathcal{L}$, there is an element y, called a *complement* of x, such that $x \vee y$ and $x \wedge y$ are the one and zero, respectively, of $\mathcal{L}$. A lattice $\mathcal{L}$ is *relatively complemented* if every interval is complemented where an *interval* $[a, b]$ of $\mathcal{L}$ is the lattice induced on $\{z : a \leq z \leq b\}$ by the order on $\mathcal{L}$. Show that:
    (a) Every geometric lattice is relatively complemented.
    (b) Every interval of a geometric lattice is a geometric lattice.

8. Let A be a matrix having $V(n+1, q)$ as its set of columns and let $\mathcal{L}$ be the lattice of flats of $M[A]$. Show that if x and y are elements of $\mathcal{L}$, then $h(x) + h(y) = h(x \vee y) + h(x \wedge y)$.

9. Find simple matroids M_1 and M_2 so that if $\mathcal{L}_1$ and $\mathcal{L}_2$ are the lattices in Figure 1.31, then $\mathcal{L}(M_1) \cong \mathcal{L}_1$ and $\mathcal{L}(M_2) \cong \mathcal{L}_2$.

10. Let $\mathcal{P}'$ be the poset obtained from the poset $\mathcal{P}$ by defining $x \le y$ in $\mathcal{P}'$ if and only if $x \ge y$ in $\mathcal{P}$. Show that:
    (a) If $\mathcal{P}$ is a lattice, then so is $\mathcal{P}'$.
    (b) If $\mathcal{P}$ is a geometric lattice, then $\mathcal{P}'$ need not be geometric.
11. Prove that two lattices $\mathcal{L}_1$ and $\mathcal{L}_2$ are isomorphic if and only if there is a bijection $\psi : \mathcal{L}_1 \to \mathcal{L}_2$ such that, for all x and y in $\mathcal{L}_1$,

$$\psi(x \vee y) = \psi(x) \vee \psi(y) \quad \text{and} \quad \psi(x \wedge y) = \psi(x) \wedge \psi(y).$$

1.8 The greedy algorithm

In this section, we consider yet another characterization of matroids. The attractive feature of this characterization is that it indicates clearly why matroids arise naturally in a number of problems in combinatorial optimization.

We begin by discussing a well-known optimization problem for graphs. Let G be a connected graph and let w be a function from $E(G)$ into $\mathbb{R}$. We call w a *weight function* on G and, for all $X \subseteq E(G)$, we define the *weight $w(X)$* of X to be $\sum_{x \in X} w(x)$. For the given G and w, the problem here is to find a spanning tree of G of minimum weight. For instance, G could be K_n where the vertices of K_n correspond to towns to be linked by a railway network, and the weight on each edge is the cost of providing a direct link between the towns corresponding to the edge's endpoints. In this case, the minimum weight of a spanning tree in G corresponds to the minimum cost of providing a railway network that will link all n towns. Kruskal's algorithm (1956) for solving this problem chooses edges one at a time. At each stage, the next edge e to be chosen is one of minimum weight such that

(i) e has not been previously chosen; and
(ii) e does not form a cycle with some set of previously chosen edges.

When no further edges can be chosen subject to these restrictions, the algorithm stops. At this stage, the set of chosen edges is the edge set of a spanning tree of minimum weight. We shall not prove the last statement directly. Instead, we shall deduce it later from the main result of this section (Theorem 1.8.5).

Example 1.8.1 Let G be the graph in Figure 1.32 with edge weights as shown. It is not difficult to check that Kruskal's algorithm will produce one of the spanning trees T_1, T_2, and T_3 shown. Precisely which of these is obtained depends upon which two edges of weight three are chosen. $\qquad\square$

The minimum-weight spanning tree problem is a special case of the following natural optimization problem (Bixby 1981). Let $\mathcal{I}$ be a collection of subsets of a set E and suppose $\mathcal{I}$ satisfies (I1) and (I2). Let w be a function from E into $\mathbb{R}$. As before, define the *weight $w(X)$* of any non-empty subset X of E by

$$w(X) = \sum_{x \in X} w(x)$$

and let $w(\emptyset) = 0$. The *optimization problem* for the pair $(\mathcal{I}, w)$ is as follows:

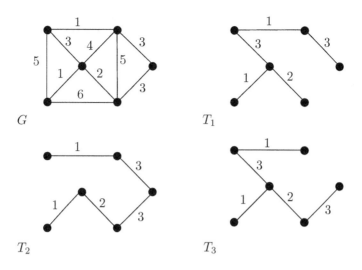

FIG. 1.32. G and its minimum-weight spanning trees.

1.8.2 *Find a maximal member B of $\mathcal{I}$ of maximum weight.*

We call the set B a *solution* to this optimization problem. If we replace the weight function w by its negative and solve the optimization problem for $(\mathcal{I}, -w)$, then we obtain a maximal member B' of $\mathcal{I}$ for which $w(B')$ is a minimum. Therefore, letting $\mathcal{I}$ be the collection of independent sets in $M(G)$ where G is a connected graph, we see that the minimum-weight spanning tree problem for graphs is a special case of 1.8.2.

The *greedy algorithm* for the pair $(\mathcal{I}, w)$ proceeds as follows:

(i) Set $X_0 = \emptyset$ and $j = 0$.

(ii) If $E - X_j$ contains an element e such that $X_j \cup e \in \mathcal{I}$, choose such an element e_{j+1} of maximum weight, let $X_{j+1} = X_j \cup e_{j+1}$, and go to (iii); otherwise, let $B_G = X_j$ and go to (iv).

(iii) Add 1 to j and go to (ii).

(iv) Stop.

Lemma 1.8.3 *If $(E, \mathcal{I})$ is a matroid M, then B_G is a solution to the optimization problem 1.8.2.*

Proof If $r(M) = r$, then $B_G = \{e_1, e_2, \ldots, e_r\}$ and B_G is a basis of M. Let B be another basis of M, say $B = \{f_1, f_2, \ldots, f_r\}$ where $w(f_1) \geq w(f_2) \geq \ldots \geq w(f_r)$. Lemma 1.8.3 follows without difficulty from the next result which shows that not only is B_G a maximum-weight basis of M, but, for all j, the jth heaviest element of B_G has weight at least that of the jth heaviest element of B.

Lemma 1.8.4 *If $1 \leq j \leq r$, then $w(e_j) \geq w(f_j)$.*

Proof Assume the contrary, taking k be the least integer for which $w(e_k) < w(f_k)$. Let $I_1 = \{e_1, e_2, \ldots, e_{k-1}\}$ and $I_2 = \{f_1, f_2, \ldots, f_k\}$. Then $|I_1| < |I_2|$ so, by (I3), there is an element f_t in $I_2 - I_1$ such that $I_1 \cup f_t \in \mathcal{I}$. But $w(f_t) \geq w(f_k) > w(e_k)$, so the greedy algorithm would have chosen f_t in preference to e_k. This contradiction completes the proof of Lemma 1.8.4 and thereby that of Lemma 1.8.3. $\square$

On combining Lemma 1.8.3 with the remarks following 1.8.2, we deduce that Kruskal's algorithm does indeed produce a spanning tree of minimum weight in a connected graph G. Observe that if G has n vertices, Kruskal's algorithm may be terminated once $n - 1$ edges have been chosen since this is the number of edges in a spanning tree of G.

We have seen that the greedy algorithm works for matroids. More strikingly, it fails for everything else.

Theorem 1.8.5 *Let $\mathcal{I}$ be a collection of subsets of a set E. Then $(E, \mathcal{I})$ is a matroid if and only if $\mathcal{I}$ has the following properties:*

(I1) $\emptyset \in \mathcal{I}$.

(I2) *If $I \in \mathcal{I}$ and $I' \subseteq I$, then $I' \in \mathcal{I}$.*

(G) *For all weight functions $w : E \to \mathbb{R}$, the greedy algorithm produces a maximal member of $\mathcal{I}$ of maximum weight.*

Proof If $(E, \mathcal{I})$ is a matroid, then (I1) and (I2) certainly hold; by Lemma 1.8.3, so too does (G). Now suppose that $\mathcal{I}$ satisfies (I1), (I2), and (G). We want to show that $\mathcal{I}$ satisfies (I3). Assume the contrary, that is, suppose that I_1, I_2 are members of $\mathcal{I}$ with $|I_1| < |I_2|$ such that $I_1 \cup e \notin \mathcal{I}$ for all e in $I_2 - I_1$.

Now $|I_1 - I_2| < |I_2 - I_1|$, so we can choose a real number ε such that

$$\frac{|I_1 - I_2|}{|I_2 - I_1|} < \varepsilon < 1. \tag{1.7}$$

Define $w : E \to \mathbb{R}$ by

$$w(e) = \begin{cases} 1, & \text{if } e \in I_1; \\ \varepsilon, & \text{if } e \in I_2 - I_1; \\ 0, & \text{otherwise.} \end{cases}$$

Then the greedy algorithm will first pick all the elements of I_1. By assumption, it cannot then pick any element of $I_2 - I_1$. Thus the remaining elements of B_G will be in $E - (I_1 \cup I_2)$. Hence

$$w(B_G) = |I_1|. \tag{1.8}$$

But, by (I2), the set I_2 is contained in a maximal member I_2' of $\mathcal{I}$ and

$$\begin{aligned} w(I_2') &\geq w(I_2) = |I_1 \cap I_2| + |I_2 - I_1|\varepsilon \\ &> |I_1 \cap I_2| + |I_2 - I_1|\frac{|I_1 - I_2|}{|I_2 - I_1|} = |I_1|. \end{aligned} \tag{1.9}$$

From (1.8) and (1.9), we deduce that $w(I_2') > w(B_G)$, that is, the greedy algorithm fails for this weight function. This contradiction completes the proof of the theorem. □

Proofs of the last result have been published by a number of different authors. Curiously, however, the first of these seems to be due to Borůvka (1926) whose paper predates even Whitney's (1935) introduction of the term 'matroid'.

We have already seen an application of the greedy algorithm to graphic matroids. The next application (Gale 1968) is to transversal matroids. Suppose that a certain set of one-worker jobs has been arranged in order of importance and that we want to fill the jobs from a pool of workers each of whom is qualified to perform some subset of the jobs. We also assume that the jobs are to be done simultaneously so that no worker can be assigned to more than one job. In general, it will not be possible to fill all the jobs, so we seek a way of choosing the set of jobs to be filled that is optimal relative to the priority order on the jobs.

Before defining optimality in this context, we describe a reformulation of the problem in terms of matroids. Let S be the set of jobs and Y the set of workers. For each y in Y, let A_y be the set of jobs that worker y is qualified to perform. Let $\mathcal{A} = (A_y : y \in Y)$. Evidently the maximum number of jobs that can be done simultaneously is the size of a largest partial transversal of $\mathcal{A}$, or equivalently, the rank of $M[\mathcal{A}]$.

Now let $p : S \to \mathbb{R}$ be the function that gives the priority order on the jobs, a lower p-value corresponding to a higher priority for the job. A *job assignment* is a basis of $M[\mathcal{A}]$. Suppose that B is the job assignment $\{x_1, x_2, \ldots, x_r\}$ where $p(x_1) \le p(x_2) \le \ldots \le p(x_r)$. We call B *optimal* if, for any other job assignment $\{z_1, z_2, \ldots, z_r\}$ where $p(z_1) \le p(z_2) \le \ldots \le p(z_r)$, we have $p(x_i) \le p(z_i)$ for all i in $\{1, 2, \ldots, r\}$. If we define $\mathcal{I}$ to be the collection of independent sets of $M[\mathcal{A}]$ and let $w = -p$, then, by Lemma 1.8.3, it follows that the greedy algorithm applied to the pair $(\mathcal{I}, w)$ will find an optimal job assignment.

Example 1.8.6 In the bipartite graph in Figure 1.33, $\{y_1, y_2, y_3, y_4\}$ corresponds to the set of workers, $\{x_1, x_2, \ldots, x_6\}$ to the set of jobs, and two vertices are joined when the corresponding worker can perform the corresponding job. Assume that the priority order on the jobs is $p(x_1) < p(x_2) < \ldots < p(x_6)$. Then it is not difficult to check that the unique optimal job assignment in this case is $\{x_1, x_2, x_3, x_6\}$. □

We remark that, in this job assignment problem, the greedy algorithm finds an optimal basis of the corresponding transversal matroid without backtracking. However, we cannot pick the workers for the jobs in this optimal basis at the same time as the jobs are added to the basis. For example, suppose Figure 1.34 represents such a problem with workers y_1 and y_2 and jobs x_1 and x_2 such that $p(x_1) < p(x_2)$. Then the greedy algorithm picks x_1 first and x_2 second. But, if x_1 is paired with y_1, then x_2 cannot be paired. In general, once an optimal basis B has been found, the problem of assigning workers to the jobs in B is precisely the problem of finding a maximum-sized matching in the bipartite graph that

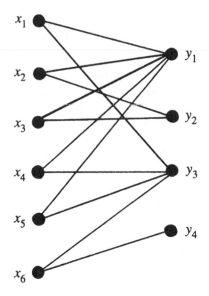

FIG. 1.33. Worker y_j can do job x_i if x_iy_j is an edge.

is obtained from $\Delta[\mathcal{A}]$ by deleting the vertices in $S - B$. Algorithms for solving the latter problem can be found in numerous books (see, for example, Cook *et al.* 1998 or Bondy and Murty 2008).

Exercises

1. (a) Find a maximum-weight spanning tree of the graph in Figure 1.32. Is this the unique such tree?
    (b) Find all maximum-weight spanning trees and all minimum-weight spanning trees of the graph in Figure 1.28(a), where the edge labels are interpreted as weights.

2. Find an optimal job assignment in Example 1.8.6 if the priority order on the jobs is changed to $p(x_6) < p(x_5) < \ldots < p(x_1)$.

3. Apply the greedy algorithm to the matroid in Figure 1.17 interpreting the element labels as weights. Find both a maximum-weight basis and a minimum-weight basis.

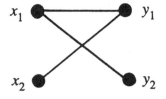

FIG. 1.34. The edge x_1y_1 is not in a maximum-sized matching.

4. Let M be a matroid and $w : E(M) \to \mathbb{R}$ be a one-to-one function. Prove that M has a unique basis of maximum weight.

5. Let M be a matroid and $w : E(M) \to \mathbb{R}$. When the greedy algorithm is applied to the pair $(\mathcal{I}(M), w)$, each application of step (ii) involves a potential choice. Thus, in general, there are a number of different sets that the algorithm can produce as solutions to the optimization problem $(\mathcal{I}(M), w)$. Let $\mathcal{B}_G$ be the set of such sets and let $\mathcal{B}_{max}$ be the set of maximum-weight bases of M. Prove that $\mathcal{B}_G = \mathcal{B}_{max}$.

6. Modify the greedy algorithm to give an algorithm for finding a maximum-weight basis B of a matroid M subject to the constraint that B contains some fixed independent set I.

7. Let M be a matroid and $\mathcal{S}$ be its set of spanning sets. Let $w : E(M) \to \mathbb{R}$ be a weight function. Prove that the set B_G produced by the following algorithm is a basis of M of minimum weight:
   (a) Set $Y_0 = E(M)$ and $j = 0$.
   (b) If Y_j contains an element e such that $Y_j - e \in \mathcal{S}$, then choose such an element e_{j+1} of maximum weight, let $Y_{j+1} = Y_j - e_{j+1}$, and go to (c); otherwise, let $Y_j = B_G$ and go to (d).
   (c) Add 1 to j and go to (b).
   (d) Stop.

8. *Let B be a maximum-weight basis of a matroid M and H be a hyperplane of M. Let X be the set of maximum-weight elements of $E(M) - H$. Prove that $X \cap B$ is non-empty.

2

DUALITY

One of the most attractive features of matroid theory is the existence of a theory of duality. This theory, which was introduced by Whitney (1935), extends both the notion of orthogonality in vector spaces and the concept of a planar dual of a plane graph. It dramatically increases the weapons at one's disposal in attacking any matroid problem. In this chapter, we define the dual of a matroid and examine its basic properties. We then discuss the duals of representable, graphic, and transversal matroids.

2.1 The definition and basic properties

In this section, we define the dual of a matroid, prove that it is also a matroid, and establish some fundamental links between matroids and their duals.

Theorem 2.1.1 *Let M be a matroid and $\mathcal{B}^*(M)$ be $\{E(M) - B : B \in \mathcal{B}(M)\}$. Then $\mathcal{B}^*(M)$ is the set of bases of a matroid on $E(M)$.*

The proof of this theorem will use the following result.

Lemma 2.1.2 *The set $\mathcal{B}$ of bases of a matroid M has the following property:*

(B2)* *If B_1 and B_2 are in $\mathcal{B}$ and $x \in B_2 - B_1$, then there is an element y of $B_1 - B_2$ such that $(B_1 - y) \cup x \in \mathcal{B}$.*

We remark here that there is a genuine difference between (B2)* and (B2) that cannot be achieved simply by relabelling.

Proof of Lemma 2.1.2 By Corollary 1.2.6, $B_1 \cup x$ contains a unique circuit, $C(x, B_1)$. As $C(x, B_1)$ is dependent and B_2 is independent, $C(x, B_1) - B_2$ is non-empty. Let y be an element of this set. Evidently $y \in B_1 - B_2$. Moreover, $(B_1 - y) \cup x$ is independent since it does not contain $C(x, B_1)$. As $|(B_1 - y) \cup x| = |B_1|$, it follows that $(B_1 - y) \cup x$ is a basis. □

Proof of Theorem 2.1.1 As $\mathcal{B}(M)$ is non-empty, so is $\mathcal{B}^*(M)$. Hence $\mathcal{B}^*(M)$ satisfies (B1). Now suppose B_1^* and B_2^* are in $\mathcal{B}^*(M)$ and $x \in B_1^* - B_2^*$. Writing E for $E(M)$, let $B_i = E - B_i^*$ for each i in $\{1, 2\}$. Then $B_i \in \mathcal{B}(M)$ and $B_1^* - B_2^* = B_1^* \cap (E - B_2^*) = (E - B_1) \cap B_2 = B_2 - B_1$. By (B2)*, as $x \in B_2 - B_1$, there is an element y of $B_1 - B_2$ such that $(B_1 - y) \cup x \in \mathcal{B}(M)$. Clearly $y \in B_2^* - B_1^*$ and $E - ((B_1 - y) \cup x) \in \mathcal{B}^*(M)$. But $E - ((B_1 - y) \cup x) = ((E - B_1) - x) \cup y = (B_1^* - x) \cup y$. Thus $\mathcal{B}^*(M)$ satisfies (B2), and $\mathcal{B}^*(M)$ is indeed the set of bases of a matroid on E. □

The matroid in the last theorem, whose ground set is $E(M)$ and whose set of bases is $\mathcal{B}^*(M)$, is called the *dual* of M and is denoted by M^*. Thus $\mathcal{B}(M^*) = \mathcal{B}^*(M)$. Moreover, it is clear that

2.1.3 $(M^*)^* = M$.

Example 2.1.4 Consider $U_{m,n}$. Its bases are all of the m-element subsets of an n-element set E. Hence $\mathcal{B}^*(U_{m,n})$ consists of all of the $(n-m)$-element subsets of E, so

$$U_{m,n}^* = U_{n-m,n}.$$ □

Corollary 1.2.5 gives one characterization of a matroid in terms of its collection of bases. A consequence of Theorem 2.1.1 is that an alternative such characterization is as follows.

Corollary 2.1.5 *Let $\mathcal{B}$ be a set of subsets of a set E. Then $\mathcal{B}$ is the collection of bases of a matroid on E if and only if $\mathcal{B}$ satisfies (B1) and (B2)*.*

Proof If $\mathcal{B}$ is the collection of bases of a matroid, then, by Corollary 1.2.5 and Lemma 2.1.2, $\mathcal{B}$ satisfies (B1) and (B2)*. Now suppose that $\mathcal{B}$ satisfies (B1) and (B2)* and consider $\mathcal{B}' = \{E - B : B \in \mathcal{B}\}$. By the proof of Theorem 2.1.1, $\mathcal{B}'$ satisfies (B1) and (B2). Hence, by Corollary 1.2.5, $\mathcal{B}'$ is the set of bases of a matroid M on E. Thus $\mathcal{B}$ is the set of bases of a matroid on E, namely M^*, and the corollary is proved. □

The exercises contain some of the many other characterizations of matroids in terms of their bases. Such characterizations are surveyed in Brylawski (1986b).

The bases of M^* are called *cobases* of M. A similar convention applies to other distinguished subsets of $E(M^*)$. Hence, for example, the circuits, hyperplanes, independent sets, and spanning sets of M^* are called *cocircuits*, *cohyperplanes*, *coindependent sets*, and *cospanning sets* of M. The next result gives some elementary relationships between these sets.

Proposition 2.1.6 *Let M be a matroid on a set E and suppose $X \subseteq E$. Then*

(i) *X is independent if and only if $E - X$ is cospanning;*

(ii) *X is spanning if and only if $E - X$ is coindependent;*

(iii) *X is a hyperplane if and only if $E - X$ is a cocircuit; and*

(iv) *X is a circuit if and only if $E - X$ is a cohyperplane.*

Proof Parts (i) and (ii) are left as exercises. For (iii), we have, by (ii) and Proposition 1.4.10(iii), that the following statements are equivalent:

(a) X is a hyperplane of M.

(b) X is a non-spanning set of M but $X \cup y$ is spanning for all $y \notin X$.

(c) $E - X$ is dependent in M^* but $(E - X) - y$ is independent in M^* for all y in $E - X$.

(d) $E - X$ is a cocircuit of M.

Part (iv) can be deduced by applying (iii) to M^*. □

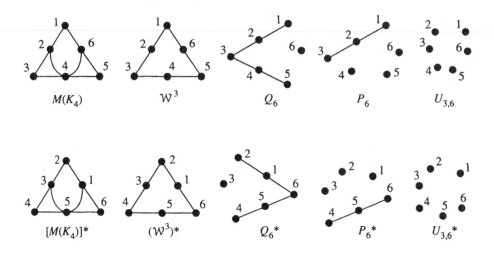

FIG. 2.1. Five matroids and their duals.

A consequence of the last result is that if X is a circuit–hyperplane of a matroid M, then $E(M) - X$ is a circuit–hyperplane of M^*. Moreover, the operation of relaxation has the following attractive property.

Proposition 2.1.7 *If M' is obtained from M by relaxing a circuit–hyperplane X of M, then $(M')^*$ can be obtained from M^* by relaxing the circuit–hyperplane $E(M) - X$ of M^*.*

Proof As $\mathcal{B}(M') = \mathcal{B}(M) \cup \{X\}$, we have $\mathcal{B}((M')^*) = \mathcal{B}(M^*) \cup \{E(M) - X\}$ and the result follows immediately. $\square$

Proposition 2.1.7 is illustrated in Figure 2.1. The rank-3 matroids in the second row are the duals of the rank-3 matroids in the first row. In both rows, each of the last four matroids is obtained from its predecessor by relaxing a circuit–hyperplane. Clearly each of $M(K_4)$, $\mathcal{W}^3$, Q_6, and P_6 is isomorphic to its dual, whereas $U_{3,6}$ is *equal* to its dual. We follow Bondy and Welsh (1971) in calling M *self-dual* if $M \cong M^*$, and *identically self-dual* if $M = M^*$. Note, however, that some authors reserve the term 'self-dual' for matroids M such that $M = M^*$.

The matroid M in Figure 2.2 is clearly simple. However, its dual M^* is not. It is mainly because the class of simple matroids is not closed under duality that matroid theorists do not, in general, confine their attention to simple matroids; duality is too important a part of matroid theory. This provides a contrast between matroid theory and graph theory, for, in the latter, many workers concentrate exclusively on simple graphs.

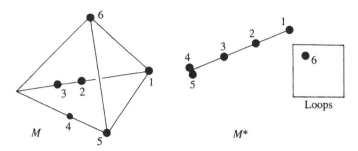

FIG. 2.2. A simple matroid with a non-simple dual.

For a matroid M, a loop of M^* is called a *coloop* of M. For example, the element 6 is a coloop of the matroid M shown in Figure 2.2. As 6 is in no basis of M^*, it is in every cobasis of M^*, that is, 6 is in every basis of M. We leave it to the reader to check that the last property actually characterizes coloops. Elements such as 4 and 5, which are in parallel in M^*, are *in series* in M.

In general, we attach an asterisk to a symbol to denote association with the dual. Thus, for example, r^* will denote the rank function of M^* while $\mathcal{C}^*$ denotes its set of circuits. Evidently

2.1.8 $r(M) + r^*(M) = |E(M)|$.

The geometric representations for matroids described in Section 1.5 apply only to matroids of rank at most four. However, such representations are also useful for matroids having rank more than four but corank at most four since, for these matroids, one can work with geometric representations for their duals.

The next result generalizes 2.1.8 to give a formula for r^*, the *corank function* of M. We caution the reader that some authors take the corank of a set X in a matroid M to be $r(M) - r(X)$ rather than $r^*(X)$ as is done here.

Proposition 2.1.9 *For all subsets X of the ground set E of a matroid M,*

$$r^*(X) = r(E - X) + |X| - r(M).$$

The proof of this will follow easily from the following augmentation result.

Lemma 2.1.10 *Let I and I^* be disjoint subsets of $E(M)$ such that I is independent and I^* is coindependent. Then M has a basis B and a cobasis B^* such that B and B^* are disjoint, $I \subseteq B$, and $I^* \subseteq B^*$.*

Proof In $M|(E - I^*)$, the set I is independent. It is therefore contained in a basis B of this restriction. So $r(B) = r(E - I^*)$. But, by Proposition 2.1.6(ii), $E - I^*$ is spanning in M. Hence $r(B) = r(M)$ and so B is a basis of M. Since $I \subseteq B$ and $I^* \subseteq E - B$, it follows that B and $E - B$ are the required basis and cobasis of M.   □

Proof of Proposition 2.1.9 Let B_X^* be a basis for X in M^*, and B_{E-X} be a basis for $E - X$ in M (see Figure 2.3). Then $r^*(X) = |B_X^*|$ and $r(E - X) =$

$|B_{E-X}|$. Now B_{E-X} and B^*_X are independent and coindependent, respectively, in M. Thus, by the last lemma, M has a basis B and a cobasis B^* with $B \cap B^* = \emptyset$ such that B contains B_{E-X}, and B^* contains B^*_X. Because B_{E-X} is a basis for $M|(E - X)$, we have $B \cap (E - X) = B_{E-X}$. Similarly, $B^* \cap X = B^*_X$, so $B \cap X = X - B^*_X$. Thus $B = B_{E-X} \dot\cup (X - B^*_X)$, so $|B| = |B_{E-X}| + |X| - |B^*_X|$. Therefore

$$r(M) = r(E - X) + |X| - r^*(X).$$

Rewriting this gives the required result. □

The following proposition expresses a fundamental link between circuits and cocircuits. This property is often called *orthogonality*.

Proposition 2.1.11 *In a matroid M, let C be a circuit and C^* be a cocircuit. Then $|C \cap C^*| \neq 1$.*

Proof Assume the contrary, letting $C \cap C^* = \{x\}$. Let $E(M) - C^* = H$. Then, by Proposition 2.1.6(iii), H is a hyperplane of M, so $\mathrm{cl}(H) = H$. But $x \in C \subseteq H \cup x$, so, by 1.4.11(ii), $x \in \mathrm{cl}(H) - H$; a contradiction. □

As we shall see in Proposition 2.1.23, the last property of circuits can even be used to characterize them. The next result can be extended to characterize when two matroids are duals of each other (see Exercise 5 of Section 2.2).

Proposition 2.1.12 *Let $(X, Y, \{z\})$ be a partition of the ground set E of a matroid M where X or Y may be empty. Then $z \in \mathrm{cl}(X)$ if and only if $z \notin \mathrm{cl}^*(Y)$.*

Proof By definition, $z \in \mathrm{cl}(X)$ if and only if $r(X \cup z) = r(X)$. By Proposition 2.1.9, the latter is true if and only if

$$r^*(Y) - |Y| + r(M) = r^*(Y \cup z) - |Y \cup z| + r(M).$$

Since $z \notin Y$, the last equation holds if and only if $r^*(Y) + 1 = r^*(Y \cup z)$, that is, if and only if $z \notin \mathrm{cl}^*(Y)$. □

A *clutter* is a collection of sets none of which is a proper subset of another. There are a number of clutters that are naturally associated with a matroid. These include the sets of bases, circuits, hyperplanes, cobases, cocircuits, and cohyperplanes. Indeed, several matroid axioms, such as (C2), are assertions that

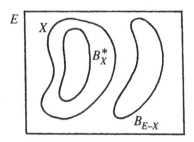

FIG. 2.3. The sets in the proof of Proposition 2.1.9.

a particular collection of sets is a clutter. If $\mathcal{A}$ is a clutter of subsets of a set S, then clearly the set $\mathcal{A}'$ of complements of members of $\mathcal{A}$ is also a clutter on S. We call $\mathcal{A}'$ the *complementary clutter* of $\mathcal{A}$. The *blocker* $b(\mathcal{A})$ of $\mathcal{A}$ consists of those minimal subsets of S that have non-empty intersection with every member of $\mathcal{A}$. Evidently $b(\mathcal{A})$ is a clutter.

Example 2.1.13 Let n be a positive integer, $S = \{1, 2, \ldots, n\}$, and, for each m in $\{0.1, \ldots, n+1\}$, let $\mathcal{A}_m$ be the collection of m-element subsets of S. Then

2.1.14 $b(\mathcal{A}_m) = \mathcal{A}_{n-m+1}$.

The only cases when this assertion may cause difficulty are when $m = 0$ and when $m = n + 1$. In these cases, we are concerned with $\mathcal{A}_0$, which equals $\{\emptyset\}$, and $\mathcal{A}_{n+1}$, which equals $\emptyset$. To verify (2.1.14) when $m = n + 1$, observe that because $\mathcal{A}_{n+1}$ is empty, it is vacuously true that every subset of S has non-empty intersection with every member of $\mathcal{A}_{n+1}$. Hence there is a unique minimal subset of S, namely $\emptyset$, that has non-empty intersection with every member of $\mathcal{A}_{n+1}$. Hence $b(\mathcal{A}_{n+1}) = \{\emptyset\} = \mathcal{A}_0$. Thus (2.1.14) holds when $m = n + 1$. It also holds when $m = 0$ since $\mathcal{A}_0 = \{\emptyset\}$, so no subset of S has non-empty intersection with the unique member, $\emptyset$, of $\mathcal{A}_0$. Thus $b(\mathcal{A}_0) = \emptyset = \mathcal{A}_{n+1}$. □

An immediate consequence of (2.1.14) is that $b(b(\mathcal{A}_m)) = \mathcal{A}_m$ for all m. This is a special case of the following attractive general result of Edmonds and Fulkerson (1970).

Proposition 2.1.15 *Let $\mathcal{A}$ be a clutter. Then $b(b(\mathcal{A})) = \mathcal{A}$.*

Proof This is left to the reader and is outlined in Exercise 16. □

Let M be a matroid and $\mathcal{B}(M)$, $\mathcal{C}(M)$, $\mathcal{H}(M)$, $\mathcal{B}^*(M)$, $\mathcal{C}^*(M)$, and $\mathcal{H}^*(M)$ be the clutters of bases, circuits, hyperplanes, cobases, cocircuits, and cohyperplanes of M. Then clearly

2.1.16 $\mathcal{B}(M)' = \mathcal{B}^*(M)$;

2.1.17 $\mathcal{H}(M)' = \mathcal{C}^*(M)$; and

2.1.18 $\mathcal{C}(M)' = \mathcal{H}^*(M)$.

Furthermore, the cocircuits of M are the minimal sets having non-empty intersection with every basis.

Proposition 2.1.19 $\mathcal{C}^*(M) = b(\mathcal{B}(M))$ *and* $b(\mathcal{C}^*(M)) = \mathcal{B}(M)$.

Proof The following statements are equivalent:
(i) $C^* \in \mathcal{C}^*(M)$.
(ii) $E(M) - C^* \in \mathcal{H}(M)$.
(iii) $E(M) - C^*$ is a maximal non-spanning set in M.
(iv) $E(M) - C^*$ does not contain a basis of M but, for all x in C^*, the set $E(M) - (C^* - x)$ does contain a basis of M.

(v) C^* meets every basis of M but, for all x in C^*, the set $C^* - x$ avoids some basis of M.

(vi) $C^* \in b(\mathcal{B}(M))$.

We conclude that $\mathcal{C}^*(M) = b(\mathcal{B}(M))$, and the rest of the proposition follows easily from this using Proposition 2.1.15.   □

If we replace M by M^* in the statement of the last proposition we obtain the statement:

$$\mathcal{C}^*(M^*) = b(\mathcal{B}(M^*)) \text{ and } b(\mathcal{C}^*(M^*)) = \mathcal{B}(M^*).$$

On rewriting this in terms of M, we get the following.

Corollary 2.1.20 $\mathcal{C}(M) = b(\mathcal{B}^*(M))$ *and* $b(\mathcal{C}(M)) = \mathcal{B}^*(M)$.   □

This corollary is called the *dual* of Proposition 2.1.19. Expressed in words, the first part of the corollary asserts that the circuits of M are the minimal sets having non-empty intersection with every cobasis. This observation is the dual of the observation that preceded the statement of 2.1.19. In general, to form the dual of a statement or proposition, one follows the procedure exemplified above. Clearly a proposition is true if and only if the dual proposition is true. The dual of 2.1.11 is the same as the original except for renaming the sets C and C^*. Propositions such as this are called *self-dual*.

Since $\mathcal{H}(M)' = \mathcal{C}^*(M)$, we can use Corollary 1.1.5 to derive the following characterization of matroids in terms of their hyperplanes. The straightforward proof is left to the reader.

Proposition 2.1.21 *Let* $\mathcal{H}$ *be a set of subsets of a set* E. *Then* $\mathcal{H}$ *is the collection of hyperplanes of a matroid on* E *if and only if* $\mathcal{H}$ *has the following properties:*

(H1) $E \notin \mathcal{H}$.

(H2) $\mathcal{H}$ *is a clutter.*

(H3) *If* H_1 *and* H_2 *are distinct members of* $\mathcal{H}$ *and* $e \in E - (H_1 \cup H_2)$, *then there is a member* H_3 *of* $\mathcal{H}$ *such that* $H_3 \supseteq (H_1 \cap H_2) \cup e$.   □

The next proposition extends 2.1.11 to give a characterization of the set of circuits of a matroid. The proof is based on the following lemma whose proof is left as an exercise.

Lemma 2.1.22 *Let* $\mathcal{X}$ *and* $\mathcal{Y}$ *be collections of subsets of a finite set* E *such that every member of* $\mathcal{X}$ *contains a member of* $\mathcal{Y}$, *and every member of* $\mathcal{Y}$ *contains a member of* $\mathcal{X}$. *Then the minimal members of* $\mathcal{X}$ *are precisely the minimal members of* $\mathcal{Y}$.   □

Proposition 2.1.23 *Let* M *be a matroid having ground set* E. *Then* D *is a circuit of* M *if and only if* D *is a minimal non-empty subset of* E *such that* $|D \cap C^*| \neq 1$ *for every cocircuit* C^* *of* M.

Proof We use the last lemma taking $\mathcal{X}$ equal to $\mathcal{C}(M)$ and $\mathcal{Y}$ equal to the set of non-empty subsets Y of E such that $|Y \cap C^*| \neq 1$ for all cocircuits C^*. By Proposition 2.1.11, if $X \in \mathcal{X}$, then $X \in \mathcal{Y}$. Now suppose Y is a minimal member of $\mathcal{Y}$. Assume Y is independent. Then M has a basis B containing Y. Choose y in Y and consider $\mathrm{cl}(B - y)$. This is a hyperplane of M. Its complement C_1^* is a cocircuit that meets Y in $\{y\}$; a contradiction. Hence Y is dependent. Thus Y contains a circuit C of M. If $Y \neq C$, then, by Proposition 2.1.11, we obtain a contradiction to the minimality of Y. We conclude that Y is a circuit of M, that is, $Y \in \mathcal{X}$. The proposition now follows immediately from Lemma 2.1.22. □

Recall from Section 1.3 that a paving matroid is a matroid M that has no circuits of size less than $r(M)$. Thus every matroid of rank less than two is paving. Next we state a characterization, in terms of their hyperplanes, of paving matroids of rank at least two (Hartmanis 1959). This result explains the reason for the name 'paving matroid', which was introduced by Welsh (1976) following a privately communicated suggestion of G.-C. Rota. We have extended Welsh's definition slightly by dropping his requirement that such matroids have rank at least two. Let k and m be integers with $k > 1$ and $m > 0$. Suppose that $\mathcal{T}$ is a set $\{T_1, T_2, \ldots, T_k\}$ of subsets of a set E such that each member of $\mathcal{T}$ has at least m elements, and each m-element subset of E is contained in a unique member of $\mathcal{T}$. Such a set $\mathcal{T}$ is called an *m-partition* of E.

Proposition 2.1.24 *If $\mathcal{T}$ is an m-partition $\{T_1, T_2, \ldots, T_k\}$ of a set E, then $\mathcal{T}$ is the set of hyperplanes of a paving matroid of rank $m + 1$ on E. Moreover, for $r \geq 2$, the set of hyperplanes of every rank-r paving matroid on E is an $(r - 1)$-partition of E.*

Proof It is straightforward to show that $\mathcal{T}$ satisfies (H1)–(H3). We leave the details of this, along with the rest of the proof, to the reader. □

Example 2.1.25 Figure 2.4 is a geometric representation of an 8-element rank-4 matroid. To see that this diagram does indeed represent a matroid, one could check that it satisfies 1.5.10–1.5.12. Alternatively, let $\mathcal{T}_1 = \{\{1, 2, 3, 4\}, \{1, 4, 5, 6\}, \{2, 3, 5, 6\}, \{1, 4, 7, 8\}, \{2, 3, 7, 8\}\}$, and $\mathcal{T} = \mathcal{T}_1 \cup \{T \subseteq \{1, 2, \ldots, 8\} :$

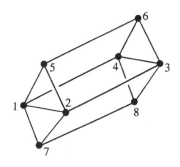

Fig. 2.4. The Vámos matroid V_8.

$|T| = 3$ and T is not contained in any member of $\mathcal{T}_1$}. Then one easily checks that $\mathcal{T}$ is a 3-partition of $\{1, 2, \ldots, 8\}$. Therefore, by the last proposition, $\mathcal{T}$ is the set of hyperplanes of a paving matroid on $\{1, 2, \ldots, 8\}$. Evidently Figure 2.4 is a geometric representation for this paving matroid. We call this matroid the *Vámos matroid* (Vámos 1968) and denote it by V_8. It occurs quite frequently throughout this book and has many interesting properties. In particular, as we shall prove Proposition 2.2.26, it is not representable over any field.

A related matroid, V_8^+, has the same ground set and is defined in the same way as V_8 but with $\{5, 6, 7, 8\}$ added to $\mathcal{T}_1$. Again one can use Proposition 2.1.24 to check that V_8^+ is actually a matroid. In fact, V_8 is obtained from V_8^+ by relaxing $\{5, 6, 7, 8\}$. It is straightforward to check that both V_8 and V_8^+ are self-dual with the latter, but not the former, being identically self-dual. $\square$

Recall from Section 1.5 that, when $r \geq 3$, a matroid M is an r-spike with tip t and legs $\{t, x_1, y_1\}, \{t, x_2, y_2\}, \ldots, \{t, x_r, y_r\}$ if its ground set E is the union of the legs and its set of circuits consists of

(i) $\{t, x_1, y_1\}, \{t, x_2, y_2\}, \ldots, \{t, x_r, y_r\}$;

(ii) all sets of the form $\{x_i, y_i, x_j, y_j\}$ with $1 \leq i < j \leq r$;

(iii) a subset $\mathcal{C}_3$ of $\{\{z_1, z_2, \ldots, z_r\} : z_i \in \{x_i, y_i\}$ for all $i\}$ such that no two members of $\mathcal{C}_3$ have more than $r - 2$ common elements; and

(iv) all $(r + 1)$-element subsets of E that contain none of the sets in (i)–(iii).

Moreover, the matroid $M \backslash t$ is a tipless r-spike. The matroid V_8^+ in the last example is isomorphic to the tipless free 4-spike, it being free because $\mathcal{C}_3$ is empty. We noted above that V_8^+ is identically self-dual. We shall show that this is true for every tipless free spike and that, in addition, every tipless spike is self-dual. The following lemma will be helpful in doing this. We use the notation above and also write L_i for each leg $\{t, x_i, y_i\}$.

Lemma 2.1.26 *The set of hyperplanes of an r-spike M consists of*

(i)   *the members of $\mathcal{C}_3$;*

(ii) *all sets of the form $E - \{x_i, y_i, x_j, y_j\}$ with $1 \leq i < j \leq r$; and*

(iii) *all $(r - 1)$-element subsets of $E - t$ that contain at most one element of each $\{x_i, y_i\}$ and are not contained in any member of $\mathcal{C}_3$.*

Proof By Proposition 1.5.18, $E - \{x_i, y_i, x_j, y_j\}$ is a hyperplane of M for all i and j with $1 \leq i < j \leq r$. Now let X be one of the sets listed in (i) or (iii). Then, from the list of circuits of M, we deduce that $r(X) = r - 1$. Moreover, if $e \in E - X$ and e lies on a leg that meets X, then $\mathrm{cl}(X \cup e)$ must contain t. Hence this closure contains all but at most one leg of M. Thus, by Proposition 1.5.18, this closure has rank r and so is E. It follows that each of the sets listed in (i)–(iii) is a hyperplane of M.

Now let H be an arbitrary hyperplane of M. Suppose first that $t \in H$. Then, for each leg L_i, either $L_i \subseteq H$ or $L_i \cap H = \{t\}$. Hence H is a union of legs so, by Proposition 1.5.18, $H = E - \{x_i, y_i, x_j, y_j\}$ for some i and j with $1 \leq i < j \leq r$.

Now suppose that $t \notin H$. Then H contains at most one element from each L_i. It follows without difficulty that H is listed under (i) or (iii). $\square$

Proposition 2.1.27 *Let e be an element of a spike M that has tip t. Then $M \backslash e$ is self-dual. Moreover, if M is a free spike, then $M \backslash t$ is identically self-dual.*

Proof Each element of M is in a circuit, so $r(M \backslash e) = r(M)$. We show first that the set of hyperplanes of $M \backslash e$ consists of all sets of the form $H - e$ where H is a hyperplane of M and $r(H - e) = r(H)$. If H is such a hyperplane of M, then, by Proposition 1.4.10(iii), H is a maximal non-spanning set of M, so $H - e$ is a maximal non-spanning set of $M \backslash e$, that is, a hyperplane of $M \backslash e$. Now suppose X is an arbitrary hyperplane of $M \backslash e$. Let H be $X \cup e$ if $r(X \cup e) = r(X)$, and H be X if $r(X \cup e) > r(X)$. Then $X = H - e$, and H is a hyperplane of M with $r(H - e) = r(H)$. We conclude that the hyperplanes of $M \backslash e$ are as specified.

Now, by Proposition 2.1.6, the cocircuits of $M \backslash e$ are the complements of its hyperplanes. The matroid $M \backslash e$ has rank r and has $2r$ elements so, by 2.1.8, its dual also has rank r. In Proposition 1.4.14, we noted that a matroid of rank r is uniquely determined by its set of non-spanning circuits, that is, by its set of circuits with at most r elements. Thus, to show that $M \backslash e$ is self-dual, we need only consider the sets of circuits and cocircuits of $M \backslash e$ with at most r elements.

First let $e = t$. The circuits of $M \backslash t$ with at most r elements consist of all members of C_3 plus, when $r > 3$, all sets of the form $\{x_i, y_i, x_j, y_j\}$. The cocircuits of $M \backslash t$ with at most r elements consist of all sets of the form $(E - t) - C$ where $C \in C_3$ plus, when $r > 3$, all sets of the form $\{x_i, y_i, x_j, y_j\}$. We deduce that if φ is the permutation of $E - t$ that interchanges x_i and y_i for all i, then φ is an isomorphism from $M \backslash t$ to $(M \backslash t)^*$. Furthermore, if M is a free spike, then C_3 is empty. Thus the sets of non-spanning circuits of $M \backslash t$ and $(M \backslash t)^*$ are empty when $r = 3$ and consist of the sets $\{x_i, y_i, x_j, y_j\}$ with $1 \le i < j \le r$ when $r \ge 4$. Hence the tipless free spike is identically self-dual.

Now suppose that $e \neq t$. Then, by symmetry, we may assume that $e = y_r$. The non-spanning circuits of $M \backslash y_r$ consist of the sets

(i) $\{t, x_1, y_1\}, \{t, x_2, y_2\}, \ldots, \{t, x_{r-1}, y_{r-1}\}$;

(ii) those members of C_3 that contain x_r; and

(iii) when $r > 3$, all sets of the form $\{x_i, y_i, x_j, y_j\}$ with $1 \le i < j \le r - 1$.

The non-spanning circuits of $(M \backslash y_r)^*$ consist of the sets

(i) $\{x_r, x_1, y_1\}, \{x_r, x_2, y_2\}, \ldots, \{x_r, x_{r-1}, y_{r-1}\}$;

(ii) the complements in $E - t$ of those members of C_3 that contain x_r; and

(iii) when $r > 3$, all sets of the form $\{x_i, y_i, x_j, y_j\}$ with $1 \le i < j \le r - 1$.

Thus the permutation of $E - y_r$ that swaps t and x_r and interchanges x_i and y_i for all i in $\{1, 2, \ldots, r - 1\}$ is an isomorphism from $M \backslash y_r$ to $(M \backslash y_r)^*$. $\square$

Next we establish a very attractive characterization of tipless spikes.

Proposition 2.1.28 *For $r \geq 3$, a rank-r matroid M on a $2r$-element set E is a tipless spike if and only if there is a partition of E into r pairs of elements such that the union of every two such pairs is both a circuit and a cocircuit of M.*

The fact that every tipless spike satisfies the specified condition is shown in the penultimate paragraph of the last proof. The next proposition proves the converse thereby finishing the proof of Proposition 2.1.28.

Proposition 2.1.29 *For some $r \geq 3$, let $E = \{x_1, y_1, x_2, y_2, \ldots, x_r, y_r\}$ and let M be a rank-r matroid on E. Suppose that $\{x_i, y_i, x_j, y_j\}$ is both a circuit and a cocircuit of M for all i and j with $1 \leq i < j \leq r$. Then there is a unique rank-r spike N with tip t and legs $\{t, x_1, y_1\}, \{t, x_2, y_2\}, \ldots, \{t, x_r, y_r\}$ such that $N \backslash t = M$. Hence M is a tipless spike.*

Proof Let $E' = E \cup \{t\}$; let $\mathcal{C}_1 = \{t, x_1, y_1\}, \{t, x_2, y_2\}, \ldots, \{t, x_r, y_r\}$; let $\mathcal{C}_2 = \{\{x_i, y_i, x_j, y_j\} : 1 \leq i < j \leq r\}$; let $\mathcal{C}_3$ be the set of non-spanning circuits of M that are not in $\mathcal{C}_1 \cup \mathcal{C}_2$; and let $\mathcal{C}_4$ be the set of $(r+1)$-element subsets of E' that contain no member of $\mathcal{C}_1 \cup \mathcal{C}_2 \cup \mathcal{C}_3$. We shall show that

(i) $\mathcal{C}_1 \cup \mathcal{C}_2 \cup \mathcal{C}_3 \cup \mathcal{C}_4$ is the set of circuits of an r-spike N with ground set E', tip t, and legs $\{t, x_1, y_1\}, \{t, x_2, y_2\}, \ldots, \{t, x_r, y_r\}$; and

(ii) $N \backslash t = M$.

Suppose that $X \in \mathcal{C}_3$. We show first that $|X| = r$ and $|X \cap \{x_i, y_i\}| = 1$ for all i. Assume not. Then, as $|X| \leq r$, we have $X \cap \{x_j, y_j\} = \emptyset$ for some j. As $\{x_j, y_j, x_k, y_k\}$ is a cocircuit of M for all $k \neq j$, it follows by orthogonality that X either contains or avoids $\{x_k, y_k\}$ for all such k. Thus we obtain the contradiction that X either contains or is contained in a member of $\mathcal{C}_2$.

Next we show that no two members of $\mathcal{C}_3$ have more than $r - 2$ common elements. Suppose X_1 and X_2 are members of $\mathcal{C}_3$ with exactly $r - 1$ common elements. Let e be one of these elements. Then $(X_1 \cup X_2) - e$ has exactly r elements and must contain a circuit of M. But this set contains no member of $\mathcal{C}_2$ and, by the last paragraph, is not in $\mathcal{C}_3$ since it avoids some $\{x_i, y_i\}$. Thus we have a contradiction.

By Proposition 1.5.17, (i) holds. The matroids $N \backslash t$ and M have the same rank, the same ground set, and the same set of non-spanning circuits. Thus, by Proposition 1.4.15, $N \backslash t = M$. The uniqueness of N follows from the fact that the only non-spanning circuits of a spike containing its tip are the legs.   □

Let M be a tipless r-spike and N be an r-spike with tip t such that $N \backslash t = M$. Let $L_1, L_2, \ldots, L_r$ be the legs of N. If $r \geq 5$, then, when two distinct 4-circuits of M meet, their intersection is $L_i - t$ for some i. Hence the legs of N are uniquely determined by the set of 4-circuits of M. We call the sets $L_1 - t, L_2 - t, \ldots, L_r - t$ the *legs* of the tipless spike M. If $r \leq 4$, then, as $AG(3, 2)$ shows, there may be different spikes N_1 and N_2 with tips t_1 and t_2 such that $N_1 \backslash t_1 = M = N_2 \backslash t_2$. Thus it may no longer be possible to unambiguously define the legs of M. However, in this case, we can still specify r pairwise disjoint 2-element subsets of $E(M)$ that we shall view as the legs of the tipless spike M.

Exercises

1. Find each of the following:
    (a) all self-dual uniform matroids;
    (b) all identically self-dual uniform matroids;
    (c) all self-dual graphic matroids on six or fewer elements;
    (d) all identically self-dual graphic matroids on six or fewer elements;
    (e) an infinite family of simple graphic self-dual matroids.

2. Let M be a matroid. Show that M^* has two disjoint circuits if and only if M has two hyperplanes whose union is $E(M)$.

3. Show that two elements are clones in a matroid if and only if they are clones in its dual.

4. In a matroid M, if X is independent and $E(M) - X$ is coindependent, show that X is a basis and $E(M) - X$ is a cobasis.

5. Show that an element e of a matroid M is a coloop of M if and only if e is in every basis of M. Now refer to Exercise 6 of Section 1.4 for a number of alternative characterizations of coloops.

6. For each of the following matroid properties, determine whether the property is necessary or sufficient (or both) for M to be identically self-dual:
    (a) Every basis of M is a cobasis of M.
    (b) Every flat of M is a flat of M^*.
    (c) The flats of M and M^* coincide.
    (d) Every circuit of M is a cocircuit of M.

7. Let e and f be distinct elements of a matroid. Prove that every circuit containing e also contains f if and only if $\{e\}$ or $\{e, f\}$ is a cocircuit.

8. Prove that a matroid is uniform if and only if every circuit meets every cocircuit.

9. Let M be a rank-r matroid.
    (a) Show that if $r^*(M) \leq r$ and every hyperplane is a circuit, then every circuit is a hyperplane.
    (b) Use duality to show that if $r^*(M) \geq r$ and every circuit is a hyperplane, then every hyperplane is a circuit.
    (c) Give examples to show that (a) and (b) can fail if the conditions on $r^*(M)$ are dropped.

10. Let B be a basis of a matroid M. If $e \in B$, denote $C_{M^*}(e, E(M) - B)$ by $C^*(e, B)$ and call it the *fundamental cocircuit of e with respect to B*.
    (a) Show that $C^*(e, B)$ is the unique cocircuit that is disjoint from $B - e$.
    (b) If $f \in E(M) - B$, prove that $f \in C^*(e, B)$ if and only if $e \in C(f, B)$.

11. (Bondy and Welsh 1971) Let $C_1^*, C_2^*, \ldots, C_r^*$ be cocircuits of a rank-r matroid M. Prove that the following are equivalent:
    (a) M has a basis B such that $C_1^*, C_2^*, \ldots, C_r^*$ are all of the fundamental cocircuits with respect to B;
    (b) for all j in $\{1, 2, \ldots, r\}$, the set C_j^* is not contained in $\bigcup_{i \neq j} C_i^*$.

12. Let M be a matroid on the set E and, for all subsets X of E, define
    $r_1(X) = r(E - X) + |X| - r(M)$. Prove directly that:
    (a) r_1 satisfies (R1)–(R3) and hence deduce that r_1 is the rank function
        of a matroid M_1 on E;
    (b) B is a basis of M_1 if and only if $E - B$ is a basis of M.

13. A *cyclic flat* of a matroid M is a flat of M that is the union of a (possibly
    empty) set of circuits. Show that:
    (a) X is a flat of M if and only if $E(M) - X$ is the union of a (possibly
        empty) set of cocircuits of M. Deduce that F is a cyclic flat of M if
        and only if $E(M) - F$ is a cyclic flat of M^*.
    (b) A flat F of M is cyclic if and only if $M|F$ has no coloops.
    (c) A matroid on a fixed set is uniquely determined by a list of its cyclic
        flats and their ranks.

14. (Brylawski 1986b) Let $\mathcal{B}$ be a collection of subsets of a set E and suppose
    $\mathcal{B}$ satisfies the following conditions:
    (B1) $\mathcal{B} \neq \emptyset$.
    (B3)   If $B_1, B_2 \in \mathcal{B}$ and $B_1 \subseteq B_2$, then $B_1 = B_2$.
    (B2)$^{*-}$ If $B_1, B_2 \in \mathcal{B}$ and $e \in B_1$, then there is an element f of B_2
            such that $(B_2 - f) \cup e \in \mathcal{B}$.

    (a) Give an example when $|E| = 5$ to show that $\mathcal{B}$ need not be the set of
        bases of a matroid on E.
    (b) Prove that the set of bases of a matroid on E satisfies (B2)$^{*-}$.

15. Find all graphs whose cycle matroids have a circuit–hyperplane.

16. (Dawson 1981) Let $\mathcal{A}$ and $\mathcal{D}$ be clutters on a set S. Prove that the following
    are equivalent by showing that each implies its successor:
    (a) $\mathcal{D} = b(\mathcal{A})$.
    (b) For all $X \subseteq S$, the set X contains a member of $\mathcal{D}$ if and only if $S - X$
        does not contain a member of $\mathcal{A}$.
    (c) For all $Y \subseteq S$, the set Y does not contain a member of $\mathcal{D}$ if and only
        if $S - Y$ contains a member of $\mathcal{A}$.
    (d) $\mathcal{A} = b(\mathcal{D})$.

17. Show that $AG(3, 2)$ is identically self-dual.

18. Let G_1 and G_2 be the graphs shown in Figure 1.20. The duals of $M(G_1)$
    and $M(G_2)$, which are denoted $M^*(G_1)$ and $M^*(G_2)$, respectively, are both
    graphic and are both transversal. Find the following for each i in $\{1, 2\}$:
    (a) a graph H_i such that $M(H_i) \cong M^*(G_i)$;
    (b) a family of sets $\mathcal{A}_i$ such that $M[\mathcal{A}_i] \cong M^*(G_i)$;
    (c) a geometric representations for $M^*(G_i)$.

19. (Brualdi 1969) Let B_1 and B_2 be bases of a matroid M. Prove that if
    $x \in B_1 - B_2$, then there is an element y of $B_2 - B_1$ such that both $(B_1 - x) \cup y$
    and $(B_2 - y) \cup x$ are bases of M.

20. Characterize paving matroids in terms of their collections of flats.

21. (a) Let I be an independent set of a matroid M. For $e \in E(M) - I$, prove that $I \cup e$ is independent if and only if M has a cocircuit C^* such that $e \in C^*$ and $C^* \cap I = \emptyset$.

   (b) (Dawson 1980) Let $\mathcal{B}$ and $\mathcal{C}^*$ be the sets of bases and cocircuits of a matroid M. Let $w : E(M) \to \mathbb{R}$ be a one-to-one weight function. Prove that the set B_D produced by the following modified greedy algorithm is the unique maximum-weight basis of M.

   (i) Set $Y_0 = \emptyset$ and $j = 0$.

   (ii) If $E(M) - Y_j$ contains no member of $\mathcal{C}^*$, then let $Y_j = B_D$ and go to (iv); otherwise, let C^*_{j+1} be a member of $\mathcal{C}^*$ contained in $E(M) - Y_j$ and choose e_{j+1} to be an element of C^*_{j+1} of maximum weight. Let $Y_{j+1} = Y_j \cup e_{j+1}$ and go to (iii).

   (iii) Add 1 to j and go to (ii).

   (iv) Stop.

   (c) Consider the Pappus matroid and the three-dimensional Desargues matroid shown in Figures 1.16 and 1.17. Taking the element labels as weights, use the above algorithm to find maximum-weight bases in the *duals* of these two matroids.

2.2 Duals of representable matroids

In this section, we prove that the dual of an $\mathbb{F}$-representable matroid M is $\mathbb{F}$-representable by constructing an explicit representation for M^* from a representation for M. We also link matroid duality and vector-space orthogonality.

Let A be an $m \times n$ matrix over a field $\mathbb{F}$, and let M be the vector matroid $M[A]$ of A. Then the ground set of M is the set E of column labels of A. In general, $M[A]$ does not uniquely determine A. Indeed, it is easily checked that M remains unchanged if one performs any of the following operations on A.

2.2.1 *Interchange two rows.*

2.2.2 *Multiply a row by a non-zero member of $\mathbb{F}$.*

2.2.3 *Replace a row by the sum of that row and another.*

2.2.4 *Adjoin or remove a zero row.*

2.2.5 *Interchange two columns (the labels moving with the columns).*

2.2.6 *Multiply a column by a non-zero member of $\mathbb{F}$.*

2.2.7 *Replace each matrix entry by its image under some automorphism of $\mathbb{F}$.*

We call operations 2.2.1–2.2.3 *elementary row operations*. Operation 2.2.7 will be examined in detail in Section 6.3. It will not be considered further in this section.

Assume that the matrix A is non-zero. It is well known and not difficult to check that, by a sequence of operations of types 2.2.1–2.2.5, one can reduce A to the form $[I_r|D]$ where I_r is the $r \times r$ identity matrix and D is some $r \times (n - r)$

matrix over $\mathbb{F}$. Evidently $r = r(M)$. Suppose that the columns of $[I_r|D]$ are labelled, in order, $e_1, e_2, \ldots, e_n$. Then $\{e_1, e_2, \ldots, e_r\}$ is a basis B of M. Moreover, it is natural to label the rows of D, in order, by $e_1, e_2, \ldots, e_r$. Thus $M[A]$ can be represented, as in Figure 2.5, both by the matrix $[I_r|D]$, whose columns are labelled $e_1, e_2, \ldots, e_n$, and by the matrix D, whose rows are labelled $e_1, e_2, \ldots, e_r$ and whose columns are labelled $e_{r+1}, e_{r+2}, \ldots, e_n$. We shall refer to both $[I_r|D]$ and D as *standard representative matrices* for M and will sometimes add *with respect to* $\{e_1, e_2, \ldots, e_r\}$ to this description. To limit possible ambiguities with this terminology, we shall also refer to D as a *reduced standard representative matrix*. Note that, in $[I_r|D]$, only the columns are labelled, whereas both the rows and columns are labelled in D.

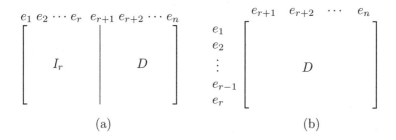

(a)   (b)

FIG. 2.5. Standard representative matrices for M.

The next proof will use the familiar notion of *rank of a matrix*. Recall that this is the maximum size of a linearly independent set of rows of the matrix or, equivalently, the maximum size of a linearly independent set of columns.

Theorem 2.2.8 *Let M be the vector matroid of the matrix $[I_r|D]$ where the columns of this matrix are labelled, in order, $e_1, e_2, \ldots, e_n$ and $1 \le r < n$. Then M^* is the vector matroid of $[-D^T|I_{n-r}]$ where its columns are also labelled $e_1, e_2, \ldots, e_n$ in that order.*

Proof We may assume that $[I_r|D]$ is as in Figure 2.5(a). Then $[-D^T|I_{n-r}]$ is

$$
\begin{array}{cc}
e_1 \;\; e_2 \cdots e_r & e_{r+1} \; e_{r+2} \cdots e_n \\
\left[\begin{array}{c|c}
 -D^T & I_{n-r}
\end{array}\right] &
\end{array}.
$$

Let B be a basis of M. We shall first show that $E - B$ is a basis of the vector matroid of $[-D^T|I_{n-r}]$. The only effect on $[-D^T|I_{n-r}]$ of rearranging rows and rearranging columns of $[I_r|D]$ is to rearrange columns and rearrange rows of $[-D^T|I_{n-r}]$. By such a rearrangement, we may assume that $B = \{e_{r-t+1}, e_{r-t+2},$

$\dots, e_r, e_{r+1}, \dots, e_{2r-t}\}$ for some t in $\{0, 1, \dots, r\}$. Thus $[I_r|D]$ can be partitioned into blocks as follows:

$$
\begin{array}{ccccc}
e_1 \cdots e_{r-t} & e_{r-t+1} \cdots e_r & e_{r+1} \cdots e_{2r-t} & e_{2r-t+1} \cdots e_n \\
\left[\begin{array}{cccc}
I_{r-t} & 0 & D_1 & D_2 \\
0 & I_t & D_3 & D_4
\end{array}\right].
\end{array}
$$

The matrix

$$
\begin{bmatrix} 0 & D_1 \\ I_t & D_3 \end{bmatrix}
$$

has rank r since B is a basis. Thus D_1, and hence $-D_1^T$, has rank $r - t$. The partition of $[I_r|D]$ induces the following partition of $[-D^T|I_{n-r}]$:

$$
\begin{array}{ccccc}
e_1 \cdots e_{r-t} & e_{r-t+1} \cdots e_r & e_{r+1} \cdots e_{2r-t} & e_{2r-t+1} \cdots e_n \\
\left[\begin{array}{cccc}
-D_1^T & -D_3^T & I_{r-t} & 0 \\
-D_2^T & -D_4^T & 0 & I_{n-(2r-t)}
\end{array}\right].
\end{array}
$$

In this matrix, the submatrix corresponding to $E - B$ is

$$
\begin{bmatrix}
-D_1^T & 0 \\
-D_2^T & I_{n-(2r-t)}
\end{bmatrix}.
$$

The rank of this submatrix is the sum of the ranks of $I_{n-(2r-t)}$ and $-D_1^T$, that is, $(n - (2r - t)) + (r - t)$, which is $n - r$. Hence $E - B$ is a basis of the vector matroid of $[-D^T|I_{n-r}]$. An easy extension of this argument shows that every basis of the last matroid arises in this way. Therefore $[-D^T|I_{n-r}]$ is indeed a representation for M^*. □

The reader may be curious why $[-D^T|I_{n-r}]$ is used in the last theorem instead of $[D^T|I_{n-r}]$ since, by 2.2.6, $M[D^T|I_{n-r}] = M[-D^T|I_{n-r}]$. The reason for this choice will be made clear in Proposition 2.2.23 where we link matroid duality and vector-space orthogonality. Another feature of the last theorem deserving comment is that the rank, r, of the matrix A is assumed to be neither 0 nor n. If $r = 0$, then $M \cong U_{0,n}$, so $M^* \cong U_{n,n} \cong M[I_n]$. If $r = n$, then $M \cong U_{n,n}$, so $M^* \cong U_{0,n} \cong M[0_n]$ where 0_n is the $n \times n$ zero matrix. On combining these observations with Theorem 2.2.8, we immediately obtain the following result.

Corollary 2.2.9 *If a matroid M is representable over a field $\mathbb{F}$, then M^* is also representable over $\mathbb{F}$.* □

In particular, the dual of a binary matroid is binary, and the dual of a ternary matroid is ternary.

Example 2.2.10 Consider the following matrix, A.

$$\left[\ I_3 \ \middle|\ \begin{array}{ccc} 0 & 1 & 1 & 1 \\ 1 & 0 & 1 & 1 \\ 1 & 1 & 0 & 1 \end{array} \right]$$

It is not difficult to show that if A is viewed as a matrix over $GF(2)$, it represents F_7, whereas, viewed as a matrix over $GF(3)$, it represents F_7^-. By Theorem 2.2.8, F_7^* and $(F_7^-)^*$ are represented by the matrix in Figure 2.6(a) viewed over $GF(2)$ and $GF(3)$, respectively, where, in the former case, $-1 = 1$. When the last matrix is viewed over $GF(3)$, scalar multiplication of columns enables us to change the signs of all negative entries. Hence F_7^* and $(F_7^-)^*$ are represented by the matrix in Figure 2.6(b) viewed over $GF(2)$ and $GF(3)$, respectively. Before leaving this example, we remark that, by Proposition 2.1.7, as F_7^- is a relaxation of F_7, it follows that $(F_7^-)^*$ is a relaxation of F_7^*.　　　　　□

$$\left[\ \begin{array}{ccc} 0 & -1 & -1 \\ -1 & 0 & -1 \\ -1 & -1 & 0 \\ -1 & -1 & -1 \end{array} \ \middle|\ I_4 \ \right] \qquad \left[\ \begin{array}{ccc} 0 & 1 & 1 \\ 1 & 0 & 1 \\ 1 & 1 & 0 \\ 1 & 1 & 1 \end{array} \ \middle|\ I_4 \ \right]$$

$$\text{(a)} \qquad\qquad\qquad\qquad \text{(b)}$$

FIG. 2.6. Both matrices represent the same matroid, which is F_7^* or $(F_7^-)^*$.

We shall show next that the dual of a regular matroid is regular. Recall that M is regular if $M \cong M[A]$ for some totally unimodular matrix A, the latter being a real matrix for which every square submatrix has determinant in $\{0, 1, -1\}$. By the last corollary, if M is regular, then M^* is certainly $\mathbb{R}$-representable, so we need to check that M^* has a totally unimodular representation. We begin by recalling for reference, both here and throughout the rest of the book, the following basic facts about determinants of matrices. The proofs of these results can be found in most linear algebra texts.

Let A be an $n \times n$ matrix $[a_{ij}]$ over a field $\mathbb{F}$. The *determinant* of A, denoted $\det A$, is defined to be

2.2.11 $\sum_\sigma \text{sgn}\,\sigma\, a_{1\sigma(1)} a_{2\sigma(2)} \cdots a_{n\sigma(n)}$

where the summation is over all permutations σ of $\{1, 2, \ldots, n\}$, and $\text{sgn}\,\sigma$ is $+1$ or -1 according to whether σ is a product of an even or an odd number of cycles of length two. The matrix A is *non-singular* if $\det A \neq 0$.

2.2.12 *For all i in $\{1, 2, \ldots, n\}$,*

$$\det A = \sum_{j=1}^{n} a_{ij}(-1)^{i+j} \det(A_{ij})$$

where A_{ij} is the matrix obtained from A by deleting row i and column j.

2.2.13 *For all j in $\{1, 2, \ldots, n\}$,*

$$\det A = \sum_{i=1}^{n} a_{ij}(-1)^{i+j} \det(A_{ij}).$$

2.2.14 $\det(A^T) = \det A$.

2.2.15 *If D is obtained from A by multiplying some row or column by a constant c, then $\det D = c \det A$.*

2.2.16 *If D is obtained from A by interchanging two rows or interchanging two columns, then $\det D = -\det A$.*

2.2.17 *If D is obtained from A by replacing row i by row $i + c$ row j for some constant c and some $j \neq i$, then $\det D = \det A$.*

On combining 2.2.15–2.2.17, it is not difficult to show that

2.2.18 $\det A = 0$ *if and only if the columns of A are linearly dependent.*

Finally,

2.2.19 $\det(AD) = \det A \cdot \det D$.

A familiar operation that forms part of the process of transforming a matrix into row–echelon form is pivoting. Given an $m \times n$ matrix X and a non-zero entry x_{st} of X, a *pivot* of X on x_{st} transforms the tth column of X into the sth standard basis vector. This is achieved by applying the following sequence of operations to X:

(i) For each i in $\{1, 2, \ldots, s - 1, s + 1, \ldots, m\}$, replace row i of X by row $i - (x_{it}/x_{st})$ row s.

(ii) Multiply row s by $1/x_{st}$.

We call x_{st} the *pivot entry* of X; its value in the transformed matrix is 1.

We need two more lemmas to show that the class of regular matroids is *closed under duality*, that is, the dual of every member of the class is also in the class.

Lemma 2.2.20 *Let X be a totally unimodular matrix. If Y is obtained from X by pivoting on the non-zero entry x_{st} of the latter, then Y is totally unimodular.*

Proof Let X' and Y' be corresponding submatrices of X and Y, each having their rows and columns indexed by the sets J_R and J_C, respectively. We want to show that $\det Y' \in \{0, 1, -1\}$. If $s \in J_R$, then it follows easily from 2.2.15 and 2.2.17 that $|\det Y'| = |\det X'|$ and so $\det Y' \in \{0, 1, -1\}$. Thus we may assume that $s \notin J_R$. In that case, if $t \in J_C$, then Y' has a zero column, so $\det Y' = 0$. Hence we may also assume that $t \notin J_C$. Now let X'' and Y'' be the submatrices of X and Y whose rows and columns are indexed by $J_R \cup \{s\}$ and $J_C \cup \{t\}$. As the only non-zero entry in the column of Y'' indexed by t is a 1, we have $|\det Y''| = |\det Y'|$. But, as above, $|\det Y''| = |\det X''|$. Hence again, $\det Y' \in \{0, 1, -1\}$ and the lemma is proved. $\qquad \square$

Lemma 2.2.21 *Let $\{e_1, e_2, \ldots, e_r\}$ be a basis of a matroid M of non-zero rank. Then M is regular if and only there is a totally unimodular matrix $[I_r|D]$ representing M over $\mathbb{R}$ whose first r columns are labelled, in order, $e_1, e_2, \ldots, e_r$.*

Proof Let M be regular. Then $M \cong M[A]$ for some totally unimodular matrix A. By 2.2.16 and Lemma 2.2.20, A can be transformed, by a sequence of pivots and row and column interchanges, into a totally unimodular matrix of the form $[I_r|D]$ in which the first r columns are labelled $e_1, e_2, \ldots, e_r$. Then $M \cong M[I_r|D]$. The converse is immediate from the definition of regularity.   □

Proposition 2.2.22 *The dual of a regular matroid is regular.*

Proof Let M be regular. If $r(M)$ or $r^*(M)$ is 0, then M^* is certainly regular, so we may assume that both $r(M)$ and $r^*(M)$ are positive. By Lemma 2.2.21, $M \cong M[I_r|D]$ for some totally unimodular matrix $[I_r|D]$. Thus, by Theorem 2.2.8, M^* is represented by $[-D^T|I_{n-r}]$ and, since the last matrix is certainly totally unimodular, the proposition follows.   □

Some authors (see, for example, White 1986) call M^* the *orthogonal matroid* of M. This is because duality for representable matroids is actually an extension of the notion of orthogonality in vector spaces. Recall that if $\underline{v}$ and $\underline{w}$ are the vectors $(v_1, v_2, \ldots, v_n)$ and $(w_1, w_2, \ldots, w_n)$ of $V(n, \mathbb{F})$, then their *dot product*, $\underline{v} \cdot \underline{w}$, is $\sum_{i=1}^{n} v_i w_i$. These vectors are *orthogonal* if $\underline{v} \cdot \underline{w} = 0$. Moreover, $\underline{v}$ is *orthogonal to a subspace* W of $V(n, \mathbb{F})$ if $\underline{v}$ is orthogonal to every member of W. One easily checks that the set $W^\perp$ of vectors in $V(n, \mathbb{F})$ that are orthogonal to W forms a subspace; we call this the *orthogonal subspace* of W. Clearly $(W^\perp)^\perp \supseteq W$. In fact, equality holds here but the proof of this fact will use the next result, which establishes the link between matroid duality and orthogonality in vector spaces. This link is stated in terms of row spaces of matrices. In general, the *row space* $\mathcal{R}(A)$ of an $m \times n$ matrix A over a field $\mathbb{F}$ is the subspace of $V(n, \mathbb{F})$ that is spanned by the rows of A. In particular, if $\mathbb{F} = GF(q)$, then $\mathcal{R}(A)$ is a subspace of $V(n, q)$. Such a subspace is called a *linear q-ary code of length n* (Hill 1986).

Proposition 2.2.23 *Let $[I_r|D]$ be an $r \times n$ matrix over a field $\mathbb{F}$ where r is in $\{1, 2 \ldots, n-1\}$. Then the orthogonal subspace of $\mathcal{R}[I_r|D]$ is $\mathcal{R}[-D^T|I_{n-r}]$.*

Proof Let $\underline{v}$ and $\underline{w}$ be rows g and h of $[I_r|D]$ and $[-D^T|I_{n-r}]$, respectively, and let $D = [d_{ij}]$. Taking δ_{ij} to be 0 unless $i = j$, in which case, $\delta_{ij} = 1$, we have

$$\underline{v} = (\delta_{g1}, \delta_{g2}, \ldots, \delta_{gg}, \ldots, \delta_{gr}, d_{g1}, d_{g2}, \ldots d_{g(n-r)})$$

and

$$\underline{w} = (-d_{1h}, -d_{2h}, \ldots, -d_{rh}, \delta_{h1}, \delta_{h2}, \ldots, \delta_{hh}, \ldots, \delta_{h(n-r)}).$$

Thus $\underline{v} \cdot \underline{w} = -d_{gh} + d_{gh} = 0$, that is, $\underline{v}$ and $\underline{w}$ are orthogonal. Therefore the orthogonal subspace of $\mathcal{R}[I_r|D]$ contains $\mathcal{R}[-D^T|I_{n-r}]$.

Now suppose $\underline{u}$ is an arbitrary member $(u_1, u_2, \ldots, u_n)$ of the orthogonal subspace of $\mathcal{R}[I_r|D]$. Then, by taking the dot product of $\underline{u}$ with each row of $[I_r|D]$, we get that $u_i = -\sum_{j=1}^{n-r} u_{r+j} d_{ij}$ for all i in $\{1, 2, \ldots, r\}$. Hence

$$\underline{u} = u_{r+1}(-d_{11}, -d_{21}, \dots, -d_{r1}, 1, 0, 0, \dots, 0)$$
$$+ u_{r+2}(-d_{12}, -d_{22}, \dots, -d_{r2}, 0, 1, 0, \dots, 0) + \dots$$
$$+ u_{r+(n-r)}(-d_{1(n-r)}, -d_{2(n-r)}, \dots, -d_{r(n-r)}, 0, 0, 0, \dots, 1).$$

Since the $n - r$ vectors on the right-hand side are the rows of $[-D^T | I_{n-r}]$, we conclude that $\underline{u} \in \mathcal{R}[-D^T | I_{n-r}]$. □

Briefly addressing this link with coding theory, we note that, when $\mathbb{F} = GF(q)$, the $r \times n$ matrix $[I_r | D]$ is a *generator matrix in standard form* for the $[n, r]$-*code* $\mathcal{R}[I_r | D]$, while $[-D^T | I_{n-r}]$ is a *parity-check matrix* for the code. Duality in coding theory corresponds to vector-space orthogonality, and $[-D^T | I_{n-r}]$ is a generator matrix for the dual code of $\mathcal{R}[I_r | D]$. The following is a straightforward consequence of the last proposition and we leave the proof to the reader.

Corollary 2.2.24 *Let W be an r-dimensional subspace of $V(n, \mathbb{F})$. Then $W^\perp$ has dimension $n - r$. Moreover, $(W^\perp)^\perp = W$.* □

By this corollary, if $[I_r | D]$ is an $r \times n$ matrix over $\mathbb{F}$ where $1 \le r \le n - 1$, then $\mathcal{R}[I_r | D]$ and $\mathcal{R}[-D^T | I_{n-r}]$ have the right dimensions to be complementary subspaces of $V(n, \mathbb{F})$. Moreover, if $\mathbb{F} = \mathbb{R}$, then these subspaces are indeed complementary. However, in general, these row spaces may have non-trivial intersection. For instance, if $\mathbb{F}$ is $GF(2)$ and $W = \mathcal{R}[I_r | I_r]$, then $W^\perp$ actually equals W.

We shall denote by $M^*[A]$ the dual of the vector matroid $M[A]$. Theorem 2.2.8 and the remarks following it give us a procedure for constructing a representation for $M^*[A]$ for any $m \times n$ matrix A of rank r. If r is 0 or n, then $M^*[A]$ is represented by I_n or 0_n, respectively. Otherwise, we reduce A to the form $[I_r | D]$. Then $M^*[A]$ is represented by $[-D^T | I_{n-r}]$. Notice that, when $0 < r < n$, the matroid $M^*[A]$ is also represented by $[D^T | I_{n-r}]$ since, as noted in 2.2.6, multiplying a column of a matrix by a non-zero scalar does not alter the associated vector matroid. In general, however, this alternative representation no longer has the property that its row space is the orthogonal subspace of $\mathcal{R}[I_r | D]$.

Example 2.2.25 Let P_8 be the vector matroid of the matrix

$$\begin{array}{cccccccc} 1 & 2 & 3 & 4 & 5 & 6 & 7 & 8 \end{array}$$
$$\left[\begin{array}{c|cccc} & & 0 & 1 & 1 & -1 \\ & & 1 & 0 & 1 & 1 \\ I_4 & & 1 & 1 & 0 & 1 \\ & & -1 & 1 & 1 & 0 \end{array} \right]$$

over $GF(3)$. Then P_8^* is the vector matroid of the matrix

$$\begin{array}{cccccccc} 5 & 6 & 7 & 8 & 1 & 2 & 3 & 4 \end{array}$$
$$\left[\begin{array}{c|cccc} & & 0 & 1 & 1 & -1 \\ & & 1 & 0 & 1 & 1 \\ I_4 & & 1 & 1 & 0 & 1 \\ & & -1 & 1 & 1 & 0 \end{array} \right].$$

Hence P_8 is self-dual. Note too that P_8 is a paving matroid.   □

This example illustrates the fact that if $M = M[I_r|D]$ where D is a symmetric matrix, then M is self-dual. The matroid P_8 will play a prominent role in Sections 6.4 and 6.5. Note that, when the matrix representing P_8 is viewed over $GF(2)$, it represents $AG(3,2)$. However, P_8 and $AG(3,2)$ have ten and fourteen circuit-hyperplanes, respectively, so they are certainly not isomorphic.

In Section 1.1, we raised the question as to whether every matroid is graphic or representable, and we saw that $U_{2,4}$ is the smallest non-graphic matroid. We shall now answer the second part of this question by proving that the Vámos matroid, introduced in Example 2.1.25 and Figure 2.4, is not representable (Vámos 1968). In fact, as we shall see in Proposition 6.4.10, V_8 is a smallest non-representable matroid. The crucial feature of V_8 that prevents it from being representable is that the points $5, 6, 7$, and 8 are not coplanar.

Proposition 2.2.26 *The Vámos matroid V_8 is not representable.*

Proof Assume that V_8 is representable over some field $\mathbb{F}$. Then there is a 4×8 matrix A over $\mathbb{F}$ whose columns are labelled by $\{1, 2, \ldots, 8\}$ so that $V_8 = M[A]$. Thus the columns of A are members of $V(4, \mathbb{F})$, the 4-dimensional vector space over $\mathbb{F}$. Moreover, for every subset X of $E(V_8)$, the rank of X in V_8 is equal to the rank of the submatrix of A whose columns are labelled by the elements of X. In turn, this is equal to the dimension of the subspace of $V(4, \mathbb{F})$ that is spanned by the columns labelled by the elements of X. If $X = \{x_1, x_2, \ldots, x_k\} \subseteq E(V_8)$, then we denote this subspace by $W(x_1, x_2, \ldots, x_k)$. Now

$$\dim(W(5,6) \cap W(1,2,3,4)) = \dim W(5,6) + \dim W(1,2,3,4)$$
$$- \dim(W(5,6) + W(1,2,3,4))$$
$$= 2 + 3 - 4 = 1.$$

Thus $W(5,6) \cap W(1,2,3,4)$ is $\langle v \rangle$, the subspace of $V(4, \mathbb{F})$ generated by some non-zero vector v. Also

$$\dim(W(1,4,5,6) \cap W(1,2,3,4)) = \dim W(1,4,5,6) + \dim W(1,2,3,4)$$
$$- \dim(W(1,4,5,6) + W(1,2,3,4))$$
$$= 3 + 3 - 4 = 2.$$

But $W(1,4)$ is a 2-dimensional subspace of $W(1,4,5,6) \cap W(1,2,3,4)$, so

$$W(1,4) = W(1,4,5,6) \cap W(1,2,3,4) \supseteq W(5,6) \cap W(1,2,3,4) = \langle v \rangle.$$

By symmetry, $\langle v \rangle \subseteq W(2,3)$.

Now
$$\dim(W(1,4) \cap W(2,3)) = 2 + 2 - 3 = 1.$$

But $\langle v \rangle$ is a subspace of $W(1,4) \cap W(2,3)$, so $\langle v \rangle = W(1,4) \cap W(2,3)$. Geometrically, if V_8 is $\mathbb{F}$-representable, then we can add a point p to the diagram so that

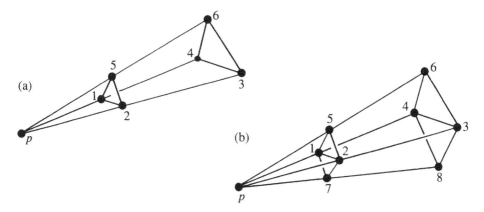

FIG. 2.7. Showing that the Vámos matroid is non-representable.

its relationship to $\{1, 2, 3, 4, 5, 6\}$ is as shown in Figure 2.7(a). Of course, p corresponds to the subspace $\langle v \rangle$ of $V(4, \mathbb{F})$. By symmetry again, as $W(1, 4) \cap W(2, 3) = W(5, 6) \cap W(1, 2, 3, 4)$, it follows that $W(1, 4) \cap W(2, 3) = W(7, 8) \cap W(1, 2, 3, 4)$. Hence $\langle v \rangle \subseteq W(7, 8)$, and so $\langle v \rangle \subseteq W(5, 6) \cap W(7, 8)$. Geometrically, this says that p is on both the line containing 5 and 6 and the line containing 7 and 8, so $\{5, 6, 7, 8\}$ is dependent (see Figure 2.7(b)); or, in terms of subspaces:

$$\dim W(5, 6, 7, 8) = \dim(W(5, 6) + W(7, 8))$$
$$= 2 + 2 - \dim(W(5, 6) \cap W(7, 8)) \leq 3.$$

This is a contradiction as $r(\{5, 6, 7, 8\}) = 4$, so V_8 is not $\mathbb{F}$-representable. □

Exercises

1. For $i = 1, 2$, let A_i be an $m_i \times n$ matrix over a field $\mathbb{F}$. If every row of A_2 is a linear combination of rows of A_1, prove that $M \begin{bmatrix} A_1 \\ A_2 \end{bmatrix} = M[A_1]$.

2. Let A be an $m \times n$ matrix over a field $\mathbb{F}$.
   (a) For each of the row operations 2.2.1–2.2.3, specify an $m \times m$ matrix L such that multiplying A on the left by L has the same effect as the row operation.
   (b) For each of the column operations 2.2.4 and 2.2.5, specify an $n \times n$ matrix R such that multiplying A on the right by R has the same effect as the column operation.

3. Let A be a non-zero matrix over a field $\mathbb{F}$ and suppose $M[A]$ has no coloops. Show that no row of A contains exactly one non-zero entry.

4. Prove that $U_{n,n+2}$ is $\mathbb{F}$-representable if and only if $|\mathbb{F}| \geq n + 1$.

5. Let M_1 and M_2 be matroids on a set E and cl_1 and cl_2 be their closure operators. Prove that $M_1 = M_2^*$ if and only if, for every partition $(X, Y, \{z\})$ of E where X or Y may be empty, the element z is in exactly one of $\text{cl}_1(X)$ and $\text{cl}_2(Y)$.

6. For matrices A_1 and A_2 over a field $\mathbb{F}$, each having n columns, show that:
   (a) If $\mathcal{R}(A_1) = \mathcal{R}(A_2)$, then $M[A_1] = M[A_2]$.
   (b) If $\mathcal{R}(A_2) = \mathcal{R}(A_1)^{\perp}$, then $M[A_2] = M^*[A_1]$.

7. Show that, in a binary matroid, a circuit and cocircuit cannot have an odd number of common elements.

8. Find an $\mathbb{R}$-representation for M^* when M is the vector matroid of the following matrix over $\mathbb{R}$:

$$
\begin{array}{ccccccccc}
1 & 2 & 3 & 4 & \cdots & n-1 & n
\end{array}
$$
$$
\begin{bmatrix}
1 & 0 & 0 & 0 & \cdots & 0 & 1 \\
0 & 1 & 1 & 0 & \cdots & 0 & 0 \\
0 & 0 & 1 & 1 & \cdots & 0 & 0 \\
\vdots & \vdots & \vdots & \vdots & \ddots & \vdots & \vdots \\
0 & 0 & 0 & 0 & \cdots & 1 & 0 \\
0 & 0 & 0 & 0 & \cdots & 1 & (-1)^n
\end{bmatrix}.
$$

9. Let T_8 and M be the vector matroids of the following matrices over $GF(3)$:

$$
\begin{array}{cccc|cccc}
1 & 2 & 3 & 4 & 5 & 6 & 7 & 8
\end{array}
$$
$$
\left[\begin{array}{cccc|cccc}
 & & & & 0 & 1 & 1 & 1 \\
 & I_4 & & & 1 & 0 & 1 & 1 \\
 & & & & 1 & 1 & 0 & 1 \\
 & & & & 1 & 1 & 1 & 0
\end{array}\right],
\qquad
\begin{array}{cccc|cccc}
1 & 2 & 3 & 4 & 5 & 6 & 7 & 8
\end{array}
$$
$$
\left[\begin{array}{cccc|cccc}
 & & & & -1 & 1 & 1 & 1 \\
 & I_4 & & & 1 & -1 & 1 & 1 \\
 & & & & 1 & 1 & -1 & 1 \\
 & & & & 1 & 1 & 1 & -1
\end{array}\right].
$$

(a) Show that $M \cong R_8$, the real affine cube from Example 1.5.8, and give a geometric representation for T_8.
(b) Show that T_8 and R_8 are both self-dual.
(c) Show that R_8 is identically self-dual but T_8 is not.
(d) Show that each of T_8 and R_8 is a tipless 4-spike by finding, for each of these matroids, a vector that corresponds to a tip.
(e) Show that if $N \in \{T_8, R_8\}$, then $(N|\{1, 2, \ldots, 7\})^* \cong F_7^-$.
(f) Consider the matrices $[I_4|D_1]$ and $[I_4|D_2]$ over $GF(3)$ where

$$
D_1 = \begin{bmatrix}
0 & 1 & 1 & 1 \\
1 & 0 & 1 & 1 \\
1 & 1 & 0 & -1 \\
1 & 1 & 1 & 0
\end{bmatrix}
\quad \text{and} \quad
D_2 = \begin{bmatrix}
0 & 1 & 1 & -1 \\
1 & 0 & 1 & -1 \\
1 & 1 & 0 & -1 \\
1 & 1 & 1 & 1
\end{bmatrix}.
$$

(g) Show, by applying a sequence of the operations 2.2.1–2.2.6 to $[I_4|D_1]$ and $[I_4|D_2]$, that $M[I_4|D_1] \cong T_8$ and $M[I_4|D_2] \cong R_8$.

10. (a) Is the matroid P_8 in Example 2.2.25 identically self-dual?
    (b) Can P_8 be obtained from $AG(3, 2)$ by relaxing circuit–hyperplanes?

11. (a) Show that if $\{b_1, b_2, \ldots, b_r\}$ is a basis of an $\mathbb{F}$-representable matroid M, then there is a matrix $[I_r|D]$ over $\mathbb{F}$ such that $M[I_r|D] = M$ and the columns of I_r are labelled, in order, by $b_1, b_2, \ldots, b_r$.

(b) Let A_1 and A_2 be matrices over a field $\mathbb{F}$, each having n columns. Assume that $\mathcal{R}(A_2)$ is orthogonal to $\mathcal{R}(A_1)$ and that the sum of the dimensions of these row spaces is n. Show that $M[A_1] \cong M^*[A_2]$.

12. Let $\{b_1, b_2, \ldots, b_r\}$ be a basis B of a binary matroid M where $r \geq 1$. Let $E(M) - B = \{e_1, e_2, \ldots, e_k\}$ and define the $r \times k$ matrix $[d_{ij}]$ by

$$d_{ij} = \begin{cases} 1, & \text{if } b_i \in C(e_j, B); \\ 0, & \text{otherwise.} \end{cases}$$

(a) Prove that M equals the vector matroid of $[I_r|D]$ where this matrix is viewed over $GF(2)$, its columns being labelled, in order, by $b_1, b_2, \ldots, b_r, e_1, e_2, \ldots, e_k$.

(b) Find binary representations for $M(K_5)$ and $M(K_{3,3})$ and their duals. (Hint: Use (a) and the fact, to be proved in Chapter 5, that graphic matroids are binary.)

(c) Use (a) to show that none of $U_{2,4}$, F_7^-, V_8, and V_8^+ is binary.

2.3 Duals of graphic matroids

For a graph G, we denote the dual of the cycle matroid of G by $M^*(G)$. This matroid is called the *bond matroid* of G or, sometimes, the *cocycle matroid* of G. An arbitrary matroid that is isomorphic to the bond matroid of some graph is called *cographic*. In this section, we investigate cographic matroids, focusing particularly on the question of when such matroids are graphic.

For a set X of edges in a graph G, we shall denote by $G\backslash X$ the subgraph of G obtained by deleting all the edges in X. If $G\backslash X$ has more connected components than G, we shall call X an *edge cut* of G. An edge e for which $\{e\}$ is an edge cut is called a *cut-edge*. A minimal edge cut will also be called a *bond* or a *cocycle* of G. In Proposition 1.4.15, we proved that the complement of a hyperplane in $M(G)$ is a bond of G. The next result follows immediately on combining that result with Proposition 2.1.6(iii).

Proposition 2.3.1 *The following statements are equivalent for a subset X of the set of edges of a graph G:*

(i) X *is a circuit of $M^*(G)$.*

(ii) X *is a cocircuit of $M(G)$.*

(iii) X *is a bond of G.* □

For an arbitrary graph G, the circuits of $M(G)$ are the edge sets of cycles of G, while the circuits of $M^*(G)$ are the edge sets of bonds. If v is a non-isolated vertex of G and X is the set of edges meeting v, then X is an edge cut. If such an X is a minimal edge cut, we call it a *vertex bond* of G.

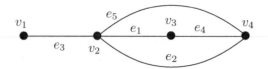

FIG. 2.8. The set of edges incident with a vertex need not be a bond.

Example 2.3.2 For the graph H shown in Figure 2.8, then the vertex bonds are $\{e_3\}$, $\{e_1, e_4\}$, and $\{e_2, e_4, e_5\}$. Evidently, although $\{e_1, e_2, e_3, e_5\}$ is the set of edges meeting the vertex v_2, it is not a bond. Moreover, it is easy to check that $M(H) \cong M^*(G)$ where G is the graph in Example 1.1.8 and Figure 1.2. $\square$

The graphic matroid in the last example has the property that its dual is also graphic. We saw in the last section that the dual of a representable matroid is always representable. We now consider the general problem of characterizing precisely when the dual of a graphic matroid is graphic. The following result indicates that this need not always occur and hints at an answer to this problem.

Proposition 2.3.3 *Neither $M^*(K_5)$ nor $M^*(K_{3,3})$ is graphic.*

Proof Let $M = M^*(K_5)$. We shall show that M is not graphic, leaving the similar argument for $M^*(K_{3,3})$ to the reader. Assume that $M \cong M(G)$ for some graph G. By Proposition 1.2.9, we may assume that G is connected. Now $M(K_5)$ has 10 elements and rank 4. Hence M has 10 elements and rank 6. Thus, by 1.3.6, G has 7 vertices and 10 edges, so its average vertex degree is $2|E(G)|/|V(G)|$, that is, $20/7$. Since this is less than three, G has a vertex of degree at most two. Hence M^* has a circuit of size one or two, so K_5 has a cycle of size one or two; a contradiction. We conclude that $M^*(K_5)$ is not graphic. $\square$

The pairing of the graphs K_5 and $K_{3,3}$ in the last result is reminiscent of Kuratowski's well-known characterization of planar graphs. This link is no coincidence and its precise nature will be stated at the end of this section. To establish the details of this link, we shall need some more definitions. A *planar embedding* of a graph G is a drawing of G in the Euclidean plane so that the vertices of G correspond to distinct points of the plane; each edge of G corresponds to a simple curve that connects the ends of the edge but meets no other vertices; and each point of intersection of two such simple curves is an end of both edges. We call a graph *planar* if it has a planar embedding. A *plane graph* consists of a graph G together with a planar embedding of G. Such a planar embedding G' of G can itself be regarded as an abstract graph and, as such, is isomorphic to G. The 5-vertex graph G shown in Figure 2.9 is not a plane graph. However, because it has a planar embedding, G', it is planar. We shall often blur the distinction between a plane graph and the associated abstract graph, just as we frequently identify an arbitrary graph H with a diagram representing H.

A plane graph G partitions the Euclidean plane into regions. Such regions, called *faces* of G, can be formally defined as follows: Suppose that P is the set of

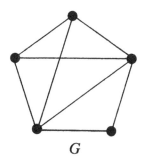
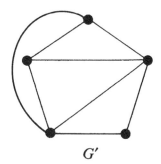

$$G \qquad\qquad\qquad G'$$

FIG. 2.9. A non-plane graph G and a planar embedding G' of it.

points of the plane that are not vertices of G and do not lie on edges of G. Two points, x and y, of P are in the same face of G if there is a simple curve joining x and y, all of whose points lie in P. A plane graph has exactly one unbounded face. We call this the *infinite face*.

Now recall that we are seeking graphs G whose bond matroids are graphic.

Theorem 2.3.4 *If G is planar, then $M^*(G)$ is graphic.*

To prove this result, we shall construct a graph whose cycle matroid is isomorphic to $M^*(G)$. We illustrate this construction by reference to the example in Figure 2.10. In that diagram, parts (a) and (c) show a plane graph G and its geometric dual, G^*. In (b), G and G^* have been superimposed to show that G^* is formed by taking a vertex in each face of G and joining two such vertices when the corresponding faces share an edge. The edges of G^* are drawn so as to cross the corresponding edges of G. The graph G' in (d) is a different planar embedding of G^*.

For an arbitrary plane graph G with at least one edge, the construction of G^*, the *geometric dual* of G, is formally described as follows: Choose a single point v_F in each face F of G. These points are to be the vertices of G^*. Suppose that the set of edges common to the boundaries of two faces F and F' is $\{e_1, e_2, \ldots, e_k\}$. Then we join v_F and $v_{F'}$ by k edges $e_1', e_2', \ldots, e_k'$, where e_i' crosses e_i but no other edge of G, and each e_i' is a simple curve. The only points common to two distinct edges of G^* can be their endpoints. If the edge e of G lies on the boundary of a single face F, we add a loop e' at v_F crossing e but no other edge of G or G^*.

To ensure that G^* is well-defined as a plane graph, we should specify when two plane graphs are equivalent. We shall not do this yet, preferring to delay a discussion of this until Section 8.6. It is not difficult to show that the graph G^* is connected. Moreover, provided G is connected,

2.3.5 $(G^*)^* = G$.

For an arbitrary planar graph G, a *geometric dual* of G is the geometric dual of some planar embedding of G. We observe that, because G may have several different planar embeddings, it may have several different geometric duals.

Example 2.3.6 In Figure 2.11, the plane graphs G_1 and G_2 are different planar embeddings of the same planar graph G. Thus G_1^* and G_2^* are geometric duals of G. Evidently $G_1^* \ncong G_2^*$.   □

Theorem 2.3.4 will follow without difficulty from the next result which shows that, although two geometric duals of a connected planar graph G need not be isomorphic, they do have isomorphic cycle matroids. As we shall show in Proposition 5.2.1, the same conclusion holds when G is disconnected.

Lemma 2.3.7 *If G^* is a geometric dual of a connected planar graph G, then*

$$M(G^*) \cong M^*(G).$$

Proof Let G^* be the geometric dual of a planar embedding G_0 of G. The construction of G^* from G_0 determines a bijection α from $E(G_0)$ to $E(G^*)$. We shall show that, under the map α,

$$M(G_0) \cong M^*(G^*).$$

Since $M(G_0) = M(G)$, the required result will then follow.

Let C be a circuit in $M(G_0)$. We want to show that $\alpha(C)$ is a bond in G^*. Now C forms a Jordan curve in the plane, and each edge in $\alpha(C)$ has one endpoint

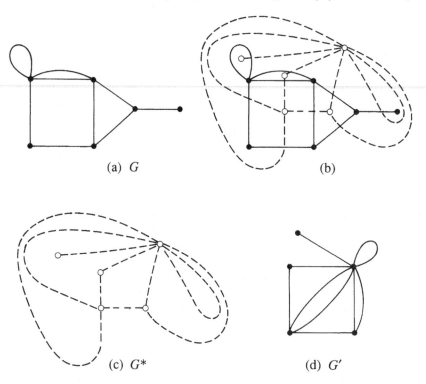

(a) G (b)

(c) G^* (d) G'

FIG. 2.10. Construction of the geometric dual of a plane graph.

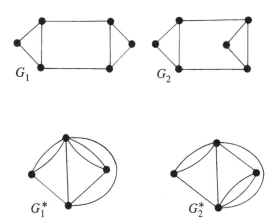

FIG. 2.11. Different embeddings giving different geometric duals.

inside and the other endpoint outside this closed curve. Thus $\alpha(C)$ is an edge cut in G^*. The fact that $\alpha(C)$ is a minimal edge cut will follow from what we shall show next, namely, that if X is a minimal edge cut in G^*, then $\alpha^{-1}(X)$ contains a cycle of G_0. Evidently, on deleting the edges of X from G^*, we obtain a graph having two components, G_1^* and G_2^*. Moreover, every edge in X has one endpoint in G_1^* and the other in G_2^*. If $|X| = 1$, then it follows from the construction of G^* that the single edge in $\alpha^{-1}(X)$ is a loop of G_0. Now suppose that $|X| > 1$ and let F be a face of G^* such that some edge, say x, of X is in the set F' of boundary edges of F. Then, as x is not a cut-edge of G^*, it is not difficult to check that x is not a cut-edge of $G^*[F']$. It is now an easy exercise in graph theory to show that F' contains a circuit C_x of $M(G^*)$ that contains x. Since X is a cocircuit of $M(G^*)$ and $x \in X \cap C_x$, it follows, by Proposition 2.1.11, that $|X \cap C_x| \geq 2$. Hence if a face of G^* meets an edge of X, it meets more than one such edge. Since G_0 is connected, we have, by (2.3.5), that $(G^*)^* = G_0$. Thus the faces of G^* correspond to vertices of G_0. Therefore, every vertex of G_0 meeting an edge of $\alpha^{-1}(X)$ meets at least two such edges. Now a graph in which every vertex has degree at least two contains a cycle (Exercise 2, Section 1.6). Thus the induced graph $G_0[\alpha^{-1}(X)]$ contains a cycle of G_0. The lemma now follows without difficulty by using Lemma 2.1.22. $\qquad\square$

Proof of Theorem 2.3.4 By Proposition 1.2.9, since G is planar, $M(G) \cong M(G_1)$ for some connected planar graph G_1. Hence $M^*(G) \cong M^*(G_1)$. But, by Lemma 2.3.7, $M^*(G_1) \cong M(G_1^*)$, so $M^*(G) \cong M(G_1^*)$ and $M^*(G)$ is graphic. $\qquad\square$

The converse of Theorem 2.3.4 is also true but we delay proving it until Section 5.2. One proof, which is outlined in Exercise 3 of that section, relies on the following famous theorem of Kuratowski (1930). A graph G' is a *subdivision* of a graph G if G' can be obtained from G by replacing non-loop edges of the latter by paths of non-zero lengths and replacing loop edges by cycles.

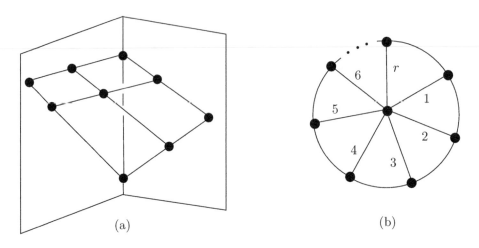

(a)   (b)

FIG. 2.12. (a) $M^*(K_{3,3})$. (b) $\mathcal{W}_r$.

Theorem 2.3.8 (Kuratowski's Theorem) *A graph is planar if and only if it has no subgraph that is a subdivision of K_5 or $K_{3,3}$.*   □

A proof of this theorem may be found, for example, in Diestel (1997).

Exercises

1. (a) Show that the geometric dual of a plane graph is connected.
   (b) Give an example of a plane graph G for which $(G^*)^* \neq G$.
2. Find geometric duals of the graphs obtained from K_5 and $K_{3,3}$ by deleting a single edge of each.
3. Find all identically self-dual graphic matroids.
4. For each positive integer n, find a planar graph G_n having at least n non-isomorphic geometric duals.
5. Complete the proof of Proposition 2.3.3 by showing that $M^*(K_{3,3})$ is not graphic.
6. Construct the geometric duals of the plane graphs in Figures 1.4(a) and 1.28(a).
7. Find all graphs for which every bond is a vertex bond.
8. If X is a set of edges in a graph G, specify the rank of X in $M^*(G)$ in terms of $G[X]$.
9. Show that $M^*(K_{3,3})$ has the geometric representation shown in Figure 2.12(a), which is a twisted 3×3 grid.
10. Let r be an integer exceeding one and $\mathcal{W}_r$ be the r-spoked wheel, that is, the graph in Figure 2.12(b). Show that:
    (a) $\mathcal{W}_3 \cong K_4$.
    (b) $\mathcal{W}_r^* \cong \mathcal{W}_r$.
    (c) $M(\mathcal{W}_4)$ is isomorphic to a restriction of $M^*(K_{3,3})$.

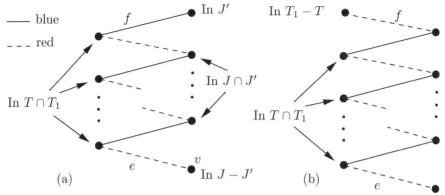

FIG. 2.13. Interchanging the colours on the edges of P.

2.4 Duals of transversal matroids

In this section, we shall identify the class of duals of transversal matroids as being a class of matroids that arise from linkings in directed graphs. While this is an important result, it is both more difficult and more specialized than the earlier results in this chapter. Consequently, after taking note of the result, the reader may wish to delay reading its proof until Chapter 11 when some of the results of this section will be generalized.

We begin with a technical result for transversal matroids. Let $\mathcal{A}$ be a family $(A_j : j \in J)$ of sets and let $M = M[\mathcal{A}]$. Evidently $r(M) \leq |J|$. We shall show next that M has a presentation having exactly $r(M)$ members.

Lemma 2.4.1 *Let $\mathcal{A}$ be $(A_1, A_2, \ldots, A_m)$, a family of subsets of a set E, and suppose that T, a maximal partial transversal of $\mathcal{A}$, is a transversal of $\mathcal{A}' = (A_1, A_2, \ldots, A_t)$ where $t < m$. Then every maximal partial transversal of $\mathcal{A}$ is a transversal of $\mathcal{A}'$.*

Proof Let $\Delta[\mathcal{A}]$ be the bipartite graph associated with $\mathcal{A}$. Let $J = \{1, 2, \ldots, m\}$ and $J' = \{1, 2, \ldots, t\}$. In $\Delta[\mathcal{A}]$, there is a matching of T onto J'. Colour the edges of this matching blue. Now suppose that there is a maximal partial transversal T_1 of $\mathcal{A}$ that is not a transversal of $\mathcal{A}'$. Then there is a matching of T_1 into J. Choose this matching so that it meets as many vertices of J' as possible, and colour the edges of this matching red. We note that some edges may now be coloured both red and blue. Now $|T_1| = |T|$ since T_1 and T are both bases of $M[\mathcal{A}]$. Moreover, T_1 is not a transversal of $\mathcal{A}'$ and therefore, for some v in $J - J'$, there is a red edge e meeting v. Let P be a path of coloured edges beginning at v that has maximum length among such paths. The first edge in this path is e. It is red but not blue. Since both the red edges and the blue edges form matchings, the edges of P alternate red and blue with no edge being both red and blue. Let the last edge of P be f. Now f is either blue or red. Assume the former (see Figure 2.13(a)). Interchange the colours on the edges of P. Then, in the new colouring of $\Delta[\mathcal{A}]$, the red edges still match T_1 into J. But this new red matching meets more vertices of J' than the original red matching; a contradiction. We conclude that f is red. Thus P has more red edges than blue (see Figure 2.13(b)). Again we interchange the colours on the edges of P, this time looking at the blue edges of the recoloured graph. They still form a matching, but this matching has $|T| + 1$ edges. This contradicts the fact that T is a maximal partial transversal of $\mathcal{A}$, thereby finishing the proof of the lemma.   □

Our approach to identifying the duals of transversal matroids is that of Ingleton and Piff (1973). It uses the bipartite-graph view of transversal matroids. We begin, however, by introducing some new ideas. Suppose that G is a *directed graph* and X and Y are subsets of the vertex set V of G. We say that X is *linked to* Y if $|X| = |Y|$ and there are $|X|$ disjoint directed paths whose initial vertex is in X and whose final vertex is in Y. Observe here that such a path may consist of just a single vertex. For $Z \subseteq V$, we say that X is *linked into* Z if X is linked to some subset of Z. In what follows, whenever we refer to a path in a directed graph, we shall mean a *directed path*. Now let B_0 be a fixed subset of V, and denote by $L(G, B_0)$ those subsets of V that are linked into B_0. As we shall see, $L(G, B_0)$ is the set of independent sets of a matroid on V. The proof of this will be indirect and will rely heavily on the following construction of a bipartite graph $\hat{G}$ from G. We begin by taking a disjoint copy $\hat{V}$ of V. For each element v of V, we shall denote by $\hat{v}$ the corresponding element of $\hat{V}$. A similar convention will apply to subsets of V. The vertex set of $\hat{G}$ is $V \cup \hat{V}$. Its edge set is $\{v\hat{v} : v \in V\} \cup \{v\hat{u} : (u, v) \in E(G)\}$.

Example 2.4.2 The above construction is illustrated in Figure 2.14, which shows a directed graph G and its corresponding bipartite graph $\hat{G}$.   □

The next result (Ingleton and Piff 1973) notes a fundamental connection between linkings in a directed graph G and matchings in the corresponding bipartite graph $\hat{G}$. This result is the key to the main theorem of the section, a characterization of the duals of transversal matroids.

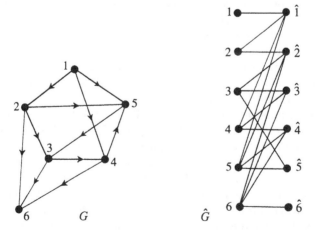

FIG. 2.14. A digraph G and the corresponding bipartite graph $\hat{G}$.

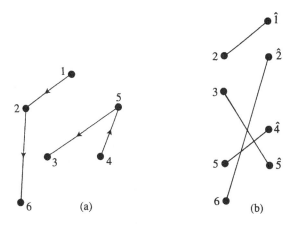

FIG. 2.15. Disjoint paths in G giving a matching in $\hat{G}$.

Lemma 2.4.3 *Let X and Y be subsets of V. Then X is linked to Y in G if and only if $V - X$ is matched to $\hat{V} - \hat{Y}$ in $\hat{G}$.*

Proof Assume first that there is a set $\{P_x : x \in X\}$ of disjoint paths in G linking X to Y. Now define a map ψ from $\hat{V} - \hat{Y}$ onto $V - X$ as follows:

$$\psi(\hat{u}) = \begin{cases} v, & \text{if } v \text{ is the successor of } u \text{ on one of the paths } \{P_x : x \in X\}; \\ u, & \text{if } u \in V - Y \text{ but } u \text{ is not on any of the paths } \{P_x : x \in X\}. \end{cases}$$

Since the paths $\{P_x : x \in X\}$ are disjoint, ψ is well-defined and one-to-one. Moreover, from the definitions of $\hat{G}$ and ψ, we see that $\{\psi(\hat{u})\hat{u} : \hat{u} \in \hat{V} - \hat{Y}\}$ is a subset of $E(\hat{G})$, and the edges of this subset match $V - X$ to $\hat{V} - \hat{Y}$.

Before proving the converse, we illustrate the above procedure on Example 2.4.2. Let $X = \{1, 4\}$ and $Y = \{6, 3\}$. Let the paths linking X to Y be as shown in Figure 2.15(a). Then $\psi(\hat{1}) = 2$, $\psi(\hat{2}) = 6$, $\psi(\hat{4}) = 5$, and $\psi(\hat{5}) = 3$, so the matching of $\hat{V} - \hat{Y}$ to $V - X$ is as shown in Figure 2.15(b).

Now, to prove the converse, suppose that $V - X$ is matched to $\hat{V} - \hat{Y}$ in $\hat{G}$. Colour the edges of this matching red. In addition, for all v in V, colour the edge $v\hat{v}$ blue. We observe that some edges may be coloured both red and blue.

We now define a set $\{P_x : x \in X\}$ of disjoint paths linking X to Y in G. This construction is illustrated with an example at the end of the proof. If a member x of X is also in Y, we let P_x consist of the single vertex x. Now suppose that $x \in X - Y$. Then x meets a blue edge but no red edge. Take a maximal path P'_x of coloured edges beginning at x.

We show next that P'_x must meet a vertex of $\hat{Y}$. Assume the contrary. The first edge in P'_x is blue and, as the red edges and the blue edges both form matchings, the edges of P'_x alternate in colour with none being coloured both red and blue. If the last edge of P'_x is blue, this edge is of the form $v\hat{v}$. Moreover, $\hat{v} \notin \hat{Y}$, as P'_x does not meet a vertex of $\hat{Y}$. Hence there is a red edge meeting $\hat{v}$. Since the other endpoint of this edge cannot already be in P'_x, we may adjoin this red edge

to P'_x thereby forming a longer path and contradicting the choice of P'_x. Thus
we may assume that the last edge in P'_x is red. If this edge is $v\hat{u}$, then $v\hat{u} \neq v\hat{v}$,
and we may adjoin the blue edge $v\hat{v}$ to P'_x to again obtain a contradiction to the
choice of P'_x. We conclude that P'_x does indeed meet a vertex in $\hat{Y}$.

Let $\hat{y}$ be the first vertex of $\hat{Y}$ met by P'_x. As $\hat{y} \notin \hat{V} - \hat{Y}$, no red edge meets $\hat{y}$,
so P'_x ends at $\hat{y}$, the last edge being the blue edge $y\hat{y}$. Clearly P'_x is the unique
maximal path of coloured edges in $\hat{G}$ beginning at x. Moreover, if $x = x_0$ and
$\hat{y} = \hat{x}_n$, then P'_x is $x_0\hat{x}_0x_1\hat{x}_1\cdots x_n\hat{x}_n$. Let P_x be $x_0x_1x_2\cdots x_n$. Then P_x is a path
in G. Now consider the set $\{P_x : x \in X\}$ of paths in G. For distinct members
z_1 and z_2 of $X - Y$, it is easily checked that P'_{z_1} and P'_{z_2} are disjoint. Hence so
are P_{z_1} and P_{z_2}. Thus all the paths in $\{P_x : x \in X\}$ are disjoint. As each links
a vertex of X to a vertex of Y and $|X| = |Y|$, these paths link X to Y.   □

To illustrate the construction of the set of paths $\{P_x : x \in X\}$ in the second
part of the last proof, we again refer to Example 2.4.2. Take the matching in $\hat{G}$
shown in Figure 2.16(a). Then $V - X = \{3, 4, 5\}$ and $\hat{V} - \hat{Y} = \{\hat{2}, \hat{3}, \hat{4}\}$. Colour
the edges in this matching red. In addition, colour blue all edges $v\hat{v}$ for v in
V. This gives the edge-coloured graph in Figure 2.16(b). As $X = \{1, 2, 6\}$ and
$Y = \{1, 5, 6\}$, both 1 and 6 are in $X \cap Y$, so P_1 and P_6 consist of the single
vertices 1 and 6, respectively. On the other hand, P'_2 is $2\hat{2}3\hat{3}4\hat{4}5\hat{5}$, so P_2 is 2345.
Figure 2.16(c) shows the three disjoint paths P_1, P_2, and P_6 in G.

Next we prove the main result of this section (Ingleton and Piff 1973), a
characterization of *cotransversal matroids*, the duals of transversal matroids.

Theorem 2.4.4 *Let G be a directed graph having vertex set V and let B_0 be
a subset of V. Then $L(G, B_0)$ is the set of independent sets of a cotransversal
matroid on V. Conversely, given any cotransversal matroid M on a set V and
any basis B_0 of M, there is a directed graph G having vertex set V such that M
has $L(G, B_0)$ as its set of independent sets.*

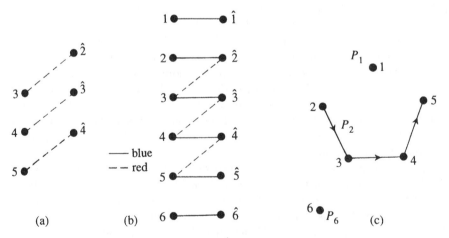

FIG. 2.16. Using a matching in $\hat{G}$ to produce disjoint paths in G.

Proof From the directed graph G, construct the corresponding bipartite graph $\hat{G}$. By the last lemma, B is a maximal member of $L(G, B_0)$ if and only if $V - B$ is matched to $\hat{V} - \hat{B}_0$ in $\hat{G}$. Let $\hat{G}_1$ be the subgraph of $\hat{G}$ obtained by deleting the vertices in $\hat{B}_0$. Then $V - B$ is matched to $\hat{V} - \hat{B}_0$ in $\hat{G}_1$ if and only if B is a maximal member of $L(G, B_0)$. But the subsets of V that are matched to $\hat{V} - \hat{B}_0$ are precisely the bases of a transversal matroid N on V, and the above assertion is that the cobases of N are the maximal members of $L(G, B_0)$. Hence $L(G, B_0)$ is the set of independent sets of the cotransversal matroid N^* on V.

Now let N be an arbitrary transversal matroid on V, and let $V - B_0$ be a basis of N. Then, by Lemma 2.4.1, $N = M[\mathcal{A}]$ for some family $\mathcal{A}$ of sets that has $V - B_0$ as a transversal. Let $\Delta[\mathcal{A}]$ be the bipartite graph associated with $\mathcal{A}$ where the latter equals $(A_j : j \in J)$. This graph contains a matching of $V - B_0$ onto J. For j in J, if j is joined to v in this matching, relabel j as $\hat{v}$. Then, for each u in B_0, add a new vertex $\hat{u}$ to $\Delta[\mathcal{A}]$ joining this vertex to u and no other. The resulting graph is of the form $\hat{G}$ for some directed graph G. It follows, by Lemma 2.4.3, that N^* has $L(G, B_0)$ as its set of independent sets. $\square$

Given a directed graph G and a subset B_0 of its vertex set, the matroid that has $L(G, B_0)$ as its collection of independent sets is called a *strict gammoid*. We shall use the same notation, $L(G, B_0)$, for this matroid as for its collection of independent sets. Such matroids were introduced by Mason (1972a). The following is an immediate consequence of the last theorem.

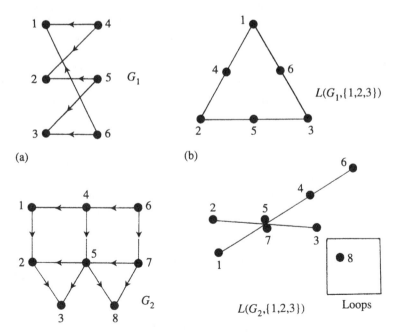

Fig. 2.17. G_1 and G_2 and the strict gammoids $L(G_i, \{1, 2, 3\})$.

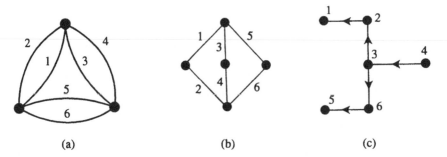

FIG. 2.18. (a) G. (b) G^*. (c) A digraph G_1 with $M(G) = L(G_1, \{1,5\})$.

Corollary 2.4.5 *A matroid is a strict gammoid if and only if its dual matroid is transversal.* □

Example 2.4.6 Let G_1 and G_2 be the directed graphs shown in Figure 2.17(a). In each case, let $B_0 = \{1, 2, 3\}$. Then it is not difficult to check that $L(G_1, B_0)$ and $L(G_2, B_0)$ have the geometric representations shown in Figure 2.17(b). Note that $L(G_1, B_0) \cong \mathcal{W}^3$ (see Figure 2.1). □

As every uniform matroid is transversal and has a uniform dual, every uniform matroid is a strict gammoid. We noted earlier that it will be proved in Chapter 11 that every transversal matroid is representable over all sufficiently large fields. By Corollary 2.2.9, the dual of an $\mathbb{F}$-representable matroid is $\mathbb{F}$-representable, so every strict gammoid is representable over all sufficiently large fields.

We showed in Example 1.6.3 that, for the planar graph G in Figure 2.18, $M(G)$ is not transversal. Thus $M(G^*)$ is not a strict gammoid where G^* is a geometric dual of G (see Figure 2.18(b)). Note, however, that $M(G) = L(G_1, B_0)$ where G_1 is the directed graph shown in Figure 2.18(c) and $B_0 = \{1, 5\}$. Thus $L(G_1, B_0)$ is a strict gammoid whose dual is not a strict gammoid. Hence $M(G^*)$ is a transversal matroid whose dual is not transversal.

Exercises

1. Prove that if M is a rank-r transversal matroid having no coloops, then every presentation of M has exactly r non-empty members.
2. Recall that a cyclic flat is a flat that is a union of circuits. Let $\mathcal{A} = (A_1, A_2, \ldots, A_m)$. Prove that if F is a cyclic flat of $M[\mathcal{A}]$ of rank k, then $|\{i : F \cap A_i \neq \emptyset\}| = k$.
3. Let G be the directed graph shown in Figure 2.19(a).
   (a) Construct the corresponding bipartite graph $\hat{G}$.
   (b) Let $V = \{1, 2, \ldots, 8\}$, $X = \{5, 7\}$, $Y = \{4, 2\}$, and the paths linking X to Y in G be 5134 and 7862. Construct the corresponding matching of $V - X$ to $\hat{V} - \hat{Y}$ in $\hat{G}$.
   (c) For the matching $\{2\hat{1}, 3\hat{5}, 4\hat{3}, 5\hat{4}, 6\hat{8}, 8\hat{7}\}$ in $\hat{G}$, construct the corresponding disjoint paths in G as in the proof of Lemma 2.4.3.

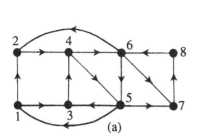 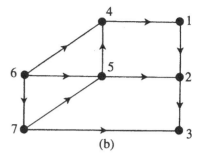

FIG. 2.19. The directed graphs in Exercises 3 and 4.

4. For the directed graph G in Figure 2.19(b), let $M = L(G, B_0)$ where $B_0 = \{1, 2, 3\}$.
    (a) Find geometric representations for M and M^*.
    (b) Give a presentation for the transversal matroid M^*.
    (c) Reverse the directions on $(5,4)$ and $(7,5)$ and repeat (a) and (b).

5. For the directed graph G_2 in Figure 2.17(a), find $L(G_2, B_0)$ when $B_0 = \{2, 3, 8\}$.

6. Let G be the directed graph in Figure 2.14.
    (a) Find $L(G, B_0)$ when $B_0 = \{2, 5, 6\}$.
    (b) Show that there is no subset B_0' of $V(G)$ such that $L(G, B_0') \cong U_{3,6}$.

7. (a) Give a presentation for the transversal matroid $U_{r,n}$.
    (b) Find a directed graph G and a subset B_0 of $V(G)$ such that $L(G, B_0) \cong U_{n-r,n}$.

8. Show that each of the matroids P_6 and Q_6 in Figure 2.1 is a strict gammoid that is also transversal.

9. Let M be a free r-spike with tip t. Show that
    (a) $M \backslash t$ is transversal if and only if $r = 3$; and
    (b) M is a strict gammoid.

10. *(Bondy and Welsh 1971) Let $(A_1, A_2, \ldots, A_r)$ be a presentation of a rank-r transversal matroid M. Show that:
    (a) If T is a transversal of $(A_2, A_3, \ldots, A_r)$ such that $A_1 \cap T$ has minimum cardinality, then $(A_1 - T, A_2, A_3, \ldots, A_r)$ is also a presentation of M.
    (b) M has distinct cocircuits $C_1^*, C_2^*, \ldots, C_r^*$ such that $(C_1^*, C_2^*, \ldots, C_r^*)$ is a presentation for M and $C_i^* \subseteq A_i$ for all i in $\{1, 2, \ldots, r\}$. Moreover, if $x \in C_j^*$ for some j in $\{1, 2, \ldots, r\}$, then $(C_1^*, C_2^*, \ldots, C_{j-1}^*, C_j^* - x, C_{j+1}^*, \ldots, C_r^*)$ is not a presentation for M.

3

MINORS

We saw in Chapter 1 that, given a matroid M on a set E and a subset T of E, there is a matroid $M \backslash T$ on $E - T$ whose independent sets are those independent sets in M that are contained in $E - T$. In this chapter, we look more closely at this operation and its dual, we consider applying a sequence of these operations to a matroid, and we examine the effects of these operations on various classes of matroids.

3.1 Contraction

In this section, we introduce the operation of contraction as the dual of the operation of deletion. We then derive a definition of contraction that does not use duality and look at how contraction affects various special sets such as independent sets, bases, and circuits.

Let M be a matroid on E, and T be a subset of E. The two fundamental operations on M that we have introduced so far are deletion and the taking of duals. The next definition combines these two operations. Let M/T, the *contraction of T from M*, be given by

3.1.1 $M/T = (M^* \backslash T)^*$.

Evidently M/T has ground set $E - T$. We shall sometimes write $M.(E - T)$ for M/T and call it the *contraction of M onto $E - T$*.

Before presenting an example of this operation, we need to note a basic property of deletion for graphic matroids. For a graph G and a subset T of $E(G)$, recall that $G \backslash T$ is the graph obtained from G by erasing or *deleting* the edges in T. Deletion for matroids extends this graph-theoretic operation, that is,

3.1.2 $M(G \backslash T) = M(G) \backslash T$.

Example 3.1.3 Let $M = M(G)$ where G is the planar graph shown in Figure 3.1. Then $M(G)/6 = (M^*(G) \backslash 6)^* = (M(G^*) \backslash 6)^*$ where G^* is a geometric dual of G and each edge of G^* has the same label as the corresponding edge of G. Thus, by 3.1.1 and 3.1.2, $M(G)/6 = (M(G^* \backslash 6))^* = M^*(G^* \backslash 6) = M((G^* \backslash 6)^*)$. But, from Figure 3.1, we see that $G^* \backslash 6$ has, as a geometric dual, the graph $G/6$ that is obtained from G by shrinking or contracting the edge 6 to a point. Therefore $M(G)/6 = M(G/6)$. This illustrates the general principle, to be proved in Section 3.2, that contraction for matroids generalizes the operation of contraction for graphs. In Section 3.3, we shall see that contraction of a non-loop element e from a matroid M corresponds geometrically to projecting the other elements of M from e onto a hyperplane avoiding e. $\square$

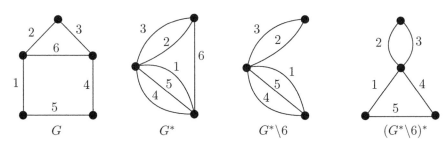

FIG. 3.1. $M(G)/6 = M(G/6)$.

The next proposition specifies the rank function of M/T. First recall from Proposition 2.1.9 that the rank function r^* of the dual, M^*, of M is given by

3.1.4 $r^*(X) = |X| + r(E - X) - r(E)$.

Evidently if $T \subseteq E$, then the rank function of $M\backslash T$ is the restriction of r_M to the set of subsets of $E - T$, that is, for all $X \subseteq E - T$,

3.1.5 $r_{M\backslash T}(X) = r_M(X)$.

Proposition 3.1.6 *If $T \subseteq E$, then, for all $X \subseteq E - T$,*

$$r_{M/T}(X) = r_M(X \cup T) - r_M(T).$$

Proof By definition, $r_{M/T}(X) = r_{(M^*\backslash T)^*}(X)$. Thus,

$$
\begin{aligned}
r_{M/T}(X) &= |X| + r_{M^*\backslash T}(E - T - X) - r_{M^*\backslash T}(E - T), &&\text{by (3.1.4),}\\
&= |X| + r^*(E - (T \cup X)) - r^*(E - T), &&\text{by (3.1.5),}\\
&= |X| + [|E - (T \cup X)| + r_M(T \cup X) - r_M(E)]\\
&\quad - [|E - T| + r_M(T) - r_M(E)], &&\text{by (3.1.4) again.}
\end{aligned}
$$

It follows, as $X \subseteq E - T$, that $r_{M/T}(X) = r_M(T \cup X) - r_M(T)$. □

The next three results determine the independent sets, bases, and circuits of M/T in terms of the corresponding sets for M. Recall that $M|T$ is $M\backslash(E - T)$.

Proposition 3.1.7 *Let B_T be a basis of $M|T$. Then*

$$
\begin{aligned}
\mathcal{I}(M/T) &= \{I \subseteq E - T : I \cup B_T \in \mathcal{I}(M)\}\\
&= \{I \subseteq E - T : M|T \text{ has a basis } B \text{ such that } B \cup I \in \mathcal{I}(M)\}.
\end{aligned}
$$

Proof Evidently $\{I \subseteq E - T : M|T \text{ has a basis } B \text{ such that } B \cup I \in \mathcal{I}\}$ contains $\{I \subseteq E - T : I \cup B_T \in \mathcal{I}\}$. Now suppose that $I \subseteq E - T$ and $B \cup I \in \mathcal{I}$ for some basis B of $M|T$. We shall show that $I \in \mathcal{I}(M/T)$. Clearly $I \cup B$ is a basis of $I \cup T$, so $r_M(I \cup B) = r_M(I \cup T)$. Therefore, as $r_{M/T}(I) = r_M(I \cup T) - r_M(T)$,

$$r_{M/T}(I) = r_M(I \cup B) - r_M(T) = |I \cup B| - |B|,$$

where the last step follows since $I \cup B \in \mathcal{I}$ and B is a basis of $M|T$. Hence $r_{M/T}(I) = |I|$, that is, $I \in \mathcal{I}(M/T)$. Therefore,

$$\{I \subseteq E - T : M|T \text{ has a basis } B \text{ such that } B \cup I \in \mathcal{I}\} \subseteq \mathcal{I}(M/T).$$

To complete the proof, we show that $\{I \subseteq E - T : I \cup B_T \in \mathcal{I}\}$ contains $\mathcal{I}(M/T)$. Thus assume $X \in \mathcal{I}(M/T)$. Then

$$|X| = r_{M/T}(X) = r_M(X \cup T) - r_M(T)$$
$$= r_M(X \cup B_T) - |B_T|,$$

since B_T is a basis of $M|T$. Hence $|X \cup B_T| = r_M(X \cup B_T)$, so $X \cup B_T \in \mathcal{I}$ and the proposition follows.   □

Corollary 3.1.8 *Let B_T be a basis of $M|T$. Then*

$$\mathcal{B}(M/T) = \{B' \subseteq E - T : B' \cup B_T \in \mathcal{B}(M)\}$$
$$= \{B' \subseteq E - T : M|T \text{ has a basis } B \text{ such that } B' \cup B \in \mathcal{B}(M)\}. \quad \square$$

From Proposition 3.1.7, it is straightforward to determine the effect of contraction on uniform matroids.

Example 3.1.9 *Let T be a t-element subset of $E = E(U_{m,n})$. Then*

$$U_{m,n}/T \cong \begin{cases} U_{0,n-t}, & \text{if } n \geq t \geq m, \\ U_{m-t,n-t}, & \text{if } t < m. \end{cases}$$

For comparison, we observe that

$$U_{m,n} \backslash T \cong \begin{cases} U_{n-t,n-t}, & \text{if } n \geq t \geq n - m, \\ U_{m,n-t}, & \text{if } t < n - m. \end{cases} \quad \square$$

Proposition 3.1.10 *The circuits of M/T consist of the minimal non-empty members of $\{C - T : C \in \mathcal{C}(M)\}$.*

Proof Evidently we may assume that T is non-empty. Since the circuits of M/T are its minimal dependent sets, it suffices to prove that

(i) all non-empty members of $\{C - T : C \in \mathcal{C}(M)\}$ are dependent in M/T; and
(ii) every circuit of M/T contains a non-empty member of $\{C - T : C \in \mathcal{C}(M)\}$.

Suppose first that $C_2 \in \mathcal{C}(M)$ and $C_2 - T \neq \emptyset$. Then $C_2 \cap T \subsetneq C_2$, so $C_2 \cap T$ is in $\mathcal{I}(M|T)$. Thus $C_2 \cap T$ is contained in a basis B_T of $M|T$. As $C_2 \cup B_T \supseteq C_2$, the set $C_2 \cup B_T$ is not in $\mathcal{I}(M)$. Hence $C_2 - T \notin \mathcal{I}(M/T)$ and (i) holds.

Now suppose $C_1 \in \mathcal{C}(M/T)$. Thus, if $B_T \in \mathcal{B}(M|T)$, then $C_1 \cup B_T \notin \mathcal{I}(M)$, but $(C_1 - e) \cup B_T \in \mathcal{I}(M)$ for all e in C_1. As $C_1 \cup B_T$ is dependent in M, it contains a circuit D of M. Moreover, $D \supseteq C_1$ since $(C_1 - e) \cup B_T \in \mathcal{I}(M)$ for all e in C_1. Hence $C_1 = D - T$. Thus (ii) holds and so does the proposition.   □

The closure operator of M/T is not difficult to derive from 1.4.1 and Proposition 3.1.6 and the details are left as an exercise.

Proposition 3.1.11 *For all* $X \subseteq E - T$,

$$\mathrm{cl}_{M/T}(X) = \mathrm{cl}_M(X \cup T) - T. \qquad \square$$

For comparison with the last four results, we note that

3.1.12 $\mathcal{I}(M\backslash T) = \{I \subseteq E - T : I \in \mathcal{I}(M)\}$;

3.1.13 $\mathcal{C}(M\backslash T) = \{C \subseteq E - T : C \in \mathcal{C}(M)\}$;

3.1.14 $\mathcal{B}(M\backslash T)$ *is the set of maximal members of* $\{B - T : B \in \mathcal{B}(M)\}$; *and*

3.1.15 $\mathrm{cl}_{M\backslash T}(X) = \mathrm{cl}_M(X) - T$.

Of these, 3.1.12 is just the definition; 3.1.13 follows immediately from 3.1.12; and 3.1.14 and 3.1.15 are straightforward and are left as exercises.

Using the above results and duality, we easily obtain the following.

3.1.16 $\mathcal{C}^*(M\backslash T)$ *is the set of minimal non-empty members of* $\{C^* - T \subseteq E - T : C^* \in \mathcal{C}^*(M)\}$;

3.1.17 $\mathcal{C}^*(M/T) = \{C^* \subseteq E - T : C^* \in \mathcal{C}^*(M)\}$;

3.1.18 $\mathcal{H}(M\backslash T)$ *is the set of maximal proper subsets of* $E - T$ *of the form* $H - T$ *where* $H \in \mathcal{H}(M)$;

3.1.19 $\mathcal{H}(M/T) = \{X \subseteq E - T : X \cup T \in \mathcal{H}(M)\}$.

Moreover, letting $\mathcal{S}(M)$ denote the set of spanning sets of M, we have

3.1.20 $\mathcal{S}(M\backslash T) = \{X \subseteq E - T : M.T \text{ has a basis } B \text{ such that } X \cup B \in \mathcal{S}(M)\}$;

3.1.21 $\mathcal{S}(M/T) = \{X \subseteq E - T : X \cup T \in \mathcal{S}(M)\} = \{Y - T : Y \in \mathcal{S}(M)\}$.

Since $M\backslash T$ and M/T are matroids on the same ground set, it is natural to try to determine when these matroids are equal. Evidently it is always true that

3.1.22 $\mathcal{I}(M\backslash T) \supseteq \mathcal{I}(M/T)$.

Proposition 3.1.23 $M\backslash T = M/T$ *if and only if* $r(T) + r(E - T) = r(M)$.

Proof Suppose $M\backslash T = M/T$ and let B be a basis of $M\backslash T$. Then B is a basis of M/T and hence $B \,\dot\cup\, B_T$ is a basis of M for some basis B_T of $M|T$. Thus $r(M) = r(T) + r(E - T)$. Conversely, suppose that $r(M) = r(T) + r(E - T)$. By 3.1.22, to prove that $M\backslash T = M/T$, we need only show that $\mathcal{I}(M\backslash T) \subseteq \mathcal{I}(M/T)$. But if $I \in \mathcal{I}(M\backslash T)$, then I is a subset of a basis B of $E - T$, and B is contained in a basis $B\,\dot\cup\,B'$ of M. Evidently $r(M) = |B \,\dot\cup\, B'| = |B| + |B'| = r(E - T) + |B'|$. Since $r(M) = r(E - T) + r(T)$, it follows that $|B'| = r(T)$, that is, B' is a basis of $M|T$. Hence $B \in \mathcal{B}(M/T)$, so $I \in \mathcal{I}(M/T)$ and $M\backslash T = M/T$. $\qquad \square$

If $T = \{e_1, e_2, \ldots, e_m\}$, then we shall often write $M\backslash e_1, e_2, \ldots, e_m$ and $M/e_1, e_2, \ldots, e_m$ as $M\backslash T$ and M/T, respectively.

Corollary 3.1.24 $M \backslash e = M/e$ *if and only if e is a loop or a coloop of M.*

Proof This is left to the reader (Exercise 10). $\qquad\qquad\qquad\qquad\qquad$ □

In a matroid M, the existence of a non-empty proper subset T of $E(M)$ for which $M \backslash T = M/T$ has important structural significance for M and this will be examined in detail in the next chapter.

Next we show that the operations of deletion and contraction commute both with each other and with themselves. This fact will be important since we shall want to consider matroids obtained from other matroids by a sequence of deletions and contractions.

Proposition 3.1.25 *In a matroid M, let T_1 and T_2 be disjoint subsets of $E(M)$. Then*

(i) $(M \backslash T_1) \backslash T_2 = M \backslash (T_1 \cup T_2) = (M \backslash T_2) \backslash T_1$;
(ii) $(M/T_1)/T_2 = M/(T_1 \cup T_2) = (M/T_2)/T_1$; *and*
(iii) $(M \backslash T_1)/T_2 = (M/T_2) \backslash T_1$.

Proof Parts (i) and (ii) are left as exercises. To prove (iii), we show that $(M \backslash T_1)/T_2$ and $(M/T_2) \backslash T_1$ have the same rank function. For $X \subseteq E - (T_1 \cup T_2)$,

$$
\begin{aligned}
r_{(M \backslash T_1)/T_2}(X) &= r_{M \backslash T_1}(X \cup T_2) - r_{M \backslash T_1}(T_2) \\
&= r_M(X \cup T_2) - r_M(T_2) \\
&= r_{M/T_2}(X) \\
&= r_{(M/T_2) \backslash T_1}(X).
\end{aligned}
$$
$\qquad\qquad$ □

In view of the last result, we may drop the parentheses in expressions such as $((M/X_1) \backslash X_2) \backslash X_3$ without introducing any ambiguity, it being understood here that X_1, X_2, and X_3 are disjoint. Another important consequence of the last result is that any sequence of deletions and contractions from M can be written in the form $M \backslash X/Y$ for some pair of disjoint sets X and Y, either of which may be empty. Matroids of this form are called *minors* of M. Such substructures were introduced by Tutte (1958) and many of the most celebrated results for matroids make reference to minors. We observe that, in view of Proposition 3.1.25(iii), a minor of M could equally well have been defined as any matroid of the form $M/Y \backslash X$ where X and Y are disjoint, possibly empty, subsets of $E(M)$. If $X \cup Y$ is non-empty, then we call $M/Y \backslash X$ a *proper minor* of M. In general, whenever we write $M \backslash X/Y$ or $M/Y \backslash X$, it will always be understood that the sets X and Y are disjoint subsets of $E(M)$.

The minors of a matroid M are its fundamental substructures and we shall give many examples in later chapters of where a knowledge of the presence of certain matroids as minors of M has important implications for the properties of M. A matroid N_1 is called an N-*minor* of M if N_1 is a minor of M that is *isomorphic* to N. More generally, if $\mathcal{N}$ is a set of matroids, then N_1 is an $\mathcal{N}$-*minor* of M if N_1 is an N-minor of M for some N in $\mathcal{N}$.

The next result follows without difficulty from 3.1.1 (Exercise 15).

Proposition 3.1.26 *A matroid N is a minor of a matroid M if and only if N^* is a minor of M^*. More particularly,*

$$N = M\backslash X/Y \text{ if and only if } N^* = M^*/X\backslash Y.$$ □

We conclude this section by showing that if e is an element of a matroid M, then the pair $(M\backslash e, M/e)$ uniquely determines M unless e is a loop or a coloop.

Proposition 3.1.27 *Let M_1 and M_2 be matroids on a common ground set E and suppose $e \in E$. Then the following statements are equivalent.*

(i) $M_1\backslash e = M_2\backslash e$ and $M_1/e = M_2/e$.

(ii) $M_1 = M_2$; or e is a loop of one of M_1 and M_2 and a coloop of the other.

Proof If (ii) holds, then so does (i). Now assume that (i) holds. As $M_1\backslash e = M_2\backslash e$, the set of circuits of M_1 avoiding e coincides with the set of circuits of M_2 avoiding e. If e is a loop or coloop of M_1, then, by Corollary 3.1.24, $M_1\backslash e = M_1/e$, so $M_2\backslash e = M_2/e$ and therefore e is a loop or a coloop of M_2. It follows that (ii) holds. We may now assume that e is neither a loop nor a coloop of M_1 or M_2. In that case, C is a circuit of M_1 containing e if and only if $C - e$ is a circuit of M_1/e but not of $M_1\backslash e$. Hence the circuits of M_1 containing e coincide with the circuits of M_2 containing e. We conclude that $M_1 = M_2$, so (ii) holds. □

Exercises

1. Show that if $T \subseteq E(M)$, then
   (a) $M\backslash T = (M^*/T)^*$;
   (b) $(M/T)^* = M^*\backslash T$; and
   (c) $M^*/T = (M\backslash T)^*$.
2. Let C be a circuit of a matroid M. For e in $E(M)$, show that:
   (a) If $e \in C$, then e is a loop of M, or $C - e$ is a circuit of M/e.
   (b) If $e \notin C$, then C is a union of circuits of M/e.
3. Show that $\text{cl}_{M/T}(X) = \text{cl}_M(X \cup T) - T$ for all $X \subseteq E - T$.
4. Let e and f be elements of a minor N of a matroid M. Show that if e and f are clones in M, they are clones in N, but that the converse of this fails.
5. Prove that elements e and f are clones in a matroid M if and only if $M/e\backslash f = M/f\backslash e$ and $r(\{e\}) = r(\{f\})$.
6. Let e, f, and g be distinct elements in a matroid M. Prove that if e and f are clones in both $M\backslash g$ and M/g, then they are clones in M.
7. Suppose that $r(M\backslash f) = r(M)$ for all elements f of a loopless matroid M, but that $r(M\backslash e, g) = r(M\backslash e/g)$ for some elements e and g. Show that $\{e, g\}$ is a cocircuit of M.
8. Let M be a matroid and T be a subset of $E(M)$. Show that:
   (a) M/T has no loops if and only if T is a flat of M.
   (b) $\text{cl}(T) = T \cup \{e \in E(M) - T : e \text{ is a loop of } M/T\}$.
   (c) $\text{cl}^*(T) = T \cup \{e \in E(M) - T : e \text{ is a coloop of } M\backslash T\}$.

9. Let B_T be a fixed basis of $M.T$. Show that $\mathcal{S}(M\backslash T) = \{X \subseteq E - T : X \cup B_T \in \mathcal{S}(M)\}$.

10. Prove that $M\backslash e = M/e$ if and only if e is a loop or a coloop of M.

11. Prove (i) and (ii) of Proposition 3.1.25.

12. Let N be a minor of a matroid M and suppose that a set X is the intersection of a circuit and a cocircuit in N. Prove that X is the intersection of a circuit and a cocircuit in M.

13. Prove that if $\{e, f, g\}$ is both a circuit and a cocircuit of a matroid M, then $M\backslash f/g = M\backslash g/f$.

14. Let M be a matroid, T be a subset of $E(M)$, and B_T be a basis of $M|T$. Prove directly that M/T is a matroid by showing that $\{B' \subseteq E - T : B' \cup B_T \in \mathcal{B}(M)\}$ is the set of bases of a matroid on $E - T$.

15. Prove Proposition 3.1.26.

16. Let T be a subset of the ground set of a matroid M. Prove that the following statements are equivalent:
    (a) $M\backslash T = M/T$.
    (b) $r(M\backslash T) \le r(M/T)$.
    (c) M has no circuits meeting both T and $E - T$.

17. Let $N = M\backslash X/Y$. Specify the rank function and the sets of bases, circuits, independent sets, hyperplanes, and cocircuits of N in terms of the corresponding objects in M.

3.2   Minors of certain classes of matroids

Some of the classes of matroids we have introduced have the property that all their minors are also in the class. We say that such classes are *closed under minors* or are *minor-closed*. Certain other classes are not minor-closed. We begin this section by noting some examples of minor-closed classes and end by showing that the classes of transversal matroids and strict gammoids are not minor-closed.

We noted in Example 3.1.9 that all deletions and all contractions of a uniform matroid are uniform. Hence the class of uniform matroids is closed under minors. This class has the further attractive property that it is closed under duality. We shall find that classes of matroids having this pair of properties are usually easier to work with than those without one or both of these properties. It was noted in 3.1.2 that all deletions of graphic matroids are graphic. Next we consider contractions. If e is an edge of a graph G, then G/e is obtained from G by deleting e and identifying its ends. Repeating this process for all the edges in a subset T of $E(G)$ gives the graph G/T. Moreover, G/T is well-defined since one can easily check that $(G/e)/f = (G/f)/e$ for all edges e and f.

Proposition 3.2.1 *If G is a graph, then $M(G)/T = M(G/T)$ for all subsets T of $E(G)$.*

Proof   We shall show that, for every edge e of G,

$$M(G)/e = M(G/e). \tag{3.1}$$

The proposition then follows by a routine induction argument on $|T|$.

If e is a loop of G, then $G/e = G\backslash e$ and $M(G)/e = M(G)\backslash e$. The result follows in this case by 3.1.2. Now suppose that e is not a loop of G. Then, for a subset I of $E(G) - e$, it is not difficult to check that $I \cup e$ contains no cycle of G if and only if I contains no cycle of G/e. Hence $\mathcal{I}(M(G)/e) = \mathcal{I}(M(G/e))$ and (3.1) holds. $\qquad\square$

On combining this proposition with 3.1.2, we deduce that the class of graphic matroids is minor-closed.

Corollary 3.2.2 *Every minor of a graphic matroid is graphic.* $\qquad\square$

The concept of a minor is useful in graphs as well as in matroids. A graph H is a *minor* of a graph G if H is a contraction of some subgraph of G, or equivalently, if H can be obtained from G by a sequence of edge deletions, edge contractions, and deletions of isolated vertices. We leave the reader to check that if G is connected, then H is a minor of G if and only if $H = G\backslash X/Y$ for some disjoint subsets X and Y of $E(G)$. Evidently if H is a minor of G, then $M(H)$ is a minor of $M(G)$. However, the converse of this is not true.

It was shown in Proposition 2.3.3 and Corollary 2.2.9 that the class of graphic matroids is not closed under duality whereas the class of $\mathbb{F}$-representable matroids is. We shall show next that the class of $\mathbb{F}$-representable matroids is also minor-closed.

Let A be a matrix over a field $\mathbb{F}$ and T be a subset of the set E of column labels of A. We shall denote by $A\backslash T$ the matrix obtained from A by deleting all the columns whose labels are in T. Evidently

3.2.3 $M[A]\backslash T = M[A\backslash T]$.

Moreover:

Proposition 3.2.4 *Every minor of an $\mathbb{F}$-representable matroid is $\mathbb{F}$-representable.*

Proof By 3.2.3, every deletion of an $\mathbb{F}$-representable matroid is $\mathbb{F}$-representable. As the dual of an $\mathbb{F}$-representable matroid is also $\mathbb{F}$-representable, we deduce, from the definition of contraction (3.1.1), that every contraction of an $\mathbb{F}$-representable matroid is $\mathbb{F}$-representable. Hence so is every minor. $\qquad\square$

In particular, the classes of binary and ternary matroids are minor-closed. Moreover, an easy modification of the last proof gives the following.

Proposition 3.2.5 *Every minor of a regular matroid is regular.* $\qquad\square$

Although we know that a contraction of an $\mathbb{F}$-representable matroid is $\mathbb{F}$-representable, our only method for obtaining an $\mathbb{F}$-representation for the contraction is a convoluted one, going via the dual. Next we describe a much more direct method for finding such a representation. The term 'minor' in matroid theory is taken from matrix theory. We have seen that deletion of elements corresponds to removing columns from the associated matrix. We shall now show that contraction essentially corresponds to removing rows.

Consider $M[A]$ and suppose that e is the label of a column of A. If this column is zero, then e is a loop of $M[A]$ and, by Corollary 3.1.24, $M[A]/e = M[A]\backslash e$. Hence, by 3.2.3, $M[A]/e$ is represented by $A\backslash e$. Now suppose that e is the label of a non-zero column of A. Then, by pivoting on a non-zero entry of e, we can transform A into a matrix A' in which the column labelled by e is a standard basis vector. In this case, A'/e will denote the matrix obtained from A' by deleting the row and column containing the unique non-zero entry in e.

Proposition 3.2.6 $M[A]/e = M[A']/e = M[A'/e]$.

Proof Since A' is obtained from A by elementary row operations, $M[A'] = M[A]$, so $M[A']/e = M[A]/e$. To prove that $M[A']/e$ and $M[A'/e]$ are equal, we first observe that, by using row and column swaps if necessary, we may assume that, in A', the unique non-zero entry of e is in row 1 and column 1. Let I be a subset of the ground set E of $M[A']$ such that $e \notin I$. Let $\{(\varepsilon_1, \underline{v}_1)^T, (\varepsilon_2, \underline{v}_2)^T, \ldots, (\varepsilon_k, \underline{v}_k)^T\}$ be the corresponding set of columns of A'. Then the set of columns labelled by $I \cup e$ is linearly independent if and only if the matrix in Figure 3.2 has rank $k+1$. The latter is true if and only if $[\underline{v}_1^T \underline{v}_2^T \cdots \underline{v}_k^T]$ has rank k, which in turn is true if and only if the columns of A'/e labelled by I are linearly independent. We conclude that $\mathcal{I}(M[A'/e]) = \mathcal{I}(M[A']/e)$. $\qquad\square$

$$
\begin{bmatrix}
1 & \varepsilon_1 & \varepsilon_2 & \cdots & \varepsilon_k \\
0 & & & & \\
\vdots & \underline{v}_1^T & \underline{v}_2^T & \cdots & \underline{v}_k^T \\
0 & & & &
\end{bmatrix}
$$

FIG. 3.2. $I \cup e$ is independent in $M[A']$ if and only if this matrix has rank $k+1$.

Example 3.2.7 Consider the vector matroid P_8 of the following matrix A over $GF(3)$. We previously considered this matroid in Example 2.2.25.

$$
\begin{array}{cccccccc}
1 & 2 & 3 & 4 & 5 & 6 & 7 & 8
\end{array}
$$
$$
\begin{bmatrix}
& & & & 0 & 1 & 1 & -1 \\
& I_4 & & & 1 & 0 & 1 & 1 \\
& & & & 1 & 1 & 0 & 1 \\
& & & & -1 & 1 & 1 & 0
\end{bmatrix}
$$

To find a $GF(3)$-representation for $P_8/8$, we first pivot on the third entry in column 8 of A. Thus, in A, we add row 3 to row 1, and subtract row 3 from row 2. This gives the matrix A' shown below.

$$A' = \begin{bmatrix} 1 & 0 & 1 & 0 & 1 & -1 & 1 & 0 \\ 0 & 1 & -1 & 0 & 0 & -1 & 1 & 0 \\ 0 & 0 & 1 & 0 & 1 & 1 & 0 & 1 \\ 0 & 0 & 0 & 1 & -1 & 1 & 1 & 0 \end{bmatrix} \begin{matrix} 1 & 2 & 3 & 4 & 5 & 6 & 7 & 8 \end{matrix}$$

Now delete row 3 and column 8 from A' to get the matrix

$$\begin{bmatrix} 1 & 0 & 1 & 0 & 1 & -1 & 1 \\ 0 & 1 & -1 & 0 & 0 & -1 & 1 \\ 0 & 0 & 0 & 1 & -1 & 1 & 1 \end{bmatrix} \begin{matrix} 1 & 2 & 3 & 4 & 5 & 6 & 7 \end{matrix} .$$

By Proposition 3.2.6, this matrix represents $P_8/8$. Of course, if we pivot on a different non-zero entry of column 8 of A, the above process will produce an alternative matrix representation for $P_8/8$. A geometric representation for $P_8/8$ is shown in Figure 3.3. □

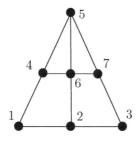

FIG. 3.3. $P_8/8$.

We have shown that the classes of uniform, graphic, and $\mathbb{F}$-representable matroids are all minor-closed. However, as we shall show next, the class of transversal matroids is not minor-closed. We first note that if $\mathcal{A}$ is a family of subsets of a set E and $T \subseteq E$, then

3.2.8 $M[\mathcal{A}]\backslash T = M[\mathcal{A}\backslash T]$

where $\mathcal{A}\backslash T$ is the family of sets obtained from $\mathcal{A}$ by deleting the elements of T from each member of $\mathcal{A}$. In contrast, a contraction of a transversal matroid need not be transversal.

Example 3.2.9 It was shown in Example 1.6.3 that, for the graphs G_1 and G_2 in Figure 1.20, $M(G_1)$ is transversal, while $M(G_2)$ is not. Since $G_2 = G_1/7$, we have $M(G_2) = M(G_1)/7$. Thus $M(G_1)$ is a transversal matroid with a non-transversal contraction. By duality, the class of strict gammoids is not closed under restriction. □

A *gammoid* is a matroid that is isomorphic to a restriction of a strict gammoid. It follows from the next two propositions (Mason 1972a) that the class

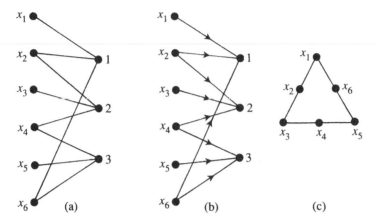

FIG. 3.4. Showing that every transversal matroid is a gammoid.

of gammoids is the smallest minor-closed class that contains all transversal matroids.

Proposition 3.2.10 *Every transversal matroid is a gammoid.*

Proof Let $M = M[\mathcal{A}]$ where $\mathcal{A}$, a family of subsets of a set S, is indexed by the set J. Now let G be the directed graph that is formed from the bipartite graph $\Delta[\mathcal{A}]$ by orienting every edge from its endpoint in S to its endpoint in J. Then it is straightforward to check that the restriction to S of the strict gammoid $L(G, J)$ is precisely the matroid $M[\mathcal{A}]$.   □

We now illustrate the construction used in the last proof.

Example 3.2.11 Let $S = \{x_1, x_2, \ldots, x_6\}$ and $\mathcal{A} = (A_1, A_2, A_3)$ where $A_1 = \{x_1, x_2, x_6\}$, $A_2 = \{x_2, x_3, x_4\}$, and $A_3 = \{x_4, x_5, x_6\}$. Then Figure 3.4(a) and (b) show $\Delta[\mathcal{A}]$ and the directed graph G constructed from it as above. The matroids $M[\mathcal{A}]$ and $L(G, \{1, 2, 3\})|S$, the equality of which was established in the last proof, are represented geometrically by Figure 3.4(c).   □

We leave the reader to prove the next result and to deduce from it that the class of gammoids coincides with the class of transversal matroids and their contractions (Exercise 8).

Proposition 3.2.12 *The class of gammoids is closed under minors and under duality.*   □

An r-spike with tip has exactly $2r+1$ elements, so the class of spikes cannot be minor-closed. However, this class behaves similarly to the class of cycle matroids of wheels in that an appropriate deletion and contraction of a member of the class gives another member of the class. We showed in Proposition 2.1.27 that every single-element deletion from a spike is self-dual. We have already discussed deleting the tip. If we delete the tip and one other element, we get a matroid that is the dual of an $(r-1)$-spike. This is one of several attractive properties of spikes that are noted in the next two results, the proofs of which are left as exercises.

Proposition 3.2.13 *Let M be an r-spike with tip t and legs $L_1, L_2, \ldots, L_r$ where $L_i = \{x_i, y_i, t\}$ for all i.*

(i) *For all $r \geq 4$ and all i, each of $M/x_i \backslash y_i$ and $M/y_i \backslash x_i$ is an $(r-1)$-spike with tip t and legs $L_1, L_2, \ldots, L_{i-1}, L_{i+1}, \ldots, L_r$. Moreover, if M is a free r-spike, then $M/x_i \backslash y_i = M/y_i \backslash x_i$ and both are free $(r-1)$-spikes.*

(ii) *For all $r \geq 4$ and all i, each of $M\backslash t/x_i$ and $(M\backslash t\backslash x_i)^*$ is an $(r-1)$-spike with tip y_i.*

(iii) *M/t can be obtained from an r-element circuit by adding r new elements, one in parallel to each existing element.* $\square$

The binary and real affine cubes, $AG(3,2)$ and R_8, are both highly symmetric tipless 4-spikes. Contracting any element from each gives a 3-spike, specifically F_7 or F_7^-, respectively. Moreover, by Proposition 2.1.27, every single-element deletion from $AG(3,2)$ or R_8 is the dual of a spike, specifically F_7^* or $(F_7^-)^*$. These observations illustrate the first part of the next result (see also Exercise 7).

Corollary 3.2.14 *Let M be a tipless r-spike with legs $L_1, L_2, \ldots, L_r$ where $L_i = \{x_i, y_i\}$ for all i.*

(i) *For all $r \geq 4$ and all i, both M/x_i and $(M\backslash x_i)^*$ are $(r-1)$-spikes with tip y_i and legs $L_1 \cup y_i, L_2 \cup y_i, \ldots, L_{i-1} \cup y_i, L_{i+1} \cup y_i, \ldots, L_r \cup y_i$. Moreover, $M/x_i \cong (M\backslash y_i)^*$.*

(ii) *For all $r \geq 4$ and all i, each of $M/x_i \backslash y_i$ and $M/y_i \backslash x_i$ is a tipless $(r-1)$-spike with legs $L_1, L_2, \ldots, L_{i-1}, L_{i+1}, \ldots, L_r$. Moreover, if M is a tipless free r-spike, then $M/x_i \backslash y_i = M/y_i \backslash x_i$ and both are tipless free $(r-1)$-spikes.*

(iii) *For all i, the matroid M/L_i can be obtained from an $(r-1)$-element circuit by replacing each element by two elements in parallel. Dually, $M\backslash L_i$ can be obtained from an $(r-1)$-element cocircuit by replacing each element by two elements in series.* $\square$

Exercises

1. Find a graph G so that the bond matroid of G is isomorphic to $M(K_5)\backslash e$.

2. (a) Show that if e and f are edges of a graph G, then $G\backslash e/f = G/f\backslash e$.
   (b) Show that if G is a connected graph, then the graph H is a minor of G if and only if there are disjoint subsets X and Y of $E(G)$ such that $H = G\backslash X/Y$, or equivalently, $H = G/Y\backslash X$.
   (c) Show that if G and H are graphs, then it is possible for $M(H)$ to be a minor of $M(G)$ without H being a minor of G.

3. Consider the wheel graph $\mathcal{W}_r$ in Exercise 10 of Section 2.3. Show that if $r \geq 3$, then $M(\mathcal{W}_r)$ has $M(K_4)$ as a minor.

4. Let M be a spike with tip t and suppose that $\{t, x, y\}$ is a leg of M. Prove that $M \backslash x, y$ is both graphic and cographic.

5. Let M be a matroid on the set $\{x_1, x_2, \ldots, x_n\}$. Show that:
   (a) The number of minors of M is at least 2^n and at most 3^n.
   (b) There are exactly 2^n matroids M with exactly 2^n minors.

6. In a matrix $[I_r | D]$ over a field $\mathbb{F}$, let e_1 label the first column of I_r. By using the construction of a representation for the dual of a matroid, give an alternative derivation of a representation for $M[I_r | D]/e_1$.

7. Let T_8 be the matroid defined in Exercise 9 of Section 2.2. Give geometric representations for each of $(T_8/8)^*$, $(T_8/4)^*$, $T_8/8$, and $T_8/4$.

8. Let $\mathcal{J}$ denote the class of gammoids. Prove the following:
   (a) $\mathcal{J}$ is minor-closed.
   (b) $\mathcal{J}$ is closed under duality.
   (c) $\mathcal{J}$ equals the class of transversal matroids and their contractions.

9. Consider the Petersen graph P_{10} shown in Figure 3.5.

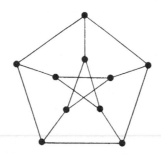

FIG. 3.5. The Petersen graph P_{10}.

   (a) Argue as in Proposition 2.3.3 to show that $M^*(P_{10})$ is not graphic.
   (b) Show that P_{10} has K_5 as a minor but has no subgraph that is a subdivision of K_5.
   (c) Show that P_{10} has a subgraph that is a subdivision of $K_{3,3}$.

10. Prove Proposition 3.2.13 and identify the legs of $(M \backslash t \backslash x_i)^*$.

11. For each of the following three possibilities for M, find all the matroids having no minor isomorphic to M:
    (a) $U_{0,1}$.
    (b) $U_{1,2}$.
    (c) $U_{0,1} \oplus U_{1,1}$ (see Exercise 7 of Section 1.1).

12. (Bondy 1972; Piff 1972)
    (a) Show that if $\{e, f\}$ is a circuit of a strict gammoid M, then $M \backslash e$ is a strict gammoid.
    (b) Deduce that if $\{e, f\}$ is a cocircuit of a transversal matroid N, then N/e is transversal.

3.3 The Scum Theorem, projection, and flats

We know from Section 3.1 that every minor N of a matroid M can be written in the form $M/Y\backslash X$. In this section, we prove the Scum Theorem, which asserts that the sets X and Y can be chosen so that M/Y has the same rank as N. We also investigate how to obtain geometric representations for $M\backslash e$ and M/e from a geometric representation for M; and we examine the effects of the operations of deletion and contraction on the flats of a matroid.

The next result (Higgs in Crapo and Rota 1970) is known as the Scum Theorem because, roughly stated, it asserts that if a matroid has a certain minor, it has such a minor hanging from the top in its lattice of flats. We shall not present this theorem in terms of lattices, preferring an approach that is more consistent with our earlier treatment.

Theorem 3.3.1 (The Scum Theorem) *Let N be a minor of a matroid M. Then there is a subset Z of $E(M) - E(N)$ such that M/Z and N have the same rank, and N is a restriction of M/Z. Moreover, if N has no loops, then Z can be chosen to be a flat of M.*

This theorem is very useful for it enables us to view the formation of a minor N as a two-stage process: a contraction to get the rank right followed by a deletion to remove the remaining elements not in N. In fact, when forming N, we frequently have a choice as to whether an element is removed by deletion or contraction. The reason for this is that if, after a sequence of deletions or contractions, an element e that is not in the minor N becomes a loop or a coloop, then, by Corollary 3.1.24, we can delete or contract e to get the same matroid.

The proof of the first part of the Scum Theorem will use the following lemma.

Lemma 3.3.2 *Every minor of a matroid M can be written in the form $M/I\backslash I^*$, where I and I^* are independent and coindependent, respectively, in M.*

Proof Every minor of M can be written as $M/Y\backslash X$ for some disjoint subsets X and Y of $E(M)$. Let Y_1 be a basis of $M|Y$. Then every element of $Y - Y_1$ is a loop in the contraction of Y_1 from $M|Y$ and so is a loop of M/Y_1. Hence

$$M/Y = M/Y_1\backslash(Y - Y_1). \tag{3.2}$$

Now let X_1 be a basis of $M^*|X$. Then, by (3.2), $M^*/X = M^*/X_1\backslash(X - X_1)$. Hence, by Proposition 3.1.26,

$$M\backslash X = M\backslash X_1/(X - X_1). \tag{3.3}$$

On combining (3.2) and (3.3), we get

$$M/Y\backslash X = M/[Y_1 \cup (X - X_1)]\backslash[X_1 \cup (Y - Y_1)].$$

As every element of $X - X_1$ is a loop of M^*/X_1, every such element is a coloop of $M\backslash X_1$. As Y_1 is independent in $M\backslash X_1$, it follows that $Y_1 \cup (X - X_1)$ is independent in $M\backslash X_1$ and hence in M. By duality, $X_1 \cup (Y - Y_1)$ is coindependent in M and the lemma follows.   □

Proof of Theorem 3.3.1   The first part is immediate from the preceding lemma. Now suppose that N has no loops. By the first part, $N = M/I\backslash I^*$ where I is independent and I^* is coindependent in M. Let $Z = \mathrm{cl}_M(I)$. Every element of $Z - I$ is a loop of M/I, so no such element is in $E(N)$. Thus $Z - I \subseteq I^*$ and $N = M/Z\backslash (I^* - Z)$.   □

The tipless free 3-spike is isomorphic to $U_{3,6}$. Hence both the 5-point line, $U_{2,5}$, and its dual, $U_{3,5}$, occur as minors of spikes. We now use the Scum Theorem to show that no longer line occurs as a minor of a spike.

Corollary 3.3.3 *Neither $U_{2,6}$ or $U_{4,6}$ is a minor of a spike.*

Proof   Take a spike M of minimal rank r such that M has a $U_{2,6}$-minor. Then, by Lemma 3.3.2, $U_{2,6} \cong M/I\backslash I^*$ for some independent set I and coindependent set I^*. If $r = 3$, then $|I| = 1$ and M has seven elements every one of which is on a 3-point line. Thus M/I has at most five rank-one flats. We deduce that $r > 3$. By Proposition 3.2.13(iii), I does not contain the tip of M. As I is non-empty, we may assume that I contains x_i where $\{x_i, y_i, t\}$ is some leg of M. Then $\{y_i, t\}$ is a circuit of M/x_i, so $U_{2,6}$ is a minor of $M/x_i\backslash t$ which, by Proposition 3.2.13(ii), is an $(r-1)$-spike. This contradicts the choice of r. We conclude that no spike has a $U_{2,6}$-minor. By duality and Proposition 2.1.27, no spike has a $U_{4,6}$-minor since every single-element deletion of a spike is self-dual.   □

We now turn our attention to obtaining geometric representations for $M\backslash e$ and M/e from a geometric representation for M. We have already implicitly used the fact that a geometric representation for $M\backslash e$ is obtained from one for M simply by deleting the point corresponding to e. Obtaining a geometric representation for the contraction of e is somewhat more complicated. We may assume that e is not a loop of M otherwise $M/e = M\backslash e$ and the geometric representation for M is as specified above. If P is the parallel class of M that contains e and P is non-trivial, then every element of $P - e$ becomes a loop of M/e. In order to determine the effect of contraction on the elements of $E(M) - P$, we shall now assume that $P = \{e\}$. To obtain a geometric representation for M/e, first note that all loops of M remain loops of M/e. Now let H be a hyperplane of M not containing e. We project every non-loop point of $E(M) - (H \cup e)$ from e onto H. Specifically, if f is such an point, we consider the line $\mathrm{cl}(\{e, f\})$. If this line meets H in g, say, then f is mapped onto g, or equivalently, onto a point touching g. If $\mathrm{cl}(\{e, f\})$ does not meet H, we add a new point on H. This point is the image of not only f but also of all other points in $\mathrm{cl}(\{e, f\}) - e$. Thus all of the non-loop points in the geometric representation for M/e are points of H. These points consist of the original non-loop points of H together with the images of the non-loop points of $E(M) - (H \cup e)$ under projection from e.

This projection from e has a predictable effect on a line L and a plane P. Specifically, if $e \in L$, then the points of $L - e$ are all mapped onto the same point or, equivalently, onto mutually touching points; if $e \notin L$, the images of distinct non-touching points of L are distinct, non-touching, and collinear in M/e. If $e \in P$, then the images of points of P are collinear in M/e; if $e \notin P$, the images of distinct non-touching points of P are distinct, non-touching, and coplanar in M/e. The verification that this construction does indeed produce a geometric representation for M/e is left to the reader with the reminder that, as e is not a loop, a subset I of $E(M) - e$ is in $\mathcal{I}(M/e)$ if and only if $I \cup e \in \mathcal{I}(M)$.

We now illustrate this construction with two examples.

Example 3.3.4 Figure 3.6(a) shows a geometric representation for F_7. To obtain a geometric representation for $F_7/1$, we can project from 1 onto the line $\{3, 4, 5\}$. The result is shown in Figure 3.6(b). Note that the same diagram is obtained by projecting from 1 onto, for example, the line $\{2, 7, 5\}$. Moreover, if we relax the circuit–hyperplane $\{2, 4, 6\}$ in F_7 to get F_7^-, and then contract 1, we see that we get the same matroid that was obtained just by contracting 1 from F_7. This last observation illustrates part of the following result (Kahn 1985), the proof of which is left to the reader (Exercise 12).

Proposition 3.3.5 *Let X be a circuit–hyperplane of a matroid M, and let M' be the matroid obtained from M by relaxing X. When $e \in E(M) - X$,*

(i) $M/e = M'/e$ *and, unless M has e as a coloop, $M'\backslash e$ is obtained from $M\backslash e$ by relaxing the circuit–hyperplane X of the latter.*

Dually, when $f \in X$,

(ii) $M\backslash f = M'\backslash f$ *and, unless M has f as a loop, M'/f is obtained from M/f by relaxing the circuit–hyperplane $X - f$ of the latter.* □

Example 3.3.6 Let M be the matroid for which a geometric representation is shown in Figure 3.7(a). To obtain a geometric representation for $M/1$, we project from 1 onto the plane $\{4, 5, 7, 8\}$. Evidently 2 must become a loop, while 3 and 6 become parallel. The resulting geometric representation for $M/1$ is shown in Figure 3.7(b) where, for example, 4, 7, and 6 are collinear in the contraction because 4, 7, 6, and 1 were coplanar in M. □

In 3.1.18 and 3.1.19, we specified the collections of hyperplanes of $M\backslash T$ and M/T. The next result extends this by determining all the flats of these matroids.

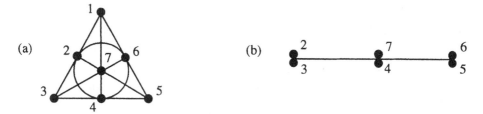

Fig. 3.6. (a) F_7. (b) $F_7/1$.

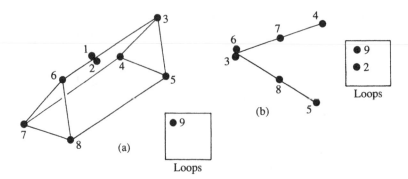

FIG. 3.7. (a) A matroid M. (b) The contraction $M/1$.

Proposition 3.3.7 *Let T be a subset of the ground set $E(M)$ of a matroid M and F be a subset of $E(M) - T$. Then*

(i)   *F is a flat of M/T if and only if $F \cup T$ is a flat of M;*

(ii)   *F is a flat of $M \backslash T$ if and only if M has a flat F' such that $F = F' - T$.*

Proof This follows easily from 3.1.15 and Proposition 3.1.11 and is left as an exercise for the reader.   □

We may use this result to relate the lattices of flats of M/T and $M \backslash T$ to that of M. If $\mathcal{L}$ is a lattice and x and y are elements of $\mathcal{L}$ such that $x \leq y$, recall that the *interval* $[x, y]$ is the lattice induced on $\{z \in \mathcal{L} : x \leq z \leq y\}$ by the order on $\mathcal{L}$.

Proposition 3.3.8 *For a subset T of the ground set of a matroid M,*

(i)   *$\mathcal{L}(M/T)$ is isomorphic to the interval $[\mathrm{cl}_M(T), E(M)]$ of $\mathcal{L}(M)$; and*

(ii)   *if $E - T$ is a flat of M, then $\mathcal{L}(M \backslash T)$ is isomorphic to the interval $[\mathrm{cl}_M(\emptyset), E - T]$ of $\mathcal{L}(M)$.*

Proof The map $\phi : \mathcal{L}(M/T) \rightarrow [\mathrm{cl}_M(T), E(M)]$ defined by $\phi(F) = F \cup T$ is easily shown to be an isomorphism. We leave the details of this and the proof of (ii) to the reader.   □

The next two results are immediate consequences of the last proposition.

Corollary 3.3.9 *If T is a flat of a matroid M, then $\mathcal{L}(M/T)$ is isomorphic to the interval $[T, E(M)]$ of $\mathcal{L}(M)$.*   □

Corollary 3.3.10 *If T_1 and $E - T_2$ are flats of a matroid M, and $T_1 \subseteq E - T_2$, then $\mathcal{L}(M/T_1 \backslash T_2)$ is isomorphic to the interval $[T_1, E - T_2]$ of $\mathcal{L}(M)$.*   □

While Proposition 3.3.8 specified $\mathcal{L}(M/T)$ for all T, it determined $\mathcal{L}(M \backslash T)$ only when $E - T$ is a flat of M. If we drop this condition on $E - T$, then $\mathcal{L}(M \backslash T)$ need not be isomorphic to an interval of $\mathcal{L}(M)$. Indeed, the elements of $\mathcal{L}(M \backslash T)$, which are the flats of $M \backslash T$, correspond precisely to those flats F of M such that $\mathrm{cl}_M(F - T) = F$.

Exercises

1. Show that the matroid $M/1$ in Example 3.3.6 is both graphic and cographic by finding graphs G_1 and G_2 such that $M/1 \cong M(G_1) \cong M^*(G_2)$.

2. (a) Show that if C is a circuit of a matroid M and $e \in C$, then
   $$M/C = M\backslash e/(C - e).$$
   (b) State the dual of the result in (a).

3. For each of the matroids $M(K_4)$, $\mathcal{W}^3$, Q_6, and P_6 shown in Figure 2.1, find geometric representations for $M/1$ and $M/6$.

4. Let X and Y be disjoint subsets of the ground set of a matroid M. Let Y_1 be a basis of $M|Y$ and X_1 be a cobasis of $M.X$. Prove that $M\backslash X/Y = M\backslash(X_1 \cup Y_2)/(Y_1 \cup X_2)$ for every partition (X_2, Y_2) of $(X - X_1) \cup (Y - Y_1)$.

5. Give necessary and sufficient conditions, in terms of flats of M, for a set F to be a flat of $M\backslash X/Y$.

6. Give an example to show that if N is a restriction of a simple matroid M, then $\mathcal{L}(N)$ need not be a sublattice of $\mathcal{L}(M)$.

7. Let M be a matroid such that $M/X \cong M/Y$ for all subsets X and Y of $E(M)$ with $|X| = |Y|$. Prove that M is uniform.

8. Prove that a matroid M has a $U_{2,n}$-minor if and only if M has a flat of rank $r(M) - 2$ that is contained in at least n distinct hyperplanes.

9. (a) Consider the matroid M for which a geometric representation is shown in Figure 1.17. Find geometric representations for $M/10$, for $M/1\backslash 4$, and for $M/3, 5\backslash 7$.
   (b) Consider the Vámos matroid V_8 (Figure 2.4). Give geometric representations for $V_8/1$ and $V_8/5$.

10. (a) Show that the non-Fano matroid F_7^- is non-binary by showing that it has a minor isomorphic to $U_{2,4}$.
    (b) Find a non-binary restriction N of F_7^- for which every proper restriction is binary.

11. Suppose that, in a matroid M, a non-empty set X is the intersection of a circuit and a cocircuit. Prove that M has a minor N such that X is a spanning circuit of both N and N^*, and $r(N) = |X| - 1 = r(N^*)$.

12. Prove Proposition 3.3.5.

13. *(Sims 1980, Bonin and de Mier 2008) Let $\mathcal{Z}$ be a collection of subsets of a set E and r be an integer-valued function on $\mathcal{Z}$. Prove that there is a matroid M on E for which $\mathcal{Z}$ is the set of cyclic flats and r is the rank function of M restricted to the sets in $\mathcal{Z}$ if and only if the following hold:
    (Z0) $\mathcal{Z}$ *is a lattice under inclusion;*
    (Z1) $r(0_{\mathcal{Z}}) = 0$ *where* $0_{\mathcal{Z}}$ *is the zero of this lattice;*
    (Z2) $0 < r(Y) - r(X) < |Y - X|$ *for all X and Y in $\mathcal{Z}$ with $X \subsetneq Y$; and*
    (Z3) *for all X and Y in $\mathcal{Z}$ having join $X \vee Y$ and meet $X \wedge Y$ in $\mathcal{Z}$,*
    $$r(X) + r(Y) \geq r(X \vee Y) + r(X \wedge Y) + |(X \cap Y) - (X \wedge Y)|.$$

4

CONNECTIVITY

We have already met several examples of basic graph-theoretic notions that have matroid-theoretic analogues. In this chapter, following Whitney (1935), we shall extend the concept of 2-connectedness from graphs to matroids. Matroid extensions of k-connectedness for $k \geq 3$ will be looked at in Chapter 8. We begin our discussion of k-connectedness with $k = 2$ rather than $k = 1$ since, by Proposition 1.2.9, every graphic matroid is isomorphic to the cycle matroid of a connected graph. In particular, if G_1 is a disconnected graph, then there is a connected graph G_2 such that $M(G_1) \cong M(G_2)$. Hence there is no matroid concept that corresponds directly to the idea of connectedness, or 1-connectedness, for graphs.

4.1 Connectivity for graphs and matroids

The purpose of this section is to introduce a notion of 2-connectedness for matroids that extends the corresponding notion for graphs. We begin by recalling the definition of k-connectedness for graphs.

A subset X of the vertex set of a graph G is called a *vertex cut* if $G - X$ has more connected components than G, where $G - X$ is obtained from G by deleting the vertices in X and all incident edges. If a vertex cut X contains a single vertex v, then v is called a *cut-vertex* of G.

Consider the graphs G_1, G_2, and G_3 in Figure 4.1. None has a cut-vertex. However, both G_1 and G_2 have 2-element vertex cuts, whereas the smallest vertex cut in G_3 has three elements. In general, for a connected graph G that has at least one pair of distinct non-adjacent vertices, the *connectivity* $\kappa(G)$ of G is the smallest integer j for which G has a j-element vertex cut. When G is connected, but has no pair of distinct non-adjacent vertices, we take $\kappa(G)$ to be

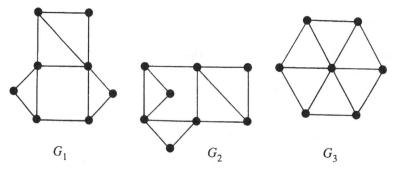

Fig. 4.1. $M(G_1)$ is isomorphic to $M(G_2)$, but $M(G_3)$ determines G_3.

$|V(G)| - 1$. Finally, if G is disconnected, we let $\kappa(G) = 0$. For a positive integer k, a graph G is said to be *k-connected* if $\kappa(G) \geq k$. Thus a graph with at least two vertices is 1-connected if and only if it is connected. For the graphs in Figure 4.1, it is easily seen that both $\kappa(G_1)$ and $\kappa(G_2)$ are 2, while $\kappa(G_3)$ is 3.

The main result of the next chapter will characterize precisely when two graphs have isomorphic cycle matroids. Connectivity will be a key tool in that discussion, and the graphs in Figure 4.1 exemplify the core of that result. The reader is urged to show that $M(G_1) \cong M(G_2)$ while, if G_4 is a graph for which $M(G_4) \cong M(G_3)$ and G_4 has no isolated vertices, then $G_4 \cong G_3$.

We have already noted that, for $k = 1$, the graph-theoretic notion of k-connectedness does not extend to matroids. Now let $k = 2$. The definition of 2-connectedness for graphs, relying as it does on deleting vertices, does not obviously extend to matroids. However, the following alternative characterization of 2-connectedness does.

Proposition 4.1.1 *Let G be a loopless graph without isolated vertices and suppose that $|V(G)| \geq 3$. Then G is 2-connected if and only if, for every pair of distinct edges of G, there is a cycle containing both.*

This proposition can be proved by a direct graph-theoretic argument. Instead, we shall give a slightly different proof involving matroids, and we postpone this until later in the section.

We now indicate how the notion of 2-connectedness for graphs can be extended to matroids. Define a relation ξ on the ground set $E(M)$ of a matroid M by $e \, \xi \, f$ if either $e = f$, or M has a circuit containing $\{e, f\}$.

Proposition 4.1.2 *For every matroid M, the relation ξ is an equivalence relation on $E(M)$.*

Proof Clearly ξ is reflexive and symmetric. To show that it is transitive, suppose that $e \, \xi \, f$ and $f \, \xi \, g$, where e, f, and g are distinct. Then M has circuits containing $\{e, f\}$ and $\{f, g\}$. Let C_1 and C_2 be circuits of M containing e and g, respectively, such that $C_1 \cap C_2$ is non-empty and that, among all such pairs of circuits, $|C_1 \cup C_2|$ is minimal. We aim to show that M has a circuit containing both e and g. Assume that this is not so. Then $C_1 \neq C_2$. Now choose an element h from $C_1 \cap C_2$. The reader may find a Venn diagram useful in keeping track of the rest of the proof. By the strong circuit elimination axiom, 1.4.12, M has a circuit C_3 such that $e \in C_3 \subseteq (C_1 \cup C_2) - h$. Moreover, by assumption, $g \notin C_3$. As $C_3 \not\subseteq C_1$, there is an element i of $C_2 - C_1$ that is in C_3. Applying the strong circuit elimination axiom to C_2 and C_3, we find that M has a circuit C_4 such that $g \in C_4 \subseteq (C_2 \cup C_3) - i$. Since $C_4 \not\subseteq C_2$, the set $C_4 \cap (C_3 - C_2)$ is non-empty, hence $C_4 \cap C_1$ is non-empty. But $C_1 \cup C_4 \subseteq (C_1 \cup C_2) - i$ and so $|C_1 \cup C_4| < |C_1 \cup C_2|$. Therefore (C_1, C_4) contradicts the choice of (C_1, C_2) as C_1 and C_4 contain e and g, respectively, and $C_1 \cap C_4 \neq \emptyset$. $\square$

The ξ-equivalence classes are called the *(connected) components* of M. Clearly every loop of M is a component; so is every coloop. If $E(M)$ is a component of

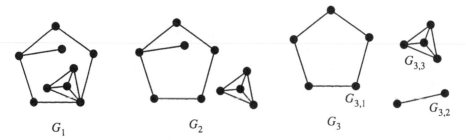

FIG. 4.2. Three different graphs with the same cycle matroid.

M or if $E(M)$ is empty, we call M *2-connected* or sometimes just *connected*; otherwise M is *disconnected*. The following is an immediate consequence of Proposition 4.1.2.

Proposition 4.1.3 *A matroid M is connected if and only if, for every pair of distinct elements of $E(M)$, there is a circuit containing both.*   □

The term 'connected component' has now been defined here for both graphs and matroids. Although this may seem undesirable, it does conform with standard usage. To highlight the difference, consider the graphs in Figure 4.2. Evidently $M(G_1) \cong M(G_2) \cong M(G_3)$. But the graphs G_1, G_2, and G_3, have one, two, and three connected components, respectively. The connected components of the matroids $M(G_1)$, $M(G_2)$, and $M(G_3)$ are isomorphic to $M(G_{3,1})$, $M(G_{3,2})$, and $M(G_{3,3})$. The graphs $G_{3,1}$, $G_{3,2}$, and $G_{3,3}$ are the blocks of each of G_1, G_2, and G_3. In general, a *block* is a connected graph whose cycle matroid is connected. Clearly a loopless graph is a block if and only if it is connected and has no cut-vertices. Hence, by Proposition 4.1.1, a block with at least three vertices is 2-connected. A *block of a graph* is a subgraph that is a block and is maximal with this property.

Example 4.1.4 The graph G in Figure 4.3(a) has eight blocks, these being the eight components of the graph in Figure 4.3(b).   □

We noted in Chapter 2 that the set of edges meeting at a vertex of a graph G is an edge cut of G but need not be a bond, a minimal edge cut. The last example

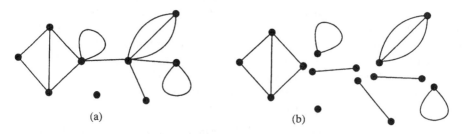

(a)   (b)

FIG. 4.3. (a) A graph G. (b) The eight blocks of G.

FIG. 4.4. $\mathcal{W}_2$.

provides another illustration of this phenomenon, which as the next result notes, cannot occur in 2-connected graphs. The proof is left as an exercise.

Proposition 4.1.5 *Let G be a connected loopless graph having at least three vertices. Then G is 2-connected if and only if, for every vertex v of G, the set of edges meeting at v is a bond.* ☐

It is not difficult to determine when a small matroid is connected.

Example 4.1.6 Table 4.1 lists all of the connected matroids on at most four elements. The graph $\mathcal{W}_2$ is shown in Figure 4.4. The reader is encouraged to check the accuracy of the table. In general, the uniform matroid $U_{m,n}$ is disconnected if and only if $n \geq 2$ and m is 0 or n. ☐

TABLE 4.1. All connected matroids on at most four elements.

Number n of elements	Connected n-element matroids
0	$U_{0,0}$
1	$U_{0,1}, U_{1,1}$
2	$U_{1,2}$
3	$U_{1,3}, U_{2,3}$
4	$U_{1,4}, U_{2,4}, U_{3,4}, M(\mathcal{W}_2)$

To conclude this section, we prove Proposition 4.1.1 by verifying the following restatement of it.

Proposition 4.1.7 *Let G be a loopless graph without isolated vertices and suppose that $|V(G)| \geq 3$. Then $M(G)$ is a connected matroid if and only if G is a 2-connected graph.*

Proof It is straightforward to show, using Proposition 4.1.3, that G is 2-connected if $M(G)$ is connected. We leave this as an exercise.

Now let G be 2-connected and let G_1 be the subgraph of G induced by some component of $M(G)$. Then, as G is loopless, $|V(G_1)| \geq 2$. Moreover, G_1 is certainly connected. As $E(G_1)$ is a component of $M(G)$, no cycle in G contains edges in both $E(G_1)$ and $E(G) - E(G_1)$. Thus, if $V(G_1) = V(G)$, then $G_1 = G$ and the result holds. Hence we may assume that $V(G_1) \neq V(G)$. Thus, for some vertex x of G_1, there is an edge xy of G such that $y \notin V(G_1)$. Then

$xy \in E(G) - E(G_1)$. Now x is not a cut-vertex of G, so $G - x$ contains a path P_{yz} that joins y to some vertex z of $G_1 - x$ but is otherwise disjoint from $G_1 - x$. Since G_1 is connected, it contains a path P_{zx} joining z and x. The union of P_{yz}, P_{zx}, and the edge xy is a cycle in G that contains xy and at least one edge of G_1. Since $xy \in E(G) - E(G_1)$, this is a contradiction.   □

Exercises

1. Show that a matroid M is connected if and only if, for every pair of distinct elements of $E(M)$, there is a hyperplane avoiding both.
2. Show that every component of a loopless matroid is a flat.
3. Find all connected matroids on a 5-element set, giving geometric representations for each.
4. If X is a subset of a matroid M and $M|X$ is connected, find necessary and sufficient conditions for $M|(\mathrm{cl}(X))$ to be connected.
5. Show that a matroid M with at least two elements is connected if and only if $\mathrm{si}(M)$ is connected and M has no loops.
6. Let X and Y be subsets of a matroid M such that both $M|(\mathrm{cl}(X))$ and $M|(\mathrm{cl}(Y))$ are connected and $X \cap Y$ is non-empty. Show that $M|(\mathrm{cl}(X \cup Y))$ is connected.
7. Let A be an $r \times n$ matrix over a field $\mathbb{F}$. Prove that $M[A]$ is connected if and only if there is no partition of the set of columns of A into non-empty sets X_1 and X_2 so that the subspaces spanned by X_1 and X_2 meet in $\{\underline{0}\}$.
8. (Harary 1969, Theorem 3.3) Let G be a loopless graph without isolated vertices and suppose that $|V(G)| \geq 3$. Prove that the following statements are equivalent:
   (a) G is 2-connected.
   (b) If u and v are vertices of G, there is a cycle meeting both.
   (c) If v is a vertex and e is an edge of G, there is a cycle containing e and meeting v.
   (d) If e and f are edges of G, there is a cycle containing both.

4.2 Properties of matroid connectivity

In this section, we indicate some of the many alternative characterizations of matroid connectivity. The first of these relates to the problem considered in the previous chapter of finding those subsets T of $E(M)$ for which $M\backslash T = M/T$. In a matroid M with ground set E, a subset X of E is a *separator* or a *1-separator* of M if X is a union of components of M. Thus both E and $\emptyset$ are separators of M. All other separators are called *non-trivial separators*. Clearly, if X is a separator of M, so is $E - X$. Next we characterize the separators of a matroid.

Proposition 4.2.1 *Let T be a subset of the ground set E of a matroid M. Then T is a separator of M if and only if*

$$r(T) + r(E - T) = r(M).$$

Proof Let T be a separator of M, and B_T and B_{E-T} be bases of $M|T$ and $M|(E-T)$, respectively. Let $B = B_T \cup B_{E-T}$. As M has no circuit meeting both T and $E-T$, there is no circuit of M contained in B. Thus B is independent in M. As B_T and B_{E-T} are maximal independent subsets of T and $E-T$, respectively, B is a maximal independent subset of M. Hence $r(T) + r(E-T) = r(M)$. The proof of the converse is similar and is left to the reader. □

We note here that, by the submodularity of r, for all subsets T of M,

$$r(T) + r(E - T) \geq r(M).$$

The last proposition characterizes when equality holds here. On combining that result with Proposition 3.1.23, we obtain the following.

Corollary 4.2.2 *If T is a subset of a matroid M, then $M\backslash T = M/T$ if and only if T is a separator of M.* □

One of the attractive features of the class of connected matroids is that it is closed under duality. To prove this, we shall use the next two results.

Lemma 4.2.3 *If T is a subset of a matroid M, then*

$$r(T) + r(E - T) - r(M) = r(T) + r^*(T) - |T|.$$

Proof This follows easily from the formula for $r^*(T)$ (Proposition 2.1.9) and is left as an exercise. □

Proposition 4.2.4 *A matroid M is disconnected if and only if, for some proper non-empty subset T of M,*

$$r(T) + r^*(T) - |T| = 0.$$

Proof By the preceding lemma, $r(T) + r^*(T) - |T| = 0$ if and only if $r(T) + r(E - T) - r(M) = 0$. By Proposition 4.2.1, the latter holds for a non-empty proper subset T of M if and only if M is disconnected. □

Corollary 4.2.5 *A matroid M is connected if and only if M^* is connected.*

Proof The equation in Proposition 4.2.4 is self-dual. □

We can obtain an alternative proof of the last corollary directly from the definition of connectedness by using the following result.

Proposition 4.2.6 *If x and y are distinct elements of a circuit C of a matroid M, then M has a cocircuit C^* such that $C \cap C^* = \{x, y\}$.*

Proof As $C - x$ is independent, it is contained in a basis B that contains y but avoids x. Now $E(M) - B$ is a cobasis of M that meets C in x. By the dual of Corollary 1.2.6, M has a unique cocircuit C^* contained in $(E(M) - B) \cup y$, and $y \in C^*$. Since C^* has the element y in common with the circuit C of M, Proposition 2.1.11 implies that C^* contains at least one other element of C. As $C \cap (y \cup (E(M) - B)) = \{x, y\}$, it follows that $x \in C^*$, so $C \cap C^* = \{x, y\}$. □

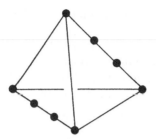

FIG. 4.5. $U_{2,4} \oplus U_{2,4}$.

The next result characterizes disconnected matroids in terms of their collections of independent sets. Its straightforward proof is omitted.

Proposition 4.2.7 *A matroid M is disconnected if and only if $E(M)$ has a proper non-empty subset T such that $\mathcal{I}(M) = \{I_1 \cup I_2 : I_1 \in \mathcal{I}(M|T)$ and $I_2 \in \mathcal{I}(M|(E - T))\}$.* $\square$

By the last result, every disconnected matroid M has the property that it can be determined from two disjoint restrictions. We now describe how to form a new matroid from two arbitrary matroids on disjoint sets.

Proposition 4.2.8 *Let M_1 and M_2 be matroids on disjoint sets E_1 and E_2. Let $E = E_1 \cup E_2$ and $\mathcal{I} = \{I_1 \cup I_2 : I_1 \in \mathcal{I}(M_1)$ and $I_2 \in \mathcal{I}(M_2)\}$. Then $(E, \mathcal{I})$ is a matroid.*

Proof   This is routine and was stated as Exercise 7 of Section 1.1. $\square$

The matroid $(E, \mathcal{I})$ in the last proposition is called the *direct sum* or *1-sum* of M_1 and M_2 and is denoted by $M_1 \oplus M_2$. Clearly $M_2 \oplus M_1 = M_1 \oplus M_2$. More generally, for n matroids, $M_1, M_2, \ldots, M_n$, on disjoint sets, $E_1, E_2, \ldots, E_n$, the direct sum $M_1 \oplus M_2 \oplus \cdots \oplus M_n$ is the pair $(E, \mathcal{I})$ where $E = E_1 \cup E_2 \cup \cdots \cup E_n$ and $\mathcal{I} = \{I_1 \cup I_2 \cup \cdots \cup I_n : I_i \in \mathcal{I}(M_i)$ for all i in $\{1, 2, \ldots, n\}\}$. It is not difficult to extend Proposition 4.2.8 to show that $M_1 \oplus M_2 \oplus \cdots \oplus M_n$ is indeed a matroid. We call $M_1, M_2, \ldots, M_n$ the *direct sum components* of $M_1 \oplus M_2 \oplus \cdots \oplus M_n$. Note that, whenever we write $M_1 \oplus M_2 \oplus \cdots \oplus M_n$, it will be implicit that $M_1, M_2, \ldots, M_n$ have disjoint ground sets.

The next result is a straightforward extension of Proposition 4.2.7.

Corollary 4.2.9 *If $T_1, T_2, \ldots, T_k$ are the connected components of a matroid M, then $M = M|T_1 \oplus M|T_2 \oplus \cdots \oplus M|T_k$. Moreover, if $M_1 \oplus M_2 \oplus \cdots \oplus M_n = N_1 \oplus N_2 \oplus \cdots \oplus N_m$, where each of $M_1, M_2, \ldots, M_n, N_1, N_2, \ldots, N_m$ is connected and non-empty, then $m = n$, and there is a permutation σ of $\{1, 2, \ldots, n\}$ such that $M_i = N_{\sigma(i)}$ for all i in $\{1, 2, \ldots, n\}$.* $\square$

Example 4.2.10 The rank-4 matroid for which a geometric representation is shown in Figure 4.5 is isomorphic to $U_{2,4} \oplus U_{2,4}$ and is clearly not uniform. Thus the direct sum of uniform matroids need not be uniform. $\square$

In contrast to the last example, we have the following result.

Proposition 4.2.11 *The classes of* $\mathbb{F}$*-representable, graphic, cographic, transversal, and regular matroids are all closed under the operation of direct sum.*

Proof This is left to the reader (Exercise 7). □

We conclude this section by noting a number of basic properties of the direct sum. These are stated for $M_1 \oplus M_2$ but can easily be extended to the direct sum of n matroids. The straightforward proofs of these facts are left as exercises.

4.2.12 $\mathcal{C}(M_1 \oplus M_2) = \mathcal{C}(M_1) \cup \mathcal{C}(M_2)$.

4.2.13 *If* $X \subseteq E(M_1 \oplus M_2)$*, then*

$$r_{M_1 \oplus M_2}(X) = r_{M_1}(X \cap E(M_1)) + r_{M_2}(X \cap E(M_2)).$$

4.2.14 $\mathcal{B}(M_1 \oplus M_2) = \{B_1 \cup B_2 : B_1 \in \mathcal{B}(M_1),\ B_2 \in \mathcal{B}(M_2)\}$.

4.2.15 $\mathcal{H}(M_1 \oplus M_2) = \{H_1 \cup E(M_2) : H_1 \in \mathcal{H}(M_1)\} \cup \{E(M_1) \cup H_2 : H_2 \in \mathcal{H}(M_2)\}$

4.2.16 F *is a flat of* $M_1 \oplus M_2$ *if and only if* $F \cap E(M_1)$ *and* $F \cap E(M_2)$ *are flats of* M_1 *and* M_2*, respectively.*

4.2.17 $\mathcal{S}(M_1 \oplus M_2) = \{S_1 \cup S_2 : S_1 \in \mathcal{S}(M_1),\ S_2 \in \mathcal{S}(M_2)\}$.

4.2.18 $\mathcal{C}^*(M_1 \oplus M_2) = \mathcal{C}^*(M_1) \cup \mathcal{C}^*(M_2)$.

4.2.19 *If* $X \subseteq E(M_1 \oplus M_2)$*, then*

$$(M_1 \oplus M_2)\backslash X = [M_1 \backslash (X \cap E(M_1))] \oplus [M_2 \backslash (X \cap E(M_2))],$$

and

$$(M_1 \oplus M_2)/X = [M_1/(X \cap E(M_1))] \oplus [M_2/(X \cap E(M_2))].$$

It is of interest to know how the components of a minor of M are related to the components of M. The following result is obtained by combining Proposition 4.1.3, Corollary 4.2.5, and 4.2.19.

Proposition 4.2.20 *For all matroids* M*, every connected component of* $M\backslash X/Y$ *is contained in a connected component of* M*. In particular, every connected matroid that is a minor of* M *is a minor of some component of* M. □

Finally, we note another attractive feature of duality.

Proposition 4.2.21 $(M_1 \oplus M_2)^* = M_1^* \oplus M_2^*$ *for all matroids* M_1 *and* M_2.

Proof
$$\begin{aligned}
\mathcal{C}((M_1 \oplus M_2)^*) &= \mathcal{C}^*(M_1 \oplus M_2) \\
&= \mathcal{C}^*(M_1) \cup \mathcal{C}^*(M_2),\ \text{by 4.2.18;} \\
&= \mathcal{C}(M_1^*) \cup \mathcal{C}(M_2^*) \\
&= \mathcal{C}(M_1^* \oplus M_2^*),\ \text{by 4.2.12.}
\end{aligned}$$
□

Exercises

1. Let T be a subset of the ground set of a matroid M. Prove that the following statements are equivalent:
    (a) T is a separator of M.
    (b) $M^*\backslash T = M^*/T$.
    (c) Every hyperplane of M contains T or $E - T$.
    (d) T is a separator of M^*.

2. Prove that the following are equivalent for a matroid M with $r(M) \geq 1$:
    (a) M is a partition matroid.
    (b) M is a direct sum of rank-one loopless matroids.
    (c) M^* is a direct sum of circuits.
    (d) $E(M)$ can be partitioned into r cocircuits.
    (e) M is loopless and has no two intersecting cocircuits.
    (f) Every circuit of M has exactly two elements.

3. Show that if $\{e, f\}$ is both a circuit and a cocircuit of a matroid M, then $\{e, f\}$ is a component of M.

4. In a matroid M, suppose $X \subseteq E(M)$, and let r^* be the rank function of M^*. Show that $r^*(X) = r((M|X)^*)$ if and only if X is a separator of M.

5. Find all disconnected paving matroids.

6. Find all matroids with at most three circuits and show that all such matroids are both graphic and cographic.

7. (a) Let A_1 and A_2 be $\mathbb{F}$-representations of matroids M_1 and M_2 where $E(M_1) \cap E(M_2) = \emptyset$. Show that $\begin{bmatrix} A_1 & 0 \\ 0 & A_2 \end{bmatrix}$ is an $\mathbb{F}$-representation of $M_1 \oplus M_2$.
    (b) Complete the proof of Proposition 4.2.11.

8. Prove directly that, in a graph G, every pair of distinct edges is in a cycle if and only if every pair of distinct edges is in a bond.

9. (a) Prove that every minor of a paving matroid is paving.
    (b) Prove that a matroid is paving if and only if it has no minor isomorphic to $U_{2,2} \oplus U_{0,1}$.

4.3   More properties of connectivity

Many matroid properties are such that a matroid M has the property if and only if all of its connected components have the property. This fact, which was exemplified in Proposition 4.2.11, means that, in many arguments, one is able to restrict attention to connected matroids. In this section, we present some more properties of connected matroids. The first result (Tutte 1966b) is particularly useful in induction arguments.

Theorem 4.3.1 *Let e be an element of a connected matroid M. Then $M\backslash e$ or M/e is connected.*

Proof Suppose $M\backslash e$ is not connected and let E_1 be one of its components. If $x \in E_1$ and $y \in E(M\backslash e) - E_1$, then M has a circuit C_{xy} containing $\{x, y\}$. Now

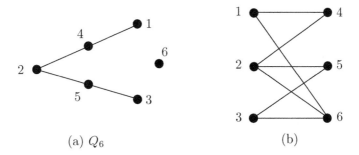

FIG. 4.6. Q_6 and its fundamental graph with respect to $\{1, 2, 3\}$.

C_{xy} is not a circuit of $M \backslash e$ as x and y are in different components of $M \backslash e$. Hence $e \in C_{xy}$ and so $C_{xy} - e$ is a circuit of M/e containing $\{x, y\}$. But x and y were arbitrarily chosen in E_1 and $E(M) - (E_1 \cup e)$, respectively. Using this and the fact that the components of M/e are $\xi_{M/e}$-equivalence classes, it is not difficult to show that $E(M) - e$ is a component of M/e; that is, M/e is connected. □

Consider the matroid Q_6 for which a geometric representation is shown in Figure 4.6(a). Clearly $\{1, 2, 3\}$ is a basis B of Q_6 and the fundamental circuits with respect to B are $\{1, 2, 4\}, \{2, 3, 5\}$, and $\{1, 2, 3, 6\}$. With $M = Q_6$, the bipartite graph in Figure 4.6(b) is that associated with the family of sets $(C(e, B) - e : e \in E(M) - B)$. Hence it has vertex classes B and $E(M) - B$ and edge set $\{(i, j) : j \in E(M) - B$ and $i \in C(j, B) - j\}$. We denote this graph by $G_B(M)$ and call it the *fundamental graph of M with respect to B*. This graph will be important in Section 6.4 when we consider constructing $\mathbb{F}$-representations of matroids. Here we observe that both the graph $G_B(Q_6)$ and the matroid Q_6 are connected. The next result (Cunningham 1973, Krogdahl 1977) establishes that, in general, we can determine whether or not M is connected from $G_B(M)$.

Proposition 4.3.2 *Let M be a non-empty matroid and B be a basis of M. Then M is connected if and only if $G_B(M)$ is connected.*

Proof The proposition clearly holds if $|E(M)| = 1$, so assume that $|E(M)| \geq 2$. Suppose first that $G_B(M)$ is connected. Then B is a non-empty proper subset of $E(M)$. Moreover, if x and y are distinct elements of M, there is an (x, y)-path in $G_B(M)$. Each edge of this path corresponds to a B-fundamental circuit, so there is a sequence $C_1, C_2, \ldots, C_n$ of B-fundamental circuits with $x \in C_1$ and $y \in C_n$ such that $C_i \cap C_{i+1}$ is non-empty for all i in $\{1, 2, \ldots, n-1\}$. Hence, by Proposition 4.1.2, M is connected.

Conversely, suppose M is connected but $G_B(M)$ is disconnected. Because $|E(M)| \geq 2$, there are no loops in M, so no vertex in $E(M) - B$ has degree 0. Hence every component of $G_B(M)$ contains some vertex of B. Let G_1 be such a component, let B_1 be the set of vertices of B that are in G_1, and let Z_1 be the remaining vertices of G_1. If $z \in Z_1$, then $C(z, B) \subseteq B_1 \cup z$. Hence $Z_1 \subseteq \mathrm{cl}(B_1)$. Similarly, $E(M) - B - Z_1 \subseteq \mathrm{cl}(B - B_1)$. Thus

$$r(B_1 \cup Z_1) + r((B - B_1) \cup (E(M) - B - Z_1)) = |B_1| + |B - B_1| = |B| = r(M),$$

so $B_1 \cup Z_1$ is a non-trivial separator of M; a contradiction. $\square$

The next result (Lehman 1964) asserts that a connected matroid is uniquely determined by the set of circuits containing any fixed element. An extension of this result for binary matroids will be noted in Exercise 11 of Section 9.4.

Theorem 4.3.3 *Let e be an element of a connected matroid M, and $\mathcal{C}_e$ be the set of circuits of M containing e. Then the circuits of M not containing e are the minimal sets of the form*

$$(C_1 \cup C_2) - \bigcap \{C : C \in \mathcal{C}_e \text{ and } C \subseteq C_1 \cup C_2\}$$

where C_1 and C_2 are distinct members of $\mathcal{C}_e$.

Proof Let C_1 and C_2 be distinct members of $\mathcal{C}_e$. We shall denote the set $\{C : C \in \mathcal{C}_e \text{ and } C \subseteq C_1 \cup C_2\}$ by $\mathcal{C}_e(C_1 \cup C_2)$. We begin by showing that

$$(C_1 \cup C_2) - \bigcap \{C : C \in \mathcal{C}_e(C_1 \cup C_2)\} \text{ is dependent.} \tag{4.1}$$

By circuit elimination, M has a circuit C_3 such that $C_3 \subseteq (C_1 \cup C_2) - e$. To prove (4.1), we shall show that

$$C_3 \subseteq (C_1 \cup C_2) - \bigcap \{C : C \in \mathcal{C}_e(C_1 \cup C_2)\}. \tag{4.2}$$

Choose f in $C_3 \cap C_1$. Then, by the strong circuit elimination axiom, there is a circuit C_4 such that $e \in C_4 \subseteq (C_1 \cup C_3) - f$. Evidently $C_4 \in \mathcal{C}_e(C_1 \cup C_2)$ and hence $f \notin \bigcap \{C : C \in \mathcal{C}_e(C_1 \cup C_2)\}$. It follows that $C_3 \cap C_1$ does not meet $\bigcap \{C : C \in \mathcal{C}_e(C_1 \cup C_2)\}$. Similarly, the last set avoids $C_3 \cap C_2$. Thus (4.2) holds and hence so does (4.1).

Now let D be an arbitrary circuit of M avoiding e. As M is connected, it has a circuit that contains e and meets D. Choose such a circuit C_1 so that $C_1 \cup D$ is minimal. As $e \in C_1 - D$, the strong circuit elimination axiom implies that M has a circuit C_2 that is distinct from C_1 such that $e \in C_2 \subseteq C_1 \cup D$. We shall show that

$$D = (C_1 \cup C_2) - \bigcap \{C : C \in \mathcal{C}_e(C_1 \cup C_2)\}. \tag{4.3}$$

If $C' \in \mathcal{C}_e(C_1 \cup C_2)$, then $C' \subseteq C_1 \cup C_2 \subseteq C_1 \cup D$. Since C' is not properly contained in C_1, it must meet D. As C' contains e, the choice of C_1 implies that $C' \cup D = C_1 \cup D$. Hence, as C' was arbitrarily chosen in $\mathcal{C}_e(C_1 \cup C_2)$,

$$\left(\bigcap \{C : C \in \mathcal{C}_e(C_1 \cup C_2)\}\right) \cup D = C_1 \cup D \supseteq C_1 \cup C_2.$$

Therefore

$$D \supseteq (C_1 \cup C_2) - \bigcap \{C : C \in \mathcal{C}_e(C_1 \cup C_2)\}. \tag{4.4}$$

But, by (4.1), the right-hand side of (4.4) contains a circuit. Hence equality must hold in (4.4), that is, (4.3) holds. We can now easily complete the proof of the theorem using, for example, Lemma 2.1.22. $\square$

The next result follows from Corollary 4.2.5 and the definition of connectivity.

Corollary 4.3.4 *A matroid M is connected if and only if, for every pair of distinct elements of M, there are a circuit and a cocircuit containing both.* □

We now present two variants of this. The first is an immediate consequence of Proposition 4.1.3 and Corollary 4.2.5. The second exemplifies the use of Theorem 4.3.1 in induction arguments.

Corollary 4.3.5 *A matroid M is connected if and only if, for every pair of distinct elements, there is a circuit or a cocircuit containing both.* □

Proposition 4.3.6 *Let M be a matroid having at least three elements. Then M is connected if and only if, for all 3-element subsets X of $E(M)$, there is a circuit or a cocircuit containing X.*

Proof If the condition on 3-element subsets of $E(M)$ holds, then M is certainly connected. Conversely, suppose that M is connected. We shall argue by induction on $|E(M)|$ to prove that the specified condition on 3-element subsets holds. If $|E(M)| = 3$, then, as M is connected, it follows from Table 4.1 (in Section 4.1) that M is isomorphic to $U_{1,3}$ or $U_{2,3}$. Hence, in this case, the result holds. Now assume the result holds for $|E(M)| < n$. Let $|E(M)| = n \geq 4$ and let X be an arbitrary 3-element subset of $E(M)$. Choose an element e from $E(M) - X$. By Theorem 4.3.1, either (i) $M \backslash e$ is connected, or (ii) M/e is connected. Assume that (i) occurs. Then, by the induction assumption, X is contained in a circuit C or a cocircuit C^* of $M \backslash e$. In the first case, C is a circuit of M containing X. In the second case, C^* or $C^* \cup e$ is a cocircuit of M containing X. Thus if (i) occurs, then the result holds for M. But if (ii) occurs, then, by Corollary 4.2.5, $(M/e)^*$ is connected, that is, $M^* \backslash e$ is connected. Applying the argument used in case (i) with M^* in place of M, we get that M^* has a circuit or a cocircuit containing X. Hence so does M and the result holds in case (ii). We conclude, by induction, that the proposition holds. □

Next we show that, given a connected minor N of a connected matroid M, we can delete or contract an element from M to both stay connected and keep N as a minor (Brylawski 1972, Seymour 1977a). In Chapter 12, we shall consider several other results of this type for matroids of higher connectivity.

Proposition 4.3.7 *Let N be a connected minor of a connected matroid M and suppose that $e \in E(M) - E(N)$. Then at least one of $M \backslash e$ and M/e is connected having N as a minor.*

Proof We may assume that N is a minor of $M \backslash e$, otherwise, in the argument that follows, we replace M and N by their duals. If $M \backslash e$ is connected, the proposition holds. Thus we may suppose that $M \backslash e$ is disconnected. It follows, by Theorem 4.3.1, that M/e is connected. To complete the proof, we shall show that N is a minor of M/e.

As N is a connected minor of $M\backslash e$, Proposition 4.2.20 implies that N is a minor of some component E_1 of $M\backslash e$. Let $E_2 = E(M\backslash e) - E_1$. Then N is a minor of $M\backslash e\backslash E_2$. Moreover, by Proposition 4.2.1 and the fact that M is connected,

$$r(E_1) + r(E_2) = r(M\backslash e) \le r(M) \tag{4.5}$$

and

$$r(E_1 \cup e) + r(E_2) > r(M). \tag{4.6}$$

By (4.5) and (4.6), $r(E_1 \cup e) > r(E_1)$, so $r(E_1 \cup e) = r(E_1) + 1$, that is, e is a coloop of $M\backslash E_2$. Thus $M\backslash e\backslash E_2 = (M\backslash E_2)\backslash e = (M\backslash E_2)/e = M/e\backslash E_2$. Since N is a minor of $M\backslash e\backslash E_2$, it follows that N is also a minor of $M/e\backslash E_2$ and hence, as required, N is a minor of M/e.   □

Corollary 4.3.8 *Let N be a connected minor of a connected matroid M. Then there is a sequence $M_0, M_1, M_2, \ldots, M_n$ of connected matroids with that $M_0 = N$ and $M_n = M$ such that, for all i in $\{0, 1, \ldots, n-1\}$, the matroid M_i is a single-element deletion or a single-element contraction of M_{i+1}*   □

Next we consider briefly what can be said about a connected matroid for which no single-element deletion is connected. Looking to graphs for guidance, we note that Dirac (1967) and Plummer (1968) proved that a 2-connected graph G for which no single-edge deletion is 2-connected must have a vertex of degree two. In particular, $M(G)$ has a 2-cocircuit, that is, a 2-element cocircuit. This result was generalized to matroids by Murty (1974). A non-empty matroid M is *minimally connected* if M is connected but $M\backslash e$ is disconnected for all elements e. Clearly $U_{2,3}$ is minimally connected and one easily checks that it is the unique minimally connected matroid of rank at most two. Murty showed that every minimally connected matroid has a 2-cocircuit and Seymour (1979) extended this as follows. A further extension will be considered in Exercise 10(d).

Proposition 4.3.9 *Let M be a minimally connected matroid. Then M has a cobasis each element of which is in a 2-element cocircuit.*

This result is a straightforward consequence of the following (Oxley 1981a).

Lemma 4.3.10 *Let M be a connected matroid having at least two elements and let $\{e_1, e_2, \ldots, e_m\}$ be a circuit of M such that $M\backslash e_i$ is disconnected for all i in $\{1, 2, \ldots, m-1\}$. Then $\{e_1, e_2, \ldots, e_{m-1}\}$ contains a 2-cocircuit of M.*

Proof As $M\backslash e_1$ is disconnected, Theorem 4.3.1 implies that M/e_1 is connected. Now suppose that $M/e_1\backslash e_i$ is connected for some i in $\{2, 3, \ldots, m-1\}$. As $M\backslash e_i$ is disconnected, it must have $\{e_1\}$ as a component. Thus e_1 is a loop or a coloop of $M\backslash e_i$. But e_1 cannot be a loop of $M\backslash e_i$, otherwise it is a loop of the connected matroid M. Thus e_1 is a coloop of $M\backslash e_i$ but not of M, so $\{e_1, e_i\}$ is a cocircuit of M. Hence the lemma holds if $M/e_1\backslash e_i$ is connected for some i in $\{2, 3, \ldots, m-1\}$ and so, in particular, it holds for $m = 3$.

Assume that the lemma is false and let k be the smallest value of m for which it fails. Since the hypotheses of the lemma are never satisfied when $m \le 2$, we must have that $k > 2$. Moreover, from above, $k \neq 3$ and $M/e_1 \backslash e_i$ is disconnected for all i in $\{2, 3, \ldots, k-1\}$. As $\{e_2, e_3, \ldots, e_k\}$ is a circuit of M/e_1, it follows by the choice of k that $\{e_2, e_3, \ldots, e_{k-1}\}$ contains a 2-cocircuit of M/e_1. Thus the last set contains a 2-cocircuit of M; a contradiction. $\qquad\square$

Proof of Proposition 4.3.9. Let X be the set of elements of M that are contained in some 2-cocircuit. Then, by the last lemma, X meets every circuit of M. Hence, by Corollary 2.1.20, X contains a cobasis B^* of M. $\qquad\square$

Generalizing another graph result of Dirac (1967) and Plummer (1968), Murty (1974) established the following sharp bound on the number of elements in a minimally connected matroid. We leave the proof as an exercise. A constructive characterization of all minimally connected matroids was given by Oxley (1981a) using an operation to be discussed in Chapter 7.

Proposition 4.3.11 *Let M be a minimally connected matroid of rank r where $r \geq 3$. Then $|E(M)| \leq 2r - 2$. Moreover, equality holds if and only if $M \cong M(K_{2,r-1})$.* $\qquad\square$

To close this chapter, we use Theorem 4.3.1 to establish a bound (Lemos and Oxley 2001) on the number of elements in a connected matroid that is polynomial in the sizes of a largest circuit and a largest cocircuit. Let M be a connected matroid with at least two elements. For e in $E(M)$, let $c_e(M)$ and $c_e^*(M)$ be the sizes of a largest circuit and a largest cocircuit containing e.

Theorem 4.3.12 *Let M be a connected matroid with at least two elements. If e is an element of M, then*

$$|E(M)| \leq (c_e(M) - 1)(c_e^*(M) - 1) + 1.$$

This theorem is an immediate consequence of the following covering result. The proof of this is a straightforward modification of a short proof of Theorem 4.3.12 given by G. Ding (private communication).

Theorem 4.3.13 *Let M be a connected matroid with at least two elements. If e is an element of M, then M has a set of at most $c_e(M) - 1$ cocircuits each containing e such that the union of these cocircuits is $E(M)$.*

Proof We argue by induction on $c_e(M)$. First observe that $c_e(M) = 2$ if and only if every element of $E(M) - e$ is parallel to e. The latter holds if and only if M is uniform of rank one, that is, if and only if $E(M)$ is a cocircuit of M. We deduce that the theorem holds when $c_e(M) = 2$. Assume it holds when $c_e(M) < n$ and let $c_e(M) = n \geq 3$.

Let C^* be a cocircuit of M that contains e. Clearly $C^* \neq E(M)$. By Theorem 4.3.1, we may remove the elements of $C^* - e$ from M one at a time by deletion or contraction so as to always maintain a connected matroid. Thus $C^* - e$ is the union of some disjoint sets X and Y such that $M \backslash X / Y$ is connected. Call this

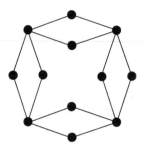

FIG. 4.7. The cycle matroid of this graph attains equality in Theorem 4.3.12.

minor N. By orthogonality (Proposition 2.1.11), every circuit of M that contains e must also meet X or Y. It follows that

$$c_e(N) < c_e(M).$$

Since $C^* \neq E(M)$, the matroid N has at least two elements. Thus, by the induction assumption, for some $k \leq c_e(N) - 1$, there are k cocircuits $C_1^*, C_2^*, \ldots, C_k^*$ of N each containing e such that the union of these cocircuits is $E(N)$. By 3.1.16 and 3.1.17, for each C_i^*, there is a cocircuit D_i^* of M such that $C_i^* = D_i^* - X$. Hence $C^*, D_1^*, D_2^*, \ldots, D_k^*$ are cocircuits of M each containing e and

$$E(M) = E(N) \cup X \cup Y \subseteq C_1^* \cup C_2^* \cup \cdots \cup C_k^* \cup C^* \subseteq D_1^* \cup D_2^* \cup \cdots \cup D_k^* \cup C^*.$$

Thus we have a family of $k + 1$ cocircuits of M whose union is $E(M)$. Since

$$k + 1 \leq (c_e(N) - 1) + 1 = c_e(N) \leq c_e(M) - 1,$$

the result follows by induction.   □

The cycle matroid of the graph in Figure 4.7 is one of a family of matroids that show that the bound in Theorem 4.3.12 is sharp. All such matroids were described in Lemos and Oxley (2001). The next result is obtained by applying Theorem 4.3.13 to graphic matroids. The proof is left as an exercise.

Corollary 4.3.14 *Let u and v be distinct vertices in a 2-connected loopless graph G. Then $|E(G)|$ cannot exceed the product of the length of a longest (u, v)-path and the size of a largest bond separating u from v.*   □

Clearly Theorem 4.3.12 implies a bound on the size of a connected matroid M in terms of the sizes, c and c^*, of a largest circuit and a largest cocircuit. This bound can be sharpened to the following best-possible bound (Lemos and Oxley 2001) whose proof is omitted. All graphic matroids attaining equality in this bound were determined by Wu (2000).

Theorem 4.3.15 *In a connected matroid M with at least two elements having largest circuit with c elements and largest cocircuit with c^* elements,*

$$|E(M)| \leq \tfrac{1}{2}cc^*.$$ □

For comparison, a lower bound on $|E(M)|$ in terms of c and c^* that holds for all matroids M having non-zero rank and corank is

4.3.16 $c + c^* - 2 \leq |E(M)|$.

To see this, note that $c \leq r(M) + 1$ and $c^* \leq r^*(M) + 1$. The bound in 4.3.16 is sharp, being attained, for example, by all uniform matroids of non-zero rank and corank.

Exercises

1. Give an example of a connected matroid M having distinct elements e_1, e_2, and e_3 such that all of $M\backslash e_1$, M/e_1, $M\backslash e_2$, and M/e_3 are connected, but M/e_2 and $M\backslash e_3$ are disconnected.

2. Let e and f be distinct elements of a connected matroid M.
   (a) Show that $M\backslash e, f$, $M\backslash e/f$, $M\backslash f/e$, or $M/e, f$ is connected.
   (b) Give examples to show that every set of three of these four matroids can be disconnected.

3. Let M_1 and M_2 be matroids on the same set E and suppose that e is an element of E that is a loop of neither M_1 nor M_2.
   (a) Show that the set of bases of M_1 containing e equals the set of bases of M_2 containing e if and only if $M_1/e = M_2/e$.
   (b) Give an example to show that a connected matroid is not uniquely determined by the set of bases containing a fixed element e.

4. Let M be a connected matroid having at least $2n - 1$ elements. Show that $E(M)$ has a subset $\{e_1, e_2, \ldots, e_n\}$ such that either $M\backslash e_i$ is connected for all i or M/e_i is connected for all i.

5. Let e be an element of a connected matroid M and $\mathcal{H}_e$ be the set of hyperplanes of M avoiding e. Specify all the hyperplanes of M in terms of the members of $\mathcal{H}_e$.

6. (Dirac 1967; Plummer 1968) Let G be a 2-connected graph such that $G\backslash e$ is not 2-connected for all edges e. Prove that G has a degree-2 vertex.

7. Consider the bound in Theorem 4.3.15.
   (a) Show that this is attained by $AG(3, 2)$.
   (b) Give an infinite family of matroids that attain the bound.
   (c) °Find a non-graphic matroid M other than $AG(3, 2)$ that attains the bound such that both M and M^* are simple.

8. (Bonin, McNulty, and Reid 1999) Let M be a loopless non-empty matroid of rank r having largest cocircuit size c^*. Prove that $|E(M)| \leq rc^*$.

9. Let M be a connected matroid with at least two elements. Prove that M has no $U_{1,3}$-minor if and only if M is a circuit.

10. (a) Show that every circuit of a minimally connected matroid is a flat.
    (b) Find a connected matroid N in which every circuit is a flat such that $|E(N)| \geq 3$ and $N\backslash x$ is connected for all x in $E(N)$.

(c) Let C^* be a cocircuit of a connected matroid M and suppose that M/e is disconnected for all e in C^*. Prove that either $E(M) = C^*$, or C^* contains at least two distinct non-trivial parallel classes of M.

(d) Let N be a matroid for which N^* is minimally connected. Prove that either $N \cong U_{1,n}$ for some $n \geq 3$, or N has at least $r(N) + 1$ non-trivial parallel classes.

(e) Prove that if N is simple non-empty connected matroid, then N has at least $r(N) + 1$ elements e such that N/e is connected.

11. Prove Proposition 4.3.11.

12. Let C be a circuit of a connected matroid M. For e in $E(M)$, prove that:
   (a) M has circuits C_1 and C_2 such that $e \in C_1 \cap C_2$ and $C \subseteq C_1 \cup C_2$.
   (b) *(Mason and Oxley 1980) If f is an element of M, then M has circuits C_e and C_f with $e \in C_e$ and $f \in C_f$ such that $C \subseteq C_e \cup C_f$.

13. Suppose $k \in \{2, 3, 4, 5\}$.
   (a) Prove that a matroid M with at least k elements is connected if and only if, for every k-element subset X of $E(M)$, there is a circuit C and a cocircuit C^* such that $C \cup C^* \supseteq X$ and $C \cap C^* \neq \emptyset$.
   (b) Show that (a) fails if $k = 6$.

14. (Seymour 1979) Let M be a connected matroid having at least two elements. Suppose that, for every element e, both $M \backslash e$ and M/e are connected but, for every pair of distinct elements, e and f, both $M \backslash e, f$ and $M/e, f$ are disconnected. Prove that $M \cong U_{2,4}$.

15. (Lemos 2004) Let M be a connected matroid having at least two elements and suppose $e \in E(M)$. Prove that if every circuit containing e is spanning and every cocircuit containing e is cospanning, then M is uniform.

16. *(Seymour in Ding, Oporowski, and Oxley 1995) Let M be a connected matroid having at least two elements and C be a largest circuit of M. Prove that all the circuits of M/C have at most $|C| - 1$ elements.

17. Let M be a connected matroid with at least two elements whose largest circuit has c elements.
   (a) °Prove that M has a collection of c not-necessarily-distinct cocircuits such that every element of M is in at least two of these cocircuits.
   (b) *(Neumann-Lara, Rivera-Campo, and Urrutia 1999) Prove that (a) holds if M is graphic.
   (c) *(McGuinness 2005) Prove that (a) holds if M is cographic.

5

GRAPHIC MATROIDS

We have already seen that graphic matroids form a fundamental class of matroids. Moreover, numerous operations and results for graphs provide the inspiration or motivation for corresponding operations and results for matroids. In this chapter, we examine graphic matroids in more detail. In particular, we shall present several proofs delayed from Chapters 1 and 2 including proofs that a graphic matroid is representable over every field, and that a cographic matroid $M^*(G)$ is graphic only if G is planar. The main result of the chapter is Whitney's 2-Isomorphism Theorem, which establishes necessary and sufficient conditions for two graphs to have isomorphic cycle matroids.

5.1 Representability

In this section, we shall prove that every graphic matroid is representable over every field and that every graphic matroid is regular. The proofs of these results use the following construction for a given graph G. Form a directed graph $D(G)$ from G by arbitrarily assigning a direction to each edge. Let $A_{D(G)}$ denote the incidence matrix of $D(G)$, that is, $A_{D(G)}$ is the matrix $[a_{ij}]$ whose rows and columns are indexed by the vertices and arcs, respectively, of $D(G)$ where

$$a_{ij} = \begin{cases} 1, & \text{if vertex } i \text{ is the tail of non-loop arc } j; \\ -1, & \text{if vertex } i \text{ is the head of non-loop arc } j; \\ 0, & \text{otherwise.} \end{cases}$$

Example 5.1.1 Let G be the graph shown in Figure 5.1(a) and $D(G)$ be as shown in Figure 5.1(b). Then $A_{D(G)}$ is as given in Figure 5.2. □

Proposition 5.1.2 *If G is a graph, then $M(G)$ is representable over every field.*

This proposition is an immediate consequence of the following result.

Lemma 5.1.3 *Let $D(G)$ be an arbitrary orientation of a graph G, and let $\mathbb{F}$ be a field. Then $A_{D(G)}$ represents $M(G)$ over $\mathbb{F}$.*

Proof It is implicit in the statement of this lemma that $A_{D(G)}$ must be interpreted as a matrix over $\mathbb{F}$. This leaves $A_{D(G)}$ unchanged unless $\mathbb{F}$ has characteristic two, in which case each negative entry changes sign. To prove the lemma, we shall show that every circuit of $M(G)$ is dependent in $M[A_{D(G)}]$, and that every circuit of $M[A_{D(G)}]$ is dependent in $M(G)$. By Lemma 2.1.22, it will follow from this that $\mathcal{C}(M(G)) = \mathcal{C}(M[A_{D(G)}])$ and so $M(G) = M[A_{D(G)}]$.

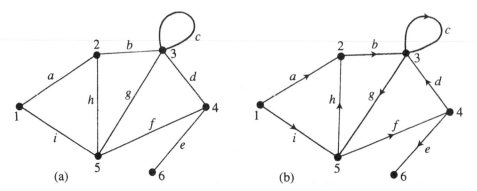

FIG. 5.1. (a) A graph G. (b) An orientation $D(G)$ of G.

Let C be a cycle of G. If C is a loop e, then the corresponding column of $A_{D(G)}$ is zero, so C is a circuit of $M[A_{D(G)}]$. Now suppose that C is not a loop and let its edges, in cyclic order, be $e_1, e_2, \ldots, e_k$. Let $\underline{v}(e_1), \underline{v}(e_2), \ldots, \underline{v}(e_k)$ be the corresponding columns of $A_{D(G)}$. Now cyclically traverse C beginning with e_1 and, for each e_j, let α_j be 1 or -1 according to whether the direction of traversal agrees or disagrees with the direction of e_j. Then $\alpha_1 \underline{v}(e_1), \alpha_2 \underline{v}(e_2), \ldots, \alpha_k \underline{v}(e_k)$ are the incidence vectors of a reorientation of C in which every arc is oriented in the direction of traversal. It follows that, in this reorientation of C, every vertex of C is the head of exactly one arc and the tail of exactly one arc. Thus $\sum_{j=1}^{k} \alpha_j \underline{v}(e_j) = \underline{0}$. Hence C is dependent in $M[A_{D(G)}]$.

Now suppose that $\{f_1, f_2, \ldots, f_m\}$ is a circuit of $M[A_{D(G)}]$. If $m = 1$, then f_1 is a loop in G, so $\{f_1\}$ is a circuit in $M(G)$. Assume, then, that $m > 1$ and consider the columns $\underline{v}(f_1), \underline{v}(f_2), \ldots, \underline{v}(f_m)$ of $A_{D(G)}$. Evidently, there are non-zero members $\varepsilon_1, \varepsilon_2, \ldots, \varepsilon_m$ of $\mathbb{F}$ such that $\sum_{i=1}^{m} \varepsilon_i \underline{v}(f_i) = \underline{0}$. Thus every row of the matrix $[\underline{v}(f_1) \ \underline{v}(f_2) \ \cdots \ \underline{v}(f_m)]$ that contains at least one non-zero entry contains at least two such entries. But the rows of this matrix correspond to vertices of G. Hence, in the subgraph G_1 of G induced by $\{f_1, f_2, \ldots, f_m\}$, every vertex has degree at least two. Thus G_1 must contain a cycle (Exercise 2, Section 1.6) and therefore $\{f_1, f_2, \ldots, f_m\}$ contains a circuit of $M(G)$. We conclude that the lemma holds.   □

$$
\begin{array}{c c}
 & \begin{array}{c c c c c c c c c} a & b & c & d & e & f & g & h & i \end{array} \\
\begin{array}{c} 1 \\ 2 \\ 3 \\ 4 \\ 5 \\ 6 \end{array} &
\left[\begin{array}{r r r r r r r r r}
1 & 0 & 0 & 0 & 0 & 0 & 0 & 0 & 1 \\
-1 & 1 & 0 & 0 & 0 & 0 & 0 & -1 & 0 \\
0 & -1 & 0 & -1 & 0 & 0 & 1 & 0 & 0 \\
0 & 0 & 0 & 1 & 1 & -1 & 0 & 0 & 0 \\
0 & 0 & 0 & 0 & 0 & 1 & -1 & 1 & -1 \\
0 & 0 & 0 & 0 & -1 & 0 & 0 & 0 & 0
\end{array} \right]
\end{array}
$$

FIG. 5.2. $A_{D(G)}$.

We shall prove in Theorem 6.6.3 that a matroid is regular if and only if it is representable over every field. Since we have not yet proved this, we shall now give a direct proof that every graphic matroid is regular. This proof will come from combining Lemma 5.1.3 with the following result of Poincaré (1900). A $(0, \pm 1)$-*matrix* is a real matrix with every entry in $\{0, 1, -1\}$.

Lemma 5.1.4 *Let A be a $(0, \pm 1)$-matrix in which every column has at most one 1 and one -1. Then A is totally unimodular.*

Proof This follows Cook *et al.* (1998). Let X be a $k \times k$ submatrix of A. We shall argue by induction on k to show that $\det X \in \{0, \pm 1\}$. This is certainly true if $k = 1$. Assume it holds for $k < n$ and let $k = n \geq 2$. If X has a column with at most one non-zero entry, then, by expanding the determinant along this column, we deduce, from the induction assumption, that $\det X \in \{0, \pm 1\}$. We may now assume that every column of X has exactly two non-zero entries. By hypothesis, one of these is 1 and the other -1, so the rows of X have zero sum and are therefore linearly dependent. Hence $\det X = 0$. □

Proposition 5.1.5 *If G is a graph, then $M(G)$ is regular.*

Proof By Lemma 5.1.3, $M(G) = M[A_{D(G)}]$ where $A_{D(G)}$ is viewed as a matrix over $\mathbb{R}$. By the last lemma, $A_{D(G)}$ is totally unimodular, so $M(G)$ is a regular matroid. □

The following is an immediate consequence of Propositions 2.2.22 and 5.1.5.

Corollary 5.1.6 *If G is a graph, then $M^*(G)$ is regular.* □

We remark that, although $M(G)$ has rank at most $|V(G)| - 1$, the matrix $A_{D(G)}$ has exactly $|V(G)|$ rows. Next we describe a direct procedure for obtaining an $\mathbb{F}$-representation A for $M(G)$ that has exactly $r(M(G))$ rows. Let $r(M(G)) = r$ where $r \geq 1$ and let $\{e_1, e_2, \ldots, e_r\}$ be a basis B of $M(G)$. Take the first r columns of A to be the matrix I_r, labelling the columns of this submatrix by $e_1, e_2, \ldots, e_r$ and letting the corresponding column vectors be $\underline{v}_1, \underline{v}_2, \ldots, \underline{v}_r$. Now take an arbitrary orientation $D(G)$ of G. For each edge e of $E(G) - B$, recall that $C(e, B)$ is the unique circuit of $M(G)$ contained in $B \cup e$. Evidently $C(e, B)$ is a cycle of G; we call it the *fundamental cycle* of e with respect to B. Let its edges in cyclic order be $e, e_{i_1}, e_{i_2}, \ldots, e_{i_k}$. Traverse this cycle in the direction determined by e letting α_{i_j} be 1 or -1 according to whether the direction of traversal agrees or disagrees with the direction of e_j. As the column of A corresponding to e, take $\sum_{j=1}^{k} \alpha_{i_j} \underline{v}_j$ (modulo $\mathbb{F}$). The proof that the matrix A is indeed an $\mathbb{F}$-representation for $M(G)$ is based on the fact that A can be obtained from $A_{D(G)}$ by a sequence of the operations 2.2.1–2.2.6 (see Exercise 4).

Example 5.1.7 For $r \geq 2$, let $\mathcal{W}_r$ denote the graph shown in Figure 5.3(a). This graph is called the r-*spoked wheel*. The edges $b_1, b_2, \ldots, b_r$ are its *spokes*, the vertex h is its *hub*, and the cycle with edge set $\{a_1, a_2, \ldots, a_r\}$ is its *rim*. Let $\{b_1, b_2, \ldots, b_r\}$ be the distinguished basis of $M(\mathcal{W}_r)$ and orient all these edges

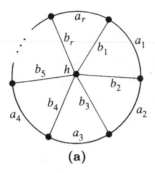

(a)

$$
\begin{array}{ccccccccc}
b_1 & b_2 & \cdots & b_r & a_1 & a_2 & a_3 & \cdots & a_{r-1} & a_r \\
\end{array}
$$

$$
\left[\quad I_r \quad \left|
\begin{array}{ccccccc}
1 & 0 & 0 & \cdots & 0 & -1 \\
-1 & 1 & 0 & \cdots & 0 & 0 \\
0 & -1 & 1 & \cdots & 0 & 0 \\
0 & 0 & -1 & \cdots & 0 & 0 \\
\vdots & \vdots & \vdots & \ddots & \vdots & \vdots \\
0 & 0 & 0 & \cdots & 1 & 0 \\
0 & 0 & 0 & \cdots & -1 & 1 \\
\end{array}
\right. \right]
$$

(b)

FIG. 5.3. (a) $\mathcal{W}_r$. (b) A representation for $M(\mathcal{W}_r)$.

away from the hub. Then, orienting the rim clockwise, we obtain the representa-
tion for $M(\mathcal{W}_r)$ in Figure 5.3(b) where, if $\mathbb{F}$ has characteristic two, all negative
entries change sign. Now let r be even and reverse the orientations on each of
$b_2, b_4, \ldots, b_r$ and each of $a_2, a_4, \ldots, a_r$. Then, we obtain the same representation
as in Figure 5.3(b) except that all negative entries have their signs changed.   □

Example 5.1.8 Orient the edges of K_5 as shown in Figure 5.4(a) and let the
distinguished basis of $M(K_5)$ be $\{1, 2, 3, 4\}$. Then the matrix in Figure 5.4(b)
represents $M(K_5)$ over all fields.   □

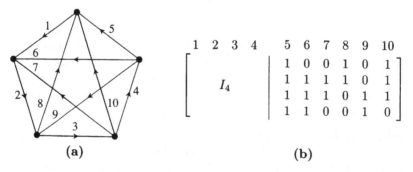

(a)

$$
\begin{array}{cccc|cccccc}
1 & 2 & 3 & 4 & 5 & 6 & 7 & 8 & 9 & 10 \\
\end{array}
$$

$$
\left[\quad I_4 \quad \left|
\begin{array}{cccccc}
1 & 0 & 0 & 1 & 0 & 1 \\
1 & 1 & 1 & 1 & 0 & 1 \\
1 & 1 & 1 & 0 & 1 & 1 \\
1 & 1 & 0 & 0 & 1 & 0 \\
\end{array}
\right. \right]
$$

(b)

FIG. 5.4. (a) An orientation of K_5. (b) A representation for $M(K_5)$.

Exercises

1. For a graph G, show that $M^*(G) \oplus M(G)$ is representable over every field.
2. (a) Construct $\mathbb{F}$-representations for $M(K_{3,3})$ by using both of the methods described in this section.
   (b) Use these representations to find representations for $M^*(K_{3,3})$.
3. Let A be a $(0, \pm 1)$-matrix over a field $\mathbb{F}$ such that every column contains at most one 1 and one -1. Let A' be obtained from A by adjoining a new row that is the negative of the sum of all the original rows. Show that:
   (a) $A' = A_{D(G)}$ for some orientation $D(G)$ of a graph G; and
   (b) $M[A] = M(G)$.
4. Let A be the matrix constructed following Corollary 5.1.6.
   (a) Show that $G[B]$ has a vertex of degree one.
   (b) Let e_1 be an edge $v_1 v_2$ of $G[B]$ where v_1 has degree one in $G[B]$. Replace row v_2 of $A_{D(G)}$ by the sum of rows v_1 and v_2 and then delete row v_1 and column e_1 from the resulting matrix. Show that the matrix thus obtained is the incidence matrix of the orientation induced on G/e_1 by $D(G)$.
   (c) Argue by induction on r using (a) and (b) to show that A can be obtained from $A_{D(G)}$ by a sequence of operations 2.2.1–2.2.6 where the only element of $\mathbb{F}$ by which a row or column is multiplied is -1.
   (d) Deduce that A is an $\mathbb{F}$-representation for $M(G)$.
   (e) Show that A is totally unimodular when viewed over $\mathbb{R}$.
5. (Kingan 1997) Let T_{12} be the vector matroid of the following matrix over $GF(2)$.

$$\left[\begin{array}{c|cccccc} & 1 & 1 & 0 & 0 & 0 & 1 \\ & 1 & 0 & 0 & 0 & 1 & 1 \\ I_6 & 0 & 0 & 0 & 1 & 1 & 1 \\ & 0 & 0 & 1 & 1 & 1 & 0 \\ & 0 & 1 & 1 & 1 & 0 & 0 \\ & 1 & 1 & 1 & 0 & 0 & 0 \end{array} \right]$$

   (a) Show that T_{12} is self-dual but not identically self-dual.
   (b) Show that all single-element deletions of T_{12} are isomorphic, and all single-element contractions of T_{12} are isomorphic.
   (c) Show that T_{12} has $M(K_5)$ and F_7 as minors but does not have $M(K_{3,3})$ as a minor.
   (d) Take the 15×12 matrix over $GF(2)$ whose rows are indexed by the edges of the Petersen graph P_{10} (see Figure 3.5) and whose columns are indexed by the 5-cycles of P_{10} with each column being the incidence vector of the corresponding 5-cycle. Prove that the vector matroid of this matrix is isomorphic to T_{12}.
   (e) Show that there is a bijection ψ from the edges of P_{10} to the four-element circuits of T_{12} such that e and f are adjacent edges of P_{10}

if and only if the symmetric difference of $\psi(e)$ and $\psi(f)$, that is, $(\psi(e) - \psi(f)) \cup (\psi(f) - \psi(e))$, is also a 4-circuit of T_{12}.

5.2 Duality in graphic matroids

In Section 2.3, we proved that if G is a planar graph, then $M^*(G)$ is graphic. In this section, we shall prove the converse of this result. Exercise 3 outlines one way to do this is by using Kuratowski's Theorem (2.3.8). Instead, we shall give a more elementary, direct proof suggested by Geelen and Gerards (private communication). First we extend Lemma 2.3.7 by showing that the cycle matroid of a geometric dual of a planar graph G is isomorphic to the bond matroid of G.

Proposition 5.2.1 *Let G^* be a geometric dual of a planar graph G. Then*

$$M(G^*) \cong M^*(G).$$

Proof We argue by induction on the number n of connected components of the graph G. Let H be a planar embedding of G of which G^* is the geometric dual. Then $M(H) \cong M(G)$ and the theorem holds by Lemma 2.3.7 when $n = 1$. Now assume that $n \geq 2$ and that the theorem holds when H has fewer than n components. Then H has a component H_1 such that every edge of $E(H) - E(H_1)$ lies in the infinite face F_1 of the plane graph H_1. Let $H_0 = H - V(H_1)$. Clearly H_0 has fewer than n components. Let F_0 be the face of H_0 that contains H_1, and let v_0 and v_1 be the vertices of H_0^* and H_1^* corresponding to F_0 and F_1, respectively. Then it is not difficult to see that G^* is obtained from H_0^* and H_1^* by identifying v_0 with v_1. Thus $M(G^*) = M(H_0^*) \oplus M(H_1^*)$. By the induction assumption, $M(H_0^*) = M^*(H_0)$ and $M(H_1^*) = M^*(H_1)$. Hence

$$M(G^*) = M^*(H_0) \oplus M^*(H_1) = M^*(H) \cong M^*(G). \qquad \square$$

By the last result, if G' is a geometric dual of a planar graph G, then there is a bijection ψ from $E(G)$ to $E(G')$ such that X is a cycle in G if and only if $\psi(X)$ is a bond in G'. Suppose now that G' is an arbitrary graph for which such a bijection exists. Then G' is called an *abstract dual* of G. Evidently G' is an abstract dual of G if and only if $M(G) \cong M^*(G')$. Thus it follows by duality that if G' is an abstract dual of a graph G, then G is an abstract dual of G'. Although every geometric dual of a graph is also an abstract dual, the converse of this statement is not true (see Exercise 1).

The next theorem is the main result of this section.

Theorem 5.2.2 *A graph G is planar if and only if $M^*(G)$ is graphic.*

The following result of Geelen and Gerards is the key to proving this theorem.

Proposition 5.2.3 *Let G and H be graphs without isolated vertices such that G is an abstract dual of H, and $M(H)$ is connected. Then there are planar embeddings of G and H such that each is the geometric dual of the other.*

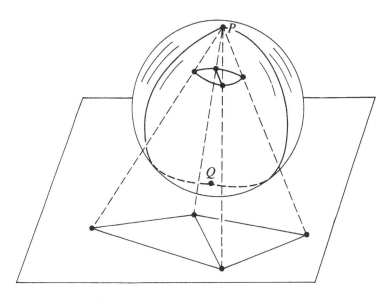

FIG. 5.5. Stereographic projection.

Proof   We may suppose that G and H are labelled so that $M(H) = M^*(G)$. By Corollary 4.2.5, a matroid is connected if and only if its dual is connected. Thus $M(G)$ is connected. We may assume that each of G and H has at least three vertices otherwise the result is easily checked. Thus the result holds if G has fewer than four edges. Assume it holds for $|E(G)| < n$ and let $|E(G)| = n \geq 4$. Now take e in $E(G)$. Since $M(G)$ is connected, it follows by Theorem 4.3.1 that either both $M(G) \backslash e$ and $(M(G) \backslash e)^*$ are connected, or both $M(G)/e$ and $(M(G)/e)^*$ are connected. But $(M(G) \backslash e)^* = M^*(G)/e = M(H)/e = M(H/e)$. Similarly, $(M(G)/e)^* = M(H \backslash e)$. Thus, by interchanging G and H if necessary, we may assume that both $M(G \backslash e)$ and $M(H/e)$ are connected. Then, by the induction assumption, $G \backslash e$ and H/e have planar embeddings so that each is the other's geometric dual.

Let w and x be the ends of e in H, and let y be the vertex of H/e that results from identifying w and x. Let Y be the face of $G \backslash e$ corresponding to y. Let u and v be the ends of e in G. By Proposition 4.1.7, as H is loopless and has at least three vertices, H is 2-connected. Thus the set E_w of edges of H meeting w is a bond in H, so E_w is a cycle in G. Hence $E_w - e$ is a (u,v)-path W in $G \backslash e$. By symmetry, $E_x - e$ is a (u,v)-path X in $G \backslash e$ where E_x is the set of edges of H meeting x. Now $(E_w - e) \cup (E_x - e)$ is the set of edges of H/e meeting y. Thus $W \cup X$ is the boundary of the face Y of $G \backslash e$, so u and v lie on the boundary of Y. Adding a simple curve in Y from u to v labelled by e gives a planar embedding G' of G. This curve splits Y into two faces, one with boundary $W \cup e$ and the other with boundary $X \cup e$. Thus the geometric dual of G' is a planar embedding, H', of H, so G' is a geometric dual of H'. The result follows by induction.   □

The proof of Theorem 5.2.2 will require some more properties of planar graphs. First we observe that a graph is planar if and only if it can be embedded on the surface of a sphere so that distinct edges do not meet, except possibly at their endpoints. To see this, suppose that we have such an embedding G' of a graph G. Choose a point P on the sphere so that P is not a vertex of G' and is not on any edge. Let Q be the point on the sphere diametrically opposite P. Now project G' from P onto the tangent plane to the sphere at Q. This operation, which is known as *stereographic projection*, gives a plane graph G'' which, as a graph, is isomorphic to G. Figure 5.5 shows an example of the operation. To establish that every planar graph can be embedded on the sphere so that distinct edges do not cross, one needs only to reverse the above construction.

Lemma 5.2.4 *Let e be an edge of a planar graph G. Then there is a planar embedding of G in which e is on the boundary of the infinite face.*

Proof Take an embedding G' of G on the surface of a sphere so that distinct edges do not meet, except at their endpoints. Now choose a point P on the sphere that is in the interior of a face containing e. The stereographic projection of G' onto the tangent plane at the point diametrically opposite P is a planar embedding of G in which e is on the boundary of the infinite face. $\square$

Proof of Theorem 5.2.2 The fact that the bond matroid of a planar graph is graphic is the content of Theorem 2.3.4. Conversely, assume that $M^*(G)$ is graphic but that G is non-planar. After deleting all loops from G, we still have a non-planar graph whose bond matroid is graphic. Thus we may assume that G is loopless and that $M^*(G) = M(H)$ for some graph H. Clearly G has a non-planar component G_1, and $M(G_1)$ is a union of components of $M(G)$. Either (i) G_1 is 2-connected and so $M(G_1)$ is a component of $M(G)$; or (ii) G_1 has a cut-vertex v. In the latter case, let $G_{1,1}, G_{1,2}, \ldots, G_{1,k}$ be the components of $G_1 - v$, and, for each i, let $G'_{1,i}$ be the subgraph of G_1 induced by $V(G'_{1,i}) \cup v$. If each $G'_{1,i}$ is planar, then, by Lemma 5.2.4, each $G'_{1,i}$ has a planar embedding having v on the boundary of its infinite face. By identifying the copies of v in these planar embeddings of $G'_{1,1}, G'_{1,2}, \ldots, G'_{1,k}$, we get a planar embedding of G_1; a contradiction. Thus some $G'_{1,i}$, which we shall call G_2, is non-planar. Moreover, $M(G_2)$ is a union of components of $M(G)$. By repeating the above process with G_2 replacing G_1, we eventually obtain, in case (ii), a 2-connected non-planar subgraph G_m of G such that $M(G_m)$ is a union of components of $M(G)$. Hence $M(G_m)$ is a component of $M(G)$. But, taking $m = 1$, we also have such a 2-connected non-planar subgraph G_m in case (i). In both cases, $M^*(G_m)$ is a component of $M(H)$. Thus $M^*(G_m) = M(H_m)$ where H_m is the subgraph of H induced by $E(G_m)$. As $M(H_m)$ is connected, it follows by Proposition 5.2.3 that G_m is planar. This contradiction completes the proof of the theorem. $\square$

The following variant of Kuratowski's Theorem is due to Wagner (1937a).

Theorem 5.2.5 *A graph is planar if and only if it has no minor isomorphic to K_5 or $K_{3,3}$.* $\square$

Matroids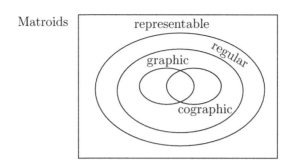

FIG. 5.6. Relationships between certain classes of matroids.

Clearly Kuratowski's Theorem implies Wagner's Theorem, and Exercise 7 indicates how to derive Kuratowski's Theorem from Wagner's Theorem. One advantage of proving Theorem 5.2.2 without using either of these theorems is that we can deduce them as corollaries of the excluded-minor characterization of graphic matroids (Theorem 10.3.2).

The Venn diagram in Figure 5.6 summarizes the relationships that have so far been proved between various classes of matroids. Although all the inclusions shown are indeed proper, we have not yet proved all of them to be so. By Theorem 5.2.2, the intersection of the classes of graphic and cographic matroids is the class of matroids that are isomorphic to the cycle matroids of planar graphs. We shall call such matroids *planar-graphic* noting that some authors call these matroids 'planar' while others use 'planar matroid' to mean a rank-3 matroid.

The classes of graphic and cographic matroids are important subclasses of the class of regular matroids. The union of the first two classes is not equal to the third for, by Proposition 2.3.3, the regular matroid $M(K_5) \oplus M^*(K_5)$ is neither graphic nor cographic. However, a difficult result of Seymour (1980b) to be proved in Chapter 13 shows that all regular matroids can be built up from graphic matroids, cographic matroids, and one special 10-element matroid.

We noted in Corollary 3.2.2 that the class of graphic matroids is minor-closed. By duality, so too is the class of cographic matroids. On combining these facts, we immediately deduce the following.

Corollary 5.2.6 *The class of planar-graphic matroids is minor-closed.* □

For any minor-closed class $\mathcal{N}$ of matroids, it is natural to seek a list of the minor-minimal matroids not in $\mathcal{N}$, that is, those matroids that are not in $\mathcal{N}$ but have all their proper minors in $\mathcal{N}$. Indeed, as we shall see in several later chapters, many of the most celebrated theorems in matroid theory are excluded-minor characterizations of various classes. In Chapter 10, we shall give such a characterization of the class of planar-graphic matroids. For the moment, we note that it is straightforward to use Kuratowski's Theorem to obtain the following characterization of the members of this class within the class of graphic matroids.

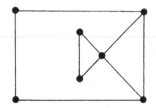

FIG. 5.7. A graph with an abstract dual that is not a geometric dual.

Proposition 5.2.7 *The following statements are equivalent for a graph* G:

(i) G *is a planar graph.*

(ii) $M(G)$ *is a planar-graphic matroid.*

(iii) $M(G)$ *has no minor isomorphic to* $M(K_5)$ *or* $M(K_{3,3})$. $\square$

Exercises

1. (D. R. Woodall, in Welsh 1976, Section 6.4) Find a connected graph that is an abstract dual but is not a geometric dual of the graph in Figure 5.7.

2. Show that $M(K_4)$ is the unique tipless spike that is graphic or cographic.

3. (Parsons 1971) Show that:
    (a) A subgraph of a graph with an abstract dual has an abstract dual.
    (b) A subdivision of a graph with an abstract dual has an abstract dual.
    (c) Kuratowski's Theorem (2.3.8) can be combined with the last two parts to show that no non-planar graph has an abstract dual.

4. (a) Show that all the inclusions indicated in Figure 5.6 are proper.
    (b) Indicate precisely how the class of binary matroids fits into Figure 5.6.

5. Show that if a graph G has a subgraph that is a subdivision of a graph H, then $M(G)$ has a minor isomorphic to $M(H)$.

6. (Whitney 1932a; Ore 1967, Section 3.3) A graph H is a *Whitney dual* of a graph G if there is a bijection $\psi : E(G) \to E(H)$ such that, for all subsets Y of $E(G)$,

$$r(M(H)) - r(M(H)\backslash\psi(Y)) = |Y| - r(M(G)|Y).$$

    (a) Show that G is a Whitney dual of H if H is a Whitney dual of G.
    (b) Determine the relationship between Whitney duals and geometric and abstract duals.
    (c) *Prove that a graph is planar if and only if it has a Whitney dual.

7. (Wagner 1937a,b; Harary and Tutte 1965) Let G be a graph. Prove that:
    (a) If G has $K_{3,3}$ as a minor, then G has a subgraph that is a subdivision of $K_{3,3}$.
    (b) If G has K_5 as a minor, then G has a subgraph that is a subdivision of K_5 or $K_{3,3}$.

8. Let G be a connected plane graph.

(a) Use matroid duality to show G satisfies Euler's polyhedron formula,

$$|V(G)| - |E(G)| + |F(G)| = 2,$$

where $F(G)$ is the set of faces of G.
(b) Prove that if G is simple and $|V(G)| \geq 3$, then $|E(G)| \leq 3|V(G)| - 6$.
(c) Characterize the graphs for which equality is attained in (b).
(d) Show that if G is simple, then it has a vertex of degree at most five.
(e) Find a simple plane graph in which every vertex has degree five.

9. (Jaeger 1979) Let M be a simple cographic matroid of rank r where $r \geq 2$. Prove that $|E(M)| \leq 3r - 3$.

5.3 Whitney's 2-Isomorphism Theorem

In this section, we shall prove Whitney's theorem characterizing when two graphs have isomorphic cycle matroids. Our proof, due to Truemper (1980), is much shorter than the original. A different short proof was given by Wagner (1985).

If G is a graph, then, evidently, adding vertices to G with no incident edges will not alter the cycle matroid of the graph. For this reason, *throughout this section, we shall consider only graphs having no isolated vertices.* The following operations on a graph G have appeared, at least implicitly, in earlier chapters. Examples of these operations are shown in Figures 5.8 and 5.9:

(a) *Vertex identification.* Let v and v' be vertices of distinct components of G. We modify G by identifying v and v' as a new vertex $\bar{v}$.

(b) *Vertex cleaving.* This is the reverse operation of vertex identification, so a graph can only be cleft at a cut-vertex or at a vertex incident with a loop.

(c) *Twisting.* Suppose that the graph G is obtained from disjoint graphs G_1 and G_2 by identifying the vertices u_1 of G_1 and u_2 of G_2 as the vertex u of G, and identifying the vertices v_1 of G_1 and v_2 of G_2 as the vertex v of G. In a *twisting*, or *Whitney twist*, of G about $\{u, v\}$, we identify, instead, u_1 with v_2, and v_1 with u_2. We call G_1 and G_2 the *pieces* of the twisting.

A graph G is 2-*isomorphic* to a graph H if H can be transformed into a graph isomorphic to G by a sequence of operations of types (a), (b), and (c). Evidently 2-isomorphism is an equivalence relation. Since none of the three operations (a)–(c) alters the edge sets of the cycles of a graph, if G is 2-isomorphic to H, then $M(G) \cong M(H)$. Most of this section will be devoted to proving the converse of this (Whitney 1933), the main result of the chapter.

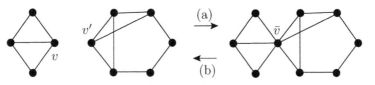

FIG. 5.8. (a) Vertex identification. (b) Vertex cleaving.

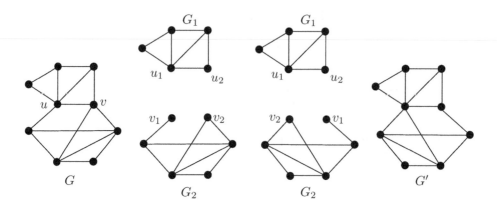

FIG. 5.9. Twisting about $\{u, v\}$.

Theorem 5.3.1 (Whitney's 2-Isomorphism Theorem) *Let G and H be graphs having no isolated vertices. Then $M(G)$ and $M(H)$ are isomorphic if and only if G and H are 2-isomorphic.*

The proof of this theorem will use the next four lemmas. Because none of operations (a)–(c) can be applied to 3-connected loopless graphs, such graphs are uniquely determined by their cycle matroids. Our proof of this is due to Edmonds (see Truemper 1980) and Greene (1971).

Lemma 5.3.2 *Let G and H be loopless graphs without isolated vertices. Suppose that $\psi : E(G) \to E(H)$ is an isomorphism from $M(G)$ to $M(H)$. If G is 3-connected, then ψ induces an isomorphism between the graphs G and H.*

Proof First note that if J is a loopless block, a hyperplane of $M(J)$ can only be connected if the complementary cocircuit is a vertex bond. Moreover, if $M(J)$ has exactly $|V(J)|$ connected hyperplanes, then, up to relabelling, $M(J)$ determines the mod-2 vertex–edge incidence matrix of J, and so uniquely determines J.

As G is 3-connected, $G - v$ is 2-connected for every vertex v of G. Thus, by Proposition 4.1.7, $M(G-v)$ is connected, so $M(G)$ has exactly $|V(G)|$ connected hyperplanes. But $M(G) \cong M(H)$, and $M(G)$ is connected and loopless, so H is a loopless block. Thus $|V(G)| = r(M(G)) + 1 = r(M(H)) + 1 = |V(H)|$. Hence $M(H)$ has precisely $|V(H)|$ connected hyperplanes and so $G \cong H$.   □

For those 2-connected graphs that are not 3-connected, the proof of Theorem 5.3.1 will use Tutte's characterization (1966a) of the structure of such graphs. To state this result, we shall need another definition. For $k \geq 2$, a connected graph G is called a *generalized cycle* with *parts* $G_1, G_2, \ldots, G_k$ if the following conditions hold:

(i) Each G_i is a connected subgraph of G having a non-empty edge set, and, if $k = 2$, both G_1 and G_2 have at least three vertices.

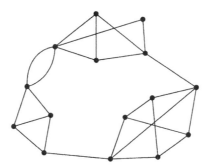

FIG. 5.10. A generalized cycle.

(ii) The edge sets of $G_1, G_2, \ldots, G_k$ partition the edge set of G, and each G_i shares exactly two vertices, its *contact vertices*, with $\bigcup_{j \neq i} G_j$.
(iii) If each G_i is replaced by an edge joining its contact vertices, the resulting graph is a cycle.

Example 5.3.3 The graph G shown in Figure 5.10 can be viewed as a generalized cycle with six parts, namely its blocks. It can also be viewed as a generalized cycle with two, three, four, or five parts. $\qquad\square$

Lemma 5.3.4 *Let G be a block having at least four vertices and suppose that G is not 3-connected. Then G has a representation as a generalized cycle, each part of which is a block.*

Proof Suppose that $\{u, v\}$ is a vertex cut of G. Choose a component H_1 of $G - \{u, v\}$ and let $H_2 = G - \{u, v\} - V(H_1)$. For each i in $\{1, 2\}$, let G_i be the subgraph of G induced by $V(H_i) \cup \{u, v\}$, and let G_2' be the graph obtained from G_2 by deleting those edges that join u and v. Then G is a generalized cycle with parts G_1 and G_2'. If G_1 is not a block, it is the union of two connected subgraphs, $G_{1,1}$ and $G_{1,2}$, that have at least two vertices each, but have only one common vertex, say x. As G is a block, $x \notin \{u, v\}$, and one of u and v is in $G_{1,1}$ while the other is in $G_{1,2}$. Therefore, G is a generalized cycle with parts $G_{1,1}, G_{1,2}$, and G_2'. If necessary, we repeat the above process finitely many times until a representation of G is obtained in which each part is a block. $\qquad\square$

The core of the proof of Whitney's Theorem is contained in the next result.

Lemma 5.3.5 *Suppose that G is a block having a representation as a generalized cycle, the parts, $G_1, G_2, \ldots, G_k$, of which are blocks. Let H be a graph for which there is an isomorphism θ from $M(G)$ to $M(H)$ and, for each i, let H_i be the subgraph of H induced by $\theta(E(G_i))$. Then H is a generalized cycle with parts $H_1, H_2, \ldots, H_k$.*

Proof For an arbitrary element j of $\{1, 2, \ldots, k\}$, let $G_{-j} = \bigcup_{i \neq j} G_i$ and $H_{-j} = \bigcup_{i \neq j} H_i$. The main part of the proof of this lemma involves showing that

$$|V(H_j) \cap V(H_{-j})| = 2. \qquad (5.1)$$

Since $G, G_1, G_2, \ldots, G_k$ are blocks, all of $H, H_1, H_2, \ldots, H_k$ are also blocks. As G has a cycle meeting all of $E(G_1), E(G_2), \ldots, E(G_k)$, the graph H_{-j} has an edge e having an endpoint x in H_j. Choose an edge f in H_j. Then, since H is a block, it has a cycle C_1 containing both e and f. Starting at x, move along C_1 using the edge e first and stop at the first vertex y of H_j that is encountered. Let P be the path from x to y thus traversed. Since C_1 contains the edge f of H_j, such a vertex y exists, and $y \in (V(H_j) \cap V(H_{-j})) - x$. Hence $|V(H_j) \cap V(H_{-j})| \geq 2$. We shall complete the proof of (5.1) by showing that

$$V(H_j) \cap V(H_{-j}) = \{x, y\}. \qquad (5.2)$$

Let u be one of the contact vertices of G_j. As G_j is a block, the set E_u of edges of G_j meeting u is a bond in G_j. Hence $\theta(E_u)$ is a bond in H_j, so $H_j \backslash \theta(E_u)$ has two components. Moreover:

Lemma 5.3.6 *Every cycle of H that contains an edge of H_j and an edge of H_{-j} also contains an edge of $\theta(E_u)$.*

Proof This is an immediate consequence of the fact that every cycle of G that contains an edge of G_j and an edge of G_{-j} also contains an edge of E_u. □

If x and y are joined by a path in $H_j \backslash \theta(E_u)$, then, by the choice of y, the union of this path with P is a cycle. Since $E(P) \subseteq E(H_{-j})$, this cycle meets both $E(H_{-j})$ and $H_j - \theta(E_u)$, and is contained in the union of the last two sets. As $\theta(E_u) \subseteq E(H_j)$, the cycle avoids $\theta(E_u)$. This contradiction to Lemma 5.3.6 implies that one of the two components of $H_j \backslash \theta(E_u)$, say H_j^x, contains x, while the other, H_j^y, contains y.

Now assume that equation (5.2) does not hold. Then there is a vertex z in $(V(H_j) \cap V(H_{-j})) - \{x, y\}$. Let g and h be edges of H_{-j} and H_j, respectively, meeting z. Since H is a block, it has a cycle C_2 that contains both g and h. Starting at z and using the edge g first, move along C_2 stopping at the first vertex w of $V(H_j) \cup V(P)$ that is encountered. Clearly $w \neq z$. Let Q be the path from z to w thus traversed.

We now distinguish three cases, showing that each leads to the conclusion that H has a cycle C_3 meeting H_j and H_{-j} but containing no edge of $\theta(E_u)$. Since this contradicts Lemma 5.3.6, it will follow that (5.2) holds and hence so does (5.1). The cases are as follows:

(i) w is a vertex of P;
(ii) $w \in V(H_j^x) - V(P)$; and
(iii) $w \in V(H_j^y) - V(P)$.

We shall describe the construction of C_3 in each case assuming that $z \in H_j^x$. We obtain C_3 similarly when $z \in H_j^y$. Recall that $E(P) \subseteq E(H_{-j})$. In case (i), C_3 is obtained by combining Q, that part of P from w to x, and an arbitrary path from x to z in H_j^x. In case (ii), we obtain C_3 by combining Q with an arbitrary path from w to z in H_j^x. Finally, in case (iii), C_3 is formed by combining P and Q with a path from x to z in H_j^x and a path from y to w in H_j^y. We conclude that (5.1) holds. Therefore, as H is a block, H is a generalized cycle having parts $H_1, H_2, \ldots, H_k$. □

We are now ready to prove Whitney's 2-Isomorphism Theorem (5.3.1).

Proof of Theorem 5.3.1 By the remarks prior to the theorem statement, it suffices to show that if there is a bijection θ from $E(G)$ onto $E(H)$ that induces an isomorphism between $M(G)$ and $M(H)$, then G and H are 2-isomorphic. Let G^+ and H^+ be the graphs that are formed from the disjoint unions of the blocks of G and H, respectively. Because $M(G^+) \cong M(G) \cong M(H) \cong M(H^+)$, it follows that there is a bijection from the set of blocks of G^+ to the set of blocks of H^+ so that corresponding blocks have isomorphic cycle matroids. Now G and G^+ are 2-isomorphic, as are H and H^+. Therefore, the required result will follow if we can show that G^+ and H^+ are 2-isomorphic. This will be done by proving, by induction on $|V(G)|$, that if G is a block and H is a graph for which $M(G)$ and $M(H)$ are isomorphic, then H can be transformed into a graph isomorphic to G by a sequence of twistings. We shall therefore assume, for the remainder of the proof, that G is a block. Clearly, if $|V(G)| \leq 4$, the required result holds. Now suppose that the result holds for $|V(G)| < n$ and let $|V(G)| = n \geq 5$. Then G is 2-connected.

By Lemma 5.3.2, if G is 3-connected, then $G \cong H$ and the required result certainly holds. It follows that we may assume that G is not 3-connected. Therefore, by Lemma 5.3.4, G has a representation as a generalized cycle in which each of the parts, $G_1, G_2, \ldots, G_k$, is a block. Hence, by Lemma 5.3.5, H has a representation as a generalized cycle, the parts of which are $H_1, H_2, \ldots, H_k$, where $H_i = H[\theta(E(G_i))]$.

For each i in $\{1, 2, \ldots, k\}$, consider the graphs $G_i + e_i$ and $H_i + f_i$ that are obtained from G_i and H_i by adding edges e_i and f_i so that, in each case, the new edge joins the contact vertices. Now C is a cycle of G meeting both $E(G_i)$ and $E(G_{-i})$ if and only if $\theta(C)$ is a cycle of H that meets $E(H_i)$ in $\theta(E(G_i) \cap C)$ and also meets $E(H_{-i})$. From this, it follows that the isomorphism between $M(G_i)$ and $M(H_i)$ can be extended to an isomorphism between $M(G_i + e_i)$ and $M(H_i + f_i)$ by mapping e_i to f_i. Now $|V(G_i + e_i)| = |V(G_i)| < |V(G)|$. Hence, by the induction assumption, $H_i + f_i$ can be transformed by a sequence of twistings into a graph isomorphic to $G_i + e_i$. Since every 2-element vertex cut in $H_i + f_i$ is also a vertex cut in H, one can perform the same sequence of twistings in H thereby transforming H_i into H_i'. By repeating this process for all i in $\{1, 2, \ldots, k\}$, one transforms H into a generalized cycle H' with parts $H_1', H_2', \ldots, H_k'$. Moreover, $M(H') \cong M(G)$ and, for each i, there is an isomorphism θ_i from G_i to H_i' under which the contact vertices of G_i are mapped to the contact vertices of H_i'.

Suppose that the cyclic order of the parts of G is $G_1, G_2, \ldots, G_k$, while the

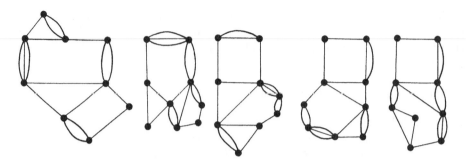

FIG. 5.11. Which of these five graphs have isomorphic cycle matroids?

cyclic order of the parts of H' is $H'_1, H'_{\sigma(2)}, \ldots, H'_{\sigma(k)}$ for some permutation σ of $\{2, 3, \ldots, k\}$. Then it is clear that, by a sequence of twistings, H' can be transformed into a generalized cycle in which the parts, in cyclic order, are $H'_1, H'_2, \ldots, H'_k$. If the resulting graph H'' is still not isomorphic to G, this last remaining difficulty can be overcome by twisting some of the parts of H'' about their contact vertices.   □

We note that in transforming H into a graph isomorphic to G, no attempt was made to minimize the number of twistings. We defer to the exercises consideration of Truemper's (1980) extension of Whitney's Theorem which determines a best-possible bound on the minimum number of twistings necessary.

Using stereographic projection and the 2-Isomorphism Theorem, one can prove the following proposition (Whitney 1932b). Instead, we shall delay the proof till Section 8.6, where various links between graph and matroid connectivity will be established that can be used to prove both this proposition and another of Whitney's results, namely that a 3-connected simple planar graph has a unique embedding on the sphere.

Proposition 5.3.7 *Let G be a 3-connected loopless planar graph. If G_1 and G_2 are abstract duals of G each having no isolated vertices, then $G_1 \cong G_2$.*   □

Exercises

1. For the five graphs in Figure 5.11,
    (a) determine which have isomorphic cycle matroids; and
    (b) for each graph G, find the number of non-isomorphic graphs H for which $M(H) \cong M(G)$.

2. (Ore 1967, Theorem 3.4.1) Let G be a planar block and G^* be a graph without isolated vertices. Prove that G^* is an abstract dual of G if and only if G^* is a geometric dual of G.

3. Let G be a 3-connected simple plane graph. Prove that the following statements are equivalent for a cycle C of G:
    (a) C is the boundary of a face of G.
    (b) $M(G/C)$ is connected.

(c) C is a vertex bond in every abstract dual of G.

4. *(Truemper 1980) Let G be a 2-connected graph having n vertices and H be a graph that is 2-isomorphic to G. Show that:

    (a) G can be transformed into a graph that is isomorphic to H by a sequence of at most $n - 2$ twistings.

    (b) For any integer m, there are such graphs G and H for which $n \geq m$ and at least $n - 2$ twistings are needed to transform G into a graph that is isomorphic to H.

5.4 Series–parallel networks

The operations of joining electrical components in series and in parallel are fundamental in electrical network theory. In this section, we shall investigate the corresponding operations for graphs. We shall show how, by beginning with a graph with just one edge and using these operations, we can build up a class of planar graphs with the property that all abstract duals of members of the class are also in the class. The graph operations considered here have been successfully extended to matroids and, in Chapter 7, a detailed discussion will be given of these matroid operations.

We begin with some examples. The graph G_1 in Figure 5.12 is obtained from G_0 by adding the edge f *in parallel* with the edge e, whereas, in G_2, the edge f has been added *in series* with e. We call G_1 a parallel extension of G_0, and G_2 a series extension of G_0. For an arbitrary graph G, these operations are defined as follows: G' is a *parallel extension* of G or, equivalently, G is a *parallel deletion* of G' if G' has a two-edge cycle $\{e, f\}$ such that $G'\backslash f = G$. If, instead, $\{e, f\}$ is a two-edge bond of G', and $G'/f = G$, then G' is a *series extension* of G, and G is a *series contraction* of G'. Thus, for example, replacing a cut-edge by a path of length two is not a series extension. The example in Figure 5.13 shows that not every series extension consists of replacing an edge by a path of length two.

The graph-theoretic operations of series and parallel extension generalize immediately to matroids. In particular, if $M\backslash f = N$ and f is in a 2-circuit of M, then M is a *parallel extension* of N, and N is a *parallel deletion* of M. If, instead, $M/f = N$ and f is in a 2-cocircuit of M, then M is a *series extension* of N, and N is a *series contraction* of M. Clearly M is a parallel extension of N if and only

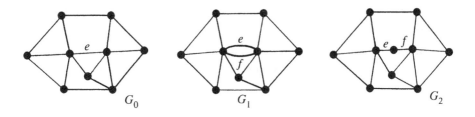

FIG. 5.12. Parallel extension and series extension.

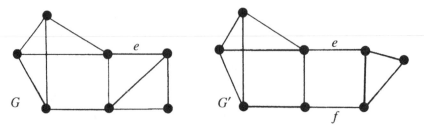

FIG. 5.13. A series extension that is not a subdivision.

if M^* is a series extension of N^*. A *series class* of M is a parallel class of M^*; it is *non-trivial* if it has at least two elements. Thus two elements are in series in M if and only if they belong to some non-trivial series class of M.

If a graph G has a subgraph that is a subdivision of a graph H, and H is 2-connected, then H can be obtained from G by a sequence of edge deletions, series contractions, and deletions of isolated vertices. Motivated by this and the form of such graph-theoretic results as Kuratowski's Theorem, Bixby (1977) defined a matroid N to be a *series minor* of a matroid M if N can be obtained from M by a sequence of deletions and series contractions. If, instead, N can be obtained from M by a sequence of contractions and parallel deletions, then N is a *parallel minor* of M. Clearly M_1 is a parallel minor of M_2 if and only if M_1^* is a series minor of M_2^*.

We noted in Proposition 3.1.25 that the operations of deletion and contraction commute. This is not the case for deletion and series contraction. For example, suppose that $\{x, y, z\}$ is a cocircuit of M but that y is not in a 2-element cocircuit of M. Then $M\backslash x$ has $\{y, z\}$ as a cocircuit, so $M\backslash x/y$ is a series minor of M. However, M/y is not a series contraction of M. In spite of the failure of these operations to commute, we do have the following result.

Lemma 5.4.1 *If $N = M/y\backslash x$ and $\{y, z\}$ is a cocircuit of M, then N is either a series contraction or a deletion of $M\backslash x$.*

Proof If $\{y, z\}$ is a cocircuit of $M\backslash x$, then N is clearly a series contraction of $M\backslash x$. Thus we may assume that $\{y, z\}$ is not a cocircuit of $M\backslash x$. It follows easily that $\{x, y\}$ must be a cocircuit of M. Therefore y is a coloop of $M\backslash x$. Hence $N = M\backslash x\backslash y$. ☐

The next result is a straightforward consequence of the last lemma. Its proof is left as an exercise.

Proposition 5.4.2 *Let M and N be matroids. If N is a series minor of M, then $N = M\backslash X/Y$ for some sets X and Y where every element of Y is in series with an element of $M\backslash X$ not in Y.* ☐

The notion of a series minor of a matroid was motivated by reference to Kuratowski's Theorem. The following variant of that theorem follows easily on combining the theorem with Proposition 5.2.7 (Bixby 1977).

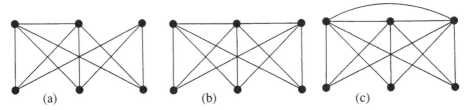

FIG. 5.14. (a) $K'_{3,3}$. (b) $K''_{3,3}$. (c) $K'''_{3,3}$.

Proposition 5.4.3 *Let G be a graph. Then G is planar if and only if $M(G)$ has no series minor isomorphic to $M(K_5)$ or $M(K_{3,3})$.* □

Wagner (1937a) showed that one can also characterize planar graphs in terms of excluded parallel minors. The proof of his result is more difficult than the proof of the last result and is omitted. The graphs $K'_{3,3}$, $K''_{3,3}$, and $K'''_{3,3}$ are shown in Figure 5.14.

Proposition 5.4.4 *A graph G is planar if and only if $M(G)$ has no parallel minor isomorphic to $M(K_5)$, $M(K_{3,3})$, $M(K'_{3,3})$, $M(K''_{3,3})$, or $M(K'''_{3,3})$.* □

Returning to the operations of series and parallel extension for graphs, we remark that these operations are special cases of the operations of series and parallel connection of graphs, which will be more closely examined in Section 7.1. Examples of these operations are illustrated in Figure 5.15. The graphs G and H are, respectively, the series and parallel connections of G_1 and G_2 relative to the directed edges e_1 and e_2. Observe that if, instead, G_2 is the graph consisting of a pair of parallel edges, then the resulting series and parallel connections are just a series and a parallel extension of G_1.

Notice that, although both the graphs G_1 and G_2 in Figure 5.15 are 3-connected, neither G nor H is. Furthermore, $G\backslash e$ and H/e are not even 2-connected. These observations foreshadow some of the important properties of series and parallel connections of matroids. These matroid operations, which are of basic structural importance, will be looked at in detail in Chapter 7.

Electrical networks that can be built by adding resistors in series or in parallel have many attractive properties. For example, the joint resistance of such a network can easily be calculated (see Duffin 1965). We now consider the corresponding graphs. A graph G is called a *series–parallel network* if it can be obtained from either of the two connected single-edge graphs by a sequence of operations each of which is either a series or a parallel extension. Thus, as is easily checked, the graphs G_1 and G_2 in Figure 5.16 are series–parallel networks, while G_3 is not. Kuratowski's Theorem characterized planar graphs by a list of excluded subgraphs, that is, a list of graphs that cannot occur as subgraphs. The main result of this section is an excluded-subgraph characterization of series–parallel networks. Some readers may wish to try to derive this result themselves. In that case, a detailed consideration of G_3 may prove helpful.

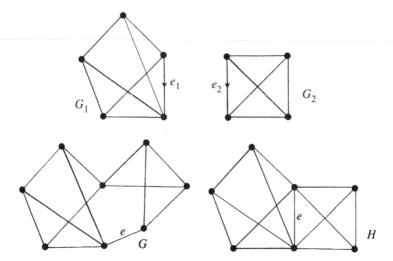

Fig. 5.15. Series and parallel connection of graphs.

The definition of series–parallel networks strongly suggests that inductive arguments can be successfully applied in determining their properties. Indeed, each of the following properties of a series–parallel network G is easily obtained using such an argument.

5.4.5 G *is a planar graph.*

5.4.6 G *is a block.*

5.4.7 *If G has at least three edges, one of which is e, then either $G\backslash e$ or G/e is not a block.*

Proposition 5.4.9 asserts that the class of series–parallel networks is closed under duality. This result is an immediate consequence of the following.

Lemma 5.4.8 *Let G be a planar graph and H be an abstract dual of G. Then, corresponding to every series extension G' of G, there is a parallel extension H' of H which is an abstract dual of G'. Moreover, every parallel extension of H is an abstract dual of some series extension of G.*

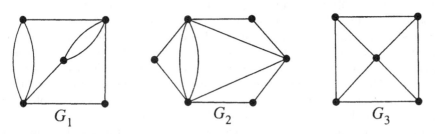

Fig. 5.16. G_1 and G_2 are series–parallel networks; G_3 is not.

Proof As H is an abstract dual of G, there is a bijection ψ from $E(G)$ onto $E(H)$ with the property that X is a cycle in G if and only if $\psi(X)$ is a bond in H. Now suppose that the graph G' is obtained by adding a new edge f in series with the edge e of G, and let H' be the graph obtained from H by adding a new edge g in parallel with $\psi(e)$. It is routine to verify that H' is an abstract dual of G', and the remainder of the proof is left as an exercise. □

Proposition 5.4.9 *Let G be a graph having no isolated vertices. If G is an abstract dual of a series–parallel network, then G is a series–parallel network.* □

The next theorem (Dirac 1952; Ádám 1957; Duffin 1965) is the main result of this section, and most of the rest of the section is devoted to proving it.

Theorem 5.4.10 *A graph G with at least one edge is a series–parallel network if and only if G is a block having no subgraph that is a subdivision of K_4.*

Proof If G is a series–parallel network, a routine induction argument shows that G is a block with no subgraph that is a subdivision of K_4. The converse of this is also proved by induction on $|E(G)|$. It clearly holds for $|E(G)| \leq 2$. Assume it holds for $|E(G)| < n$ and let $|E(G)| = n \geq 3$. If G has two edges in series or two edges in parallel, the result follows by the induction assumption. Thus we may assume that G is simple with every vertex degree exceeding two. The next lemma (Dirac 1952) completes the proof of the theorem by establishing the contradiction that G has a subgraph that is a subdivision of K_4.

A *chordal path* of a cycle in a graph is a path that joins two non-consecutive vertices of the cycle but is otherwise disjoint from the cycle. An *internal vertex* of such a path is a vertex of the path other than one of its end-vertices.

Lemma 5.4.11 *A simple 2-connected graph G in which the degree of every vertex is at least three has a subgraph that is a subdivision of K_4.*

Proof Let C be a cycle of G with the maximum number of edges. Let the vertices of C, in cyclic order, be $a_1, a_2, \ldots, a_m$. We shall show next that every vertex of C meets a chordal path of C. Since every vertex of G has degree at least three, for each i in $\{1, 2, \ldots, m\}$, there is an edge e_i that meets a_i but is not in C. As G is 2-connected, e_i is the first edge of a path P_i in G that joins a_i to some other vertex a_j of C but is otherwise disjoint from C. If this path is not a chordal path of C, then a_i and a_j are consecutive vertices of C. But then, either P_i has more than one edge, in which case the choice of C is contradicted, or P_i has exactly one edge, and the simplicity of G is contradicted. We conclude that every vertex of C does indeed meet a chordal path of C.

We now distinguish two cases:

(i) C has two chordal paths, R_1 and R_2, that join different pairs of vertices and have a common internal vertex.

(ii) Whenever two chordal paths of C have a common internal vertex, they have the same end-vertices.

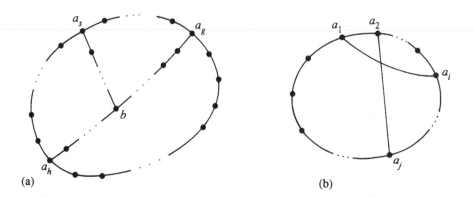

FIG. 5.17. Chordal paths of C giving subdivisions of K_4.

In the first case, suppose that R_1 joins a_g and a_h, and R_2 joins a_s and a_t where $a_s, a_g,$ and a_h are distinct, although a_t may equal a_g or a_h. Beginning at a_s, follow R_2 towards a_t until the first time a vertex b of R_1 is encountered. Then the subgraph of G induced by C, the chordal path R_1, and that part of the chordal path R_2 joining a_s and b is a subdivision of K_4 (see Figure 5.17(a)).

In case (ii), for every pair $\{a_k, a_l\}$ of distinct vertices of C that are joined by a chordal path, arbitrarily choose such a path P_{kl}, and let Q_{kl} be a shortest (a_k, a_l)-path in the cycle C. Let G' be the subgraph of G induced by C and the set of chosen paths P_{kl}. By assumption, no two chordal paths of C in G' share a common internal vertex. From among the paths of the form Q_{kl}, choose a shortest path, say Q, and assume it joins a_1 and a_i and contains a_2. Evidently $i \neq 2$. Also, since C has a chordal path that meets a_2, there is a chordal path in G' joining a_2 to a_j for some j not in $\{1, 2, 3\}$. As a shortest path joining a_2 and a_j in C cannot be shorter than Q, we must have that $j > i$. Thus the subgraph of G' induced by the union of C, the chordal path joining a_1 and a_i, and the chordal path joining a_2 and a_j is a subdivision of K_4 (see Figure 5.17(b)).

This completes the proof of the lemma and thereby finishes the proof of Theorem 5.4.10.   □

The next result essentially restates Theorem 5.4.10 as an excluded-minor result. The proof is left as an exercise. In Chapter 12, we shall give an excluded-minor characterization of series–parallel networks within the class of all matroids.

Corollary 5.4.12 *The following statements are equivalent for a graph G having at least one edge and no isolated vertices:*

(i)   *G is a series–parallel network.*

(ii) *$M(G)$ is connected and has no series minor isomorphic to $M(K_4)$.*

(iii) *$M(G)$ is connected and has no parallel minor isomorphic to $M(K_4)$.*

(iv) *$M(G)$ is connected and has no minor isomorphic to $M(K_4)$.*   □

Exercises

1. For each of the graphs in Figures 5.11, 5.12, 5.13, and 5.18, determine whether the graph is a series–parallel network and, if not, find the least number of edges that must be deleted to give such a graph.

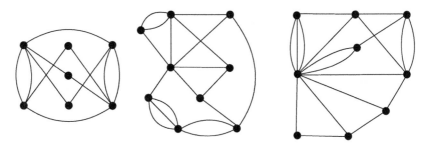

FIG. 5.18. Which of these three graphs are series–parallel networks?

2. Prove that if e and f are parallel edges in a series–parallel network G, then $G \backslash e$ is a series–parallel network.

3. Let G be a series–parallel network with no parallel edges. For a cycle C, find the maximum number of chordal paths C can have in terms of $|C|$.

4. Show that if a graph G has a subgraph that is a subdivision of a graph H, then $M(G)$ has $M(H)$ as a series minor.

5. Prove Proposition 5.4.2.

6. Find matroids M and N such that M has an N-minor, but M has no series minor and no parallel minor isomorphic to N.

7. Prove Corollary 5.4.12.

8. Show that a graph is a series–parallel network if and only if it is connected having a single edge, or it can be obtained from a cycle with two edges by a sequence of subdivisions and parallel extensions.

9. (Dirac 1960) Show that a simple n-vertex series–parallel network has at most $2n - 3$ edges and characterize the graphs attaining this bound.

10. (Oxley 1984a) Prove that a block with at least one edge is a series–parallel network if and only if it does not have a cycle and a bond that meet in exactly four edges.

11. (Duffin 1965) Prove that if a 2-connected graph G has a planar embedding in which every vertex lies on the boundary of the infinite face, then G is a series–parallel network. Is the converse of this true?

12. *(Duffin 1965) Let G be a block and assign a direction to every edge of G. Prove that G is a series–parallel network if and only if, whenever a cycle C of G meets a pair of edges, x and y, that are oppositely directed relative to C, the edges x and y are oppositely directed relative to every cycle containing them both.

6

REPRESENTABLE MATROIDS

Ever since Whitney (1935) used graphic and vector matroids as the two fundamental examples of matroids, there have been numerous basic problems associated with these classes, one of the most prominent of which has been to characterize the classes. In the case of vector matroids, this problem is usually specialized to that of characterizing the class of $\mathbb{F}$-representable matroids for some particular field $\mathbb{F}$. Since the classes of graphic and $\mathbb{F}$-representable matroids are minor-closed, one of the most commonly sought ways to characterize these classes has been via a list of *excluded minors*, that is, minor-minimal matroids that are not in the class. In Section 6.6, we shall present Tutte's excluded-minor characterization of the class of graphic matroids, a result that will be proved in Chapter 10. In Section 6.5, we state the excluded-minor characterizations for the only fields for which the excluded minors are known explicitly, namely, the classes of binary, ternary, and quaternary ($GF(4)$-representable) matroids. The first two of these results will be proved in Chapters 9 and 10. A focal point for research in this area has been Rota's Conjecture (1971) that, for all prime powers q, the class of $GF(q)$-representable matroids has a finite set of excluded minors. At present, this conjecture is open for all values of q except for the three for which the complete sets of excluded minors are known.

There are two main goals of this long chapter. The first is to provide an overview of the basic questions associated with matroid representability. The second is to indicate how one actually goes about constructing representations. In view of the length of this chapter, perhaps a word of guidance to the reader is in order. The key ideas are presented in Sections 6.1 and 6.3–6.6. The other five sections discuss topics that are less central. Of these, probably the most important is Section 6.10, which discusses an important class of matroids introduced by Dowling (1973a, 1973b). That section can essentially be read independently of the rest of the chapter as very little of it depends on earlier sections. Section 6.2 looks at affine geometries, a class of highly symmetric structures that are closely linked to the projective geometries of Section 6.1; Section 6.7 discusses algebraic matroids, a class of matroids that properly contains the class of representable matroids and arises from algebraic dependence over a field; Section 6.8 focuses on characteristic sets, its main idea being concerned with how one can capture geometrically certain algebraic properties of a field; and Section 6.9 examines modularity, a special property of flats that is important in several contexts including matroid constructions.

6.1 Projective geometries

This section introduces projective geometries, which play a role among representable matroids analogous to that of complete graphs in graph theory. Most of this section will discuss algebraic and geometric properties of projective geometries. But, for the reader who would prefer a quicker introduction to matroid representations, we begin by discussing these in terms of vector spaces.

Let M be an n-element rank-r matroid and $\mathbb{F}$ be a field. Then, by definition, M is $\mathbb{F}$-representable if and only if M is isomorphic to $M[A]$ for some $m \times n$ matrix A over $\mathbb{F}$ with $m \geq r$. Associated with this isomorphism, there is a natural map ψ_A from $E(M)$ into $V(m, \mathbb{F})$.

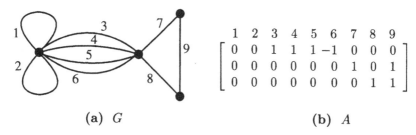

(a) G **(b)** A

FIG. 6.1. A graph G and a matrix A representing $M(G)$ over $\mathbb{R}$.

Example 6.1.1 Let $E = \{1, 2, \ldots, 9\}$ and $M = M(G)$ where G is the graph shown in Figure 6.1(a). Let A be the matrix over $\mathbb{R}$ shown in Figure 6.1(b). Then, under the identity map on E, it is easily checked that $M(G) \cong M[A]$. The corresponding map ψ_A from E into $V(3, \mathbb{R})$ is $\psi_A(1) = \psi_A(2) = (0, 0, 0)^T$, $\psi_A(3) = \psi_A(4) = \psi_A(5) = (1, 0, 0)^T$, $\psi_A(6) = (-1, 0, 0)^T$, $\psi_A(7) = (0, 1, 0)^T$, $\psi_A(8) = (0, 0, 1)^T$, and $\psi_A(9) = (0, 1, 1)^T$. □

It is clear that, in general, a rank-r matroid M is $\mathbb{F}$-representable if and only if, for some $m \geq r$, there is a function $\psi : E(M) \to V(m, \mathbb{F})$ such that $r_M(X) = \dim\langle\psi(X)\rangle$ for all $X \subseteq E(M)$. Such a map ψ will be called a *coordinatization* of M over $\mathbb{F}$ or an $\mathbb{F}$-*coordinatization* of M. Clearly if ψ is a coordinatization of M, then $r = \dim\langle\psi(E(M))\rangle$. Hence the rank-$r$ matroid M is $\mathbb{F}$-representable if and only if M has a coordinatization ϕ over $\mathbb{F}$ such that $\phi : E(M) \to V(r, \mathbb{F})$.

For a matrix A over a field $\mathbb{F}$, an element e of $M[A]$ is a loop of this matroid if and only if the corresponding column of A is the zero vector. Two elements f and g are parallel in $M[A]$ if and only if the corresponding columns of A are scalar multiples of each other and neither is zero. It follows from these observations that a matroid M is $\mathbb{F}$-representable if and only if its associated simple matroid $si(M)$ is $\mathbb{F}$-representable. Hence, when we discuss representability questions, we shall usually concentrate on simple matroids. Now suppose that the matroid M is simple and that $\psi : E(M) \to V(m, \mathbb{F})$ is a coordinatization of M over $\mathbb{F}$. Then ψ is one-to-one. Moreover, $\psi(E(M))$ does not contain $\underline{0}$; nor does it contain more than one element of any 1-dimensional subspace of $V(m, \mathbb{F})$. Thus, when $\mathbb{F}$ is finite, we can view $\psi(E(M))$ as being a restriction of the simple matroid associated with $V(m, \mathbb{F})$. As we shall see, this simple matroid is the matroid associated with the projective geometry $PG(m-1, \mathbb{F})$. Our definition of a projective geometry will be valid for both finite and infinite fields. We required $\mathbb{F}$ to be finite above only to ensure that $V(m, \mathbb{F})$ had finitely many elements and so could be viewed as a matroid.

Let V be a vector space over a field $\mathbb{F}$. The *projective geometry $PG(V)$ associated with V* consists of a set of *points*, a disjoint set of *lines*, and an *incidence relation* between points and lines. The points and lines are precisely the 1- and 2-dimensional subspaces of V, and incidence is determined by set inclusion. Evidently the construction of $PG(V)$ from V is analogous to the construction of the simple matroid si(M) from a matroid M. Indeed, if V is finite and we view both V and $PG(V)$ as matroids, these constructions are identical. An alternative way to construct $PG(V)$ from V is to first delete the zero vector and then, from each 1-dimensional subspace, delete all but one of the remaining elements. The elements that are left become the points of the projective geometry. Frequently these representatives for the 1-dimensional subspaces are chosen according to some pattern. For example, if $V = V(m, \mathbb{F})$, we often choose from each 1-dimensional subspace that vector whose first non-zero coordinate is a one. When $PG(V)$ is constructed in this way, the lines are the intersections of the 2-dimensional subspaces of V with the chosen set of points.

Having defined projective geometries associated with vector spaces, we now define projective geometries in general. It is not difficult to check that $PG(V)$ has the following properties:

(i) *Every two distinct points, a and b, are on exactly one line ab.*

(ii) *Every line contains at least three points.*

(iii) *If a, b, c, and d are four distinct points, no three of which are collinear, and if the line ab intersects the line cd, then the line ac intersects the line bd.*

In general, suppose P and L are disjoint sets of points and lines, respectively, and ι is an incidence relation such that (i)–(iii) hold. Then the triple (P, L, ι) is called a *projective geometry* or a *projective space*. In view of (i) and (ii), we can interpret lines as subsets of P. When this is done, it is natural to indicate incidence by using set-theoretic notation. Two projective geometries (P, L, ι) and (P', L', ι') are *isomorphic* if there is a bijection $\zeta : P \to P'$ such that a subset X of P is in L if and only if $\zeta(X)$ is in L'. As we shall see in the next theorem, most finite projective geometries are actually isomorphic to projective geometries associated with vector spaces.

A *subspace* of a projective geometry (P, L, ι) is a subset P_1 of P such that if a and b are distinct elements of P_1, then all points on the line ab are in P_1. Thus, for example, $\emptyset$ and P are subspaces, as are $\{p\}$ and X for all points p and all lines X. A subspace is a *hyperplane* if the only subspace properly containing it is P. The subspaces of (P, L, ι), ordered by inclusion, form a partially ordered set having a zero. The *(projective) dimension* of a subspace is one less than its height in this poset provided this height is finite; otherwise the dimension of the subspace is ∞. Thus $\emptyset$ has dimension -1, while points and lines have dimensions 0 and 1, respectively. A finite projective geometry of dimension one consists of a single line containing k points where k is an arbitrary integer exceeding two. If $V = V(n+1, \mathbb{F})$, then $PG(V)$ has dimension n and it is customary to denote this projective geometry by $PG(n, \mathbb{F})$. When $\mathbb{F}$ is the finite field $GF(q)$, we shall usually write $PG(n, q)$ for $PG(n, \mathbb{F})$.

Theorem 6.1.2 *Every finite projective geometry of dimension greater than two is isomorphic to $PG(n, q)$ for some integer n exceeding two and some prime power q.* $\qquad\square$

In fact, a natural extension of this result holds for infinite projective geometries. The reader is referred to Dembowski's book (1968, pp. 27–28) for a discussion of this result and for several references for proofs of it. A word of warning here concerning terminology: Dembowski calls the projective geometry $PG(n, q)$ a 'finite desarguesian projective geometry'. A projective geometry of dimension 2 is usually called a *projective plane*. Figure 1.11 shows the projective planes $PG(2, 2)$ and $PG(2, 3)$, both of which are certainly associated with vector spaces. It is known, however, that there are finite projective planes that are not associated with any vector space (see, for example, Hughes and Piper 1973). Clearly any finite projective plane gives rise to a rank-3 paving matroid on the set of points. Using this fact and Theorem 6.1.2, we deduce that every finite projective geometry can be regarded as a matroid.

We now know that the projective geometry $PG(n, \mathbb{F})$ can be viewed algebraically, using its derivation from $V(n+1, \mathbb{F})$, or geometrically, using (i)-(iii) above. Moreover, for $n \le 3$, we can draw a diagram representing $PG(n, q)$, this diagram being just a geometric representation of the corresponding matroid; that is, the points, lines, and planes of the projective space coincide with the points, lines, and planes of the matroid.

For every finite subset S of $PG(m-1, \mathbb{F})$, there is a matroid induced on S by linear independence over $\mathbb{F}$. We shall denote this matroid by $PG(m-1, \mathbb{F})|S$. Summarizing the above discussion, we have the following theorem, the main result of this section.

Theorem 6.1.3 *Let M be a simple rank-r matroid and $\mathbb{F}$ be a field. The following statements are equivalent:*

(i) M *is $\mathbb{F}$-representable.*
(ii) $PG(r-1, \mathbb{F})$ *has a finite subset T such that $M \cong PG(r-1, \mathbb{F})|T$.*
(iii) *For some $m \ge r$, there is a finite subset S of $PG(m-1, \mathbb{F})$ such that $M \cong PG(m-1, \mathbb{F})|S$.* $\qquad\square$

In view of this result, we see that a study of $\mathbb{F}$-representable simple matroids is a study of the restrictions of projective geometries over $\mathbb{F}$. Just as every n-vertex simple graph can be obtained from K_n by deleting edges, so too can every rank-r simple $\mathbb{F}$-representable matroid be obtained from $PG(r-1, \mathbb{F})$ by deleting elements. We remark here that, in a geometric representation of $PG(r-1, \mathbb{F})|S$, the points of the representation are just points of the projective geometry; the lines and planes of the representation come from intersecting S with lines and planes of the projective geometry that contain, respectively, at least two points and at least three non-collinear points of S.

The remainder of this section will discuss some properties of finite projective geometries. We begin by noting some statistics for $PG(r-1, q)$. Featured prominently here is the *Gaussian coefficient* $\begin{bmatrix} m \\ k \end{bmatrix}_q$, which is defined for all integers m and k with $0 \le k \le m$ by

$$\begin{bmatrix} m \\ k \end{bmatrix}_q = \frac{(q^m - 1)(q^m - q) \cdots (q^m - q^{k-1})}{(q^k - 1)(q^k - q) \cdots (q^k - q^{k-1})}$$

$$= \frac{(q^m - 1)(q^{m-1} - 1) \cdots (q^{m-k+1} - 1)}{(q^k - 1)(q^{k-1} - 1) \cdots (q - 1)}.$$

Note that $\begin{bmatrix} m \\ 0 \end{bmatrix}_q = 1$ for all $m \ge 0$ since, by convention, an empty product is 1.

We have seen that $PG(r-1, q)$ can be looked at in several different ways. For most of the rest of this book, we shall view it as the simple matroid associated with $V(r, q)$, where the latter is also being thought of as a matroid here. Evidently the matroid $PG(r-1, q)$ has rank r. Moreover, every flat of $PG(r-1, q)$ corresponds to a flat of $V(r, q)$. Since every rank-k flat of $V(r, q)$ is isomorphic to $V(k, q)$, every rank-k flat of $PG(r-1, q)$ is isomorphic to $PG(k-1, q)$. The next result specifies the number of such flats. From this, we get immediately that $PG(r-1, q)$ has exactly $(q^r - 1)/(q-1)$ elements.

Proposition 6.1.4 *Let k be a non-negative integer.*

(i) *The number of rank-k flats in $PG(r-1, q)$ is $\begin{bmatrix} r \\ k \end{bmatrix}_q$.*

(ii) *The number of k-element independent sets in $PG(r-1, q)$ is*

$$\frac{1}{k!} \left(\frac{q^r - 1}{q - 1} \right) \left(\frac{q^r - q}{q - 1} \right) \cdots \left(\frac{q^r - q^{k-1}}{q - 1} \right).$$

(iii) *The number of k-element circuits in $PG(r-1, q)$ is 0 for $k < 3$ and is*

$$\frac{1}{k!(q-1)} (q^r - 1)(q^r - q) \cdots (q^r - q^{k-2}) \quad \text{for } k \ge 3.$$

Proof We shall use the following result.

Lemma 6.1.5 *The number of ordered k-tuples $(\underline{v}_1, \underline{v}_2, \ldots, \underline{v}_k)$ of distinct members of $V(r, q)$ such that $\{\underline{v}_1, \underline{v}_2, \ldots, \underline{v}_k\}$ is linearly independent is*

$$(q^r - 1)(q^r - q) \cdots (q^r - q^{k-1}).$$

Proof Evidently $V(r, q)$ has q^r members. As $\underline{v}_1$ must be non-zero, there are $q^r - 1$ choices for it. Once distinct vectors $\underline{v}_1, \underline{v}_2, \ldots, \underline{v}_j$ have been chosen so that $\{\underline{v}_1, \underline{v}_2, \ldots, \underline{v}_j\}$ is linearly independent, $\underline{v}_{j+1}$ can be any of the q^r elements of $V(r, q)$ except for the q^j elements of $\langle \underline{v}_1, \underline{v}_2, \ldots, \underline{v}_j \rangle$. Hence there are $q^r - q^j$ choices for $\underline{v}_{j+1}$. □

From this lemma, we deduce that the number of k-element independent sets in $V(r, q)$ is

$$\frac{1}{k!}(q^r - 1)(q^r - q) \cdots (q^r - q^{k-1}). \tag{6.1}$$

Since every non-zero member of $V(r, q)$ is in a parallel class of size $q - 1$, part (ii) follows.

The number of rank-k flats of $PG(r - 1, q)$ is the same as the number n_k of rank-k flats of $V(r, q)$. But every rank-k flat of $V(r, q)$ is isomorphic to $V(k, q)$ and so the total number of k-element independent sets in $V(r, q)$ is the product of n_k and the number of k-element independent sets in $V(k, q)$. It follows from (6.1) that

$$n_k = \frac{(1/k!)(q^r - 1)(q^r - q) \cdots (q^r - q^{k-1})}{(1/k!)(q^k - 1)(q^k - q) \cdots (q^k - q^{k-1})} = \begin{bmatrix} r \\ k \end{bmatrix}_q ,$$

that is, (i) holds.

To prove (iii), we use the next lemma whose proof is left as an exercise.

Lemma 6.1.6 *If $r \geq 2$ and B is a basis of $PG(r - 1, q)$, then there are precisely $(q - 1)^{r-1}$ elements x of $PG(r - 1, q)$ such that $B \cup x$ is a circuit.* □

Let $\mathcal{C}_{r,k}$ denote the set of k-element circuits of $PG(r - 1, q)$. Evidently $|\mathcal{C}_{r,k}| = 0$ for $k < 3$, so suppose $k \geq 3$. Then, as every k-element circuit is in a unique flat of rank $k - 1$,

$$|\mathcal{C}_{r,k}| = \begin{bmatrix} r \\ k - 1 \end{bmatrix}_q |\mathcal{C}_{k-1,k}|. \tag{6.2}$$

Now, in $PG(k - 2, q)$, consider the set of ordered pairs (B, C) where B is a basis and C is a circuit containing B. By counting the number of such pairs in two different ways, first over circuits and then over bases, we get, using Lemma 6.1.6, that

$$k|\mathcal{C}_{k-1,k}| = (q - 1)^{k-2}|\mathcal{B}(PG(k - 2, q))|.$$

Thus, by (ii),

$$|\mathcal{C}_{k-1,k}| = \frac{(q - 1)^{k-2}}{k(k - 1)!} \left(\frac{q^{k-1} - 1}{q - 1} \right) \left(\frac{q^{k-1} - q}{q - 1} \right) \cdots \left(\frac{q^{k-1} - q^{k-2}}{q - 1} \right)$$

$$= \frac{1}{k!(q - 1)} \left(q^{k-1} - 1 \right) \left(q^{k-1} - q \right) \cdots \left(q^{k-1} - q^{k-2} \right).$$

Therefore, by (6.2), for $k \geq 3$,

$$|\mathcal{C}_{r,k}| = \frac{1}{k!(q - 1)} \left(q^r - 1 \right) \left(q^r - q \right) \cdots \left(q^r - q^{k-2} \right). \qquad □$$

We shall show in Proposition 14.10.4 that the bound on $|E(M)|$ in the next result also holds for all simple rank-r matroids M with no $U_{2,q+2}$-minor.

Corollary 6.1.7 *A simple rank-r matroid M that is representable over $GF(q)$ has at most $\frac{q^r-1}{q-1}$ elements. Moreover, if $|E(M)| = \frac{q^r-1}{q-1}$, then M is isomorphic to $PG(r-1,q)$.*

Proof By Theorem 6.1.3, as M is representable over $GF(q)$ and has rank r, there is a subset S of $E(PG(r-1,q))$ such that M is isomorphic to $PG(r-1,q)|S$. Since, by Proposition 6.1.4(i), $PG(r-1,q)$ has $\frac{q^r-1}{q-1}$ elements, both parts of the result now follow easily. $\square$

The lattice of flats of $PG(r-1,q)$ is particularly well-behaved. Clearly it is isomorphic to the lattice of flats of $V(r,q)$. Using this fact and Corollary 3.3.10, it is straightforward to prove the following.

Proposition 6.1.8 *Let X and Y be flats of $PG(r-1,q)$ with $X \subseteq Y$ and $r(Y) - r(X) = k$. Then the interval $[X,Y]$ of $\mathcal{L}(PG(r-1,q))$ is isomorphic to $\mathcal{L}(PG(k-1,q))$.* $\square$

The next corollary follows immediately from this proposition and the fact that every line of $PG(r-1,q)$ contains exactly $q+1$ points.

Corollary 6.1.9 *If $r \geq 2$ and $e \in PG(r-1,q)$, then $PG(r-1,q)/e$ is the matroid that is obtained from $PG(r-2,q)$ by replacing each element by q elements in parallel.* $\square$

From elementary properties of $V(r,q)$, it is not difficult to see that the lattice obtained from $\mathcal{L}(PG(r-1,q))$ by reversing all order relations is actually isomorphic to $\mathcal{L}(PG(r-1,q))$. Informally, $\mathcal{L}(PG(r-1,q))$ looks the same upside down as it does right side up. A geometric lattice with this property is an example of a modular lattice, an arbitrary lattice $\mathcal{L}$ being *modular* if it satisfies the Jordan–Dedekind chain condition and, for all x, y in $\mathcal{L}$,

$$h(x) + h(y) = h(x \vee y) + h(x \wedge y).$$

Modularity of lattices can be defined in several other equivalent ways (see, for example, Birkhoff 1967). In Section 6.9, we shall examine modularity in matroids. For the moment, we note that the fact that $\mathcal{L}(PG(r-1,q))$ is modular is an immediate consequence of the following familiar identity for subspaces U and W of a finite-dimensional vector space:

$$\dim U + \dim W = \dim(U + W) + \dim(U \cap W). \tag{6.3}$$

In Figure 6.2, we summarize several basic statistics for $\mathcal{L}(PG(r-1,q))$ including the numbers of flats of each rank, the number of rank-$(j-1)$ flats contained in each rank-j flat, and the number of rank-$(j+1)$ flats containing each rank-j flat. These numbers are not difficult to derive using Proposition 6.1.4.

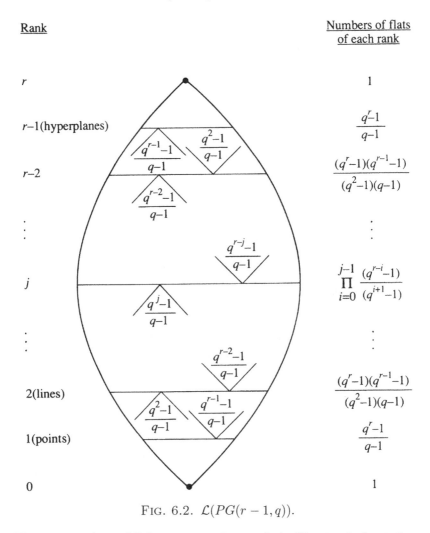

Rank

Numbers of flats
of each rank

FIG. 6.2. $\mathcal{L}(PG(r-1,q))$.

The next result establishes an assertion made in Chapter 1 about the non-representability of the non-Pappus matroid.

Proposition 6.1.10 *If $\mathbb{F}$ is a field, then the non-Pappus matroid is not $\mathbb{F}$-representable.*

Proof Assume the non-Pappus matroid N is $\mathbb{F}$-representable for some field $\mathbb{F}$. Then, by Theorem 6.1.3, there is a 9-element subset S of $PG(2,\mathbb{F})$ such that the matroid $PG(2,\mathbb{F})|S$ has the geometric representation shown in Figure 6.3. In particular, $7, 8,$ and 9 are not collinear in this diagram, that is, $\{7,8,9\}$ is a basis of N. But the next result precludes the occurrence of such a configuration in $PG(2,\mathbb{F})$, thereby completing the proof of the proposition. $\qquad\square$

Theorem 6.1.11 (Pappus's Theorem) *Let* $\{1, 2, 3\}$ *and* $\{4, 5, 6\}$ *be triples of distinct points that lie on the lines* L *and* L', *respectively, of* $PG(2, \mathbb{F})$ *such that none of these six points is on both* L *and* L'. *Let* $7, 8$, *and* 9 *be the points of intersection of the pairs of lines,* 15 *and* 24, 16 *and* 34, *and* 26 *and* 35, *respectively. Then* $7, 8$, *and* 9 *are collinear.*

Proof This can be found in Hughes and Piper (1973, Theorem 2.6).   □

We remark here that the pairs of lines whose intersections give the points $7, 8$, and 9 in the last theorem are actually opposite sides of the twisted hexagon whose vertices are, in cyclic order, $1, 5, 3, 4, 2$, and 6.

If we modify Figure 6.3 to make $7, 8$, and 9 collinear, the resulting configuration is called the *Pappus configuration*. As noted in Chapter 1, the corresponding matroid is called the Pappus matroid.

We now look briefly at projective planes, that is, projective spaces of dimension two. From Figure 6.2, we know that both the number of points and the number of lines of $PG(2, q)$ is $\frac{q^3 - 1}{q - 1}$, that is, $q^2 + q + 1$. Moreover, each line contains $\frac{q^2 - 1}{q - 1}$, that is, $q + 1$ points, and each point is on $q + 1$ lines. The projective planes that are not isomorphic to $PG(2, q)$ for any q also have a very uniform structure. In particular, if π is a finite projective plane, then one can show that all lines of π contain the same number of points. If this number is $m + 1$, then π is said to have *order* m. The next proposition is not difficult to prove.

Proposition 6.1.12 *In a projective plane of order* m,

(i)   *every point is on exactly* $m + 1$ *lines;*
(ii)   *there are exactly* $m^2 + m + 1$ *points; and*
(iii) *there are exactly* $m^2 + m + 1$ *lines.*   □

Clearly the plane $PG(2, q)$ has order q. It is known that there are unique planes of orders $2, 3, 4, 5, 7$, and 8, and it follows from the theorem of Bruck and Ryser stated below that there is no plane of order 6. Moreover, there are at least four non-isomorphic planes of order 9 and at least two non-isomorphic planes of order m for all integers m exceeding 8 that are of the form p^k for p a prime

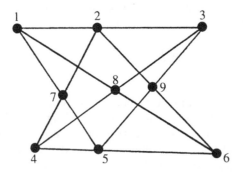

FIG. 6.3. The non-Pappus matroid.

and k an integer exceeding 1. Among the many interesting unsolved questions associated with projective planes are the following, both of which are very hard:

(i) If p is a prime exceeding seven, is $PG(2,p)$ the only projective plane of order p?

(ii) Is there a finite projective plane whose order is not a power of a prime?

The following important result of Bruck and Ryser (1949) eliminates many potential orders for projective planes. The first order left open by this theorem is 10, and Lam, Thiel, and Swiercz (1989) used a computer search to show that there is no plane of this order.

Proposition 6.1.13 *Let m be an integer of the form $4k + 1$ or $4k + 2$ where $k \geq 1$. If there is a projective plane of order m, then there are integers a and b such that $m = a^2 + b^2$.*

Proof This can be found in Hall (1986). □

From a matroid-theoretic point of view, projective planes are of particular interest in connection with the problem of matroid reconstruction, an analogue of the well-known reconstruction problem for graphs. Let M_1 and M_2 be non-isomorphic projective planes of order m. Then every single-element contraction of each of M_1 and M_2 is isomorphic to the matroid obtained from $U_{2,m+1}$ by replacing each element by m parallel elements. Equivalently, the multiset that consists of the isomorphism classes of all of the single-element deletions of M_1^* is identical to the corresponding multiset for M_2^*; yet $M_1^* \not\cong M_2^*$. This shows that we may not be able to uniquely determine a matroid, even if we know all of its single-element deletions up to isomorphism. For further discussion of this interesting topic of matroid reconstruction, the reader is referred to Chapter 15 and the work of Brylawski (1974, 1975a).

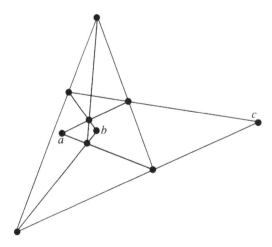

FIG. 6.4. The non-Desargues matroid.

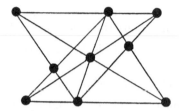

FIG. 6.5. A matroid that is not representable over any division ring.

To close this section, we comment briefly on a more general type of representability than we have considered so far. Let A be an $m \times n$ matrix over a division ring R. Let E be the set of column labels of A and $\mathcal{I}$ be the set of subsets I of E for which the multiset of columns labelled by I is a linearly independent set in the left vector space of m-tuples over R. Then the argument that was used in Proposition 1.1.1 to show that $(E, \mathcal{I})$ is a matroid when R is a field also establishes that $(E, \mathcal{I})$ is a matroid when R is an arbitrary division ring. As before, we shall denote this matroid by $M[A]$. An arbitrary matroid that is isomorphic to $M[A]$ is said to be *representable over R*. The reader who is unfamiliar with left vector spaces should consult Jacobson's *Lectures in Abstract Algebra, Volume II* (1953) where most of the treatment of linear algebra is in terms of one-sided vector spaces. We note here that the definition given earlier for the projective geometry $PG(V)$ extends without difficulty to the more general case when V is an arbitrary left vector space. Moreover, the extension of Theorem 6.1.2 referred to earlier is that every projective geometry of dimension greater than two is isomorphic to $PG(V)$ for some left vector space V.

A good reference for matroids representable over division rings is Ingleton (1971a). It is noted there that the non-Pappus matroid, while it is not representable over any field, is representable over some division ring. By contrast, the Vámos matroid is not representable over any division ring. This follows from the proof of Proposition 2.2.26 because the dimension identity (6.3), upon which that proof is based, holds for all vector spaces over division rings. Ingleton's paper gives several other examples of matroids that are not representable over any division ring including the 10-point rank-3 *non-Desargues matroid* for which a geometric representation is shown in Figure 6.4. The fact that this diagram does actually represent a matroid follows by 1.5.9; the fact that this matroid is not representable over any division ring follows by Desargues's Theorem in projective geometry (see, for example, Dembowski 1968, p. 26) which asserts that if the configuration of points and lines in Figure 6.4 occurs in a projective geometry over a division ring, then the points a, b, and c must be collinear.

Exercises

1. What is the maximum number of bases in a simple rank-r matroid that is representable over $GF(q)$, and which matroids attain this maximum?
2. Prove that the Gaussian coefficients have the following properties:

(a) $\left[\begin{smallmatrix} m \\ k \end{smallmatrix}\right]_q = \left[\begin{smallmatrix} m \\ m-k \end{smallmatrix}\right]_q.$

(b) If $q > 1$, then $q^{k(m-k)} \leq \left[\begin{smallmatrix} m \\ k \end{smallmatrix}\right]_q \leq q^{k(m-k+1)}.$

(c) $\left[\begin{smallmatrix} m \\ k \end{smallmatrix}\right]_q = \left[\begin{smallmatrix} m-1 \\ k \end{smallmatrix}\right]_q + q^{m-k}\left[\begin{smallmatrix} m-1 \\ k-1 \end{smallmatrix}\right]_q.$

(d) $\lim_{q \to 1} \left[\begin{smallmatrix} m \\ k \end{smallmatrix}\right]_q = \binom{m}{k}.$

3. Show that $\left[\begin{smallmatrix} r-t \\ s-u \end{smallmatrix}\right]_q \left[\begin{smallmatrix} t \\ u \end{smallmatrix}\right]_q q^{(s-u)(t-u)}$ is the number of rank-s subspaces of $PG(r-1, q)$ whose intersection with a fixed rank-t subspace has rank u.

4. Describe $PG(r-1, q)/I$ for a k-element independent set I in $PG(r-1, q)$.

5. Let M be a simple rank-r matroid that is representable over $GF(q)$ and has exactly $\left[\begin{smallmatrix} r \\ k \end{smallmatrix}\right]_q$ flats of rank k for some k with $0 \leq k \leq r$. For which values of k must it be true that $M \cong PG(r-1, q)$?

6. (Ingleton 1971a) Show that the matroid for which a geometric representation is shown in Figure 6.5 is not representable over any division ring.

7. *(Ingleton 1971a) Let V_1, V_2, V_3, and V_4 be subspaces of a vector space V.

   (a) Show that

   $$\dim(V_1 \cap V_2 \cap V_3) \geq \dim(V_1 \cap V_2) + \dim V_3 - \dim(V_1 + V_3) \\ - \dim(V_2 + V_3) + \dim(V_1 + V_2 + V_3)$$

   and

   $$\dim(V_1 \cap V_2 \cap V_3 \cap V_4) \geq \dim(V_1 \cap V_2 \cap V_3) \\ + \dim(V_1 \cap V_2 \cap V_4) - \dim(V_1 \cap V_2).$$

   (b) Deduce from (a) that

   $$\dim(V_1 \cap V_2 \cap V_3 \cap V_4) \geq \dim(V_1 \cap V_2) + \dim V_3 + \dim V_4 \\ - \dim(V_1 + V_3) - \dim(V_2 + V_3) \\ - \dim(V_1 + V_4) - \dim(V_2 + V_4) \\ + \dim(V_1 + V_2 + V_3) + \dim(V_1 + V_2 + V_4).$$

   (c) Prove that

   $$\dim V_1 + \dim V_2 + \dim(V_1 + V_2 + V_3) + \dim(V_1 + V_2 + V_4) + \dim(V_3 + V_4) \\ \leq \dim(V_1 + V_2) + \dim(V_1 + V_3) + \dim(V_1 + V_4) \\ + \dim(V_2 + V_3) + \dim(V_2 + V_4).$$

   (d) Deduce from (c) that if M is a representable matroid and X_1, X_2, X_3, and X_4 are subsets of $E(M)$, then

   $$r(X_1) + r(X_2) + r(X_1 \cup X_2 \cup X_3) + r(X_1 \cup X_2 \cup X_4) + r(X_3 \cup X_4) \\ \leq r(X_1 \cup X_2) + r(X_1 \cup X_3) + r(X_1 \cup X_4) + r(X_2 \cup X_3) + r(X_2 \cup X_4).$$

   (e) Deduce from (d) that V_8 (see Figure 2.4) is non-representable.

6.2 Affine geometries

Affine matroids were introduced in Section 1.5. In this section, we introduce affine geometries and indicate their relationship to both affine matroids and projective geometries. Finite affine geometries form an important class of highly symmetric matroids and play a fundamental role in geometry. However, they are not as central to a discussion of matroid representability as projective geometries. This section is independent of the remaining sections of this chapter and so could be omitted by a reader whose main interest is in representability.

Let n be an integer exceeding -2. The *affine geometry* $AG(n, \mathbb{F})$ is obtained from $PG(n, \mathbb{F})$ by deleting all the points of a hyperplane of the latter. Equivalently, we can construct $AG(n, \mathbb{F})$ directly from $V(n+1, \mathbb{F})$ as follows. First delete a hyperplane H of $V(n+1, \mathbb{F})$. Then, for each 1-dimensional subspace X of $V(n+1, \mathbb{F})$ that is not contained in H, choose a single representative from $X - H$. These representatives are the points of $AG(n, \mathbb{F})$. Since, for any two hyperplanes H_1 and H_2 of $V(n+1, \mathbb{F})$, there is a non-singular linear transformation of $V(n+1, \mathbb{F})$ mapping H_1 onto H_2, it does not matter which hyperplane H of $V(n+1, \mathbb{F})$ is deleted in the above construction; that is, $AG(n, \mathbb{F})$ is well-defined. Now every hyperplane of $V(n+1, \mathbb{F})$ is the kernel of a linear functional, the latter being a linear transformation from $V(n+1, \mathbb{F})$ onto $V(1, \mathbb{F})$. In particular, as $V(n+1, \mathbb{F}) = \{(x_1, x_2, \ldots, x_{n+1}) : x_i \in \mathbb{F} \text{ for all } i\}$, the set of vectors $(x_1, x_2, \ldots, x_{n+1})$ with $x_1 = 0$ is a hyperplane of $V(n+1, \mathbb{F})$. If this is the hyperplane H that is deleted, then all the remaining vectors have non-zero first coordinates. If we pick, as the representative for each 1-dimensional subspace, that vector whose first non-zero coordinate is a one, then we see that the point set of $AG(n, \mathbb{F})$ can be viewed as $\{(1, x_2, x_3, \ldots, x_{n+1}) : x_i \in \mathbb{F} \text{ for all } i\}$. Now let S be a finite subset $\{(1, x_{i2}, x_{i3}, \ldots, x_{i(n+1)}) : 1 \leq i \leq m\}$ of $AG(n, \mathbb{F})$. Then there is a matroid, $AG(n, \mathbb{F})|S$, induced on the points of S by linear dependence over $\mathbb{F}$. This matroid is precisely the affine matroid on $\{(x_{i2}, x_{i3}, \ldots, x_{i(n+1)}) : 1 \leq i \leq m\}$ identified in Section 1.5. It is now not difficult to see that a matroid M is affine over $\mathbb{F}$ if and only if M has no loops and $si(M) \cong AG(n, \mathbb{F})|S$ for some $n \geq -1$ and some finite subset S of $AG(n, \mathbb{F})$.

We now consider the affine geometry $AG(r-1, q)$, that is, $AG(r-1, GF(q))$, viewing this as a matroid. Clearly this matroid has rank r and, when $r \geq 1$, it has q^{r-1} elements. Moreover, all of its rank-k flats are isomorphic to $AG(k-1, q)$.

Example 6.2.1 The affine matroid $AG(3, 2)$ was looked at several times in

$$
\begin{bmatrix}
1 & 1 & 1 & 1 & 1 & 1 & 1 & 1 \\
1 & 0 & 0 & 1 & 1 & 0 & 1 & 0 \\
0 & 1 & 0 & 1 & 0 & 1 & 1 & 0 \\
0 & 0 & 1 & 0 & 1 & 1 & 1 & 0
\end{bmatrix}
\qquad
\left[
\begin{array}{c|cccc}
 & 0 & 1 & 1 & 1 \\
I_4 & 1 & 0 & 1 & 1 \\
 & 1 & 1 & 0 & 1 \\
 & 1 & 1 & 1 & 0
\end{array}
\right]
$$

(a)   (b)

FIG. 6.6. Two $GF(2)$-representations for $AG(3, 2)$.

Chapters 1 and 2. With $V(4,2) = \{(x_1, x_2, x_3, x_4)^T : x_i \in \{0,1\}$ for all $i\}$, we have, from above, that $AG(3,2)$ is represented by the matrix in Figure 6.6(a), this representation being obtained from $V(4,2)$ by deleting all the points of the hyperplane $x_1 = 0$. Another representation for $AG(3,2)$, which was noted following Example 2.2.25, is given in Figure 6.6(b). It is obtained from $V(4,2)$ by deleting all the points of the hyperplane $x_1 + x_2 + x_3 + x_4 = 0$. □

Example 6.2.2 Consider $AG(2,2)$ and $AG(2,3)$. Viewed as the matroids induced on $V(2,2)$ and $V(2,3)$ by affine dependence, these affine geometries have the geometric representations shown in Figure 6.7. Notice that each point has been labelled by the corresponding element of $V(2,q)$, and dependence among these points must be determined affinely rather than linearly. To obtain linear representations for these matroids, recall, from Section 1.5, that if a point has affine coordinates (v_1, v_2), we take its corresponding linear coordinates to be $(1, v_1, v_2)$. We observe that $AG(2,2) \cong U_{3,4}$. Moreover, by comparing Figure 6.7 with Figure 1.11, we can see, in these two cases, how $AG(2,q)$ has been obtained from $PG(2,q)$ by deleting a line. □

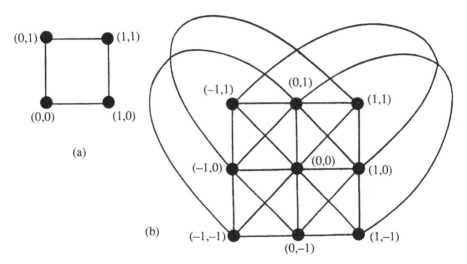

FIG. 6.7. (a) $AG(2,2)$ and (b) $AG(2,3)$.

The next result is the analogue of Proposition 6.1.4 for $AG(r-1, q)$.

Proposition 6.2.3 *Let k be a non-negative integer.*

(i)   *The number of rank-k flats of $AG(r-1, q)$ is*

$$q^{r-k} \begin{bmatrix} r-1 \\ k-1 \end{bmatrix}_q = q^{r-k} \frac{(q^{r-1} - 1)(q^{r-1} - q) \cdots (q^{r-1} - q^{k-2})}{(q^{k-1} - 1)(q^{k-1} - q) \cdots (q^{k-1} - q^{k-2})}.$$

(ii) *The number of k-element independent sets of $AG(r-1,q)$ is*

$$\frac{1}{k!}q^{r-1}(q^{r-1}-1)(q^{r-1}-q)\cdots(q^{r-1}-q^{k-2}).$$

(iii) *$AG(r-1,q)$ has no circuits with fewer than three elements. For $k \geq 3$, its number of k-element circuits is*

$$\frac{1}{k!}\,q^{r-2}(q^{r-1}-1)(q^{r-1}-q)\cdots(q^{r-1}-q^{k-3})\left[(q-1)^{k-1}-(-1)^{k-1}\right].$$

Proof The matroid $AG(r-1,q)$ is obtained from $PG(r-1,q)$ by deleting the points of a hyperplane H of the latter. Thus every flat of $AG(r-1,q)$ is of the form $X - H$ where X is a flat of $PG(r-1,q)$. It follows without difficulty, using the modularity of $\mathcal{L}(PG(r-1,q))$, that the rank-k flats of $AG(r-1,q)$ are all the non-empty sets of the form $X - H$ where X is a rank-k flat of $PG(r-1,q)$. Thus the number of rank-k flats of $AG(r-1,q)$ is the difference between the number of such flats in $PG(r-1,q)$ and the number in a hyperplane of $PG(r-1,q)$. Since every hyperplane of $PG(r-1,q)$ is isomorphic to $PG(r-2,q)$, we deduce from Proposition 6.1.4(i) that the number of rank-k flats in $AG(r-1,q)$ is $\left[{r \atop k}\right]_q - \left[{r-1 \atop k}\right]_q$, which simplifies to $q^{r-k}\left[{r-1 \atop k-1}\right]_q$.

To prove (ii), we first count the number of ordered k-tuples $(\underline{v}_1, \underline{v}_2, \ldots, \underline{v}_k)$ of distinct members of $\{(1, x_2, \ldots, x_r) : \text{each } x_i \in GF(q)\}$ such that $\{\underline{v}_1, \underline{v}_2, \ldots, \underline{v}_k\}$ is linearly independent. Evidently there are q^{r-1} choices for $\underline{v}_1$. If distinct vectors $\underline{v}_1, \underline{v}_2, \ldots, \underline{v}_j$ have been chosen so that $\{\underline{v}_1, \underline{v}_2, \ldots, \underline{v}_j\}$ is linearly independent, then $\{\underline{v}_1, \underline{v}_2, \ldots, \underline{v}_j\}$ spans a rank-j flat of $AG(r-1,q)$ and this flat is isomorphic to $AG(j-1,q)$. Hence there are $q^{r-1} - q^{j-1}$ choices for $\underline{v}_{j+1}$. We conclude that the number of k-element independent sets of $AG(r-1,q)$ is

$$\frac{1}{k!}\,q^{r-1}(q^{r-1}-1)(q^{r-1}-q)\cdots(q^{r-1}-q^{k-2}).$$

The proof of (iii) mimics the proof of (iii) of Proposition 6.1.4. It uses the following lemma in which, for convenience, $AG(r-1,q)$ is viewed as a restriction of $V(r,q)$.

Lemma 6.2.4 *If $r \geq 2$ and B is a basis of $AG(r-1,q)$, then the number of elements $\underline{x}$ of $AG(r-1,q)$ such that $B \cup \underline{x}$ is a circuit equals*

$$q^{-1}[(q-1)^r - (-1)^r].$$

Proof Let $B = \{(1, \underline{v}_i) : 1 \leq i \leq r\}$ and $\underline{x} = (1, \underline{w})$. Then $B \cup \underline{x}$ is a circuit of $AG(r-1,q)$ if and only if there are non-zero elements $\alpha_1, \alpha_2, \ldots, \alpha_r$ of $GF(q)$ such that

$$\sum_{i=1}^{r} \alpha_i(1, \underline{v}_i) = (1, \underline{w}).$$

By considering the first coordinates here, we get that, for $r \geq 2$, the number of circuits of $AG(r-1,q)$ containing B equals the number, $m(r)$, of r-tuples

$(\alpha_1, \alpha_2, \ldots, \alpha_r)$ of non-zero elements of $GF(q)$ such that $\sum_{i=1}^{r} \alpha_i = 1$. Clearly $m(r)$ is defined for all $r \geq 1$, and $m(1) = 1$. Moreover, for $r \geq 2$,

$$m(r) = |\{(\alpha_1, \alpha_2, \ldots, \alpha_r) : \sum_{i=1}^{r} \alpha_i = 1 \text{ and } \alpha_i \in GF(q) - \{0\} \text{ for } 1 \leq i \leq r\}|$$

$$= |\{(\alpha_1, \alpha_2, \ldots, \alpha_{r-1}, \alpha_r) : \sum_{i=1}^{r-1} \alpha_i \neq 1 \text{ and } \alpha_r = 1 - \sum_{i=1}^{r-1} \alpha_i,$$

$$\text{where } \alpha_i \in GF(q) - \{0\} \text{ for } 1 \leq i \leq r - 1\}|.$$

Thus

$$m(r) = |\{(\alpha_1, \alpha_2, \ldots, \alpha_{r-1}) : \alpha_i \in GF(q) - \{0\} \quad \text{for } 1 \leq i \leq r - 1\}|$$

$$- |\{(\alpha_1, \alpha_2, \ldots, \alpha_{r-1}) : \sum_{i=1}^{r-1} \alpha_i = 1 \text{ and each } \alpha_i \in GF(q) - \{0\}|$$

$$= (q-1)^{r-1} - m(r-1).$$

An easy induction argument now completes the proof of the lemma. □

Let $\mathcal{C}'_{r,k}$ denote the set of k-element circuits in $AG(r-1, q)$. Clearly $|\mathcal{C}'_{r,k}| = 0$ for $k < 3$, while, if $k \geq 3$, then $|\mathcal{C}'_{r,k}|$ is the product of $|\mathcal{C}'_{k-1,k}|$ and the number of rank-$(k-1)$ flats in $AG(r-1, q)$. The rest of the argument closely follows the proof of (iii) of Proposition 6.1.4 and the details are left to the reader. □

We now consider $\mathcal{L}(AG(r-1, q))$. As the next result shows, this lattice is somewhat like $\mathcal{L}(PG(r-1, q))$ in that every interval of it that does not include the zero of the lattice is isomorphic to $\mathcal{L}(PG(k-1, q))$ for some k.

Proposition 6.2.5 *Let X and Y be flats of $AG(r-1, q)$ with $X \subseteq Y$ and $r(Y) - r(X) = k$. Then*

$$[X, Y] \cong \begin{cases} \mathcal{L}(AG(k-1, q)), \text{ if } X = \emptyset; \\ \mathcal{L}(PG(k-1, q)), \text{ otherwise.} \end{cases}$$

Proof As noted earlier, every flat of $AG(r-1, q)$ of rank k is isomorphic to $AG(k-1, q)$. This proves the result when $X = \emptyset$. Now suppose $X \neq \emptyset$. We shall establish the result in the case that $Y = E(AG(r-1, q))$. The general result will then follow since, by Proposition 6.1.8, every interval in the lattice of flats of a finite projective geometry is also the lattice of flats of a projective geometry.

Clearly X has rank $r - k$. Let m denote the number of rank-$(r-k+1)$ flats of $AG(r-1, q)$ containing X. Now, for all x in $E(AG(r-1, q)) - X$, the set $X \cup x$ is contained in a unique rank-$(r-k+1)$ flat. As each rank-$(r-k+1)$ flat containing X has q^{r-k} elements, each such flat has $q^{r-k} - q^{r-k-1}$ elements not in X. Therefore

$$m = \left(\frac{1}{q^{r-k} - q^{r-k-1}}\right) |E(AG(r-1, q)) - X| = \frac{q^{r-1} - q^{r-k-1}}{q^{r-k} - q^{r-k-1}} = \frac{q^k - 1}{q - 1}.$$

Thus the simple matroid M associated with $AG(r-1,q)/X$ has $\frac{q^k-1}{q-1}$ elements. But $r(M) = k$ and M is representable over $GF(q)$. Hence, by Corollary 6.1.7, $M \cong PG(k-1,q)$ and so $\mathcal{L}(AG(r-1,q)/X) \cong \mathcal{L}(PG(k-1,q))$. $\qquad \square$

Using Propositions 6.2.3 and 6.2.5, and arguing as in their proofs, it is not difficult to check that the various numbers shown in Figure 6.8 are correct.

Every line of $AG(r-1,q)$ has exactly q elements. Using this, together with Proposition 6.2.5, we immediately obtain the following:

Corollary 6.2.6 *Let e be an element of $AG(r-1,q)$. Then $AG(r-1,q)/e$ is isomorphic to the matroid obtained from $PG(r-2,q)$ by replacing every element by $q-1$ elements in parallel.* $\qquad \square$

Rank   Numbers of flats of each rank

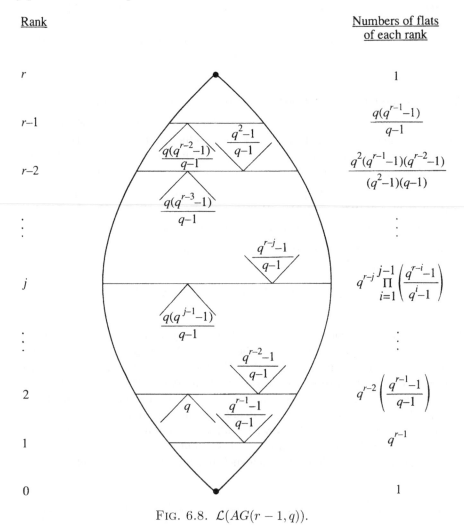

FIG. 6.8. $\mathcal{L}(AG(r-1,q))$.

In the final result of this section, the matroid $AG(r-1,q)$ will be identified with $PG(r-1,q)\backslash H$ where H is a hyperplane of $PG(r-1,q)$. We then consider the effect of adding an element of H to $AG(r-1,q)$ and contracting out this added element. This operation is an example of the quotient operation, which we shall examine in more detail in the next chapter.

Proposition 6.2.7 *For an element e of a hyperplane H of $PG(r-1,q)$, let $M = PG(r-1,q)\backslash(H-e)/e$. Then M is isomorphic to the matroid that is obtained from $AG(r-2,q)$ by replacing every element by q elements in parallel.*

Proof This is not difficult and is left to the reader. $\square$

Brylawski and Kelly (1980, p. 80) have noted that the last result corresponds intuitively to the fact that photographs from an infinite distance give perfect perspective.

Exercises

1. Show that $AG(1,q) \cong U_{2,q}$.
2. Show that $M(K_4)$ is not affine over $GF(2)$ or $GF(3)$ but that it is affine over all fields with more than three elements.
3. Show that $AG(r-1,q)$ has the following properties:
   (a) The intersection of two distinct hyperplanes is either empty or has rank $r-2$.
   (b) There is a partition of the ground set into q sets, each of which is a hyperplane.
4. Let X and Y be flats of $AG(n,q)$ of ranks s and t, respectively. If $s \le m \le t$, find the number of rank-m flats Z for which $X \subseteq Z \subseteq Y$.
5. Prove Proposition 6.2.7.
6. Which of the affine geometries $AG(n,q)$ are self-dual matroids?
7. Let M be a binary matroid. Show that M is affine over $GF(2)$ if and only if $E(M)$ is a disjoint union of cocircuits.
8. Recall that if W is a subspace of a vector space V, then a *coset* of W is a subset of V of the form $\{\underline{v} + \underline{w} : \underline{w} \in W\}$ where $\underline{v}$ is some fixed member of V. View $AG(n,q)$ as the matroid on $V(n,q)$ whose independent sets are the affinely independent subsets of $V(n,q)$.
   (a) Show that, for all $k \ge 1$, a set X is a rank-k flat of $AG(n,q)$ if and only if X is a coset of a $(k-1)$-dimensional subspace of $V(n,q)$.
   (b) Show that H is a hyperplane of $AG(n,q)$ if and only if there are a non-zero vector $\underline{w}$ of $V(n,q)$ and an element c of $GF(q)$ such that $H = \{\underline{v} \in V(n,q) : \underline{v} \cdot \underline{w} = c\}$.
9. *(Oxley 1978a) Let M be a $GF(q)$-representable matroid whose ground set is a disjoint union of cocircuits. Prove that M is affine over $GF(q)$.

6.3   Different matroid representations

A key issue in the consideration of matrix representations of a matroid M is to resolve when two representations are genuinely different. It was observed in Chapter 2 that, when certain operations are performed on a matrix A, the matroid $M[A]$ is unchanged. These matrix operations will be the basis of the definition of equivalence of matroid representations to be given in this section. One of the appealing features of this definition is that it has a geometric reformulation in terms of automorphisms of projective geometries.

For a non-zero matrix A over a field $\mathbb{F}$, Section 2.2 noted seven operations on A that do not alter $M[A]$. Six of these operations are standard matrix operations.

6.3.1 *Interchange two rows.*

6.3.2 *Multiply a row by a non-zero member of* $\mathbb{F}$.

6.3.3 *Replace a row by the sum of that row and another.*

6.3.4 *Adjoin or remove a zero row.*

6.3.5 *Interchange two columns (moving their labels with the columns).*

6.3.6 *Multiply a column by a non-zero member of* $\mathbb{F}$.

The seventh operation differs from the first six in that it is based on a property of the field $\mathbb{F}$. It will be discussed in more detail below.

6.3.7 *Replace each matrix entry by its image under some automorphism of* $\mathbb{F}$.

Operations 6.3.2 and 6.3.6 are known, respectively, as *row scaling* and *column scaling*. We shall define two notions of equivalence of representations based on these seven operations. Let M be a rank-r matroid on the set $\{e_1, e_2, \ldots, e_n\}$ where $r \geq 1$. For each i in $\{1, 2\}$, let A_i be an $r_i \times n$ matrix over the field $\mathbb{F}$ whose columns are labelled, in some order, by $e_1, e_2, \ldots, e_n$. Assume that the identity map on $\{e_1, e_2, \ldots, e_n\}$ is an isomorphism between M and $M[A_i]$ for each i. We define A_1 and A_2 to be *equivalent representations* of M if A_2 can be obtained from A_1 by a sequence of the operations 6.3.1–6.3.7. The two representations are *projectively equivalent* if A_2 can be obtained from A_1 by a sequence of these operations that does not use a field automorphism, that is, by a sequence of the matrix operations 6.3.1–6.3.6. The term *strongly equivalent* has sometimes been used in place of 'projectively equivalent'. Both equivalence and projective equivalence are easily seen to be equivalence relations on the set of non-zero matrices over $\mathbb{F}$ whose columns are labelled by $e_1, e_2, \ldots, e_n$. Two representations that are not projectively equivalent will be called *projectively inequivalent*.

We remark here that equivalence of representations for rank-0 matroids has not been defined. In this trivial case, the representation problem is completely solved: every such matroid is representable over every field, its representations consisting of all of the zero matrices with the appropriate number of columns.

When $\mathbb{F} = GF(p)$, for p prime, A_1 and A_2 are projectively equivalent if and only if they are equivalent. This is because the automorphism group of the field $GF(p^k)$ is $\{x \to x^{p^i} : i \in \{0, 1, 2, \ldots, k-1\}\}$ (see, for example, Hungerford 1974). This group is cyclic of order k and is generated by the map $x \to x^p$, which is sometimes called the *Frobenius automorphism*. In particular, if k is 1, the automorphism group is trivial. Thus operation 6.3.7 can be ignored in any discussion of binary and ternary matroids. It can also be ignored for representations over $\mathbb{R}$ or $\mathbb{Q}$, since these two infinite fields have no non-trivial automorphisms. By contrast, complex conjugation is an example of a non-trivial automorphism of $\mathbb{C}$, while $GF(4)$ has a unique non-trivial automorphism.

Example 6.3.8 Recall from the Preliminaries that $GF(4)$ can be viewed as having elements $0, 1, \omega$, and $1 + \omega$ where $\omega^2 = \omega + 1$ and the sum of every element with itself is 0. Over $GF(4)$, the two matrices

$$
\begin{array}{cccc}
1 & 2 & 3 & 4
\end{array}
\begin{bmatrix}
1 & 0 & 1 & 1 \\
0 & 1 & 1 & \omega
\end{bmatrix}
\quad \text{and} \quad
\begin{array}{cccc}
1 & 2 & 3 & 4
\end{array}
\begin{bmatrix}
1 & 0 & 1 & 1 \\
0 & 1 & 1 & \omega + 1
\end{bmatrix}
$$

are representations for $U_{2,4}$ on the ground set $\{1, 2, 3, 4\}$. These representations are equivalent as the non-identity automorphism of $GF(4)$ maps ω to $\omega + 1$. But the representations are not projectively equivalent. The reader may wish to show this last fact directly. It will follow from Proposition 6.3.12 below. □

Example 6.3.9 As the reader can check, over $GF(5)$, the matrix

$$
\begin{array}{ccccc}
1 & 2 & 3 & 4 & 5
\end{array}
\begin{bmatrix}
1 & 0 & 0 & 1 & 1 \\
0 & 1 & 0 & 1 & a \\
0 & 0 & 1 & 1 & b
\end{bmatrix}
$$

represents $U_{3,5}$ provided $\{a, b\} \cap \{0, 1\} = \emptyset$ and $a \neq b$. Thus there are six choices for the pair (a, b). No two of these six representations are projectively equivalent for, as will follow from Proposition 6.3.12, none can be obtained from another by a sequence of operations 6.3.1–6.3.6. Because $GF(5)$ has no non-trivial automorphisms, it follows that all of these representations are inequivalent. □

Next we note that if we need to apply a field automorphism to establish the equivalence of two representations, then we can ensure that we only need to do this once.

Lemma 6.3.10 *Let $\mathbb{F}$ be a field and A_1 and A_2 be equivalent $\mathbb{F}$-representations of a matroid of positive rank. Then either*

(i) A_1 *and A_2 are projectively equivalent; or*

(ii) A_1 *and A_2 are not projectively equivalent, and A_2 can be obtained from A_1 by a sequence of operations 6.3.1–6.3.7 that involves exactly one operation of type 6.3.7. Moreover, this operation can be done either first or last in the sequence.*

Proof As A_1 and A_2 are equivalent, A_2 can be obtained from A_1 by a sequence
of the operations 6.3.1–6.3.7. If two operations of type 6.3.7 occur consecutively
in such a sequence, they can be combined into a single operation of that type.
It now suffices to show that, when an operation of type 6.3.7 occurs in such
a sequence, it can be interchanged both with the operation that precedes it
and with the operation that succeeds it. This is immediate if the neighbouring
operation is of type 6.3.1, 6.3.4, or 6.3.5. If the neighbouring operation is of
type 6.3.3, it follows from the fact that a field automorphism preserves addition.
Finally, suppose a row or column is multiplied by a non-zero field element c and
then a field automorphism σ is applied to the entries of the resulting matrix. We
can achieve the same result by first applying σ and then scaling the relevant row
or column by $\sigma(c)$. The lemma now follows without difficulty. $\square$

Next we examine equivalence and projective equivalence of representations
more closely with a view to relating these definitions to matrix multiplications
and automorphisms of projective geometries. First we recall that a sequence of
elementary row operations (6.3.1–6.3.3) can be achieved by left multiplying by
a non-singular matrix. Indeed, we have the following result (see, for example,
Hungerford 1974).

Lemma 6.3.11 *Let A and A' be $r \times n$ matrices over a field $\mathbb{F}$. Then the following
statements are equivalent:*

(i)   *A' can be obtained from A by a sequence of elementary row operations.*

(ii) *A' can be obtained by multiplying A on the left by a non-singular $r \times r$
      matrix.*

(iii) *There is a non-singular linear transformation τ on $V(r, \mathbb{F})$ such that, for
      all i in $\{1, 2, \ldots, n\}$, the ith column of A' is the image under τ of the ith
      column of A.* $\square$

The following characterization of projectively equivalent representations (Brylawski and Lucas 1976) comes from combining the last lemma with the observation that multiplying a matrix on the right by a non-singular diagonal matrix
has the same effect as performing a sequence of column scalings.

Proposition 6.3.12 *Suppose that $r \geq 1$ and A_1 and A_2 are $r \times n$ matrices over
a field $\mathbb{F}$ with the columns of each matrix being labelled, in order, by $e_1, e_2, \ldots, e_n$.
Then A_1 and A_2 are projectively equivalent representations of a matroid on
$\{e_1, e_2, \ldots, e_n\}$ if and only if there are a non-singular $r \times r$ matrix X and a
non-singular $n \times n$ diagonal matrix Y such that $A_2 = X A_1 Y$.* $\square$

It is straightforward to use the last proposition to verify the assertions about
the representations in Examples 6.3.8 and 6.3.9 being projectively inequivalent.

Equivalence of representations has a very useful geometric interpretation.
Let M be an $\mathbb{F}$-representable matroid of rank r where $r \geq 3$. Roughly speaking,
two representations of M are equivalent if one is the image of the other under
an automorphism of $PG(r - 1, \mathbb{F})$ where an *automorphism* of this projective

geometry is a permutation of its set of points that maps lines onto lines. These automorphisms are often referred to as *collineations*. It is not difficult to check that, under an automorphism of $PG(r-1,\mathbb{F})$, every subspace is mapped onto one of the same dimension (Exercise 2).

Lemma 6.3.11 relates a sequence of elementary row operations to a linear transformation. When one is also allowed to apply a field automorphism, semi-linear transformations arise. Formally, a permutation σ of $V(r,\mathbb{F})$ is a *non-singular semilinear transformation* if there are a non-singular linear transformation τ of $V(r,\mathbb{F})$ and an automorphism α of $\mathbb{F}$ such that if $\underline{v} \in V(r,\mathbb{F})$ and $\tau(\underline{v}) = (w_1, w_2, \ldots, w_r)$, then $\sigma(\underline{v}) = (\alpha(w_1), \alpha(w_2), \ldots, \alpha(w_r))$. It is not difficult to characterize non-singular semilinear transformations in terms of a weakened linearity condition (see Exercise 1).

We now reformulate the definition of equivalence of representations in terms of non-singular semilinear transformations.

Proposition 6.3.13 *Let M be a matroid on the set $\{e_1, e_2, \ldots, e_n\}$. Suppose that $r \geq 1$ and let A_1 and A_2 be $r \times n$ matrices over a field $\mathbb{F}$ such that, for each i in $\{1, 2\}$, the map taking e_j to the jth column $\underline{v}_j^{(i)}$ of A_i is an isomorphism from M to $M[A_i]$. Then A_1 and A_2 are equivalent representations of M if and only if there are a non-singular semilinear transformation σ of $V(r,\mathbb{F})$ and a sequence $c_1, c_2, \ldots, c_n$ of non-zero elements of $\mathbb{F}$ such that $\underline{v}_j^{(2)} = c_j \sigma(\underline{v}_j^{(1)})$ for all j in $\{1, 2, \ldots, n\}$.*

Proof This follows easily from Lemma 6.3.11 and the preceding discussion. The details are left to the reader. $\square$

The following result indicates the importance of semilinear transformations.

Theorem 6.3.14 *If $r \geq 3$, then every automorphism of $PG(r-1,\mathbb{F})$ is induced by a non-singular semilinear transformation of $V(r,\mathbb{F})$.*

Proof This can be found in Artin (1957) and Baer (1952). $\square$

We can now indicate the precise relationship between equivalence of representations and automorphisms of projective geometries, thereby justifying the definition of the former. Let A_1 and A_2 be $r \times n$ matrices representing a matroid M over a field $\mathbb{F}$ and suppose that $r(M) \geq 1$. Let ψ_1 and ψ_2 be the $\mathbb{F}$-coordinatizations ψ_{A_1} and ψ_{A_2} that correspond to A_1 and A_2, respectively (see Example 6.1.1). Then $\psi_i : E(M) \to V(r,\mathbb{F})$ for $i = 1, 2$. Moreover, ψ_i maps every loop of M to $\underline{0}$. Now let $T_i = \{\langle \psi_i(e) \rangle : e$ is a non-loop element of $M\}$. Define $\theta : T_1 \to T_2$ by $\theta(\langle \psi_1(e) \rangle) = \langle \psi_2(e) \rangle$ for all non-loop elements e of M. Then θ is well-defined and can be viewed as a map between two subsets of $PG(r-1,\mathbb{F})$.

Corollary 6.3.15 *When $r \neq 2$, the matrices A_1 and A_2 are equivalent representations of M if and only if θ is the restriction to T_1 of an automorphism of $PG(r-1,\mathbb{F})$.*

Proof The result is trivial when $r = 1$ and follows immediately on combining Proposition 6.3.13 and Theorem 6.3.14 when $r \geq 3$. $\square$

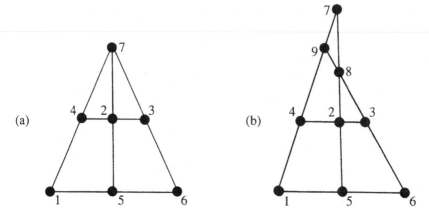

FIG. 6.9. Different embeddings of $\{1, 2, \ldots, 6\}$ in $PG(2, \mathbb{R})$.

The requirement that $r \neq 2$ is necessary in the last corollary since, when $r = 2$, every permutation of the point set of $PG(r - 1, \mathbb{F})$ is an automorphism of the projective geometry. But not every such permutation is induced by a sequence of the operations 6.3.1–6.3.7.

The next example illustrates the use of the last corollary to show that two particular representations are inequivalent.

Example 6.3.16 Consider the following matrices over $\mathbb{R}$:

$$A_1 = \left[\begin{array}{ccc|ccc} & & & 0 & 1 & 1 \\ & I_3 & & 1 & 2 & 1 \\ & & & 2 & 2 & 1 \end{array} \right], \qquad A_2 = \left[\begin{array}{ccc|ccc} & & & 0 & 1 & 1 \\ & I_3 & & 1 & 2 & 3 \\ & & & 2 & 2 & 3 \end{array} \right].$$

with column labels $1\ 2\ 3\ \ 4\ 5\ 6$ over both matrices.

We shall view the geometric representations for $M[A_1]$ and $M[A_2]$ shown in Figure 6.9 as subsets of $PG(2, \mathbb{R})$ where certain additional points have been included in these diagrams. Clearly the identity map on the set $\{1, 2, \ldots, 6\}$ is an isomorphism between $M[A_1]$ and $M[A_2]$, so A_1 and A_2 are both representations for $M[A_1]$. However, they are not equivalent representations for, as we can see with the aid of the point 7, which corresponds to $(1, 1, 2)^T$, there is no collineation of $PG(2, \mathbb{R})$ that acts as the identity on $\{1, 2, \ldots, 6\}$.   □

A rank-r matroid M on an n-element set is called *uniquely $\mathbb{F}$-representable* if all of the $r \times n$ matrices representing M over $\mathbb{F}$ are equivalent. As we shall see in the next section and in Section 14.6, all binary matroids are uniquely $GF(2)$-representable, and all ternary matroids are uniquely $GF(3)$-representable.

Exercises

1. Consider the set S of non-singular semilinear transformations on $V(r, \mathbb{F})$.

(a) Show that S consists of the permutations η of $V(r, \mathbb{F})$ for which $\mathbb{F}$ has an automorphism α such that $\eta(\underline{x} + \underline{y}) = \eta(\underline{x}) + \eta(\underline{y})$ for all $\underline{x}$, $\underline{y}$ in $V(r, \mathbb{F})$, and $\eta(c\underline{x}) = \alpha(c)\eta(\underline{x})$ for all $\underline{x}$ in $V(r, \mathbb{F})$ and all c in $\mathbb{F}$.

(b) Prove that S forms a group under composition.

2. Let ζ be a permutation of the set of points of $PG(r-1, \mathbb{F})$. Show that:

   (a) if ζ is an automorphism of $PG(r-1, \mathbb{F})$ and X is a subspace of $PG(r-1, \mathbb{F})$, then both $\zeta(X)$ and $\zeta^{-1}(X)$ are subspaces having the same dimension as X;

   (b) ζ is an automorphism of $PG(r-1, \mathbb{F})$ if and only if it maps subspaces onto subspaces.

3. Show that the following matrices are equivalent $GF(4)$-representations for Q_6, this matroid being as shown in Figure 2.1:

$$
\begin{array}{ccc}
6 \quad 3 \quad 5 & 2 \quad 4 \quad 1 \\
\left[\begin{array}{c|ccc}
& 1 & 0 & 1 \\
I_3 & 1 & 1 & \omega \\
& \omega+1 & 1 & \omega+1
\end{array}\right], &
\end{array}
\qquad
\begin{array}{ccc}
1 \quad 4 \quad 3 & 6 \quad 2 \quad 5 \\
\left[\begin{array}{c|ccc}
& 1 & 1 & 0 \\
I_3 & \omega & 0 & 1 \\
& 1 & \omega & 1
\end{array}\right].
\end{array}
$$

4. (a) Show that the matroid in Figure 6.9(a) is a ternary affine matroid by exhibiting it as a restriction of $AG(2,3)$.

   (b) Show that the matroid in Figure 6.9(b) is non-ternary by finding a non-ternary proper minor of it.

5. (a) Show that the following matrices are equivalent $GF(2)$-representations of the same matroid, S_8:

$$
\begin{array}{cccc}
1 \quad 2 \quad 3 \quad 4 & 5 \quad 6 \quad 7 \quad 8 \\
\left[\begin{array}{c|cccc}
& 0 & 1 & 1 & 0 \\
I_4 & 1 & 0 & 1 & 1 \\
& 1 & 1 & 0 & 1 \\
& 1 & 1 & 1 & 0
\end{array}\right], &
\end{array}
\qquad
\begin{array}{cccc}
4 \quad 7 \quad 3 \quad 1 & 5 \quad 6 \quad 2 \quad 8 \\
\left[\begin{array}{c|cccc}
& 0 & 1 & 1 & 1 \\
I_4 & 1 & 0 & 1 & 1 \\
& 1 & 1 & 0 & 1 \\
& 1 & 1 & 1 & 1
\end{array}\right].
\end{array}
$$

   (b) Show that S_8 is self-dual.

   (c) Show that S_8 has a unique element e such that $S_8/e \cong F_7$ and a unique element f such that $S_8 \backslash f \cong F_7^*$.

6.4 Constructing representations for matroids

In this section, we describe explicitly how to construct representations for matroids, and we illustrate this technique on a number of examples. In general, finding whether a matroid is representable over some field will require a lot of computation. The question of the complexity of this problem will not be addressed here, although in Section 9.4, we shall prove a result of Seymour (1981a) that there is no polynomial-time algorithm to decide whether or not a given matroid is binary. We shall also note there that Seymour's result extends to $GF(q)$-representable matroids for all prime powers q.

Recall from Chapter 2 that if M is a rank-r matroid with $r \geq 1$, then every matrix that represents M over a field $\mathbb{F}$ is projectively equivalent to a standard representative matrix $[I_r|D]$ for M. In view of this fact, we shall now concentrate on matrices of the latter type. Given such a matrix, let its columns be labelled, in order, $e_1, e_2, \ldots, e_n$. If B is the basis $\{e_1, e_2, \ldots, e_r\}$ of M, then it is natural to label the rows of $[I_r|D]$ by $e_1, e_2, \ldots, e_r$. Hence D has its rows labelled by $e_1, e_2, \ldots, e_r$ and its columns labelled by $e_{r+1}, e_{r+2}, \ldots, e_n$. Evidently, for all k in $\{r+1, r+2, \ldots, n\}$, the fundamental circuit $C(e_k, B)$ is $e_k \cup \{e_i : e_i \in B$ and D has a non-zero entry in row e_i and column $e_k\}$. Thus, if we let $D^\#$ be the matrix obtained from D by replacing each non-zero entry of D by a 1, then the columns of $D^\#$ are precisely the incidence vectors of the sets $C(e_k, B) - e_k$ where the rows and columns of $D^\#$ inherit their labels from D. This matrix $D^\#$ is called the *B-fundamental-circuit incidence matrix* of M (Brylawski and Lucas 1976). In addition, $[I_r|D^\#]$ is sometimes called a *partial representation* for M (Truemper 1984). This name is justified by the following result, which is easily proved using fundamental circuits.

Proposition 6.4.1 *Let $[I_r|D_1]$ and $[I_r|D_2]$ be matrices over the fields $\mathbb{K}_1$ and $\mathbb{K}_2$ with the columns of each labelled, in order, by $e_1, e_2, \ldots, e_n$. If the identity map on $\{e_1, e_2, \ldots, e_n\}$ is an isomorphism from $M[I_r|D_1]$ to $M[I_r|D_2]$, then $D_1^\# = D_2^\#$.*   □

Note that an immediate consequence of this is that a binary matroid is uniquely $GF(2)$-representable. In fact, as we shall see in Proposition 6.6.5, if $\mathbb{F}$ is an arbitrary field and M is an $\mathbb{F}$-representable matroid that is also binary, then M is uniquely $\mathbb{F}$-representable.

Partial representations share many of the properties of representations. For example, using Proposition 2.1.11, it is straightforward to prove the following analogue of Theorem 2.2.8. Some other properties of partial representations will be looked at in the exercises.

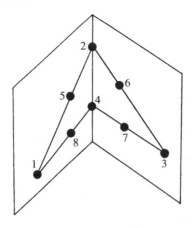

FIG. 6.10. $\mathcal{W}^4$.

$$X = \begin{array}{c} \\ 1 \\ 2 \\ 3 \\ 4 \end{array} \begin{array}{cccc} 5 & 6 & 7 & 8 \\ \left[\begin{array}{cccc} 1 & 0 & 0 & 1 \\ 1 & 1 & 0 & 0 \\ 0 & 1 & 1 & 0 \\ 0 & 0 & 1 & 1 \end{array}\right] \end{array} \qquad Y = \begin{array}{c} \\ 1 \\ 2 \\ 3 \\ 4 \end{array} \begin{array}{cccc} 5 & 6 & 7 & 8 \\ \left[\begin{array}{cccc} 1 & 1 & 0 & 0 \\ 0 & 1 & 1 & 0 \\ 0 & 0 & 1 & 1 \\ 1 & 0 & 0 & 1 \end{array}\right] \end{array}$$

FIG. 6.11. Fundamental-circuit incidence matrices for $\mathcal{W}^4$ and its dual.

Proposition 6.4.2 *Let M be a matroid on a set E and B be a basis of M. If X is the B-fundamental-circuit incidence matrix of M, then X^T is the $(E - B)$-fundamental-circuit incidence matrix of M^*.* □

Example 6.4.3 Let M be the matroid $\mathcal{W}^4$ for which a geometric representation is shown in Figure 6.10 and let $B = \{1, 2, 3, 4\}$. Then B is a basis for $\mathcal{W}^4$. The B-fundamental-circuit incidence matrix of M and the $(E - B)$-fundamental-circuit incidence matrix of M^* are the matrices X and Y in Figure 6.11. Evidently $Y = X^T$. We remark that if we modify $\mathcal{W}^4$ by requiring that $5, 6, 7$, and 8 be coplanar, then the matroid we obtain is graphic. We leave it to the reader to identify exactly which graphic matroid it is. □

The Fano and non-Fano matroids, F_7 and F_7^-, are of fundamental importance in matroid theory. Figure 6.12(a) gives geometric representations for both F_7 and F_7^-. In the former, $\{4, 5, 6\}$ is a line; in the latter, it is not. For each of these matroids, we aim to determine exactly the fields over which it is representable. As a first step towards this, take M in $\{F_7, F_7^-\}$ and let $B = \{1, 2, 3\}$. Then B is a basis of M, and the B-fundamental-circuit incidence matrix X of M is as in Figure 6.12(b). In Example 2.2.10, it was noted, though not proved, that $[I_3|X]$ can represent F_7 or F_7^- depending on the field over which it is viewed. This follows immediately from the next result, the proof of which *will* be given.

Lemma 6.4.4 *Let $A = [I_3|X]$ where X is the matrix in Figure 6.12(b). View A over a field $\mathbb{F}$. If the characteristic of $\mathbb{F}$ is two, then $M[A] = F_7$, while if the characteristic of $\mathbb{F}$ is not two, then $M[A] = F_7^-$.*

Proof Let M be F_7 if $\mathbb{F}$ has characteristic two, and F_7^- otherwise. To prove that $M = M[A]$, we first perform the routine check that each 3-point line of M is a circuit of $M[A]$. It then remains to show that $M[A]$ has no other circuits of size less than four. To achieve this, we note that, as each of the unbroken 3-point lines in Figure 6.12(a) is certainly a circuit of $M[A]$, and $M[A]$ has rank three, there are very few other possibilities for circuits of $M[A]$ of size less than four. For instance, if $\{4, 6, 7\}$ is a circuit, then $1, 4, 6$, and 7 are all collinear. But 3 is collinear with 6 and 7, and 2 is collinear with 1 and 6, so $1, 4, 6, 7, 3$, and 2 are collinear. Hence, as 5 is collinear with 1 and 3, all seven elements are collinear; that is, $M[A]$ has collapsed to rank two; a contradiction. Using this idea, it is not difficult to show that if $\mathbb{F}$ has characteristic two, then $M[A]$ has no other circuits of size less than four and so $M[A] = F_7$. Similarly, if $\mathbb{F}$ has characteristic

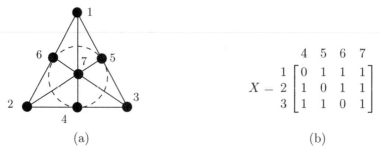

$$X - \begin{matrix} 1 \\ 2 \\ 3 \end{matrix} \begin{matrix} 4 & 5 & 6 & 7 \\ \begin{bmatrix} 0 & 1 & 1 & 1 \\ 1 & 0 & 1 & 1 \\ 1 & 1 & 0 & 1 \end{bmatrix} \end{matrix}$$

(a)   (b)

FIG. 6.12. Illustrating the proof of Lemma 6.4.4.

other than two, then the only other possible circuit of $M[A]$ of size less than four is $\{4, 5, 6\}$. But the 3×3 submatrix of X whose columns are $4, 5$, and 6 has determinant 2. Thus $\{4, 5, 6\}$ is independent in $M[A]$, and $M[A] = F_7^-$. □

Later in the section, we shall complete the determination, for each of F_7 and F_7^-, of all the fields over which the matroid is representable. In the last example, we used determinants to decide about the independence of a set of vectors. The next result uses this elementary idea again to get an extension of Proposition 6.4.1. The proof follows Brylawski and Lucas (1976).

Proposition 6.4.5 *Let* $[I_r|D_1]$ *and* $[I_r|D_2]$ *be* $r \times n$ *matrices over the fields* $\mathbb{K}_1$ *and* $\mathbb{K}_2$ *with the columns of each being labelled, in order, by* $e_1, e_2, \ldots, e_n$. *Then* $M[I_r|D_1]$ *and* $M[I_r|D_2]$ *are isomorphic under the identity map on* $\{e_1, e_2, \ldots, e_n\}$ *if and only if, whenever* D_1' *and* D_2' *are corresponding square submatrices of* D_1 *and* D_2, *respectively,* $\det D_1' = 0$ *exactly when* $\det D_2' = 0$.

Proof We shall prove the result in one direction. The proof for the converse can be obtained by reversing this argument. Assume that the identity map on $\{e_1, e_2, \ldots, e_n\}$ is an isomorphism from $M[I_r|D_1]$ to $M[I_r|D_2]$. Let D_1' and D_2' be the square submatrices of D_1 and D_2 whose rows and columns are labelled by $\{e_{i_1}, e_{i_2}, \ldots, e_{i_k}\}$ and $\{e_{j_1}, e_{j_2}, \ldots, e_{j_k}\}$, respectively. Let $\{e_1, e_2, \ldots, e_r\} = B$. Then the following statements are equivalent:

(i) $\det D_1' \neq 0$;
(ii) $(B - \{e_{i_1}, e_{i_2}, \ldots, e_{i_k}\}) \cup \{e_{j_1}, e_{j_2}, \ldots, e_{j_k}\}$ is a basis of $M[I_r|D_1]$;
(iii) $(B - \{e_{i_1}, e_{i_2}, \ldots, e_{i_k}\}) \cup \{e_{j_1}, e_{j_2}, \ldots, e_{j_k}\}$ is a basis of $M[I_r|D_2]$;
(iv) $\det D_2' \neq 0$.   □

Next we describe how to construct representations. Let M be a rank-r matroid and $\{e_1, e_2, \ldots, e_r\}$ be a basis B for M. Let X be the B-fundamental-circuit incidence matrix of M, the columns of this matrix being labelled by $e_{r+1}, e_{r+2}, \ldots, e_n$. From Proposition 6.4.1, if $[I_r|D]$ is an $\mathbb{F}$-representation of M with columns labelled, in order, $e_1, e_2, \ldots, e_n$, then $[I_r|D^\#] = [I_r|X]$. Thus the task of finding an $\mathbb{F}$-representation for M can be viewed as being one of finding the specific elements of $\mathbb{F}$ that correspond to the non-zero elements of $D^\#$. By column scaling, the first non-zero entry in each column of D can be taken to be one. Moreover, by row scaling, the first non-zero entry in each row can also be taken to be one. The next theorem (Brylawski and Lucas 1976) gives a best-possible lower bound on the number of entries of D whose values can be predetermined by row and column scaling. It also provides an easy way to see whether or not two representations are projectively equivalent. Before stating this theorem, we shall require some preliminaries.

Consider the matrix $D^{\#}$. Its rows are indexed by $e_1, e_2, \ldots, e_r$ and its columns by $e_{r+1}, e_{r+2}, \ldots, e_n$. Let $G(D^{\#})$ denote the *associated* simple *bipartite graph*, that is, $G(D^{\#})$ has vertex classes $\{e_1, e_2, ..., e_r\}$ and $\{e_{r+1}, e_{r+2}, \ldots, e_n\}$, and two vertices e_i and e_j are adjacent if and only if the entry in row e_i and column e_j of $D^{\#}$ is 1. Equivalently, $G(D^{\#})$ is the fundamental graph $G_B(M)$ associated with the basis $B = \{e_1, e_2, \ldots, e_r\}$ of M and considered in Proposition 4.3.2.

Example 6.4.6 If $D^{\#}$ is the matrix in Figure 6.13(a), then $G(D^{\#})$ is as shown in Figure 6.13(b) where, for example, e_2 and e_5 are adjacent because the entry in the row indexed by e_2 and the column indexed by e_5 is 1. Similarly, e_3 and e_8 are non-adjacent because the (e_3, e_8)-entry is 0. □

Theorem 6.4.7 *For a field $\mathbb{F}$, let the $r \times n$ matrix $[I_r|D_1]$ be an $\mathbb{F}$-representation of a matroid M. Let $\{b_1, b_2, \ldots, b_k\}$ be a basis of the cycle matroid of $G(D_1^{\#})$. Then $k = n - \omega(G(D_1^{\#}))$. Moreover, if $(\theta_1, \theta_2, \ldots, \theta_k)$ is an ordered k-tuple of non-zero elements of $\mathbb{F}$, then M has a unique $\mathbb{F}$-representation $[I_r|D_2]$ that is projectively equivalent to $[I_r|D_1]$ such that, for each i in $\{1, 2, \ldots, k\}$, the entry of D_2 corresponding to b_i is θ_i. Indeed, $[I_r|D_2]$ can be obtained from $[I_r|D_1]$ by a sequence of row and column scalings.*

$$
\begin{array}{c}
 & \begin{array}{ccccc} e_4 & e_5 & e_6 & e_7 & e_8 \end{array} \\
\begin{array}{c} e_1 \\ e_2 \\ e_3 \end{array}
& \left[\begin{array}{ccccc}
1 & 0 & 0 & 1 & 1 \\
0 & 1 & 1 & 1 & 0 \\
1 & 1 & 0 & 1 & 0
\end{array} \right]
\end{array}
$$

(a)   (b)

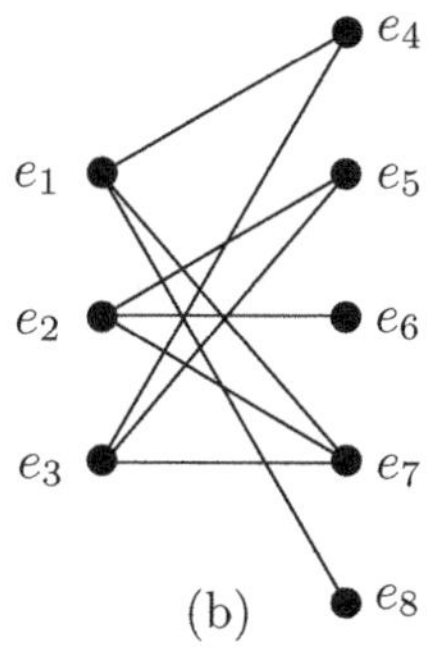

FIG. 6.13. (a) A matrix and (b) its associated bipartite graph.

Before proving this, consider Example 6.4.6 again. The boxed entries in $D^{\#}$ below correspond to a basis of the cycle matroid of $G(D^{\#})$.

$$
\begin{array}{c}
 & \begin{array}{ccccc} e_4 & e_5 & e_6 & e_7 & e_8 \end{array} \\
\begin{array}{c} e_1 \\ e_2 \\ e_3 \end{array}
& \left[\begin{array}{ccccc}
\boxed{1} & 0 & 0 & \boxed{1} & \boxed{1} \\
0 & \boxed{1} & \boxed{1} & \boxed{1} & 0 \\
\boxed{1} & 1 & 0 & 1 & 0
\end{array} \right]
\end{array}
$$

Hence if $[I_3|D]$ is an $\mathbb{F}$-representation of a matroid M, then, by Theorem 6.4.7, we can independently assign arbitrary non-zero elements of $\mathbb{F}$ to each of these boxed entries. In particular, we can assume that, in D, each of these entries is 1. In this case, then, there are only two non-zero entries of D whose values we are not free to arbitrarily assign.

Proof of Theorem 6.4.7 The fact that $k = n - \omega(G(D_1^\#))$ is immediate from 1.3.7. Next, suppose that G_1 is a forest that is a subgraph of $G(D_1^\#)$ and, for each edge x of G_1, let $\theta(x)$ be a non-zero element of $\mathbb{F}$. We shall show, by induction on $|E(G_1)|$, that, by a sequence of row and column scalings of $[I_r|D_1]$, we can obtain a matrix $[I_r|D_2]$ such that, for all edges x of G_1, the entry in D_2 corresponding to x is $\theta(x)$. This is trivially true for $|E(G_1)| = 0$. Assume it true for $|E(G_1)| < m$ and let $|E(G_1)| = m \geq 1$. As G_1 is a forest with at least one edge, it has a vertex v of degree one. Let y be the edge of G_1 incident with v. On applying the induction assumption to $G_1 \backslash y$, it follows that, by appropriate row and column scaling of $[I_r|D_1]$, we obtain a matrix $[I_r|D_2]$ such that, for all edges x of $G_1 \backslash y$, the entry of D_2 corresponding to x is $\theta(x)$.

Now the vertex v of $G(D_1^\#)$ corresponds to a row or a column of D_2. In the former case, as y is the only edge of G_1 meeting v, none of the entries of D_2 corresponding to edges of $G_1 \backslash y$ is in this row. We can therefore multiply this row by an appropriate non-zero scalar t to make the y-entry equal to $\theta(y)$ without changing any of the $(G_1 \backslash y)$-entries. This multiplication may alter the entry in row v of I_r. But multiplying the corresponding column of I_r by t^{-1} will fix this without affecting any other entries. If the vertex v corresponds to a column of D_2, then, by an appropriate scaling of this column, we can make the y-entry equal to $\theta(y)$ without altering any of the $(G_1 \backslash y)$-entries.

We conclude by induction that, when $\{b_1, b_2, \ldots, b_k\}$ is a basis of the cycle matroid of $G(D_1^\#)$, there is a matrix $[I_r|D_2]$ that is projectively equivalent to $[I_r|D_1]$ so that, for each edge b_i, the corresponding entry of D_2 is θ_i. To complete the proof of the theorem, it remains to show that $[I_r|D_2]$ is unique. Assume that $[I_r|D_3]$ is also projectively equivalent to $[I_r|D_1]$ and that, for each edge b_i, the corresponding entry of D_3 is θ_i. Then $[I_r|D_2]$ and $[I_r|D_3]$ are projectively equivalent so, by Theorem 6.3.12, $[I_r|D_3] = X[I_r|D_2]Y$ for some non-singular $r \times r$ matrix X and some non-singular diagonal matrix Y. Then $Y = \begin{bmatrix} Y_1 & 0 \\ 0 & Y_2 \end{bmatrix}$ where Y_1 is $r \times r$ and Y_2 is $(n-r) \times (n-r)$. Thus

$$[I_r|D_3] = X[I_r|D_2]\begin{bmatrix} Y_1 & 0 \\ 0 & Y_2 \end{bmatrix}$$

$$= [X|XD_2]\begin{bmatrix} Y_1 & 0 \\ 0 & Y_2 \end{bmatrix}$$

$$= [XY_1|XD_2Y_2].$$

Hence $XY_1 = I_r$ and so X is a non-singular diagonal matrix. Therefore D_3 is obtained from D_2 by scaling rows and scaling columns.

Let G' be an arbitrary component of the graph $G(D_1^\#)$. To complete the proof that $D_3 = D_2$, it suffices to show that, for each edge of G', the corresponding entries of D_2 and D_3 are equal. Since $\{b_1, b_2, \ldots, b_k\}$ is a basis of the cycle matroid of $G(D_1^\#)$, the component G' of this graph has a spanning tree T' with edge set contained in $\{b_1, b_2, \ldots, b_k\}$ and, for each edge of T', the corresponding entries of D_2 and D_3 coincide. Now let e be an edge of $E(G') - T'$ and let C_e be the unique cycle of G' contained in $T' \cup e$. Let $r_1, c_1, r_2, c_2, \ldots, r_m, c_m, r_1$ be the cyclic order of the vertices of C_e where $e = c_m r_1$ and $\{r_1, r_2, \ldots, r_m\}$ and $\{c_1, c_2, \ldots, c_m\}$ correspond to subsets of the rows and columns of $D_1^\#$. The scaling of rows and columns that produces D_3 from D_2 does not alter the values of the entries corresponding to each of the edges $r_1 c_1, c_1 r_2, r_2 c_2, \ldots, r_m c_m$. Assume that row r_1 is scaled by a. Then column c_1 is scaled by a^{-1}, so row r_2 is scaled by a, and hence column c_2 is scaled by a^{-1}. Repeating this, we see that each of rows $r_1, r_2, \ldots, r_m$ is scaled by a, while each of columns $c_1, c_2, \ldots, c_m$ is scaled by a^{-1}. Hence the entry corresponding to the edge $c_m r_1$, that is to e, is multiplied by both a^{-1} and a and so is unchanged. Therefore $D_2 = D_3$, as required. $\square$

The last result enables us to complete our determination of the fields over which F_7 is representable and the fields over which F_7^- is representable.

Proposition 6.4.8 *Let $\mathbb{F}$ be a field. Then*

(i) *F_7 is $\mathbb{F}$-representable if and only if the characteristic of $\mathbb{F}$ is two; and*

(ii) *F_7^- is $\mathbb{F}$-representable if and only if the characteristic of $\mathbb{F}$ is not two.*

By Lemma 6.4.4, to prove this result it suffices to establish the following.

Lemma 6.4.9 *If $M \in \{F_7, F_7^-\}$ and M is $\mathbb{F}$-representable, then $M = M[I_3|X]$ where X is the matrix in Figure 6.12(b).*

Proof Clearly X is the B-fundamental-circuit incidence matrix for M where B is the basis $\{1, 2, 3\}$ of M. Thus if $[I_3|D]$ is an $\mathbb{F}$-representation for M, then

$$D^\# = X = \begin{array}{c} \\ 1 \\ 2 \\ 3 \end{array} \begin{array}{cccc} 4 & 5 & 6 & 7 \\ \left[\begin{array}{cccc} 0 & \boxed{1} & \boxed{1} & \boxed{1} \\ \boxed{1} & 0 & 1 & \boxed{1} \\ 1 & 1 & 0 & \boxed{1} \end{array} \right] \end{array}$$

where the boxed entries correspond to a basis for $M(G(D^\#))$. Hence there are non-zero elements $a, b,$ and c of $\mathbb{F}$ such that $M = M[I_3|D_1]$ where D_1 is as shown in Figure 6.14(a). As $\{1, 4, 7\}$ is a circuit of M, the determinant of the submatrix of $[I_3|D_1]$ in Figure 6.14(b) is 0. Thus $a = 1$. Similarly, as $\{2, 5, 7\}$ and $\{3, 6, 7\}$ are circuits of M, both b and c equal one. We conclude that, as required, $D_1 = X$. $\square$

We showed in Proposition 2.2.26 that the Vámos matroid V_8 is not representable. In fact, it is a consequence of the following result of Fournier (1971) that V_8 is a smallest non-representable matroid.

$$
\begin{array}{c}
\begin{array}{cccc} 4 & 5 & 6 & 7 \end{array} \\
\begin{array}{c} 1 \\ 2 \\ 3 \end{array}
\begin{bmatrix}
0 & 1 & 1 & 1 \\
1 & 0 & c & 1 \\
a & b & 0 & 1
\end{bmatrix}
\end{array}
\qquad\qquad
\begin{array}{c}
\begin{array}{ccc} 5 & 6 & 7 \end{array} \\
\begin{bmatrix}
1 & 0 & 1 \\
0 & 1 & 1 \\
0 & a & 1
\end{bmatrix}
\end{array}
$$

(a) D_1 (b)

FIG. 6.14. Two matrices from the proof of Lemma 6.4.9.

Proposition 6.4.10 *Let M be a matroid having fewer than eight elements. Then M is representable.* □

The Vámos matroid is not the only non-representable 8-element matroid, though as Fournier (1971) also showed, all such matroids have rank four.

Example 6.4.11 (Blackburn, Crapo, and Higgs 1973) The following is a reduced standard matrix representation for $AG(3,2)$:

$$
\begin{array}{c}
\begin{array}{cccc} 5 & 6 & 7 & 8 \end{array} \\
\begin{array}{c} 1 \\ 2 \\ 3 \\ 4 \end{array}
\begin{bmatrix}
0 & 1 & 1 & 1 \\
1 & 0 & 1 & 1 \\
1 & 1 & 0 & 1 \\
1 & 1 & 1 & 0
\end{bmatrix}
\end{array}
$$

Evidently $\{1,2,3,8\}$ is a circuit of this matroid. From the geometric representation of $AG(3,2)$ in Figure 1.14, we know that every circuit of $AG(3,2)$ is also a hyperplane. Moreover, it is straightforward to check that, up to isomorphism, $AG(3,2)$ has a unique relaxation. Let $AG(3,2)'$ be the matroid obtained from $AG(3,2)$ by relaxing $\{1,2,3,8\}$. By Proposition 3.3.5(ii), $AG(3,2)'\backslash 8 = AG(3,2)\backslash 8$. Clearly the last matroid is isomorphic to F_7^*. Hence if $AG(3,2)'$ is $\mathbb{F}$-representable, then $\mathbb{F}$ has characteristic two. But, by Proposition 3.3.5(ii), $AG(3,2)'/1$ can be obtained from $AG(3,2)/1$ by relaxing the circuit–hyperplane $\{2,3,8\}$ of the latter. Moreover, by Proposition 6.2.5, $AG(3,2)/1 \cong F_7$. Thus $AG(3,2)'/1 \cong F_7^-$. Hence $AG(3,2)'$ is not representable over any characteristic-two fields. We conclude that $AG(3,2)'$ is not representable. Since $AG(3,2)'$ has thirteen 4-circuits, but V_8 has only five such circuits, $AG(3,2)'$ is a non-representable matroid that is different from V_8. □

In Example 2.2.25, we considered the matroid P_8 that has the following matrix as a standard representation over $GF(3)$:

$$
\begin{array}{c}
\begin{array}{cccc} 5 & 6 & 7 & 8 \end{array} \\
\begin{array}{c} 1 \\ 2 \\ 3 \\ 4 \end{array}
\begin{bmatrix}
0 & 1 & 1 & -1 \\
1 & 0 & 1 & 1 \\
1 & 1 & 0 & 1 \\
-1 & 1 & 1 & 0
\end{bmatrix}.
\end{array}
$$

It is not difficult to check that P_8 has no circuits of size less than four and that its 4-circuits are $\{1,2,3,8\}, \{1,2,4,7\}, \{1,3,4,6\}, \{2,3,4,5\}, \{1,4,5,8\}, \{2,3,6,7\}$,

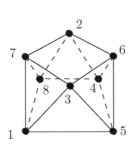
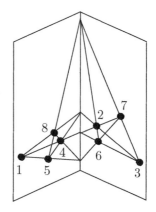

FIG. 6.15. Two geometric representations for P_8.

$\{1,5,6,7\}$, $\{2,5,6,8\}$, $\{3,5,7,8\}$, and $\{4,6,7,8\}$. Geelen, Gerards, and Kapoor (2000) noted that P_8 has an attractive geometric representation which is derived as follows. Begin with an ordinary 3-dimensional cube. Take its top face and rotate that face $45°$ in its plane about the centre of the face (see the left-hand side of Figure 6.15). After this is done, each diagonal of the top face becomes parallel to and hence coplanar with two of the edges of the bottom face. By symmetry, each diagonal of the bottom face is parallel to two of the edges of the top face. This accounts for eight of the ten 4-point planes in P_8. The remaining two are the top and bottom faces of the cube, which are the only 4-point planes from the cube that are retained in this construction. It is easy to see that these two faces, $\{1,4,5,8\}$ and $\{2,3,6,7\}$, are the only pair of disjoint circuit–hyperplanes in P_8. Moreover, from the geometric construction for P_8, we deduce that the two matroids obtained by relaxing one of these circuit-hyperplanes are isomorphic. Let P_8^- be the matroid that is obtained from P_8 by relaxing $\{2,3,6,7\}$, and let $P_8^=$ be obtained from P_8 by relaxing both $\{1,4,5,8\}$ and $\{2,3,6,7\}$. The matroids P_8 and $P_8^=$ have the following important representability properties (Oxley 1986a, Geelen, Gerards, and Kapoor 2000).

Proposition 6.4.12 *Let k be an integer exceeding one. Then P_8 is a minor-minimal matroid that is not representable over $GF(2^k)$. Moreover, $P_8^=$ is a minor-minimal matroid that is not representable over $GF(4)$.*

To prove this, we first establish several other attractive properties of P_8, P_8^-, and $P_8^=$. For a matroid M, an *automorphism* is a permutation σ of $E(M)$ such that $r(X) = r(\sigma(X))$ for all $X \subseteq E(M)$. The set of automorphisms of M forms a group under composition. This automorphism group is *transitive* if, for every two elements x and y of M, there is an automorphism that maps x to y.

Lemma 6.4.13 *Each of the matroids P_8, P_8^-, and $P_8^=$ is self-dual. Moreover, both P_8 and $P_8^=$ have transitive automorphism groups.*

Proof From Theorem 2.2.8, the permutation $(1,5)(2,6)(3,7)(4,8)$ of $E(P_8)$ is an isomorphism between P_8 and its dual. Under this permutation, the complementary sets $\{1,4,5,8\}$ and $\{2,3,6,7\}$ are both fixed. It follows that the same permutation shows that both P_8^- and $P_8^=$ are also isomorphic to their duals.

From the lists of 4-circuits of P_8 and $P_8^=$, it is easy to check that the permutations $(1,8,4,5)(2,7,3,6)$ and $(1,2,4,3)(5,6,8,7)$ are automorphisms of both matroids. By using products of these two automorphisms, it is straightforward to show that both P_8 and $P_8^=$ have transitive automorphism groups.   □

The proof of Proposition 6.4.12 will use the next two lemmas. One can prove this proposition by beginning with the original matrix representation for P_8, but the following argument of Geelen, Gerards, and Kapoor (2000) is cleaner.

Lemma 6.4.14 *Suppose that $M \in \{P_8, P_8^-, P_8^=\}$. If M is representable over a field $\mathbb{F}$, then there are elements b and c of $\mathbb{F} - \{0,1\}$ such that M is represented by the matrix $[I_4|D]$ where D is*

$$
\begin{array}{c}
\;\;8\;\;\;5\;\;\;\;6\;\;\;\;2 \\
\begin{array}{c}1\\4\\7\\3\end{array}
\left[
\begin{array}{cccc}
1 & 1 & 1 & 1 \\
1 & 1 & c^{-1} & b \\
1 & b & 0 & b \\
1 & c & 1 & 0
\end{array}
\right].
\end{array}
$$

Proof Let $[I_r|D]$ be an $\mathbb{F}$-representation for M. By using Theorem 6.4.7 and the list of 4-circuits of M, we can assume that D is as shown in Figure 6.16(a) where a, b, c, d, e, f, and g are non-zero elements of $\mathbb{F}$. As $\{3,5,7,8\}$ is a circuit of

$$
\begin{array}{c}
\;\;8\;\;\;5\;\;\;6\;\;\;2 \\
\begin{array}{c}1\\4\\7\\3\end{array}
\left[
\begin{array}{cccc}
1 & 1 & 1 & 1 \\
1 & a & d & f \\
1 & b & 0 & g \\
1 & c & e & 0
\end{array}
\right]
\end{array}
\qquad
\begin{array}{c}
\;\;7\;\;\;3\;\;\;8\;\;\;5 \\
\left[
\begin{array}{cccc}
0 & 0 & 1 & 1 \\
0 & 0 & 1 & a \\
1 & 0 & 1 & b \\
0 & 1 & 1 & c
\end{array}
\right]
\end{array}
$$

$$
\qquad\qquad\quad (a) \qquad\qquad\qquad\qquad\qquad (b)
$$

FIG. 6.16. Two matrices from the proof of Lemma 6.4.14.

M, the submatrix of $[I_r|D]$ shown in Figure 6.16(b) has determinant 0. Hence so does $\left[\begin{smallmatrix}1&1\\1&a\end{smallmatrix}\right]$. Thus $a = 1$. Similarly, the circuits $\{4,6,7,8\}, \{2,3,4,5\}, \{1,2,3,8\}$, and $\{1,5,6,7\}$ imply, respectively, that $e = 1$, that $g = b$, that $f = g$, and that $ae = cd$. Also, as neither $\{3,4,5,8\}$ nor $\{4,5,7,8\}$ is a circuit of M, neither b nor c is 1. Combining this with the five equations above, we get the lemma.   □

Lemma 6.4.15 *Let $\mathbb{F}$ be a field.*

(i)   *P_8 is $\mathbb{F}$-representable if and only if the characteristic of $\mathbb{F}$ is not two;*
(ii)   *P_8^- is $\mathbb{F}$-representable if and only if $|\mathbb{F}| \geq 4$; and*
(iii) *$P_8^=$ is $\mathbb{F}$-representable if and only if $|\mathbb{F}| \geq 5$.*

Proof Let $A_{b,c} = [I_4|D]$, where D is as specified in Lemma 6.4.14 and b and c are in $\mathbb{F} - \{0, 1\}$. Let $N_{b,c}$ be the matroid that is represented over $\mathbb{F}$ by $A_{b,c}$. By Lemma 6.4.14, to prove (i)–(iii), we need to determine, for each M in $\{P_8, P_8^-, P_8^=\}$, precisely when b and c can be found in $\mathbb{F}$ so that $N_{b,c} = M$. One easily checks that each 4-circuit of $P_8^=$ and, in particular, $\{2, 5, 6, 8\}$ is a circuit of $N_{b,c}$. It follows easily from this and the fact that neither b nor c is 1 that $N_{b,c}$ has no 3-circuits. Similarly, by considering D^T, we deduce that $N_{b,c}$ has no 3-cocircuits. It follows that $N_{b,c}$ has no 4-circuits meeting $\{2, 5, 6, 8\}$ in exactly three elements. The only 4-circuits of $N_{b,c}$ meeting $\{2, 5, 6, 8\}$ in exactly one element are $\{1, 3, 4, 6\}$ and $\{1, 2, 4, 7\}$. Now consider the possible 4-circuits of $N_{b,c}$ that meet $\{2, 5, 6, 8\}$ in exactly two elements. Apart from the 4-circuits of $P_8^=$, there are two other possibilities, $\{1, 4, 5, 8\}$ and $\{2, 3, 6, 7\}$. Clearly,

6.4.16 $\{1, 4, 5, 8\}$ *is a circuit of $N_{b,c}$ if and only if $b = c$;*

and

6.4.17 $\{2, 3, 6, 7\}$ *is a circuit of $N_{b,c}$ if and only if $b = c^{-1}$.*

As $P_8^=$ has neither $\{1, 4, 5, 8\}$ nor $\{2, 3, 6, 7\}$ as a circuit, $A_{b,c}$ represents $P_8^=$ if and only if $\mathbb{F} - \{0, 1\}$ contains elements b and c such that $b \notin \{c, c^{-1}\}$. The last condition fails if $|\mathbb{F}| \leq 4$ but holds if $|\mathbb{F}| \geq 5$. Hence (iii) holds.

Next consider P_8. It has both $\{1, 4, 5, 8\}$ and $\{2, 3, 6, 7\}$ as circuits. Hence $A_{b,c}$ represents P_8 if and only if $\mathbb{F} - \{0, 1\}$ contains elements b and c such that $c = b = c^{-1}$. The last condition holds if and only if $c = -1$ but $c \neq 1$, that is, if and only if the characteristic of $\mathbb{F}$ is not two. Thus (i) holds.

Finally, consider P_8^-. It is represented by $A_{b,c}$ if and only if $\mathbb{F} - \{0, 1\}$ contains elements b and c such that $b = c \neq c^{-1}$. The last condition holds if and only if $|\mathbb{F}| \geq 4$. Hence (ii) holds. □

Proof of Proposition 6.4.12 By the last lemma, to prove the proposition, we need to show that every proper minor of $P_8^=$ is $GF(4)$-representable, while every proper minor of P_8 is $GF(2^k)$-representable for all $k > 1$. By Lemma 6.4.13, both P_8 and $P_8^=$ are self-dual having transitive automorphism groups. Hence it suffices to consider exactly one single-element deletion of each matroid. Recall that P_8^- and $P_8^=$ are obtained from P_8 and P_8^- by relaxing $\{2, 3, 6, 7\}$ and $\{1, 4, 5, 8\}$, respectively. Thus, by Proposition 3.3.5, $P_8 \backslash 2 = P_8^- \backslash 2$ and $P_8^= \backslash 1 = P_8^- \backslash 1$. As P_8^- is $\mathbb{F}$-representable for all $|\mathbb{F}| \geq 4$, we deduce that the same is true for every proper minor of each of P_8 and $P_8^=$ and the proposition follows. □

Exercises

1. Let X be a matrix all of whose entries are in $\{0, 1\}$. Show that the associated bipartite graph $G(X)$ has a cycle if and only if X has a submatrix in which every row and every column contains at least two ones.

2. Show that P_8 has no minor isomorphic to $M(K_4)$.

3. Show that $AG(3, 2)$ is representable only over fields of characteristic two.

4. Suppose that the columns of the matrix $[I_r|X]$ are labelled, in order, $e_1, e_2, \ldots, e_n$ and let $[I_r|X]$ be a partial representation for the matroid M. Show that:

   (a) If $r + 1 \leq i \leq n$, then the matrix obtained from $[I_r|X]$ by deleting column i is a partial representation for $M \backslash e_i$.

   (b) If $1 \leq i \leq r$, then the matrix obtained from $[I_r|X]$ by deleting row i and column i is a partial representation for M / e_i.

5. Let F_8 be the 8-element rank-4 matroid shown in Figure 6.17. Show that F_8 is a non-representable matroid that is not isomorphic to either V_8 or the matroid $AG(3,2)'$ in Example 6.4.11.

6. By attempting to construct a representation, show directly that the non-Pappus matroid is non-representable.

7. Determine all fields over which the matroid Q_6 in Figure 2.1 is representable.

8. (a) Show that $\mathcal{W}^4$ is $\mathbb{F}$-representable if and only if $|\mathbb{F}| \geq 3$.

    (b) Show that $U_{3,6}$ is $\mathbb{F}$-representable if and only if $|\mathbb{F}| \geq 4$.

9. (Brylawski and Kelly 1980)

   (a) Show that $AG(2,3)$ is $\mathbb{F}$-representable if and only if $\mathbb{F}$ contains a root of the equation $x^2 - x + 1 = 0$.

   (b) Using quadratic residues, deduce from (a) that $AG(2,3)$ is representable over $GF(q)$ if and only if q is a square, q is a power of 3, or $q \equiv 1 \pmod 3$, that is, if and only if $q \not\equiv 2 \pmod 3$.

10. Show that $AG(2,4)$ is $\mathbb{F}$-representable if and only if $\mathbb{F}$ has $GF(4)$ as a subfield.

6.5 Representability over finite fields

This section will be primarily concerned with surveying various results on matroid representability over finite fields. The main results characterize the classes of binary, ternary, and $GF(4)$-representable matroids. Proofs of the first two of these results will be presented here and in Chapter 10 respectively. It is worth noting here that, as will be shown in Corollary 6.8.13, every matroid that is representable over some field is representable over some finite field.

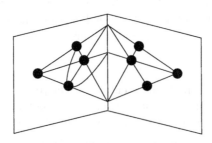

FIG. 6.17. The non-representable matroid F_8.

Probably the most popular approach to finding necessary and sufficient conditions for a matroid to be representable over some fixed field $\mathbb{F}$ has been to find the minimal obstructions to $\mathbb{F}$-representability. Since a matroid M is $\mathbb{F}$-representable if and only if all its minors are $\mathbb{F}$-representable, one way to characterize the class of $\mathbb{F}$-representable matroids is by listing the minor-minimal matroids that are not $\mathbb{F}$-representable. These matroids, which are called the *excluded* or *forbidden minors* for $\mathbb{F}$-representability, are just the non-$\mathbb{F}$-representable matroids for which every proper minor is $\mathbb{F}$-representable. Finding the complete set of excluded minors for representability over a particular field is a notoriously difficult problem and has, in fact, only been solved for the 2-, 3-, and 4-element fields. Nevertheless, we can still make some general observations about excluded minors. For example, since the class of $\mathbb{F}$-representable matroids is closed under duality, we have the following.

Lemma 6.5.1 *If a matroid M is an excluded minor for representability over a field $\mathbb{F}$, then so is its dual M^*.* □

Clearly all excluded minors for $\mathbb{F}$-representability are simple and, therefore, so are their duals. These excluded minors fall loosely into two categories: those that are excluded because the field is too small, and those that are excluded for structural reasons, be they geometric or algebraic. A class of matroids of the first type is the class of rank-two uniform matroids. The proof of the next result is straightforward (see Exercise 5 of Section 1.2).

Proposition 6.5.2 *Let $\mathbb{F}$ be a field and k be an integer exceeding one. Then $U_{2,k}$ is $\mathbb{F}$-representable if and only if $|\mathbb{F}| \geq k - 1$.* □

From this proposition and Lemma 6.5.1, it follows that, for all finite fields $GF(q)$, some of the excluded minors for $GF(q)$-representability are excluded because of the size of the field.

Corollary 6.5.3 *The matroids $U_{2,q+2}$ and $U_{q,q+2}$ are excluded minors for $GF(q)$-representability.* □

In fact, Tutte (1958) established that $U_{2,4}$ is the unique excluded minor for $GF(2)$-representability.

Theorem 6.5.4 *A matroid is binary if and only if it has no $U_{2,4}$-minor.*

Proof By Proposition 6.5.2, $U_{2,4}$ is an excluded minor for the class of binary matroids. Now let M be an arbitrary such excluded minor. Then M has no 1- or 2-element cocircuits. Thus if x and y are distinct elements of M, then $r(M \backslash \{x, y\}) = r(M)$. Suppose $[I_r | D]$ represents $M \backslash \{x, y\}$ over $GF(2)$. As $M \backslash x$ and $M \backslash y$ are binary, there are column vectors $\underline{v}_x$ and $\underline{v}_y$ such that $[I_r | D | \underline{v}_x]$ and $[I_r | D | \underline{v}_y]$ represent $M \backslash y$ and $M \backslash x$, respectively, over $GF(2)$. Now let M' be the matroid that is represented over $GF(2)$ by $[I_r | D | \underline{v}_x | \underline{v}_y]$. Then $M \backslash x = M' \backslash x$ and $M \backslash y = M' \backslash y$. Since $M \neq M'$, there is a set that is independent in one of M and M' and dependent in the other. Take Z to be a minimal such set. Then Z is an independent set in one of M and M', say M_I, and a circuit in the other, M_C. As $M_I \backslash x = M_C \backslash x$ and $M_I \backslash y = M_C \backslash y$, we have that $\{x, y\} \subseteq Z$.

We shall show that

(i) $Z = \{x, y\}$; and

(ii) $r(M_I) = 2$.

As we show below, both these assertions are consequences of the following:

(iii) If J is an independent set of M_I containing Z, then $J = \{x, y\}$.

To prove (iii), suppose that $J - \{x, y\}$ is non-empty. This set is independent in $M_I \backslash \{x, y\}$ and so in $M_C \backslash \{x, y\}$. Hence if we contract $J - \{x, y\}$ from M_I and M_C, we get matroids N_I and N_C of the same rank. Since one of M_I and M_C is binary and the other is an excluded minor for the class of binary matroids, both N_I and N_C are binary. Clearly,

$$N_I \neq N_C$$

since $\{x, y\}$ is independent in N_I and dependent in N_C. But $N_I \backslash x = N_C \backslash x$ and $N_I \backslash y = N_C \backslash y$. Consider $N_I \backslash \{x, y\}$, which equals $N_C \backslash \{x, y\}$. Take a basis B of this matroid. Because N_I and N_C have the same rank, either B is a basis of both these matroids, or it is a basis of neither. In the latter case, $r(N_I \backslash \{x, y\}) < r(N_I)$, so $\{x, y\}$ contains a cocircuit of both N_I and N_C. Thus, by 3.1.16, $\{x, y\}$ contains a cocircuit of both M_I and M_C. This contradicts the fact that M has no 1- or 2-cocircuits.

We now know that B is a basis of both N_I and N_C. Since $N_I \backslash x = N_C \backslash x$ and $N_I \backslash y = N_C \backslash y$, we have $C_{N_I}(e, B) = C_{N_C}(e, B)$ for all e in $E(N_I) - B$. Thus, as N_I and N_C are both binary and their B-fundamental-circuit incidence matrices are equal, $N_I = N_C$. This contradiction implies that (iii) holds.

Clearly (i) follows from (iii) as $\{x, y\} \subseteq Z \subseteq J = \{x, y\}$. To get (ii), note that, as Z is independent in M_I, it is contained in a basis J of M_I. By (iii), $J = \{x, y\}$ so $r(M_I) = 2$.

We now know that all of $M_I, M_C, M_I \backslash \{x, y\}$, and $M_C \backslash \{x, y\}$ have rank 2. Thus M, which is M_I or M_C, has rank 2, has at least four elements, and is simple. Hence $U_{2,4}$ is a deletion of M and we conclude that $M \cong U_{2,4}$. $\square$

A number of other characterizations of binary matroids will be given in Chapter 9. The next two results show that the Fano and non-Fano matroids and their duals are examples of structural excluded minors for representability.

Proposition 6.5.5 *Let $\mathbb{F}$ be a field of characteristic other than two. Then F_7 and F_7^* are excluded minors for $\mathbb{F}$-representability.*

Proof By Proposition 6.4.8, F_7 is not $\mathbb{F}$-representable. In Example 1.5.7, we noted that $F_7 \backslash e \cong M(K_4)$ for all elements e. Hence, by Proposition 5.1.2, every single-element deletion of F_7 is $\mathbb{F}$-representable. On the other hand, by extending the argument in Example 3.3.4 and using Proposition 5.1.2 again, we get that every single-element contraction of F_7 is graphic and so is $\mathbb{F}$-representable. Thus F_7 is an excluded-minor for $\mathbb{F}$-representability. By Lemma 6.5.1, so is F_7^*. $\square$

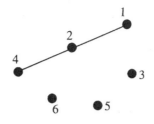

FIG. 6.18. P_6.

The matroid P_8 in the next result was last discussed after Example 6.4.11.

Proposition 6.5.6 *Let $\mathbb{F}$ be a field of characteristic two. If $\mathbb{F} \neq GF(2)$, then P_8, F_7^-, and $(F_7^-)^*$ are excluded minors for $\mathbb{F}$-representability.*

Proof The assertion for P_8 was proved in Proposition 6.4.12. The proofs for F_7^- and $(F_7^-)^*$ are similar to the last proof and are left as exercises. □

The reason why $GF(2)$ is an exception in the last proposition is that each of P_8, F_7^-, and $(F_7^-)^*$ has the non-binary matroid $U_{2,4}$ as a proper minor. Hence none of P_8, F_7^-, and $(F_7^-)^*$ is a *minor-minimal* non-binary matroid.

On combining Corollary 6.5.3 and Proposition 6.5.5, we get that $U_{2,5}, U_{3,5}, F_7$, and F_7^* are excluded minors for $GF(3)$-representability. The fact that they are the only such excluded minors was announced in 1971 by Ralph Reid but was never published by him. The first published proofs of this result were due independently to Bixby (1979) and Seymour (1979). In Chapter 10, we shall present a proof that is essentially due to Gerards (1989).

Theorem 6.5.7 *A matroid is ternary if and only if it has no minor isomorphic to any of the matroids $U_{2,5}, U_{3,5}, F_7$, and F_7^*.* □

Given the excluded-minor characterizations of the classes of binary and ternary matroids stated above, it is natural to consider next the class of *quaternary matroids*, that is, matroids representable over $GF(4)$. By Corollary 6.5.3 and Propositions 6.5.6 and 6.4.12, all of the matroids $U_{2,6}, U_{4,6}, F_7^-, (F_7^-)^*, P_8$, and $P_8^=$ are excluded minors for $GF(4)$-representability. Another such matroid is P_6, a geometric representation of which is shown in Figure 6.18. It is easy to check that every proper minor of the self-dual matroid P_6 is quaternary. However, P_6 itself is not quaternary because the field is too small.

Proposition 6.5.8 *The matroid P_6 is $\mathbb{F}$-representable if and only if $|\mathbb{F}| \geq 5$.*

Proof Suppose that P_6 is $GF(q)$-representable. Then P_6 is a restriction of $PG(2, q)$. Let L be the line of $PG(2, q)$ containing $1, 2$, and 4. Then the three lines of $PG(2, q)$ spanned by $\{3, 5\}, \{3, 6\}$, and $\{5, 6\}$ meet L in distinct points of $PG(2, q)$, and none of these points of intersection is in $\{1, 2, 4\}$. Thus L has at least six points, so $q \geq 5$. The proof of the converse is left to the reader. □

We have now identified seven excluded minors for the class of quaternary matroids. An important paper of Geelen, Gerards, and Kapoor (2000) proved that there are no others.

Theorem 6.5.9 *A matroid is quaternary if and only if it has no minor isomorphic to any of the matroids $U_{2,6}, U_{4,6}, P_6, F_7^-, (F_7^-)^*, P_8$, and $P_8^=$.*   □

For the fields $GF(q)$ with $q \geq 5$, there do not seem to have been any conjectures made as to what a complete list of excluded minors for $GF(q)$-representability might be. When $q = 5$, we know that $U_{2,7}, U_{5,7}, F_7$, and F_7^* are excluded minors. Moreover, it is not difficult to check that each of V_8, the non-Pappus matroid, and the dual of the non-Pappus matroid is an excluded minor for $GF(5)$-representability. We leave it to the reader to check that the matroid in the next example is another such excluded minor.

Example 6.5.10 Let Q_8 be the rank-4 paving matroid whose circuits of size less than five are the faces of the cube in Figure 6.19 together with exactly five of the six diagonal planes such as $\{1, 2, 7, 8\}$. Clearly Q_8 is self-dual.   □

As we can check, the list of excluded minors for $GF(5)$-representability also includes $U_{3,7}$ and $U_{4,7}$ as well as two matroids noted in Exercises 3 and 4. These examples provide some indication of how the complexity of finding the complete set of excluded minors for $GF(q)$-representability grows with q. For large q, finding such a complete list appears impossible with current methods. Instead, the focus of research for $q \geq 5$ has been on resolving the following conjecture of Rota (1971), one of the major unsolved problems in matroid theory.

Conjecture 6.5.11 (Rota's Conjecture) *For all prime powers q, the set of excluded minors for $GF(q)$-representability is finite.*   □

Theorems 6.5.4, 6.5.7, and 6.5.9 prove the special cases of this conjecture when q is $2, 3$, and 4. Interestingly, the first two of these results predated the conjecture with only the last being more recent. Although no other case of the conjecture has yet been resolved, there has been interesting progress made on it. This is reviewed in Chapter 14.

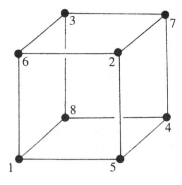

Fɪɢ. 6.19. The 4-circuits of Q_8 are all of the faces and five diagonal planes.

In Chapter 11, we shall introduce matroid extensions of the graph operations of delta-wye and wye-delta exchange which, when applied to excluded minors for $GF(q)$-representability, produce additional excluded minors. By applying sequences of these operations beginning with the matroid $U_{2,q+2}$, Oxley, Semple, and Vertigan (2000) proved the following exponential lower bound on the number of excluded minors for $GF(q)$-representability.

Theorem 6.5.12 *For every prime power q, the class of $GF(q)$-representable matroids has at least 2^{q-4} excluded minors.*   □

Rota's Conjecture asserts that if a field $\mathbb{F}$ is finite, then so is the set of excluded minors for $\mathbb{F}$-representability. This raises the natural question as to whether the converse of this assertion holds. We shall follow Geelen and Whittle (private communication) in using spikes to answer this question. We begin with a general discussion of the representability of spikes noting that we have already investigated the representability of certain specific tipped and tipless spikes including $F_7, F_7^-, P_6, AG(3,2)$, and Q_8.

Lemma 6.5.13 *Let r be an integer exceeding three and $\mathbb{F}$ be a field. Let M be a rank-r spike with legs $\{t, x_1, y_1\}, \{t, x_2, y_2\}, \ldots, \{t, x_r, y_r\}$ and let $N = M\backslash t$. Then M is $\mathbb{F}$-representable if and only if N is $\mathbb{F}$-representable. Indeed, if A is a matrix representing N over $\mathbb{F}$, then A can be extended by adjoining a column labelled by t to give an $\mathbb{F}$-representation for M.*

Proof Certainly if M is $\mathbb{F}$-representable, then N is too. Now let A be a matrix representing N over $\mathbb{F}$. This determines a subset Z of $PG(r-1, \mathbb{F})$ labelled $\{x_1, x_2, \ldots, x_r, y_1, y_2, \ldots, y_r\}$ such that $PG(r-1, \mathbb{F})|Z = N$. As $\{x_1, y_1, x_2, y_2\}$ is a circuit of N, these four points are coplanar. Let t be the point of $PG(r-1, \mathbb{F})$ that lies on the intersection of the lines spanned by $\{x_1, y_1\}$ and $\{x_2, y_2\}$. Let M_t be $PG(r-1, \mathbb{F})|(Z \cup t)$.

In M_t, for each i with $3 \le i \le r$, the union of the rank-3 sets $\{t, x_1, y_1, x_i, y_i\}$ and $\{t, x_2, y_2, x_i, y_i\}$ has rank 4 and their intersection is $\{t, x_i, y_i\}$. By the submodularity of the rank function, this intersection has rank at most 2, so $\{t, x_i, y_i\}$ is a circuit. Therefore, by Proposition 1.5.18, M_t is a spike with legs $\{t, x_1, y_1\}$, $\{t, x_2, y_2\}, \ldots, \{t, x_r, y_r\}$, and $M_t\backslash t = N$. By Proposition 2.1.29, there is a unique such spike. Hence $M_t = M$, so M is $\mathbb{F}$-representable and the lemma holds.   □

We note here that the last lemma does not hold if $r = 3$. For example, if t is an element of the spike F_7, then we may view t as the tip of the spike. Deleting it gives a matroid isomorphic to $M(K_4)$. Thus, by Proposition 5.1.2, $F_7\backslash t$ is representable over every field whereas, by Proposition 6.4.8, F_7 is representable only over fields of characteristic two.

Let $\mathbb{F}$ be a field and $\alpha_1, \alpha_2, \ldots, \alpha_r$ be non-zero elements of $\mathbb{F}$. Let $\mathbf{1}$ be the $r \times 1$ matrix of all ones, and let

$$A_r = \begin{bmatrix} 1+\alpha_1^{-1} & 1 & 1 & \cdots & 1 & 1 \\ 1 & 1+\alpha_2^{-1} & 1 & \cdots & 1 & 1 \\ 1 & 1 & 1+\alpha_3^{-1} & \cdots & 1 & 1 \\ \vdots & \vdots & \vdots & \ddots & \vdots & \vdots \\ 1 & 1 & 1 & \cdots & 1+\alpha_{r-1}^{-1} & 1 \\ 1 & 1 & 1 & \cdots & 1 & 1+\alpha_r^{-1} \end{bmatrix}. \tag{6.4}$$

The following lemma of Geelen, Gerards, and Whittle (2002) is the key result for describing the representability of spikes.

Lemma 6.5.14 *Suppose that $r \geq 3$ and $\mathbb{F}$ is a field. Let M be an $\mathbb{F}$-representable rank-r spike with legs $\{t, x_1, y_1\}, \{t, x_2, y_2\}, \ldots, \{t, x_r, y_r\}$ such that $\{x_1, x_2, \ldots, x_r\}$ is independent. Then every $\mathbb{F}$-representation of M is projectively equivalent to a matrix of the form $[I_r|A_r|1]$ whose columns are labelled, in order, $x_1, x_2, \ldots, x_r, y_1, y_2, \ldots, y_r, t$, where A_r is as in (6.4). Moreover, for $K \subseteq \{1, 2, \ldots, r\}$, the set $\{x_k : k \notin K\} \cup \{y_k : k \in K\}$ is a circuit of M if and only if*

$$\sum_{k \in K} \alpha_k = -1.$$

Finally, if $r \geq 4$ and N is an $\mathbb{F}$-representable rank-r tipless spike with legs $\{x_1, y_1\}, \{x_2, y_2\}, \ldots, \{x_r, y_r\}$ such that $\{x_1, x_2, \ldots, x_r\}$ is independent, then every $\mathbb{F}$-representation of N is projectively equivalent to a matrix of the form $[I_r|A_r]$ whose columns are labelled, in order, $x_1, x_2, \ldots, x_r, y_1, y_2, \ldots, y_r$.

In the last lemma, the assumption that $\{x_1, x_2, \ldots, x_r\}$ is independent does not impose a genuine constraint since, by the definition of a spike (see Proposition 1.5.17), at least one of $\{x_1, x_2, \ldots, x_r\}$ and $\{y_1, x_2, x_3, \ldots, x_r\}$ is independent. Our proof of this lemma uses the following result from Mirsky (1955, Exercise I.13). The matrix A_r' is obtained from A_r by replacing its bottom right-hand entry by 1.

Lemma 6.5.15 *For all positive integers r,*

$$\det A_r = [1 + \sum_{i=1}^{r} \alpha_i] \cdot \prod_{i=1}^{r} \alpha_i^{-1}.$$

Proof We shall use a single induction argument to prove both this formula and the assertion that $\det A_r' = \prod_{i=1}^{r-1} \alpha_i^{-1}$. Noting that $A_1' = [1]$ and that an empty product takes the value 1, we see that both determinants are as specified when $r = 1$. Assume both formulas hold for $r < n$ and let $r = n$. By (2.2.17), the determinants of A_n and A_n' are unchanged by subtracting row $n-1$ from row n. Thus A_n' and A_n have the same determinants as the matrices B_n' and B_n that are obtained by replacing the last rows of A_n' and A_n by $(0, 0, \ldots, 0, -\alpha_{n-1}^{-1}, 0)$

and $(0, 0, \ldots, 0, -\alpha_{n-1}^{-1}, \alpha_n^{-1})$. By expanding $\det B_n'$ and $\det B_n$ along the last rows of these matrices, we get $\det A_n' = \det B_n' = \alpha_{n-1}^{-1} \det A_{n-1}'$ and

$$\det A_n = \det B_n = \alpha_{n-1}^{-1} \det A_{n-1}' + \alpha_n^{-1} \det A_{n-1}.$$

Thus, by the induction assumption, $\det A_n' = \alpha_{n-1}^{-1} \prod_{i=1}^{n-2} \alpha_i^{-1} = \prod_{i=1}^{n-1} \alpha_i^{-1}$ and

$$\det A_n = \alpha_{n-1}^{-1} \prod_{i=1}^{n-2} \alpha_i^{-1} + \alpha_n^{-1} [1 + \sum_{i=1}^{n-1} \alpha_i] \cdot \prod_{i=1}^{n-1} \alpha_i^{-1}$$

$$= \alpha_n \prod_{i=1}^{n} \alpha_i^{-1} + [1 + \sum_{i=1}^{n-1} \alpha_i] \cdot \prod_{i=1}^{n} \alpha_i^{-1}$$

$$= [1 + \sum_{i=1}^{n} \alpha_i] \cdot \prod_{i=1}^{n} \alpha_i^{-1}.$$

We conclude, by induction, that the lemma holds. □

Proof of Lemma 6.5.14 Let $B = \{x_1, x_2, \ldots, x_r\}$. Since M is $\mathbb{F}$-representable having B as a basis, we may assume that an $\mathbb{F}$-representation of M has the form $[I_r | D]$ where the columns are labelled, in order, $x_1, x_2, \ldots, x_r, y_1, y_2, \ldots, y_r, t$ and $D = [d_{ij}]$. As $\{x_1, x_2, \ldots, x_r, t\}$ is a circuit of M, we may assume, by scaling rows, that $d_{i(r+1)} = 1$ for all i in $\{1, 2, \ldots, r\}$. Moreover, for each such i, the set $\{t, x_i, y_i\}$ is a circuit, so $d_{ij} \neq 0$ for all j in $\{1, 2, \ldots, r\} - i$. By scaling columns, we may assume that $d_{21} = 1$ and that $d_{1k} = 1$ for all k in $\{2, 3, \ldots, r\}$. Using the leg $\{t, x_i, y_i\}$ again, we get that $d_{ij} = 1$ for all j in $\{1, 2, \ldots, r\} - i$. Finally, as M is simple, no d_{ii} is 1 so we can take α_i to be $(d_{ii} - 1)^{-1}$ for all i. We conclude that $[I_r | D]$ is indeed projectively equivalent to $[I_r | A_r | \mathbf{1}]$.

For $K \subseteq \{1, 2, \ldots, r\}$, the set $\{x_k : k \notin K\} \cup \{y_k : k \in K\}$ is a circuit of M if and only if the determinant of a matrix of the form of $A_{|K|}$ is zero. It follows from Lemma 6.5.15 that this occurs if and only if $\sum_{k \in K} \alpha_k = -1$.

Finally, for some $r \geq 4$, let N be an $\mathbb{F}$-representable rank-r tipless spike with legs $\{x_1, y_1\}, \{x_2, y_2\}, \ldots, \{x_r, y_r\}$ such that $\{x_1, x_2, \ldots, x_r\}$ is independent. By Proposition 2.1.29, there is a unique rank-r spike M with legs $\{t, x_1, y_1\}, \{t, x_2, y_2\}, \ldots, \{t, x_r, y_r\}$ such that $N = M \backslash t$. Moreover, by Lemma 6.5.13, M is $\mathbb{F}$-representable and every $\mathbb{F}$-representation for M can be obtained from one for N by adjoining a column. Thus, by the first part of the lemma, every $\mathbb{F}$-representation of N is projectively equivalent to a matrix of the form $[I_r | A_r]$. □

The following immediate consequence of Lemma 6.5.14 allows us to specify precisely when a spike if representable over a field. The corresponding result for tipless spikes of rank at least four also holds.

Corollary 6.5.16 *Suppose that $r \geq 3$, that $\mathbb{F}$ is a field, and that M is a rank-r spike with legs $\{t, x_1, y_1\}, \{t, x_2, y_2\}, \ldots, \{t, x_r, y_r\}$ such that $\{x_1, x_2, \ldots, x_r\}$ is independent. Then M is $\mathbb{F}$-representable if and only if there are non-zero* *elements* $\alpha_1, \alpha_2, \ldots, \alpha_r$ *of* $\mathbb{F}$ *such that, for all* $K \subseteq \{1, 2, \ldots, r\}$, *the set* $\{x_k : k \notin K\} \cup \{y_k : k \in K\}$ *is a circuit of* M *if and only if* $\sum_{k \in K} \alpha_k = -1$. $\square$

The next result was proved by Geelen and Whittle (private communication). It verifies the converse of Rota's Conjecture (6.5.11).

Theorem 6.5.17 *Let* $\mathbb{F}$ *be a field. If* $\mathbb{F}$ *is infinite, then the set of excluded minors for* $\mathbb{F}$-*representability is infinite.*

We prove this theorem by showing that, when $\mathbb{F}$ is infinite, the set of excluded minors for $\mathbb{F}$-representability contains infinitely many tipless spikes.

Lemma 6.5.18 *For each* $r \geq 4$ *and each* k *in* $\{2, 3, \ldots, r - 2\}$, *let* $M_r(k)$ *be the tipless rank-r spike with legs* $\{x_1, y_1\}, \{x_2, y_2\}, \ldots, \{x_r, y_r\}$ *whose only non-spanning circuits meeting each* $\{x_i, y_i\}$ *are* $\{y_1, y_2, \ldots, y_r\}$, $\{x_1, x_2, \ldots, x_k, y_{k+1}, y_{k+2}, \ldots, y_r\}$, *and* $\{y_1, y_2, \ldots, y_k, x_{k+1}, x_{k+2}, \ldots, x_r\}$. *Then* $M_r(k)$ *is a non-representable matroid that is an excluded minor for* $\mathbb{F}$-*representability for every infinite field* $\mathbb{F}$.

Proof By assumption, $\{x_1, x_2, \ldots, x_r\}$ is not a circuit, so it is a basis. Let $\mathbb{F}$ be a field and assume that $M_r(k)$ is $\mathbb{F}$-representable. Then, by Lemma 6.5.14, $M_r(k)$ has an $\mathbb{F}$-representation of the form $[I_r | A_r]$ where the columns are labelled, in order, $x_1, x_2, \ldots, x_r, y_1, y_2, \ldots, y_r$, and A_r is as in (6.4). Moreover, the three specified circuits of $M_r(k)$ imply that $\sum_{i=1}^{r} \alpha_i = -1$, that $\sum_{i=k+1}^{r} \alpha_i = -1$, and that $\sum_{i=1}^{k} \alpha_i = -1$. Adding the last two of these equations, we get $\sum_{i=1}^{r} \alpha_i = -2$, which contradicts the first equation. We conclude that $M_r(k)$ is non-representable.

As $M_r(k)$ is a tipless spike, it is isomorphic to its dual by Proposition 2.1.27. Thus, to prove that $M_r(k)$ is an excluded minor for $\mathbb{F}$-representability, it suffices to show that every single-element contraction of $M_r(k)$ is $\mathbb{F}$-representable. Now let $\mathbb{F}$ be infinite. By symmetry, we need only show that $M_r(k)/y_r$ and $M_r(k)/x_r$ are $\mathbb{F}$-representable for all k. The first of these is a rank-$(r-1)$ spike with tip x_r and legs $\{x_r, x_1, y_1\}, \{x_r, x_2, y_2\}, \ldots, \{x_r, x_{r-1}, y_{r-1}\}$ whose only non-spanning circuits meeting each $\{x_i, y_i\}$ with $1 \leq i \leq r - 1$ are $\{y_1, y_2, \ldots, y_{r-1}\}$ and $\{x_1, x_2, \ldots, x_k, y_{k+1}, y_{k+2}, \ldots, y_{r-1}\}$. By Corollary 6.5.16, $M_r(k)/y_r$ is $\mathbb{F}$-representable provided there are non-zero elements $\alpha_1, \alpha_2, \ldots, \alpha_{r-1}$ of $\mathbb{F}$ such that $\sum_{i \in I} \alpha_i = -1$ if and only if I is $\{1, 2, \ldots, r-1\}$ or $\{k+1, k+2, \ldots, r-1\}$. We describe a recursive procedure for choosing these elements. First choose α_2 not in $\{0, 1, -1\}$. Assume $\alpha_2, \alpha_3, \ldots, \alpha_i$ have been chosen for some i with $2 \leq i \leq r-3$. Choose α_{i+1} not in $\{\varepsilon_0 + \sum_{h=2}^{i} \varepsilon_h \alpha_h : \varepsilon_0, \varepsilon_1, \ldots, \varepsilon_i \in \{0, 1, -1\}\}$. Finally, choose $\alpha_1 = -(\alpha_2 + \alpha_3 + \cdots + \alpha_k)$ and $\alpha_{r-1} = -1 - (\alpha_{k+1} + \alpha_{k+2} + \cdots + \alpha_{r-2})$. These choices mean that if I is $\{1, 2, \ldots, r-1\}$ or $\{k+1, k+2, \ldots, r-1\}$, then $\sum_{i \in I} \alpha_i = -1$. It remains to prove that the converse of this holds.

Assume that $\sum_{i \in I} \alpha_i = -1$ for some subset I of $\{1, 2, \ldots, r-1\}$ other than $\{1, 2, \ldots, r-1\}$ or $\{k+1, k+2, \ldots, r-1\}$. Then I is a non-empty proper subset of $\{1, 2, \ldots, r-1\}$. Suppose first that $r - 1 \notin I$. If $1 \notin I$, let j be the largest element of I. Then $\alpha_j = -1 - \sum_{i \in I - \{j\}} \alpha_i$, violating the choice of α_j. If $1 \in I$, then $-1 = \sum_{i \in I - \{1\}} \alpha_i + \alpha_1 = \sum_{i \in I - \{1\}} \alpha_i - \sum_{i=2}^{k} \alpha_i$. This means that the choice of α_j is violated where j is now the largest element of the symmetric difference of the sets $I - \{1\}$ and $\{2, 3, \ldots, k\}$.

We may now assume that $r - 1 \in I$. If $1 \notin I$, then $-1 = \sum_{i \in I - \{r-1\}} \alpha_i - 1 - \sum_{i=k+1}^{r-2} \alpha_i$ and the choice of α_j is violated where j is the largest member of the symmetric difference of $I - \{r - 1\}$ and $\{k + 1, k + 2, \ldots, r - 2\}$. If $1 \in I$, then $-1 = \sum_{i \in I - \{1, r-1\}} \alpha_i - \sum_{i=2}^{k} \alpha_i - 1 - \sum_{i=k+1}^{r-2} \alpha_i$ and the choice of α_j is violated where j is the largest member of $\{2, 3, \ldots, r - 2\}$ not in I. We conclude that $M_r(k)/y_r$ is $\mathbb{F}$-representable. We leave it to the reader to provide the similar argument that establishes that $M_r(k)/x_r$ is $\mathbb{F}$-representable. We conclude that $M_r(k)$ is an excluded minor for $\mathbb{F}$-representability for all infinite fields $\mathbb{F}$. $\qquad \square$

We now turn to another unsolved problem that concerns minors which are excluded for size reasons.

Problem 6.5.19 *Find all the fields over which the uniform matroid $U_{r,n}$ is representable.* $\qquad \square$

If $r \leq 1$, then evidently $U_{r,n}$ is $\mathbb{F}$-representable for every field $\mathbb{F}$. Moreover, Proposition 6.5.2 solves 6.5.19 for $r = 2$. For larger values of r, the problem has received considerable attention from both projective geometers and coding theorists. In projective geometry, an *n-arc* is a set S of points of $PG(r-1, q)$ such that $PG(r - 1, q)|S \cong U_{r,n}$ (see, for example, Hirschfeld 1997), and Problem 6.5.19 is equivalent to a problem enunciated by Segre (1955). In coding theory, the equivalent notion is that of a *maximum distance separable (MDS) code* of length n, dimension r, and minimum distance $n - r + 1$. For all r in $\{2, 3, 4, 5\}$, the matroid $U_{r,n}$ is $GF(q)$-representable if and only if n does not exceed the value of k specified in Table 6.1. This table is obtained by combining Proposition 6.5.2 with results of Bose (1947), Bush (1952), Segre (1955), Casse (1969), and Gulati and Kounias (1970). For a more detailed consideration of n-arcs, MDS codes, and their properties, the reader is referred to Hirschfeld's (1997) survey paper.

TABLE 6.1. $U_{r,n}$ is $GF(q)$-representable if and only if $n \leq k$.

r	k	Restriction on q
3	$q + 1$	q odd
	$q + 2$	q even
4	5	$q \leq 3$
	$q + 1$	$q \geq 4$
5	6	$q \leq 4$
	$q + 1$	$q \geq 5$

For r exceeding five, although partial results are known, there is no complete answer. The solution of Problem 6.5.19 would be completed by settling the following conjecture, a restatement of the main conjecture for MDS codes.

Conjecture 6.5.20 *For all prime powers q, the matroid $U_{r,q+2}$ is an excluded minor for the class of $GF(q)$-representable matroids for all r in $\{2, 3, \ldots, q\}$ when q is odd and for all r in $\{2, 3, \ldots, q\} - \{3, q-1\}$ when q is even.* □

Hirschfeld (1997, Theorem 3.17) summarized progress towards this conjecture. From that result, we get immediately that the conjecture holds for all $q \leq 27$. Note that the exceptions that arise when q is even are genuine as, by the second line of Table 6.1 and duality, $U_{3,q+2}$ and $U_{q-1,q+2}$ are $GF(q)$-representable. Recently, Ball (2010) proved Conjecture 6.5.20 when q is prime. More generally, he established the following.

Theorem 6.5.21 *Let q be a power of a prime p. When $2 \leq r \leq p$, the matroid $U_{r,n}$ is $GF(q)$-representable if and only if $n \leq q+1$.* □

Corollary 6.5.22 *For all primes p, the matroid $U_{r,p+2}$ is an excluded minor for the class of $GF(p)$-representable matroids if and only if $2 \leq r \leq p$.* □

To close this section, we note that there are various known necessary and sufficient conditions for representability. Among these are results due to Vámos (1971a) and White (1980a). In general, these conditions amount to giving an algebraic reformulation of what it means for a matroid to be representable. For a survey of these results and how they have been used, we refer the reader to White (1987b) where a proof can also be found of the following proposition of Tutte (1958), the first result of this kind.

Proposition 6.5.23 *Let M be a matroid and $\mathbb{F}$ be a field. A necessary and sufficient condition for M to be $\mathbb{F}$-representable is that, for all hyperplanes H of M, there is a function $f_H : E(M) \to \mathbb{F}$ such that*

(i) $H = \{x \in E(M) : f_H(x) = 0\}$; *and*

(ii) *if H_1, H_2, and H_3 are distinct hyperplanes of M for which the rank of $H_1 \cap H_2 \cap H_3$ is $r(M) - 2$, then there are non-zero elements c_1, c_2, and c_3 of $\mathbb{F}$ such that*

$$c_1 f_{H_1} + c_2 f_{H_2} + c_3 f_{H_3} = 0.$$ □

Exercises

1. Let $\mathbb{F}$ be a field. Show that every excluded minor for $\mathbb{F}$-representability is connected and simple.
2. Find all fields $\mathbb{F}$ for which
   (a) Q_8 is $\mathbb{F}$-representable;
   (b) Q_8 is an excluded minor for $\mathbb{F}$-representability.
3. (Brylawski 1975d) Prove that:
   (a) Figure 6.20(a) shows an excluded minor for $GF(5)$-representability.

(b) For all $m \geq 5$, the matroid in Figure 6.20(b) is $\mathbb{F}$-representable if and only if $|\mathbb{F}| \geq m$.

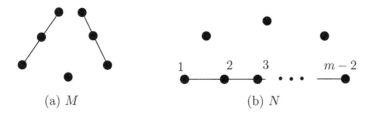

(a) M (b) N

FIG. 6.20. Geometric representations of the matroids in Exercise 3.

4. Let T_8 and R_8 be as in Exercise 9 of Section 2.2. Show that:

   (a) R_8 is $\mathbb{F}$-representable if and only if the characteristic of $\mathbb{F}$ is not two.
   (b) T_8 is $\mathbb{F}$-representable if and only if $\mathbb{F}$ has characteristic three.
   (c) R_8 has a transitive automorphism group.
   (d) T_8 is an excluded minor for $\mathbb{F}$-representability for all fields $\mathbb{F}$ whose characteristic is not two or three.
   (e) Q_8 is obtained from R_8 by relaxing a circuit–hyperplane.

5. (Lazarson 1958) Let p be a prime number exceeding two and L_p be the vector matroid of $[I_{p+1}|J_{p+1} - I_{p+1}]$ over $GF(p)$, where J_n is the $n \times n$ matrix of all ones. Then $L_3 = T_8$. Let the columns of $[I_{p+1}|J_{p+1} - I_{p+1}]$ be $x_1, x_2, \ldots, x_{p+1}, y_1, y_2, \ldots, y_{p+1}$. Show the following:

   (a) L_p is a tipless spike of rank $p + 1$.
   (b) When evaluated over $\mathbb{R}$, the determinant of $[J_{n+1} - I_{n+1}]$ is $(-1)^n n$.
   (c) The non-spanning circuits of L_p that meet $\{x_i, y_i\}$ for all i are $\{y_1, y_2, \ldots, y_{p+1}\}$ and all the sets $\{x_1, x_2, \ldots, x_{k-1}, y_k, x_{k+1}, \ldots, x_{p+1}\}$ for $1 \leq k \leq p + 1$.
   (d) L_p is $\mathbb{F}$-representable if and only if $\mathbb{F}$ has characteristic p.
   (e) $L_p \backslash y_{p+1}$ is represented over $\mathbb{Q}$ and over $GF(p')$ for all primes $p' \geq p$ by the matrix obtained from $[I_{p+1}|J_{p+1} - I_{p+1}]$ by deleting the last column.
   (f) The matrix in Figure 6.21 is a standard matrix representation for L_p / y_{p+1} over $\mathbb{Q}$ and over $GF(p')$ for all $p' \geq p$.
   (g) If $\mathbb{F}$ is a field of characteristic p', then L_p is an excluded minor for $\mathbb{F}$-representability if $p < p'$, or if $p' = 0$.

6. (Geelen, Oxley, Vertigan, and Whittle 2002) For a prime p exceeding 4:

   (a) Prove that the free $(p-1)$-spike with tip is not $GF(p)$-representable.
   (b) Deduce that, for $r \geq p - 1$, the free tipless r-spike is not $GF(p)$-representable.

$$\begin{array}{c} \\ x_2 \\ x_3 \\ x_4 \\ \vdots \\ x_{p-1} \\ x_p \\ x_{p+1} \end{array}\begin{array}{c} \begin{array}{cccccccc} x_1 & y_1 & y_2 & y_3 & y_4 & \cdots & y_{p-1} & y_p \end{array} \\ \left[\begin{array}{cccccccc} 1 & 1 & p-1 & 0 & 0 & \cdots & 0 & 0 \\ 1 & 1 & 0 & p-1 & 0 & \cdots & 0 & 0 \\ 1 & 1 & 0 & 0 & p-1 & \cdots & 0 & 0 \\ \vdots & \vdots & \vdots & \vdots & \vdots & \ddots & \vdots & \vdots \\ 1 & 1 & 0 & 0 & 0 & \cdots & p-1 & 0 \\ 1 & 1 & 0 & 0 & 0 & \cdots & 0 & p-1 \\ 0 & 1 & 1 & 1 & 1 & \cdots & 1 & 1 \end{array}\right] \end{array}.$$

FIG. 6.21. A standard matrix representation for L_p/y_{p+1}.

7. (Geelen, Gerards, and Whittle 2002) For $r \geq 4$, let M_r be the rank-r tipless spike with legs $\{x_1, y_1\}, \{x_2, y_2\}, \dots, \{x_r, y_r\}$ and exactly two circuit-hyperplanes, $\{y_1, x_2, x_3, \dots, x_r\}$ and $\{x_1, y_2, y_3, \dots, y_r\}$. Show that:
   (a) M_r is representable over every infinite field.
   (b) If $5 \leq s < r$, then M_r has no M_s-minor.
8. *(Z. Wu 2003) Prove that, for all $r \geq 3$, there are exactly
   (a) two non-isomorphic rank-r ternary spikes; and
   (b) $\lfloor (r^2 + 6r + 24)/12 \rfloor$ non-isomorphic rank-r quaternary spikes.
9. *(Lindström 1987a) Use Proposition 6.5.23 to prove that a matroid is both binary and ternary if and only if it is $\mathbb{F}$-representable for every field $\mathbb{F}$.

6.6 Regular matroids

The last section focused on matroids that are representable over some fixed finite field. We now consider the class of matroids that are representable over every field. By Proposition 5.1.2 and duality, this class includes the classes of graphic and cographic matroids. In this section, we shall show that it coincides with the class of regular matroids. In addition, we shall state Tutte's important excluded-minor theorems for the classes of regular, graphic, and cographic matroids.

First recall that we defined a matroid to be regular if it has a totally unimodular representation, that is, an $\mathbb{R}$-representation in which every square submatrix has determinant in $\{0, 1, -1\}$. The proof that the class of regular matroids equals the class of matroids representable over every field will use a modification of the pivot operation described in Section 2.2. In a standard representative matrix $[I_r|D]$, suppose that a pivot is performed on a non-zero entry d_{st} of D where $D = [d_{ij}]$. Then the resulting matrix need no longer have the form $[I_r|X]$. However, one obtains a matrix in this form simply by interchanging columns s and $r+t$ of the transformed matrix. Thus, in order to maintain the form of the matrix, whenever a pivot is performed in $[I_r|D]$ on an entry of D, we shall always follow this by the appropriate column interchange; that is, we absorb the natural column interchange into the pivoting operation.

$$\begin{bmatrix} & & & & \Big| & 0 & 1 & 1 & 1 \\ & & I_4 & & \Big| & 1 & 0 & 1 & 1 \\ & & & & \Big| & 1 & 1 & 0 & 1 \\ & & & & \Big| & 1 & 1 & 1 & 0 \end{bmatrix} \qquad \begin{bmatrix} & & & & \Big| & 0 & 1 & 1 & 1 \\ & & I_4 & & \Big| & 1 & -1 & 0 & -1 \\ & & & & \Big| & 1 & 0 & -1 & -1 \\ & & & & \Big| & 1 & 1 & 1 & 0 \end{bmatrix}$$

with column labels:
(a): 1 2 3 4 | 5 6 7 8
(b): 8 2 3 4 | 5 6 7 1

(a)   (b)

FIG. 6.22. Pivoting on the first entry of column 8 of (a) gives (b).

Example 6.6.1 Let A be the matrix over $\mathbb{Q}$ shown in Figure 6.22(a). Pivoting on the entry in the top right corner, we get the matrix shown in Figure 6.22(b). Note that the cost of preserving the matrix in the form $[I_4|X]$ is that the ordering on the column labels has changed. We remark that $M[A]$ can be obtained from $AG(3,2)$ by a sequence of circuit–hyperplane relaxations (see Exercise 1). □

The following lemma plays an important role in the proof of the next theorem, which characterizes regular matroids in terms of representability (Tutte 1965a).

Lemma 6.6.2 *Let $[d_{ij}]$ be a matrix D_1 all of whose entries are in $\{0, 1, -1\}$. Suppose $[I_r|D_1]$ is an $\mathbb{F}$-representation of a binary matroid M where $\mathbb{F}$ has characteristic different from two. Assume that $[I_r|D_2]$ is obtained from $[I_r|D_1]$ by pivoting on a non-zero entry d_{st} of D_1. Then every entry of D_2 is in $\{0, 1, -1\}$. Moreover, $[I_r|D_2]$ is also obtained if $[I_r|D_1]$ is viewed as a $(0, \pm 1)$-matrix over $\mathbb{R}$ and the pivot is done over $\mathbb{R}$.*

Proof It is straightforward to check that all the entries in row s of D_2 and all those in column t are in $\{0, 1, -1\}$. Now suppose $j \neq t$ and $i \neq s$. Then the pivot replaces d_{ij} by $d_{ij} - d_{it}d_{sj}d_{st}^{-1}$, which equals $d_{st}^{-1}(d_{st}d_{ij} - d_{it}d_{sj})$. As all the entries of D_1 are in $\{0, 1, -1\}$ and d_{st} is non-zero, $d_{st}^{-1}(d_{st}d_{ij} - d_{it}d_{sj})$ is in $\{0, 1, -1\}$ unless $d_{st}d_{ij} - d_{it}d_{sj} = \pm 2$. Hence assume that this equation holds. Then $\begin{bmatrix} d_{st} & d_{sj} \\ d_{it} & d_{ij} \end{bmatrix}$, or some row or column permutation thereof, is a submatrix D_1' of D_1 whose determinant is ± 2. But, as every non-zero entry of D_1 is in $\{1, -1\}$, the matrix $[I_r|D_1^\#]$ is a $GF(2)$-representation for the binary matroid M. Hence, when $[I_r|D_1]$ is viewed over $GF(2)$, it represents M. Thus, by Proposition 6.4.5, as $\det D_1'$ is 0 over $GF(2)$, it must also be 0 over $\mathbb{F}$; a contradiction.

Finally, suppose that we view $[I_r|D_1]$ as a $(0, \pm 1)$-matrix over $\mathbb{R}$ and perform the pivot over $\mathbb{R}$. Then, arguing as in the previous paragraph, we get the required result unless some $d_{st}d_{ij} - d_{it}d_{sj}$ is ± 2 when calculated over $\mathbb{R}$. In the exceptional case, the form of D_1 implies that $d_{st}d_{ij} - d_{it}d_{sj}$ is ± 2 when calculated over $\mathbb{F}$. But this possibility was eliminated above. □

Theorem 6.6.3 *The following statements are equivalent for a matroid M:*
(i)   *M is regular.*
(ii) *M is representable over every field.*

(iii) *M is binary and, for some field* $\mathbb{F}$ *of characteristic other than two, M is* $\mathbb{F}$-*representable.*

Before proving this theorem, we make two important observations. Firstly, an immediate consequence of the theorem is that the class of regular matroids is the intersection of the classes of binary and ternary matroids. Secondly, from the first part of the proof, it will follow that if $[I_r|D]$ is a totally unimodular matrix that represents the matroid M over $\mathbb{R}$, then, when $[I_r|D]$ is viewed over an arbitrary field $\mathbb{F}$, it is an $\mathbb{F}$-representation for M.

The following rather technical lemma contains the core of the proof that (iii) implies (i) in Theorem 6.6.3. As we shall see, this lemma also implies that an $\mathbb{F}$-representable binary matroid is uniquely $\mathbb{F}$-representable. Recall from Section 6.3 that two representations are projectively equivalent if one can be obtained from the other by a sequence of the matrix operations 6.3.1–6.3.6. In particular, no field automorphism is used in this process. A *diagonal* of a cycle C in a graph G is an edge of G that joins two distinct vertices of C but is not in C.

Lemma 6.6.4 *For a field* $\mathbb{F}$, *let* $[I_r|D]$ *be an* $\mathbb{F}$-*representation for a binary matroid M. Let* B_D *be a basis for the cycle matroid of* $G(D^\#)$. *If every entry of D corresponding to an edge of* B_D *is 1, then every other non-zero entry of D has a uniquely determined value in* $\{1, -1\}$.

Proof This is due to Brylawski (1975c) (see also White 1987b). Let d be any non-zero entry of D that does not correspond to an edge of B_D, and let e_d be the corresponding edge of $G(D^\#)$. Then there is a unique cycle C_d in $G(D^\#)$ that contains e_d and is contained in $B_D \cup e_d$. Of course, C_d is the fundamental cycle of e_d with respect to B_D. We prove the lemma by induction on $|C_d|$.

Before getting into the details of the proof, we consider an example. Assume $D^\#$ is the matrix shown in Figure 6.23(a) with the boxed entries corresponding to the edges in B_D. Then (b) and (c) of Figure 6.23 show $G(D^\#)$ and the subgraph

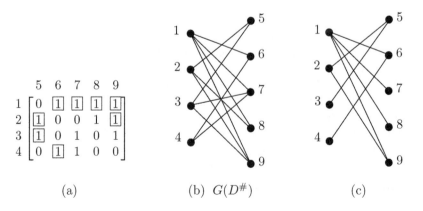

(a)   (b) $G(D^\#)$ (c)

FIG. 6.23. The bipartite graphs associated with (a) and its boxed entries.

of $G(D^{\#})$ induced by B_D. Now, for some non-zero elements d_1, d_2, d_3, and d_4 of $\mathbb{F}$, the matrix D is as shown in Figure 6.24(a). Thus C_{d_1} and C_{d_2} are as shown in Figure 6.24(b) and (c). The submatrices of D whose rows and columns are indexed by the vertices of C_{d_1} and C_{d_2} are

$$
\begin{array}{c}
\begin{array}{cc} 8 & 9 \end{array} \\
\begin{array}{c} 1 \\ 2 \end{array}
\begin{bmatrix} 1 & 1 \\ d_1 & 1 \end{bmatrix}
\end{array}
\quad \text{and} \quad
\begin{array}{c}
\begin{array}{ccc} 5 & 7 & 9 \end{array} \\
\begin{array}{c} 1 \\ 2 \\ 3 \end{array}
\begin{bmatrix} 0 & 1 & 1 \\ 1 & 0 & 1 \\ 1 & d_2 & d_3 \end{bmatrix}
\end{array},
$$

respectively. We call these submatrices D_{d_1} and D_{d_2}.

Returning to the proof of the lemma, we have that, in general, for some $k \geq 2$, there are exactly k rows and exactly k columns of D that contain entries corresponding to edges of C_d. Let D_d be the $k \times k$ submatrix of D induced by these sets of k rows and k columns. In D_d, each row and each column contains exactly two entries that correspond to edges of C_d. Now either (a) D_d contains some other non-zero entries, or (b) D_d contains no other non-zero entries. In case (a), for every other non-zero entry d' of D_d, the corresponding edge $e_{d'}$ of $G(D^{\#})$ is a diagonal of C_d. Hence $|C_{d'}| < |C_d|$ and so, by the induction assumption, $d' \in \{1, -1\}$. Thus every entry of D_d, except possibly d, is in $\{0, 1, -1\}$. Now let $G(D_d^{\#})$ be the subgraph of $G(D^{\#})$ induced by the vertices of C_d and, among the cycles of $G(D_d^{\#})$ containing e_d, let C'_d be one of shortest length. The submatrix D'_d of D_d induced by vertices of C'_d has j rows and j columns for some $j \leq k$. Moreover, each row and each column of D'_d contains exactly two entries corresponding to edges of C'_d and has no other non-zero entries. Thus, in case (a), D has a $j \times j$ submatrix D'_d having d as an entry such that each row and column contains exactly two non-zero entries and all these entries, except

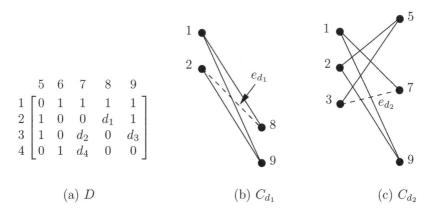

$$
\begin{array}{c}
\begin{array}{ccccc} 5 & 6 & 7 & 8 & 9 \end{array} \\
\begin{array}{c} 1 \\ 2 \\ 3 \\ 4 \end{array}
\begin{bmatrix} 0 & 1 & 1 & 1 & 1 \\ 1 & 0 & 0 & d_1 & 1 \\ 1 & 0 & d_2 & 0 & d_3 \\ 0 & 1 & d_4 & 0 & 0 \end{bmatrix}
\end{array}
$$

(a) D \qquad\qquad (b) C_{d_1} \qquad\qquad (c) C_{d_2}

FIG. 6.24. The fundamental cycles of d_1 and d_2 in $G(D^{\#})$.

possibly d, have uniquely determined values in $\{1, -1\}$. But this also holds in case (b) for, in that case, we take $j = k$ and $D'_d = D_d$.

Evaluating $\det D'_d$ using the summation formula 2.2.11, we see that there are exactly two non-zero terms in this summation and so $\det D'_d \in \{1+d, 1-d, -1+d, -1-d\}$. Since M is binary, $[I_r|D^\#]$ is a $GF(2)$-representation for M. Hence d is 1 over $GF(2)$. Therefore $\det D'_d$ is 0 when evaluated over $GF(2)$. Thus, by Proposition 6.4.5, $\det D'_d$ is also 0 when evaluated over $\mathbb{F}$, and so d has a uniquely determined value in $\{1, -1\}$. The lemma follows by induction. □

Proof of Theorem 6.6.3 Evidently we may assume that $r(M) > 0$. Now suppose that (i) holds. Then it follows by Lemma 2.2.21 that M has a totally unimodular representation of the form $[I_r|D]$. Let $\mathbb{F}$ be an arbitrary field and X be a set of r columns of $[I_r|D]$. Then X is a basis of M if and only if the determinant over $\mathbb{R}$ of the submatrix corresponding to X is a non-zero member of $\{0, 1, -1\}$. But the latter holds if and only if this determinant, when evaluated over $\mathbb{F}$, is non-zero. In turn, this holds if and only if X is a basis of the vector matroid on $[I_r|D]$ when the latter is viewed as a matrix over $\mathbb{F}$. We conclude that $[I_r|D]$ represents M over $\mathbb{F}$, so (i) implies (ii). Since (ii) clearly implies (iii), it remains to show that (iii) implies (i).

Assume that (iii) holds and let $[I_r|D]$ be an $\mathbb{F}$-representation for M. By Theorem 6.4.7, if B_D is a basis of the cycle matroid of $G(D^\#)$, then, we may assume that every entry in D corresponding to an edge of this basis is equal to 1. Then, by Lemma 6.6.4, every non-zero entry of D is in $\{1, -1\}$. We know that $[I_r|D]$ represents M over $\mathbb{F}$. To complete the proof, we shall show that, when viewed over $\mathbb{R}$, the matrix $[I_r|D]$ is a totally unimodular representation of M. To establish that $[I_r|D]$ represents M over $\mathbb{R}$, we note that, by Proposition 6.4.5, it suffices to show that, for every square submatrix D' of D, the determinant of D' is 0 over $\mathbb{F}$ if and only if it is 0 over $\mathbb{R}$. Clearly if $\det D'$ is 0 over $\mathbb{R}$, it is 0 over $\mathbb{F}$. Moreover, the converse of this will follow if we can show that if $\det D'$ is non-zero over $\mathbb{R}$, then it is in $\{1, -1\}$. But establishing this will also verify that M is regular and thereby complete the proof of the theorem.

Now let D' have k columns and suppose that $\det D' \neq 0$ when calculated over $\mathbb{R}$. Take a non-zero entry d_{st} of D that is also an entry of D'. Over $\mathbb{F}$, we can pivot in D on d_{st} to reduce column t of D to a standard basis vector of length r. This makes the corresponding column of D' a standard basis vector of length k. The natural column interchange with which we follow the pivot will not change D' only altering its location in the matrix representing M. By Lemma 6.6.2, the pivot produces another matrix with every entry in $\{0, 1, -1\}$ and, if this pivot is done over $\mathbb{R}$ instead of $\mathbb{F}$, it produces the same matrix. By repeated pivots, we eventually obtain a matrix representing M over $\mathbb{F}$ in which every column of D' is a standard basis vector. Now exactly the same matrix is obtained when these pivots are done over $\mathbb{R}$, and these operations will either leave $\det D'$ unchanged, or will multiply it by -1. Therefore, in the final matrix, $\det D'$ is still non-zero when calculated over $\mathbb{R}$. But the form of this matrix now implies that $\det D' \in \{1, -1\}$, as required. □

The next result is due to Brylawski and Lucas (1976).

Proposition 6.6.5 *Let M be a binary matroid and $\mathbb{F}$ be a field. Then all $\mathbb{F}$-representations of M are projectively equivalent, so M is uniquely $\mathbb{F}$-representable.*

Proof This follows easily by combining Theorem 6.4.7 and Lemma 6.6.4. □

From Theorem 6.6.3, Corollary 6.5.3, and the proof of Proposition 6.5.5, we deduce that the excluded minors for regularity must include $U_{2,4}, F_7$, and F_7^*. The next result (Tutte 1958) establishes that this list is complete. We shall prove this theorem in Chapter 10.

Theorem 6.6.6 *A matroid is regular if and only if it has no minor isomorphic to any of the matroids $U_{2,4}$, F_7, and F_7^*.* □

In Chapter 10, we shall also prove the following excluded-minor character-izations of the classes of graphic and cographic matroids. These results, which are again due to Tutte (1959), may be viewed as matroid generalizations of Kuratowski's Theorem (2.3.8).

Theorem 6.6.7 **(i)** *A matroid is graphic if and only if it has no minor isomor-phic to any of the matroids $U_{2,4}, F_7, F_7^*, M^*(K_5)$, and $M^*(K_{3,3})$.*

(ii) *A matroid is cographic if and only if it has no minor isomorphic to any of the matroids $U_{2,4}, F_7, F_7^*, M(K_5)$, and $M(K_{3,3})$.* □

Corollary 6.6.8 *The following statements are equivalent for a matroid M.*

(i) *$M \cong M(G)$ for some planar graph G.*

(ii) *M is both graphic and cographic.*

(iii) *M has no minor isomorphic to any of the matroids $U_{2,4}, F_7, F_7^*, M(K_5)$, $M^*(K_5), M(K_{3,3})$, and $M^*(K_{3,3})$.* □

In contrast to the excluded-minor characterization of Theorem 6.6.6, Sey-mour (1980b) gave a constructive characterization of regular matroids, the pre-cise statement of which will be given in Chapter 12. This theorem shows that every regular matroid can be built from graphic matroids, cographic matroids, and one special 10-element matroid. This matroid, R_{10}, is the vector matroid of the matrix A_{10} over $GF(2)$, where

$$
A_{10} = \left[\begin{array}{ccccc|ccccc}
 & & & & & 1 & 1 & 0 & 0 & 1 \\
 & & & & & 1 & 1 & 1 & 0 & 0 \\
 & & I_5 & & & 0 & 1 & 1 & 1 & 0 \\
 & & & & & 0 & 0 & 1 & 1 & 1 \\
 & & & & & 1 & 0 & 0 & 1 & 1
\end{array} \right].
$$

with column labels $1\ 2\ 3\ 4\ 5\ 6\ 7\ 8\ 9\ 10$.

The matroid R_{10} has many attractive features (Bixby 1977, Seymour 1980b). It is clearly self-dual, although it is not identically self-dual. Moreover:

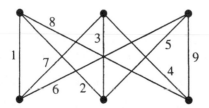

FIG. 6.25. $R_{10}\backslash 10$ is isomorphic to $M(K_{3,3})$.

Lemma 6.6.9 *Let e be an element of R_{10}. Then $R_{10}\backslash e \cong M(K_{3,3})$ and $R_{10}/e \cong M^*(K_{3,3})$.*

Proof When Z is $\{1,2,3,4,5\}$ or $\{6,7,8,9,10\}$, there is an automorphism of R_{10} that interchanges any two elements of Z. By pivoting on the first entry in column 6 of A_{10} and then swapping row 3 with row 4, and column 3 with column 4, we get

$$
\begin{array}{ccccccccccc}
 & 6 & 2 & 4 & 3 & 5 & 1 & 7 & 8 & 9 & 10 \\
\left[\begin{array}{ccccc} & & & & \\ & & & & \\ & & I_5 & & \\ & & & & \\ & & & & \end{array}\right. & & & & & & \begin{array}{ccccc} 1 & 1 & 0 & 0 & 1 \\ 1 & 0 & 1 & 0 & 1 \\ 0 & 0 & 1 & 1 & 1 \\ 0 & 1 & 1 & 1 & 0 \\ 1 & 1 & 0 & 1 & 0 \end{array} & & & & \left.\begin{array}{c} \\ \\ \\ \\ \\ \end{array}\right]
\end{array}.
$$

Interchanging columns 7 and 10 now gives the matrix A_{10} with its columns relabelled. Hence R_{10} has an automorphism that interchanges 1 and 6, so R_{10} has a transitive automorphism group. Thus, to complete the proof of the lemma, it suffices to show that $R_{10}\backslash 10 \cong M(K_{3,3})$. The matrix $A_{10}\backslash 10$ has the form $[I_5|X]$, where X is the fundamental-circuit incidence matrix of $M(K_{3,3})$ with respect to the basis $\{1,2,3,4,5\}$, the labelling of $K_{3,3}$ being as in Figure 6.25. As $M(K_{3,3})$ is binary, $M(K_{3,3}) \cong M[I_5|X] = M[A_{10}\backslash 10] = R_{10}\backslash 10$. $\square$

Corollary 6.6.10 *The matroid R_{10} is regular.*

Proof The corollary follows by combining Lemma 6.6.9 with Theorem 6.6.6. Alternatively, since we have yet to prove the latter, let $A_{10} = [I_5|D]$ and replace each entry on the main diagonal of D by -1. It is straightforward to check that the resulting matrix is totally unimodular. Hence R_{10} is regular. $\square$

The results of this section enable us to update the Venn diagram in Figure 5.6 (see Figure 6.26). Moreover, the reader should now have no difficulty in showing that each of the indicated inclusions is proper (Exercise 2).

Exercises

1. (a) Show that the matroid $M[A]$ in Example 6.6.1 is isomorphic to neither R_8 nor T_8 from Exercise 8 of Section 2.2.
   (b) How many circuit–hyperplane relaxations are needed to get $M[A]$ from $AG(3,2)$?

Matroids

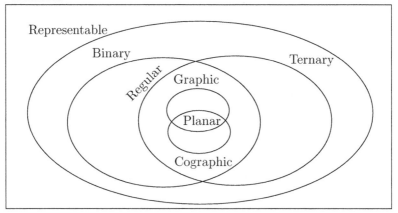

FIG. 6.26. Relationships between various classes of matroids.

2. Show that each of the inclusions in Figure 6.26 is proper.

3. A simple rank-r matroid M is *maximal regular* if M is regular and there is no simple rank-r regular matroid N of which M is a proper restriction. Show that:

   (a) $M(K_n)$ is maximal regular for all $n \geq 2$.

   (b) $M^*(K_{3,3})$ is maximal regular.

4. Let M be a matroid and A be a matrix $[I_r|D]$ such that, for every field $\mathbb{F}$, when A is viewed over $\mathbb{F}$, it is an $\mathbb{F}$-representation for M. Prove that, for every square submatrix X of A, the determinant of X, calculated over $\mathbb{F}$, is $0, 1$, or -1.

5. (Seymour 1980b) Show that:

   (a) R_{10} can be represented over $GF(2)$ by the ten 5-tuples that have exactly three entries equal to one.

   (b) If a, b, c, and d are elements of R_{10} with $a \neq b$ and $c \neq d$, there is an automorphism of R_{10} mapping a to c and b to d.

   (c) With $A_{10} = [I_5|D]$, if each entry on the main diagonal of D is replaced by -1, the resulting matrix is totally unimodular.

6. (Aigner 1979) Let A be a rank-r totally unimodular matrix with r rows and $M = M[A]$. Use the Cauchy–Binet formula to show that $\det(AA^T)$ is the number of bases of the regular matroid M.

6.7 Algebraic matroids

The focus of this chapter thus far has been on matroids that arise via linear dependence. There is another type of dependence that occurs in algebra, specifically in the study of fields. In this section, we shall see how this notion of algebraic dependence also gives rise to a class of matroids. This observation was first made by Van der Waerden (1937) when he noted several features common to both linear and algebraic dependence. Soon after, the properties of these algebraic matroids were investigated by Mac Lane (1938). Despite the proximity of this work to Whitney's (1935) founding paper in matroid theory, algebraic matroids have received far less attention over the years than representable matroids. Undoubtedly, an important reason for this is that algebraic matroids are not nearly as easy to deal with as representable matroids. Probably the most striking illustration of this fact is that it is still not known whether or not the class of algebraic matroids is closed under duality.

Before giving the precise definition of an algebraic matroid, we recall some of the basics of field theory. While a detailed development of this theory can be found in most graduate algebra texts, we shall not assume this background, attempting to make the treatment here as self-contained as possible. Much of our discussion will follow Van der Waerden (1937). For a field $\mathbb{F}$, we denote by $\mathbb{F}[x_1, x_2, \ldots, x_n]$ and $\mathbb{F}(x_1, x_2, \ldots, x_n)$, respectively, the ring of polynomials over $\mathbb{F}$ in the n indeterminates $x_1, x_2, \ldots, x_n$, and the associated field of rational functions in $x_1, x_2, \ldots, x_n$. Suppose $\mathbb{K}$ is a field such that $\mathbb{F} \subseteq \mathbb{K}$ and the addition and multiplication operations in $\mathbb{F}$ are restrictions of those in $\mathbb{K}$. Then $\mathbb{F}$ is a *subfield* of $\mathbb{K}$, and $\mathbb{K}$ is an *extension field* of $\mathbb{F}$. So, for instance, $GF(4)$ and $GF(8)$ are extension fields of $GF(2)$, although $GF(8)$ is not an extension field of $GF(4)$. The extension field $\mathbb{K}$ is a vector space over the field $\mathbb{F}$. Clearly the dimension of this $\mathbb{F}$-vector space $\mathbb{K}$ may be finite or infinite. An element u of $\mathbb{K}$ is said to be *algebraic* over $\mathbb{F}$ if u is a root of some non-zero polynomial in $\mathbb{F}[x]$. If u is not algebraic over $\mathbb{F}$, it is *transcendental* over $\mathbb{F}$. If every element of $\mathbb{K}$ is algebraic over $\mathbb{F}$, then $\mathbb{K}$ is an *algebraic extension of* $\mathbb{F}$; otherwise $\mathbb{K}$ is a *transcendental extension of* $\mathbb{F}$.

Now let $\{t_1, t_2, \ldots, t_n\}$ be a subset of $\mathbb{K}$. Then $\mathbb{F}(t_1, t_2, \ldots, t_n)$ denotes the subfield of $\mathbb{K}$ generated by $\{t_1, t_2, \ldots, t_n\}$, that is, $\mathbb{F}(t_1, t_2, \ldots, t_n)$ consists of all elements of the form $h(t_1, t_2, \ldots, t_n)/k(t_1, t_2, \ldots, t_n)$ where h and k are in $\mathbb{F}[x_1, x_2, \ldots, x_n]$. An element s of $\mathbb{K}$ is *algebraically dependent* on $\{t_1, t_2, \ldots, t_n\}$ over $\mathbb{F}$ if s is algebraic over $\mathbb{F}(t_1, t_2, \ldots, t_n)$. The latter occurs if and only if s is a root of an equation of the form

$$a_0(t_1, t_2, \ldots, t_n)x^m + a_1(t_1, t_2, \ldots, t_n)x^{m-1} + \cdots + a_m(t_1, t_2, \ldots, t_n) = 0$$

where each $a_i(t_1, t_2, \ldots, t_n)$ is a polynomial in $t_1, t_2, \ldots, t_n$ with coefficients in $\mathbb{F}$, and at least one of these polynomials is non-zero.

A finite subset T of $\mathbb{K}$ is *algebraically dependent* over $\mathbb{F}$ if, for some t in T, the element t is algebraically dependent on $T - t$. If T is not algebraically dependent over $\mathbb{F}$, it is called *algebraically independent* over $\mathbb{F}$. Reference to the particular field $\mathbb{F}$ is often omitted when one is discussing algebraic dependence.

We show next that the abstract notion of dependence that is captured in the definition of a matroid has algebraic dependence as a special case.

Theorem 6.7.1 *Suppose that $\mathbb{K}$ is an extension field of a field $\mathbb{F}$ and E is a finite subset of $\mathbb{K}$. Then the collection $\mathcal{I}$ of subsets of E that are algebraically independent over $\mathbb{F}$ is the set of independent sets of a matroid on E.*

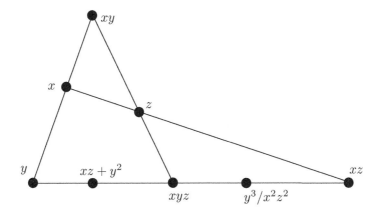

FIG. 6.27. A geometric representation of an algebraic matroid.

Before proving this theorem, we note the following.

Example 6.7.2 Let $\mathbb{F} = GF(2)$ and $\mathbb{K} = \mathbb{F}(x, y, z)$ where x, y, and z are indeterminates. If $E = \{x, y, z, xy, xz, xyz, xz + y^2, y^3/x^2z^2\}$, then it is not difficult to check that Figure 6.27 gives a geometric representation for the matroid $(E, \mathcal{I})$ whose existence is asserted by the theorem. $\square$

Following Van der Waerden (1937), we shall use a sequence of three elementary lemmas to prove Theorem 6.7.1.

Lemma 6.7.3 *If s is algebraically dependent on $\{t_1, t_2, \ldots, t_n\}$ but not on $\{t_1, t_2, \ldots, t_{n-1}\}$, then t_n is algebraically dependent on $\{t_1, t_2, \ldots, t_{n-1}, s\}$.*

Proof In $\mathbb{F}(t_1, t_2, \ldots, t_{n-1})$, the element s is algebraically dependent on $\{t_n\}$. Thus, for some polynomials $a_0(t_n), a_1(t_n), \ldots, a_m(t_n)$ that have coefficients in $\mathbb{F}(t_1, t_2, \ldots, t_{n-1})$ and are not all zero,

$$a_0(t_n)s^m + a_1(t_n)s^{m-1} + \cdots + a_m(t_n) = 0.$$

Grouping the terms in this equation by powers of t_n, we get the equation

$$b_0(s)t_n^k + b_1(s)t_n^{k-1} + \cdots + b_k(s) = 0 \tag{6.5}$$

where each $b_i(s)$ is a polynomial having coefficients in $\mathbb{F}(t_1, t_2, \ldots, t_{n-1})$. By hypothesis, the element s is not algebraic over $\mathbb{F}(t_1, t_2, \ldots, t_{n-1})$. Therefore each of $b_0(s), b_1(s), \ldots, b_k(s)$ is either identically zero in s or is non-zero. But if all of these polynomials are identically zero in s, then all of $a_0(t_n), a_1(t_n), \ldots, a_m(t_n)$ are also identically zero; a contradiction. Thus $b_i(s)$ is non-zero for some i. Hence, by (6.5), t_n is algebraically dependent on $\{s\}$ over $\mathbb{F}(t_1, t_2, \ldots, t_{n-1})$, so t_n is algebraically dependent on $\{t_1, t_2, \ldots, t_{n-1}, s\}$ over $\mathbb{F}$. $\square$

Lemma 6.7.4 *Suppose that s is algebraic over $\mathbb{F}(t)$, and t is algebraic over $\mathbb{F}$. Then s is algebraic over $\mathbb{F}$.*

Proof As t is algebraic over $\mathbb{F}$, and s is algebraic over $\mathbb{F}(t)$, the vector spaces $\mathbb{F}(t)$ and $\mathbb{F}(t)(s)$ are finite dimensional over $\mathbb{F}$ and $\mathbb{F}(t)$, respectively. From this, it follows easily that, for some positive integer k, the vector space $\mathbb{F}(t)(s)$ has dimension k over $\mathbb{F}$. Therefore $\{1, s, s^2, \ldots, s^k\}$ cannot be linearly independent over $\mathbb{F}$ and so s is algebraic over $\mathbb{F}$. $\square$

Lemma 6.7.5 *If s is algebraically dependent on $\{t_1, t_2, \ldots, t_n\}$ and each t_i is algebraically dependent on $\{u_1, u_2, \ldots, u_k\}$, then s is algebraically dependent on $\{u_1, u_2, \ldots, u_k\}$.*

Proof We argue by induction on n. If $n = 1$, the result is immediate by the last lemma. Assume the result true for $n < m$ and let $n = m$. Now $\mathbb{F}(t_1, t_2, \ldots, t_m) = \mathbb{F}(t_m)(t_1, t_2, \ldots, t_{m-1})$. Thus s is algebraic over $\mathbb{F}(t_m)(t_1, t_2, \ldots, t_{m-1})$. Moreover, for i in $\{1, 2, \ldots, m-1\}$, each t_i is algebraic over $\mathbb{F}(u_1, u_2, \ldots, u_k)$ and hence over $\mathbb{F}(t_m)(u_1, u_2, \ldots, u_k)$. Thus, by the induction assumption, s is algebraic over $\mathbb{F}(t_m)(u_1, u_2, \ldots, u_k)$. As the last field equals $\mathbb{F}(u_1, u_2, \ldots, u_k, t_m)$, and t_m is algebraic over $\mathbb{F}(u_1, u_2, \ldots, u_k)$, it follows, by Lemma 6.7.4, that s is algebraic over $\mathbb{F}(u_1, u_2, \ldots, u_k)$. Hence, by induction, the lemma holds. $\square$

The next result is an easy consequence of the last three lemmas and its proof is left to the reader.

Corollary 6.7.6 *Let T and U be finite subsets of an extension field $\mathbb{K}$ of a field $\mathbb{F}$ and suppose that U is algebraically independent over $\mathbb{F}$.*

(i)   *If every element of T is algebraic over $\mathbb{F}$, then U is algebraically independent over $\mathbb{F}(T)$.*

(ii) *$T \cup U$ is algebraically independent over $\mathbb{F}$ if and only if T is algebraically independent over $\mathbb{F}(U)$.* $\square$

With these preliminaries, we are now ready to prove that algebraic dependence does indeed give rise to a matroid.

Proof of Theorem 6.7.1 Evidently $\mathcal{I}$ satisfies (I1) and (I2). Assume that $\mathcal{I}$ does not satisfy (I3), and let I_1 and I_2 be algebraically independent subsets of E with $|I_1| < |I_2|$ such that, for every element e of $I_2 - I_1$, the set $I_1 \cup e$ is algebraically dependent. It follows easily by Lemma 6.7.3 that each such e depends algebraically on some subset of I_1. Hence every element of I_2 depends algebraically on I_1. Now, from among the minimum-cardinality subsets of $I_1 \cup I_2$ on which every element of I_2 depends algebraically, choose one, say I_1', such that $I_1' - I_2$ is as small as possible. Evidently $|I_1'| \leq |I_1|$. As $|I_2| > |I_1|$, the set $I_2 - I_1'$ is non-empty. Let e be an element of this set. Then e depends algebraically on I_1'. Let I_1'' be a minimal subset of I_1' on which e is algebraically dependent. As $I_1'' \cup e$ is not contained in the algebraically independent set I_2, there is an element f in $I_1'' - I_2$. Now e depends algebraically on I_1'' but not on $I_1'' - f$. Thus, by Lemma 6.7.3, f depends algebraically on $(I_1'' - f) \cup e$ and hence on $(I_1' - f) \cup e$. Hence every element of I_1' depends algebraically on $(I_1' - f) \cup e$. But every element of I_2 depends algebraically on I_1'. Thus, by Lemma 6.7.5, every element of I_2

depends algebraically on $(I_1' - f) \cup e$. As $|((I_1' - f) \cup e) - I_2| < |I_1' - I_2|$, the choice of I_1' is contradicted. Hence $\mathcal{I}$ satisfies (I3), so $(E, \mathcal{I})$ is a matroid. □

Let M be an arbitrary matroid and let $(E, \mathcal{I})$ be the matroid in the last theorem. Suppose that there is a map φ from $E(M)$ into E such that, for all subsets T of $E(M)$, the set T is independent in M if and only if $|\varphi(T)| = |T|$ and $\varphi(T)$ is independent in $(E, \mathcal{I})$. Then the matroid M is said to be *algebraic over* $\mathbb{F}$ or *algebraically representable over* $\mathbb{F}$, and the map φ is called an *algebraic representation* of M over $\mathbb{F}$. An *algebraic matroid* is one that is algebraic over some field. To emphasize the distinction between algebraic representability and the type of representability discussed earlier, we shall sometimes refer to an $\mathbb{F}$-representable matroid as being *linearly representable* over $\mathbb{F}$. Before giving some examples of algebraic matroids, we prove a characterization of the independent sets of such a matroid. This result will facilitate the determination of the collection of independent sets from a list of the elements.

Proposition 6.7.7 *Let* $\{t_1, t_2, \ldots, t_n\}$ *be a non-empty subset of an extension field* $\mathbb{K}$ *of a field* $\mathbb{F}$. *Then* $\{t_1, t_2, \ldots, t_n\}$ *is algebraically independent over* $\mathbb{F}$ *if and only if there is no non-zero polynomial* $f(x_1, x_2, \ldots, x_n)$ *with coefficients in* $\mathbb{F}$ *such that* $f(t_1, t_2, \ldots, t_n) = 0$.

Proof If $\{t_1, t_2, \ldots, t_n\}$ is algebraically dependent over $\mathbb{F}$, then some t_i is algebraically dependent on $\{t_1, t_2, \ldots, t_{i-1}, t_{i+1}, \ldots, t_n\}$. Therefore there is a non-zero member f of $\mathbb{F}[x_1, x_2, \ldots, x_n]$ such that $f(t_1, t_2, \ldots, t_n) = 0$.

Now let $\{t_1, t_2, \ldots, t_n\}$ be algebraically independent. We shall argue by induction on n to show that there is no non-zero member f of $\mathbb{F}[x_1, x_2, \ldots, x_n]$ such that $f(t_1, t_2, \ldots, t_n) = 0$. This is immediate if $n = 1$. Assume it true for $n < m$ and suppose that $n = m \geq 2$. Let f be a member of $\mathbb{F}[x_1, x_2, \ldots, x_m]$ for which $f(t_1, t_2, \ldots, t_m) = 0$. Group the terms in the polynomial f according to the power of t_m involved. Since $\{t_1, t_2, \ldots, t_m\}$ is algebraically independent, it follows that, for all i, the coefficient $f_i(t_1, t_2, \ldots, t_{m-1})$ of t_m^i is zero. But $\{t_1, t_2, \ldots, t_{m-1}\}$ is also algebraically independent and therefore, by the induction assumption, each f_i is identically zero. Hence f itself is identically zero and the proposition follows by induction. □

Example 6.7.8 We noted in Proposition 6.5.2 that the uniform matroid $U_{2,n}$ is $\mathbb{F}$-representable if and only if $|\mathbb{F}| \geq n - 1$. In contrast, $U_{2,n}$ is algebraically representable over $\mathbb{F}$ for all fields $\mathbb{F}$. To see this, let E be the subset $\{xy, x^2y, \ldots, x^ny\}$ of $\mathbb{F}(x, y)$ where x and y are indeterminates. It is straightforward to check that if $\mathcal{I}$ is the set of algebraically independent subsets of E over $\mathbb{F}$, then $(E, \mathcal{I}) \cong U_{2,n}$. Similarly, if $U_{3,n}$ has ground set $\{1, 2, \ldots, n\}$ and φ is the map from this set into $\mathbb{F}(x, y, z)$ defined by $\varphi(i) = xy^iz^{i^2}$, then one can show that φ is an algebraic representation for $U_{3,n}$ over $\mathbb{F}$. Indeed, as we shall see in Chapter 11, all transversal matroids are algebraic over all fields, hence so are all uniform matroids. □

Let E be a finite subset of an extension field $\mathbb{K}$ of a field $\mathbb{F}$ and let M be the matroid induced on E by algebraic dependence over $\mathbb{F}$. Then an element

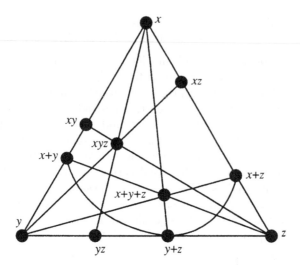

FIG. 6.28. An algebraic matroid that is not linearly representable.

t of E is a loop of M if and only if t is algebraic over $\mathbb{F}$. Equivalently, t is transcendental over $\mathbb{F}$ if and only if $\{t\}$ is algebraically independent over $\mathbb{F}$. More generally, suppose that $\{t_1, t_2, \dots, t_n\}$ is algebraically independent over $\mathbb{F}$. Then $t_1, t_2, \dots, t_n$ are also called *independent transcendentals* over $\mathbb{F}$. Now let $x_1, x_2, \dots, x_n$ be indeterminates over $\mathbb{F}$. Then it is not difficult to see that the polynomial rings $\mathbb{F}[t_1, t_2, \dots, t_n]$ and $\mathbb{F}[x_1, x_2, \dots, x_n]$ are isomorphic and hence that the quotient fields $\mathbb{F}(t_1, t_2, \dots, t_n)$ and $\mathbb{F}(x_1, x_2, \dots, x_n)$ are also isomorphic. Thus, algebraically, the n independent transcendentals $t_1, t_2, \dots, t_n$ behave just like the n indeterminates $x_1, x_2, \dots, x_n$.

We saw in Example 6.7.8 that a matroid can be algebraic over a field but be non-representable over that field. The next example (Ingleton 1971a) is of an algebraic matroid that is not linearly representable over any field.

Example 6.7.9 Let E be the subset $\{x, y, z, xy, xz, yz, x + y, x + z, y + z, xyz, x + y + z\}$ of $GF(2)(x, y, z)$ where x, y, and z are independent transcendentals. Let M be the matroid induced on E by algebraic dependence over $GF(2)$. Then Figure 6.28 is a geometric representation of M. To see this, note first that, since $(x + y) + (y + z) + (x + z) = 0$ over $GF(2)$, the set $\{x + y, y + z, x + z\}$ is algebraically dependent. The algebraic equations that correspond to each of the other 3-element circuits shown are easily derived. Therefore Figure 6.28 is indeed a geometric representation of M unless some 3-element subset of $\{xy, xz, yz, x + y + z\}$ is a circuit. But it is straightforward to show that there is no non-zero polynomial f in $GF(2)[x_1, x_2, x_3]$ such that $f(xy, xz, yz) = 0$. Hence $\{xy, xz, yz\}$ is not a circuit. By symmetry, either all or none of the remaining 3-element subsets of $\{xy, xz, yz, x + y + z\}$ is a circuit. But, in the former case, circuit elimination implies that $\{xy, xz, yz\}$ is a circuit. We conclude that no 3-element subset of $\{xy, xz, yz, x+y+z\}$ is a circuit and this completes the proof that Figure

6.28 is a geometric representation of M. Now M is certainly algebraic. But, since it clearly has both F_7 and F_7^- as restrictions, it is not linearly representable. $\square$

We show next that the class of matroids that are linearly representable over a field $\mathbb{F}$ is a subset of the class of matroids that are algebraic over $\mathbb{F}$. The proof of this follows Piff (1969) (see also Welsh 1976).

Proposition 6.7.10 *If a matroid M is linearly representable over a field $\mathbb{F}$, then M is algebraic over $\mathbb{F}$.*

Proof Let $r(M) = r$. Evidently we may assume that $r > 0$. Take an $\mathbb{F}$-coordinatization $\psi : E(M) \rightarrow V(r, \mathbb{F})$ of M where the members of $V(r, \mathbb{F})$ are viewed as column vectors. Let $\{e_1, e_2, \ldots, e_r\}$ be a basis of M, and, for all i in $\{1, 2, \ldots, r\}$, let $\psi(e_i) = \underline{v}_i$. Then $\{\underline{v}_1, \underline{v}_2, \ldots, \underline{v}_r\}$ is a basis for $V(r, \mathbb{F})$. Now let $t_1, t_2, \ldots, t_r$ be independent transcendentals over $\mathbb{F}$. Then, for every element e of $E(M)$, the vector $\psi(e)$ can be uniquely written in the form $\sum_{i=1}^{r} a_i \underline{v}_i$ where $a_1, a_2, \ldots, a_r \in \mathbb{F}$. For each such e, let $\phi(e)$ be $\sum_{i=1}^{r} a_i t_i$. Then ϕ is a well-defined function from $E(M)$ into $\mathbb{F}(t_1, t_2, \ldots, t_r)$. To complete the proof of the proposition, we shall show that ϕ is an algebraic representation of M over $\mathbb{F}$.

First let $\{g_1, g_2, \ldots, g_r\}$ be an arbitrary basis of M. Then, for some non-singular $r \times r$ matrix A over $\mathbb{F}$,

$$(\psi(g_1)\,\psi(g_2) \cdots \psi(g_r)) = (\underline{v}_1\,\underline{v}_2 \cdots \underline{v}_r)A.$$

Thus $(\phi(g_1), \phi(g_2), \ldots, \phi(g_r)) = (t_1, t_2, \ldots, t_r)A$, and so,

$$(\phi(g_1), \phi(g_2), \ldots, \phi(g_r))A^{-1} = (t_1, t_2, \ldots, t_r).$$

Hence each of $t_1, t_2, \ldots, t_r$ is algebraically dependent on $\{\phi(g_1), \phi(g_2), \ldots, \phi(g_r)\}$. But $\{t_1, t_2, \ldots, t_r\}$ is a basis for the algebraic matroid on $\phi(E(M))$ and therefore so too is $\{\phi(g_1), \phi(g_2), \ldots, \phi(g_r)\}$.

Now let $\{c_1, c_2, \ldots, c_k\}$ be a circuit C of M. Then $\psi(C)$ is a minimal linearly dependent subset of $V(r, \mathbb{F})$. Hence $\sum_{i=1}^{k} a_i \psi(c_i) = \underline{0}$ for some non-zero elements $a_1, a_2, \ldots, a_k$ of $\mathbb{F}$. Thus $\sum_{i=1}^{k} a_i \phi(c_i) = 0$, so $\phi(C)$ is algebraically dependent over $\mathbb{F}$. We conclude that ϕ is an algebraic representation for M over $\mathbb{F}$. $\square$

Example 6.7.8 shows that the converse of Proposition 6.7.10 fails for all finite fields. However, for fields of characteristic zero, we have the following result whose proof we omit (see, for example, Lang 1965, Chapter 10, Proposition 10).

Proposition 6.7.11 *If a matroid M is algebraic over a field $\mathbb{F}$ of characteristic zero, then M is linearly representable over $\mathbb{F}(T)$ for some finite set T of transcendentals over $\mathbb{F}$.* $\square$

We know from Example 6.7.9 and Proposition 6.7.10 that the class of algebraic matroids is large. Indeed, Ingleton (1971a) asked whether this class coincides with the class of all matroids. Subsequently, he and Main (1975) proved that the Vámos matroid is non-algebraic (Exercise 10). Their result was used

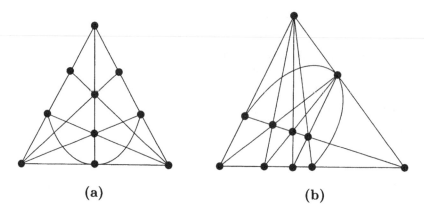

(a)   **(b)**

FIG. 6.29. Two non-algebraic matroids.

by Lindström (1984a, 1986b) to prove that the matroid for which a geometric representation is shown in Figure 6.29(a) is also non-algebraic. By generalizing the construction of the Vámos matroid, Lindström (1988a) defined a class of non-algebraic matroids consisting of one matroid of each rank exceeding three. In addition, Lindström (1987c) found an infinite class of rank-3 non-algebraic matroids, each having every proper minor algebraic. A geometric representation for the first member of this class is shown in Figure 6.29(b).

Next we consider the effect on algebraic matroids of the three fundamental operations of deletion, contraction, and the taking of duals. Evidently if a matroid M is algebraic over $\mathbb{F}$ and $T \subseteq E(M)$, then $M|T$ is algebraic over $\mathbb{F}$. The behaviour of M under contraction is more difficult to determine and relies on the following result of Lindström (1989) which proves a conjecture of Piff (1972). We remark that Lindström's paper notes that two earlier papers (Lindström 1987b, Shameeva 1985) claiming to prove Piff's conjecture were, in fact, incomplete.

Proposition 6.7.12 *If a matroid M is algebraic over an extension field $\mathbb{F}(t)$ of a field $\mathbb{F}$, then M is algebraic over $\mathbb{F}$.*   □

The real substance of the last result lies in the case when t is transcendental over $\mathbb{F}$, for Piff (1972) showed that the result is straightforward when t is algebraic over $\mathbb{F}$ (Exercise 3). From the last proposition, Lindström (1989) was able to deduce that the class of algebraic matroids is closed under contraction.

Proposition 6.7.13 *If a matroid M is algebraic over a field $\mathbb{F}$ and $T \subseteq E(M)$, then $M.T$ is algebraic over $\mathbb{F}$.*

Proof By Proposition 6.7.12, it suffices to show that $M.T$ is algebraic over $\mathbb{F}(U)$ for some finite set U. Our proof of this follows Welsh (1976). Let $\phi : E(M) \to \mathbb{K}$ be an algebraic representation of M over $\mathbb{F}$. Then $\mathbb{K}$ is an extension field of $\mathbb{F}(\phi(E(M) - T))$. Denote by ϕ' the restriction of the map ϕ to the set T. Let I be a subset of T and B be a basis of $E(M) - T$. The following sequence of equivalent statements shows that ϕ' is an algebraic representation of $M.T$ over $\mathbb{F}(\phi(E(M) - T))$.

(i) I is independent in $M.T$.

(ii) $I \cup B$ is independent in M.

(iii) $\phi(I \cup B)$ is algebraically independent over $\mathbb{F}$, and $|\phi(I \cup B)| = |I \cup B|$.

(iv) $\phi(I)$ is algebraically independent over $\mathbb{F}(\phi(B))$, and $|\phi(I)| = |I|$.

(v) $\phi(I)$ is algebraically independent over $\mathbb{F}(\phi(E(M) - T))$, and $|\phi(I)| = |I|$.

The verification that (iii) is equivalent to (iv) uses Corollary 6.7.6(ii), while the equivalence of (iv) and (v) follows by Corollary 6.7.6(i). □

Corollary 6.7.14 *If a matroid M is algebraic over a field $\mathbb{F}$, then every minor of M is algebraic over $\mathbb{F}$.* □

In view of this corollary and the characterizations of binary and ternary matroids noted in Section 6.5, it is natural to try to determine the excluded minors for algebraic representability over certain specific fields. However, as Lemos (1988) has noted, one can use one of Lindström's (1987c) classes of non-algebraic matroids or another class of matroids considered by Gordon (1984) to show that, for every field $\mathbb{F}$, the set of such excluded minors is infinite.

Probably the most basic unsolved problem in the study of algebraic matroids is the following.

Problem 6.7.15 *Is the dual of an algebraic matroid also algebraic?* □

Lindström (1985b) showed that, to answer this question, it suffices to solve the following problem, although he noted that this problem also seems very hard.

Problem 6.7.16 *If a matroid is algebraic over a field of a certain characteristic, is its dual also algebraic over a field of this characteristic?* □

For a discussion of various other unsolved problems related to algebraic matroids, the reader is referred to Lindström (1988b).

Exercises

1. Show that all graphic matroids are algebraic over every field.
2. By giving an explicit algebraic representation, show that $U_{r,n}$ is algebraic over all fields for all r and n with $0 \leq r \leq n$.
3. Show that if M is algebraic over $\mathbb{F}(t)$ where t is algebraic over $\mathbb{F}$, then M is algebraic over $\mathbb{F}$.
4. Suppose that $M \cong M(G)$ for some graph G and let $\mathbb{F}$ be a field. Describe how to obtain explicit algebraic representations for M and M^* over $\mathbb{F}$.
5. (Lindström 1988b) Prove that if M is algebraic over a field $\mathbb{F}$, then M is algebraic over the prime field of $\mathbb{F}$. Deduce that if n is a positive integer and M is algebraic over all fields with at least n elements, then M is algebraic over all fields.
6. (Lindström 1983, 1986a) Let p be a prime number and λ be an element of $GF(p^2)$ such that $\lambda^p \neq \lambda$.

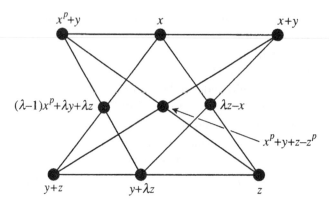

$$x^p+y \qquad x \qquad x+y$$

$$(\lambda-1)x^p+\lambda y+\lambda z \qquad \lambda z-x$$

$$x^p+y+z-z^p$$

$$y+z \qquad y+\lambda z \qquad z$$

FIG. 6.30. The non-Pappus matroid is algebraic over $GF(p^2)$.

(a) Show that Figure 6.30 depicts an algebraic representation for the non-Pappus matroid over $GF(p^2)$.

(b) Deduce that the non-Pappus matroid is algebraic over $GF(p)$ for all primes p.

7. (a) *(Lindström 1985d) Prove that if a matroid is linearly representable over $\mathbb{Q}$, then it is algebraic over all fields.

(b) Deduce that a matroid that is algebraic over some field of characteristic 0 is algebraic over all fields.

8. (a) *(Lindström 1985c) Prove that the Fano matroid is algebraic only over fields of characteristic two.

(b) Show that the non-Fano matroid is algebraic over all fields.

9. *(Lindström 1985a) Show that the non-Desargues matroid (see Figure 6.4) is non-algebraic.

10. *(Ingleton and Main 1975) Let $\{1, 1', 2, 2', 3, 3', 4, 4'\}$ be a set E of elements of an extension field $\mathbb{K}$ of a field $\mathbb{F}$, and let M be the matroid induced on E by algebraic dependence over $\mathbb{F}$. Suppose that every subset of E with fewer than four elements is independent, while every subset of E with more than four elements has rank 4.

(a) Show that if $\{1, 1', 2, 2'\}$, $\{1, 1', 3, 3'\}$, $\{1, 1', 4, 4'\}$, $\{2, 2', 3, 3'\}$, and $\{2, 2', 4, 4'\}$ are dependent, then $\{3, 3', 4, 4'\}$ is also dependent.

(b) Deduce that the Vámos matroid, the matroid Q_8 from Example 6.5.10, and the relaxation of $AG(3, 2)$ are all non-algebraic.

6.8 Characteristic sets and decidability

For a matroid M, the *characteristic set* $\mathcal{K}(M)$ of M is the set $\{k : M$ is representable over some field of characteristic $k\}$. In this section, we consider the possibilities for the characteristic set of a matroid M. We also consider the questions of whether there are finite algorithms to determine whether a matroid is representable and to find its characteristic set. Ingleton (1971a) was the first to define characteristic sets although his definition was somewhat broader than ours, since it was in terms of representability over division rings. We have seen a number of examples of where a matroid M fails to be representable over some prime field $GF(p)$ simply because the field is too small. In such circumstances, p is not excluded from $\mathcal{K}(M)$ since M will be representable over some suitable extension of $GF(p)$. In general, one can view the fact that p is not in $\mathcal{K}(M)$ as a reflection of some non-trivial structural property of M.

If $\mathbb{P}$ is the set of prime numbers, then $\mathcal{K}(M) \subseteq \mathbb{P} \cup \{0\}$ for all matroids M. The following question was asked in the division-ring case by Ingleton (1971a).

Problem 6.8.1 *For which subsets T of $\mathbb{P} \cup \{0\}$, is there a matroid M having T as its characteristic set?* □

This question has now been completely answered and, in this section, we survey the results that lead to the answer along with various results about characteristic sets of certain classes of matroids. We shall assume familiarity with the basics of field theory as sketched in the Preliminaries and at the beginning of the previous section. In particular, a field $\mathbb{F}$ has characteristic p if and only if $\mathbb{F}$ has $GF(p)$ as a subfield; and $\mathbb{F}$ has $\mathbb{Q}$ as a subfield if and only if it has characteristic 0. Part of the motivation for Problem 6.8.1 derived from the following result of Rado (1957). Some comments on its proof appear at the end of the section.

Proposition 6.8.2 *If $0 \in \mathcal{K}(M)$, then $\mathcal{K}(M)$ contains all sufficiently large primes. Moreover, M is representable over infinitely many prime fields.* □

Vámos (1971b) showed that the converse of this proposition is also true. Indeed, he established the following stronger result.

Proposition 6.8.3 *If $\mathcal{K}(M)$ is infinite, then $0 \in \mathcal{K}(M)$.* □

On combining the last two results, we deduce that, if T is the characteristic set of some matroid, then either

(i) $0 \in T$ and $\mathbb{P} - T$ is finite; or
(ii) $0 \notin T$ and T is finite.

The question as to which sets satisfying (i) do actually occur as characteristic sets was resolved by R. Reid (see Brylawski and Kelly 1980, pp. 101–2) who discovered the following.

Example 6.8.4 Let k be an integer exceeding one and α be a primitive kth root of unity in $\mathbb{C}$, that is, α is an element of $\mathbb{C}$ for which $\alpha^k = 1$ and $1, \alpha, \alpha^2, \ldots, \alpha^{k-1}$ are distinct. Let A_k be the following matrix over $\mathbb{C}$:

$$
\begin{array}{ccccccccccc}
e_1 & e_2 & e_3 & e_4 & e_5 & a_0 & b_0 & a_1 & b_1 & \cdots & a_{k-2} & b_{k-2}
\end{array}
$$
$$
\begin{bmatrix}
1 & 0 & 0 & \boxed{1} & \boxed{1} & \boxed{1} & 0 & \boxed{1} & 0 & \cdots & \boxed{1} & 0 \\
0 & 1 & 0 & 1-\alpha & 1 & \boxed{1} & \boxed{1} & 1 & \boxed{1} & \cdots & 1 & \boxed{1} \\
0 & 0 & 1 & 1 & 0 & \boxed{1} & 1 & f_1(\alpha) & f_1(\alpha) & \cdots & f_{k-2}(\alpha) & f_{k-2}(\alpha)
\end{bmatrix}
$$

where $f_i(\alpha) = \alpha^i + \alpha^{i-1} + \cdots + 1$ for all i in $\{1, 2, \ldots, k-1\}$. Then it is not difficult to check that Figure 6.31 is a geometric representation for $M[A_k]$. Evidently

$0 \in \mathcal{K}(M[A_k])$. In what follows, we shall be viewing A_k as a matrix over fields other than $\mathbb{C}$. We emphasize here that, throughout this discussion, $M[A_k]$ will always denote the vector matroid of the *complex* matrix A_k.

Lemma 6.8.5 *Let p be a prime number. Then $p \in \mathcal{K}(M[A_k])$ if and only if k is not divisible by p.*

Proof. Let $\mathbb{F}$ be a field. If $M[A_k]$ is $\mathbb{F}$-representable, then, by Theorem 6.4.7, we may assume that $M[A_k]$ has an $\mathbb{F}$-representation $[I_3|X]$ so that, in X, each entry corresponding to a boxed entry of A_k equals 1. Since $\{e_3, e_5, a_0, a_1, a_2, \ldots, a_{k-2}\}$ is a line of $M[A_k]$, the second entries in the e_5-, a_1-, a_2-, $\ldots$, a_{k-2}-columns of X are all 1. The circuit $\{e_2, a_0, e_4\}$ implies that the third entry in the e_4-column of X is a 1, while the circuits $\{e_1, a_i, b_i\}$, for i in $\{0, 1, \ldots, k-2\}$, imply that the third entries of the a_i- and b_i-columns of X are equal. Thus, for some non-zero elements $x_0, x_1, \ldots, x_{k-2}$ of $\mathbb{F}$, the matrix X is

$$
\begin{array}{c}
\begin{array}{ccccccccccc}
e_4 & e_5 & a_0 & b_0 & a_1 & b_1 & a_2 & b_2 & \cdots & a_{k-2} & b_{k-2}
\end{array} \\
\left[
\begin{array}{ccccccccccc}
1 & 1 & 1 & 0 & 1 & 0 & 1 & 0 & \cdots & 1 & 0 \\
x_0 & 1 & 1 & 1 & 1 & 1 & 1 & 1 & \cdots & 1 & 1 \\
1 & 0 & 1 & 1 & x_1 & x_1 & x_2 & x_2 & \cdots & x_{k-2} & x_{k-2}
\end{array}
\right].
\end{array}
$$

Now writing $1 - x$ for x_0, we deduce from the circuit $\{e_4, b_0, a_1\}$ of $M[A_k]$ that $x_1 = 1 + x$. In general, since $\{e_4, b_i, a_{i+1}\}$ is a circuit of $M[A_k]$ for all i in $\{1, 2, \ldots, k-3\}$, we get that $x_{i+1} = xx_i + 1$. It follows that $x_{i+1} = f_{i+1}(x)$.

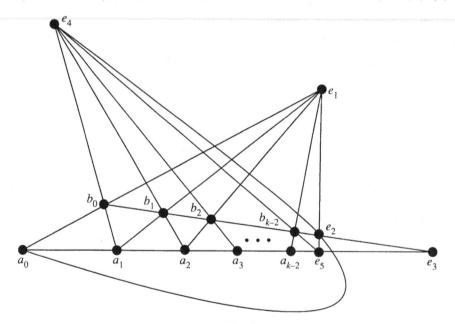

FIG. 6.31. $M[A_k]$.

Finally, as $\{e_4, b_{k-2}, e_5\}$ is a circuit, $xx_{k-2} + 1 = 0$, that is, $f_{k-1}(x) = 0$. Hence $(x-1)f_{k-1}(x) = 0$, so $x^k - 1 = 0$.

As x_0 is non-zero, $x \neq 1$. Moreover, as none of $x_1, x_2, \ldots, x_{k-2}$ is 0, none of $f_1(x), f_2(x), \ldots, f_{k-2}(x)$ is 0, that is, $x^2 - 1, x^3 - 1, \ldots, x^{k-1} - 1$ are all non-zero. We conclude that if $M[A_k]$ is $\mathbb{F}$-representable, then $\mathbb{F}$ contains x, a primitive kth root of unity, and an $\mathbb{F}$-representation for $M[A_k]$ is $[I_3|X]$ which equals

$$
\begin{array}{ccccccccccccc}
e_1 & e_2 & e_3 & e_4 & e_5 & a_0 & b_0 & a_1 & b_1 & \cdots & a_{k-2} & b_{k-2} \\
\left[\begin{array}{cccccccccccc}
1 & 0 & 0 & 1 & 1 & 1 & 0 & 1 & 0 & \cdots & 1 & 0 \\
0 & 1 & 0 & 1-x & 1 & 1 & 1 & 1 & 1 & \cdots & 1 & 1 \\
0 & 0 & 1 & 1 & 0 & 1 & 1 & f_1(x) & f_1(x) & \cdots & f_{k-2}(x) & f_{k-2}(x)
\end{array}\right].
\end{array}
$$

But, as every line of $M[A_k]$ is also a line of $M[I_3|X]$, it follows without difficulty that $[I_3|X]$ is indeed an $\mathbb{F}$-representation for $M[A_k]$. Thus $M[A_k]$ is $\mathbb{F}$-representable if and only if, when A_k is viewed as a matrix over $\mathbb{F}$, it is an $\mathbb{F}$-representation for $M[A_k]$ where α is a primitive kth root of unity in $\mathbb{F}$. Now suppose $\mathbb{F}$ has characteristic p. Then, viewed over $\mathbb{F}$, every entry in A_k is in $GF(p)(\alpha)$. Hence $M[A_k]$ is $\mathbb{F}$-representable if and only if it is represented over $GF(p)(\alpha)$ by A_k. The result now follows by using the next lemma, a standard result from field theory. □

We recall that every field $\mathbb{F}$ has an algebraic extension $\mathbb{K}$ such that every non-zero polynomial in $\mathbb{F}[x]$ has a root in $\mathbb{K}$. This field $\mathbb{K}$ is unique up to isomorphism and is called the *algebraic closure of* $\mathbb{F}$.

Lemma 6.8.6 *The algebraic closure of $GF(p)$ contains a primitive kth root of unity if and only if k is not divisible by p.*

Proof This follows, for example, from Hungerford (1974, p. 295). □

Proposition 6.8.7 *If $0 \in T$ and $\mathbb{P} - T$ is finite, then there is a matroid M for which $\mathcal{K}(M) = T$.*

Proof Let $k = \prod_{p \in \mathbb{P} - T} p$ and $M = M[A_k]$. The result follows immediately from Lemma 6.8.5. □

The following result of Kahn (1982) completes the answer to Problem 6.8.1.

Proposition 6.8.8 *If T is a finite set of prime numbers, then there is a matroid whose characteristic set is T.* □

A sketch of the main ideas in the proof of the last result can be found in White (1987b). This proof uses several transcendental field extensions in its construction of a matroid with characteristic set T, so the matroids formed are, in general, not representable over the corresponding prime fields. By extending the ideas in Example 6.8.4 and some other unpublished work of R. Reid, Brylawski (1982b) constructed large numbers of rank-3 matroids that are only representable over a small number of prime fields. For example, one such matroid is representable over $GF(23)$ and $GF(59)$ but over no other characteristics.

In contrast to the classes of matroids with small characteristic sets just considered, many of the matroids M we have already met have characteristic set equal to $\mathbb{P} \cup \{0\}$. This is true, for example, if M is regular, and so, in particular, if M is graphic or cographic. Moreover, it follows from a result of Piff and Welsh (1970), to be proved in Chapter 11, that if M is transversal, and hence if M is uniform, then $\mathcal{K}(M) = \mathbb{P} \cup \{0\}$.

One important issue that has yet to be discussed in this chapter is whether there is a finite algorithm to determine if a given matroid is representable. More precisely, let M be a matroid $(E, \mathcal{I})$. Is there a finitely terminating algorithm that will determine whether or not there is a field over which M is representable? Note that this question makes no mention of a specific field. After we have answered the question, we shall briefly discuss the natural variant of it when the field is specified. The answer to the original question will require some algebraic machinery that is closely linked to what is needed to prove, for example, Proposition 6.8.2. Some of the theory underlying this machinery will be omitted but references will be provided to help the reader close the gaps.

Vámos (1971a) showed that the problem of determining whether a given matroid is representable is decidable.

Theorem 6.8.9 *Let M be a matroid $(E, \mathcal{I})$. There is a finite algorithm for determining whether or not M is representable.*

Proof We can certainly determine, in finitely many steps, the rank r of M and its set $\mathcal{B}$ of bases. Let $|E| = n$. Let X be an $r \times n$ matrix $[x_{ij}]$ whose columns are labelled by the elements of E. Of course, M is representable if and only if there is some field $\mathbb{F}$ so that we can replace each x_{ij} by an element of $\mathbb{F}$ to give a matrix $X_{\mathbb{F}}$ such that $M = M[X_{\mathbb{F}}]$. We shall denote by $\mathbb{Z}[x_{ij}]$ the polynomial ring over $\mathbb{Z}$ with rn variables x_{ij}. For each r-element subset Y of $E(M)$, let p_Y be the determinant of the $r \times r$ submatrix of X whose set of columns is labelled by Y. Then $p_Y \in \mathbb{Z}[x_{ij}]$. The set Y is dependent in M if and only if $p_Y = 0$. If Y is independent in M, then it is a basis of M, and $p_Y \neq 0$. Since we are seeking a representation over some field, we can encapsulate all of these inequations corresponding to bases into a single equation by introducing a new variable y that will be the inverse of $\prod_{B \in \mathcal{B}} p_B$ in any field, that is,

$$y \prod_{B \in \mathcal{B}} p_B - 1 = 0.$$

Thus if S is the polynomial ring $\mathbb{Z}[x_{ij}; y]$ in the $rn + 1$ variables x_{ij} and y, we have a set of equations in S that precisely capture what it means for M to be representable over some field. The corresponding ideal I in S is generated by $\{p_Y : Y \text{ is a dependent } r\text{-element set in } M\} \cup \{y \prod_{B \in \mathcal{B}} p_B - 1\}$.

Next we show that

6.8.10 *M is representable if and only if $1 \notin I$.*

Suppose that $1 \in I$. For a representation to exist, we must be able to choose values for each x_{ij} and y in some field $\mathbb{F}$ so that the evaluation in $\mathbb{F}$ of each generator of I and hence of each member of I is zero. But this is impossible since $1 \in I$, so M is not representable. Conversely, suppose that $1 \notin I$. Then $I \subsetneq S$, so I is contained in a maximal ideal T of S. As factoring out a maximal ideal from a commutative ring with unity gives a field (see, for example, Hungerford 1974, Theorem 2.20), S/T is a field. It is over this field that M is representable. To see this, observe that there is a natural homomorphism φ from S into S/T determined by $\varphi(a) = a + T$. We replace each entry x_{ij} of X by $\varphi(x_{ij})$. Let the resulting matrix be X_φ. To see that $M = M[X_\varphi]$, we observe that if Y is an r-element dependent set in M, then $p_Y \in I$, so $p_Y \in T$. Thus the determinant of the submatrix of X_φ whose columns are labelled by Y is zero in S/T, so Y is dependent in $M[X_\varphi]$. On the other hand, let B' be a basis of M. We should like to show that $p_{B'}$ maps to a non-zero element in S/T. Suppose not. Then $p_{B'} \in T$. Since $y \prod_{B \in \mathcal{B}} p_B - 1 \in I \subseteq T$, it follows that $1 \in T$ contradicting the fact that T is a proper ideal. Hence $p_{B'} \notin T$. Thus the determinant of the submatrix of X_φ whose set of columns is labelled by B' is non-zero in S/T. Hence X_φ does indeed represent M over the field S/T, and 6.8.10 holds.

By 6.8.10, to complete the proof of the theorem, we need to establish that there is a finite algorithm for determining whether or not $1 \in I$. Here we rely on the theory of Gröbner bases. Although this theory is more usually applied to polynomial rings over fields, it extends, for example, to polynomial rings over $\mathbb{Z}$ (see, for example, Adams and Loustaunau 1994, Chapter 4). Following the last reference (p. x), we now briefly outline the main ideas of this theory. Let $\mathbb{F}$ be a field and I be an ideal in $\mathbb{F}[x]$ that is generated by $f_1, f_2, \ldots, f_n$. Then, by using the Euclidean algorithm, we can compute a single polynomial d, the greatest common divisor of $f_1, f_2, \ldots, f_n$, such that d generates I. Then an arbitrary polynomial g in $\mathbb{F}[x]$ is in I if and only if the remainder, when divided by d, is zero. Analogously, a Gröbner basis G for an ideal I in $\mathbb{Z}[y_1, y_2, \ldots, y_m]$ generates I; and a member f of $\mathbb{Z}[y_1, y_2, \ldots, y_m]$ is in I if and only if the remainder of the division of f by the polynomials in G is zero. An important aspect of the theory is formulating an appropriate concept of division and Adams and Loustaunau illustrate this with numerous helpful examples. From our perspective, the two key points of the theory are that, for every ideal I in $\mathbb{Z}[y_1, y_2, \ldots, y_m]$,

(i)   there is a finite algorithm to determine a Gröbner basis for I; and

(ii) given a Gröbner basis, there is a finite algorithm to determine whether any element, specifically 1, is in I.

By applying (i) and (ii), we conclude that the theorem holds. □

The next proposition is one of the two results that Rado used in proving Proposition 6.8.2.

Proposition 6.8.11 *Suppose that a matroid M is representable over a field $\mathbb{F}$. If $\mathbb{F}$ has characteristic p, then M is representable over $GF(p^k)$ for some positive integer k. If $\mathbb{F}$ has characteristic 0, then M is representable over $\mathbb{Q}(\alpha)$ for some α that is algebraic over $\mathbb{Q}$ and so M is representable over $\mathbb{C}$.* □

We sketch a proof of this result using the ideas in the last proof together with the following weak version of Hilbert's Nullstellensatz (Adams and Loustaunau, 1994, Theorem 2.2.3), which is proved in Atiyah and MacDonald (1969).

Theorem 6.8.12 *Let $\mathbb{K}$ be a field and $\mathbb{L}$ be its algebraic closure. Let I be an ideal in $\mathbb{K}[x_1, x_2, \ldots, x_n]$. Then there are elements $a_1, a_2, \ldots, a_n$ of $\mathbb{L}$ such that $f(a_1, a_2, \ldots, a_n) = 0$ for all f in I if and only if $I \neq \mathbb{K}[x_1, x_2, \ldots, x_n]$.* □

Proof of Proposition 6.8.11   We prove the result when $\mathbb{F}$ has characteristic zero leaving the characteristic-p case as an exercise. We argue as in the proof of Theorem 6.8.9. Let M have n elements and rank r. Beginning with an $r \times n$ matrix $X = [x_{ij}]$ whose columns are labelled by the elements of M, we can replace each x_{ij} by an element of $\mathbb{F}$ such that the resulting matrix $X_{\mathbb{F}}$ represents M over $\mathbb{F}$. Let $I_{\mathbb{F}}$ be the ideal in $\mathbb{F}[x_{ij}; y]$ generated by $\{p_Y :$ Y is a dependent r-element set in $M\} \cup \{y \prod_{B \in \mathcal{B}} p_B - 1\}$. Since M is representable, Theorem 6.8.12 implies that $I_{\mathbb{F}} \neq \mathbb{F}[x_{ij}; y]$. Now each generator for $I_{\mathbb{F}}$ is in $\mathbb{Z}[x_{ij}; y]$ and so is in $\mathbb{Q}[x_{ij}; y]$. Let $I_{\mathbb{Q}}$ be the ideal in $\mathbb{Q}[x_{ij}; y]$ generated by $\{p_Y :$ Y is a dependent r-element set in $M\} \cup \{y \prod_{B \in \mathcal{B}} p_B - 1\}$.. As $I_{\mathbb{F}} \neq \mathbb{F}[x_{ij}; y]$, it follows that $I_{\mathbb{Q}} \neq \mathbb{Q}[x_{ij}; y]$. Thus, by Theorem 6.8.12 again, there are elements a_{ij} in the algebraic closure $\mathbb{A}$ of $\mathbb{Q}$ so that $[a_{ij}]$ represents M over $\mathbb{A}$. As every entry of $[a_{ij}]$ is algebraic over $\mathbb{Q}$, it is in some finite algebraic extension of $\mathbb{Q}$. Hence there is some finite algebraic extension $\mathbb{F}'$ of $\mathbb{Q}$ that contains every element of $[a_{ij}]$, so M is representable over $\mathbb{F}'$. Since every finite extension of a field of characteristic zero is simple (see, for example, Herstein, 1975, p. 236), we deduce that $\mathbb{F}' = \mathbb{Q}(\alpha)$ for some α that is algebraic over $\mathbb{Q}$. Thus M is representable over $\mathbb{Q}(\alpha)$ and hence over $\mathbb{C}$. □

By combining Proposition 6.8.11 with Proposition 6.8.2, we obtain the following interesting corollary. The proof is left as an exercise.

Corollary 6.8.13 *Every representable matroid is representable over infinitely many finite fields.* □

Again using Gröbner bases over the integers, Baines and Vámos (2003) proved the following result, which blends the two themes of this section.

Theorem 6.8.14 *Let M be a matroid $(E, \mathcal{I})$. There is a finite algorithm for determining its characteristic set, $\mathcal{K}(M)$.* □

To conclude this section, we consider, for a fixed field $\mathbb{F}$, whether there is a finite algorithm to determine if a given matroid M is $\mathbb{F}$-representable. When $\mathbb{F}$ is finite, such a finite algorithm certainly exists. Now let $\mathbb{F}$ be infinite. We may proceed as before to consider the ideal $I_{\mathbb{F}}$ in $\mathbb{F}[x_{ij}; y]$ generated by $\{p_Y :$ Y is a dependent r-element set in $M\} \cup \{y \prod_{B \in \mathcal{B}} p_B - 1\}$. By using Gröbner bases again, this time over the field $\mathbb{F}$, we get a finite algorithm to determine whether $I_{\mathbb{F}} \neq \mathbb{F}[x_{ij}; y]$. However, knowing this need not settle the question of whether M is $\mathbb{F}$-representable since it follows, by Theorem 6.8.12, that $I_{\mathbb{F}} \neq \mathbb{F}[x_{ij}; y]$ if and only if M is representable over the algebraic closure of $\mathbb{F}$.

Thus when $\mathbb{F}$ is algebraically closed, the problem of determining whether a given matroid if $\mathbb{F}$-representable is decidable. But what if $\mathbb{F}$ is not algebraically closed, say $\mathbb{F}$ is $\mathbb{R}$ or $\mathbb{Q}$? By a result of Tarski (1951) (see also Jacobson 1974, Theorem 5.6), the problem of determining whether a given matroid is $\mathbb{R}$-representable is decidable. On the other hand, Sturmfels (1987) showed that the problem of determining whether a given matroid is $\mathbb{Q}$-representable is equivalent to Hilbert's Tenth Problem for $\mathbb{Q}$. The latter asks whether it is decidable if a single multivariate polynomial with integer coefficients has a solution in $\mathbb{Q}$. Although it was proved by Matiyasevic (1993) that Hilbert's Tenth Problem for $\mathbb{Z}$ is undecidable, the problem remains open for $\mathbb{Q}$.

Problem 6.8.15 *Is there a finite algorithm to determine whether a given matroid is $\mathbb{Q}$-representable?*   □

Sturmfels also showed that a conjecture of Grünbaum (1972, Conjecture 2.14) is equivalent to the assertion that the answer to the last problem is negative.

Exercises

1. With A_k as in Example 6.8.4, show that $M[A_2] \cong F_7^-$.
2. Let M be a matroid that is representable over $\mathbb{Q}$. Show that there is an integer $n(M)$ such that M is representable over $GF(p)$ for all primes p exceeding $n(M)$.
3. (Reid in Brylawski and Kelly 1980) Let p be a prime number exceeding two and M_p be the vector matroid of the following matrix over $GF(p)$:

$$
\begin{bmatrix}
1 & 0 & 0 & 1 & 1 & 0 & 0 & \cdots & 0 & 1 & 1 & \cdots & 1 \\
0 & 1 & 0 & 1 & 1 & 1 & 1 & \cdots & 1 & 0 & 0 & \cdots & 0 \\
0 & 0 & 1 & 1 & 0 & 1 & 2 & \cdots & p-1 & 1 & 2 & \cdots & p-1
\end{bmatrix}.
$$

   (a) Give a geometric representation for M_p.
   (b) Show that $M[A_p]$ is isomorphic to a relaxation of M_p.
   (c) Prove that M_p has characteristic set $\{p\}$.
   (d) *Delete the third element from M_p. Prove that the resulting matroid is an excluded minor for $\mathbb{R}$-representability.

4. (Brylawski 1982b) Let p be an odd prime number and let $m = \lfloor \log_2(p+1) \rfloor$. For $0 \le i \le m$, define x_i to be $\lfloor (p+1)/2^{m-i+1} \rfloor$. Let N_p be the vector matroid of the following matrix over $GF(p)$:

$$
\begin{array}{ccccccccccccc}
e_1 & e_2 & e_3 & e_4 & e_5 & e_6 & f_1 & g_1 & f_2 & g_2 & \cdots & f_m & g_m \\
\begin{bmatrix}
1 & 0 & 0 & 1 & 1 & 1 & 1 & 0 & 1 & 0 & \cdots & 1 & 0 \\
0 & 1 & 0 & 1 & 1 & 0 & 2 & 1 & 2 & 1 & \cdots & 2 & 1 \\
0 & 0 & 1 & 1 & 0 & 1 & x_1 & x_1 & x_2 & x_2 & \cdots & x_m & x_m
\end{bmatrix}
\end{array}.
$$

   (a) Show that N_{13} is the vector matroid of the following matrix over $GF(13)$:

$$\begin{bmatrix} 1 & 0 & 0 & 1 & 1 & 1 & 1 & 0 & 1 & 0 & 1 & 0 \\ 0 & 1 & 0 & 1 & 1 & 0 & 2 & 1 & 2 & 1 & 2 & 1 \\ 0 & 0 & 1 & 1 & 0 & 1 & 1 & 1 & 3 & 3 & 7 & 7 \end{bmatrix}.$$

(b) Suppose that $y_1 y_2 \cdots y_{m+1}$ is $p+1$ written in binary, that is, $p+1 = y_1 2^m + y_2 2^{m-1} + \cdots + y_{m+1} 2^0$. Show that x_i is the number which, when written in binary, is $y_1 y_2 \cdots y_i$.

(c) *Show that N_p has characteristic set $\{p\}$.

(d) Suppose that $p \geq 5$. Prove that $AG(2,p)$ has characteristic set $\{p\}$ by showing that it has N_p as a restriction.

6.9 Modularity

Let X and Y be flats in a matroid M. Then (X, Y) is a *modular pair of flats* if

$$r(X) + r(Y) = r(X \cup Y) + r(X \cap Y).$$

If Z is a flat of M such that (Z, Y) is a modular pair for all flats Y, then Z is called a *modular flat* of M. Evidently $E(M)$, $\mathrm{cl}(\emptyset)$, and all rank-1 flats of M are modular flats; so are all separators of M. In this section, we shall study some of the properties of modular flats. Such flats are, in general, relatively well-behaved and the property of modularity will be particularly important in certain matroid constructions that will be considered in Chapters 7 and 11. We noted in Section 6.1 that every projective geometry $PG(n, q)$ has the property that all its flats are modular. Another important result of this section will note that there are relatively few other matroids with this property.

Because modularity is a property of flats, it is basically a lattice property. Indeed, it is clear that a flat X is modular in a matroid M if and only if, in the simple matroid $\mathrm{si}(M)$ associated with M, the flat corresponding to X is modular. For this reason, much of the present discussion of modularity will concentrate on simple matroids. The reader should have little difficulty deriving the corresponding results for non-simple matroids.

We begin by looking at *modular matroids*, those matroids in which every flat is modular. Every free matroid, $U_{n,n}$, is easily seen to be modular; so too are every rank-2 matroid and every finite projective plane. Since, by Theorem 6.1.2, a finite projective geometry of dimension greater than two is isomorphic to some $PG(n, q)$, we deduce that every finite projective geometry is a modular matroid. Moreover, it is easy to show that the direct sum of two modular matroids is also modular. Thus, by taking direct sums of free matroids and finite projective geometries, one can construct more examples of modular matroids. The next result asserts that every simple modular matroid is obtainable in this way.

Proposition 6.9.1 *A matroid M is modular if and only if, for every connected component N of M, the simple matroid associated with N is either a free matroid or a finite projective geometry.*

Proof A proof, in the context of geometric lattices, can be found in Birkhoff (1967, pp. 90–93). □

A flat Y in a matroid M is a *complement* of a flat X if $X \cap Y = \mathrm{cl}(\emptyset)$ and $\mathrm{cl}(X \cup Y) = E(M)$. Thus Y is a complement of X if and only if, in $\mathcal{L}(M)$, the meet and join of X and Y are the zero and one, respectively, of the lattice.

The remaining results in this section are taken from Brylawski (1975b). The first such result gives two alternative characterizations of modular flats which appear to be progressively weaker than the definition.

Proposition 6.9.2 *The following statements are equivalent for a flat X in a matroid M:*

(i) X *is modular.*
(ii) $r(X) + r(Y) = r(X \cup Y)$ *for all flats Y that meet X in* $\mathrm{cl}(\emptyset)$.
(iii) $r(X) + r(Y) = r(M)$ *for all complements Y of X.*

Proof Evidently (i) implies (ii), and (ii) implies (iii). We shall show that (iii) implies (ii) and omit the similar proof that (ii) implies (i).

Assume that (iii) holds, but (ii) does not, and let Y be a flat for which $X \cap Y = \mathrm{cl}(\emptyset)$ and $r(X) + r(Y) \neq r(X \cup Y)$. Because (iii) holds, Y is not a complement of X, so $r(X \cup Y) < r(M)$. Let Z be a basis for $M/(X \cup Y)$. We show next that $\mathrm{cl}(Y \cup Z)$ is a complement of X. Clearly $\mathrm{cl}(X \cup \mathrm{cl}(Y \cup Z)) = E(M)$. Now suppose that a is a non-loop element of $X \cap \mathrm{cl}(Y \cup Z)$. Then $a \notin Y$ since $X \cap Y = \mathrm{cl}(\emptyset)$. Thus if B is a basis of Y, then $B \cup a$ is independent in $M|(X \cup Y)$. But Z is independent in $M/(X \cup Y)$. Therefore $Z \cup B \cup a$ is independent in M, a contradiction to the fact that $a \in \mathrm{cl}(Y \cup Z)$. Hence $X \cap \mathrm{cl}(Y \cup Z) = \mathrm{cl}(\emptyset)$, so $\mathrm{cl}(Y \cup Z)$ is indeed a complement of X. By (iii),

$$r(M) = r(X) + r(\mathrm{cl}(Y \cup Z)) = r(X) + r(Y) + |Z|$$
$$= r(X) + r(Y) + r(M) - r(X \cup Y).$$

Therefore $r(X \cup Y) = r(X) + r(Y)$. This contradiction completes the proof that (iii) implies (ii). □

This following is an immediate consequence of the last proposition. Recall that a *line* in a matroid is a rank-2 flat.

Corollary 6.9.3 *In a loopless matroid, a hyperplane is modular if and only if it intersects every line; and a line is modular if and only if it intersects every hyperplane.* □

Example 6.9.4 Consider the matroids $\mathcal{W}^3$ and M for which geometric representations are shown in Figure 6.32. From the last corollary, we deduce that $\mathcal{W}^3$ has no modular hyperplanes, whereas M has exactly two modular hyperplanes, $\{2, 3, 4, 8\}$ and $\{5, 6, 7, 8\}$. □

The next result enables us to identify certain particular flats as being modular. Its straightforward proof is left to the reader (Exercise 6).

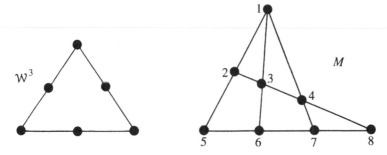

FIG. 6.32. M has two modular lines; $\mathcal{W}^3$ has none.

Proposition 6.9.5 *Let X be a modular flat in a matroid M and T be a subset of $E(M)$ containing X. Then X is a modular flat of $M|T$.*   □

Using this result, the reader can easily prove the following.

Corollary 6.9.6 *If M is a simple matroid representable over $GF(q)$, and X is a subset of $E(M)$ such that $M|X \cong PG(k-1,q)$ for some k, then X is a modular flat of M.*   □

From this corollary, we deduce that 3-point and 4-point lines are always modular flats in simple binary and simple ternary matroids, respectively. In particular, a 3-circuit is always a modular flat in a simple regular matroid.

The proofs of the next two results are based on the characterization of modular flats in (iii) of Proposition 6.9.2. Each adds further support to the earlier assertion that modular flats are relatively well-behaved.

Proposition 6.9.7 *Let X be a modular flat in a matroid M and Y be a modular flat in $M|X$. Then Y is a modular flat in M.*

Proof Clearly Y is a flat of M. Now let Z be a complement of Y in M. Then $\mathrm{cl}(Y \cup Z) = E(M)$. Thus, as $X \supseteq Y$, it follows that $\mathrm{cl}(X \cup Z) = E(M)$. Moreover, $X \cap Z \subseteq \mathrm{cl}((X \cap Z) \cup Y) \subseteq X$, so $X \cap Z \subseteq \mathrm{cl}((X \cap Z) \cup Y) \cap Z \subseteq X \cap Z$, that is,

$$\mathrm{cl}((X \cap Z) \cup Y) \cap Z = X \cap Z.$$

Therefore, since $Y \cap Z = \mathrm{cl}(\emptyset)$, the Hasse diagram in Figure 6.33 represents a sublattice of $\mathcal{L}(M)$ where, possibly, $\mathrm{cl}((X \cap Z) \cup Y) = X$.

As X is a modular flat and $\mathrm{cl}(X \cup Z) = E(M)$,

$$r(M) = r(X \cup Z) = r(X) + r(Z) - r(X \cap Z). \tag{6.6}$$

Thus

$$
\begin{aligned}
r(M) &\geq r(\mathrm{cl}((X \cap Z) \cup Y)) + r(Z) - r(X \cap Z) \\
&= r(\mathrm{cl}((X \cap Z) \cup Y)) + r(Z) - r(\mathrm{cl}((X \cap Z) \cap Y) \cap Z) \\
&\geq r(\mathrm{cl}((X \cap Z) \cup Y \cup Z)) \\
&\geq r(Y \cup Z) = r(M).
\end{aligned}
$$

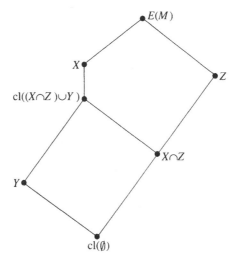

FIG. 6.33. A sublattice of $\mathcal{L}(M)$.

Clearly equality must hold throughout the last sequence of inequalities and so

$$r(M) = r(\mathrm{cl}((X \cap Z) \cup Y)) + r(Z) - r(X \cap Z). \qquad (6.7)$$

Comparing (6.6) and (6.7), we deduce, since $\mathrm{cl}((X \cap Z) \cup Y) \subseteq X$, that $X = \mathrm{cl}((X \cap Z) \cup Y)$. Thus, in the Hasse diagram in Figure 6.33, the vertices corresponding to $\mathrm{cl}((X \cap Z) \cup Y)$ and X coincide. This diagram now suggests how to finish the proof, for Y is modular in $M|X$, so

$$\begin{aligned} r(Y) + r(X \cap Z) &= r((X \cap Z) \cup Y) + r((X \cap Z) \cap Y) \\ &= r(X) + r(\mathrm{cl}(\emptyset)). \end{aligned}$$

Substituting for $r(X \cap Z)$ from (6.6) and simplifying, we get that

$$r(Y) + r(Z) = r(M),$$

and the proposition follows by Proposition 6.9.2(iii). □

Corollary 6.9.8 *If X and Y are modular flats in a matroid M, then $X \cap Y$ is a modular flat.*

Proof By the last result, it suffices to show that $X \cap Y$ is modular in $M|X$. This can be done using an argument similar to the last one. We leave it to the reader to complete the details. □

Brylawski (1975b) called the next result the *modular short-circuit axiom*. It provides a useful characterization of modular flats in terms of circuits.

Theorem 6.9.9 *The following statements are equivalent for a non-empty set X of elements in a simple matroid M:*

(i)   *X is a modular flat of M.*
(ii) *For every circuit C such that C − X is non-empty, there is an element x of X such that $(C - X) \cup x$ is dependent.*
(iii) *For every circuit C and every element e of C − X, there is an element f of X and a circuit C′ such that $e \in C' \subseteq f \cup (C - X)$.*

Proof Assume that X satisfies (iii). We shall show that X satisfies (i). As M is simple and $\mathrm{cl}(X) = X \cup \{x \colon M \text{ has a circuit } D \text{ with } x \in D \subseteq X \cup x\}$, it follows easily that $\mathrm{cl}(X) = X$. Now assume that X is not modular. Then, by Proposition 6.9.2(ii) and submodularity, M has a flat Y disjoint from X such that $r(X) + r(Y) > r(X \cup Y)$. Let B_X and B_Y be bases for X and Y, respectively. Then, as $r(B_X \cup B_Y) < |B_X \cup B_Y|$, the set $B_X \cup B_Y$ must contain a circuit C, and C must meet both X and Y. Choose e in $C - X$. Then there is an element f in X and a circuit $C′$ such that $e \in C' \subseteq f \cup (C - X)$. Hence $f \in X \cap \mathrm{cl}(Y) = X \cap Y = \emptyset$; a contradiction. Thus (iii) implies (i).

Next we show that (i) implies (ii). Suppose that X is a modular flat and let C be a circuit such that $C - X$ is non-empty. If $C \cap X$ is empty, then (ii) certainly holds. Thus we may assume that $C \cap X$ is non-empty. Then both $C \cap X$ and $C - X$ are independent. Now (ii) will hold if we can show that $X \cap \mathrm{cl}(C - X)$ is non-empty. Assume the contrary. Then, as X is modular,

$$r(X \cup \mathrm{cl}(C - X)) = r(X) + r(\mathrm{cl}(C - X))$$
$$= r(X) + r(C - X)$$
$$= r(X) + |C - X|.$$

Therefore, as $r(X \cup \mathrm{cl}(C - X)) = r(X \cup \mathrm{cl}(C))$,

$$r(X \cup \mathrm{cl}(C)) = r(X) + |C - X|. \tag{6.8}$$

Again, as X is modular,

$$r(X \cup \mathrm{cl}(C)) = r(X) + r(\mathrm{cl}(C)) - r(X \cap \mathrm{cl}(C))$$
$$\leq r(X) + r(C) - r(X \cap C). \tag{6.9}$$

Therefore, as $r(X \cap C) = |X \cap C|$, we get, on combining (6.8) and (6.9), that $|C| \leq r(C)$, a contradiction to the fact that C is a circuit. Hence $X \cap \mathrm{cl}(C - X)$ is non-empty, and (i) implies (ii).

Finally, we show that (ii) implies (iii). Let C be a circuit of M and e be an element of $C - X$. By (ii), there is an element x of X such that $(C - X) \cup x$ is dependent. We shall argue by induction on $|C - X|$ to show that the required circuit $C′$ exists. Since $x \in X \cap \mathrm{cl}(C - X)$, there is a circuit C_1 such that $x \in C_1 \subseteq (C - X) \cup x$. As M is simple, $|C_1| \geq 3$. Thus $|C - X| \geq 2$, and if $|C - X| = 2$, then $e \in C_1$ and we can take $C' = C_1$ and $f = x$. Thus (iii) holds when $|C - X| = 2$. Assume it holds for $|C - X| < k$ and let $|C - X| = k$. If $e \in C_1$, we again take $C' = C_1$ and $f = x$. Thus assume $e \notin C_1$. Now take g in

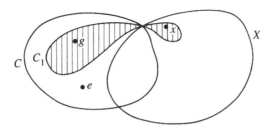

FIG. 6.34. Applying strong circuit elimination.

$C_1 \cap C$ (see Figure 6.34). Then, by the strong circuit elimination axiom, there is a circuit C_2 such that $x \in C_2 \subseteq (C \cup C_1) - g$. Since $(C_1 \cup C_2) - x \subseteq C$ and $(C_1 \cup C_2) - x$ contains a circuit, we must have that $(C_1 \cup C_2) - x = C$. Hence $C_2 \supseteq C - C_1$, so $e \in C_2$. As $C_2 - X$ is a subset of $(C - g) - X$, it follows that $|C_2 - X| < |C - X|$. Thus, by the induction assumption, as $e \in C_2$, there is a circuit C_3 and an element h of X such that $e \in C_3 \subseteq h \cup (C_2 - X)$. But $h \cup (C_2 - X) \subseteq h \cup (C - X)$ so, letting $C' = C$ and $f = h$, we have a circuit and an element with the required properties, and (iii) follows by induction. $\square$

We close this section by noting two consequences of the last theorem, neither of which has a difficult proof. Some further consequences of the theorem will be considered in the exercises. The first application of modularity we shall meet will be in Section 7.2 where we consider the reverse of the deletion operation.

Corollary 6.9.10 *In a simple matroid M, let X be a flat and $X_1, X_2, \ldots, X_k$ be the components of $M|X$. Then X is modular if and only if M has distinct components $M_1, M_2, \ldots, M_k$ such that X_i is a modular flat of M_i for all i. In particular, modular flats of simple connected matroids are connected.* $\square$

One can determine when a set is a modular flat in a graphic matroid by combining the last corollary with the following result.

Proposition 6.9.11 *Let X be a set of edges of a simple graph G and suppose that $G[X]$ is connected. Then X is a modular flat in $M(G)$ if and only if, whenever a pair $\{u, v\}$ of distinct vertices of $G[X]$ is joined by a path that meets $G[X]$ in $\{u, v\}$, there is an edge of X joining u and v.* $\square$

An immediate consequence of this proposition is that if G is a simple graph and $G[X] \cong K_n$ for some $n \geq 2$, then X is a modular flat of $M(G)$. However, not every modular flat in a graphic matroid comes from a complete subgraph.

Example 6.9.12 Consider the graph G in Figure 6.35. By Proposition 6.9.11, $M(G)$ has several modular flats including $\{1, 2, 3, 4, 5\}, \{6, 7, 8, 9, 10, 11\}, \{1, 2, 3, 4, 5, 11, 12, 13\}$, and $\{3, 6, 7, 8, 9, 10, 11, 12, 13\}$. $\square$

Exercises

1. Prove that the direct sum of two modular matroids is modular.

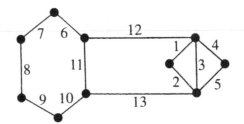

FIG. 6.35. Not every modular flat comes from a complete subgraph.

2. Show that a matroid is modular if and only if every hyperplane is modular.

3. Prove that (ii) implies (i) in Proposition 6.9.2.

4. Find all simple $GF(q)$-representable matroids N with the property that if M is a simple $GF(q)$-representable matroid of which N is a restriction, and $E(N)$ is a flat of M, then $E(N)$ is a modular flat of M.

5. Prove that a line X in a connected, simple, binary matroid is a modular flat if and only if $|X| = 3$.

6. Prove Proposition 6.9.5.

7. (Brylawski 1975b)

  (a) For distinct elements x and y in a simple matroid, prove that $\{x, y\}$ is a modular flat if and only if x and y are in different components.
  (b) Use (a) and Theorem 6.9.9 to prove Corollary 6.9.10.

8. (Brylawski 1975b) Suppose that X is a flat of a matroid M and assume that $M/X = M_1 \oplus M_2$. Prove that $E(M_1) \cup X$ is a modular flat of M if and only if X is a modular flat of $M|(E(M_2) \cup X)$.

9. (Brylawski (1975b) Prove that the following statements are equivalent for a set X in a simple matroid M:

  (a) X is a modular flat.
  (b) If I is an independent subset of $E(M) - X$ and $I \cup e$ is in $\mathcal{I}(M)$ for all e in X, then $I \cup I' \in \mathcal{I}(M)$ for all independent subsets I' of X.

10. (White 1971) Let C_1 and C_2 be distinct circuits in a matroid M such that $(E(M) - C_1, E(M) - C_2)$ is a modular pair of flats of M^*. Show that:

  (a) $r(C_1) + r(C_2) = r(C_1 \cup C_2) + r(C_1 \cap C_2)$.
  (b) $r(C_1 \cup C_2) = |C_1 \cup C_2| - 2$.
  (c) If $e \in C_1 \cap C_2$, then M has a unique circuit C_3 contained in the set $(C_1 \cup C_2) - e$. Moreover, $C_3 \supseteq (C_1 - C_2) \cup (C_2 - C_1)$.
  (d) (Fournier 1987) If B is a basis and C is a circuit of M such that $|C - B| \geq 2$, then M has two distinct circuits D_1 and D_2 such that $(D_1 - D_2) \cup (D_2 - D_1) \subseteq C$ and $(E(M) - D_1, E(M) - D_2)$ is a modular pair of flats of M^*.

11. *Let M be a simple rank-r matroid having W_1 points and W_{r-1} hyperplanes. Prove that:

(a) (Motzkin 1951; Basterfield and Kelly 1968; Greene 1970; Heron 1973) $W_1 \leq W_{r-1}$.

(b) (Greene 1970) $W_1 = W_{r-1}$ if and only if M is modular.

(c) (Mason 1973) $\mathcal{L}(M)$ has W_1 disjoint maximal chains each of which is from a point to a hyperplane.

6.10 Dowling geometries

Dowling (1973a, 1973b) introduced a class of matroids that play a similar role to that of projective geometries. In this section, we define these matroids, note some of their important properties, and consider some generalizations of them. We begin by describing a particular class of Dowling geometries which will help provide some intuition for such matroids in general. Recall that, for a field $\mathbb{F}$, we denote by $\mathbb{F}^\times$ the multiplicative group of non-zero elements of $\mathbb{F}$.

Example 6.10.1 Let $\{\underline{v}_1, \underline{v}_2, \ldots, \underline{v}_r\}$ be a basis of $V(r,q)$. The *Dowling geometry* $Q_r(GF(q)^\times)$ is the rank-r simple matroid obtained by restricting the matroid $V(r,q)$ to $\{\underline{v}_1, \underline{v}_2, \ldots, \underline{v}_r\} \cup \{\underline{v}_i - \alpha \underline{v}_j : 1 \leq i < j \leq r \text{ and } \alpha \in GF(q)^\times\}$, that is, to $\{\underline{v}_1, \underline{v}_2, \ldots, \underline{v}_r\} \cup \{\underline{v}_i + \alpha \underline{v}_j : 1 \leq i < j \leq r \text{ and } \alpha \in GF(q)^\times\}$. Evidently, $Q_2(GF(q)^\times) \cong U_{2,q+1}$. Moreover, $Q_r(GF(2)^\times) \cong M(K_{r+1})$. In general, $Q_r(GF(q)^\times)$ has exactly $r + (q-1)\binom{r}{2}$ elements. This matroid has numerous other attractive properties about which more will be said later. □

Next, we introduce a generalization of the matroids considered above, although the link may not be immediately apparent. Let A be a finite (multiplicative) group with identity element 1, and let r be a non-negative integer. The *Dowling geometry* $Q_r(A)$, a rank-r matroid which we describe next, was introduced by Dowling (1973a) using a lattice-theoretic approach. Our approach to these matroids is a blend of those taken by Kahn and Kung (1982), Whittle (1989b), and Zaslavsky (1989, 1991). First, we define $Q_0(A)$ to be $U_{0,0}$. For $r \geq 1$, we begin by constructing a graph K_r^A on the vertex set $\{1, 2, \ldots, r\}$ as follows:

(i) for each pair (i,j) of vertices with $i < j$ and each α in A, add an edge joining i and j labelled α_{ij}; and

(ii) for each vertex i, add a loop at i giving it a label, say b_i, that has nothing to do with A.

As an example, note that when A is the multiplicative group of order 2 whose elements are 1 and -1, the graph K_3^A is as shown in Figure 6.36(a). In general, the ground set of $Q_r(A)$ is $\{b_1, b_2, \ldots, b_r\} \cup \{\alpha_{ij} : 1 \leq i < j \leq r \text{ and } \alpha \in A\}$. For $1 \leq j < i \leq r$, it will be convenient to define α_{ij} to be $(\alpha)_{ji}^{-1}$ and to abbreviate the latter as α_{ji}^{-1}. To define the circuits of $Q_r(A)$, we first distinguish certain cycles of K_r^A. Let C be such a cycle with at least two edges and arbitrarily assign an orientation to it. Let the vertices and edges of C, beginning with a vertex, be $i_1, e_1, i_2, e_2, \ldots, i_k, e_k, i_1$. If e_j is labelled by the element β of A, we let $\phi(e_j)$ be β if $i_j < i_{j+1}$, and we let $\phi(e_j)$ be β^{-1} otherwise. We call C *balanced* if $\phi(e_1)\phi(e_2)\ldots\phi(e_k) = 1$. A cycle is *unbalanced* if either it has a single edge, or it has at least two edges but is not balanced. The reader can easily check that neither the orientation of C nor the vertex at which one begins the cycle will affect whether or not C is balanced. Thus this concept is well-defined. Consider the three graphs shown in Figure 6.37. A subdivision of such a graph is called, respectively, a Θ-*graph*, a *loose handcuff*, or a *tight handcuff*. The circuits of $Q_r(A)$ consist of the edge sets of all of the balanced cycles together with the edge sets of all of the Θ-graphs, loose handcuffs, and tight handcuffs in which, in all three of these cases, none of the cycles is balanced. We delay, until after the following example, the proof that $Q_r(A)$ is actually a matroid.

Example 6.10.2 As noted above, if A is the multiplicative group on $\{1, -1\}$, then K_3^A is as shown in Figure 6.36(a). Clearly K_3^A has exactly four balanced cycles: $\{1_{12}, 1_{23}, 1_{13}\}$, $\{1_{12}, -1_{23}, -1_{13}\}$, $\{-1_{12}, 1_{23}, -1_{13}\}$, and $\{-1_{12}, -1_{23}, 1_{13}\}$. Using this, it is not difficult to show that $Q_3(A)$ is the matroid having the geometric representation shown in Figure 6.36(b), where the four balanced cycles correspond to the four curved lines. Note that, in this case, as $A = \{1, -1\}$, every Θ-graph in K_3^A contains a balanced cycle. The following matrix represents $Q_3(A)$ over $GF(3)$ where, for example, α_{23} is mapped to $(0, 1, -\alpha)^T$:

$$
\begin{array}{cccccccccc}
& b_1 & b_2 & b_3 & 1_{12} & -1_{12} & 1_{23} & -1_{23} & 1_{13} & -1_{13} \\
\left[\begin{array}{c}\\\\\\\end{array}\right. & \begin{array}{c}1\\0\\0\end{array} & \begin{array}{c}0\\1\\0\end{array} & \begin{array}{c}0\\0\\1\end{array} & \begin{array}{c}1\\-1\\0\end{array} & \begin{array}{c}1\\1\\0\end{array} & \begin{array}{c}0\\1\\-1\end{array} & \begin{array}{c}0\\1\\1\end{array} & \begin{array}{c}1\\0\\-1\end{array} & \begin{array}{c}1\\0\\1\end{array} & \left.\begin{array}{c}\\\\\\\end{array}\right].
\end{array}
$$

Thus $Q_3(A) \cong Q_3(GF(3)^\times)$.   □

The elements $b_1, b_2, \ldots, b_r$ are called the *joints* of $Q_r(A)$. This terminology derives from viewing $Q_r(A)$ geometrically by taking $b_1, b_2, \ldots, b_r$ labelling the vertices of an r-simplex, that is, r affinely independent points. The remaining points of $Q_r(A)$ lie on edges of this simplex, that is, on lines between two joints.

The isomorphism between $Q_3(A)$ and $Q_3(GF(3)^\times)$ noted above is reassuring, for we have yet to reconcile the definition for $Q_r(GF(q)^\times)$ given in Example 6.10.1 with the circuit definition of $Q_r(A)$. Moreover, we still need to

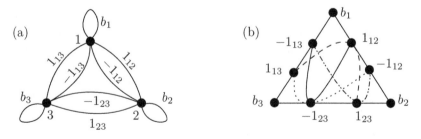

FIG. 6.36. (a) The graph K_3^A. (b) The matroid $Q_3(A)$ when $A = \{1, -1\}$.

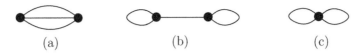

FIG. 6.37. Subdivide to get (a) a Θ-graph, a (b) loose or (c) tight handcuff.

prove that the specified collection of subsets of $E(Q_r(A))$ is indeed the set of circuits of a matroid. We shall delay remedying the first of these deficits until Lemma 6.10.11. To remedy the second, we show next that $Q_r(A)$ is indeed a matroid. Rather than doing this directly, we shall deduce it from a more general result. A *biased graph* (G, Ψ) consists of a graph G and a set Ψ of cycles of G such that if C_1 and C_2 are in Ψ and $G[C_1 \cup C_2]$ is a Θ-graph, then the third cycle in $G[C_1 \cup C_2]$ is also in Ψ. The next lemma shows that (K_r^A, Ψ) is a biased graph when Ψ is the set of balanced cycles of K_r^A.

Lemma 6.10.3 *In K_r^A, if two of the cycles in a Θ-subgraph are balanced, then so is the third.*

Proof Suppose that the Θ-graph consists of three paths P_1, P_2, and P_3 joining the vertices u and v and that the cycles C_2 and C_3 formed by P_1 and P_2 and by P_1 and P_3 are both balanced. We may assume that these cycles are directed from u to v along P_1, and from v to u along P_2 and P_3. Let δ_1 be the product of the adjusted edge labels along P_1 where, as before, an edge labelled α_{ij} has *adjusted edge label* α or α^{-1} depending on whether it is traversed from i to j, or from j to i. Similarly, let δ_2 and δ_3 be the products of the adjusted edge labels from v to u along P_2 and P_3, respectively. Then $\delta_1 \delta_2 = 1 = \delta_1 \delta_3$, so $1 = \delta_2^{-1} \delta_3$. If we now take the cycle formed by directing P_2 from u to v, and P_3 from v to u, then the product of the adjusted edge labels in this cycle is $\delta_2^{-1} \delta_3$, so this cycle is also balanced. $\square$

In general, in a biased graph, we call the cycles in Ψ *balanced*. A Θ-graph, loose handcuff, or tight handcuff is *unbalanced* if none of the cycles it contains is balanced. The next theorem establishes that biased graphs give rise to a class of matroids and, using this and the last lemma, we get that each $Q_r(A)$ is a matroid. The proof of this theorem will use the following elementary result.

Lemma 6.10.4 *A connected graph G with at least two cycles has a Θ-graph, a loose handcuff, or a tight handcuff as a subgraph.*

Proof Two distinct cycles of G are either vertex disjoint, have a unique common vertex, or have more than one common vertex. Thus G contains, respectively, a loose handcuff, a tight handcuff, or a Θ-graph. $\square$

Theorem 6.10.5 *Let (G, Ψ) be a biased graph. Then there is a matroid on $E(G)$ whose collection $\mathcal{C}$ of circuits consists of the edge sets of all balanced cycles along with all unbalanced Θ-graphs, all unbalanced loose handcuffs, and all unbalanced tight handcuffs.*

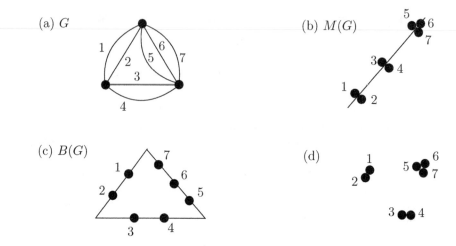

FIG. 6.38. (a) G; (b) its cycle matroid, $M(G)$; (c) its bicircular matroid, $B(G)$; and (d) its even-cycle matroid.

Before proving this theorem, we note that the matroid whose existence it shows is called the *bias matroid* or *frame matroid* of (G, Ψ) and is denoted by $M(G, \Psi)$. When Ψ is the set of all cycles of G, it is clear that $M(G, \Psi) = M(G)$, the usual cycle matroid of G. At the other extreme, when Ψ is empty, $M(G, \Psi)$ is the *bicircular matroid* $B(G)$ of G, that is, the matroid on $E(G)$ whose circuits consist precisely of the edge sets of all Θ-graphs, all loose handcuffs, and all tight handcuffs. When Ψ consists of all cycles of G with an even number of edges, the bias matroid $M(G, \Psi)$ is called the *even-cycle matroid* of G (Doob 1973) or the *factor matroid* of G (Wagner 1988).

As an example, let G be the graph in (a) of Figure 6.38. Geometric representations for the cycle matroid, the bicircular matroid, and the even-cycle matroid of G are shown in (b), (c), and (d) of the figure.

Proof of Theorem 6.10.5 Evidently the empty set is not in $\mathcal{C}$, and no member of $\mathcal{C}$ is a proper subset of another. Hence (C1) and (C2) hold.

Now suppose that C_1 and C_2 are distinct members of $\mathcal{C}$ and $e \in C_1 \cap C_2$. We need to show that $(C_1 \cup C_2) - e$ contains a member of $\mathcal{C}$. Assume not. Clearly $G[C_1 \cup C_2]$ is connected. Let $H = G[C_1 \cup C_2] \backslash e$. Then, by assumption, H contains no balanced cycles. Next we show that:

6.10.6 H *is connected.*

Suppose H is disconnected. Then e is not in a cycle of $G[C_1]$ or of $G[C_2]$. Hence each of $G[C_1]$ and $G[C_2]$ is a loose handcuff in which e appears on the path connecting the two cycles. Now H has a component H_1 whose edge set is not contained in C_1. For each i in $\{1, 2\}$, the edges of C_i in H_1 induce an unbalanced cycle D_i together with a path P_i from a vertex of $G[D_i]$ to some end

v of e. Since, by Lemma 6.10.4, H_1 contains a unique cycle, this cycle is D_1 so $D_2 = D_1$. As $E(H_1) - C_1$ is non-empty, P_2 must contain an edge f that is not in C_1. Clearly, $G[C_2 \cap E(H_1)] \setminus f$ contains paths from the endpoints of f to v and to $V(D_1)$. Using these paths and P_1, we see that $H_1 \setminus f$ is connected. Thus f is in a cycle of H_1 different from D_1; a contradiction. Therefore 6.10.6 holds.

Suppose H has no cycles. Then each of C_1 and C_2 must be a cycle, and e is in both these cycles. Then, by (C3) applied to $M(G)$, we deduce that $(C_1 \cup C_2) - e$ contains a cycle. Thus H has a cycle; a contradiction.

By Lemma 6.10.4 and 6.10.6, we may now assume that H is connected having a unique cycle, this cycle being unbalanced. As neither $G[C_1]$ nor $G[C_2]$ has a vertex of degree one, every degree-one vertex in H must be an end of e. Thus H consists of one of the following, where each path described has non-zero length:

(i)   a single cycle having e as a diagonal;

(ii)   a cycle C together with two paths P_1 and P_2 that are attached to distinct vertices u_1 and u_2 of the cycle such that the only vertices common to two of P_1, P_2, and C are u_1 and u_2; in this case, e joins the ends of P_1 and P_2 that are not in C;

(iii) a cycle C, a (u, v)-path P that has one end u on C but is otherwise vertex disjoint from C, and a path Q that has one end w in $V(P) - v$ but is otherwise vertex disjoint from both P and C; in this case, possibly $u = w$, and e must join v to the end of Q that differs from w; or

(iv)   a cycle C and a path P that has one end u on C, but which is otherwise vertex disjoint from C; in this case, e joins the other end of P to either (a) a vertex of C other than u; or (b) a vertex of P, possibly u.

In the first case, C_1 and C_2 must both be balanced cycles and their union is a Θ-graph, the third cycle of which is H. Since (G, Ψ) is a biased graph, we deduce that H is a balanced cycle; a contradiction. In the second case, $G[C_1 \cup C_2]$ is a Θ-graph, and C_1 and C_2 must both contain P_1, P_2, and e. As neither C_1 nor C_2 contains the other, each of C_1 and C_2 is a cycle and hence is balanced. Again, we obtain the contradiction that the cycle in H is also balanced. Cases (iii) and (iv)(b) cannot arise since neither C_1 nor C_2 contains the other. In case (iv)(a), C_1 and C_2 must both contain P and e, and one can argue as in case (ii) to again obtain a contradiction and thereby finish the proof of the theorem. $\square$

We show next that all minors of bias matroids are also bias matroids.

Proposition 6.10.7 *The class of bias matroids is minor-closed.*

Proof Let (G, Ψ) be a biased graph. For e in $E(G)$, it is easy to check that $M(G, \Psi) \setminus e = M(G \setminus e, \Psi \setminus e)$ where $\Psi \setminus e$ is the collection of cycles in Ψ that do not contain e. Contraction is more complex and depends on what type of edge e is. The simplest case is when $\{e\}$ is a balanced loop. Then $M(G, \Psi)/e = M(G, \Psi) \setminus e$.

Suppose next that $\{e\}$ is an unbalanced loop at the vertex v. We shall describe a graph G' and a collection Ψ' of cycles such that $M(G, \Psi)/e = M(G', \Psi')$. To construct G' from G, first delete e and declare all remaining loops at v to be in Ψ'. Then, for every edge joining v to some other vertex u, replace the edge by a loop at u, and declare that this loop is not in Ψ'. Finally, take each cycle of G that is in Ψ and avoids v and add it to Ψ'. Evidently, a Θ-graph in G' does not use the vertex v and so is a Θ-graph in G. It follows that if two of the cycles in this Θ-graph are in Ψ', then so is the third. We now need to show that the circuits of $M(G, \Psi)/e$ coincide with those of $M(G', \Psi')$. This involves straightforward case-checking and we leave it to the reader to supply the details.

Finally, suppose that e joins distinct vertices u and v in G. In this case, it is not difficult to show that $(G/e, \Psi'')$ is a biased graph and that $M(G, \Psi)/e = M(G/e, \Psi'')$, where Ψ'' consists of the collection of minimal sets of the form $C - e$ where $C \in \Psi$.   □

Descriptions of bias matroids in terms of their bases, independent sets, and so on will be given in the exercises. Although there is an extensive literature on biased graphs and bias matroids, the main focus of this section is on Dowling geometries, and the rest of the section is primarily devoted to describing some of their interesting properties. We refer the reader interested in a more detailed discussion of bias matroids to the sequence of papers on this topic by Zaslavsky (1982a, 1987b, 1989, 1990, 1991, 1994) and to his extensive bibliography of papers in this and related areas (Zaslavsky 1998a).

We show next that the matroid $Q_r(A)$ determines the group A up to isomorphism, provided $r \geq 3$.

Proposition 6.10.8 (i) *For all groups A,*

$$Q_0(A) = U_{0,0}, \; Q_1(A) \cong U_{1,1}, \; and \; Q_2(A) \cong U_{2,|A|+2}.$$

(ii) *For $r \geq 3$, if $Q_r(A) \cong Q_r(A')$ for groups A and A', then $A \cong A'$.*

The following elementary lemma will be useful in proving both this proposition and the next theorem.

Lemma 6.10.9 *Let A and A' be groups and, for all i in $\{1, 2, 3\}$, let $\phi_i : A \to A'$ be a bijection. If*

$$\phi_3(\alpha\beta) = \phi_1(\alpha)\phi_2(\beta) \tag{6.10}$$

for all α, β in A, then $A \cong A'$.

Proof By taking $\beta = 1$ and then $\alpha = 1$, we get $\phi_3(\alpha) = \phi_1(\alpha)\phi_2(1)$ and $\phi_3(\beta) = \phi_1(1)\phi_2(\beta)$. Thus

$$\phi_1(\alpha) = \phi_3(\alpha)[\phi_2(1)]^{-1} \tag{6.11}$$

and

$$\phi_2(\beta) = [\phi_1(1)]^{-1}\phi_3(\beta). \tag{6.12}$$

Let $\phi = [\phi_1(1)]^{-1}\phi_3[\phi_2(1)]^{-1}$. Then ϕ is a bijection from A to A' and

$$\phi(\alpha\beta) = [\phi_1(1)]^{-1}\phi_3(\alpha\beta)[\phi_2(1)]^{-1}$$
$$= [\phi_1(1)]^{-1}\phi_1(\alpha)\phi_2(\beta)[\phi_2(1)]^{-1} \text{ by (6.10);}$$
$$= [\phi_1(1)]^{-1}\phi_3(\alpha)[\phi_2(1)]^{-1}[\phi_1(1)]^{-1}\phi_3(\beta)[\phi_2(1)]^{-1} \text{ by (6.11) and (6.12);}$$
$$= \phi(\alpha)\phi(\beta).$$

Hence ϕ is an isomorphism from A to A'. □

Proof of Proposition 6.10.8 For all groups A, we defined $Q_0(A)$ to be $U_{0,0}$, and it is clear that $Q_1(A) \cong U_{1,1}$. For $r \geq 2$, since $Q_r(A)$ has $r + \binom{r}{2}|A|$ elements, it follows that $Q_2(A) \cong U_{2,|A|+2}$. Now assume that $r \geq 3$ and $Q_r(A) \cong Q_r(A')$. Then $|A| = |A'|$. If $|A| < 4$, then A is the unique group of that order, so $A \cong A'$. Hence we may assume that $|A| \geq 4$. Each joint of $Q_r(A)$ is on exactly $r - 1$ non-trivial lines each of which has $|A| + 2$ elements. Each non-joint element of $Q_r(A)$ is on a unique line with $|A| + 2$ elements. Hence we can distinguish between the joints and non-joints in $Q_r(A)$. Let ϕ be an isomorphism between $Q_r(A)$ and $Q_r(A')$. Take three distinct joints, b_1, b_2, and b_3, of $Q_r(A)$ and consider the copy of $Q_3(A)$ that is obtained by restricting $Q_r(A)$ to the flat spanned by $\{b_1, b_2, b_3\}$. If we restrict the map ϕ to this copy of $Q_3(A)$, then ϕ induces an isomorphism between $Q_3(A)$ and $Q_3(A')$. Hence we may assume that $r = 3$ and that ϕ maps the joints b_1, b_2, b_3 of $Q_3(A)$ to the joints b'_1, b'_2, b'_3 of $Q_3(A')$ so that $\phi(b_i) = b'_i$. We may also assume that, for all (i, j) in $\{(1, 2), (2, 3), (1, 3)\}$, the isomorphism ϕ maps $\{\alpha_{ij} : \alpha \in A\}$ to $\{\alpha'_{ij} : \alpha' \in A'\}$. Hence ϕ induces three bijections ϕ_{12}, ϕ_{23}, and ϕ_{13} from A to A', where $\phi_{ij}(\alpha) = \alpha'$ if $\phi(\alpha_{ij}) = \alpha'_{ij}$. Now the following statements are equivalent for all α, β, and γ in A.

(i) $\alpha\beta = \gamma$.
(ii) $\alpha\beta\gamma^{-1} = 1$.
(iii) $\{\alpha_{12}, \beta_{23}, \gamma_{13}\}$ is a line of $Q_3(A)$.
(iv) $\{\phi(\alpha_{12}), \phi(\beta_{23}), \phi(\gamma_{13})\}$ is a line of $Q_3(A')$.
(v) $\phi_{12}(\alpha)\phi_{23}(\beta)[\phi_{13}(\gamma)]^{-1} = 1$.
(vi) $\phi_{12}(\alpha)\phi_{23}(\beta) = \phi_{13}(\gamma)$.

We conclude that $\phi_{12}(\alpha)\phi_{23}(\beta) = \phi_{13}(\alpha\beta)$ for all α, β in A. It follows, by Lemma 6.10.9, that $A \cong A'$. □

Next we present Dowling's (1973a) very attractive solution to the representability problem for $Q_r(A)$. Clearly $Q_1(A)$, which is isomorphic to $U_{1,1}$, is representable over all fields, irrespective of the group A. If $r = 2$, then $Q_r(A) \cong U_{2,|A|+2}$, so $Q_r(A)$ is $\mathbb{F}$-representable if and only if $|\mathbb{F}| \geq |A| + 1$.

Theorem 6.10.10 *Suppose $r \geq 3$. Then $Q_r(A)$ is representable over a field $\mathbb{F}$ if and only if A is isomorphic to a subgroup of $\mathbb{F}^\times$.*

We prove the easier part of the theorem first. For notational convenience, when A is isomorphic to a subgroup of $\mathbb{F}^\times$, we shall assume that this isomorphism is the identity. Note that the next lemma reconciles the definition of $Q_r(GF(q)^\times)$ given in Example 6.10.1 with the definition of $Q_r(A)$ given in terms of K_r^A.

Lemma 6.10.11 *For a field* $\mathbb{F}$, *let* A *be a finite subgroup of* $\mathbb{F}^\times$. *For* $r \geq 1$, *let* B *be a basis* $\{\underline{v}_1, \underline{v}_2, \ldots, \underline{v}_r\}$ *of* $V(r, \mathbb{F})$ *and let* M *be the matroid induced on* $B \cup \{\underline{v}_i - \alpha \underline{v}_j : 1 \leq i < j \leq r$ *and* $\alpha \in A\}$ *by linear independence over* $\mathbb{F}$. *Then the map* ψ *that takes* $\underline{v}_i$ *to* b_i *and takes* $\underline{v}_i - \alpha \underline{v}_j$ *to* α_{ij} *is an isomorphism between* M *and* $Q_r(A)$.

Proof We view each $\underline{v}_i$ as a column vector. First we show that:

6.10.12 *Every circuit of* M *maps to a dependent set of* $Q_r(A)$.

Let C be a circuit of M and consider the subgraph G_C of K_r^A induced by the set of edges in $\psi(C)$. If G_C is not connected, then the set of vectors corresponding to one of its components is linearly dependent, contradicting the fact that C is a circuit. Hence G_C is connected. If G_C contains at least two cycles, then, by Lemma 6.10.4, it has a subgraph that is a Θ-graph or a loose or tight handcuff. Such a subgraph is either unbalanced or contains a balanced cycle. In each case, $\psi(C)$ is dependent in $Q_r(A)$. We may now assume that G_C contains at most one cycle and that this cycle is unbalanced. Now G_C has no vertices of degree one, otherwise no linear combination of the vectors in C with all non-zero coefficients is $\underline{0}$. We deduce that G_C is an unbalanced cycle. Clearly this cycle has at least two edges. By shuffling rows and columns and rescaling some columns, replacing $\underline{v}_i - \alpha \underline{v}_j$ by $\underline{v}_j - \alpha^{-1} \underline{v}_i$, we may assume that the columns corresponding to the circuit C yield a matrix of the form $\left[\begin{smallmatrix} D \\ 0 \end{smallmatrix}\right]$, where

$$
D = \begin{bmatrix}
1 & 0 & 0 \cdots & 0 & 0 & 1 \\
-\alpha_1 & 1 & 0 \cdots & 0 & 0 & 0 \\
0 & -\alpha_2 & 1 \cdots & 0 & 0 & 0 \\
\vdots & \vdots & \vdots \ddots & \vdots & \vdots & \vdots \\
0 & 0 & 0 \cdots & 1 & 0 & 0 \\
0 & 0 & 0 \cdots -\alpha_{k-2} & 1 & 0 \\
0 & 0 & 0 \cdots & 0 & -\alpha_{k-1} & -\alpha_k
\end{bmatrix}.
$$

Since G_C is an unbalanced cycle, it follows that $\alpha_1 \alpha_2 \ldots \alpha_{k-1} \alpha_k^{-1} \neq 1$, that is, $\alpha_1 \alpha_2 \ldots \alpha_{k-1} \neq \alpha_k$. But, expanding the determinant of D along the first row, we get $-\alpha_k + (-1)^{k+1}(-\alpha_1)(-\alpha_2) \ldots (-\alpha_{k-1})$, which equals $-\alpha_k + \alpha_1 \alpha_2 \ldots \alpha_{k-1}$ and so is non-zero. This contradicts the fact that C is a circuit, so 6.10.12 holds.

To complete the proof of the lemma, it suffices to show that:

6.10.13 *If* C *is a circuit in* $Q_r(A)$, *then* $\psi^{-1}(C)$ *is dependent in* M.

If C is a balanced cycle in K_r^A, then we may assume, by shuffling rows and columns and rescaling, that the corresponding matrix has the form $\left[\begin{smallmatrix} D \\ 0 \end{smallmatrix}\right]$ where D is as above. Since C is balanced, $\alpha_1 \alpha_2 \ldots \alpha_{k-1} \alpha_k^{-1} = 1$. We showed above that $\det D = -\alpha_k + \alpha_1 \alpha_2 \ldots \alpha_{k-1}$. Since this determinant is zero, we deduce that $\psi^{-1}(C)$ is dependent in M.

We may now assume that the subgraph G_C of K_r^A induced by C is a a Θ-graph or a loose or tight handcuff. Then one can obtain a spanning tree for G_C by deleting exactly two edges. Hence C meets precisely $|C| - 1$ vertices, and the vectors corresponding to these vertices span C. Thus $\psi^{-1}(C)$ consists of $|C|$ elements in rank $|C| - 1$ and so is dependent in M. □

We shall finish the proof of Theorem 6.10.10 by proving the converse of the last lemma.

Proof of Theorem 6.10.10 Suppose that $Q_r(A)$ is representable over $\mathbb{F}$. As $Q_3(A)$ is a restriction of $Q_r(A)$, it suffices to prove that A is a subgroup of $\mathbb{F}^\times$ when $r = 3$. We may also assume that the columns corresponding to the joints $b_1, b_2,$ and b_3 of $Q_3(A)$ are the three natural basis vectors $\underline{v}_1, \underline{v}_2,$ and $\underline{v}_3$, respectively. Now, for all (i, j) in $\{(1,2), (2,3), (3,1)\}$, the set $\{b_i, \alpha_{ij}, b_j\}$ labels an unbalanced loose handcuff in K_r^A where we recall that $\alpha_{31} = \alpha_{13}^{-1}$. Hence, by scaling, we may assume that the column corresponding to the edge α_{ij} is $\underline{v}_i - \sigma_{ij}(\alpha)\underline{v}_j$. Clearly, if $\alpha \neq \alpha'$, then $\sigma_{ij}(\alpha) \neq \sigma_{ij}(\alpha')$. Now the following statements are equivalent for all $\alpha, \beta,$ and γ in A.

(i) $\alpha\beta\gamma = 1$

(ii) $\{\alpha_{12}, \beta_{23}, \gamma_{31}\}$ is a circuit of $Q_3(A)$.

(iii) The matrix

$$\begin{bmatrix} 1 & 0 & -\sigma_{31}(\gamma) \\ -\sigma_{12}(\alpha) & 1 & 0 \\ 0 & -\sigma_{23}(\beta) & 1 \end{bmatrix}$$

has zero determinant.

(iv) $\sigma_{12}(\alpha)\sigma_{23}(\beta)\sigma_{31}(\gamma) = 1$.

Hence

$$[\sigma_{31}((\alpha\beta)^{-1})]^{-1} = \sigma_{12}(\alpha)\sigma_{23}(\beta). \tag{6.13}$$

Thus

$$\sigma_{12}(\alpha)\sigma_{23}(1) = [\sigma_{31}((\alpha \cdot 1)^{-1})]^{-1} = [\sigma_{31}((1 \cdot \alpha)^{-1})]^{-1} = \sigma_{12}(1)\sigma_{23}(\alpha).$$

So $\sigma_{12}(\alpha) = \sigma_{12}(1)\sigma_{23}(\alpha)[\sigma_{23}(1)]^{-1}$. Hence $\sigma_{23}(\alpha) = [\sigma_{12}(1)]^{-1}\sigma_{12}(\alpha)\sigma_{23}(1)$. Thus, as $\mathbb{F}^\times$ is abelian,

$$\sigma_{12}(\alpha)\sigma_{23}(\beta) = \sigma_{12}(1)\sigma_{23}(\alpha)[\sigma_{23}(1)]^{-1}[\sigma_{12}(1)]^{-1}\sigma_{12}(\beta)\sigma_{23}(1)$$
$$= \sigma_{12}(\beta)\sigma_{23}(\alpha).$$

It follows by (6.13) that $\sigma_{31}((\alpha\beta)^{-1}) = \sigma_{31}((\beta\alpha)^{-1})$. Thus, since σ_{31} is one-to-one, $(\alpha\beta)^{-1} = (\beta\alpha)^{-1}$, so $\alpha\beta = \beta\alpha$. Therefore, A is abelian.

Let

$$H_3 = \{\sigma_{12}(\alpha) : \alpha \in A\},$$
$$H_1 = \{\sigma_{23}(\beta) : \beta \in A\}, \text{ and}$$
$$H_2 = \{\sigma_{31}(\gamma) : \gamma \in A\}.$$

Clearly, $|H_i| = |A|$ for all i.

Our next aim is to show that $\mathbb{F}^\times$ has a subgroup A_0 such that each H_i is a coset of A_0. We shall then use Lemma 6.10.9 to finish the proof of the theorem. Using (6.13), it is straightforward to check that, for every permutation (i, j, k) of $(1, 2, 3)$, an element a_i is in H_i if and only if there are elements a_j and a_k of H_j and H_k, respectively, such that

$$a_i a_j a_k = 1. \tag{6.14}$$

Thus, for example, as $\mathbb{F}^\times$ is an abelian group containing $H_1 \cup H_2 \cup H_3$, if $d_j \in H_j$ and $d_k \in H_k$, then $d_j^{-1} d_k^{-1} \in H_i$. This fact will be used repeatedly in establishing the following:

6.10.14 *If $d_i, e_i, c_i \in H_i$, then $d_i e_i c_i^{-1} \in H_i$.*

To see this, choose d_j in H_j. Then, as $d_i \in H_i$, it follows that $d_i^{-1} d_j^{-1} \in H_k$. Thus, as $c_i \in H_i$, we deduce that $d_i d_j c_i^{-1} \in H_j$. Hence, as $e_i \in H_i$, it follows that $d_i^{-1} d_j^{-1} c_i e_i^{-1} \in H_k$. Finally, since $d_j \in H_j$, we get that $d_i d_j c_i^{-1} e_i d_j^{-1} \in H_i$, that is, 6.10.14 holds.

For each i in $\{1, 2, 3\}$, choose a fixed element c_i in H_i, and let $A_i = \{a_i c_i^{-1} : a_i \in H_i\}$. We show next that:

6.10.15 *Each A_i is a subgroup of $\mathbb{F}^\times$.*

Clearly $1 \in A_i$. Since A_i is a finite subset of $\mathbb{F}^\times$, it suffices to show that A_i is closed. But, if $d_i c_i^{-1}$ and $e_i c_i^{-1}$ are in A_i, then $d_i, e_i \in H_i$, so, by 6.10.14, $d_i e_i c_i^{-1} \in H_i$. Hence $(d_i c_i^{-1})(e_i c_i^{-1}) \in A_i$, so A_i is a subgroup of $\mathbb{F}^\times$.

Clearly, $H_i = A_i c_i$ for all i, so $|A_i| = |H_i| = |A|$.

6.10.16 $A_1 = A_2 = A_3$ *and this set contains $c_1 c_2 c_3$.*

Suppose $d_1 \in A_1$ and $d_2 \in A_2$. Then $d_1 c_1 \in H_1$ and $d_2 c_2 \in H_2$, so by (6.14), $(d_1 c_1 d_2 c_2)^{-1} \in H_3$. Hence, as $\mathbb{F}^\times$ is abelian, $(d_1 d_2 c_1 c_2 c_3)^{-1} \in A_3$. Let $c = c_1 c_2 c_3$. Then $(d_1 d_2 c)^{-1} \in A_3$ for all $d_1 \in A_1$ and all $d_2 \in A_2$. In particular, if $d_1 = d_2 = 1$, then c^{-1} and hence c is in A_3. Thus, $(d_1 d_2)^{-1} c \in A_3$, so $(d_1 d_2)^{-1} \in A_3$ and hence $d_1 d_2$ in A_3. We conclude that if $d_1 \in A_1$ and $d_2 \in A_2$, then $d_1 d_2 \in A_3$. Putting $d_1 = 1$, we deduce that $A_2 \subseteq A_3$. Similarly, $A_1 \subseteq A_3$. Since $|A_1| = |A_2| = |A_3|$, we conclude that $A_1 = A_2 = A_3$. Moreover, as $c = c_1 c_2 c_3$ and $c \in A_3$, we have $c_1 c_2 c_3 \in A_3$.

We may now rename each of A_1, A_2, and A_3 as A_0, observing that each of H_1, H_2, and H_3 is a coset of A_0. Now choose three elements z_1, z_2, and z_3 in A_0 such that $z_1 z_2 z_3 = c_1 c_2 c_3$. For all α, β, and γ in A, let $\tau_{12}(\alpha) = \sigma_{12}(\alpha) z_3 c_3^{-1}$, let $\tau_{23}(\beta) = \sigma_{23}(\beta) z_1 c_1^{-1}$, and let $\tau_{31}(\gamma) = \sigma_{31}(\gamma) z_2 c_2^{-1}$. Then each of τ_{12}, τ_{23}, and τ_{31} is a bijection from A to A_0. Moreover, since $z_1 z_2 z_3 = c_1 c_2 c_3$, it follows that $\tau_{12}(\alpha) \tau_{23}(\beta) \tau_{31}(\gamma) = \sigma_{12}(\alpha) \sigma_{23}(\beta) \sigma_{31}(\gamma)$. The equivalence of (i)-(iv) established above gives that the right-hand side of the last equation is 1 if and only if $\alpha \beta \gamma = 1$. Hence, for all α, β in A,

$$[\tau_{31}((\alpha\beta)^{-1})]^{-1} = \tau_{12}(\alpha)\tau_{23}(\beta).$$

Let $\tau'_{13}(\delta) = [\tau_{31}(\delta^{-1})]^{-1}$ for all δ in A. Then τ'_{13} is a bijection from A to A_0, and $\tau'_{13}(\alpha\beta) = \tau_{12}(\alpha)\tau_{23}(\beta)$ for all α, β in A. Therefore, by Lemma 6.10.9, $A \cong A_0$; that is, A is isomorphic to a subgroup of $\mathbb{F}^\times$. On combining this with Lemma 6.10.11, we conclude that Theorem 6.10.10 holds. □

A glossary of terminology used in the area of biased graphs appears in Zaslavsky (1998b). Because we have been focused here on matroids rather than graphs, we have been able to ignore Zaslavsky's 'half-edges' and 'free loops', replacing these by unbalanced and balanced loops, respectively. Some further brief remarks may help the reader interested in delving into the literature in this area. A *signed graph* is obtained from a graph by placing a positive or negative sign on each edge. If G is a signed graph, then it is easy to see that one obtains a biased graph (G, Ψ) by taking a cycle of G to be balanced if it contains an even number of negative edges. A matroid that is isomorphic to the bias matroid of a signed graph, that is, to the bias matroid of such a biased graph, is called a *signed-graphic matroid*. Letting A be the 2-element group, we observe that $Q_3(A)$ from Example 6.10.1 is such a matroid. More generally, the simple matroid associated with every signed-graphic matroid is isomorphic to a restriction of the Dowling geometry $Q_r(A)$, where A is the 2-element group. As $A \cong GF(3)^\times$, it follows that a matroid is signed-graphic if and only if it can be represented over $GF(3)$ by a matrix every column of which has at most two non-zero entries.

A *gain graph* is obtained from a graph G by assigning to each edge both a direction and a *gain*, an element α from a group A, where A need not be finite. A cycle in a gain graph is *balanced* if, when the cycle is traversed, the product of the adjusted edge labels is 1. As before, an edge labelled α has adjusted edge label α^{-1} if the direction in which it is traversed is opposite to that assigned. If Ψ is the set of balanced cycles, then the argument in Lemma 6.10.3 establishes that (G, Ψ) is a biased graph. Evidently, when A is finite, the simple graph associated with $M(G, \Psi)$ is a restriction of $Q_r(A)$ for some r. Zaslavsky (1982b) originally called such restrictions of Dowling geometries *voltage-graphic matroids* but this terminology is no longer current.

Lemma 6.10.11 showed that the definition of $Q_r(GF(q)^\times)$ given in Example 6.10.1 is consistent with the definition of $Q_r(A)$ in terms of $K_r(A)$ when $A = GF(q)^\times$. This fact will be central to the argument that follows, which shows how to obtain a representation for the bicircular matroid $B(G)$ of a graph G over all sufficiently large fields. Recall that $B(G)$ is the matroid with ground set $E(G)$ whose circuits consist of the edge sets of all Θ-graphs, all loose handcuffs, and all tight handcuffs in G.

Proposition 6.10.17 *For every graph G, there is an integer $n(G)$ such that the bicircular matroid $B(G)$ of G is representable over all fields with at least $n(G)$ elements.*

Proof If e and f are loops at the same vertex of G, then $\{e, f\}$ is the edge set of a tight handcuff, so $\{e, f\}$ is a circuit of $B(G)$. Thus it suffices to prove the

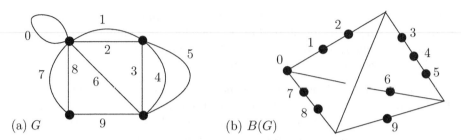

FIG. 6.39. A graph G and its bicircular matroid $B(G)$.

proposition in the case that G has at most one loop at each vertex.

Label $V(G)$ as $\{1, 2, \ldots, r\}$. Then, provided $|A|$ is sufficiently large, G is isomorphic to a subgraph H of K_r^A where this isomorphism ignores the group labels on the edges of H. Now let us consider these edge labels on H. If H contains no balanced cycles of K_r^A, then the circuits of $Q_r(A)|E(H)$ consist of all Θ-graphs and all loose and tight handcuffs in K_r^A whose edge sets are contained in $E(H)$. Hence $Q_r(A)|E(H) = B(H)$. Thus it suffices to show that the isomorphism σ between G and H can be chosen so that H contains no balanced cycles of K_r^A.

We shall assume that A is a cyclic group having γ as a generator. It is not difficult to show that, provided the order of A is chosen to be sufficiently large, H can be chosen so that it contains no balanced cycles of K_r^A. One way to do this, for example, is to start with an ordering $g_1, g_2, \ldots, g_m$ of the non-loop edges of G. Then take A to have order at least 2^m and, for each k, let $\sigma(g_{k+1}) = (\gamma^{2^k})_{ij}$ where g_{k+1} joins i to j and $i < j$. Then it is straightforward to check that every cycle of H is unbalanced. We conclude that if q is a prime power with $q \geq 2^m + 1$, then $B(G)$ is isomorphic to a restriction of $Q_r(GF(q)^\times)$. Explicitly, by using Lemma 6.10.11, we have that if $\{\underline{v}_1, \underline{v}_2, \ldots, \underline{v}_r\}$ is a basis of $V(r, q)$ and $\{g_1, g_2, \ldots, g_m\}$ is the set of non-loop edges of G, then $B(G)$ is isomorphic to the restriction of $V(r, q)$ to

$$\{\underline{v}_i : G \text{ has a loop at } i\} \cup \{\underline{v}_i - \gamma^{2^k} \underline{v}_j : g_{k+1} \text{ joins } i \text{ to } j \text{ where } i < j\}.$$

The argument above shows that $B(G)$ is representable over every sufficiently large finite field $\mathbb{F}$. When $\mathbb{F}$ is an infinite field, we can assign a gain γ_k from $\mathbb{F}^\times$ to each non-loop edge g_{k+1} of G so that, in the resulting gain graph, (G, Ψ), the set Ψ of balanced cycles is empty. Then $M(G, \Psi) = B(G)$. Moreover, if we replace γ^{2^k} by γ_k in the last displayed set, it follows by the argument for Lemma 6.10.11 that $B(G)$ is $\mathbb{F}$-representable. $\square$

An alternative proof of the last result will be possible when we reach Chapter 11 by combining two results from that chapter: Corollary 11.2.10, which shows that every bicircular matroid is transversal; and Corollary 11.2.17, which shows that every transversal matroid is representable over every sufficiently large field.

For the graph G in Figure 6.39(a), a geometric representation of its bicircular matroid is shown in Figure 6.39(b). This exemplifies a natural general procedure for constructing a geometric representation for $B(G)$ for any graph G. Let $V(G) = \{1, 2, \ldots, r\}$. We first modify G, adding a loop b_i at each vertex i to give the graph H. It suffices to describe a geometric representation for $B(H)$. Let $\{b_1, b_2, \ldots, b_r\}$ label r affinely independent points in $\mathbb{R}^n$ where $n \geq r-1$. Let e be an edge of the original graph G. If e is a loop and e meets the vertex i, add a point corresponding to e touching b_i. If e joins distinct vertices i and j of G, add a point corresponding to e freely on the line joining b_i and b_j. We defer to the exercises the proof that this procedure does indeed give a geometric representation for $B(H)$. We shall consider bicircular matroids again in Proposition 11.1.6 when we consider matroids derived from submodular functions.

One of the reasons why Dowling geometries are of such importance in matroid theory is that they are linked to projective geometries and free matroids through a very attractive result of Kahn and Kung (1982). As a preliminary to stating this theorem, we now note some more properties of $Q_r(A)$. The first of these shows that every simple minor of a Dowling geometry is isomorphic to a restriction of a Dowling geometry. Recall that $Q_0(A)$ is $U_{0,0}$.

Proposition 6.10.18 *Let A be a group of order m.*

(i)   *For every element e of $Q_r(A)$, the simplification of $Q_r(A)/e$ is isomorphic to $Q_{r-1}(A)$. In particular, for $r \geq 3$, the contraction $Q_r(A)/e$ can be obtained from $Q_{r-1}(A)$ either (a) by adding m elements in parallel with each joint when e is a joint of $Q_r(A)$, or (b) by adding m elements in parallel with exactly one joint f of $Q_{r-1}(A)$ and, for all non-trivial lines through f and every non-joint element g on such a line, adding one element in parallel with g.*

(ii) *A set X is a flat of $Q_r(A)$ if and only if the components of K_r^A induced by X consist of complete graphs in which every cycle is balanced along with at most one component that is isomorphic to K_k^A for some k in $\{0, 1, \ldots, r\}$. Hence, when X is a flat, $Q_r(A)|X$ is isomorphic to the direct sum of the cycle matroids of a family of complete graphs and $Q_k(A)$ for some k in $\{0, 1, \ldots, r\}$.*

(iii) *A rank-k flat X of $Q_r(A)$ is modular if and only if $Q_r(A)|X \cong Q_k(A)$.*

(iv) *$Q_r(A)$ has exactly $((m+1)^r - 1)/m$ hyperplanes.*

Proof   The proof of (i) follows without difficulty from considering the 3-element circuits in $Q_r(A)$. Next, we consider the flats of $Q_r(A)$ showing first that:

6.10.19 *If T is a tree in K_r^A, then the closure of T in $Q_r(A)$ induces a complete graph on $V(T)$ and all cycles in this complete graph are balanced.*

For all distinct i and j in $V(T)$ that are not adjacent in T, there is a unique edge e of K_r^A joining i and j such that the unique cycle contained in $T \cup e$ is balanced. Let T' be the set of such edges e. Evidently $T' \subseteq \mathrm{cl}(T)$. To show that

$\text{cl}(T) = T \cup T'$, it suffices to show that every cycle C contained in $T \cup T'$ is balanced. This is not difficult to establish by arguing by induction on $|C - T|$ using the fact that a Θ-graph cannot contain exactly two balanced cycles. We leave completion of the details as an exercise and conclude that 6.10.19 holds.

Let G_X be the subgraph of K_r^A induced by a flat X of $Q_r(A)$. If at least two components of G_X contain unbalanced cycles, then, by using loose handcuffs, we see that all edges joining these components are in the closure of X; a contradiction. Thus at most one component of G_X contains an unbalanced cycle.

6.10.20 *If G_X has a component H containing an unbalanced cycle, then*

$$Q_r(A)|E(H) \cong Q_{|V(H)|}(A).$$

To see this, observe, by considering tight and loose handcuffs, that X contains all loops at vertices of H. These loops are joints of $Q_r(A)$ that span all the edges of the subgraph of K_r^A induced by $V(H)$. Thus 6.10.20 holds.

6.10.21 *If H is a component of G_X containing a balanced cycle but no unbalanced cycles, then $Q_r(A)|E(H) \cong M(K_{|V(H)|})$.*

Let T be a spanning tree of H. As X is a flat, $\text{cl}(T) \subseteq E(H)$. As H contains no unbalanced cycles, H must be simple. It follows by 6.10.19 that $\text{cl}(T) = E(H)$ and $Q_r(A)|E(H)$ is the cycle matroid of H and so is isomorphic to $M(K_{|V(H)|})$.

The following is an immediate consequence of 6.10.19.

6.10.22 *If H is a component of G_X having no cycles, then $Q_r(A)|E(H) \cong U_{1,1}$.*

On combining 6.10.20–6.10.22, we obtain (ii). Part (iii) follows from (ii) by constructing examples to show that the only flats of $Q_r(A)$ that can be modular are those that are isomorphic to $Q_k(A)$ for some k in $\{0, 1 \ldots, r\}$; moreover, every such flat is indeed modular. We omit the straightforward details.

From (ii), we see that, to obtain a hyperplane of $Q_r(A)$, we first partition the vertex set of K_r^A into sets V_1 and V_2 where V_1 may be empty but V_2 is not. Then we take the union of the subgraph of K_r^A induced by V_1 with a complete graph G' on V_2 in which every cycle is balanced. Now, by 6.10.19, G' is uniquely determined by any one of its spanning trees. There are $|V_2| - 1$ edges in such a spanning tree and, since every two vertices of K_r^A are joined by m edges, the number of choices for G' once V_2 is chosen is $m^{|V_2|-1}$. Now, letting $|V_1| = k$, we have $\binom{r}{k}$ choices for V_1, and $|V_2| = r - k$. Therefore, the number of hyperplanes of $Q_r(A)$ is $\sum_{k=0}^{r-1} \binom{r}{k} m^{r-k-1}$, which equals

$$\frac{1}{m} \left[\sum_{k=0}^{r} \binom{r}{k} m^{r-k} - 1 \right].$$

By the binomial theorem, this is $((m+1)^r - 1)/m$, and (iv) is proved. $\qquad\square$

The class of simple matroids representable over $GF(q)$ and the class of simple graphic matroids are both closed under direct sums and have the property that every simple minor of a member of the class is also in the class. In addition, a rank-r member of one of these classes is a restriction of $PG(r-1, q)$ or $M(K_{r+1})$, respectively. Kahn and Kung's (1982) theorem identifies exactly which other classes of matroids share these properties. The following definitions will be helpful in giving a precise statement of their theorem. A collection $\mathcal{T}$ of non-empty simple matroids is a *hereditary class* if every non-empty simple minor of a member of $\mathcal{T}$ is also in $\mathcal{T}$, and the direct sum of two members of $\mathcal{T}$ is also in $\mathcal{T}$. Given a hereditary class $\mathcal{T}$, a sequence $T_1, T_2, \ldots$ of simple matroids in $\mathcal{T}$ is a *sequence of universal models* for $\mathcal{T}$ if each T_r has rank r, and every rank-r member of $\mathcal{T}$ is a restriction of T_r. Finally, a *variety* of simple matroids is a hereditary class of simple matroids with a sequence of universal models. Thus the class of simple matroids representable over $GF(q)$ and the class of free matroids $U_{r,r}$ are examples of varieties. So too is the class of simple graphic matroids; it is the special case, when A is the trivial group, of the variety in which the sequence of universal models is $Q_1(A), Q_2(A), \ldots$. There are only two more families of examples of varieties, and the members of each have a very uncomplicated structure. The *variety of matchstick geometries of order n* is the set of all restrictions of the matroids $M_r(n)$ where r and n are positive integers and $M_1(n) = U_{1,1}$ while, for all $k \geq 1$, the matroid $M_{2k}(n)$ is the direct sum of k copies of $U_{2,n+1}$; and $M_{2k+1}(n)$ is the direct sum of $M_{2k}(n)$ and $U_{1,1}$. For all positive integers n, the *variety of origami geometries of order n* is the set of all restrictions of the matroids $O_r(n)$, where $O_1(n) = U_{1,1}$ and, for all $r \geq 2$, if B is a basis $\{\underline{v}_1, \underline{v}_2, \ldots, \underline{v}_r\}$ of $V(r, \mathbb{R})$, then $O_r(n)$ is the matroid induced on $B \cup \{\underline{v}_i + j\underline{v}_{i+1} : 1 \leq i \leq r-1 \text{ and } 1 \leq j \leq n-1\}$ by linear dependence. Geometrically, $O_r(n)$ is formed from a free matroid on $\{1, 2, \ldots, r\}$ by, for all i in $\{1, 2, \ldots, r-1\}$, adding $n-1$ points freely on the line joining i and $i+1$. As we shall see in the next chapter, this means that $O_r(n)$ is formed from $r-1$ lines by a sequence of $r-1$ parallel connections.

Theorem 6.10.23 *Let $\mathcal{V}$ be a variety of simple matroids with a sequence (T_r) of universal models. Then this sequence is $(U_{r,r}), (M_r(n)), (O_r(n)), (PG(r-1, q))$, or $(Q_r(A))$ where, in the second and third sequences, n is some positive integer; in the fourth, q is some prime power; and, in the last, A is some finite group.*□

The proof of this theorem can be found in Kahn and Kung's (1982) paper.

Exercises

1. Let X be a rank-k flat of $Q_r(A)$. Show that:
    (a) The simplification of $Q_r(A)/X$ is isomorphic to $Q_{r-k}(A)$.
    (b) $Q_r(A)$ has $(r-k) + |A|\binom{r-k}{2}$ rank-$(k+1)$ flats containing X.
2. Let B be the set of joints of $Q_r(A)$ and let $Q'_r(A)$ be $Q_r(A) \backslash B$, the *jointless Dowling geometry*. Show that:
    (a) If $|A| = 1$, then $Q'_r(A) \cong M(K_r)$.
    (b) If $|A| > 1$ and $r > 1$, then $r(Q'_r(A)) = r$.
    (c) If $|A| = 2$, then $Q'_3(A) \cong M(K_4)$.

(d) If $|A| = 3$, then $Q'_3(A) \cong AG(2,3)$.

(e) If D is a subgroup of A, then $Q_r(D)$ and $Q'_r(D)$ are isomorphic to restrictions of $Q_r(A)$ and $Q'_r(A)$, respectively.

3. (Zaslavsky 1991) Let (G, Ψ) be a biased graph. Show that:

(a) $\mathcal{I}(M(G, \Psi)) = \{I \subseteq E(G) : G[I]$ contains no balanced cycles and no component with more than one cycle$\}$.

(b) When G is connected, a set B is a basis in $M(G, \Psi)$ if and only if B is a spanning tree of G when G has no unbalanced cycles and, otherwise, each component of the graph $(V(G), B)$ is the union of a spanning tree T of that component and an edge e for which the unique cycle contained in $T \cup e$ is unbalanced.

(c) If X is a subset of $E(G)$, then the rank of X in $M(G, \Psi)$ is the difference between $|V(G[X])|$ and the number of components of $G[X]$ in which every cycle is balanced.

4. Let H be a graph having a distinguished loop at every vertex and let N_H be the matroid having the geometric representation described following Proposition 6.10.17. Show that:

(a) Every circuit of $B(H)$ is dependent in N_H.

(b) Every circuit of N_H corresponds to a connected subgraph of H that contains at least two cycles and so is dependent in $B(H)$.

(c) $B(H) = N_H$.

(d) This procedure gives another way to verify that $B(H)$ is representable over every infinite field.

5. (Matthews 1978, Zaslavsky 1989) For n in $\mathbb{Z}^+$, let Ψ be the set of cycles of a directed graph in which the number of edges directed one way is congruent modulo n to the number directed the other way. Show that:

(a) (G, Ψ) is a biased graph.

(b) (G, Ψ) arises as a gain graph over an n-element cyclic group with generator α where the gain assigned to an edge is α if it is traversed in its direction in G and is α^{-1} otherwise.

6. (Matthews 1978, Zaslavsky 1989) Let G be a directed graph and Ψ be the set of cycles in which no two consecutive edges are the same way.

(a) Show that (G, Ψ) is a biased graph.

(b) Prove that (G, Ψ) arises as a gain graph over the free abelian group generated by the vertex labels of G in which the gain of an edge directed from v to w is $v + w$.

7. (Zaslavsky 1991) Show that:

(a) The unique biased graph (G, Ψ) for which the corresponding bias matroid is $U_{3,6}$ has Ψ being empty and has G consisting of K_3 with each edge replaced by two parallel edges.

(b) None of $U_{3,7}$, F_7, or F_7^* arises as a bias matroid.

7

CONSTRUCTIONS

This chapter discusses several different ways of using some given set of matroids to form a new matroid. The five most basic examples of such constructions are deletion, contraction, and the formation of duals, minors, and direct sums. Each of these operations has been looked at in detail in an earlier chapter. In this chapter and in Chapter 11, we shall consider some other matroid operations. The discussion here of matroid constructions will be far from complete. The reader seeking a more detailed treatment of this subject is referred to the paper of Brylawski (1986a), which has been used as a source for much of this chapter. Another strong influence on the present treatment of constructions is Mason's (1977) expository paper.

The direct-sum operation provides a way to join matroids on disjoint sets. In Section 7.1, we look at three closely related ways of joining two matroids with exactly one common element. In Chapter 11, we shall discuss some ways of joining two matroids with more than one common element. Section 7.2 investigates the operations of extension and coextension, which reverse the operations of deletion and contraction, respectively. In Section 7.3. we use the techniques of Section 7.2 to examine the relationship between two matroids such that every flat of one is a flat of the other. Finally, Section 7.4 discusses an interesting non-commutative operation for joining two matroids. While the results in Section 7.1 will be used quite frequently in later chapters, the reader could delay reading Sections 7.2 and 7.3 until Chapter 11 when various other constructions will be considered. There are no results later in the book that depend on Section 7.4.

7.1   Series and parallel connection and 2-sums

In the last section of Chapter 5, we looked briefly at the operations of series and parallel connection of graphs. We now formally define these operations and show how they extend naturally to matroids. For each i in $\{1, 2\}$, let p_i be an edge of a graph G_i. Arbitrarily assign a direction to p_i and label its tail by u_i and its head by v_i. To form the *series* and *parallel connections* of G_1 and G_2 with respect to the directed edges p_1 and p_2, we begin by deleting p_1 from G_1 and p_2 from G_2; we then identify u_1 and u_2 as the vertex u. To complete the series connection, we add a new edge p joining v_1 and v_2. The parallel connection is completed by identifying v_1 and v_2 as the vertex v and then adding a new edge p joining u and v. Thus, unless exactly one of p_1 and p_2 is a loop, the parallel connection is obtained by simply identifying p_1 and p_2 so that their directions agree.

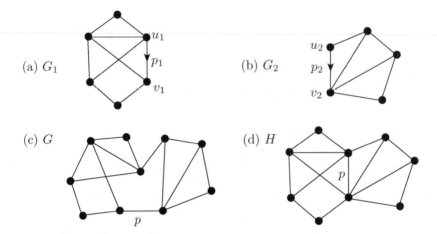

FIG. 7.1. Series and parallel connections of two graphs.

Example 7.1.1 The graphs G and H in Figure 7.1(c) and 7.1(d) are, respectively, the series and parallel connections of the graphs G_1 and G_2 in (a) and (b) with respect to the directed edges p_1 and p_2. Observe that reversing the direction of p_2 will alter the series and parallel connections of the graphs. However, by Whitney's 2-Isomorphism Theorem (5.3.1), this reversal will not alter the cycle matroids of the series and parallel connections. □

Let $\mathcal{C}_S$ and $\mathcal{C}_P$ denote the collections of circuits of the cycle matroids of the series and parallel connections of the graphs G_1 and G_2. Then, in the last example, and indeed in general, it is not difficult to specify $\mathcal{C}_S$ and $\mathcal{C}_P$ in terms of $\mathcal{C}(M(G_1))$ and $\mathcal{C}(M(G_2))$. Writing M_1 for $M(G_1)$ and M_2 for $M(G_2)$ and assuming neither p_1 nor p_2 is a loop or a cut edge, we have

7.1.2 $\mathcal{C}_S = \mathcal{C}(M_1 \backslash p_1) \cup \mathcal{C}(M_2 \backslash p_2)$
$$\cup \{(C_1 - p_1) \cup (C_2 - p_2) \cup p : p_i \in C_i \in \mathcal{C}(M_i) \text{ for each } i\}$$

and

7.1.3 $\mathcal{C}_P = \mathcal{C}(M_1 \backslash p_1) \cup \{(C_1 - p_1) \cup p : p_1 \in C_1 \in \mathcal{C}(M_1)\}$
$$\cup\, \mathcal{C}(M_2 \backslash p_2) \cup \{(C_2 - p_2) \cup p : p_2 \in C_2 \in \mathcal{C}(M_2)\}$$
$$\cup \{(C_1 - p_1) \cup (C_2 - p_2) : p_i \in C_i \in \mathcal{C}(M_i) \text{ for each } i\}.$$

Now suppose that M_1 and M_2 are arbitrary matroids on disjoint sets. Let p_1 and p_2 be elements of M_1 and M_2, respectively, such that neither p_1 nor p_2 is a loop or a coloop. Take p to be an element that is not in $E(M_1)$ or $E(M_2)$ and let $E = E(M_1 \backslash p_1) \cup E(M_2 \backslash p_2) \cup p$. Then, with $\mathcal{C}_S$ and $\mathcal{C}_P$ as specified in 7.1.2 and 7.1.3, respectively, we have the following result.

Proposition 7.1.4 *Each of $\mathcal{C}_S$ and $\mathcal{C}_P$ is the collection of circuits of a matroid on E.*

Proof It is straightforward to check that both $\mathcal{C}_S$ and $\mathcal{C}_P$ satisfy (C1) and (C2). We shall show that $\mathcal{C}_S$ satisfies (C3), leaving the proof that $\mathcal{C}_P$ satisfies this condition to the reader.

Let C and D be distinct members of $\mathcal{C}_S$ having a common element e. If $\mathcal{C}(M_1\backslash p_1)$ or $\mathcal{C}(M_2\backslash p_2)$ contains both C and D, then $\mathcal{C}_S$ certainly has a member contained in $(C \cup D) - e$. Thus we may assume that $p \in C$ or $p \in D$.

Suppose that $p \in C \cap D$. Then each M_i has circuits C_i and D_i both containing p_i such that $C = (C_1 - p_1) \cup (C_2 - p_2) \cup p$ and $D = (D_1 - p_1) \cup (D_2 - p_2) \cup p$. As $C \neq D$, we may assume, without loss of generality, that $C_1 \neq D_1$. If $e = p$ or if $e \in (C_2 - p_2) \cap (D_2 - p_2)$, then, on applying (C3) to the circuits C_1 and D_1 of M_1, eliminating the common element p_1, we deduce that M_1 has a circuit C_1' contained in $(C_1 \cup D_1) - p_1$. Evidently $C_1' \in \mathcal{C}_S$ and $C_1' \subseteq (C \cup D) - e$ so, in this case, (C3) holds for $\mathcal{C}_S$. We may now assume that $e \in (C_1 - p_1) \cap (D_1 - p_1)$. Then, applying (C3) in M_1 again, we get that M_1 has a circuit C_1'' contained in $(C_1 \cup D_1) - e$. If $p_1 \notin C_1''$, then C_1'' is a member of $\mathcal{C}_S$ contained in $(C \cup D) - e$. If $p_1 \in C_1''$, then $(C_1'' - p_1) \cup (C_2 - p_2) \cup p$ is a member of $\mathcal{C}_S$ contained in $(C \cup D) - e$. We conclude that (C3) holds when $p \in C \cap D$. By symmetry, the only case that remains to be checked is when $p \in C$ and $p \notin D$. We leave it to the reader to provide the straightforward details of this argument. □

The matroids on E that have $\mathcal{C}_S$ and $\mathcal{C}_P$ as their sets of circuits will be denoted by $S((M_1; p_1), (M_2; p_2))$ and $P((M_1; p_1), (M_2; p_2))$, or briefly, $S(M_1, M_2)$ and $P(M_1, M_2)$. These matroids are called the *series* and *parallel connections* of M_1 and M_2 with respect to the *basepoints* p_1 and p_2 (Brylawski 1971).

It is often convenient to view $S(M_1, M_2)$ and $P(M_1, M_2)$ as being formed from two matroids M_1 and M_2 whose ground sets meet in a single element p. In this context, p is called the *basepoint* of the connection and we take $E = E(M_1) \cup E(M_2)$. Moreover, with $p_1 = p_2 = p$, the sets $\mathcal{C}_S$ and $\mathcal{C}_P$ are defined as in 7.1.2 and 7.1.3 provided neither M_1 nor M_2 has p as a loop or a coloop. When p is a loop of M_1, we define

7.1.5 $P(M_1, M_2) = P(M_2, M_1) = M_1 \oplus (M_2/p)$ and

7.1.6 $S(M_1, M_2) = S(M_2, M_1) = (M_1/p) \oplus M_2$.

When p is a coloop of M_1, we define

7.1.7 $P(M_1, M_2) = P(M_2, M_1) = (M_1\backslash p) \oplus M_2$ and

7.1.8 $S(M_1, M_2) = S(M_2, M_1) = M_1 \oplus (M_2\backslash p)$.

Note that if $\{p\}$ is a separator of both M_1 and M_2, then we have given two definitions of each of $P(M_1, M_2)$ and $S(M_1, M_2)$. To show that these definitions are consistent, one uses the fact that if $\{e\}$ is a separator of a matroid N, then $N\backslash e = N/e$ (Corollary 3.1.24). Furthermore, on combining 7.1.2, 7.1.3, and 7.1.5–7.1.8, we observe that, in all cases, $P(M_1, M_2) = P(M_2, M_1)$ and $S(M_1, M_2) = S(M_2, M_1)$.

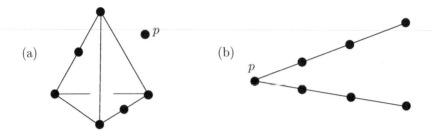

FIG. 7.2. (a) $S(U_{2,4}, U_{2,4})$. (b) $P(U_{2,4}, U_{2,4})$.

Now let $\{p\}$ be a separator of M_1 or M_2 and suppose that $M_1 \cong M(G_1)$ and $M_2 \cong M(G_2)$ for some graphs G_1 and G_2. Then it is routine to verify that the matroids $S(M_1, M_2)$ and $P(M_1, M_2)$, defined as in 7.1.5–7.1.8, are isomorphic to the cycle matroids of graphs that are series and parallel connections, respectively, of G_1 and G_2.

For matroids, just as for graphs, the operations of series and parallel connection generalize the operations of series and parallel extension. Indeed, if N is a series or parallel extension of M, then N is isomorphic to a series or parallel connection of M and $U_{1,2}$.

The next two examples illustrate the fact that, when M_1 and M_2 have small rank, it is easy to find a geometric representation for $P(M_1, M_2)$ from geometric representations for M_1 and M_2.

Example 7.1.9 Let both M_1 and M_2 be isomorphic to $U_{2,4}$. Then geometric representations for $S(M_1, M_2)$ and $P(M_1, M_2)$ are given in Figure 7.2. In the former, the basepoint p is free in space, that is, p is in no circuits of size less than five. In the latter, M_1 and M_2 have been glued together at p.   □

In the last example, the symmetry of M_1 and M_2 meant that, up to isomorphism, $S(M_1, M_2)$ and $P(M_1, M_2)$ did not depend on the choice of basepoints. In the next example, changing the basepoint in M_2 changes the isomorphism class of the parallel connection.

Example 7.1.10 Let M_1 and M_2 be the matroids for which geometric representations are shown in Figure 7.3(a) and 7.3(b). Then geometric representations for $P((M_1; p_1), (M_2; x))$ and $P((M_1; p_1), (M_2; y))$ are as shown in Figure 7.3 (c) and (d).   □

The operations of series and parallel connection are actually duals of each other. To show this, we shall determine the sets of bases of the series and parallel connections of two matroids. We first note that:

7.1.11 *The basepoint p is a coloop of $S(M_1, M_2)$ if and only if p is a coloop of M_1 or M_2.*

7.1.12 *The basepoint p is a loop of $P(M_1, M_2)$ if and only if p is a loop of M_1 or M_2.*

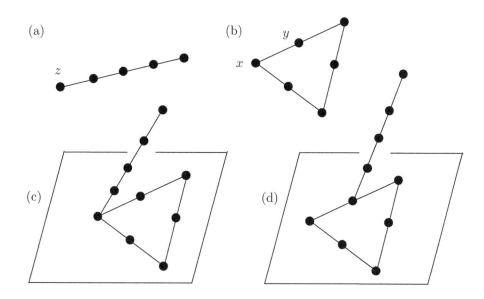

FIG. 7.3. (a) M_1. (b) M_2. (c) $P((M_1; z), (M_2; x))$. (d) $P((M_1; z), (M_2; y))$.

Proposition 7.1.13 *Let M_1 and M_2 be matroids with $E(M_1) \cap E(M_2) = \{p\}$. Let $E = E(M_1) \cup E(M_2)$ and B be a subset of E.*

(i) *Assume that M_1 and M_2 do not both have p as a coloop. Then B is a basis of $S(M_1, M_2)$ if and only if M_1 and M_2 have disjoint bases B_1 and B_2 such that $B = B_1 \cup B_2$.*

(ii) *Assume that M_1 and M_2 do not both have p as a loop. Then B is a basis of $P(M_1, M_2)$ if and only if M_1 and M_2 have bases B_1 and B_2 such that*

    (a) *p is in both B_1 and B_2, and $B = B_1 \cup B_2$; or*

    (b) *p is in exactly one of B_1 and B_2, and $B = (B_1 \cup B_2) - p$.*

Proof If p is a loop or a coloop of M_1 or M_2, the result is not difficult to check using 7.1.5–7.1.8 Thus assume that neither M_1 nor M_2 has p as a loop or as a coloop. We shall prove (i) and leave the proof of (ii) to the reader.

Let B be a basis of $S(M_1, M_2)$. For each i in $\{1, 2\}$, let $B_i = B \cap E(M_i)$. First suppose that $p \notin B$. Then B_1 does not contain a member of $\mathcal{C}(M_1 \backslash p)$. However, if $e \in E(M_1) - (B_1 \cup p)$, then $B \cup e$ contains a circuit of $S(M_1, M_2)$. Hence $B_1 \cup e$ contains a member of $\mathcal{C}(M_1 \backslash p)$. Thus B_1 or $B_1 \cup p$ is a basis of M_1. But $B \cup p$ contains a circuit of $S(M_1, M_2)$ containing p. Hence $B_1 \cup p$ contains a circuit of M_1. We conclude that B_1 is a basis of M_1. Similarly, B_2 is a basis of M_2 and so B is a disjoint union of a basis of M_1 and a basis of M_2.

Now suppose that $p \in B$. Again no B_i contains a member of $\mathcal{C}(M_i \backslash p)$. Moreover, either B_1 does not contain a circuit of M_1 containing p, or B_2 does not contain a circuit of M_2 containing p, or both. Suppose that the third possibility occurs. Then, for all e in $E(M_1) - B_1$, the set $B \cup e$ contains a member $C(e)$ of $\mathcal{C}_S$. Evidently $C(e)$ cannot be of the form $C_1 \cup C_2$ where each C_i is a circuit of M_i containing p. Thus $C(e) \in \mathcal{C}(M_1 \backslash p)$ for all e in $E(M_1) - B_1$. Hence $B_1 - p$ spans $M_1 \backslash p$. As $B_1 - p$ does not span M_1, it follows that p is a coloop of M_1; a contradiction. We conclude that exactly one of B_1 and B_2, say B_1, contains a circuit containing p. It is now straightforward to show that $B_1 - p$ is a basis of M_1, and B_2 is a basis of M_2.

   To complete the proof of (i), we need to show that if B is the union of disjoint bases A_1 of M_1 and A_2 of M_2, then B is a basis of $S(M_1, M_2)$. Without loss of generality, we may assume that $p \notin A_2$. Evidently B is independent in $S(M_1, M_2)$. We show next that it is a maximal independent set. If $e \in E(M_2) - (A_2 \cup p)$, then $C_{M_2}(e, A_2)$ is a circuit of $S(M_1, M_2)$ contained in $B \cup e$, so $B \cup e$ is dependent in $S(M_1, M_2)$. If $e \in E(M_1) - A_1$, then either $C_{M_1}(e, A_1)$ contains p or not. In the first case, $B \cup e$ contains $C_{M_1}(e, A_1) \cup C_{M_2}(p, A_2)$, a circuit of $S(M_1, M_2)$. In the second case, $B \cup e$ contains $C_{M_1}(e, A_1)$, a circuit of $S(M_1, M_2)$. We conclude that B is indeed a maximal independent set of $S(M_1, M_2)$. $\qquad\square$

   We can now show that series and parallel connection are dual operations.

Proposition 7.1.14 *Let M_1 and M_2 be matroids such that $E(M_1) \cap E(M_2) = \{p\}$. Then*
$$S(M_1, M_2) = [P(M_1^*, M_2^*)]^*$$
and
$$P(M_1, M_2) = [S(M_1^*, M_2^*)]^*.$$

Proof   As the first equation implies the second by duality, we need only prove the former. Assume initially that p is a loop of M_1. Then p is a coloop of M_1^*, so
$$\begin{aligned}
S(M_1, M_2) &= (M_1/p) \oplus M_2 \\
&= (M_1^* \backslash p)^* \oplus (M_2^*)^* \\
&= [(M_1^* \backslash p) \oplus M_2^*]^* = [P(M_1^*, M_2^*)]^*.
\end{aligned}$$

A similar argument establishes the result when p is a coloop of M_1.

   We may now assume that neither M_1 nor M_2 has $\{p\}$ as a separator. Suppose that $E - B$ is a basis of $[P(M_1^*, M_2^*)]^*$ where $E = E(M_1) \cup E(M_2)$. We shall show that $E - B$ is a basis of $S(M_1, M_2)$. Evidently B is a basis of $P(M_1^*, M_2^*)$. We now distinguish two cases: (a) $p \in B$, and (b) $p \notin B$. In case (a), by Proposition 7.1.13(ii), $B \cap E(M_1^*)$ and $B \cap E(M_2^*)$ contain p and are bases of M_1^* and M_2^*, respectively. Thus $(E - B) \cap E(M_1)$ and $(E - B) \cap E(M_2)$ avoid p and are bases of M_1 and M_2, respectively. Hence $E - B$ can be written as a disjoint union of bases of M_1 and M_2. Therefore, by Proposition 7.1.13(i), $E - B$ is a basis of $S(M_1, M_2)$.

   In case (b), Proposition 7.1.13(ii) implies that, for distinct i and j in $\{1, 2\}$, the sets $[B \cap E(M_i^*)] \cup p$ and $B \cap E(M_j^*)$ are bases of M_i^* and M_j^*. Therefore

$[(E-B)\cap E(M_i)]-p$ and $(E-B)\cap E(M_j)$ are bases of M_i and M_j, respectively, the first avoiding p and the second containing p. Hence $E-B$ can be written as a disjoint union of bases of M_1 and M_2 and so is a basis of $S(M_1, M_2)$.

We conclude that $\mathcal{B}([P(M_1^*, M_2^*)]^*) \subseteq \mathcal{B}(S(M_1, M_2))$. A similar argument gives the reverse inclusion thereby finishing the proof □

Brylawski (1971) noted numerous attractive properties of the operations of series and parallel connection. Some of these are summarized below, while others appear in the exercises. Several of these results are stated in terms of M_1, but, since $S(M_2, M_1) = S(M_1, M_2)$, and $P(M_2, M_1) = P(M_1, M_2)$, the corresponding results for M_2 also hold.

Proposition 7.1.15 *Let M_1 and M_2 be matroids with $E(M_1) \cap E(M_2) = \{p\}$. Then*

(i)
$$r(S(M_1, M_2)) = \begin{cases} r(M_1) + r(M_2) - 1, & \text{if } p \text{ is a coloop of both} \\ & M_1 \text{ and } M_2; \\ r(M_1) + r(M_2), & \text{otherwise.} \end{cases}$$

$$r(P(M_1, M_2)) = \begin{cases} r(M_1) + r(M_2), & \text{if } p \text{ is a loop of both} \\ & M_1 \text{ and } M_2; \\ r(M_1) + r(M_2) - 1, & \text{otherwise.} \end{cases}$$

(ii) $M_1 = S(M_1, M_2)/(E(M_2) - p)$ *unless p is a loop of M_1 and a coloop of M_2; and*

$M_1 = P(M_1, M_2)\backslash(E(M_2) - p)$ *unless p is a coloop of M_1 and a loop of M_2.*

(iii) $S(M_1, M_2)\backslash p = (M_1\backslash p) \oplus (M_2\backslash p)$ *and* $P(M_1, M_2)/p = (M_1/p) \oplus (M_2/p)$.

(iv) $S(M_1, M_2)/p = P(M_1, M_2)\backslash p$.

(v) *For e in $E(M_1) - p$,*

$S(M_1, M_2)\backslash e = S(M_1\backslash e, M_2)$; $\quad S(M_1, M_2)/e = S(M_1/e, M_2)$;

$P(M_1, M_2)\backslash e = P(M_1\backslash e, M_2)$; *and* $P(M_1, M_2)/e = P(M_1/e, M_2)$.

Proof The proofs of these results use Propositions 7.1.13 and 7.1.14 along with 7.1.2, 7.1.3, and 7.1.5–7.1.8. The straightforward details are omitted. □

The operations of series and parallel connection arise naturally in the consideration of matroid structure because of the following fundamental link with matroid connectivity.

Theorem 7.1.16 *Let p be an element of a connected matroid M.*

(i) *If $M\backslash p = M_1 \oplus M_2$ where both M_1 and M_2 are non-empty, then*

$$M = S(M/E(M_1), M/E(M_2)).$$

Dually,

(ii)   *if $M/p = M_1 \oplus M_2$ where both M_1 and M_2 are non-empty, then*

$$M = P(M \backslash E(M_1), M \backslash E(M_2)).$$

Proof   Evidently we need only prove (i). We shall show that $\mathcal{B}(M) = \mathcal{B}(N)$ where $N = S(M/E(M_1), M/E(M_2))$. First, we observe that, as $|E(M)| \geq 2$ and M is connected, $r(M) = r(M\backslash p)$, so p is not a coloop of M. Thus, by 3.1.17, p is not a coloop of $M/E(M_1)$ or of $M/E(M_2)$.

Now, by Proposition 7.1.15(iii),

$$\begin{aligned} N\backslash p &= (M/E(M_1)\backslash p) \oplus (M/E(M_2)\backslash p) \\ &= (M\backslash p/E(M_1)) \oplus (M\backslash p/E(M_2)) \\ &= M_2 \oplus M_1 = M\backslash p. \end{aligned}$$

Hence

$$\mathcal{B}(N\backslash p) = \mathcal{B}(M\backslash p).$$

Suppose next that B is a basis of N containing p. Then, without loss of generality, we may assume that $B = B_1 \,\dot\cup\, p \,\dot\cup\, B_2$ where $B_1 \cup p$ is a basis of $M/E(M_1)$ and B_2 is a basis of $M/E(M_2)$. As $p \notin B_2$, it follows that B_2 is a basis of $M/E(M_2)\backslash p$. But $M/E(M_2)\backslash p = M\backslash p/E(M_2) = M_1 = M|E(M_1)$. Therefore, as $B_1 \cup p$ is a basis of $M/E(M_1)$, it follows that $B_1 \cup p \cup B_2$ is a basis of M. Hence every basis of N containing p is a basis of M.

Finally, suppose that B is a basis of M containing p. Then $B - p$ is independent in $M\backslash p$. Hence, as $M\backslash p = M_1 \oplus M_2$,

$$|B - p| = r(M) - 1 = r(M_1) + r(M_2) - 1.$$

Thus $B - p$ contains a basis of M_1 or of M_2. Without loss of generality, assume the latter, letting B_2 be such a basis. Then B_2 is a basis of $M|E(M_2)$ and so is a basis of $M\backslash p/E(M_1)$. As B is a basis of M, it follows that $B - B_2$ is a basis of $M/E(M_2)$. As p is not a coloop of $M/E(M_1)$, the set B_2, which is a basis of $M\backslash p/E(M_1)$, is also a basis of $M/E(M_1)$. Hence, by Proposition 7.1.13(i), B is a basis of N. We conclude that $\mathcal{B}(N) = \mathcal{B}(M)$, so $N = M$. $\square$

By the last result, a connected matroid is a series or parallel connection if and only if it can be disconnected by deleting or contracting a single element. Next we show that connectivity is preserved under both of the operations of series and parallel connection.

Proposition 7.1.17   *If the matroids M_1 and M_2 each have at least two elements and $E(M_1) \cap E(M_2) = \{p\}$, then the following statements are equivalent:*

(i)   *Both M_1 and M_2 are connected.*
(ii)   *$S(M_1, M_2)$ is connected.*
(iii) *$P(M_1, M_2)$ is connected.*

Proof Since the dual of a matroid is connected if and only if the matroid itself is connected, it suffices to establish the equivalence of (i) and (iii). Assume that (i) holds. Then it follows by Proposition 4.1.2 that $P(M_1, M_2)$ is connected since $E(M_1)$ and $E(M_2)$ meet, and $M_1 = P(M_1, M_2)|E(M_1)$ and $M_2 = P(M_1, M_2)|E(M_2)$.

Now assume that $P(M_1, M_2)$ is connected but M_1 is not. Then, by 7.1.5 and 7.1.7, neither M_1 nor M_2 has $\{p\}$ as a separator. As M_1 is disconnected, there is a pair of distinct elements e and f of $E(M_1) - p$ such that no circuit of M_1 contains $\{e, f\}$. It follows from 7.1.3 that no circuit of $P(M_1, M_2)$ contains $\{e, f\}$; a contradiction. $\square$

It is not difficult to generalize the definitions of series and parallel connection so that more than two matroids are joined. Indeed, the next proposition can be proved using Proposition 7.1.4 and a straightforward induction argument.

Proposition 7.1.18 *Let n be an integer exceeding one and $M_1, M_2, \ldots, M_n$ be matroids such that $E(M_i) \cap E(M_j) = \{p\}$ for all distinct i and j, and no M_k has $\{p\}$ as a separator. Then there are matroids $P(M_1, M_2, \ldots, M_n)$ and $S(M_1, M_2, \ldots, M_n)$ each with ground set $E(M_1) \cup E(M_2) \cup \cdots \cup E(M_n)$ whose collections of circuits are $\mathcal{C}_P$ and $\mathcal{C}_S$, respectively, where*

$$\mathcal{C}_P = (\bigcup_{i=1}^{n} \mathcal{C}(M_i)) \cup \{(C_i \cup C_j) - p : 1 \le i < j \le n \text{ and } p \in C_k \in \mathcal{C}(M_k) \text{ for } k = i, j\}$$

and

$$\mathcal{C}_S = (\bigcup_{i=1}^{n} \mathcal{C}(M_i \backslash p)) \cup \{\cup_{i=1}^{n} C_i : p \in C_i \in \mathcal{C}(M_i) \text{ for all } i\}. \quad \square$$

It is clear that the definition given in the last proposition is consistent with that given earlier. Observe that, because no M_k has $\{p\}$ as a separator, each M_k has at least two elements. We call $P(M_1, M_2, \ldots, M_n)$ and $S(M_1, M_2, \ldots, M_n)$ the *parallel connection* and *series connection*, respectively, of $M_1, M_2, \ldots, M_n$. The properties of these matroids that were proved above when $n = 2$ can be extended to larger values of n. Some of these are considered in the exercises.

The assertion in 7.1.15(iv) that $P(M_1, M_2)\backslash p = S(M_1, M_2)/p$ is an important one. The next example illustrates this for graphic matroids.

Example 7.1.19 Consider Figure 7.4. Clearly the matroids $P(M(G_1), M(G_2))$ and $S(M(G_1), M(G_2))$ are isomorphic to the cycle matroids of the graphs in (c) and (d). The graph in (e) can be obtained both from the graph in (c) by deleting p and from the graph in (d) by contracting p. The cycle matroid of this graph is isomorphic to both $P(M(G_1), M(G_2))\backslash p$ and $S(M(G_1), M(G_2))/p$.

The graph in Figure 7.4(e) can be obtained directly from G_1 and G_2 by identifying the edges labelled p and then deleting the identified edge. Following Robertson and Seymour (1984), we call a graph obtained in this way a *2-sum* of G_1 and G_2. To ensure that this operation is well-defined, we insist that if the edge p is a loop in one of G_1 and G_2, then it is a loop in the other. $\square$

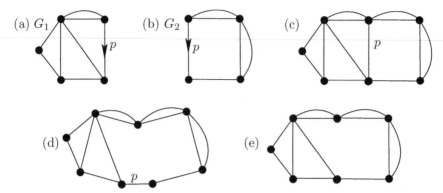

FIG. 7.4. The parallel and series connections and 2-sum of two graphs.

By analogy with the above situation for graphs, we have the following definition for matroids. Let M and N be matroids, each with at least two elements.
Let $E(M) \cap E(N) = \{p\}$ and suppose that neither M nor N has $\{p\}$ as a separator. Then the *2-sum* $M \oplus_2 N$ of M and N is $S(M, N)/p$ or, equivalently,
$P(M, N)\backslash p$. Clearly $N \oplus_2 M = M \oplus_2 N$. The element p is called the *basepoint*
of the 2-sum, and M and N are called the *parts* of the 2-sum. We remark that
sometimes, to ensure that the 2-sum has more elements than its parts, the definition of $M \oplus_2 N$ requires that each of M and N has at least three elements.
Although this restriction was imposed in the first edition of this book, it will not
be used in this edition.

The next proposition, whose proof is an easy modification of the proof of
Proposition 7.1.14, can be used as a definition of 2-sum that is independent of
the definitions of series and parallel connection.

Proposition 7.1.20 *Let M and N be matroids each with at least two elements.
Let $E(M) \cap E(N) = \{p\}$ and suppose that neither M nor N has $\{p\}$ as a
separator. Then $M \oplus_2 N$ is the matroid with ground set $(E(M) \cup E(N)) - p$ and
set of circuits*

$$\mathcal{C}(M\backslash p) \cup \mathcal{C}(N\backslash p) \cup \{(C \cup D) - p : p \in C \in \mathcal{C}(M) \text{ and } p \in D \in \mathcal{C}(N)\}. \quad \square$$

The 2-sum operation is of fundamental importance. We noted in Chapter 4
that a matroid is not 2-connected if and only if it can be written as a direct
sum. In Chapter 8, where higher connectivity for matroids is discussed, we shall
see that a 2-connected matroid M is not 3-connected if and only if M can be
written as a 2-sum of two matroids each with fewer elements than M. The next
result contains a basic and frequently used property of 2-sums.

Proposition 7.1.21 *Both M and N are isomorphic to minors of $M \oplus_2 N$. In
particular, if p' is an element of $E(N) - p$ that is in the same component of N as
p, then there is a minor of $M \oplus_2 N$ from which M can be obtained by relabelling
p' by p. Moreover, if N has at least three elements, then M is isomorphic to a
proper minor of $M \oplus_2 N$.*

Proof By the choice of p', there is a circuit C of N that contains $\{p, p'\}$. Now consider the matroid obtained from $P(M, N)$ by deleting $E(N) - C$ and then contracting $C - \{p, p'\}$. By Proposition 7.1.15(v), this matroid is the parallel connection of M and $N\backslash(E(N) - C)/(C - \{p, p'\})$. But the last matroid is a circuit with elements p and p'. It follows that M can be obtained from the matroid $(M \oplus_2 N)\backslash(E(N) - C)/(C - \{p, p'\})$ by relabelling p' by p. This proves the first part of the lemma and the last part is immediate. $\square$

Two of the many other attractive properties of the 2-sum operation are contained in the following proposition, the proof of which is left to the reader.

Proposition 7.1.22 (i) $(M \oplus_2 N)^* = M^* \oplus_2 N^*$.

(ii) *Suppose $|E(M)| \geq 2$ and $|E(N)| \geq 2$. Then $P(M, N)\backslash p$ is connected if and only if both M and N are connected. In particular, $M \oplus_2 N$ is connected if and only if both M and N are connected.* $\square$

The next result establishes that the operation of 2-sum is associative.

Proposition 7.1.23 *Let N_1, N_2, and N_3 be matroids each having at least two elements. Suppose that $E(N_1)$ and $E(N_3)$ are disjoint, that $E(N_1) \cap E(N_2) = \{p_{12}\}$, and that $E(N_2) \cap E(N_3) = \{p_{23}\}$. If $\{p_{12}\}$ is not a separator of N_1 or N_2, and $\{p_{23}\}$ is not a separator of N_2 or N_3, then*

$$N_1 \oplus_2 (N_2 \oplus_2 N_3) = (N_1 \oplus_2 N_2) \oplus_2 N_3.$$

Proof The matroids $N_1 \oplus_2 (N_2 \oplus_2 N_3)$ and $(N_1 \oplus_2 N_2) \oplus_2 N_3$ have equal ground sets and can be shown to be the same matroid by, for example, calculating the collection of circuits of each. We omit the straightforward details. $\square$

We have seen that both the series and parallel connections of two graphic matroids are graphic. Moreover, Example 7.1.9 shows that, not surprisingly, neither the series nor the parallel connection of two uniform matroids need be uniform. We now consider the effect of the operations of series and parallel connection on various other classes of matroids.

$$
\begin{array}{cc}
E(M_1) - p \quad p & \quad p \quad E(M_2) - p \\
\begin{bmatrix} & \begin{matrix} 0 \\ 0 \\ \vdots \\ 0 \\ 1 \end{matrix} \\ A_1 & \end{bmatrix} &
\begin{bmatrix} \begin{matrix} 1 \\ 0 \\ \vdots \\ 0 \\ 0 \end{matrix} & \\ & A_2 \end{bmatrix} \\
\text{(a)} & \text{(b)}
\end{array}
$$

FIG. 7.5. Representations for (a) M_1 and (b) M_2.

Proposition 7.1.24 *Let $\mathbb{F}$ be a field. If M_1 and M_2 are $\mathbb{F}$-representable matroids such that $E(M_1) \cap E(M_2) = \{p\}$, then both $P(M_1, M_2)$ and $S(M_1, M_2)$ are $\mathbb{F}$-representable.*

Proof If p is a loop or a coloop of M_1 or M_2, then, since $S(M_1, M_2)$ and $P(M_1, M_2)$ are direct sums of minors of M_1 and M_2, each is $\mathbb{F}$-representable. Now assume that neither M_1 nor M_2 has $\{p\}$ as a separator. Then we may assume that M_1 and M_2 are represented over $\mathbb{F}$ by the matrices in Figure 7.5(a) and 7.5(b), respectively. It is now not difficult to check that $P(M_1, M_2)$ and $S(M_1, M_2)$ are represented over $\mathbb{F}$ by the matrices in Figure 7.6(a) and 7.6(b). The details of this argument are left to the reader. □

We remark that once we had shown $P(M_1, M_2)$ to be $\mathbb{F}$-representable in the last proof, duality could have been used to deduce that $S(M_1, M_2)$ is $\mathbb{F}$-representable. However, this approach does not give the explicit $\mathbb{F}$-representation in Figure 7.6(b).

$$
\begin{array}{c}
\begin{array}{ccc} E(M_1)-p & p & E(M_2)-p \end{array}\\
\left[\begin{array}{c|c|c}
 & 0 & \\
 & 0 & \\
A_1 & \vdots & 0 \\
 & 0 & \\
\hline
 & 1 & \\
\hline
 & 0 & \\
0 & \vdots & A_2 \\
 & 0 & \\
 & 0 &
\end{array}\right]\\
\text{(a)}
\end{array}
\qquad
\begin{array}{c}
\begin{array}{ccc} E(M_1)-p & p & E(M_2)-p \end{array}\\
\left[\begin{array}{c|c|c}
 & 0 & \\
 & 0 & \\
A_1 & \vdots & 0 \\
 & 0 & \\
 & 1 & \\
\hline
 & 1 & \\
 & 0 & \\
0 & \vdots & A_2 \\
 & 0 & \\
 & 0 &
\end{array}\right]\\
\text{(b)}
\end{array}
$$

FIG. 7.6. Representations for (a) $P(M_1, M_2)$ and (b) $S(M_1, M_2)$.

Corollary 7.1.25 *If M_1 and M_2 are regular matroids with $E(M_1) \cap E(M_2) = \{p\}$, then both $P(M_1, M_2)$ and $S(M_1, M_2)$ are regular.*

Proof By Theorem 6.6.3, a matroid is regular if and only if it is both binary and ternary. The corollary follows from this and the last proposition. □

We leave it to the reader to show that the converses of the last two results hold. Now suppose, in Corollary 7.1.25, that p is a loop of neither M_1 nor M_2. Then M_1 and M_2 are represented by totally unimodular matrices of the form shown in Figure 7.5(a) and 7.5(b). Moreover, it is not difficult to check that the matrices in Figure 7.6(a) and 7.6(b) are totally unimodular representations for $P(M_1, M_2)$ and $S(M_1, M_2)$.

Since the 2-sum of M_1 and M_2 is a minor of their parallel connection, the following corollary is a straightforward consequence of earlier results. A class of matroids is *closed under 2-sums* if every 2-sum of two members of the class is also in the class.

Corollary 7.1.26 *The classes of graphic, cographic, $\mathbb{F}$-representable, and regular matroids are all closed under 2-sums.*   □

We saw in the proof of Proposition 7.1.24 how to obtain an $\mathbb{F}$-representation for $P(M_1, M_2)$ from $\mathbb{F}$-representations for M_1 and M_2. Because $M_1 \oplus_2 M_2 = P(M_1, M_2)\backslash p$, we get an $\mathbb{F}$-representation for $M_1 \oplus_2 M_2$ simply by deleting the column in Figure 7.6(a) labelled by p.

In contrast to the last corollary, the next result shows that the class of transversal matroids is not closed under 2-sums or parallel connections. However, it is closed under series connections. A proof of the last fact will be given in Chapter 11, while the last exercise of this section characterizes precisely when the parallel connection of two transversal matroids is transversal.

Example 7.1.27 We noted in Example 1.6.3 that the cycle matroid of the graph G in Figure 7.7 is not transversal. But $M(G)$ is isomorphic to both $P(M(G_2), M(G_3))$ and $M(G_1) \oplus_2 M(G_3)$, and each of $M(G_1)$, $M(G_2)$, and $M(G_3)$ is easily shown to be transversal.   □

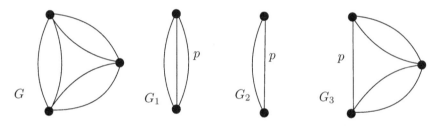

FIG. 7.7. The cycle matroid of the first graph, G, is not transversal.

Exercises

1. Let M_1 and M_2 be matroids such that $E(M_1) \cap E(M_2) = \{p\}$.
   (a) (Jaeger, Vertigan, and Welsh 1990) Let $M = P(M_1, M_2)\backslash p$. Show that if $A_1 \subseteq E(M_1)$ and $A_2 \subseteq E(M_2)$, then

   $$r_M(A_1 \cup A_2) = r_1(A_1) + r_2(A_2) - \theta(A_1, A_2) + \theta(\emptyset, \emptyset)$$

   where $\theta(X, Y)$ is 1 if $r_1(X \cup p) = r_1(X)$ and $r_2(Y \cup p) = r_2(Y)$, and $\theta(X, Y)$ is 0 otherwise.
   (b) For $N = P(M_1, M_2)$, determine the rank function, closure operator, hyperplanes, cocircuits, flats, and independent sets of N.
   (c) Repeat (b) with $N = S(M_1, M_2)$.
2. Let M_1, M_2, and M_3 be matroids such that $E(M_1) \cap E(M_2) = \{p\}$ and $E(M_3) \cap (E(M_1) \cup E(M_2)) = \emptyset$. Show that $S(M_1, M_2) \oplus M_3 = S(M_1, M_2 \oplus M_3)$, and $P(M_1, M_2) \oplus M_3 = P(M_1, M_2 \oplus M_3)$.
3. Prove Proposition 7.1.15.

4. Show that the 2-sum of two identically self-dual matroids is identically self-dual.

5. Find all fields $\mathbb{F}$ for which $P(AG(2,3), AG(3,2))$ is $\mathbb{F}$-representable (see Exercises 3 and 9 of Section 6.4).

6. Does Theorem 7.1.16 hold for all matroids M?

7. Prove Proposition 7.1.22.

8. Let $M = M_1 \oplus_2 M_2$ and suppose that $E(M_2) - p$ has no element e such that M_1 is isomorphic to a minor of both $M \backslash e$ and M/e. Prove that M_2 is a uniform matroid of rank or corank one.

9. For $n \geq 2$, let $M_1, M_2, \ldots, M_n$ be matroids such that $E(M_i) \cap E(M_j) = \{p\}$ for all distinct i and j, and no M_k has $\{p\}$ as a separator. Prove that:

   (a) $S(M_1, M_2, \ldots, M_k) = P(M_1^*, M_2^*, \ldots, M_k^*)$.

   (b) If σ is a permutation of $\{1, 2, \ldots, n\}$, then

   $$P(M_{\sigma(1)}, M_{\sigma(2)}, \ldots, M_{\sigma(n)}) = P(M_1, M_2, \ldots, M_n) \text{ and}$$
   $$S(M_{\sigma(1)}, M_{\sigma(2)}, \ldots, M_{\sigma(n)}) = S(M_1, M_2, \ldots, M_n).$$

   (c) If $3 \leq k \leq n$, then

   $$P(M_1, M_2, \ldots, M_k) = P(P(M_1, M_2, \ldots, M_{k-1}), M_k) \text{ and}$$
   $$S(M_1, M_2, \ldots, M_k) = S(S(M_1, M_2, \ldots, M_{k-1}), M_k).$$

   (d) The following are equivalent:
       (i) All of $M_1, M_2, \ldots, M_n$ are connected.
       (ii) $P(M_1, M_2, \ldots, M_n)$ is connected.
       (iii) $S(M_1, M_2, \ldots, M_n)$ is connected.

10. *Suppose that N_1 and N_2 are connected matroids such that $|E(N_1)|$, $|E(N_2)| \geq 2$. Prove that $P((N_1; p), (N_2; p))$ is transversal if and only if N_1 and N_2 are both transversal and each has a presentation in which p occurs in exactly one of the sets of the presentation.

7.2 Single-element extensions

If a matroid M is obtained from a matroid N by deleting a non-empty subset T of $E(N)$, then N is called an *extension* of M. In particular, if $|T| = 1$, then N is a *single-element extension* of M. In this section, following Crapo (1965), we shall investigate how to find all of the single-element extensions of a given matroid. The term 'single-element extension' is often abbreviated in the literature to 'extension', and we shall follow this practice here when it should not cause confusion. Another term that is sometimes used instead of 'single-element extension' is 'addition' (see, for example, Truemper 1985).

Two obvious ways to extend a matroid M are to adjoin a loop or to adjoin a coloop. In these cases, the resulting matroids are isomorphic to $M \oplus U_{0,1}$ and $M \oplus U_{1,1}$, respectively. Another type of extension that we have already met is a parallel extension where we add a new element in parallel to some existing element of M.

If N^* is an extension of M^*, then N is called a *coextension* of M. In this case, $M = N/T$ for some subset T of $E(N)$. A single-element coextension of M has also been called a 'lift' of M (see, for example, Mason 1977), but we shall reserve this term for a slightly different construction to be considered towards the end of Section 7.3. We remark that if N is a series extension of M, then N is actually a coextension of M rather than an extension. Nevertheless, we shall continue to use the well-established term 'series extension' instead of the more accurate 'series coextension'.

Now suppose that the matroid M is given and we wish to adjoin an element e to M to form an extension N. By Proposition 3.3.7, the flats of M are of the form $X - e$ where X is a flat of N. Thus to specify N in terms of M, we consider the effect on the flats of M of the addition of e. Clearly, there are three possibilities for each flat F of M:

(i) $F \cup e$ is a flat of N and $r(F \cup e) = r(F)$;

(ii) $F \cup e$ is a flat of N and $r(F \cup e) = r(F) + 1$; or

(iii) $F \cup e$ is not a flat of N.

We shall be particularly interested in the flats satisfying (i). The following example revisits the Escher matroid from Example 1.5.5.

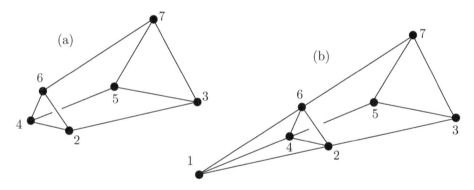

FIG. 7.8. (a) A matroid M. (b) An extension of M by the element 1.

Example 7.2.1 Let M be the matroid for which a geometric representation is shown in Figure 7.8(a). Suppose we wish to adjoin the element 1 to M so that it is on both the line through 2 and 3 and the line through 4 and 5. Assume that such an extension N exists. If $\mathcal{M}$ is the set of flats F of M such that $F \cup 1$ is a flat of N having the same rank as F, then we have insisted that both $\{2, 3\}$ and $\{4, 5\}$ are in $\mathcal{M}$. Therefore every flat of M containing $\{2, 3\}$ or $\{4, 5\}$ must also be in $\mathcal{M}$. In particular, $\{2, 3, 6, 7\}$ and $\{4, 5, 6, 7\}$ are in $\mathcal{M}$. As N is a matroid, r_N is submodular. Thus

$$3 + 3 = r_M(\{2, 3, 6, 7\}) + r_M(\{4, 5, 6, 7\}) = r_N(\{1, 2, 3, 6, 7\}) + r_N(\{1, 4, 5, 6, 7\})$$
$$\geq r_N(\{1, 2, 3, 4, 5, 6, 7\}) + r_N(\{1, 6, 7\})$$
$$\geq r_M(\{2, 3, 4, 5, 6, 7\}) + r_M(\{6, 7\})$$
$$= 4 + 2.$$

We deduce that $r_N(\{1, 6, 7\}) = 2 = r_M(\{6, 7\})$, so $\{6, 7\}$ must also be in $\mathcal{M}$.

One choice for the extension N has the geometric representation shown in Figure 7.8(b). As we shall see in Example 7.2.7, the only other choice for N adds 1 as a loop.   □

Examining the last example, we see that the reason that $\{6, 7\}$ is forced to be in $\mathcal{M}$ is that $\{6, 7\}$ is the intersection of two flats, $\{2, 3, 6, 7\}$ and $\{4, 5, 6, 7\}$, in $\mathcal{M}$ that form a modular pair, that is, the sum of their ranks equals the sum of the rank of their union and the rank of their intersection. The next lemma generalizes the ideas in this example.

Lemma 7.2.2 *Let N be an extension of a matroid M by an element e and let $\mathcal{M}$ be the set of flats F of M such that $F \cup e$ is a flat of N having the same rank as F. Then $\mathcal{M}$ has the following properties:*

(i)   *If $F \in \mathcal{M}$ and F' is a flat of M containing F, then $F' \in \mathcal{M}$.*

(ii)   *If $F_1, F_2 \in \mathcal{M}$ and (F_1, F_2) is a modular pair, then $F_1 \cap F_2 \in \mathcal{M}$.*

Proof (i) As $F \in \mathcal{M}$, we have $r(F \cup e) = r(F)$. Thus $e \in \mathrm{cl}_N(F) \subseteq \mathrm{cl}_N(F')$. Hence $r(F' \cup e) = r(F')$ and so $F' \cup e$ is a flat of N. Therefore $F' \in \mathcal{M}$.

(ii) Clearly $r(F_i \cup e) = r(F_i)$ for each i. Thus, as (F_1, F_2) is a modular pair, $r(F_1 \cap F_2) + r(F_1 \cup F_2) = r(F_1) + r(F_2) = r(F_1 \cup e) + r(F_2 \cup e)$. Applying the submodularity of r to this equation, we get that

$$r(F_1 \cap F_2) + r(F_1 \cup F_2) \geq r((F_1 \cap F_2) \cup e) + r(F_1 \cup F_2 \cup e). \qquad (7.1)$$

As $r(F_1 \cup e) = r(F_1)$, it follows that $r(F_1 \cup F_2 \cup e) = r(F_1 \cup F_2)$. Thus (7.1) implies that $r(F_1 \cap F_2) \geq r((F_1 \cap F_2) \cup e)$. Hence $r(F_1 \cap F_2) = r((F_1 \cap F_2) \cup e)$ and so $F_1 \cap F_2 \in \mathcal{M}$.   □

An arbitrary set $\mathcal{M}$ of flats of a matroid M is called a *modular cut* if it satisfies (i) and (ii) of Lemma 7.2.2. By that lemma, every single-element extension of a matroid gives rise to a modular cut. So, for example, the modular cut corresponding to the extension $M \oplus U_{0,1}$ of M consists of the set of all flats of M, while the modular cut corresponding to $M \oplus U_{1,1}$ is empty. The next result establishes that every modular cut gives rise to a unique extension.

Theorem 7.2.3 *Let $\mathcal{M}$ be a modular cut of a matroid M on a set E. Then there is a unique extension N of M on $E \dot\cup e$ such that $\mathcal{M}$ consists of those flats F of M for which $F \cup e$ is a flat of N having the same rank as F. Moreover, for all subsets X of E,*

$$r_N(X) = r_M(X) \quad and$$

$$r_N(X \cup e) = \begin{cases} r_M(X), & \text{if } \text{cl}_M(X) \in \mathcal{M}, \\ r_M(X) + 1, & \text{if } \text{cl}_M(X) \notin \mathcal{M}. \end{cases}$$

Proof It is straightforward to check that if N is an extension of M satisfying the specified condition on flats, then N must have rank function r_N. Hence if such an extension exists, it is unique.

To complete the proof of the theorem, we must show that r_N is the rank function of a matroid on $E \cup e$. One can easily check that r_N satisfies (R1) and (R2). Since r_M satisfies (R3), to show that r_N satisfies (R3), we need only check that the submodular inequality holds for all pairs of sets having one of the forms $(X \cup e, Y)$ and $(X \cup e, Y \cup e)$ where $X, Y \subseteq E$.

In the first case,

$$r_N(X \cup e) + r_N(Y) = r_M(X) + r_M(Y) + \delta_X$$

where δ_X is 0 or 1 according to whether or not $\text{cl}_M(X)$ is in $\mathcal{M}$. Thus

$$\begin{aligned} r_N(X \cup e) + r_N(Y) &\geq r_M(X \cup Y) + r_M(X \cap Y) + \delta_X \\ &= r_M(X \cup Y) + r_N((X \cup e) \cap Y) + \delta_X. \end{aligned} \tag{7.2}$$

But if $\text{cl}_M(X) \in \mathcal{M}$, then $\text{cl}_M(X \cup Y) \in \mathcal{M}$, so

$$r_M(X \cup Y) + \delta_X \geq r_N(X \cup Y \cup e). \tag{7.3}$$

If $\text{cl}_M(X) \notin \mathcal{M}$, then $\delta_X = 1$ and again (7.3) holds. On substituting from (7.3) into (7.2), we get that (R3) holds for the pair $(X \cup e, Y)$.

Finally, we show that (R3) holds for the pair $(X \cup e, Y \cup e)$. Clearly

$$\begin{aligned} r_N(X \cup e) + r_N(Y \cup e) &= r_M(X) + \delta_X + r_M(Y) + \delta_Y \\ &= r_M(\text{cl}_M(X)) + \delta_X + r_M(\text{cl}_M(Y)) + \delta_Y \\ &\geq r_M(\text{cl}_M(X) \cup \text{cl}_M(Y)) + \delta_X \\ &\quad + r_M(\text{cl}_M(X) \cap \text{cl}_M(Y)) + \delta_Y. \end{aligned} \tag{7.4}$$

If $\delta_X = \delta_Y = 1$, then

$$r_N(X \cup e) + r_N(Y \cup e) \geq r_M(X \cup Y) + 1 + r_M(X \cap Y) + 1, \quad \text{so}$$

$$\begin{aligned} r_N(X \cup e) + r_N(Y \cup e) &\geq r_N((X \cup e) \cup (Y \cup e)) \\ &\quad + r_N((X \cup e) \cap (Y \cup e)). \end{aligned} \tag{7.5}$$

If one of δ_X and δ_Y is 0 and the other is 1, then exactly one of $\text{cl}_M(X)$ and $\text{cl}_M(Y)$ is in $\mathcal{M}$. Hence $\text{cl}_M(\text{cl}_M(X) \cup \text{cl}_M(Y))$ is in $\mathcal{M}$ but $\text{cl}_M(X) \cap \text{cl}_M(Y)$ is not. Thus, from (7.4),

$$r_N(X \cup e) + r_N(Y \cup e) \geq r_N(\mathrm{cl}_M(X) \cup \mathrm{cl}_M(Y) \cup e)$$
$$+ r_N((\mathrm{cl}_M(X) \cup e) \cap (\mathrm{cl}_M(Y) \cup e)) \qquad (7.6)$$

and again (7.5) holds. Similarly, (7.5) holds if $\delta_X = \delta_Y = 0$ unless equality holds in (7.4). But, in the exceptional case, $(\mathrm{cl}_M(X), \mathrm{cl}_M(Y))$ is a modular pair of flats of M, so $\mathrm{cl}_M(X) \cap \mathrm{cl}_M(Y) \in \mathcal{M}$. As $\mathrm{cl}_M(\mathrm{cl}_M(X) \cup \mathrm{cl}_M(Y))$ is also in $\mathcal{M}$, (7.6) follows and hence so does (7.5). We conclude that r_N satisfies (R3).   □

On combining Theorem 7.2.3 and Lemma 7.2.2, we get that there is a one-to-one correspondence between single-element extensions of a matroid M and modular cuts of M. In view of this, we shall often refer to a particular extension as being *determined* by the corresponding modular cut. If $M = N\backslash e$ and $\mathcal{M}$ is the modular cut corresponding to the extension N, we shall often write N as $M +_{\mathcal{M}} e$.

By finding all modular cuts of a matroid M, we can determine all possible single-element extensions of M. This gives a systematic technique by which, theoretically, one can determine all matroids on some fixed number of elements. Although this technique quickly becomes computationally impractical, it was used, for example, by Blackburn, Crapo, and Higgs (1973) to find all simple matroids on at most eight elements.

A modular cut of a matroid M provides one way to specify an extension of M. Crapo (1965) showed that a more compact way to do this is by listing the hyperplanes in the modular cut, for these hyperplanes determine the modular cut (see Exercise 6). Another, less compact, way to specify an extension of M by e is by a list $\mathcal{U}$ of those subsets of $E(M)$ whose rank is unchanged by the addition of e. Again we postpone to the exercises consideration of $\mathcal{U}$ and its properties.

Example 7.2.4 In the matroid Q_6, whose geometric representation is shown in Figure 7.9(a), let $\mathcal{M}_1 = \{\{1,4\}, \{2,5\}, E(Q_6)\}$ and $\mathcal{M}_2 = \{\{1,4\}, \{2,5\}, \{3,6\}, E(Q_6)\}$. Each of $\mathcal{M}_1$ and $\mathcal{M}_2$ is easily shown to be a modular cut of Q_6, so each determines an extension. Geometric representations for these extensions

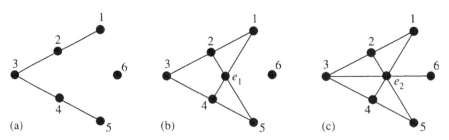

(a)   (b)   (c)

FIG. 7.9. (a) Q_6. (b) N_1 and (c) N_2 are incompatible extensions.

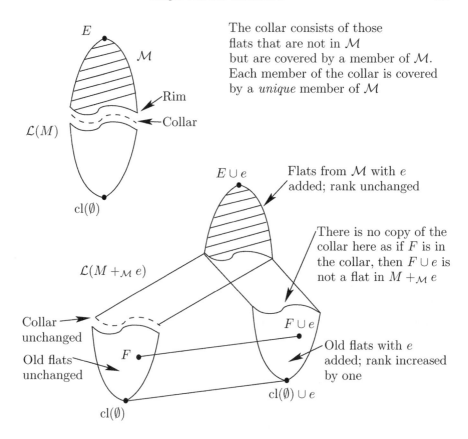

The collar consists of those
flats that are not in $\mathcal{M}$
but are covered by a member of $\mathcal{M}$.
Each member of the collar is covered
by a *unique* member of $\mathcal{M}$

Flats from $\mathcal{M}$ with e
added; rank unchanged

There is no copy of the
collar here as if F is in
the collar, then $F \cup e$ is
not a flat in $M +_{\mathcal{M}} e$

Old flats with e
added; rank increased
by one

FIG. 7.10. $\mathcal{L}(M)$ and $\mathcal{L}(M +_{\mathcal{M}} e)$.

are shown in Figure 7.9(b) and 7.9(c), the added elements being e_1 and e_2,
respectively. The extensions N_1 and N_2 are examples of *incompatible single-
element extensions* of Q_6, for both are extensions of Q_6, yet there is no matroid
N such that $N\backslash e_1 = N_2$ and $N\backslash e_2 = N_1$. We defer to the exercises consideration
of the problem of determining precisely when two single-element extensions are
compatible.   □

The rank function of $M +_{\mathcal{M}} e$ is given in Theorem 7.2.3. Using this, it follows
that $r(M +_{\mathcal{M}} e) = r(M) + 1$ if and only if $\mathcal{M} = \emptyset$.

Corollary 7.2.5 *The flats of $M +_{\mathcal{M}} e$ fall into three disjoint classes:*

(i)   *flats F of M that are not in $\mathcal{M}$;*

(ii)   *sets $F \cup e$ where F is a flat of M that is in $\mathcal{M}$; and*

(iii)   *sets $F \cup e$ where F is a flat of M that is not in $\mathcal{M}$, and F is not contained
in a member F' of $\mathcal{M}$ of rank $r(F) + 1$.*   □

In terms of the lattice $\mathcal{L}(M)$ of flats of M, an alternative description of the condition governing a flat F of M in (iii) of the last corollary is that F is not in $\mathcal{M}$ and F is not covered in $\mathcal{L}(M)$ by any member of $\mathcal{M}$. This corollary is illustrated in Figure 7.10 where a schematic representation of an arbitrary modular cut in $\mathcal{L}(M)$ is shown along with the lattice $\mathcal{L}(M +_{\mathcal{M}} e)$ of flats of the corresponding extension (see Mason 1977).

If F is a flat of a matroid M, then the set $\mathcal{M}_F$ of flats containing F is easily shown to be a modular cut. We call $\mathcal{M}_F$ a *principal modular cut* and the extension determined by $\mathcal{M}_F$ a *principal extension*. We shall denote this principal extension by $M +_F e$ and we shall also say, in this case, that the element e has been *freely added* to the flat F or, if $F = E(M)$, that e has been freely added to M. In the last case, e is said to be *free* in $M +_{E(M)} e$ and this matroid is called the *free extension* of M. When adding the element e freely to the flat F, we add e to F so that the only new circuits created are those which are forced by the fact that e has been placed on F. Every such circuit spans F.

The next proposition, a straightforward consequence of Corollary 7.2.5, specifies the independent sets of a principal extension of M. Descriptions of principal extensions in terms of circuits, bases and so on will be considered in the exercises.

Proposition 7.2.6 *Let F be a flat of a matroid M and $N = M +_F e$. Then*

$$\mathcal{I}(N) = \mathcal{I}(M) \cup \{I \cup e : I \in \mathcal{I}(M) \text{ and } \operatorname{cl}_M(I) \not\supseteq F\}. \qquad \square$$

One attractive feature of modular cuts is that the intersection of two modular cuts is a modular cut. Suppose that we wish to add the element e to each of the flats $F_1, F_2, \ldots, F_m$. Then the modular cut corresponding to such an extension must contain $\{F_1, F_2, \ldots, F_m\}$. The smallest modular cut containing this set, which is the intersection of all modular cuts containing $\{F_1, F_2, \ldots, F_m\}$, is called the modular cut *generated* by $\{F_1, F_2, \ldots, F_m\}$. Frequently this modular cut is equal to the set of all flats of the matroid, in which case the only way that e can be added to all of $F_1, F_2, \ldots, F_m$ is as a loop.

Example 7.2.7 Let M be the matroid for which a geometric representation is given in Figure 7.8(a) and let $\mathcal{M}$ be the modular cut generated by $\{\{2, 3\}, \{4, 5\}\}$. We argued in Example 7.2.1 that $\mathcal{M}$ contains $\{\{2, 3\}, \{4, 5\}, \{6, 7\}, \{2, 3, 4, 5\}, \{2, 3, 6, 7\}, \{4, 5, 6, 7\}, E(M)\}$. It is not difficult to check that the last collection is actually a modular cut, so this modular cut equals $\mathcal{M}$. Hence $M +_{\mathcal{M}} 1$ is the matroid that is represented geometrically by Figure 7.8(b). By symmetry, any other modular cut $\mathcal{M}'$ containing $\{\{2, 3\}, \{4, 5\}\}$ may be assumed to contain $\{2, 5\}$. Then $\mathcal{M}'$ contains each of $\{2, 5\} \cap \{2, 3\}$ and $\{2, 5\} \cap \{4, 5\}$, and so also contains $\{2\} \cap \{5\}$. Hence $\mathcal{M}'$ consists of the set of all flats of M. Thus, as asserted in Example 7.2.1, there are exactly two distinct extensions of M by the element 1 in which both $\{1, 2, 3\}$ and $\{1, 4, 5\}$ are rank-2 flats. $\qquad \square$

Exercises

1. Prove Proposition 7.2.6.

2. Find all non-isomorphic single-element extensions of Q_6 and specify the corresponding modular cuts.

3. Suppose that $N \backslash e = M$. Show that e is free in N if and only if $r(N) = r(M)$ and every circuit of N containing e has $r(N) + 1$ elements.

4. Suppose that F is a flat of a matroid M. Let $E(M) = E$ and let $N = M +_F e$. Show the following:
   (a) $\mathcal{B}(N) = \mathcal{B}(M) \cup \{(B - f) \cup e : B \in \mathcal{B}(M) \text{ and } f \in B \cap F\}$.
   (b) $\mathcal{C}(N) = \mathcal{C}(M) \cup \{X \cup e : \operatorname{cl}_M(X) \supseteq F \text{ and } \operatorname{cl}_M(X - x) \not\supseteq F \text{ for all } x \text{ in } X\}$.
   (c) For all $X \subseteq E$, $r_N(X) = r_M(X)$, and

   $$r_N(X \cup e) = \begin{cases} r_M(X), & \text{if } r_M(X \cup F) = r_M(X), \\ r_M(X) + 1, & \text{otherwise.} \end{cases}$$

   (d) For all $X \subseteq E$,

   $$\operatorname{cl}_N(X) = \begin{cases} \operatorname{cl}_M(X) \cup e, & \text{if } \operatorname{cl}_M(X) \supseteq F, \\ \operatorname{cl}_M(X), & \text{otherwise,} \end{cases}$$

   and

   $$\operatorname{cl}_N(X \cup e) = \begin{cases} \operatorname{cl}_M(X \cup F) \cup e, & \text{if } r_M(X \cup F) = r_M(X) + 1, \\ \operatorname{cl}_M(X) \cup e, & \text{otherwise.} \end{cases}$$

5. Let N be an extension of a matroid M by an element e. Let $\mathcal{U}$ be the set of subsets of $E(M)$ whose rank is unchanged by the addition of e.
   (a) Let $\mathcal{M}$ be the modular cut of M corresponding to the extension N. Specify how to obtain each of $\mathcal{M}$ and $\mathcal{U}$ from the other.
   (b) Show that $\mathcal{U}$ has the following properties:
      (i) If $X \in \mathcal{U}$ and Y is a subset of $E(M)$ such that $\operatorname{cl}_M(Y) \supseteq X$, then $Y \in \mathcal{U}$.
      (ii) If $X_1, X_2 \in \mathcal{U}$ and $(\operatorname{cl}_M(X_1), \operatorname{cl}_M(X_2))$ is a modular pair, then $\operatorname{cl}_M(X_1) \cap \operatorname{cl}_M(X_2) \in \mathcal{U}$.
   (c) Let M be a matroid and $\mathcal{U}$ be any set of subsets of $E(M)$ satisfying (i) and (ii). Prove that there is a unique extension N' of M by an element e' in which $\mathcal{U}$ consists of those subsets X of $E(M)$ for which $r(X \cup e') = r(X)$.

6. (Crapo 1965) A *linear subclass* of a matroid M is a subset $\mathcal{H}'$ of the set of hyperplanes of M such that if H_1 and H_2 are members of $\mathcal{H}'$ for which $r(H_1 \cap H_2) = r(M) - 2$, and H_3 is a hyperplane containing $H_1 \cap H_2$, then $H_3 \in \mathcal{H}'$. Show the following:
   (a) If $\mathcal{M}$ is a modular cut of M, then the hyperplanes of M in $\mathcal{M}$ form a linear subclass.
   (b) If $\mathcal{H}'$ is a linear subclass of M, and $\mathcal{M}$ consists of all flats X of M for which every hyperplane containing X is in $\mathcal{H}'$, then $\mathcal{M}$ is a modular cut of M.

(c) If $\mathcal{H}'$ is a linear subclass of M, then there is a unique extension N of M by e such that $\mathcal{H}'$ is the set of hyperplanes H of M for which $H \cup e$ is a hyperplane of N.

7. (Oxley 1984b) Verify the following: Let $\{e_1, e_2, \ldots, e_k\}$ be a circuit in a matroid M where $k \geq 3$ and suppose that e_1 is in every flat of M that is a dependent set. Then a flat F of M is in the modular cut generated by $\mathrm{cl}(\{e_1, e_2\})$ and $\mathrm{cl}(\{e_3, e_4, \ldots, e_k\})$ if and only if F contains one of the two generating flats. Moreover, the generating flats are disjoint.

8. (Cheung 1974) Let M, N_1, and N_2 be matroids such that $N_1 \backslash e_1 = M = N_2 \backslash e_2$ where $e_1 \neq e_2$. Let $\mathcal{M}_1$ and $\mathcal{M}_2$ be the modular cuts of M corresponding to N_1 and N_2, respectively. We call N_1 and N_2 *compatible extensions* of M if there is a matroid N on the set $E(M) \cup \{e_1, e_2\}$ such that $N \backslash e_2 = N_1$ and $N \backslash e_1 = N_2$.

   (a) Prove that N_1 and N_2 are compatible extensions of M if and only if $N_{12} = N_{21}$, where N_{12} is the extension of N_1 by e_2 that is determined by the modular cut generated by $\{\mathrm{cl}_{N_1}(X) : X \in \mathcal{M}_2\}$, and N_{21} is defined analogously.

   (b) Show that if N_1 and N_2 are both principal extensions, then N_1 and N_2 are compatible extensions of M.

7.3 Quotients and related operations

What is the precise relationship between two matroids M_1 and M_2 on a common ground set E if every flat of M_2 is a flat of M_1? In this section, we shall answer this question by introducing a new operation which uses the operation of extension. As in the last section, modular cuts will play an important role here.

A modular cut of a matroid M will be called *proper* if it is not equal to the set of all flats of M. Let $\mathcal{M}$ be such a modular cut and assume also that $\mathcal{M}$ is non-empty. The *elementary quotient* of M with respect to $\mathcal{M}$ is the matroid $(M +_{\mathcal{M}} e)/e$. Evidently its rank is one less than that of M and its ground set is the same as that of M. The reason for not allowing $\mathcal{M}$ to be empty or non-proper is that otherwise e is a coloop or a loop of $M +_{\mathcal{M}} e$ and so the matroid $(M +_{\mathcal{M}} e)/e$ is just M. Hence the effect of these constraints is only to exclude two trivial cases. In general, a matroid Q is a *quotient* of the matroid M if there is a matroid N and a subset X of $E(N)$ such that $M = N \backslash X$ and $Q = N/X$. Thus if Q is a quotient of M, then $E(Q) = E(M)$. Another easy consequence of the definition is the following.

Proposition 7.3.1 *Let M_1 and M_2 be matroids. Then M_1 is a quotient of M_2 if and only if M_2^* is a quotient of M_1^*.* $\square$

Example 7.3.2 In Figure 7.11, M_1 and M_2 are two of the many elementary quotients of $M(K_4)$. Observe that every flat of each of M_1 and M_2 is also a flat of $M(K_4)$. $\square$

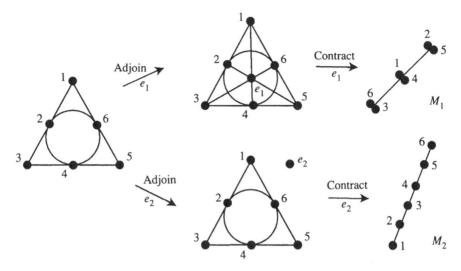

FIG. 7.11. M_1 and M_2 are two of the quotients of $M(K_4)$.

The next proposition shows that an arbitrary quotient can be formed by taking a (possibly empty) sequence of elementary quotients. The proof will use the following result.

Lemma 7.3.3 *A matroid Q is a quotient of M if and only if there is a matroid N of rank $r(M)$ having an independent set I such that $M = N\backslash I$ and $Q = N/I$.*

Proof If N and I exist satisfying the specified conditions, then Q is certainly a quotient of M. Conversely, let Q be a quotient of M. Then there is a matroid N_1 and a subset X of $E(N_1)$ such that $M = N_1\backslash X$ and $Q = N_1/X$. Choose such a matroid N_1 of minimum rank. Let I be a basis of X in N_1 and let $N = N_1\backslash(X - I)$. Then $N\backslash I = M$ and $N/I = N_1/I\backslash(X - I)$. Moreover, the last matroid equals Q since every element of $X - I$ is a loop of N_1/I. Also the choice of N_1 means that $r(N) = r(N_1)$. Assume that $r(N) \neq r(M)$. Then $r(E(M) \cup x) > r(M)$ for some x in I. Thus $N\backslash(I - x)/x = M$ and $N/x/(I - x) = Q$. Since N/x contradicts the choice of N_1, we conclude that $r(N) = r(M)$ and this completes the proof of the lemma. □

An immediate consequence of this lemma is the following.

Corollary 7.3.4 *If Q is a quotient of M and $r(Q) = r(M)$, then $Q = M$.* □

Proposition 7.3.5 *A matroid Q is a quotient of M with $r(M) - r(Q) = k$ if and only if there is a sequence $M_0, M_1, M_2, \ldots, M_k$ of matroids with $M_0 = M$ and $M_k = Q$ such that, for all i in $\{1, 2, \ldots, k\}$, the matroid M_i is an elementary quotient of M_{i-1}.*

Proof Let Q be a quotient of M with $r(M) - r(Q) = k$. Then, by Lemma 7.3.3, there is a matroid N of rank $r(M)$ having an independent set I such that $M = N \backslash I$ and $Q = N/I$. Evidently $|I| = k$, so let $I = \{e_1, e_2, \ldots, e_k\}$. For each i in $\{0, 1, \ldots, k\}$, let M_i be the matroid $N \backslash (I - \{e_1, e_2, \ldots, e_i\})/\{e_1, e_2, \ldots, e_i\}$. Clearly $M_0 = M$ and $M_k = Q$. Moreover, it is straightforward to check that if $1 \le i \le k$, then M_i is an elementary quotient of M_{i-1}.

Conversely, assume that a sequence $M_0, M_1, \ldots, M_k$ of matroids exists with the specified properties. Then, for each i in $\{1, 2, \ldots, k\}$, there is an extension N_i of M_{i-1} on $E(M_{i-1}) \cup e_i$ with $r(N_i) = r(M_{i-1})$ such that $N_i/e_i = M_i$ and $r(M_i) = r(N_i) - 1$. To complete the proof we show, by induction on i, that there is an extension N'_i of M on $E(M) \cup \{e_1, e_2, \ldots, e_i\}$ such that $\{e_1, e_2, \ldots, e_i\}$ is independent in N'_i, and $N'_i/\{e_1, e_2, \ldots, e_i\} = M_i$. This is certainly true if $i = 1$. Assume it true for $i = j$ and let $i = j+1$. Then, by the induction assumption, M has an extension N'_j such that $M_j = N'_j/\{e_1, e_2, \ldots, e_j\}$ and $\{e_1, e_2, \ldots, e_j\}$ is independent in N'_j. Now $N_{j+1} \backslash e_{j+1} = M_j$ and $r(N_{j+1}) = r(M_j)$. Also $M_{j+1} = N_{j+1}/e_{j+1}$ and $r(M_{j+1}) = r(N_{j+1}) - 1$. Thus $N'_j/\{e_1, e_2, \ldots, e_j\}$ has a proper non-empty modular cut corresponding to the extension N_{j+1}. Let $\mathcal{M}' = \{F \cup \{e_1, e_2, \ldots, e_j\} : F \in \mathcal{M}\}$. Then it is routine to check that $\mathcal{M}'$ is a proper non-empty modular cut of N'_j. Let the corresponding extension of N'_j have ground set $E(N'_j) \cup e_{j+1}$ and call this extension N'_{j+1}. By comparing their rank functions, one can show that N_{j+1} and $N'_{j+1}/\{e_1, e_2, \ldots, e_j\}$ are equal. Therefore, as $N_{j+1}/e_{j+1} = M_{j+1}$, we get that $M_{j+1} = N'_{j+1}/\{e_1, e_2, \ldots, e_{j+1}\}$. Finally, as M_{j+1} is an elementary quotient of M_j, the set $\{e_{j+1}\}$ is independent in $N'_{j+1}/\{e_1, e_2, \ldots, e_j\}$. Thus $\{e_1, e_2, \ldots, e_{j+1}\}$ is independent in N'_{j+1}, so N'_{j+1} is the required extension of M. The proposition now follows by induction. □

We began this section seeking a characterization of the link between matroids M_1 and M_2 such that every flat of M_2 is a flat of M_1. The next result provides several such characterizations including one in terms of quotients. Some further such characterizations will be considered in the exercises.

Proposition 7.3.6 *Let M_1 and M_2 be matroids having rank functions r_1 and r_2, closure operators cl_1 and cl_2, and a common ground set E. The following statements are equivalent:*

(i)   *M_2 is a quotient of M_1.*
(ii) *Every flat of M_2 is a flat of M_1.*
(iii) *If $X \subseteq Y \subseteq E$, then $r_1(Y) - r_1(X) \ge r_2(Y) - r_2(X)$.*
(iv) *Every circuit of M_1 is a union of circuits of M_2.*
(v)   *If $X \subseteq E$, then $\mathrm{cl}_1(X) \subseteq \mathrm{cl}_2(X)$.*

Proof It is straightforward to show using Proposition 3.3.7 and the definition of a quotient that (i) implies (ii). Next we establish the equivalence of (ii)–(v) by showing that each of (ii)–(iv) implies its successor and that (v) implies (ii).

Assume (ii) holds. We argue by induction on $|Y - X|$ to show that (iii) holds. This is certainly true if $|Y - X| = 0$. Now assume (iii) holds for $|Y - X| < k$ and let $|Y - X| = k \geq 1$. Choose y in $Y - X$. If $r_1(Y) - r_1(X) = r_1(Y-y)+1-r_1(X)$, then, by the induction assumption, $r_1(Y) - r_1(X) \geq r_2(Y-y)+1-r_2(X) \geq r_2(Y) - r_2(X)$. Thus we may suppose that $r_1(Y) - r_1(X) = r_1(Y-y)-r_1(X)$. Hence $r_1(Y) = r_1(Y-y)$, so $y \in \mathrm{cl}_1(Y-y)$. If $r_2(Y) = r_2(Y-y)$, then the required result follows by the induction assumption. Thus we may assume that $r_2(Y) \neq r_2(Y-y)$, that is, $y \notin \mathrm{cl}_2(Y-y)$. Then $\mathrm{cl}_2(Y-y)$ and $\mathrm{cl}_2(Y)$ are distinct flats of M_2 and hence of M_1. Therefore $\mathrm{cl}_2(Y-y) = \mathrm{cl}_1(\mathrm{cl}_2(Y-y)) \supseteq \mathrm{cl}_1(Y-y) \supseteq Y$, so $\mathrm{cl}_2(Y-y) = \mathrm{cl}_2(Y)$; a contradiction. Hence (ii) implies (iii).

Now assume (iii) holds and let C be a circuit of M_1. If C is not a union of circuits of M_2, then C has an element x that is not in a circuit of $M_2|C$. Thus $r_2(C) - r_2(C - x) = 1$. Hence, by (iii), $r_1(C) - r_1(C - x) \geq 1$, contradicting the fact that C is a circuit. Therefore (iii) implies (iv).

Next assume that (iv) holds and suppose $X \subseteq E$. Then

$$\mathrm{cl}_1(X) = X \cup \{x : M_1 \text{ has a circuit } C \text{ such that } x \in C \subseteq X \cup x\}$$
$$\subseteq X \cup \{x : M_2 \text{ has a circuit } D \text{ such that } x \in D \subseteq X \cup x\}$$
$$= \mathrm{cl}_2(X),$$

that is, (iv) implies (v).

We now show that (v) implies (ii), so suppose that (v) holds. Let F be a flat of M_2. Then $F = \mathrm{cl}_2(F) \supseteq \mathrm{cl}_1(F) \supseteq F$, so F is a flat of M_1, and (ii) holds. Hence (v) implies (ii) and we conclude that (ii)–(v) are equivalent.

Now, to finish the proof of Proposition 7.3.6, we shall show that (ii) implies (i), the proof of this being quite long. Thus suppose that every flat of M_2 is a flat of M_1. Then, from the above, each of (iii)–(v) holds. By (iii), $r_1(M_1) \geq r_2(M_2)$. We shall show, by induction on $r_1(M_1) - r_2(M_2)$, that M_2 is a quotient of M_1. The following lemma will be useful in this proof. Recall that, in a matroid M, a flat F_1 covers a flat F_2 in $\mathcal{L}(M)$ if $F_1 \supseteq F_2$ and $r(F_1) = r(F_2) + 1$.

Lemma 7.3.7 *Let X be a flat of M_2 for which $r_1(X) - r_2(X) = r_1(M_1) - r_2(M_2)$, and let Y be a flat of M_1 that covers X in $\mathcal{L}(M_1)$. Then Y is a flat of M_2 covering X in $\mathcal{L}(M_2)$, and $r_1(Y) - r_2(Y) = r_1(M_1) - r_2(M_2)$.*

Proof Choose y in $Y - X$. Then, as X is a flat of M_2, the flat $\mathrm{cl}_2(X \cup y)$ of M_2 covers X in $\mathcal{L}(M_2)$. Thus $\mathrm{cl}_2(X \cup y)$ is a flat of M_1. By (iii),

$$r_1(E) - r_1(\mathrm{cl}_2(X \cup y)) \geq r_2(E) - r_2(\mathrm{cl}_2(X \cup y)).$$

Hence $[r_1(M_1) - r_2(M_2)] + r_2(\mathrm{cl}_2(X \cup y)) \geq r_1(\mathrm{cl}_2(X \cup y))$, so

$$[r_1(X) - r_2(X)] + r_2(\mathrm{cl}_2(X \cup y)) \geq r_1(\mathrm{cl}_2(X \cup y)).$$

Therefore $r_1(X) + [r_2(\mathrm{cl}_2(X \cup y)) - r_2(X)] \geq r_1(\mathrm{cl}_2(X \cup y))$, that is,

$$r_1(X) + 1 \geq r_1(\mathrm{cl}_2(X \cup y)). \tag{7.7}$$

As $\mathrm{cl}_2(X \cup y) \neq X$, equality must hold in (7.7). Thus both $\mathrm{cl}_2(X \cup y)$ and Y cover X in $\mathcal{L}(M_1)$. Since the intersection of $\mathrm{cl}_2(X \cup y)$ and Y properly contains X, we conclude that $\mathrm{cl}_2(X \cup y) = Y$ and the lemma follows. $\square$

Now to show that M_2 is a quotient of M_1, suppose first that $r_1(M_1) - r_2(M_2) = 0$. We shall show that the sets of flats of M_1 and M_2 coincide. By assumption, every flat of M_2 is a flat of M_1. If M_1 has a flat that is not a flat of M_2, take Y to be such a flat of smallest rank. By (iii), $r_1(M_1) - r_1(\mathrm{cl}_2(\emptyset)) \geq r_2(M_2) - r_2(\mathrm{cl}_2(\emptyset))$. Thus $r_1(\mathrm{cl}_2(\emptyset)) = 0$, so, by (ii), $\mathrm{cl}_2(\emptyset) = \mathrm{cl}_1(\emptyset)$. Hence $Y \neq \mathrm{cl}_1(\emptyset)$ so $r_1(Y) > 0$. By the choice of Y, there is a flat X of M_2 that is covered by Y in $\mathcal{L}(M_1)$, and the sets of flats of M_1 and M_2 of rank less than $r_1(Y)$ coincide. Thus $r_1(X) = r_2(X)$. Hence, by Lemma 7.3.7, Y is a flat of M_2. We conclude that if $r_1(M_1) - r_2(M_2) = 0$, then $M_1 = M_2$, so M_2 is certainly a quotient of M_1.

Now assume that if $r_1(M_1) - r_2(M_2) < k$, then M_2 is a quotient of M_1 and let

$$r_1(M_1) - r_2(M_2) = k \geq 1.$$

Let $\mathcal{M}$ be the set of flats F of M_1 such that F is a flat of M_2 with

$$r_2(F) = r_1(F) - k.$$

Then $\mathcal{M}$ consists of those flats X for which equality holds in (iii) when $Y = E$.

Lemma 7.3.8 *The set $\mathcal{M}$ is a modular cut of M_1.*

Proof Suppose $F \in \mathcal{M}$. By Lemma 7.3.7, every flat covering F in $\mathcal{L}(M_1)$ is in $\mathcal{M}$. It follows easily that every flat in M_1 containing F is in $\mathcal{M}$. Using this and (iii), it is not difficult to complete the proof that $\mathcal{M}$ is a modular cut. □

As $\mathcal{M}$ is neither empty nor equal to the set of all flats of M_1, we can form the elementary quotient $(M_1 +_\mathcal{M} e)/e$ of M_1. Call this elementary quotient N. We show next that every flat of M_2 is a flat of N. This will enable us to apply the induction assumption to M_2 and N.

Let F be a flat of M_2. Certainly F is a flat of M_1. If $F \in \mathcal{M}$, then $F \cup e$ is a flat of $M_1 +_\mathcal{M} e$, so F is a flat of $(M_1 +_\mathcal{M} e)/e$, that is, F is a flat of N. Now suppose that $F \notin \mathcal{M}$ and assume that F is covered in $\mathcal{L}(M_1)$ by a flat F' in $\mathcal{M}$. Then, by (iii),

$$1 = r_1(F') - r_1(F) \geq r_2(F') - r_2(F). \tag{7.8}$$

Since $r_2(F') - r_2(F) \geq 1$, equality must hold in (7.8), so $r_1(F) - r_2(F) = r_1(F') - r_2(F') = k$. Thus $F \in \mathcal{M}$; a contradiction. We conclude that F is not covered in $\mathcal{L}(M_1)$ by a flat in $\mathcal{M}$. Hence $F \cup e$ is a flat of $M_1 +_\mathcal{M} e$, so F is a flat of N.

We have now shown that every flat of M_2 is a flat of N. As $r(N) - r(M_2) < r(M_1) - r(M_2)$, we can apply the induction assumption to get that M_2 is a quotient of N. But N is a quotient of M_1, and so, by Proposition 7.3.5, M_2 is a quotient of M_1. This completes the proof of Proposition 7.3.6. □

Next we shall look briefly at the reverse of the quotient operation. A matroid M_1 is a *lift* of a matroid M_2 if M_2 is a quotient of M_1. If M_2 is an elementary quotient of M_1, then M_1 is an *elementary lift* of M_2. Thus M_1 is a lift of M_2

if there is a matroid N and a subset Y of $E(N)$ such that $N \backslash Y = M_1$ and $N/Y = M_2$. Hence M_1 is a lift of M_2 if and only if M_1^* is a quotient of M_2^*.

To form an elementary lift of the matroid M, we first coextend M by an element e so that e is neither a loop nor a coloop in the resulting matroid N. The elementary lift is then obtained from N by deleting e.

Now let N be a coextension of M by e. We consider next how to represent N geometrically when e is not a loop. As contraction of e corresponds geometrically to projection from e, to coextend M by e, one first adds a new point to correspond to e, this point being added in a new dimension. Then, each non-loop element f of M is slid up along the line joining e and f. More formally, the point in the coextension corresponding to f is some point that is on the line joining e and f but is distinct from e. For each loop g of M, either g remains a loop of N, or not, in which case we add g as a point touching e in N. Three examples of this construction being applied to loopless matroids are shown in Figures 7.12–7.14. The corresponding elementary lifts of M are obtained in each case by deleting e. Clearly these elementary lifts are isomorphic to $U_{3,6}$, $\mathcal{W}^3$, and $U_{4,6}$, respectively.

It is clear that an arbitrary matroid will have many different single-element extensions and many different single-element coextensions. The particular coextensions shown in Figures 7.12 and 7.14 are special because, in each case, every dependent flat of the coextension contains e, where a *dependent flat* is a flat that is also a dependent set. In general, if N is the coextension of M by e in which e is a non-loop element that is in every dependent flat, then N is called the *free coextension* of M. An immediate consequence of the next result is that N is the free coextension of M if and only if N^* is the free extension of M^*.

Proposition 7.3.9 *Let e be an element of a matroid N and suppose that e is not a loop. Then e is in every dependent flat of N if and only if e is free in N^*.*

Proof As e is not a loop of N, each of statements (ii)–(vi) below is equivalent to its predecessor. The fact that (ii) implies (iii) follows because a circuit and a cocircuit cannot have exactly one common element (Proposition 2.1.11).

(i) The element e is in every dependent flat of N.

(ii) The element e is in the closure of every circuit of N.

(iii) N does not have a circuit D and a cocircuit C such that $e \in C$ and $C \cap D$ is empty.

(iv) N^* does not have a cocircuit D and a circuit C such that $e \in C$ and $C \cap D$ is empty.

(v) Every circuit of N^* containing e is spanning.

(vi) The element e is free in N^*. □

An important special case of the quotient operation is the operation of truncation. For $r(M) > 0$, the *truncation* $T(M)$ of M is the elementary quotient corresponding to the free extension, that is, $T(M) = (M +_{E(M)} e)/e$. In other words, we freely add an element to M and then contract it out. When $r(M) = 0$, we take the truncation $T(M)$ to be M. Evidently $r(T(M))$ is $r(M) - 1$ unless

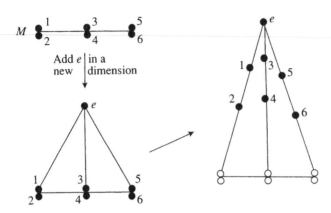

FIG. 7.12. Constructing the free coextension of M.

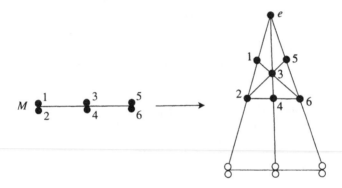

FIG. 7.13. A non-free coextension of M.

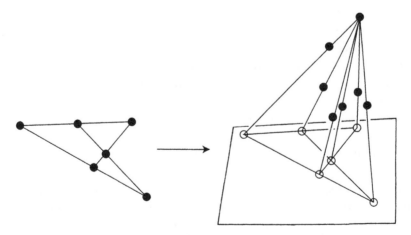

FIG. 7.14. Constructing the free coextension of $M(K_4)$.

M has rank 0, in which case so does $T(M)$. For all positive integers i, the *ith truncation* $T^i(M)$ of M is defined inductively by $T^i(M) = T(T^{i-1}(M))$ where $T^0(M) = M$. Thus, for example, if $k \leq n$, then $T^k(U_{n,n}) = U_{n-k,n}$.

It is very easy to describe the lattice of flats and the set of independent sets of the ith truncation $T^i(M)$ in terms of those of M. In particular, if $i \geq r(M)$, then $T^i(M)$ is the rank-zero matroid on $E(M)$.

Proposition 7.3.10 *Let M be a matroid of non-zero rank and let i be a non-negative integer not exceeding $r(M)$. Then*

$$\mathcal{I}(T^i(M)) = \{X \in \mathcal{I}(M) : |X| \leq r(M) - i\}.$$

Moreover, $\mathcal{L}(T^i(M))$ is obtained from $\mathcal{L}(M)$ by removing all flats of the latter of rank exceeding $r(M) - i - 1$, and making $E(M)$ the unique flat of rank $r(M) - i$.

Proof This is straightforward by induction and is omitted. ☐

A generalization of the truncation operation was introduced by Brown (1971). For a flat F of M of positive rank, the *principal truncation* $T_F(M)$ is $(M +_F e)/e$. Geometrically, $T_F(M)$ is obtained from M by freely adding e to the flat F and then contracting e. This operation can be iterated up to $r(F)$ times by using the following definition: $T_F^i(M) = T_{T^{i-1}(M|F)}(T_F^{i-1}(M))$. The *complete principal truncation* is $T_F^i(M)$ where $i = r(F) - 1$. Geometrically, it is obtained by freely adding an $(r(F) - 1)$-element independent set I to F and then contracting I.

The theory of quotients of matroids is part of an extensive theory of maps between matroids developed by Crapo (1965, 1967) and Higgs (1966a, 1966b, 1968). There are two types of such maps: strong and weak, and, in general, these maps are defined from a matroid M_1 on a set E_1 to a matroid M_2 on a set E_2. We now look briefly at such maps in the case when $|E_1| = |E_2|$ and the map from E_1 to E_2 is a bijection φ. In that case, φ is a *strong map* if $\varphi^{-1}(F)$ is a flat of M_1 for every flat F of M_2. Hence, by Proposition 7.3.6, when $E_1 = E_2$, the identity map ι is a strong map from M_1 to M_2 if and only if M_2 is a quotient of M_1. When considering strong and weak maps, it is common, as in the last statement, to assume that the bijection φ is the identity map.

We say that a bijection φ is a *weak map* from M_1 to M_2 if $\varphi^{-1}(I)$ is independent in M_1 for every independent set I of M_2. In this case, M_2 is called a *weak-map image* of M_1. The following result contains a number of alternative characterizations of weak maps. Its straightforward proof is left as an exercise.

Proposition 7.3.11 *Let M_1 and M_2 be matroids having rank functions r_1 and r_2 and ground sets E_1 and E_2. Let $\varphi : E_1 \to E_2$ be a bijection. The following statements are equivalent:*

(i) *The function φ is a weak map from M_1 to M_2.*

(ii) *If D is dependent in M_1, then $\varphi(D)$ is dependent in M_2.*

(iii) *If C is a circuit of M_1, then $\varphi(C)$ contains a circuit of M_2.*

(iv) *If $X \subseteq E_1$, then $r_1(X) \geq r_2(\varphi(X))$.* ☐

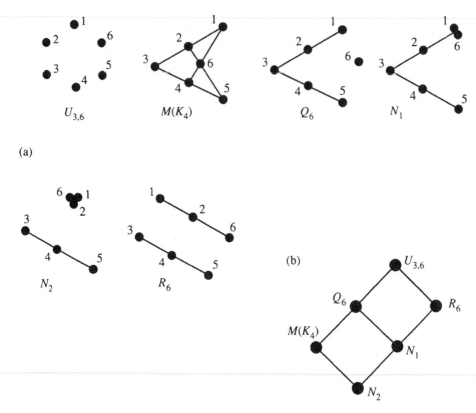

FIG. 7.15. (a) Six matroids and (b) the weak order on them.

The next two results are easy consequences of this proposition.

Corollary 7.3.12 *Suppose that M_1 and M_2 are matroids on E_1 and E_2 and that $\varphi : E_1 \to E_2$ is a bijection. If φ is a strong map from M_1 to M_2, then φ is a weak map from M_1 to M_2.*   □

Corollary 7.3.13 *Let M_1 and M_2 be matroids of the same rank on E_1 and E_2 and let $\varphi : E_1 \to E_2$ be a bijection. If φ is a weak map from M_1 to M_2, then φ is a weak map from M_1^* to M_2^*.*   □

The next corollary will be used in the proof of the main result of Section 7.4.

Corollary 7.3.14 *Suppose that M_1 and M_2 are matroids on E_1 and E_2 and that $\varphi : E_1 \to E_2$ and $\psi : E_2 \to E_1$ are bijections. If both φ and ψ are weak maps, then $M_1 \cong M_2$.*

Proof We show first that $\psi\varphi : E(M_1) \to E(M_1)$ is an automorphism of M_1. Let C be a circuit of M_1. Then, by Proposition 7.3.11, $\varphi(C)$ contains a circuit of M_2, so $\psi\varphi(C)$ contains a circuit C_1 of M_1. Thus $(\psi\varphi)^2(C)$ contains $\psi\varphi(C_1)$ which contains a circuit C_2 of M_1. As $\psi\varphi$ is a permutation of $E(M_1)$, some power of it is the identity map, so $(\psi\varphi)^n(C) = C$ for some positive integer n. It follows that $|C| = |C_1|$, that is, $\psi\varphi(C)$ contains a unique circuit, namely $\psi\varphi(C)$ itself. We deduce that $\varphi(C)$ is a circuit of M_2 for all circuits C of M_1. Since φ and ψ are both weak maps, by Proposition 7.3.11, $r(M_1) \geq r(M_2)$ and $r(M_2) \geq r(M_1)$ so $r(M_1) = r(M_2)$. By supposing that $E(M_1) = E(M_2)$, we may apply Proposition 7.3.6 and then Corollary 7.3.4 to give that $M_1 \cong M_2$. $\square$

The collection $\mathcal{E}$ of matroids on a fixed set E can be partially ordered by taking $M_1 \geq M_2$ if the identity map on E is a weak map from M_1 to M_2. Under this partial order, the *weak order* on $\mathcal{E}$, the set $\mathcal{E}$ has both a one and a zero, namely the free matroid on E and the rank-0 matroid on E. We say that M_1 is *freer* than M_2 if $M_1 \geq M_2$. Making a matroid freer corresponds to destroying small circuits, that is, circuits of size less than the rank. Clearly a relaxation of a matroid M is freer than M itself. In Figure 7.15, six matroids on six elements are shown along with the poset induced on them by the weak order.

We have taken a brief glimpse at the theory of strong and weak maps. For a much more detailed exposition of this theory, the reader is referred to White (1986) where there are chapters devoted to both types of maps.

Exercises

1. Show that $T(AG(3,2)) \cong U_{3,8}$.

2. Find all non-isomorphic elementary quotients of $M(K_4)$.

3. Let M be the free coextension of F_7 by the element e. If f is in $E(F_7) - e$, give a geometric representation for M/f.

4. Is a matroid of rank at least one uniquely determined by its collection of elementary quotients?

5. Prove the following are equivalent for matroids M_1 and M_2 on a set E:
   (a) M_2 is a quotient of M_1.
   (b) M_1^* is a quotient of M_2^*.
   (c) Every cocircuit of M_2 is a union of cocircuits of M_1.
   (d) Every hyperplane of M_2 is an intersection of hyperplanes of M_1.

6. The matroids P_6 and $\mathcal{W}^3$ can be formed by relaxing $\{1, 2, 3\}$ in the matroids Q_6 and $M(K_4)$, respectively, in Figure 7.15(a). Find how these matroids fit into the weak order in Figure 7.15(b).

7. For $n \geq 3$, consider the bipartite graph $K_{2,n}$. Show that:
   (a) $T(M(K_{2,n}))$ is the (tipless) free spike of rank n;
   (b) this construction can be modified to obtain the free spike of rank n as a truncation of a graphic matroid.

8. Give necessary and sufficient conditions on a matroid M for $T(M)$ to be connected.

9. Let M_1 and M_2 be matroids having rank functions r_1 and r_2 and a common ground set E. Suppose every flat of M_2 is a flat of M_1. Prove that if X is a flat of M_2 with $r_1(X) - r_2(X) = r_1(M_1) - r_2(M_2)$, then $M_1/X = M_2/X$.

10. Let $\mathcal{M}$ be a proper non-empty modular cut of a matroid M. If $A \subseteq E(M)$, let

$$r'(A) = \begin{cases} r_M(A), & \text{if } \mathrm{cl}_M(A) \notin \mathcal{M}, \\ r_M(A) - 1, & \text{otherwise.} \end{cases}$$

    Show that r' is the rank function of an elementary quotient of M and every elementary quotient of M arises in this way.

11. (Brylawski 1986a) If M_1 and M_2 are matroids on the same set E, prove that M_2 is a quotient of M_1 if and only if, for all subsets X of E, the matroid M_2/X is a weak-map image of M_1/X.

12. (Lucas 1975) Suppose that M_2 is a weak-map image of M_1 and $r(M_2) = r(M_1)$. Prove that every separator of M_1 is a separator of M_2.

13. Let F be a flat of a matroid M and let N be $T_F(M)$.
    (a) Specify $\mathcal{L}(N)$ in terms of $\mathcal{L}(M)$.
    (b) Show that

$$r_N(X) = \begin{cases} r_M(X) - 1, & \text{if } r_M(X) = r_M(X \cup F), \\ r_M(X), & \text{otherwise.} \end{cases}$$

    (c) Find $\mathcal{I}(N)$.
    (d) Show that

$$\mathrm{cl}_N(X) = \begin{cases} \mathrm{cl}_M(X \cup F), & \text{if } r_M(X \cup F) = r_M(X) + 1, \\ \mathrm{cl}_M(X), & \text{otherwise.} \end{cases}$$

14. (Oxley and Whittle 1991) Let $\mathcal{C}'$ be the set of non-spanning circuits of a rank-r matroid M on a set E. Show that there is a matroid N and a non-negative integer k such that the set of matroids on E having $\mathcal{C}'$ as its set of non-spanning circuits is $\{N, T(N), T^2(N), \ldots, T^k(N)\}$.

15. (Kung 1977) Let M_1 and M_2 be matroids on a set E and cl_1 denote the closure operator of M_1. Prove that the following are equivalent:
    (a) M_2 is a quotient of M_1.
    (b) For all pairs $\{I_1, I_2\}$ of independent sets in M_1 such that $\mathrm{cl}_1(I_1) = \mathrm{cl}_1(I_2)$, if I_1 is dependent in M_2, then I_2 is dependent in M_2.

16. (Lucas 1975) Let M_2 be a weak-map image of a binary matroid M_1 and suppose that $r(M_2) = r(M_1)$. Prove that:
    (a) M_2 is binary;
    (b) if $M_2 \neq M_1$, then M_2 is disconnected.

7.4 A non-commutative operation

In this section, we consider a relatively new operation which, unlike the other operations discussed in this chapter for joining two matroids, is non-commutative.

This operation was introduced by Crapo and Schmitt (2005a) and they used it to prove a natural bound on the number of non-isomorphic matroids on an n-element set. The results in this section are interesting but they will not be used elsewhere in the book.

Theorem 7.4.1 *Let M_1 and M_2 be matroids on disjoint sets E_1 and E_2. Let $\mathcal{B}$ consist of those subsets B of $E_1 \cup E_2$ having exactly $r(M_1) + r(M_2)$ elements such that $B \cap E_1$ is independent in M_1, and $B \cap E_2$ is spanning in M_2. Then $\mathcal{B}$ is the set of bases of a matroid $M_1 \square M_2$ with ground set $E_1 \cup E_2$.*

Proof Clearly all the bases of $M_1 \oplus M_2$ are in $\mathcal{B}$, so $\mathcal{B}$ is non-empty. To check that $\mathcal{B}$ satisfies (B2), suppose that B_1 and B_2 are in $\mathcal{B}$ and $x \in B_1 - B_2$. Then $|B_1| = |B_2|$. We shall write r_2 for the rank function of M_2.

Assume that $x \in E_1$. Suppose first that $(B_2 - B_1) \cap E_2$ is non-empty. Then, choosing y arbitrarily in this set, we see that $[(B_1 - x) \cup y] \cap E_2$ spans M_2, while $[(B_1 - x) \cup y] \cap E_1$, which equals $(B_1 - x) \cap E_1$, is independent in M_1. Hence $(B_1 - x) \cup y \in \mathcal{B}$. We may now assume that $(B_2 - B_1) \cap E_2$ is empty. Then $|B_1 \cap E_2| \geq |B_2 \cap E_2|$, so $|B_1 \cap E_1| \leq |B_2 \cap E_1|$. By applying (I3) in M_1, we deduce that $[B_2 - (B_1 - x)] \cap E_1$ contains an element y such that $[(B_1 - x) \cup y] \cap E_1$ is in $\mathcal{I}(M_1)$. Hence, in this case too, $(B_1 - x) \cup y \in \mathcal{B}$.

We may now assume that $x \in E_2$. Suppose first that $r_2((B_1 - x) \cap E_2) < r_2(B_2 \cap E_2)$. Then $r_2((B_1 - x) \cap E_2) = r(M_2) - 1$ and there is an element y of $(B_2 - B_1) \cap E_2$ such that $r_2([(B_1 - x) \cup y] \cap E_2) > r_2((B_1 - x) \cap E_2)$. Hence $[(B_1 - x) \cup y] \cap E_2$ spans M_2 and so, as $[(B_1 - x) \cup y] \cap E_1$ equals $B_1 \cap E_1$ and is therefore independent in M_1, we conclude that $(B_1 - x) \cup y \in \mathcal{B}$. We may now assume that $r_2((B_1 - x) \cap E_2) = r_2(B_2 \cap E_2)$. In that case, we can choose y to be an arbitrary element of $(B_2 - B_1) \cap E_2$ to get that $(B_1 - x) \cup y \in \mathcal{B}$ unless $(B_2 - B_1) \cap E_2$ is empty. In the exceptional case, as $x \in (B_1 - B_2) \cap E_2$, we have $|B_1 \cap E_2| > |B_2 \cap E_2|$. Thus $|B_2 \cap E_1| > |B_1 \cap E_1|$, so we can choose y in $(B_2 - B_1) \cap E_1$ such that $[(B_1 - x) \cup y] \cap E_1 \in \mathcal{I}(M_1)$ and again $(B_1 - x) \cup y \in \mathcal{B}$. We conclude that $\mathcal{B}$ satisfies (B2) and so $M_1 \square M_2$ is indeed a matroid. $\square$

We call $M_1 \square M_2$ the *free product* of M_1 and M_2. When we write this, it will be implicit that M_1 and M_2 have disjoint ground sets.

Example 7.4.2 Suppose that M_1 and M_2 are both isomorphic to $U_{2,4}$ with the first having ground set $\{1, 2, 3, 4\}$ and the second $\{a, b, c, d\}$. Then $M_1 \square M_2$ has rank 4 and can be represented geometrically by taking a, b, c, and d as the vertices of a tetrahedron and then adding a line $\{1, 2, 3, 4\}$ that is freely placed relative to this tetrahedron, that is, so that none of a, b, c, or d is on any of the faces of the tetrahedron.

We note that, while $M_1 \square M_2$ is clearly isomorphic to $M_2 \square M_1$, the two matroids are not equal since $\{a, b, c, d\}$ has rank 4 in the first and rank 2 in the second. The reader can easily check that $U_{1,3} \square U_{2,3}$ and $U_{2,3} \square U_{1,3}$ are not even isomorphic, for the second is simple but the first is not. $\square$

Extending the last example, we show next that $M_1 \square M_2$ and $M_2 \square M_1$ are almost never equal. The proof is left to the reader (Exercise 2).

Proposition 7.4.3 *Let M_1 and M_2 be matroids. Then $M_1 \square M_2 = M_2 \square M_1$ if and only if both M_1 and M_2 have rank 0, or both M_1^* and M_2^* have rank 0.* □

The next two results prove attractive properties of the free product. The ground set and rank function of each M_i will be written as E_i and r_i, respectively.

Corollary 7.4.4 $(M_1 \square M_2)^* = M_2^* \square M_1^*$.

Proof The following statements are equivalent.

(i) B is a basis of $(M_1 \square M_2)^*$.

(ii) $(E_1 \cup E_2) - B$ is a basis of $M_1 \square M_2$.

(iii) $E_1 - B$ is independent in M_1 and $E_2 - B$ is spanning in M_2, and
      $|E_1 - B| + |E_2 - B| = r(M_1) + r(M_2)$.

(iv) $E_1 \cap B$ is spanning in M_1^* and $E_2 \cap B$ is independent in M_2^*, and
      $|B| = r(M_1^*) + r(M_2^*)$.

(v) B is a basis of $M_2^* \square M_1^*$.   □

Proposition 7.4.5 $(M_1 \square M_2)|E_1 = M_1$ and $(M_1 \square M_2)/E_1 = M_2$.

Proof The first equation follows easily from the definition of $M_1 \square M_2$; the second follows from the first by duality using the last corollary.   □

Next we identify the independent sets of $M_1 \square M_2$.

Proposition 7.4.6 *A subset I of $E_1 \cup E_2$ is independent in $M_1 \square M_2$ if and only if $I \cap E_1$ is independent in M_1 and*

$$r(M_1) - r_1(I \cap E_1) \geq |I \cap E_2| - r_2(I \cap E_2).$$

Proof Let I be a subset of $E_1 \cup E_2$ for which $I \cap E_1 \in \mathcal{I}(M_1)$ and the specified rank condition holds. Then I is contained in a maximal subset B of $E_1 \cup E_2$ for which $B \cap E_1 \in \mathcal{I}(M_1)$ and $r(M_1) - r_1(B \cap E_1) \geq |B \cap E_2| - r_2(B \cap E_2)$. As B is maximal, if $e \in E_2 - B$, then $|(B \cup e) \cap E_2| - r_2((B \cup e) \cap E_2)$ exceeds both $|B \cap E_2| - r_2(B \cap E_2)$ and $r(M_1) - r_1(B \cap E_1)$. From the first of these, we deduce that $r_2((B \cup e) \cap E_2) = r_2(B \cap E_2)$, so $B \cap E_2$ spans M_2. From the second, we see that $r(M_1) - r_1(B \cap E_1) = |B \cap E_2| - r_2(B \cap E_2)$. As $B \cap E_1$ is independent in M_1, and $B \cap E_2$ spans M_2, the last equation reduces to

$$r(M_1) - |B \cap E_1| = |B \cap E_2| - r(M_2),$$

that is, $|B| = r(M_1) + r(M_2)$. Thus $B \in \mathcal{B}(M_1 \square M_2)$, so $I \in \mathcal{I}(M_1 \square M_2)$.

Now suppose $I \in \mathcal{I}(M_1 \square M_2)$. Then $I \subseteq B$ for some B in $\mathcal{B}(M_1 \square M_2)$. Thus $B \cap E_1$ is independent in M_1 and $B \cap E_2$ is spanning in M_2, and $r(M_1) + r(M_2) = |B|$. Hence $|B \cap E_1)| = r_1(B \cap E_1)$ and $r_2(B \cap E_2) = r(M_2)$. Thus, as above, $r(M_1) - r_1(B \cap E_1) = |B \cap E_2| - r_2(B \cap E_2)$. Now $r_1(I \cap E_1) \leq r_1(B \cap E_1)$, so $r(M_1) - r_1(I \cap E_1) \geq |B \cap E_2| - r_2(B \cap E_2) \geq |I \cap E_2| - r_2(I \cap E_2)$, where the last inequality holds since $|B \cap E_2| - |I \cap E_2| \geq r_2(B \cap E_2) - r_2(I \cap E_2)$. Hence I satisfies the desired rank condition. Moreover, $I \cap E_1 \subseteq B \cap E_1$, so $I \cap E_1$ is independent in M_1, and the proposition follows. □

Suppose we are given $M_1 \square M_2$ without being told what E_1 and E_2 are. The goal of the rest of this section is to show that we can still recover M_1 and M_2, up to isomorphism, provided we know $|E_1|$. We shall need several preliminaries.

Lemma 7.4.7 *In $M_1 \square M_2$, let U be a subset of $E_1 \cup E_2$ having $|E_1|$ elements. Then $r(M_1) - r_1(U \cap E_1) \leq |U \cap E_2|$ and equality holds if and only if $E_1 - U$ is a set of coloops of M_1.*

Proof Since $|U| = |E_1|$, we have $|U \cap E_2| = |E_1 - U|$. Let Z be a basis for $M_1|(U \cap E_1)$. Then the following inequalities are equivalent:

$$|E_1 - U| + |Z| \geq r(M_1).$$
$$|U \cap E_2| + |Z| \geq r(M_1).$$
$$|U \cap E_2| + r_1(U \cap E_1) \geq r(M_1).$$
$$|U \cap E_2| \geq r(M_1) - r_1(U \cap E_1).$$

As the first inequality holds, so does the last. Also, equality holds in the first, and so in the last, if and only if all elements of $E_1 - U$ are coloops of M_1. □

The next lemma shows that we can determine the rank of M_1 from $M_1 \square M_2$ provided we know $|E_1|$.

Lemma 7.4.8 *Let $M = M_1 \square M_2$. Then*

$$r(M_1) = \min\{r_M(U) : U \subseteq E_1 \cup E_2 \text{ and } |U| = |E_1|\}.$$

Proof By Proposition 7.4.5, $r(M_1) = r_M(E_1)$. Now let U be a subset of $E_1 \cup E_2$ with $|U| = |E_1|$. It remains to show that $r(M_1) \leq r_M(U)$. Let Z be a basis of $M_1|(U \cap E_1)$. By Lemma 7.4.7, $r(M_1) - r_1(U \cap E_1) \leq |U \cap E_2|$. Thus there is a subset A of $U \cap E_2$ such that $r(M_1) - r_1(U \cap E_1) = |A|$. Then, as $A \cup Z \subseteq U$ and $Z \subseteq E_1$, we have $(A \cup Z) \cap E_2 = A$ and

$$r(M_1) - r_1((A \cup Z) \cap E_1) \geq r(M_1) - r_1(U \cap E_1)$$
$$= |A|$$
$$\geq |(A \cup Z) \cap E_2| - r_2((A \cup Z) \cap E_2).$$

Thus, by Proposition 7.4.6, $A \cup Z$ is independent in M. Now U contains $A \cup Z$, and Z is a basis of $M_1|(U \cap E_1)$. Also $|A| = r(M_1) - r_1(U \cap E_1)$. Therefore

$$r_M(U) \geq |A| + |Z| = [r(M_1) - r_1(U \cap E_1)] + r_1(U \cap E_1) = r(M_1),$$

so the lemma holds. □

The proof of the main result, Theorem 7.4.11, will require two more preliminaries. The second contains the core of the proof of the theorem; the first implies
that if M_1 has no coloops and M_2 has no loops, then E_1, and hence M_1 itself,
can be recovered from $M_1 \square M_2$ provided we know $|E_1|$.

Lemma 7.4.9 *Let* $M = M_1 \square M_2$ *and* U *be a subset of* $E_1 \cup E_2$ *having* $|E_1|$
elements. If $U \cap E_2$ *does not consist entirely of loops of* M_2, *and* $E_1 - U$ *does
not consist entirely of coloops of* M_1, *then* $r_M(U) > r(M_1)$.

Proof Let Z be a basis of $M_1|(U \cap E_1)$. By Lemma 7.4.7, since $E_1 - U$ does
not consist entirely of coloops of M_1, we have $r(M_1) - r_1(Z) \le |U \cap E_2| - 1$. Let
Y be a subset of $U \cap E_2$ having $r(M_1) - r_1(Z) + 1$ elements including a non-loop
element of M_2. Then

$$
\begin{aligned}
r(M_1) - r_1((Y \cup Z) \cap E_1) &= r(M_1) - r_1(Z) \\
&= |Y| - 1 \\
&\ge |(Y \cup Z) \cap E_2| - r_2((Y \cup Z) \cap E_2).
\end{aligned}
$$

Hence, by Proposition 7.4.6, $Z \cup Y$ is independent in M, so

$$r_M(U) \ge |Z| + |Y| = |Z| + [r(M_1) - r_1(Z) + 1] = r(M_1) + 1 > r(M_1). \qquad \square$$

Proposition 7.4.10 *Let* $M = M_1 \square M_2$. *If* $U \subseteq E_1 \cup E_2$ *such that* $|U| = |E_1|$
and $r_M(U) = r(M_1)$, *then there are bijective weak maps from* $M|U$ *and* M/U
onto M_1 *and* M_2, *respectively.*

Proof Let V be the complement of U in $E_1 \cup E_2$ and let g be an arbitrary
bijection from $V \cap E_1$ onto $U \cap E_2$. Define $\varphi : E_1 \cup E_2 \to E_1 \cup E_2$ by

$$
\varphi(x) = \begin{cases}
g(x), & \text{if } x \in V \cap E_1; \\
g^{-1}(x), & \text{if } x \in U \cap E_2; \\
x, & \text{otherwise.}
\end{cases}
$$

The restrictions of φ to U and to V are bijections mapping U and V to E_1 and
E_2, respectively. We denote these bijections by φ_1 and φ_2. The rest of the proof
will show that these bijections are the desired weak maps. By the last lemma,
since $r_M(U) = r(M_1)$, there are two cases:

(i) every element of $V \cap E_1$ is a coloop of M_1; and
(ii) every element of $U \cap E_2$ is a loop of M_2.

Suppose that (i) holds. Then the bases of M_1 are precisely the sets of the form
$(V \cap E_1) \cup B_1$ where B_1 is a basis of $M_1|(U \cap E_1)$. Now recall that $M_1 = M|E_1$,
so $M_1|(U \cap E_1) = M|(U \cap E_1)$. Let B_1 be a basis of $M_1|(U \cap E_1)$. Then

$$
\begin{aligned}
r(M_1) - r_1(B_1) &= r(M_1) - r_1(U \cap E_1) \\
&= |V \cap E_1| \\
&= |U \cap E_2| \\
&\ge |U \cap E_2| - r_2(U \cap E_2).
\end{aligned}
$$

It follows by Lemma 7.4.6 that $B_1 \cup (U \cap E_2)$ is independent in M. Moreover, $B_1 \cup (U \cap E_2)$ must be a basis of $M|U$ and every basis of $M|U$ must have this form. Hence every element of $U \cap E_2$ is a coloop of $M|U$. Since φ_1 acts as the identity map on $U \cap E_1$, and maps $V \cap E_1$ onto $U \cap E_2$, it follows that φ_1 is an isomorphism from $M|U$ onto M_1. Thus, in case (i), φ_1 is certainly a bijective weak map from $M|U$ onto M_1.

Next we consider φ_2. Let $B = B_1 \cup (U \cap E_2)$. From above, B is a basis for $M|U$. Let Z be a basis for M_2. To show that $\varphi_2 : M/U \to M_2$ is a weak map, we need to show that $\varphi^{-1}(Z)$ is independent in M/U or, equivalently, that $\varphi^{-1}(Z) \cup B$ is independent in M. Now $\varphi^{-1}(Z) = \varphi^{-1}(Z \cap U) \cup (Z \cap V)$. Hence $\varphi^{-1}(Z) \cup B = \varphi^{-1}(Z \cap U) \cup (Z \cap V) \cup B$. Recall that $B = B_1 \cup (U \cap E_2)$, so $\varphi^{-1}(Z) \cup B = \varphi^{-1}(Z \cap U) \cup (Z \cap V) \cup B_1 \cup (U \cap E_2)$. Thus $\varphi^{-1}(Z) \cup B$ meets E_1 in $\varphi^{-1}(Z \cap U) \cup B_1$ and meets E_2 in $(Z \cap V) \cup (U \cap E_2)$. Therefore, by Lemma 7.4.6, $\varphi^{-1}(Z) \cup B$ is independent in M if and only if

$$r(M_1) - r_1(\varphi^{-1}(Z \cap U) \cup B_1) \geq |(Z \cap V) \cup (U \cap E_2)| - r_2((Z \cap V) \cup (U \cap E_2)). \quad (7.9)$$

Now $\varphi^{-1}(Z \cap U) \subseteq E_1 - U$, so every element of $\varphi^{-1}(Z \cap U)$ is a coloop of M_1. Moreover, B_1 is a basis for $M_1|(U \cap E_1)$, so

$$\begin{aligned}
r(M_1) - r_1(\varphi^{-1}(Z \cap U) \cup B_1) &= |E_1 - U| - |\varphi^{-1}(Z \cap U)| \\
&= |E_1 - U| - |Z \cap U|. \quad (7.10)
\end{aligned}$$

On the other hand, $(Z \cap V) \cup (U \cap E_2)$ contains Z, a basis of M_2, so

$$\begin{aligned}
|(Z \cap V) \cup (U \cap E_2)| & \\
- r_2((Z \cap V) \cup (U \cap E_2)) &= |Z \cap V| + |U \cap E_2| - r(M_2) \\
&= |Z \cap V| + |U \cap E_2| - |Z| \\
&= |Z \cap V| + |U \cap E_2| - |Z \cap V| - |Z \cap U| \\
&= |U \cap E_2| - |Z \cap U| \\
&= |E_1 - U| - |Z \cap U|. \quad (7.11)
\end{aligned}$$

By comparing (7.10) and (7.11), we see that (7.9) holds. We conclude that, in case (i), φ_2 is a weak map from M/U onto M_2.

Now assume that (ii) holds. Then every element of $U \cap E_2$ is a loop of M_2 and so is a coloop of M_2^*. By Corollary 7.4.4, $M^* = M_2^* \square M_1^*$ and, since $|U| = |E_1|$ and $r_M(U) = r(M_1)$, it follows that $|V| = |E_2|$ and

$$r_{M^*}(V) = |V| - r(M) + r_M(U) = |E_2| - r(M_1) - r(M_2) + r(M_1) = r(M_2^*).$$

By interchanging the roles of M_1, M_2, and U with, respectively, M_2^*, M_1^*, and V in case (i), we obtain that the restrictions of φ to V and U induce weak maps from $M^*|V$ and M^*/V onto M_2^* and M_1^*. Since $M^*|V = (M/U)^*$ and $M^*/V = (M|U)^*$, it follows that φ_2 is a weak map from $(M/U)^*$ onto M_2^*

while φ_1 is a weak map from $(M/U)^*$ onto M_1^*. As $r((M/U)^*) = r(M_2^*)$ and $r((M|U)^*) = r(M_1^*)$, it follows by Corollary 7.3.13 that $\varphi_2 : M/U \to M_2$ and $\varphi_1 : M|U \to M_1$ are weak maps. □

We are now ready to show that, given $|E_1|$, the matroid $M_1 \square M_2$ determines M_1 and M_2 up to isomorphism.

Theorem 7.4.11 *For matroids M_1, M_2, N_1, and N_2, if $M_1 \square M_2 \cong N_1 \square N_2$ and $|E(M_1)| = |E(N_1)|$, then $M_1 \cong N_1$ and $M_2 \cong N_2$.*

Proof Let $M_1 \square M_2$ and $N_1 \square N_2$ be M and N, respectively. By choosing an isomorphism from M to N and relabelling if necessary, we may assume that $M = N$. Now $r(M_1) = r_N(E(M_1))$ and $r(N_1) = r_M(E(N_1))$. Moreover, as $|E(N_1)| = |E(M_1)|$, Lemma 7.4.8 implies that $r(M_1) \leq r_M(E(N_1))$. Thus $r(M_1) \leq r(N_1)$ and, by symmetry, equality holds. Then, by applying Proposition 7.4.10 to M, we obtain bijective weak maps $\varphi_1 : M|E(N_1) \to M_1$ and $\varphi_2 : M/E(N_1) \to M_2$. By Proposition 7.4.5, φ_1 maps N_1 onto M_1, and φ_2 maps N_2 onto M_2. Applying Proposition 7.4.10 to N, we obtain bijective weak maps $\psi_1 : M_1 \to N_1$ and $\psi_2 : M_2 \to N_2$. Corollary 7.3.14 now gives that $M_1 \cong N_1$ and $M_2 \cong N_2$. □

Crapo and Schmitt (2005a) used this theorem to prove the following result thereby verifying a conjecture of Welsh (1969a). An alternative proof of this conjecture was given by Lemos (2004). We leave the straightforward derivation of this corollary from the theorem as an exercise for the reader.

Corollary 7.4.12 *Let $f(n)$ denote the number of non-isomorphic matroids on an n-element set. Then, for all non-negative integers n and m,*

$$f(n + m) \geq f(n)f(m).$$ □

Exercises

1. Give an example of four non-isomorphic matroids M_1, N_1, M_2, and N_2 such that $M_1 \square N_1 = M_2 \square N_2$.
2. Prove that:
   (a) if $M_1 \square M_2 = M_2 \square M_1$, then $M_1 \square M_2 = M_1 \oplus M_2$;
   (b) if $M_1 \square M_2 = M_1 \oplus M_2$, then M_2 or M_1^* is free;
   (c) $M_1 \square M_2 = M_2 \square M_1$ if and only if either both M_1 and M_2 are free, or both M_1^* and M_2^* are free.
3. (Crapo and Schmitt 2005b) Let $M = M_1 \square M_2$. For $X \subseteq E(M)$, show that:
   (a) $r_M(X) = r_1(X \cap E_1) + r_2(X \cap E_2) + \min\{r(M_1) - r_1(X \cap E_1), |X \cap E_2| - r_2(X \cap E_2)\}$.
   (b)

$$\mathrm{cl}_M(X) = \begin{cases} \mathrm{cl}_1(X \cap E_1) \cup (X \cap E_2) & \text{if } r(M_1) - r_1(X \cap E_1) \\ & \qquad > |X \cap E_2| - r_2(X \cap E_2); \\ E_1 \cup \mathrm{cl}_2(X \cap E_2) & \text{otherwise.} \end{cases}$$

(c) If $r(M_1) - r_1(X \cap E_1) > |X \cap E_2| - r_2(X \cap E_2)$, then X is a flat of M if and only if $X \cap E_1$ is a flat of M_1; if $r(M_1) - r_1(X \cap E_1) \leq |X \cap E_2| - r_2(X \cap E_2)$, then X is a flat of M if and only if $X \supseteq E_1$ and $X \cap E_2$ is a flat of M_2.

(d) X is a circuit of M if and only if either $X \subseteq E_1$ and X is a circuit of M_1, or $X \cap E_1$ is independent in M_1, the matroid $M_2|(X \cap E_2)$ has no coloops, and $r(M_1) - r_1(X \cap E_1) + 1 = |X \cap E_2| - r_2(X \cap E_2)$.

4. (Crapo and Schmitt 2005b) For a matroid M, let $E(M) = E_1 \dot\cup E_2$. Let $M_1 = M|E_1$ and $M_2 = M/E_1$. Prove that:

(a) if I is independent in M, then

$$r(M_1) - r_1(I \cap E_1) \geq |I \cap E_2| - r_2(I \cap E_2);$$

(b) if $X \subseteq E(M)$, then the identity map on $E(M)$ is a rank-preserving weak map from $(M|X)\square(M/X)$ to M.

5. (Higgs 1968) For a matroid M, define the *Higgs lift* $L(M)$ to be the elementary lift of M that is obtained by taking the free coextension of M by e and then deleting e. For all positive integers i, let $L^i(M)$ be $L(L^{i-1}(M))$ where $L^0(M) = M$. Show that:

(a) $L(M)$ is the matroid on $E(M)$ whose independent sets are those sets X of $E(M)$ for which $r_M(X) \geq |X| - 1$;

(b) $L(M) = (T(M^*))^*$.

(c) $L^j(M^*) = (T^j(M))^*$ for all $j \geq 0$.

(d) if $i \geq 0$, then $T^i(M)/X = T^i(M/X)$ and $L^i(M)|X = L^i(M|X)$.

6. (Crapo and Schmitt 2005b) Let $M = M_1\square M_2$. For $X \subseteq E(M)$, show that:

(a) $M|X = [M_1|(X \cap E_1)]\square L^i(M_2|(X \cap E_2))$ where $i = r(M_1) - r_1(X \cap E_1)$; and

(b) $M/X = T^j(M_1/(X \cap E_1))\square(M_2/(X \cap E_2))$ where $j = |X \cap E_2| - r_2(X \cap E_2)$.

(c)
$$T(M_1\square M_2) = \begin{cases} M_1\square T(M_2), & \text{if } r(M_2) > 0; \\ T(M_1)\square M_2, & \text{if } r(M_2) = 0. \end{cases}$$

(d)
$$L(M_1\square M_2) = \begin{cases} L(M_1)\square M_2, & \text{if } r(M_1) < |E_1|; \\ M_1\square L(M_2), & \text{if } r(M_1) = |E_1|. \end{cases}$$

7. (Crapo and Schmitt 2005b) Let $M_1, M_2, \ldots, M_k$ be matroids on disjoint sets $E_1, E_2, \ldots, E_k$. If $k = 1$, let $M_1\square M_2\square \cdots \square M_k$ be M_1. Prove that:

(a) if $k \geq 3$, then $(M_1\square M_2)\square M_3 = M_1\square(M_2\square M_3)$;

(b) if $M = M_1\square M_2\square \cdots \square M_k$ and $X \subseteq E(M)$, then X is independent in M if and only if, for all j such that $1 \leq j \leq k$,

$$\sum_{i=1}^{j-1}[r(M_i) - r_i(X \cap E_i)] \geq \sum_{i=1}^{j}[|X \cap E_i| - r_i(X \cap E_i)].$$

8. (Crapo and Schmitt 2005b) Let M_1 and M_2 be matroids on disjoint ground sets. Prove that:
   (a) if M_1 and M_2 are representable over a field $\mathbb{F}$ and $\mathbb{F}$ is sufficiently large, then $M_1 \square M_2$ is $\mathbb{F}$-representable;
   (b) if M_1 and M_2 are transversal matroids, then so is $M_1 \square M_2$.
9. (Crapo and Schmitt 2005b) Let $M = M_1 \square M_2$. For $X \subseteq E(M)$, prove that:
   (a) E_1 is a cyclic flat of M if and only if M_1 has no coloops and M_2 has no loops;
   (b) if $X \neq E_1$, then X is a cyclic flat of M if and only if X is a cyclic flat of M_1, or $X = E_1 \cup Y$ for some non-empty cyclic flat Y of M_2.
10. (Lemos 2004) Let M_1 and M_2 be matroids with disjoint ground sets E_1 and E_2. Suppose both M_1 and M_2^* have at least two non-loop elements. For some element e not in $E_1 \cup E_2$, let N_1 and N_2 be obtained by freely extending M_1 and M_2^* by e. Define $M_1 \triangledown M_2$ to be $N_1 \oplus_2 N_2^*$. Show that:
    (a) $(M_1 \triangledown M_2)^* = M_2^* \triangledown M_1^*$;
    (b) *if $M_1 \triangledown M_2 = M_1' \triangledown M_2'$ and $|E(M_1)| = |E(M_1')|$, and neither M_1 nor M_2^* is the direct sum of a uniform matroid and a rank-0 matroid, then $M_1 \cong M_1'$ and $M_2 \cong M_2'$.
11. (Lemos 2004) Let $\mathcal{F}$ be a non-empty class of matroids that is closed under duality, direct sums, 2-sums, and free extensions. Show that:
    (a) If $\mathcal{F}$ is closed under minors, then $\mathcal{F}$ contains all uniform matroids.
    (b) *If $\mathcal{F}$ contains $U_{0,1}$ and m and n are non-negative integers, then $g(m + n) \geq g(m)g(n)$, where $g(k)$ denotes the number of non-isomorphic k-element matroids belonging to $\mathcal{F}$.
12. (Crapo and Schmitt 2005b) Let M be a matroid and suppose $X \subseteq E(M)$.
    (a) Prove that the following are equivalent:
        (i) $M = (M|X)\square(M/X)$;
        (ii) every cyclic flat of M is comparable to X by inclusion.
    (b) Deduce that the following are equivalent when M is non-empty:
        (i) M is irreducible with respect to free product, that is, every factorization of M as a free product has M as a factor;
        (ii) M has no proper non-empty subset X such that every cyclic flat is comparable to X by inclusion.
    (c) *If $M_1 \square M_2 \square \cdots \square M_m = N_1 \square N_2 \square \cdots \square N_n$ where all the M_i and N_j are irreducible, then $m = n$ and $M_i \cong N_i$ for all i in $\{1, 2, \ldots, m\}$.

8

HIGHER CONNECTIVITY

The property of 2-connectedness for matroids has already proved to be of basic structural importance. When we introduced this property in Chapter 4, we noted that it is closely related to the property of 2-connectedness for graphs. In view of this, it is natural to ask whether, for arbitrary n, the property of n-connectedness for graphs can be extended to matroids. This chapter addresses this question focusing particularly on the case when $n = 3$. Some of the tools developed here will be used later to obtain important decomposition results.

In Sections 8.1 and 8.2, we present Tutte's definition of n-connectedness for matroids and establish some basic properties of the matroid connectivity function and the related local connectivity function. In Section 8.3, we show that a matroid is 3-connected if and only if it cannot be decomposed as a direct sum or 2-sum. In addition, we prove a theorem of Cunningham and Edmonds that gives a unique 2-sum decomposition of a connected matroid into 3-connected matroids, circuits, and cocircuits. Section 8.4 discusses some properties of the cycle matroids of wheels and of their relaxations, whirls. In Section 8.5, we prove Tutte's Linking Theorem, a matroid generalization of Menger's Theorem. Section 8.6 considers the differences between matroid n-connectedness and graph n-connectedness and introduces a matroid generalization of the latter. Section 8.7 proves some extremal connectivity results including Tutte's Triangle Lemma. This lemma is a basic tool in the proof of Tutte's Wheels-and-Whirls Theorem, which is given in Section 8.8. That theorem establishes that wheels and whirls are the only 3-connected matroids for which no single-element deletion or contraction is 3-connected.

8.1 Tutte's definition

The concept of n-connectedness for matroids was introduced by W. T. Tutte (1966b). The motivation for Tutte's definition appears to derive from two sources: a desire to generalize the corresponding concept for graphs, and a wish to incorporate duality into the theory. These two aims are not totally compatible.

Example 8.1.1 It follows from the definition of graph connectivity in Chapter 4 that the planar graph G in Figure 8.1 is 3-connected. However, the graph G^*, a geometric dual of G, has a degree-2 vertex and so is not 3-connected. Thus if we make the matroid connectivities of $M(G)$ and its dual $M(G^*)$ equal, then, for at least one of G and G^*, the graph connectivity and the connectivity of the corresponding cycle matroid will be different. □

FIG. 8.1. The connectivities of G and G^* are different.

The problems that occur in the last example are predictable and, indeed, when $n = 3$, the case most studied by Tutte, his matroid definition of n-connectedness is preserved under duality and, as Proposition 8.1.9 below shows, it essentially succeeds in generalizing graphical n-connectedness.

In Section 4.1, to extend the notion of 2-connectedness from graphs to matroids, we needed to find a reformulation of the definition that did not mention vertices since, in an arbitrary matroid, there is no direct generalization of the concept of a vertex. This approach led to a matroid being defined to be 2-connected if, for every pair of distinct elements, there is a circuit containing both. For $n \geq 3$, the matroid definition of n-connectedness is less straightforward. It generalizes the rank formulation of 2-connectedness for matroids. If X is a subset of the ground set E of a matroid M, then, by the submodularity of the rank function,

$$r(X) + r(E - X) - r(M) \geq 0. \tag{8.1}$$

Moreover, by Proposition 4.2.1, M is not 2-connected if and only if, for some proper non-empty subset X of E, equality holds in (8.1). Before formally defining n-connectedness for matroids, we consider the following.

Example 8.1.2 Let G be the graph shown in Figure 8.2. Evidently G is 2-connected. As G is loopless and has at least three vertices, it follows by Proposition 4.1.1 that $M(G)$ is 2-connected. Thus, by the remarks preceding this example, for all partitions (X, Y) of $E(G)$ such that

$$\min\{|X|, |Y|\} \geq 1, \tag{8.2}$$

we have

$$r(X) + r(Y) - r(M(G)) > 0. \tag{8.3}$$

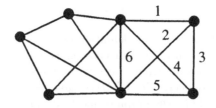

FIG. 8.2. A graph with connectivity 2.

Now let $X = \{1, 2, \ldots, 6\}$ and $Y = E(G) - X$. Then

$$r(X) + r(Y) - r(M(G)) = 1,$$

that is, with this choice of X and Y, the value of $r(X) + r(Y) - r(M(G))$ is minimized over all partitions of $E(G)$ satisfying (8.2) and (8.3). $\qquad\square$

Let M be a matroid with ground set E. If $X \subseteq E$, let

$$\lambda_M(X) = r(X) + r(E - X) - r(M).$$

We call λ_M the *connectivity function* of M and will often abbreviate it as λ. Clearly,

8.1.3 $\lambda(E - X) = \lambda(X)$.

Moreover, we have the following direct consequence of Lemma 4.2.3.

Lemma 8.1.4 *Let M be a matroid with ground set E. If $X \subseteq E$, then*

$$\lambda_M(X) = r(X) + r^*(X) - |X|. \qquad\square$$

Let k be a positive integer and M be a matroid with ground set E. For $X \subseteq E(M)$, if $\lambda_M(X) < k$, then both X and $(X, E - X)$ are called *k-separating*. Thus a 1-separating set is what we have been calling a separator. A k-separating pair $(X, E - X)$ for which $\min\{|X|, |E - X|\} \geq k$ is called a *k-separation* of M with *sides* X and $E - X$. Thus M is 2-connected if and only if it has no 1-separations. More generally, for all integers $n \geq 2$, Tutte defined M to be *n-connected* if, for all k in $\{1, 2, \ldots, n - 1\}$, it has no k-separations. For example, the graphic matroid $M(G)$ in Example 8.1.2, has a 2-separation $(\{1, 2, 3, 4, 5\}, E(G) - \{1, 2, 3, 4, 5\})$, so it is not 3-connected. Evidently, G is not a 3-connected graph.

Lemma 8.1.4 immediately implies one of the most frequently used properties of the connectivity function, that it is invariant under duality.

Corollary 8.1.5 *Let M be a matroid with ground set E. If $X \subseteq E$, then*

$$\lambda_M(X) = \lambda_{M^*}(X).$$

Moreover, M is n-connected if and only if M^* is n-connected. $\qquad\square$

A link between 2-connected graphs and 2-connected matroids was noted in Proposition 4.1.7. The following slight strengthening of that result is easy to deduce from it.

Corollary 8.1.6 *Let G be a graph having no isolated vertices. If $|V(G)| \geq 3$, then $M(G)$ is 2-connected if and only if G is 2-connected and loopless.* $\qquad\square$

The exact relationship between matroid connectivity and graph connectivity will be specified in Section 8.6. To aid the reader's intuition for higher connectivity, we begin our discussion of that relationship here. First, we note a lemma, which is a straightforward consequence of the fact observed in 1.3.8 that if X is a set of edges of a graph G and $G[X]$, the subgraph of G induced by X, has $\omega(G[X])$ components, then

$$r(X) = |V(G[X])| - \omega(G[X]).$$

The details of the proof of this lemma are left as an exercise.

Lemma 8.1.7 *In graph G without isolated vertices, for every subset X of $E(G)$,*

$$\lambda_{M(G)}(X) = |V(G[X]) \cap V(G[E - X])| - \omega(G[X]) - \omega(G[E - X]) + \omega(G).$$

Hence, when G is connected and X and $E - X$ are non-empty,

$$\lambda_{M(G)}(X) \leq |V(G[X]) \cap V(G[E - X])| - 1$$

and equality holds if and only if both $G[X]$ and $G[E - X]$ are connected. □

We can think of $V(G[X]) \cap V(G[E - X])$ as the *vertex boundary* of X. The last lemma notes that, when $G[X]$ and $G[E - X]$ are both connected, this vertex boundary has exactly $\lambda_{M(G)}(X) + 1$ members. We now apply this lemma.

Lemma 8.1.8 *Let G be a connected graph and suppose that G has a k-element minimal vertex cut V'. Then $M(G)$ has a k-separation (X, Y) such that*

$$\min\{r(X), r(Y)\} \geq k.$$

Proof Let H_1 and H_2 be distinct components of $G - V'$ and, for $i = 1, 2$, let $E_i = E(G[V' \cup V(H_i)]) - E(G[V'])$. Since V' is a minimal vertex cut, $G[E_i]$ is connected. As this graph has at least $k + 1$ vertices, it follows that $r(E_i) \geq k$. Let $E_2' = E(G) - E_1$. As $E_2' \supseteq E_2$, we have $r(E_2') \geq k$. Moreover, $V(G[E_1])$ and $V(G[E_2'])$ meet in V'. Therefore, by Lemma 8.1.7, $\lambda_{M(G)}(E_1) \leq |V'| - 1 = k - 1$, and (E_1, E_2') is the required k-separation of $M(G)$. □

As noted earlier, most of Tutte's work on n-connected matroids concentrated on the case $n = 3$. In that case, we have the following attractive link.

Proposition 8.1.9 *Let G be a graph without isolated vertices and suppose that $|E(G)| \geq 4$. Then $M(G)$ is 3-connected if and only if G is 3-connected and simple.*

Proof Assume that $M(G)$ is 3-connected. Then G is connected and, by Lemma 8.1.8, G has no 1- or 2-element vertex cuts, so G is 3-connected. If X is the edge set of a cycle of G with k edges and $k \in \{1, 2\}$, then $(X, E - X)$ is a k-separation of $M(G)$; a contradiction. Hence G is simple.

Conversely, let G be simple and 3-connected. Then, by Corollary 8.1.6, $M(G)$ has no 1-separations. Suppose that $M(G)$ has a 2-separation (X, Y). Then, it is not difficult to show by moving edges that $M(G)$ has a 2-separation (X', Y') such that both $G[X']$ and $G[Y']$ are connected having at least three vertices. Then Lemma 8.1.7 implies that $G[X']$ and $G[Y']$ have exactly two common vertices. The deletion of these two vertices from G disconnects it, contradicting the fact that G is 3-connected. □

For $n \geq 4$, the relationship between n-connectedness in graphs and n-connectedness in matroids will be examined in Section 8.6. Based on Corollary 8.1.6 and Proposition 8.1.9, the reader may wish to predict this relationship. We show next that spikes of rank at least four are 3-connected.

Example 8.1.10 For $r \geq 4$, let M_r be a tipless rank-r spike with legs $L_1, L_2, \ldots,$ L_r, where $L_i = \{a_i, b_i\}$ for all i. Two rank-4 examples of such spikes are the real affine cube R_8 and the binary affine space $AG(3,2)$. We shall show that M_r is 3-connected but not 4-connected. It is straightforward to establish the latter since the union of any two distinct legs L_i and L_j is both a circuit and a cocircuit. Thus, by Lemma 8.1.4,

$$\lambda_{M_r}(L_i \cup L_j) = r(L_i \cup L_j) + r^*(L_i \cup L_j) - |L_i \cup L_j| = 2.$$

Hence M_r has numerous 3-separations. The 4-circuits of M_r ensure that it is 2-connected and so has no 1-separations. Suppose (X, Y) is a 2-separation of M_r. Then $|X|, |Y| \geq 2$ and $r(X) + r(Y) = r + 1$. Thus $r(X), r(Y) \leq r - 1$. If each of X and Y meets every leg, then $r(X), r(Y) \geq r - 1$, so $r + 1 = r(X) + r(Y) \geq 2(r-1)$. Hence $r \leq 3$; a contradiction. Thus X or Y contains a leg. But if Y contains a leg, so does X, otherwise $r(Y) = r$. Hence we may assume that both X and Y contain legs. Suppose that X and Y meet x and y legs, respectively. Then $x + y \geq r$. Since both X and Y avoid legs, $r + 1 = r(X) + r(Y) \geq (x + 1) + (y + 1)$, so $x + y \leq r - 1$; a contradiction. We conclude that M_r is 3-connected. $\square$

Exercises

1. Show that F_7 and F_7^- are 3-connected.
2. Let M be an n-connected matroid with $|E(M)| < 2n$. Show that M is m-connected for all $m \geq 2$.
3. Let M be a rank-3 matroid having at least four elements. Give necessary and sufficient conditions, in terms of its geometric representation, for M to be 3-connected.
4. For all positive integers r, give an example of a rank-r matroid that is n-connected for all $n \geq 2$.
5. Let M be a 3-connected matroid having at least four elements and let N be a single-element extension of M. Prove that N is 3-connected if and only if N is simple and has the same rank as M.
6. Prove Lemma 8.1.7.
7. Prove that if M is an n-connected matroid having at least $2n - 1$ elements, then $E(M)$ has no n-element subset that is both a circuit and a cocircuit.
8. Let G be a simple 2-connected graph and (X, Y) be a 2-separation of $M(G)$. Prove that $M(G)$ has a 2-separation (X', Y') such that both $G[X']$ and $G[Y']$ are connected having at least three vertices.
9. (Akkari 1988) Let M be an n-connected matroid having at least $2(n-1)$ elements and suppose that X is an independent subset of $E(M)$. Show that if $Y \subseteq X$ and M/X is n-connected, then M/Y is n-connected.

10. (Oxley 1981b) Let x and y be distinct elements of an n-connected matroid M where $n \geq 2$ and $|E(M)| \geq 2(n-1)$. Suppose $M \backslash x/y$ is n-connected but $M \backslash x$ is not. Prove that M has a cocircuit of size n containing $\{x, y\}$.

8.2 Properties of the connectivity function

This section will present some basic properties of the matroid connectivity function and of n-connected matroids. It is easy to see that an n-connected graph has no vertices of degree less than n and, more generally, that such a graph has no bonds of size less than n. This prompts one to ask what can be said about the size of cocircuits in an n-connected matroid. The next result answers this question. Note that, since n-connectedness is preserved under duality, the result also gives information about circuit sizes.

Proposition 8.2.1 *If M is an n-connected matroid and $|E(M)| \geq 2(n-1)$, then all circuits and all cocircuits of M have at least n elements.*

Proof If, for some $j < n$, the matroid M has a j-element subset X that is a circuit or a cocircuit, then it is easy to check that $(X, E(M)-X)$ is a j-separation of M, contradicting the fact that M is n-connected. $\square$

A k-separation (X, Y) of a matroid M is *minimal* if $\min\{|X|, |Y|\} = k$. For example, if X consists of two parallel elements in a 2-connected matroid M with $|E(M)| \geq 4$, then $(X, E - X)$ is a minimal 2-separation of M. Likewise, every 3-element circuit and every 3-element cocircuit induces a minimal 3-separation in a 3-connected matroid with at least six elements. These observations are special cases of the following result, which is a straightforward consequence of the preceding proposition.

Corollary 8.2.2 *Let (X, Y) be a k-separation of a k-connected matroid and suppose that $|X| = k$. Then X is either a coindependent circuit or an independent cocircuit.* $\square$

The last proposition begins to reveal the limitations of Tutte's definition of n-connectedness. For a graph G, let $M = M(G)$. When G has at least four edges, for M to be a 3-connected matroid, the graph G must be simple. This is a mild restriction. However, when $|E(G)| \geq 6$, for the matroid M to be 4-connected, the graph G must have no 3-cycles. This is a severe restriction implying, for example, that, for all $m \geq 4$, the matroid $M(K_m)$ is not 4-connected. Likewise, for all $r \geq 3$, the projective geometry $PG(r-1, q)$ is not 4-connected. On the other hand, the uniform matroid $U_{r,n}$ is 4-connected provided $r, n - r \geq 3$. In Section 8.6, we shall show how Tutte's definition of n-connectedness can be modified so that it directly generalizes graph connectivity.

The notion of a connectivity function was introduced by Tutte (1966b), although the connectivity function he defined added one to the value of λ defined above. Tutte did this to ensure agreement between graph and matroid connectivities (see Lemma 8.1.7). Here we have followed the trend of recent work and discarded this added one. This translation has minimal effect on the properties of λ and has the benefit of simplifying many calculations. However, caution is required when reading the literature as many papers follow Tutte's original definition.

Next we describe a useful companion to the function λ (Geelen, Gerards, and Whittle 2003). For sets X and Y in a matroid M, the *local connectivity* between X and Y, denoted $\sqcap(X, Y)$ or $\sqcap_M(X, Y)$, is defined by

$$\sqcap(X, Y) = r(X) + r(Y) - r(X \cup Y).$$

Evidently, $\sqcap(Y, X) = \sqcap(X, Y)$. Moreover,

$$\sqcap(X, E - X) = \lambda(X) \tag{8.4}$$

and, when X and Y are disjoint,

$$\sqcap(X, Y) = r_M(X) - r_{M/Y}(X).$$

If M is a representable matroid and we view its ground set as labelling a multiset of vectors in a vector space V, then the modularity of the dimension function in V implies that $\sqcap(X, Y)$ equals the rank of the subspace that is the intersection of the closures in V of X and Y. We can think of this subspace as being the *subspace boundary* of X and Y in V. Two lines in space are skew if their union has rank four. More generally, we define sets X and Y in a matroid M to be *skew* if the rank of their union is the sum of their ranks or, equivalently, if $\sqcap(X, Y) = 0$.

The following elementary property of $\sqcap$ will be used frequently.

Lemma 8.2.3 *Let $X_1, X_2, Y_1,$ and Y_2 be subsets of the ground set of a matroid M. If $X_1 \supseteq Y_1$ and $X_2 \supseteq Y_2$, then*

$$\sqcap(X_1, X_2) \geq \sqcap(Y_1, Y_2)$$

or, equivalently,

$$r(X_1) + r(X_2) - r(X_1 \cup X_2) \geq r(Y_1) + r(Y_2) - r(Y_1 \cup Y_2).$$

Proof Let $n = |X_1 - Y_1| + |X_2 - Y_2|$. The result is immediate if $n = 0$. Assume it holds when $n < k$ and let $n = k > 0$. We may assume that $X_1 - Y_1$ contains an element e. Then, by the induction assumption,

$$r(X_1 - e) + r(X_2) - r((X_1 - e) \cup X_2) \geq r(Y_1) + r(Y_2) - r(Y_1 \cup Y_2).$$

Hence the result follows provided that

$$r(X_1) - r(X_1 \cup X_2) \geq r(X_1 - e) - r((X_1 - e) \cup X_2). \tag{8.5}$$

But, since r is a submodular, increasing function,

$$r(X_1) + r((X_1 - e) \cup X_2) \geq r(X_1 \cup [(X_1 - e) \cup X_2]) + r(X_1 \cap [(X_1 - e) \cup X_2])$$
$$\geq r(X_1 \cup X_2) + r(X_1 - e).$$

Thus (8.5) holds and hence, by induction, so does the lemma. □

We show next that λ does not increase under taking minors and we characterize when its value is preserved (Geelen, Gerards, and Whittle 2002).

Lemma 8.2.4 *Let X, C, and D be disjoint subsets of the ground set E of a matroid M. Then*

(i) $\lambda_{M \backslash D/C}(X) \leq \lambda_M(X)$ *with equality if and only if $r(X \cup C) = r(X) + r(C)$ and $r(E - X) + r(E - D) = r(E) + r(E - (X \cup D))$);*

(ii) *when C is independent and D is coindependent,*

$$\lambda_{M \backslash D/C}(X) \geq \lambda_M(X) - |C| - |D|$$

with equality if and only if $\mathrm{cl}_M(X)$ contains C and $\mathrm{cl}^*_M(X)$ contains D.

Proof For $Z \subseteq E - (D \cup C)$, we have $r_{M \backslash D/C}(Z) = r_M(Z \cup C) - r_M(C)$. Thus

$$\lambda_M(X) - \lambda_{M \backslash D/C}(X)$$
$$= [r_M(X) + r_M(E - X) - r_M(E)]$$
$$\quad - [r_{M \backslash D/C}(X) + r_{M \backslash D/C}(E - (X \cup D \cup C)) - r_{M \backslash D/C}(E - (D \cup C))]$$
$$= [r(X) + r(E - X) - r(E)]$$
$$\quad - [r(X \cup C) - r(C) + r(E - (X \cup D)) - r(C) - r(E - D) + r(C)]$$
$$= [r(X) + r(C) - r(X \cup C)]$$
$$\quad + [r(E - X) + r(E - D) - r(E - (X \cup D)) - r(E)].$$

By the submodularity of r, each of the last two square-bracketed terms is non-negative and (i) follows.

To prove (ii), we observe that, since C is independent and D is coindependent, $r(C) = |C|$ and $r(E - D) = r(E)$. From Proposition 2.1.9,

$$r(E - X) - r(E - (X \cup D)) = [r^*(X) - |X| + r(M)]$$
$$\quad - [r^*(X \cup D) - |X \cup D| + r(M)]$$
$$= r^*(X) + |D| - r^*(X \cup D).$$

Hence

$$\lambda_M(X) - \lambda_{M \backslash D/C}(X) = [r(X) + |C| - r(X \cup C)] + [r^*(X) + |D| - r^*(X \cup D)].$$

Part (ii) now follows immediately. $\square$

Next we prove the natural extension of the inequality in (i) above when the sets C and D are allowed to meet X.

Corollary 8.2.5 *Let X, C, and D be subsets of the ground set E of a matroid M where C and D are disjoint. Then*

$$\lambda_{M \backslash D/C}(X - (D \cup C)) \leq \lambda_M(X).$$

Equivalently, for $Y = E - X$, if (X, Y) is k-separating in M, and N is a minor of M, then $(X \cap E(N), Y \cap E(N))$ is k-separating in N.

Proof We have the following, where the inequalities follow by Lemma 8.2.4(i), and the equalities hold by (8.1.3):

$$\lambda_M(X) \geq \lambda_{M\backslash(D\cap Y)/(C\cap Y)}(X)$$
$$= \lambda_{M\backslash(D\cap Y)/(C\cap Y)}(Y - (D\cup C))$$
$$\geq \lambda_{M\backslash(D\cap Y)/(C\cap Y)\backslash(D\cap X)/(C\cap X)}(Y - (D\cup C))$$
$$= \lambda_{M\backslash D/C}(X - (D\cup C)).$$ $\square$

To maximize its usefulness, the next result contains a lot of redundancy.

Corollary 8.2.6 *Let z be an element of a matroid M and let (X,Y) be a partition of $E(M) - z$ into possibly empty sets. Then*

$$\lambda_M(X) - 1 \leq \lambda_{M\backslash z}(X) \leq \lambda_M(X)$$

and

$$\lambda_M(X) - 1 \leq \lambda_{M/z}(X) \leq \lambda_M(X).$$

Moreover, in each of (i)–(iv), parts (a)–(c) are equivalent.

(i) (a) $\lambda_{M\backslash z}(X) = \lambda_M(X)$;
     (b) $z \in \mathrm{cl}_M(Y)$ *or z is a coloop of M;*
     (c) $z \notin \mathrm{cl}^*{}_M(X)$ *or z is a coloop of M;*

(ii) (a) $\lambda_{M\backslash z}(X) = \lambda_M(X) - 1$;
      (b) $z \notin \mathrm{cl}_M(Y)$ *and z is not a coloop of M;*
      (c) $z \in \mathrm{cl}^*{}_M(X)$ *and z is not a coloop of M;*

(iii) (a) $\lambda_{M/z}(X) = \lambda_M(X)$;
       (b) $z \in \mathrm{cl}^*{}_M(Y)$ *or z is a loop of M;*
       (c) $z \notin \mathrm{cl}_M(X)$ *or z is a loop of M;*

(iv) (a) $\lambda_{M/z}(X) = \lambda_M(X) - 1$;
      (b) $z \notin \mathrm{cl}^*{}_M(Y)$ *and z is not a loop of M;*
      (c) $z \in \mathrm{cl}_M(X)$ *and z is not a loop of M.*

Proof First we prove (i) and (ii). In each, the equivalence of (b) and (c) is immediate from Proposition 2.1.12, which implies that z is in exactly one of $\mathrm{cl}_M(Y)$ and $\mathrm{cl}^*{}_M(X)$, and z is in exactly one of $\mathrm{cl}_M(X)$ and $\mathrm{cl}^*{}_M(Y)$. By Lemma 8.2.4(i), $\lambda_{M\backslash z}(X) = \lambda_M(X)$ if and only if $r(Y\cup z)+r(E-z) = r(E)+r(Y)$. The last equation certainly holds if z is a coloop of M. If z is not a coloop, the equation holds if and only if $r(Y\cup z) = r(Y)$, that is, if and only if $z \in \mathrm{cl}_M(Y)$. Thus (i) holds. Moreover, we may assume that z is not a coloop. Then, by Lemma 8.2.4(ii), $\lambda_{M\backslash z}(X) \geq \lambda_M(X) - 1$. Thus (ii) follows from (i). The rest of the corollary follows by duality. $\square$

We show next that the only ways to destroy n-connectedness by a single-element extension are by adding a coloop or creating a small circuit.

Proposition 8.2.7 *Let e be an element of a matroid M. Suppose that $M\backslash e$ is n-connected but M is not. Then either e is a coloop of M, or M has a circuit that contains e and has fewer than n elements.*

Proof As M is not n-connected, it has an $(n-j)$-separation (X,Y) for some positive integer j. Thus $\lambda_M(X) < n-j$ and $\min\{|X|,|Y|\} \geq n-j$. Without loss of generality, we may assume that $e \in Y$. By Lemma 8.2.4, $\lambda_{M\backslash e}(X) \leq \lambda_M(X)$. If this inequality is strict, then $(X, Y-e)$ is an $(n-j-1)$-separation of the n-connected matroid $M\backslash e$; a contradiction. Thus $\lambda_{M\backslash e}(X) = \lambda_M(X)$, so, by Corollary 8.2.6(i), either e is a coloop of M, or $e \in \text{cl}_M(Y-e)$. In the latter case, as $(X, Y-e)$ is not an $(n-j)$-separation of $M\backslash e$, we have that $|Y-e| < n-j$. Thus Y contains a circuit containing e and having at most $n-j$ elements. □

Another elementary property of matroid connectivity that is similar to a property of graph connectivity is that deletion or contraction of a single element drops the connectivity by at most one provided the number of elements is not too small. This is a straightforward consequence of Lemma 8.2.4(ii) and we leave the proof as an exercise.

Proposition 8.2.8 *If e is an element of an n-connected matroid M, then, provided $|E(M)| \geq 2(n-1)$, both $M\backslash e$ and M/e are $(n-1)$-connected.* □

One of the most useful properties of λ_M is that it is a submodular function.

Lemma 8.2.9 *If X and Y are subsets of the ground set E of a matroid, then*

$$\lambda(X) + \lambda(Y) \geq \lambda(X \cup Y) + \lambda(X \cap Y).$$

Proof The proof relies only on the submodularity of the rank function r. We have

$$\begin{aligned}
&\lambda(X) + \lambda(Y) + 2r(E) \\
&= r(X) + r(E-X) + r(Y) + r(E-Y) \\
&= [r(X) + r(Y)] + [r(E-X) + r(E-Y)] \\
&\geq [r(X \cup Y) + r(X \cap Y)] + [r(E-(X\cap Y)) + r(E-(X\cup Y))] \\
&= [r(X \cup Y) + r(E-(X\cup Y))] + [r(X \cap Y) + r(E-(X\cap Y))] \\
&= \lambda(X \cup Y) + \lambda(X \cap Y) + 2r(E).
\end{aligned}$$
□

The following is an immediate consequence of the last lemma.

Corollary 8.2.10 *Let X and Y be k-separating sets in a matroid. If one of $X \cup Y$ and $X \cap Y$ is not $(k-1)$-separating, then the other is k-separating.* □

The next result (Oxley, Semple, and Whittle 2004) provides another useful link between the local connectivity and connectivity functions.

Lemma 8.2.11 *Let X and Y be disjoint subsets of the ground set of a matroid M. Then*

$$\sqcap_M(X,Y) + \sqcap_{M^*}(X,Y) = \lambda(X) + \lambda(Y) - \lambda(X \cup Y).$$

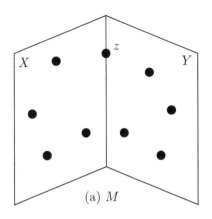 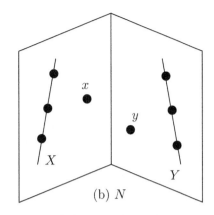

FIG. 8.3. $z \in \mathrm{cl}_M(X) \cap \mathrm{cl}_M(Y)$ and $y \in \mathrm{cl}^*{}_N(X) \cap \mathrm{cl}^*{}_N(Y \cup x)$.

Proof The result follows easily by substitution using the facts that X and Y are disjoint and $\lambda(Z) = r_M(Z) + r_{M^*}(Z) - |Z|$. $\qquad\square$

In a matroid M, a set X or a pair $(X, E - X)$ is *exactly k-separating* if $\lambda_M(X) = k - 1$. Similarly, a k-separation (X, Y) is *exact* if $\lambda_M(X) = k - 1$.

Example 8.2.12 Consider the rank-4 matroid M for which a geometric representation is shown in Figure 8.3(a). Let z be as labelled, let X be the set of four points on the left-hand plane other than z, and let Y be the corresponding set of four points on the right-hand plane. It is not difficult to verify that M has no 1- or 2-separations and that $(X, Y \cup z)$ and $(X \cup z, Y)$ are exact 3-separations. Thus the element z can be moved from one side of the 3-separation to the other. We observe that $z \in \mathrm{cl}(X) \cap \mathrm{cl}(Y)$.

Next let N be the rank-4 matroid shown in Figure 8.3(b), where X and Y are 3-point lines as indicated. Then N has no 1- or 2-separations and has each of $(X, Y \cup \{x, y\}), (X \cup x, Y \cup y)$, and $(X \cup \{x, y\}, Y)$ as an exact 3-separation. Moreover, although it is perhaps less obvious from the figure, $(X \cup y, Y \cup x)$ is also an exact 3-separation. In this case, we have, for example, that $y \in \mathrm{cl}^*(X) \cap \mathrm{cl}^*(Y \cup x)$, where we recall that $e \in \mathrm{cl}^*(S) - S$ if and only if e is a coloop of $M \backslash S$. $\qquad\square$

The phenomenon that was just exemplified can be generalized.

Lemma 8.2.13 *Let z be an element of a matroid M and (X, Y) be a partition of $E(M) - z$ into possibly empty sets. Then there is an integer k such that both $(X \cup z, Y)$ and $(X, Y \cup z)$ are exactly k-separating if and only if $z \in \mathrm{cl}(X) \cap \mathrm{cl}(Y)$ or $z \in \mathrm{cl}^*(X) \cap \mathrm{cl}^*(Y)$.*

This lemma is an immediate consequence of the following more explicit result.

Proposition 8.2.14 *Let z be an element of a matroid M and let (X,Y) be a partition of $E(M) - z$ into possibly empty sets. Then*

$$\lambda_M(X) - 1 \le \lambda_M(X \cup z) \le \lambda_M(X) + 1.$$

Moreover, in each of (i)–(iii), parts (a)–(c) are equivalent.

(i) (a) $\lambda_M(X \cup z) = \lambda_M(X)$;
    (b) z is in exactly one of $\mathrm{cl}_M(X)$ and $\mathrm{cl}^*{}_M(X)$;
    (c) z is in exactly one of $\mathrm{cl}_M(X) \cap \mathrm{cl}_M(Y)$ and $\mathrm{cl}^*{}_M(X) \cap \mathrm{cl}^*{}_M(Y)$.

(ii) (a) $\lambda_M(X \cup z) = \lambda_M(X) - 1$;
    (b) $z \in \mathrm{cl}_M(X) \cap \mathrm{cl}^*{}_M(X)$;
    (c) $z \notin \mathrm{cl}_M(Y) \cup \mathrm{cl}^*{}_M(Y)$.

(iii) (a) $\lambda_M(X \cup z) = \lambda_M(X) + 1$;
    (b) $z \notin \mathrm{cl}_M(X) \cup \mathrm{cl}^*{}_M(X)$;
    (c) $z \in \mathrm{cl}_M(Y) \cap \mathrm{cl}^*{}_M(Y)$.

Proof By Lemma 8.1.4, $\lambda_M(X) = r(X) + r^*(X) - |X|$. Thus

$$\lambda_M(X \cup z) - \lambda_M(X) = [r_M(X \cup z) - r_M(X)] + [r^*_M(X \cup z) - r^*_M(X)] - 1.$$

Hence $\lambda_M(X \cup z) - \lambda_M(X) \in \{-1, 0, 1\}$.

Now $r_M(X \cup z) = r_M(X)$ if and only if $z \in \mathrm{cl}_M(X)$. Thus, in each of (i)–(iii), the equivalence of (a) and (b) follows easily. The equivalence of (b) and (c) follows from Proposition 2.1.12, which implies that z is in exactly one of $\mathrm{cl}_M(X)$ and $\mathrm{cl}^*{}_M(Y)$. $\qquad\square$

How much information about a matroid does its connectivity function convey? We have already noted that the connectivity functions of a matroid and its dual coincide. Moreover, it is straightforward to check that if M_1 and M_2 are matroids on disjoint sets, then $M_1 \oplus M_2$ and $M_1 \oplus M_2^*$ have the same connectivity function. An attractive unpublished conjecture of Cunningham asserted that the two situations just described are the only ways in which two distinct matroids

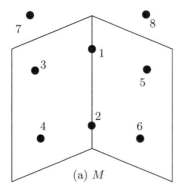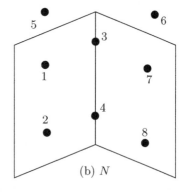

FIG. 8.4. Seymour's counterexample to Cunningham's conjecture.

can have equal connectivity functions. However, Seymour (1988) showed that this conjecture is false.

Example 8.2.15 Let $E = \{1, 2, \ldots, 8\}$ and let M and N be the rank-4 matroids on E for which geometric representations are shown in Figure 8.4, where we note that the elements 7 and 8 are free in M, while 5 and 6 are free in N. Evidently, M and N are connected, and M is not equal to N or N^*, although M is isomorphic to both these matroids. The reader can easily check that the connectivity functions of both M and N are given by

$$\lambda(X) = \begin{cases} |X|, & \text{if } |X| \leq 3; \\ |E - X|, & \text{if } |X| \geq 5; \\ 3, & \text{if } X \text{ or } E - X \text{ is } \{1,2,3,4\} \text{ or } \{1,2,5,6\}; \\ 4, & \text{otherwise.} \end{cases}$$

□

Despite its failure in general, Cunningham's conjecture does hold in some important special cases. The next theorem combines results of Seymour (1988) and Lemos (1994, 2002).

Theorem 8.2.16 *Let M and N be connected matroids with the same ground set and the same connectivity function. If*

(i) *M is binary, or*
(ii) *$r(M) \neq r^*(M)$,*
then $M = N$, or $M = N^*$.

□

Exercises

1. Let e and f be distinct elements of a matroid M, and X and Y be subsets of $E(M) - \{e, f\}$. Prove that
$$\lambda_{M\backslash e}(X) + \lambda_{M/f}(Y) \geq \lambda_M(X \cap Y) + \lambda_{M\backslash e/f}(X \cup Y).$$

2. In a matroid M, suppose $X_1 \supseteq Y_1$ and $X_2 \supseteq Y_2$. Prove that $\sqcap(X_1, X_2) = \sqcap(Y_1, Y_2)$ if and only if
$$r_{M/Y_1}(X_1 - Y_1) + r_{M/Y_2}(X_2 - Y_2) = r_{M/(Y_1 \cup Y_2)}([X_1 \cup X_2] - [Y_1 \cup Y_2]).$$

3. (Geelen, Gerards, and Whittle 2006a) Let N be a minor of a matroid M and X be a subset of $E(N)$. Prove that if $\lambda_M(X) = \lambda_N(X)$, then $M|X = N|X$.

4. Let X, C, and D be disjoint subsets of the ground set of a matroid M. Prove that $\lambda_{M\backslash D/C}(X) = \lambda_M(X)$ if and only if $\sqcap_M(X, C) = 0 = \sqcap_{M^*}(X, D)$.

5. (Geelen, Gerards, and Whittle 2007a) Let X, Y, and Z be subsets of the ground set of a matroid M with X disjoint from $Y \cup Z$. Prove that $\sqcap_{M/X}(Y, Z) = \sqcap_M(X \cup Y, Z) - \sqcap_M(X, Z)$.

6. (Oxley, Semple, and Whittle 2004) Let P, Q, R, and S be subsets of the ground set of a matroid M. Prove the following:

(a) $\sqcap(P \cup Q, R \cup S) + \sqcap(P, Q) + \sqcap(R, S) = \sqcap(P \cup R, Q \cup S) + \sqcap(P, R) + \sqcap(Q, S)$.

(b) $\sqcap(P \cup Q, R) + \sqcap(P, Q) = \sqcap(P \cup R, Q) + \sqcap(P, R)$.

(c) $\sqcap(P \cup Q, R) + \sqcap(P, Q) \geq \sqcap(P, R) + \sqcap(Q, R)$.

(d) If (P, Q, R) is a partition of $E(M)$ into possibly empty sets, then $\lambda(P) + \sqcap(Q, R) = \lambda(R) + \sqcap(P, Q)$.

7. For some $n \geq 2$, let M be an n-connected matroid and X and Y be n-separating subsets of $E(M)$. Prove that

(a) if $|X \cap Y| \geq n - 1$, then $X \cup Y$ is n-separating; and

(b) if $|E(M) - (X \cup Y)| \geq n - 1$, then $X \cap Y$ is n-separating.

8. Let X_1, X_2, X_3, and X_4 be subsets of $E(M)$ where M is a representable matroid. Use Exercise 7(d) of Section 6.1 to prove that

(a) $\sqcap(X_1, X_3) + \sqcap(X_2, X_3) + \sqcap(X_2, X_4) \leq \sqcap(X_1 \cup X_2, X_3) + \sqcap(X_1 \cup X_4, X_2) + \sqcap(X_3, X_4)$; and

(b) $\sqcap(X_1, X_3) + \sqcap(X_2, X_3) + \sqcap(X_2, X_4) + \sqcap(X_1, X_4) \leq \sqcap(X_1 \cup X_2, X_3) + \sqcap(X_1 \cup X_2, X_4) + \sqcap(X_1, X_2) + \sqcap(X_3, X_4)$.

8.3   3-connected matroids and 2-sums

In this section, we shall establish a basic link between the operation of 2-sum and the property of 3-connectedness. This link will then be used to associate a unique labelled tree with every 2-connected matroid M. The vertices of this tree are labelled by 3-connected matroids, circuits, and cocircuits; M can be reconstructed from it; and every 2-separation of M can be determined from it.

To begin, recall that a matroid M fails to be 2-connected if and only if it can be written as a direct sum of two non-empty matroids, each of which is isomorphic to a minor of M. The following analogue of this result for 3-connectedness was proved independently by Bixby (1972), Cunningham (1973), and Seymour (1980b). The proof given here is Seymour's. The theorem uses the operation of 2-sum, a quick definition of which is given in Proposition 7.1.20.

Theorem 8.3.1 *A 2-connected matroid M is not 3-connected if and only if $M = M_1 \oplus_2 M_2$ for some matroids M_1 and M_2, each of which has at least three elements and is isomorphic to a proper minor of M.*

Proof It is an easy consequence of Proposition 7.1.15(i) that if $M = M_1 \oplus_2 M_2$ and $|E(M_1)|, |E(M_2)| \geq 3$, then $(E(M_1) - E(M_2), E(M_2) - E(M_1))$ is a 2-separation of M and so M is not 3-connected. The proof of the converse is not quite as straightforward. Suppose that M is not 3-connected. Then M has an exact k-separation (X_1, X_2) for some $k < 3$. As M is 2-connected, $k = 2$, so

$$\sqcap(X_1, X_2) = 1. \tag{8.6}$$

The next two lemmas are fundamental to the construction of matroids M_1 and M_2 of which M is the 2-sum.

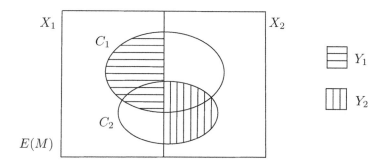

FIG. 8.5. The sets in the proof of Lemma 8.3.3.

Lemma 8.3.2 *Let C_1 and C_2 be circuits of M each of which meets both X_1 and X_2. Then $C_1 \cap X_1$ is not a proper subset of $C_2 \cap X_1$.*

Proof Assume the contrary and choose x_1 in $C_1 \cap X_1$ and x_2 in $(C_2 - C_1) \cap X_1$. Thus $\{x_1, x_2\} \subseteq C_2 \cap X_1$. Moreover, $C_2 \cap X_1$ is independent in M, so it is contained in a basis B_1 of X_1. Clearly $\mathrm{cl}((B_1 \cup X_2) - \{x_1, x_2\})$ contains x_1 and hence x_2; that is, $(B_1 \cup X_2) - \{x_1, x_2\}$ spans M. Hence $r(M) \le |B_1| + r(X_2) - 2 = r(X_1) + r(X_2) - 2$. Thus $\sqcap(X_1, X_2) \ge 2$, a contradiction to (8.6). $\square$

Lemma 8.3.3 *Let Y_1 and Y_2 be non-empty subsets of X_1 and X_2, respectively. Suppose that M has circuits C_1 and C_2 with $C_1 \cap X_1 = Y_1$ and $C_2 \cap X_2 = Y_2$ such that $C_1 \cap X_2$ and $C_2 \cap X_1$ are non-empty. Then $Y_1 \cup Y_2$ is a circuit of M.*

Proof Choose circuits C_1 and C_2 with the properties described such that $C_1 \cup C_2$ is minimal (see Figure 8.5). If $C_1 = C_2$, then the lemma holds so assume that $C_1 \ne C_2$. We show next that $(C_1 \cup C_2) \cap X_1$ is independent. Assume not and let C be a circuit contained in it. Now $C \ne C_1$ since C_1 meets X_2 but C does not. Thus there is an element x in $C - C_1$. Since $C \subseteq C_1 \cup C_2$, it follows that $x \in C_2$. Choose y in $C_2 \cap X_2$. Then $y \in C_2 - C$ and $x \in C_2 \cap C$. Thus, by the strong circuit elimination axiom, there is a circuit C_2' that contains y such that $C_2' \subseteq (C_2 \cup C) - x$. Clearly $C_2' \cap X_2 \subseteq (C_2 \cup C) \cap X_2 = C_2 \cap X_2$. It follows, since $C_2' \ne C_2$, that $C_2' \cap X_1$ is non-empty. Moreover, $C_2' \cap X_2$ is non-empty since it contains y. Applying Lemma 8.3.2 to C_2 and C_2', we deduce, since $C_2' \cap X_2 \subseteq C_2 \cap X_2$, that $C_2' \cap X_2 = C_2 \cap X_2 = Y_2$. But $C_1 \cup C_2' \subseteq (C_1 \cup C_2) - x$. This contradicts the minimality of $C_1 \cup C_2$. We conclude that $(C_1 \cup C_2) \cap X_1$ is independent. Similarly, $(C_1 \cup C_2) \cap X_2$ is independent.

By Lemma 8.2.3,

$$\sqcap((C_1 \cup C_2) \cap X_1, (C_1 \cup C_2) \cap X_2) \le \sqcap(X_1, X_2) = 1.$$

Since $(C_1 \cup C_2) \cap X_1$ and $(C_1 \cup C_2) \cap X_2$ are both independent, it follows that $|(C_1 \cup C_2) \cap X_1| + |(C_1 \cup C_2) \cap X_2| - r(C_1 \cup C_2) \le 1$, so $r(C_1 \cup C_2) \ge |C_1 \cup C_2| - 1$. As $C_1 \cup C_2$ contains two distinct circuits, its rank is at most $|C_1 \cup C_2| - 2$; a contradiction. $\square$

To finish the proof of Theorem 8.3.1, we now construct matroids M_1 and M_2 whose 2-sum is M. Let p be an element that is not in $E(M)$ and let $\mathcal{C}_1 = \mathcal{C}(M|X_1) \cup \{(C \cap X_1) \cup p : C$ is a circuit of M that meets both X_1 and $X_2\}$. It is not difficult to check using the last lemma that $\mathcal{C}_1$ is the set of circuits of a matroid on $X_1 \cup p$. We call this matroid M_1. Define M_2 similarly as the matroid on $X_2 \cup p$ whose set of circuits is $\mathcal{C}(M|X_2) \cup \{(C \cap X_2) \cup p : C$ is a circuit of M that meets both X_1 and $X_2\}$. It now follows easily that $M = M_1 \oplus_2 M_2$. Finally, since Proposition 7.1.21 showed that the parts of a 2-sum are isomorphic to proper minors of the 2-sum itself provided these parts have at least three elements, we deduce that M_1 and M_2 are isomorphic to proper minors of M. $\square$

The next result, a straightforward combination of Theorem 8.3.1 and Proposition 4.2.7, emphasizes the structural importance of 3-connected matroids within the class of all matroids.

Corollary 8.3.4 *Every matroid that is not 3-connected can be constructed from 3-connected proper minors of itself by a sequence of the operations of direct sum and 2-sum.* $\square$

This result will be used frequently in Chapter 12 where we shall specify the structure of many classes of matroids. We already know, by virtue of a host of results in Chapter 4, that in many matroid arguments the general result can be obtained by restricting attention to the 2-connected case. It follows, on combining Corollary 8.3.4 with the results on 2-sums at the end of Section 7.1, that one can frequently be even more selective and concentrate just on 3-connected matroids.

We noted in Proposition 4.2.20 that if a 2-connected matroid N is a minor of a matroid M, then N is a minor of some connected component of M. The next result generalizes this.

Proposition 8.3.5 *Let M, N, M_1, and M_2 be matroids such that $M = M_1 \oplus_2 M_2$ and N is 3-connected. If M has an N-minor, then M_1 or M_2 has an N-minor.*

Proof Assume that the proposition fails and let M be a counterexample that minimizes $|E(M)|$. If $|E(M_2)| = 2$, then $M \cong M_1$; a contradiction. Thus we may assume that $\min\{|E(M_1)|, |E(M_2)|\} \geq 3$. Let p be the basepoint of the 2-sum. Then, as in the first sentence of the proof of Theorem 8.3.1, $(E(M_1) - p, E(M_2) - p)$ is a 2-separation of M. As N is 3-connected, it follows that $|E(M)| > |E(N)|$.

We shall show next that M is connected. Suppose not. Then, by Proposition 7.1.22(ii), M_1 or M_2, say M_1, is disconnected. Thus M can be written as $(M_{1,1} \oplus M_{1,2}) \oplus_2 M_2$ where p is an element of $M_{1,2}$. Then $M = M_{1,1} \oplus (M_{1,2} \oplus_2 M_2)$. By Proposition 4.2.20, $M_{1,1}$ or $M_{1,2} \oplus_2 M_2$ has an N-minor. By the choice of M, we deduce that $M_{1,1}, M_{1,2}$, or M_2 has an N-minor. But each of $M_{1,1}$ and $M_{1,2}$ is a minor of M_1, so we have a contradiction. Thus M is indeed connected.

Now let N_0 be an N-minor of M and suppose $e \in E(M) - E(N_0)$. Then, by Proposition 4.3.7, either $M \backslash e$ or M/e is connected and has N_0 as a minor. We may assume the former otherwise we replace each of M, M_1, M_2, and N_0 by its dual in the argument that follows. Without loss of generality, we may assume that $e \in E(M_1) - p$. As $M = M_1 \oplus_2 M_2 = P(M_1, M_2) \backslash p$ and e is in $E(M_1) - p$, we have, by Proposition 7.1.15(v), that $M \backslash e = P(M_1 \backslash e, M_2) \backslash p$. Then, by Proposition 7.1.22(ii) again, since $M \backslash e$ is connected, so too are $M_1 \backslash e$ and M_2. Thus $M_1 \backslash e \oplus_2 M_2$ is well-defined and equals $M \backslash e$. The choice of M now implies that $M_1 \backslash e$ or M_2 has an N-minor; a contradiction. $\square$

By Theorem 8.3.1, a 2-connected matroid M that is not 3-connected can be written as a 2-sum, say $M = M_1 \oplus_2 M_2$. By Proposition 7.1.22(ii), both M_1 and M_2 are 2-connected. If either of these matroids fails to be 3-connected, then it too can be written as a 2-sum, and this process can continue. In order to keep track of this decomposition, we introduce the following concept.

A *matroid-labelled tree* is a tree T with vertex set $\{M_1, M_2, \ldots, M_k\}$ for some positive integer k such that

(i)   each M_i is a matroid;

(ii)   if M_{j_1} and M_{j_2} are joined by an edge e_i of T, then $E(M_{j_1}) \cap E(M_{j_2}) = \{e_i\}$, and $\{e_i\}$ is not a separator of M_{j_1} or M_{j_2}; and

(iii) if M_{j_1} and M_{j_2} are non-adjacent, then $E(M_{j_1}) \cap E(M_{j_2})$ is empty.

We call $M_1, M_2, \ldots, M_k$ the *vertex labels* of T.

The reader may feel that the idea of two matroid-labelled trees being 'equal to within relabelling of their edges' is self-explanatory. If not, we observe that if we change the label on an edge of a matroid-labelled tree T from e to f, then, to recover a matroid-labelled tree, we must relabel e by f in the two matroids that had labelled the endpoints of e. Two matroid-labelled trees are *equal to within relabelling of their edges* if one can be obtained from the other by a sequence of such relabelling moves associated with single edges.

Let e be an edge of a matroid-labelled tree T and suppose that e joins vertices labelled by N_1 and N_2. Suppose that we contract the edge e of the tree T and relabel by $N_1 \oplus_2 N_2$ the vertex that results by identifying the endpoints of e, leaving all other edge and vertex labels unchanged. Then it is not difficult to see that we retain a matroid-labelled tree and it is natural to denote this tree by T/e. This process can be repeated and we note that, since the operation of 2-sum is associative (Proposition 7.1.23), for every subset $\{e_{i_1}, e_{i_2}, \ldots, e_{i_m}\}$ of $E(T)$, the matroid-labelled tree $T/e_{i_1}, e_{i_2}, \ldots, e_{i_m}$ is well-defined.

We shall be interested in associating with a 2-connected matroid a certain type of matroid-labelled tree that will describe the matroid's structure. Before defining such a tree, we present a technical result that is a straightforward consequence of Proposition 7.1.20.

Corollary 8.3.6 *Let* $M_1 \oplus_2 N_1 = M_2 \oplus_2 N_2$ *where* $E(M_i) \cap E(N_i) = \{p_i\}$ *for each* i. *If* $E(M_1) - p_1 = E(M_2) - p_2$, *then* M_2 *and* N_2 *can be obtained from* M_1 *and* N_1 *by relabelling* p_1 *by* p_2 *in each.*

Proof Let $M = M_1 \oplus_2 N_1$. By Proposition 7.1.20, $M|(E(M_1) - p_1) = M_1 \backslash p_1$ since these two matroids have the same ground sets and the same sets of circuits. Moreover, the collection of circuits of M_1 containing p_1 is

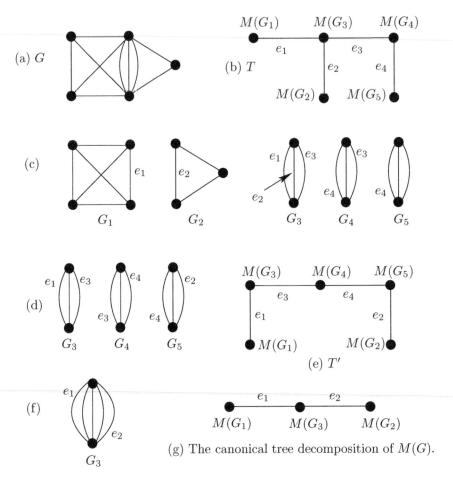

FIG. 8.6. A graph G and three tree decompositions of $M(G)$.

$$\{(C \cap E(M_1)) \cup p_1 : C \in \mathcal{C}(M) \text{ with } C \cap E(M_1) \neq \emptyset \neq C \cap E(N_1)\}.$$

Since $E(M_1) - p_1 = E(M_2) - p_2$, it follows that M_2 can be obtained from M_1 by relabelling p_1 by p_2. By symmetry, we deduce that the lemma holds. $\qquad\square$

A *tree decomposition* of a 2-connected matroid M is a matroid-labelled tree T such that if $V(T) = \{M_1, M_2, \ldots, M_k\}$ and $E(T) = \{e_1, e_2, \ldots, e_{k-1}\}$, then

(i) $E(M) = (E(M_1) \cup E(M_2) \cup \cdots \cup E(M_k)) - \{e_1, e_2, \ldots, e_{k-1}\}$;

(ii) $|E(M_i)| \geq 3$ for all i unless $|E(M)| < 3$, in which case $k = 1$ and $M_1 = M$; and

(iii) M is the matroid that labels the single vertex of $T/e_1, e_2, \ldots, e_{k-1}$.

Example 8.3.7 Let M be the cycle matroid of the graph G shown in Figure 8.6(a). If $G_1, G_2, \ldots, G_5$ are as shown in Figure 8.6(c), then the tree T in Figure 8.6(b) is a tree decomposition for M in which every vertex label is 3-connected. However, T is not the only such tree decomposition for M, another being the matroid-labelled tree T' in Figure 8.6(e) where G_3, G_4, and G_5 are relabelled as in Figure 8.6(d). □

The following is an immediate consequence of the fact that the dual of a 2-sum of two matroids is the 2-sum of their duals (Proposition 7.1.22(ii)).

Lemma 8.3.8 *If T is a tree decomposition for a 2-connected matroid M and every vertex label is replaced by its dual, then the resulting matroid-labelled tree T^* is a tree decomposition for M^*.* □

The next result (Cunningham and Edmonds 1980, Seymour 1981b) describes explicitly how every 2-connected matroid can be written in terms of 2-sums of 3-connected matroids.

Proposition 8.3.9 *Let M be a 2-connected matroid. Then M has a tree decomposition T in which every vertex label is 3-connected.*

Before proving this result, we make some observations. Example 8.3.7 showed that requiring every vertex label of a tree decomposition to be 3-connected does not guarantee its uniqueness. Indeed, the non-uniqueness arose there because we were able to decompose a cocircuit in more than one way. Dually, a circuit can be decomposed in more than one way. By restricting the ways in which circuits and cocircuits can occur in the decomposition, Cunningham and Edmonds (in Cunningham 1973) were able to guarantee uniqueness of the decomposition.

Theorem 8.3.10 *Let M be a 2-connected matroid. Then M has a tree decomposition T in which every vertex label is 3-connected, a circuit, or a cocircuit, and there are no two adjacent vertices that are both labelled by circuits or are both labelled by cocircuits. Moreover, T is unique to within relabelling of its edges.*

The tree decomposition of M whose existence and uniqueness are asserted in the last theorem will be called the *canonical tree decomposition* for M. For the cycle matroid of the graph G in Figure 8.6(a), the canonical tree decomposition is the tree in Figure 8.6(g) where G_1 and G_2 are labelled as in Figure 8.6(c) while G_3 is labelled as in Figure 8.6(f). We noted in Corollary 4.2.9 that every matroid has a unique decomposition into 2-connected components. The canonical tree decomposition of a 2-connected matroid M provides a way to break M into a unique collection of pieces, each of which is a 3-connected matroid, a circuit, or a cocircuit. Moreover, M can be reconstructed from these pieces using the operation of 2-sum, with the common elements of the pieces guiding how this reconstruction is to be done.

Because every circuit or cocircuit with at most three elements is 3-connected, a useful first step in the proof of Theorem 8.3.10 will be to prove Proposition 8.3.9. The following preliminaries will assist in the proof. The reader may prefer to skip the details of these proofs. In that case, there is one further result in this section that should be noted, namely Proposition 8.3.16. The definitions needed to make sense of that are contained in the next paragraph and in the paragraph preceding Lemma 8.3.15.

Let T be a tree decomposition of a matroid M. Then, for every element x of M, there is a unique vertex of T that is labelled by a matroid whose ground set contains x. Thus every edge e_i of T induces a partition (X_i, Y_i) of $E(M)$. We shall say that (X_i, Y_i) is *displayed by* e_i. Similarly, if G is a connected subgraph of T, then the subset of $E(M)$ *corresponding to* G is the set of elements of M that lie in the ground set of some vertex label of G.

By Proposition 7.1.21, if the 2-connected matroid M is the 2-sum of M_1 and N_1 with basepoint p, and q is an element of $E(N_1) - p$, then M_1 can be obtained from some minor of M by relabelling q by p. The following is a straightforward extension of this result and we omit the proof.

Lemma 8.3.11 *Let the matroid N label a vertex v of a tree decomposition T of a 2-connected matroid M. Then M has a minor isomorphic to N. More precisely, let $f_1, f_2, \ldots, f_m$ be the edges of T meeting v and let G_i be the component of $T - v$ containing an endpoint of f_i. Then, for each i, if x_i is in the subset of $E(M)$ corresponding to G_i, there are subsets U and V of $E(M) - \{x_1, x_2, \ldots, x_m\}$ such that N can be obtained from $M \backslash U / V$ by relabelling each x_i by f_i.* $\quad\square$

Proof of Proposition 8.3.9 Beginning with a tree consisting of a single vertex labelled by M, we iterate the process of finding a vertex N of the tree that is not 3-connected and replacing that vertex by two adjacent vertices labelled by two matroids of which N is the 2-sum. Formally, if N is the 2-sum, with basepoint p_N, of N_1 and N_2, we form a new matroid-labelled tree T' from the current matroid-labelled tree T as follows:

(i) delete N;

(ii) add two new vertices N_1 and N_2 joined by a new edge labelled by p_N; and

(iii) for each edge p of T meeting N, if p joins N and N', say, add an edge to T' labelled by p joining N' to the member of $\{N_1, N_2\}$ that contains p.

This iterative process ends when every vertex label is 3-connected at which stage, it is not hard to see that we have the required tree decomposition of M. $\quad\square$

We shall show in the next proof that we can easily modify the tree decomposition of M obtained in the last proof to obtain the canonical tree decomposition of M. This proof is quite long but the result is an important one and will lead, for example, to a description of all the 2-separations of a 2-connected matroid.

Proof of Theorem 8.3.10 By Proposition 8.3.9, there is a tree decomposition T' for M in which every vertex is labelled by a 3-connected matroid. Now T' may have an edge e joining two vertices that are both labelled by circuits or are both labelled by cocircuits. If this occurs, we can contract e from T' to get a new tree decomposition for M. We repeat this process until we obtain a tree decomposition T for M in which there are no two adjacent vertices labelled by circuits and no two adjacent vertices labelled by cocircuits. We shall call T a *canonical tree decomposition* for M. To complete the proof of the theorem, we need to show that T is unique to within relabelling of its edges.

We shall prove this uniqueness by induction on $|E(M)|$. If M is 3-connected, a circuit, or a cocircuit, then T must consist of a single vertex labelled by M. We may now assume that M is not 3-connected and is neither a circuit nor a cocircuit and that all 2-connected matroids with fewer elements than M have a unique canonical tree decomposition. Evidently $|E(M)| \geq 4$. Let T_1 and T_2 be canonical tree decompositions for M. Let e be an edge of T_i for some i in $\{1, 2\}$. Then the partition of $E(M)$ displayed by e is a 2-separation of M. To see this, note that if T_i' is the matroid-labelled tree obtained by contracting every edge of T_i except e, then M is the 2-sum of the two matroids that label the vertices of T_i'. Thus the partition of $E(M)$ displayed by the edge e of T_i', which equals the partition of $E(M)$ displayed by the edge e of T_i, is a 2-separation of M.

As we are arguing by induction on $|E(M)|$, it is natural to consider a degree-one vertex M_1 of T_1. Let p_1 be the unique edge incident with M_1 and let $X_1 = E(M_1) - p_1$. The rest of the proof of Theorem 8.3.10 will use three lemmas.

Lemma 8.3.12 *If M_1 is a circuit or a cocircuit, then T_1 and T_2 are equal to within relabelling their edges.*

Proof Clearly $E(M_1) = X_1 \cup p_1$ and $X_1 \subseteq E(M)$. Assume first that M_1 is a cocircuit and choose x in X_1. The elements of X_1 are parallel in M. We want to know where these elements occur in T_2. If f_1 and f_2 are parallel in $N_1 \oplus_2 N_2$ and $f_i \in E(N_i)$ for each i, then, by Proposition 7.1.20, each f_i is parallel in N_i to the basepoint p of the 2-sum. Since no two adjacent vertices of T_1 or T_2 label cocircuits, it follows that X_1 is a parallel class of M, and T_2 has a vertex M_2 labelled by a cocircuit such that $E(M_2) \cap E(M) = X_1$.

Suppose first that $|X_1| \geq 3$. Then $M \backslash x$ is 2-connected and, by the induction assumption, it has a unique canonical tree decomposition. Evidently two such tree decompositions are T_1' and T_2' where T_i' is obtained from T_i by relabelling the vertex M_i by $M_i \backslash x$. By the induction assumption, T_1' and T_2' are equal to within relabelling their edges. Hence so too are T_1 and T_2.

We may now assume that $|X_1| = 2$, say $X_1 = \{x, y\}$. Let J_1 be the unique neighbour of M_1 in T_1. Then we can obtain a canonical tree decomposition T_1' for $M \backslash x$ by deleting the vertex M_1 from T_1 and relabelling the element p_1 of J_1 by y. Suppose the vertex M_2 of T_2 has degree one, being adjacent to J_2 say. Then, by constructing T_2' from T_2 just as T_1' was constructed from T_1, we deduce, since T_1' and T_2' are equal to within relabelling their edges, that the same is true for T_1 and T_2. If M_2 has degree exceeding one, then $|E(M_2)| \geq 4$. In this case, let T_2' be obtained from T_2 by relabelling the vertex M_2 of the latter by $M_2 \backslash x$. By the induction assumption, T_1' and T_2' are equal to within edge relabelling. But the vertex of T_2' containing y is the cocircuit $M_2 \backslash x$, whereas the vertex J_1 of T_1' containing y is not a cocircuit otherwise T_1 has M_1 and J_1 as adjacent vertices

that are both labelled by cocircuits. This contradiction implies that the lemma
holds if M_1 is a cocircuit. By duality, it also holds if M_1 is a circuit.   □

By symmetry, we may now assume that, in both T_1 and T_2, the label on every
degree-one vertex is a 3-connected matroid with at least four elements. Hence
$|X_1| \geq 3$ where we recall that $X_1 = E(M_1) \cap E(M)$. Next we prove the following.

Lemma 8.3.13 *Let* (X_2, Y_2) *be a 2-separation of* M. *Then* $X_1 \cap X_2$ *or* $X_1 \cap Y_2$
is empty.

Proof Assume that both $X_1 \cap X_2$ and $X_1 \cap Y_2$ are non-empty. Suppose first that
$X_2 \subseteq X_1$. By Lemma 8.3.11, there are subsets U and V of $E(M)$ and an element
z of $E(M) - X_1$ such that M_1 is obtained from $M \backslash U / V$ by relabelling z by p_1.
Since $\lambda_M(X_2) = 1$, Lemma 8.2.4 implies that $\lambda_{M \backslash U / V}(X_2) \leq 1$, so $\lambda_{M_1}(X_2) \leq 1$.
Since $|Y_2 \cap X_1| \geq 1$ and $p_1 \in E(M_1) - X_2$, it follows that $|E(M_1) - X_2| \geq 2$,
so $(X_2, E(M_1) - X_2)$ is a 2-separation of M_1; a contradiction. Hence $X_2 \not\subseteq X_1$
and, similarly, $Y_2 \not\subseteq X_1$. Thus all four of the sets $X_1 \cap X_2, X_1 \cap Y_2, X_2 - X_1$,
and $Y_2 - X_1$ are non-empty. As $X_2 - X_1$ and $Y_2 - X_1$ are the complements of
$X_1 \cup Y_2$ and $X_1 \cup X_2$, respectively, and M is 2-connected, we have, by (8.1.3),
that $\lambda(X_1 \cup Y_2) \geq 1$ and $\lambda(X_1 \cup X_2) \geq 1$.
Now each of $\lambda(X_2), \lambda(Y_2)$, and $\lambda(X_1)$ is 1. Hence, by Lemma 8.2.9,

$$2 = \lambda(X_1) + \lambda(X_2) \geq \lambda(X_1 \cap X_2) + \lambda(X_1 \cup X_2) \geq \lambda(X_1 \cap X_2) + 1.$$

As $X_1 \cap X_2 \neq \emptyset$, it follows that $\lambda(X_1 \cap X_2) = 1$. Similarly, $\lambda(X_1 \cap Y_2) = 1$. Then
$\lambda(X_1 \cap X_2) + \lambda(X_1 \cap Y_2) - \lambda(X_1) = 1 + 1 - 1 = 1$. Thus, by Lemma 8.2.11,

$$\sqcap_M(X_1 \cap X_2, X_1 \cap Y_2) + \sqcap_{M^*}(X_1 \cap X_2, X_1 \cap Y_2) = 1.$$

Hence, for some N in $\{M, M^*\}$, we have $\sqcap_N(X_1 \cap X_2, X_1 \cap Y_2) = 0$. Since $|X_1| \geq 3$
and $X_1 \cap X_2$ and $X_1 \cap Y_2$ are non-empty, we may assume that $|X_1 \cap X_2| \geq 1$
and $|X_1 \cap Y_2| \geq 2$.
The matroid M is the 2-sum, with basepoint p_1, of M_1 and some matroid
N_1. Thus, by Lemma 7.1.22(i), M^* is the 2-sum, with basepoint p_1, of M_1^*
and N_1^*. Therefore $M_1 \backslash p_1$ and $M_1^* \backslash p_1$ are restrictions of M and M^*, respectively. Since $\sqcap_N(X_1 \cap X_2, X_1 \cap Y_2) = 0$, either $\sqcap_{M_1 \backslash p_1}(X_1 \cap X_2, X_1 \cap Y_2) = 0$
or $\sqcap_{M_1^* \backslash p_1}(X_1 \cap X_2, X_1 \cap Y_2) = 0$. Thus $\sqcap_{M_1}((X_1 \cap X_2) \cup p_1, X_1 \cap Y_2) \leq 1$ or
$\sqcap_{M_1^*}((X_1 \cap X_2) \cup p_1, X_1 \cap Y_2) \leq 1$. By Corollary 8.1.5 and (8.4), it follows that
$\lambda_{M_1}(X_1 \cap Y_2) = \lambda_{M_1^*}(X_1 \cap Y_2) \leq 1$. This contradiction to the fact that M_1 is
3-connected completes the proof of Lemma 8.3.13.   □

We know that every edge of T_2 induces a 2-separation (X_2, Y_2) of M. Moreover, from the last lemma, either $X_1 \subseteq Y_2$ or $X_1 \subseteq X_2$. The next lemma essentially finishes the proof of Theorem 8.3.10.

Lemma 8.3.14 *The tree decomposition* T_2 *has a degree-one vertex* M_2 *such that*
$E(M_2) - q_1 = X_1$ *where* q_1 *is the unique edge of* T_2 *meeting* M_2.

Proof Suppose that the lemma fails. Assume first that T_2 has a degree-one vertex M_2 such that $X_1 \subseteq E(M_2)$. Then, because $q_1 \notin X_1$, we deduce that $X_1 \subsetneq E(M_2) - q_1$. As $\lambda_M(X_1) = 1$, Lemmas 8.3.11 and 8.2.4 imply that $\lambda_{M_2}(X_1) \leq 1$. Since $|E(M_2) - X_1| \geq 2$, this contradicts the fact that M_2 is 3-connected. Hence we may assume that T_2 has no degree-one vertex M_2 for which $X_1 \subseteq E(M_2)$. By the last lemma, all the elements of X_1 lie on one side or the other of each 2-separation induced by an edge of T_2. Hence there is a vertex N_2 of T_2 of degree at least two such that $X_1 \subseteq E(N_2)$.

We show next that N_2 is 3-connected. If not, then N_2 is a circuit or a cocircuit. It follows that all the elements of X_1 are in series or in parallel in N_2 and hence that the same is true in M and also in M_1. But this contradicts the fact that the last matroid is 3-connected having at least four elements.

By Lemma 8.3.11, there are subsets U and V of $E(M) - X_1$ such that N_2 can be obtained from $M \backslash U/V$ by relabelling some of the elements of $E(M) - X_1$. Hence, by Lemma 8.2.4, $1 = \lambda_M(X_1) \geq \lambda_{M \backslash U/V}(X_1) = \lambda_{N_2}(X_1)$. Since N_2 has degree at least two, $|E(N_2) - X_1| \geq 2$ and so we have a contradiction to the fact that N_2 is 3-connected. We conclude that Lemma 8.3.14 holds. $\square$

By the last lemma, $M = M_1 \oplus_2 M_1'$ and $M = M_2 \oplus_2 M_2'$ where $E(M_1) - E(M_1') = X_1 = E(M_2) - E(M_2')$. Recall that p_1 is the common element of M_1 and M_1'. Let p_2 be the common element of M_2 and M_2'. If we rename p_2 by p_1, then, by Corollary 8.3.6, $M_1 = M_2$ and $M_1' = M_2'$. Let T_1' be obtained from T_1 by deleting the vertex M_1 of the latter, and let T_2' be obtained from T_2 by deleting M_2 recalling that we have relabelled the element p_2 of M_2' by p_1. Then T_1' and T_2' are both canonical tree decompositions of M_1'. Thus, by the induction assumption, T_1' and T_2' are equal to within edge relabelling. It follows that the same is true for T_1 and T_2 and this completes the proof of Theorem 8.3.10. $\square$

Let T be the canonical tree decomposition of a 2-connected matroid M. We know that every edge of T displays a 2-separation of M. The final result of the section, Proposition 8.3.16, shows that every 2-separation of M that is not displayed by an edge of T is, in a very natural sense, displayed by a special vertex of T. Let v be a vertex of T that is labelled by a circuit or cocircuit, N say. Let $G_1, G_2, \ldots, G_k$ be the components of $T - v$ and, for each i, let Z_i be the subset of $E(M)$ corresponding to G_i. We say that a partition (X, Y) of $E(M)$ into subsets with at least two elements is *displayed by the vertex v* if, for all i, either $Z_i \subseteq X$ or $Z_i \subseteq Y$. For example, in the canonical tree decomposition for $M(G)$ in Figure 8.6(g), let v be the vertex labelled by $M(G_3)$ and let a, b, and c be the edges of G_3 other than e_1 and e_2. Then two of the partitions displayed by v are $(\{a, b, c\}, E(G) - \{a, b, c\})$ and $((E(G_1) - e_1) \cup a, (E(G_2) - e_2) \cup \{b, c\})$.

Lemma 8.3.15 *If (X, Y) is a partition of $E(M)$ that is displayed by a vertex of T, then (X, Y) is a 2-separation of M.*

Proof By duality, it suffices to prove the result when N is a cocircuit. Let T' be the matroid-labelled tree obtained by contracting all the edges of T that do not meet v. If e_i is the edge of T joining v to G_i, let Q_i label the endpoint of e_i in T' that is different from v. Then $E(Q_i) = Z_i \cup e_i$ and M can be obtained from N by adjoining $Q_1, Q_2, \ldots, Q_k$, one by one, via 2-sums.

To see that (X, Y) is a 2-separation of M, we begin by defining a partition (X', Y') of $E(N)$. Let X' be $X \cap E(N)$ together with all elements e_i for which $Z_i \subseteq X$. Let $Y' = E(N) - X'$. Since N is a cocircuit and neither X' nor Y' is empty, $r(X') = r(Y') = r(N) = 1$, so $r(X') + r(Y') - r(N) = 1$. Now suppose, without loss of generality, that $e_1 \in X'$. Then

$$r((X' \cup E(Q_1)) - e_1) + r(Y') - r(N \oplus_2 Q_1) = r(Q_1) + 1 - (r(N) + r(Q_1) - 1)$$
$$= 2 - r(N) = 1.$$

By repeating this process, adjoining Q_2, then Q_3, and so on through Q_k, we eventually obtain that (X, Y) is a 2-separation of M as required. $\qquad\square$

The next result (Cunningham and Edmonds in Cunningham (1973)) establishes that every 2-separation of a 2-connected matroid is displayed by its canonical tree decomposition.

Proposition 8.3.16 *Let T be the canonical tree decomposition of a 2-connected matroid M. A partition (X, Y) of $E(M)$ into sets with at least two elements is a 2-separation of M if and only if (X, Y) is displayed by an edge of T or by a vertex of T that is labelled by a circuit or a cocircuit.*

Proof We have already shown that every partition that is displayed by an edge or by a vertex labelled by a circuit or cocircuit is a 2-separation. Now let (X, Y) be an arbitrary 2-separation of M. Following the proof of Proposition 8.3.9, we can obtain a tree decomposition T' for M in which every vertex label is 3-connected by first writing M as the 2-sum $N_1 \oplus_2 N_2$ where $E(N_1) = X \cup p$ and $E(N_2) = Y \cup p$ for some $p \notin E(M)$. If we continue this process to completion as described in that proof, the resulting tree decomposition T' has p as an edge and this edge displays (X, Y).

The tree decomposition T' need not be the canonical tree decomposition for M. The latter, T, is obtained from T' by contracting every edge that joins two vertices that are both labelled by circuits or are both labelled by cocircuits. If p is an edge of T, then it displays (X, Y). Using this and duality, we may assume that, in T', the edge p joins two vertices that are both labelled by cocircuits, say P_1 and P_2. Now, for every edge e of T' other than p, let $(Z_e, E(M) - Z_e)$ be the partition of $E(M)$ displayed by e where Z_e corresponds to the component of $T' \backslash e$ avoiding p. Because (X, Y) is displayed by p, it follows that, for each e, either $Z_e \subseteq X$ or $Z_e \subseteq Y$. Hence (X, Y) is displayed in T'/p by the vertex labelled by $P_1 \oplus_2 P_2$. If this vertex is joined by an edge q to another vertex labelled by a cocircuit, then the argument just given establishes that (X, Y) is displayed by a vertex of $T'/p, q$. By continuing this process of edge contraction until we eventually obtain the canonical tree decomposition T of M, we deduce that (X, Y) is displayed by a vertex of T. $\qquad\square$

In closing this section, we recall two attractive features of the canonical tree decomposition of a 2-connected matroid M. Specifically, it provides

(i) a description of all of the 2-separations of M; and

(ii) a unique decomposition of M into a collection of 3-connected matroids, circuits, and cocircuits from which M can be reconstructed.

Oxley, Semple, and Whittle (2004, 2007) considered the problem of describing all of the 3-separations of a 3-connected matroid M. In particular, they showed that one can associate a tree with M so that, up to a certain natural equivalence, every non-minimal 3-separation of M is displayed by an edge or a special vertex of the tree. This feature of the 3-separation tree is analogous to one of the two key features of Cunningham and Edmonds's tree decomposition. By contrast, however, the 3-separation tree does not, in general, encapsulate a way to uniquely reconstruct M. In Chapter 11, we shall look in more detail at the idea of sticking matroids together across subsets with more than one element and consider the difficulties that arise in doing this.

Exercises

1. Show that the matroid R_6, for which a geometric representation is shown in Figure 8.7, is not 3-connected.

2. Let $\mathbb{F}$ be an arbitrary field and M be a single-element extension of a 3-connected $\mathbb{F}$-representable matroid N. Prove that if M is not $\mathbb{F}$-representable, then M is 3-connected.

3. Prove that M_1 and M_2 are series minors and parallel minors of $M_1 \oplus_2 M_2$.

4. Find the canonical tree decompositions for the cycle matroids of each of the graphs in Figure 5.18.

5. Let G be a graph whose cycle matroid is 2-connected and non-empty. Prove that G is a series-parallel network if and only if every vertex in the canonical tree decomposition of $M(G)$ is labelled by a circuit or a cocircuit.

6. Prove Lemma 8.3.11.

7. Let M be a 2-connected matroid with $|E(M)| \geq 3$. For e in $E(M)$, prove that $M \backslash e$ is disconnected if and only if, in the canonical tree decomposition of M, the vertex label whose ground set contains e is a circuit.

8. Let M be a 2-connected matroid and T_1 and T_2 be tree decompositions for M in which every vertex label is 3-connected. Let M_1 label a vertex of T_1. Prove that

   (a) $|V(T_1)| = |V(T_2)|$;

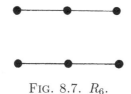

FIG. 8.7. R_6.

(b) if $|E(M_1)| \neq 3$, then T_2 has a vertex label M_2 that is isomorphic to M_1 such that $E(M_2) \cap E(M) = E(M_1) \cap E(M)$ and if M_1' is adjacent to M_1 in T_1, then T_2 has a vertex M_2' adjacent to M_2 such that $M_2' \cong M_1'$;

(c) if $|E(M_1)| = 3$ and $e \in E(M_1) \cap E(M)$, then T_2 has a vertex label M_2 such that $M_2 \cong M_1$ and $e \in E(M_2) \cap E(M)$.

8.4 Wheels and whirls

In this section, we introduce two fundamental families of 3-connected matroids and begin to discuss the significance of these families.

Table 4.1 listed all 2-connected matroids with at most four elements. Every such matroid with at most three elements has no 2-separations and so is 3-connected. Table 8.1 lists all 3-connected matroids with at most five elements. The reader is urged to check the completeness of this table. One false impression that the table may convey is that all 3-connected matroids are uniform. But we recall from Proposition 8.1.9 that, for every simple 3-connected graph G with at least four vertices, $M(G)$ is a 3-connected matroid, and none of these matroids is uniform. In fact, the table reflects the fact that that all 3-connected matroids having rank or corank at most two are uniform (Exercise 1).

Example 8.4.1 For $r \geq 2$, the r-spoked wheel graph $\mathcal{W}_r$ shown in Figure 8.8 was introduced in Example 5.1.7. Clearly $\mathcal{W}_3 \cong K_4$ and it is easy to check that, for $r \geq 3$, the graph $\mathcal{W}_r$ is 3-connected, so $M(\mathcal{W}_r)$ is a 3-connected matroid. By contrast, $M(\mathcal{W}_2)$ is not 3-connected since it has $(\{a_1, a_2\}, \{b_1, b_2\})$ as a 2-separation. Geometric representations for the matroids $M(\mathcal{W}_2)$, $M(\mathcal{W}_3)$, and $M(\mathcal{W}_4)$ are shown in Figure 8.9. It is commonplace to use the term *wheel* for both the graph $\mathcal{W}_r$ and its cycle matroid $M(\mathcal{W}_r)$. □

The family $\{M(\mathcal{W}_r) : r \geq 3\}$ is the first of two basic families of 3-connected matroids. The second basic family is related to the first by the operation of relaxation of a circuit–hyperplane, which was introduced in Section 1.5. We shall use the following result of Kahn (1985), the proof of which is straightforward.

TABLE 8.1. All 3-connected matroids with at most five elements.

Number n of elements	3-connected n-element matroids
0	$U_{0,0}$
1	$U_{0,1}, U_{1,1}$
2	$U_{1,2}$
3	$U_{1,3}, U_{2,3}$
4	$U_{2,4}$
5	$U_{2,5}, U_{3,5}$

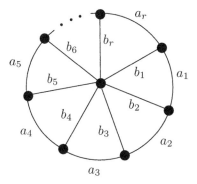

FIG. 8.8. The r-spoked wheel graph $\mathcal{W}_r$.

Proposition 8.4.2 *Let M' be a matroid that is obtained by relaxing a circuit–hyperplane of a matroid M. If M is n-connected, then so is M'.* □

Example 8.4.3 For $\mathcal{W}_r$ labelled as in Figure 8.8, the rim, $\{a_1, a_2, \ldots, a_r\}$, of the wheel is the unique circuit–hyperplane of $M(\mathcal{W}_r)$. We define $\mathcal{W}^r$, the *rank-r whirl*, to be the matroid obtained from $M(\mathcal{W}_r)$ by relaxing this circuit–hyperplane. Thus $E(\mathcal{W}^r) = E(\mathcal{W}_r)$, while the set of bases of $\mathcal{W}^r$ consists of the rim together with all edge sets of spanning trees of $\mathcal{W}_r$. By Proposition 1.5.14, the set of circuits of $\mathcal{W}^r$ consists of $\{C : C$ is the edge set of a cycle of $\mathcal{W}_r$ other than the rim$\} \cup \{\{a_1, a_2, \ldots, a_r\} \cup \{b_i\} : 1 \le i \le r\}$. In Figure 8.10, we give geometric representations for $\mathcal{W}^2$, $\mathcal{W}^3$, and $\mathcal{W}^4$. All three of these matroids have been considered earlier in the book (see, for example, 6.5.4 and Figures 2.1 and 6.10). Evidently $\mathcal{W}^2 \cong U_{2,4}$, so $\mathcal{W}^2$ is 3-connected. Moreover, for $r \ge 3$, by Proposition 8.4.2, since $M(\mathcal{W}_r)$ is 3-connected, so is $\mathcal{W}^r$. Hence, for all $r \ge 2$, the matroid $\mathcal{W}^r$ is 3-connected. □

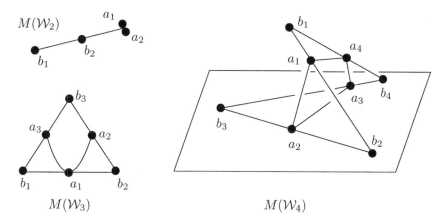

FIG. 8.9. Geometric representations for the three smallest wheels.

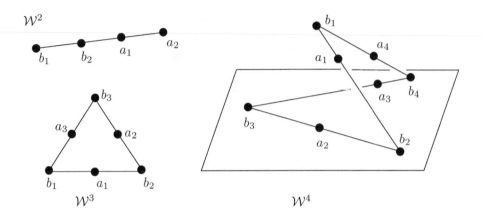

$\mathcal{W}^2$

$\mathcal{W}^3$

$\mathcal{W}^4$

FIG. 8.10. The three smallest whirls.

What distinguishes $\{M(\mathcal{W}_r) : r \geq 3\}$ and $\{\mathcal{W}^r : r \geq 2\}$ within the class of 3-connected matroids? One obvious property of these matroids is that every element occurs in both a *triangle*, a 3-element circuit, and a *triad*, a 3-element cocircuit. In fact, wheels and whirls are the only 3-connected matroids with this property. Indeed, a much stronger result is true. This result, Tutte's Wheels-and-Whirls Theorem (1966b) will be stated and proved in Section 8.8. The next proposition notes one of the many attractive properties of wheels and whirls.

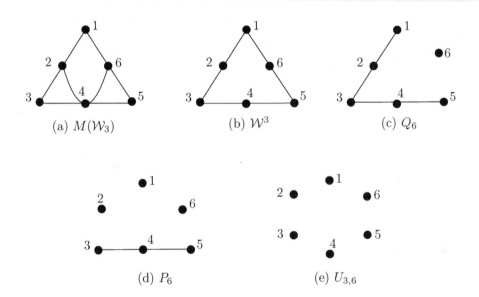

(a) $M(\mathcal{W}_3)$

(b) $\mathcal{W}^3$

(c) Q_6

(d) P_6

(e) $U_{3,6}$

FIG. 8.11. $M(\mathcal{W}_3)$ and four successive relaxations of it.

Proposition 8.4.4 *For all $r \geq 2$, both $M(\mathcal{W}_r)$ and $\mathcal{W}^r$ are self-dual.*

Proof The geometric dual of the planar embedding of $\mathcal{W}_r$ in Figure 8.8 is clearly isomorphic to $\mathcal{W}_r$ (Exercise 10 of Section 2.3). Hence $M^*(\mathcal{W}_r) \cong M(\mathcal{W}_r)$. Furthermore, one easily checks that $\mathcal{W}^r$ is self-dual since the map that interchanges a_i and b_i for all i is an isomorphism between $\mathcal{W}^r$ and its dual. □

The terms 'rim' and 'spoke' are used in the obvious way in whirls of rank at least three. Hence the spokes of a whirl are the elements that are in more than one triangle, while the rim consists of all elements that are not spokes. Although $\mathcal{W}^2$ has been excluded here because of its symmetry, it is sometimes convenient to view two of its elements as spokes and the other two as rim elements.

Example 8.4.5 Each of the last four matroids in Figure 8.11 can be obtained from its predecessor by relaxing a circuit–hyperplane. Since the first matroid in the list, $M(\mathcal{W}_3)$, is certainly 3-connected, all five of these matroids are 3-connected. Moreover, as noted in Section 2.1, all of these matroids are self-dual. We shall see in Chapter 12 that these five matroids are the only rank-3, 6-element, 3-connected matroids. The only other 6-element 3-connected matroids are $U_{2,6}$ and $U_{4,6}$. □

Exercises

1. Prove that all 3-connected matroids of rank at most two are uniform.
2. Prove Proposition 8.4.2.
3. (a) Let M be a self-dual matroid on a set E and M' be the matroid that is obtained by relaxing a circuit–hyperplane X of M. Prove that M' is self-dual if and only if there is an isomorphism ψ from M to M^* such that $\psi(X) = E - X$.
   (b) Find a relaxation of a self-dual matroid that is not self-dual.

8.5 Tutte's Linking Theorem and its applications

One of the most useful connectivity results for graphs is Menger's Theorem (1927). Tutte (1965b) proved a matroid generalization of the vertex form of this theorem but the significance of his result was largely ignored until recently. In this section, we state and prove what is now called Tutte's Linking Theorem and we show how it implies Menger's Theorem. The reader may be curious about matroid analogues of the edge form of Menger's Theorem. Such a result was proved by Seymour (1977b); we delay consideration of it until Chapter 12. This section is quite technical and the reader may prefer to focus on the highlights. In that case, the results that deserve particular attention are Theorem 8.5.7, which is an attractive extension of Tutte's Linking Theorem, and Lemma 8.5.3. The latter is the only result from this section that will be used again in this chapter, although Tutte's Linking Theorem will be applied in the next chapter and again in Chapter 13.

We begin by stating Menger's Theorem and showing how this result is implied by a certain property of the connectivity function. Tutte's Linking Theorem establishes that the connectivity function of every matroid has this property. First we introduce some more notation. For disjoint subsets X and Y of the ground set E of a matroid M, let

$$\kappa_M(X,Y) = \min\{\lambda_M(S) : X \subseteq S \subseteq E - Y\}.$$

If S is a set for which this minimum is attained where $X \subseteq S \subseteq E - Y$, then clearly

$$\kappa_M(X,Y) = \lambda_M(S) = \kappa_M(S, E - S). \tag{8.7}$$

In general, by (8.1.3) and Corollary 8.1.5, respectively, we have

$$\kappa_M(Y, X) = \kappa_M(X, Y) \tag{8.8}$$

and

$$\kappa_{M^*}(X, Y) = \kappa_M(X, Y). \tag{8.9}$$

If $\{P_1, P_2, \ldots, P_n\}$ is a set of paths in a graph G each of which joins the distinct vertices s and t, then these paths are *internally disjoint* if no vertex of $V(G) - \{s, t\}$ is in more than one of these paths. A subset V' of $V(G) - \{s, t\}$ *separates s from t* if s and t are in different components of $G - V'$.

Theorem 8.5.1 (Menger's Theorem) *Let s and t be distinct non-adjacent vertices in a graph G. Then the maximum number of internally disjoint (s, t)-paths equals the minimum number of vertices needed to separate s from t.*

Proof We may assume that G has at least one (s, t)-path otherwise the result is immediate. We may also assume that G is connected. Let S and T be the sets of edges incident with s and t, respectively.

Now suppose that the minimum number of vertices needed to separate s from t is k, and the maximum number of internally disjoint (s, t)-paths is h. Clearly,

$$k \geq h. \tag{8.10}$$

We shall show next that

$$\kappa_{M(G)}(S, T) \geq k - 1 \tag{8.11}$$

and that

$$h - 1 \geq \max\{\kappa_{M(H)}(S, T) : M(H) \text{ is a minor of } M(G) \text{ on } S \cup T\}. \tag{8.12}$$

To show (8.11), suppose that $\kappa_{M(G)}(S, T) = \kappa_{M(G)}(S'', T'')$ where (S'', T'') is a partition of $E(G)$ with $S \subseteq S''$ and $T \subseteq T''$. If $G[S'']$ has a component Z that does not contain S, then $V(Z)$ must meet $V(G[T''])$. Thus, by moving the edges of Z from S'' to T'', we decrease $r(S'')$ by $|V(Z)| - 1$, we increase $r(T'')$ by at most $|V(Z)| - 1$, but we do not increase the number of components of $G[T'']$. The minimality of $\kappa_{M(G)}(S'', T'')$ means that its value is unaltered. By repeating this procedure, we may assume that $G[S'']$ is connected. Then, applying the same procedure to $G[T'']$, we ensure that $G[T'']$ is connected and that $G[S'']$ remains connected. Thus, by (8.7) and Lemma 8.1.7,

$$\kappa_{M(G)}(S,T) = \kappa_{M(G)}(S'',T'') = |V(G[S'']) \cap V(G[T''])| - 1.$$

Since $V(G[S'']) \cap V(G[T''])$ separates s from t, we deduce that (8.11) holds.

To prove (8.12), let $M(H)$ be a minor of $M(G)$ with edge set $S \cup T$ such that $\kappa_{M(H)}(S,T)$ attains the maximum on the right-hand side of (8.12). Since the value of $\kappa_{M(H)}(S,T)$ is unaffected by isolated vertices in H, we may assume that H has none. Certainly $H[S]$ and $H[T]$ are connected. If H is disconnected, then, since $E(H) = S \cup T$, we have $\kappa_{M(H)}(S,T) = 0$ so $\kappa_{M(H)}(S,T) \leq h - 1$. To show that the last inequality also holds if H is connected, first note that, in that case, by Lemma 8.1.7, $\kappa_{M(H)}(S,T) = |V(H[S]) \cap V(H[T])| - 1$. If $|V(H[S]) \cap V(H[T])| > h$, then H, and hence G, has more than h internally disjoint (s,t)-paths. This contradiction completes the proof that (8.12) holds.

On combining (8.11), (8.10), and (8.12), we see that Menger's Theorem holds provided that $\kappa_{M(G)}(S,T)$ equals

$$\max\{\kappa_{M(H)}(S,T) : M(H) \text{ is a minor of } M(G) \text{ with edge set } S \cup T\}.$$

Hence Menger's Theorem follows immediately from the next theorem. □

Theorem 8.5.2 (Tutte's Linking Theorem) *Let X and Y be disjoint subsets of the ground set of a matroid M. Then $\kappa_M(X,Y)$ equals the maximum value of $\kappa_N(X,Y)$ over all minors N of M with $E(N) = X \cup Y$.*

As we shall see in Theorem 8.5.7, an even stronger result is true: we can find such a minor N with $\kappa_N(X,Y) = \kappa_M(X,Y)$ so that $N|X = M|X$ and $N|Y = M|Y$. Before proving Theorem 8.5.2, we note three useful lemmas. The first of these (Geelen, Gerards, and Whittle 2002) has several other significant applications, some of which will be considered in Section 8.7.

Lemma 8.5.3 *Let e be an element of a matroid M, and X and Y be subsets of $E(M) - e$. Then*

$$\lambda_{M\backslash e}(X) + \lambda_{M/e}(Y) \geq \lambda_M(X \cap Y) + \lambda_M(X \cup Y \cup e) - 1.$$

Proof The key steps in the proof involve applying the submodularity of the rank function. Writing E for $E(M)$, we have

$$\lambda_{M\backslash e}(X) + \lambda_{M/e}(Y)$$
$$= [r_M(X) + r_M(E - e - X) - r_M(E - e)]$$
$$\quad + [r_{M/e}(Y) + r_{M/e}(E - e - Y) - r_{M/e}(E - e)]$$
$$= r(X) + r(Y \cup e) + r(E - (X \cup e)) + r(E - Y) - r(E - e) - r(E) - r(e)$$
$$\geq [r(X) + r(Y \cup e)] + [r(E - (X \cup e)) + r(E - Y)] - 2r(E) - 1$$
$$\geq [r(X \cup Y \cup e) + r(X \cap Y)] + [r(E - (X \cap Y)) + r(E - (X \cup Y \cup e))]$$
$$\quad - 2r(E) - 1$$

Hence

$$\lambda_{M\backslash e}(X) + \lambda_{M/e}(Y) = [r(X \cap Y) + r(E - (X \cap Y)) - r(E)]$$
$$+ [r(X \cup Y \cup e) + r(E - (X \cup Y \cup e)) - r(E)] - 1$$
$$= \lambda_M(X \cap Y) + \lambda_M(X \cup Y \cup e) - 1. \qquad \sqcup$$

We leave the straightforward proof of the next lemma as an exercise.

Lemma 8.5.4 *Let X' and Y' be disjoint subsets of the ground set of a matroid M. If $X \subseteq X'$ and $Y \subseteq Y'$, then $\kappa_M(X,Y) \leq \kappa_M(X',Y')$.* □

Lemma 8.5.5 *Let N be a minor of a matroid M, and let X' and Y' be disjoint subsets of $E(M)$. If X and Y are subsets of $E(N)$ with $X \subseteq X'$ and $Y \subseteq Y'$, then*

$$\kappa_N(X,Y) \leq \kappa_M(X',Y').$$

Proof By the last lemma, $\kappa_M(X,Y) \leq \kappa_M(X',Y')$, so we need only show that $\kappa_N(X,Y) \leq \kappa_M(X,Y)$. Let (X_1,Y_1) be a partition of $E(M)$ with $X_1 \supseteq X$ and $Y_1 \supseteq Y$ such that $\kappa_M(X_1,Y_1) = \kappa_M(X,Y)$. Let $N = M\backslash D/C$. We show that

$$\kappa_N(X_1 - (D \cup C), Y_1 - (D \cup C)) \leq \kappa_M(X_1,Y_1). \qquad (8.13)$$

By (8.9), it suffices to show that if $d \in D$, then

$$\kappa_{M\backslash d}(X_1 - d, Y_1 - d) \leq \kappa_M(X_1,Y_1). \qquad (8.14)$$

Without loss of generality, $d \in Y_1$. Thus

$$\kappa_{M\backslash d}(X_1 - d, Y_1 - d) = \kappa_{M\backslash d}(X_1, Y_1 - d)$$
$$= \lambda_{M\backslash d}(X_1)$$
$$\leq \lambda_M(X_1) \text{ by Lemma 8.2.4(i);}$$
$$= \kappa_M(X_1,Y_1).$$

Therefore, (8.14) holds and so does (8.13). Hence, by Lemma 8.5.4,

$$\kappa_N(X,Y) \leq \kappa_N(X_1 - (D \cup C), Y_1 - (D \cup C)) \leq \kappa_M(X_1,Y_1) = \kappa_M(X,Y). \quad \square$$

Proof of Theorem 8.5.2 We shall establish the following assertion, which, by using the last lemma (8.5.5), is easily seen to be equivalent to the theorem.

8.5.6 $\kappa_M(X,Y) \geq k$ *if and only if M has a minor N with ground set $X \cup Y$ such that $\kappa_N(X,Y) \geq k$.*

Suppose that $\kappa_M(X,Y) \geq k$. We show, by induction on $|E - (X \cup Y)|$, that M has a minor N on $X \cup Y$ such that $\kappa_N(X,Y) \geq k$. This is immediate if $|E - (X \cup Y)| = 0$. Assume it holds for $|E - (X \cup Y)| < n$ and let $|E - (X \cup Y)| = n \geq 1$. Let $e \in E - (X \cup Y)$. If $\kappa_{M\backslash e}(X,Y) \geq k$ or $\kappa_{M/e}(X,Y) \geq k$, then the induction assumption implies that M has a minor N with ground set $X \cup Y$ such that $\kappa_N(X, Y) \geq k$. Thus there are sets S and T both containing X and avoiding $Y \cup e$ such that $\lambda_{M \setminus e}(S) \leq k - 1$ and $\lambda_{M/e}(T) \leq k - 1$. Then, by Lemma 8.5.3, either $\lambda_M(S \cap T)$ or $\lambda_M(S \cup T \cup e)$ is at most $k - 1$. Thus $\kappa_M(X, Y) < k$. This contradiction implies that the required minor N of M exists.

Conversely, assume that M has a minor N with ground set $X \cup Y$ such that $\kappa_N(X, Y) \geq k$. Then, by Lemma 8.5.5, $\kappa_M(X, Y) \geq k$. $\square$

Tutte's Linking Theorem has several alternative statements. When X and Y are disjoint subsets of the ground set of a matroid M, it follows immediately from the definitions that if $J \subseteq E(M) - (X \cup Y)$, then

$$\sqcap_{M/J}(X, Y) = \lambda_{(M/J)|(X \cup Y)}(X) = \kappa_{(M/J)|(X \cup Y)}(X, Y). \qquad (8.15)$$

Tutte's Linking Theorem asserts that there is such a set J for which

$$\sqcap_{M/J}(X, Y) = \kappa_M(X, Y). \qquad (8.16)$$

Geelen, Gerards, and Whittle (2007a) proved the following attractive generalization of Tutte's Theorem. Recall that sets S and T in M are skew if $\sqcap(S, T) = 0$ or, equivalently, if $M|(S \cup T) = M|S \oplus M|T$.

Theorem 8.5.7 *Let X and Y be disjoint sets of elements in a matroid M. Then M has a minor N with ground set $X \cup Y$ for which $\kappa_N(X, Y) = \kappa_M(X, Y)$ such that $N|X = M|X$ and $N|Y = M|Y$.*

Proof By Theorem 8.5.2, there is a subset J of $E(M) - (X \cup Y)$ such that $\sqcap_{M/J}(X, Y) = \kappa_M(X, Y)$. Choose J to be a minimal such set. If J contains an element e such that $r(J - e) = r(J)$, then $\sqcap_{M/(J-e)}(X, Y) = \sqcap_{M/J}(X, Y)$; a contradiction. Hence J is independent.

We shall show next that J is skew to both X and Y. Assume that J is not skew to X. Then J contains an element f such that $r(X \cup J) = r(X \cup (J - f))$. Thus, writing J' for $J - f$, we have $r_{M/J}(X) = r_{M/J'}(X) - 1$ and $r_{M/J}(X \cup Y) = r_{M/J'}(X \cup Y) - 1$, while $r_{M/J}(Y) = r_{(M/J')/f}(Y) \leq r_{M/J'}(Y)$. Hence

$$\begin{aligned}
\sqcap_{M/J'}(X, Y) &= r_{M/J'}(X) + r_{M/J'}(Y) - r_{M/J'}(X \cup Y) \\
&\geq [r_{M/J}(X) + 1] + r_{M/J}(Y) - [r_{M/J}(X \cup Y) + 1] \\
&= r_{M/J}(X) + r_{M/J}(Y) - r_{M/J}(X \cup Y) \\
&= \sqcap_{M/J}(X, Y) \\
&= \kappa_M(X, Y).
\end{aligned}$$

But, by (8.15) and Lemma 8.5.5,

$$\sqcap_{M/J'}(X, Y) = \kappa_{(M/J')|(X \cup Y)}(X, Y) \leq \kappa_M(X, Y).$$

Hence $\sqcap_{M/J'}(X, Y) = \kappa_M(X, Y)$ so J' contradicts the minimality of J. We conclude that J is skew to X. By symmetry, it is also skew to Y.

Let $K = E(M) - (X \cup Y \cup J)$ and $N = M/J\backslash K$. Then $\kappa_N(X,Y) = \sqcap_{M/J}(X,Y) = \kappa_M(X,Y)$. Moreover, as J and X are skew in M, we have

$$N|X = [M|(J \cup X)]/J = [M|J \oplus M|X]/J = M|X.$$

By symmetry, $N|Y = M|Y$. □

In the following list of exercises, all but the first four are results from Geelen, Gerards, and Whittle (2007a).

Exercises

1. For disjoint sets X and Y in a matroid M, prove that M has a minor N with ground set $X \cup Y$ for which $\kappa_N(X,Y) = \kappa_M(X,Y)$ such that $N.X = M.X$ and $N.Y = M.Y$.

2. (Seymour 1980b) Let X and Y be disjoint subsets of the ground set of a matroid M and suppose $z \in E(M) - (X \cup Y)$. Prove that if $\kappa_M(X,Y) = k$, then $\kappa_{M\backslash z}(X,Y) = k$ or $\kappa_{M/z}(X,Y) = k$.

3. Prove Lemma 8.5.4.

4. Let M be a 3-connected binary matroid having a pair of disjoint triangles, X and Y. Prove that M has an $M^*(K_{2,3})$-minor with ground set $X \cup Y$.

5. Let T be a set of elements of a matroid M and, for each subset X of $E(M) - T$, let $s(X) = \kappa_M(X,T)$. Prove that s is the rank function of a matroid on $E(M) - T$.

6. Let S and T be disjoint subsets of the ground set of a matroid M. Prove that S and T have subsets S_1 and T_1, respectively, such that $|S_1| = |T_1| = \kappa_M(S_1, T_1) = \kappa_M(S,T)$.

7. Let S and T be disjoint subsets of the ground set of a matroid M such that $|S| = |T| = \kappa_M(S,T)$. Prove that $E(M) - (S \cup T)$ has a subset J such that $J \cup S$ and $J \cup T$ are both bases of M.

8. Let S and T be disjoint subsets of the ground set of a matroid M such that $\kappa_M(S,T) \geq k$ and $S \cup T$ is independent. Prove that S and T have subsets S' and T', respectively, such that $\kappa_{M/[(S \cup T)-(S' \cup T')]}(S',T') = |S'| = |T'| = k$.

9. For a matroid M, let X and Y be disjoint subsets of $E(M)$ such that $\kappa_M(X,Y) \geq k$. Suppose $E(M) - (X \cup Y)$ is non-empty and, for every element e of this set, $\kappa_{M\backslash e}(X,Y) < k$ or $\kappa_{M/e}(X,Y) < k$. Prove that $\lambda_M(X) = k$ and $E(M) - (X \cup Y)$ has an ordering $(z_1, z_2, \ldots, z_m)$ such that $\lambda_M(X \cup \{z_1, z_2, \ldots, z_i\}) = k$ for all i in $\{1, 2, \ldots, m\}$.

8.6 Matroid versus graph connectivity

In this section, we indicate the precise relationship between matroid connectivity and graph connectivity. We also show how Tutte's definition of matroid connectivity can be modified to give a direct generalization of graph connectivity.

If G is a connected graph, then the addition of isolated vertices to G produces a disconnected graph with the same cycle matroid as G. For this reason, many of the results in this section will be stated for graphs that have no isolated vertices.

We saw in Section 8.1 how 2- and 3-connected graphs and matroids are related and Lemma 8.1.8 hinted at how the matroid definition of n-connectedness could be modified to agree with the graph definition. We now clarify this modification.

For a positive integer k, a partition (X, Y) of the ground set of a matroid M is a *vertical k-separation* of M if $\lambda_M(X) < k$ and $\min\{r(X), r(Y)\} \geq k$. We remark here that, in this context, the word 'vertical' is the adjective corresponding to the noun 'vertex'. If M has a vertical k-separation (X, Y), then

$$k + k - r(M) \leq r(X) + r(Y) - r(M) \leq k - 1,$$

so $k \leq r(M) - 1$ and $\max\{r(X), r(Y)\} \leq r(M) - 1$. Hence X and Y contain cocircuits of M. Conversely, if M has two disjoint cocircuits, then it is straightforward to check that M has a vertical k-separation for some positive integer $k \leq r(M) - 1$. Thus, provided a matroid has two disjoint cocircuits, we define its *vertical connectivity* $\kappa(M)$ of M to be the least positive integer j such that M has a vertical j-separation; otherwise we let $\kappa(M) = r(M)$. Hence $\kappa(M) = r(M)$ if and only if M has no two disjoint cocircuits. As an example,

$$\kappa(U_{r,n}) = \begin{cases} n - r + 1, & \text{if } n \leq 2r - 2; \\ r, & \text{otherwise.} \end{cases}$$

In general, a matroid M is called *vertically n-connected* if n is an integer for which $2 \leq n \leq \kappa(M)$.

The next theorem shows that vertical n-connectedness for matroids directly generalizes the notion of n-connectedness for graphs. This fact motivates the use of the notation $\kappa(M)$ for the vertical connectivity of M. This usage of $\kappa(M)$ is confined to this section. Our predominant usage of κ in this book is within expressions such as $\kappa_M(X, Y)$ as in the last section.

Theorem 8.6.1 *Let G be a connected graph. Then*

$$\kappa(M(G)) = \kappa(G).$$

Proof Suppose first that the simple graph associated with G is complete. Then $\kappa(G) = |V(G)| - 1$ and, since $M(G)$ has no two disjoint cocircuits, $\kappa(M(G)) = r(M(G)) = |V(G)| - 1$. Thus we may now assume that the simple graph associated with G is not complete. Let $\kappa(G) = n$. Then G has an n-element minimal vertex cut. Hence, by Lemma 8.1.8, $M(G)$ has a vertical n-separation and so $\kappa(M(G)) \leq n$. Thus $\kappa(M(G)) \leq \kappa(G)$.

It remains to establish the reverse inequality. For the rest of this proof, we shall refer to two vertices as being *connected* in a graph if they lie in the same component of the graph. Suppose that G is n-connected but that, for some $k < n$, there is a vertical k-separation (E_1, E_2) of $M(G)$. Let $G_1 = G[E_1]$ and $G_2 = G[E_2]$. Now assume that G has two vertices u and v that are not connected in either G_1 or G_2. Then u and v are not adjacent in G. As G is n-connected, Menger's Theorem (8.5.1) implies that there are n internally disjoint paths $P_1, P_2, \ldots, P_n$ in G joining u to v. By assumption, none of these paths lies entirely in E_1 or entirely in E_2. Let $Y_i = E_i \cap (\cup_{j=1}^n E(P_j))$ for each i in $\{1,2\}$. Then clearly $r(Y_i) = |Y_i|$ and $r(Y_1 \cup Y_2) = |Y_1 \cup Y_2| - n + 1$. Thus $\sqcap(Y_1, Y_2) = n - 1$. Since $E_1 \supseteq Y_1$ and $E_2 \supseteq Y_2$, Lemma 8.2.3 implies that $\sqcap(E_1, E_2) \geq n - 1 \geq k$. But $\sqcap(E_1, E_2) = \lambda_{M(G)}(E_1)$, so we have a contradiction to the fact that (E_1, E_2) is a vertical k-separation of $M(G)$. We conclude that every two vertices of G are connected in at least one of G_1 and G_2.

As shown before the theorem, $\max\{r(E_1), r(E_2)\} < r(M(G))$. Hence neither G_1 nor G_2 is a connected graph whose edge set is spanning in $M(G)$. Thus some component H_1 of G_1 has distinct vertices u and v that are not connected in G_2. If not, then every two vertices that are connected in G_1 are also connected in G_2, so G_2 is connected and E_2 is spanning in $M(G)$; a contradiction. Choose a vertex w of G that is not in $V(H_1)$. Then v and w are not connected in G_1, so must be connected in G_2. Likewise, u and w must be connected in G_2. Hence u and v are connected in G_2; a contradiction. We conclude that if $\kappa(G) \geq n$, then $\kappa(M(G)) \geq n$, so $\kappa(M(G)) \geq \kappa(G)$ and the theorem holds.   □

The notion of vertical connectivity for matroids was introduced independently by Inukai and Weinberg (1981), Cunningham (1981), and Oxley (1981c); all three papers prove results equivalent to Theorem 8.6.1. Our proof is Cunningham's. We follow Geelen, Gerards, and Whittle (2003) and call a matroid *round* if it has no two disjoint cocircuits. We remark that both Cunningham (1981) and Inukai and Weinberg (1981) take $\kappa(M)$ to be ∞ instead of $r(M)$ when M is round.

In the first part of the last proof, we noted that, for a connected graph G, if the associated simple graph is complete, then $M(G)$ is round. The converse of this is easily verified. Thus round matroids are an analogue of complete graphs. We omit the routine proof of the next result which gives several characterizations of round matroids. The reader can use this lemma to show that not only is every $PG(r-1, q)$ round, but so too is every matroid we get by deleting at most $q^{r-1} - q^{r-2} - 1$ elements from it. This follows from Figure 6.2 as the union of two hyperplanes of $PG(r-1, q)$ misses exactly $q^{r-1} - q^{r-2}$ elements of $PG(r-1, q)$.

Lemma 8.6.2 *The following statements are equivalent for a matroid M.*

(i) $\kappa(M) = r(M)$.

(ii) *M has no two disjoint cocircuits.*

(iii) *Every cocircuit of M is spanning.*

(iv) *M cannot be written as the union of two proper flats.*   □

The analogue of vertical connectivity for n-connectedness is not often used. We introduce it now to help make precise the relationship between graph and matroid connectivity. If a matroid M has a k-separation for some k, we define the *Tutte connectivity* $\tau(M)$ of M to be $\min\{j : M \text{ has a } j\text{-separation}\}$; otherwise we take $\tau(M)$ to be ∞. Hence, for example, $\tau(U_{2,4}) = \infty$. Clearly, for an integer n exceeding one, M is n-connected if and only if $\tau(M) \geq n$. Moreover, it is immediate from Corollary 8.1.5 that, for all matroids M,

$$\tau(M^*) = \tau(M). \tag{8.17}$$

Richardson (1973) and Inukai and Weinberg (1978) showed that the only matroids with infinite Tutte connectivity are uniform, proving the following result, which is not difficult to derive from Proposition 8.2.1.

Corollary 8.6.3 *A matroid with infinite Tutte connectivity is uniform, and*

$$\tau(U_{r,n}) = \begin{cases} r + 1, & \text{if } n \geq 2r + 2; \\ n - r + 1, & \text{if } n \leq 2r - 2; \\ \infty, & \text{otherwise.} \end{cases} \qquad \square$$

The precise relationship between Tutte connectivity and vertical connectivity involves a matroid generalization of a familiar graph notion. The *girth* $g(M)$ of a matroid M is the minimum circuit size of M unless M has no circuits, in which case, $g(M) = \infty$. Thus the *girth* $g(G)$ of a graph G is the girth $g(M(G))$ of its cycle matroid.

Theorem 8.6.4 *Let M be a matroid and suppose that M is not isomorphic to any uniform matroid $U_{r,n}$ with $n \geq 2r - 1$. Then*

$$\tau(M) = \min\{\kappa(M), g(M)\}.$$

Proof By Corollary 8.6.3, $\tau(M)$ is finite taking the value k, say. Thus M has a k-separation (X, Y). Hence $|E(M)| \geq 2k$ and so, by Proposition 8.2.1,

$$g(M) \geq k.$$

If $g(M) > k$, then, as $\min\{|X|, |Y|\} \geq k$, we have $\min\{r(X), r(Y)\} \geq k$. Hence (X, Y) is a vertical k-separation of M, so $\kappa(M) \leq k$. But

$$\kappa(M) = \min\{j : M \text{ is vertically } j\text{-separated}\}$$
$$\geq \min\{j : M \text{ is } j\text{-separated}\} = k = \tau(M). \tag{8.18}$$

We conclude that if $g(M) > k$, then the theorem holds.

We may now suppose that $g(M) = k$. If M has two disjoint cocircuits, then the theorem follows by (8.18). Thus we may assume that M does not have two disjoint cocircuits. Hence $\kappa(M) = r(M)$. Thus if $r(M) \geq g(M) = k$, then $\tau(M) = k = \min\{\kappa(M), g(M)\}$. Hence we may suppose that $r(M) < g(M) = k$. Therefore $g(M) = r(M) + 1$ and so $M \cong U_{r,n}$ for some non-negative integer n. As M does not have two disjoint cocircuits, $2(n - r + 1) > n$, that is, $n \geq 2r - 1$. This contradiction completes the proof of the theorem. $\qquad \square$

For the uniform matroids excluded in Theorem 8.6.4, the Tutte connectivity, τ, is either ∞ or $r + 1$, but the vertical connectivity, κ, is r. On combining that theorem with Theorem 8.6.1, it is not difficult to deduce the corresponding result for graphic matroids, a generalization of Corollary 8.1.6 and Proposition 8.1.9.

Corollary 8.6.5 *Let G be a connected graph having at least three vertices and suppose that $G \not\cong K_3$. Then $\tau(M(G)) = \min\{\kappa(G), g(G)\}$.*

Proof As G is connected and $|V(G)| \geq 3$, we have $r(M(G)) \geq 2$. Since $U_{2,4}$ is not graphic, $U_{r,n}$ is not graphic if both r and $n - r$ exceed one. It follows, since $G \not\cong K_3$, that $M(G)$ is not isomorphic to any $U_{r,n}$ with $n \geq 2r - 1$. The corollary now follows easily from Theorems 8.6.1 and 8.6.4.   □

In addition to examining both Tutte connectivity and vertical connectivity, Cunningham (1981) discussed the dual of vertical connectivity calling it *cyclic connectivity*. If k is a positive integer, a partition (X, Y) of the ground set $E(M)$ of a matroid M is a *cyclic k-separation* of M if $\lambda_M(X) < k$ and both X and Y contain circuits of M. Provided M has a pair of disjoint circuits, we define its *cyclic connectivity* $\kappa^*(M)$ to be the least positive integer j for which M has a cyclic j-separation; otherwise, we let $\kappa^*(M) = r^*(M)$. Thus, for all matroids M,

$$\kappa^*(M) = \kappa(M^*). \tag{8.19}$$

We call M *cyclically n-connected* if n is an integer such that $2 \leq n \leq \kappa^*(M)$. By using Lemma 8.6.2 and duality, we get several characterizations of when $\kappa^*(M) = r^*(M)$.

It should, perhaps, be emphasized here that, although we now have three different types of n-connectedness for matroids, when we refer to an 'n-connected' matroid, we mean one that is n-connected in Tutte's sense. The next result shows that Tutte n-connectedness is basically the conjunction of vertical n-connectedness and cyclic n-connectedness. The proof is left to the reader.

Proposition 8.6.6 *Suppose that M is a matroid for which $\tau(M)$ is finite and let n be a positive integer. Then M is n-connected if and only if M is both vertically and cyclically n-connected.*   □

The next proposition (Oxley and Wu 1995) generalizes the well-known result that if G and H are n-connected graphs having at least n common vertices, then $G \cup H$ is n-connected. Recall that $\sqcap(X, Y) = r(X) + r(Y) - r(X \cup Y)$.

Proposition 8.6.7 *Let n be an integer exceeding one and X and Y be subsets of the ground set of a matroid M. Suppose that $M|X$ and $M|Y$ are both vertically n-connected and that $\sqcap(X, Y) \geq n - 1$. Then $M|(X \cup Y)$ is vertically n-connected.*

Proof It suffices to show that M is vertically n-connected under the assumption that $E(M) = X \cup Y$. Assume that M has a vertical k-separation for some $k < n$. Then there is a partition $\{S, T\}$ of $E(M)$ such that $k \leq \min\{r(S), r(T)\}$ and

$$r(S) + r(T) - r(M) \leq k - 1 \leq n - 2. \tag{8.20}$$

By Lemma 8.2.3,

$$r(S) + r(T) - r(M) \geq r(S \cap X) + r(T \cap X) - r(X), \tag{8.21}$$

so $r(S \cap X) + r(T \cap X) - r(X) = j$ for some $j \leq n - 2$. As $M|X$ is vertically n-connected, it has no vertical $(j+1)$-separations, so $\min\{r(S \cap X), r(T \cap X)\} \leq j$.

Thus $\min\{r(S \cap X), r(T \cap X)\} \leq r(S \cap X) + r(T \cap X) - r(X)$ and the reverse inequality is immediate. Hence

$$\min\{r(S \cap X), r(T \cap X)\} = r(S \cap X) + r(T \cap X) - r(X),$$

so $S \cap X$ or $T \cap X$ spans X. By symmetry, $S \cap Y$ or $T \cap Y$ spans Y.

Suppose $S \cap X$ spans X, and $S \cap Y$ spans Y. Then S spans M so, by (8.20), $r(T) \leq k - 1$ which contradicts the fact that $r(T) \geq k$. Hence, by symmetry, we may assume that $S \cap X$ spans X, and $T \cap Y$ spans Y. Thus

$$\sqcap(X, Y) = r(X) + r(Y) - r(M) \leq r(S) + r(T) - r(M) \leq n - 2$$

and this contradiction completes the proof. □

By combining the last proposition with results from earlier in the section, it is straightforward to prove the following corollary. We omit the details.

Corollary 8.6.8 *Let n be an integer exceeding one and M be a matroid having no circuits with fewer than n elements. If $M|X$ and $M|Y$ are n-connected and $\sqcap(X, Y) \geq n - 1$, then $M|(X \cup Y)$ is n-connected.* □

One way to ensure that $\sqcap(X, Y) \geq n - 1$ is for the closures of X and Y to have at least $n - 1$ common elements. In particular, if X and Y are subsets of a loopless matroid M such that $M|X$ and $M|Y$ are connected and $X \cap Y$ is non-empty, then $M|(X \cup Y)$ is connected. The last observation is also easily deduced from the definition of connectedness in terms of circuits (Proposition 4.1.2).

To close this section, we use the links established here between graph and matroid connectivity, together with some results from Section 5.3, to prove two of Whitney's (1932b) results: a 3-connected simple planar graph has a unique planar embedding, and a 3-connected loopless planar graph has a unique abstract dual. In order to formulate the first of these results precisely, we shall need some more definitions. Let G be a connected planar graph. Two planar embeddings G_1 and G_2 of G are *equivalent* provided that every walk that is obtained by traversing the boundary of a face of G_1 can also be obtained by traversing the boundary of a face of G_2. Observe that, when G is 2-connected and loopless, all faces of G_1 and G_2 are bounded by cycles. It follows that, in this case, G_1 and G_2 are equivalent if and only if every set of edges that bounds a face of G_1 also bounds a face of G_2. A connected planar graph is *uniquely embeddable in*

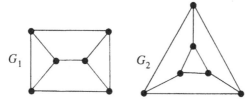

FIG. 8.12. Two equivalent plane graphs.

Higher connectivity

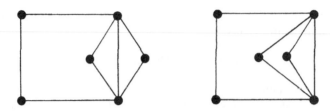

FIG. 8.13. Two isomorphic but inequivalent plane graphs.

the plane if any two planar embeddings of it are equivalent. Since stereographic projection allows us to transform a plane graph into a graph embedded on the surface of a sphere and vice versa, we shall also say that a connected planar graph is *uniquely embeddable on the sphere* if it is uniquely embeddable in the plane. The plane graphs G_1 and G_2 in Figure 8.12 are equivalent. Each can be obtained from the same graph on the sphere by stereographic projection: for G_1, the point P of projection is taken inside a face bounded by four edges; for G_2, we take P inside a face bounded by three edges. It is not difficult to see that the isomorphic plane graphs in Figure 8.13 are not equivalent. The essential difference between the pairs of graphs in these two diagrams is their connectivities.

Theorem 8.6.9 *Let G be a simple 3-connected planar graph. Then G is uniquely embeddable on the sphere.*

The proof will use the following. Recall from Section 5.2 that a graph G' is an abstract dual of a graph G if $M(G) \cong M^*(G')$. In particular, every geometric dual of G is an abstract dual of it.

Lemma 8.6.10 *Let G' be an abstract dual of a simple 3-connected planar graph G and suppose G' has no isolated vertices. Then G' is 3-connected and simple.*

Proof This is a straightforward combination of Proposition 8.1.9 and (8.17), and the details are left to the reader. □

Proof of Theorem 8.6.9 Let G_1 and G_2 be inequivalent planar embeddings of G, and let G_1^* and G_2^* be their geometric duals. Then, by the last lemma, both G_1^* and G_2^* are 3-connected. If we label the edges of G by the set E, then there is a natural labelling of the edge sets of all of G_1, G_2, G_1^*, and G_2^* by E. The identity map ι on E induces an isomorphism between $M(G_1)$ and $M(G_2)$ and so between $M^*(G_1)$ and $M^*(G_2)$. Thus, by Lemmas 5.2.1 and 5.3.2, ι induces an isomorphism between $M(G_1^*)$ and $M(G_2^*)$ and hence between G_1^* and G_2^*.

As G_1 and G_2 are inequivalent, we may assume that there is a set X of edges that bounds a face in G_1 but not in G_2. Then, since G_1^* and G_2^* are 3-connected, X is a vertex bond in G_1^* but not in G_2^*; a contradiction. □

Now, as promised in Chapter 5, we give a proof of Proposition 5.3.7, first restating the proposition for convenient reference.

Proposition 8.6.11 *Let G be a 3-connected loopless planar graph. If G_1 and G_2 are abstract duals of G, each having no isolated vertices, then $G_1 \cong G_2$.*

Proof We first observe that if G is simple, the result follows easily on combining Lemmas 8.6.10 and 5.3.2. Now suppose that G is non-simple and let X be a subset of $E(G)$ that is formed by choosing all but one edge from each parallel class of G. For each i in $\{1, 2\}$, there is a bijection ψ_i from $E(G)$ onto $E(G_i)$ that induces an isomorphism between $M^*(G)$ and $M(G_i)$. Evidently $G \backslash X$ is simple and has both $G_1/\psi_1(X)$ and $G_2/\psi_2(X)$ as abstract duals. Moreover, by Lemma 8.6.10, the last two graphs are 3-connected, so, by Lemma 5.3.2, the restriction of $\psi_2 \circ \psi_1^{-1}$ to $E(G_1) - \psi_1(X)$ induces an isomorphism between these two graphs.

For each i in $\{1, 2\}$, every edge of $\psi_i(X)$ is in series in G_i with some edge of $E(G_i) - \psi_i(X)$. Furthermore, as $G_i/\psi_i(X)$ is 3-connected, G_i is a subdivision of $G_i/\psi_i(X)$. Indeed, G_i can be constructed from $G_i/\psi_i(X)$ by replacing each edge e by a path whose length equals the size of the parallel class in G that contains $\psi_i^{-1}(e)$. It now follows easily that $\psi_2 \circ \psi_1^{-1}$ induces an isomorphism between G_1 and G_2. $\qquad\square$

Exercises

1. Find $\kappa^*(U_{r,n})$.
2. Give examples of a matroid M and a sequence $M_1, M_2, \ldots$ of matroids such that $\kappa^*(M) \neq \kappa(M)$ and $\kappa(M_n) - \tau(M_n) \geq n$ for all positive integers n.
3. (Seymour 1980b) Prove that the following are equivalent for a connected matroid M:
   (a) M is cyclically 3-connected.
   (b) M can be obtained from a 3-connected matroid by a sequence of series extensions.
   (c) For every partition (X, Y) of $E(M)$ with $r(X) + r(Y) \leq r(M) + 1$, either X or Y is contained in a series class of M.
4. (Inukai and Weinberg 1981) Prove that the following are equivalent for a matroid M:
   (a) M is round.
   (b) Either X or $E(M) - X$ is spanning for all X with $\emptyset \subsetneq X \subsetneq E(M)$.
   (c) Either $r(M.X) = 0$ or $r(M/X) = 0$ for all X with $\emptyset \subsetneq X \subsetneq E(M)$.
5. (Geelen, Gerards, and Whittle 2003) Prove the following for a matroid M:
   (a) If M is round and $e \in E(M)$, then M/e is round.
   (b) If $M|X$ is round and $r(X) = r(M)$, then M is round.
   (c) If M is round, then $si(M)$ is round.
   (d) If $N = M \backslash D/C$ where D is coindependent and N is round, then $si(M/C)$ is round.
6. (Hausmann and Korte 1981) If M is a matroid and $X \subseteq E$, denote the girth of $M|X$ by $g(X)$. This defines a function from 2^E into $\mathbb{Z}^+ \cup \{0, \infty\}$. Prove that an arbitrary such function is the girth function of a matroid on E if and only if g satisfies the following conditions:
   (G1) If $X \subseteq E$ and $g(X) < \infty$, then X has a subset Y such that $g(X) = g(Y) = |Y|$.

(G2) If $X \subseteq Y \subseteq E$, then $g(X) \geq g(Y)$.

(G3) If X and Y are distinct subsets of E with $g(X) = |X|$ and $g(Y) = |Y|$, and $e \in X \cap Y$, then $g((X \cup Y) - e) < \infty$.

7. (Cunningham 1981) Prove that if $M \backslash e$ is vertically n-connected and e is not a coloop of M, then M is vertically n-connected.

8. In the following classes of matroids, find all the round matroids:
   (a) cographic matroids;
   (b) *(Li Weixuan 1983) simple binary matroids.

8.7 Some extremal connectivity results

We saw in Section 8.5 how Lemma 8.5.3 was useful in proving Tutte's Linking Theorem (8.5.2). We begin this section with some more applications of this lemma. The section concludes with a proof of Tutte's Triangle Lemma, an extremal connectivity result that is in frequent use when one is dealing with 3-connected matroids. For sets X_1, X_2, Y_1, and Y_2, the pairs (X_1, X_2) and (Y_1, Y_2) *cross* if all four of the sets $X_1 \cap Y_1$, $X_1 \cap Y_2$, $X_2 \cap Y_1$, and $X_2 \cap Y_2$ are non-empty. The next result appears somewhat technical but the two subsequent results indicate how useful it is.

Lemma 8.7.1 *Let M be a $(k+1)$-connected matroid for some $k \geq 1$ and suppose $e \in E(M)$. Assume that (X_1, X_2) and (Y_1, Y_2) are k-separations of $M \backslash e$ and M/e, respectively. Then (X_1, X_2) and (Y_1, Y_2) cross. Moreover,*

$$|X_1 \cap Y_1| < k \ or \ |X_2 \cap Y_2| < k.$$

Proof Evidently, $\lambda_{M \backslash e}(X_1) \leq k-1$ and $\lambda_{M/e}(Y_1) \leq k-1$. Thus, by Lemma 8.5.3, $\lambda_M(X_1 \cap Y_1) + \lambda_M(X_1 \cup Y_1 \cup e) - 1 \leq 2(k-1)$. Hence, by (8.1.3),

$$\lambda_M(X_1 \cap Y_1) + \lambda_M(X_2 \cap Y_2) \leq 2(k-1) + 1.$$

Thus $\lambda_M(X_1 \cap Y_1) \leq k-1$ or $\lambda_M(X_2 \cap Y_2) \leq k-1$. Since M is $(k+1)$-connected, we deduce that

$$|X_1 \cap Y_1| < k \ or \ |X_2 \cap Y_2| < k. \tag{8.22}$$

By interchanging Y_1 and Y_2 above, we also get that $|X_1 \cap Y_2| < k$ or $|X_2 \cap Y_1| < k$.

If (X_1, X_2) and (Y_1, Y_2) do not cross, then, by symmetry, we may assume that $X_1 \cap Y_2 = \emptyset$. Then $X_1 \subseteq Y_1$, so $X_2 \supseteq Y_2$. Hence $|X_1 \cap Y_1| = |X_1| \geq k$ and $|X_2 \cap Y_2| = |Y_2| \geq k$. Thus (8.22) fails. We conclude that (X_1, X_2) and (Y_1, Y_2) do indeed cross. $\square$

We show next that the last lemma yields another proof of Theorem 4.3.1.

Corollary 8.7.2 *Let e be an element of a 2-connected matroid M. Then $M \backslash e$ or M/e is 2-connected.*

Proof Taking $k = 1$ in Lemma 8.7.1, we have that (X_1, X_2) and (Y_1, Y_2) cross yet $X_1 \cap Y_1$ or $X_2 \cap Y_2$ is empty. This contradiction implies that (X_1, X_2) or (Y_1, Y_2) does not exist, so the corollary holds. $\square$

As wheels and whirls exemplify, a 3-connected matroid M can have an element e for which neither $M \backslash e$ nor M/e is 3-connected. Bixby (1982) showed that, when this occurs, $M \backslash e$ or M/e is, in a natural sense, close to being 3-connected. His result, which is frequently called Bixby's Lemma, is stated and proved next. The statement of this result involves $\mathrm{si}(M/e)$, the simple matroid associated with M/e, and $\mathrm{co}(M \backslash e)$, the cosimple matroid associated with $M \backslash e$. In general, a matroid is *cosimple* if its dual is simple. For an arbitrary matroid N, its *cosimplification*, $\mathrm{co}(N)$, is $(\mathrm{si}(N^*))^*$. Thus $\mathrm{co}(N)$ is obtained from N by contracting all coloops and contracting all but one element from each non-trivial series class. Bixby's Lemma has been used extensively in the study of 3-connected matroids. Recall that a 2-separation (X, Y) is minimal if $|X|$ or $|Y|$ is 2.

Lemma 8.7.3 (Bixby's Lemma) *Let e be an element of a 3-connected matroid M. Then either $M \backslash e$ or M/e has no non-minimal 2-separations. Moreover, in the first case, $\mathrm{co}(M \backslash e)$ is 3-connected, while, in the second case, $\mathrm{si}(M/e)$ is 3-connected.*

Proof Take $k = 2$ in Lemma 8.7.1 and suppose that the 2-separations (X_1, X_2) and (Y_1, Y_2) of $M \backslash e$ and M/e are both non-minimal. Then

$$|X_1 \cap Y_1| \leq 1 \text{ or } |X_2 \cap Y_2| \leq 1$$

and, by symmetry,

$$|X_1 \cap Y_2| \leq 1 \text{ or } |X_2 \cap Y_1| \leq 1.$$

It follows that at least one of $|X_1|, |X_2|, |Y_1|$, or $|Y_2|$ is at most 2. This contradiction implies that $M \backslash e$ or M/e has no non-minimal 2-separations. We shall assume the latter; if the former holds, we dualize the argument that follows.

Assume that $|E(M)| \geq 4$ and consider $\mathrm{si}(M/e)$. By Proposition 8.2.8, M/e is 2-connected. Hence so is $\mathrm{si}(M/e)$. Assume that the last matroid is not 3-connected. Then it has a 2-separation (Z_1, Z_2). Each element of M/e that is not in $\mathrm{si}(M/e)$ is parallel to some element of Z_1 or Z_2. Thus M/e has a 2-separation (Z_1', Z_2') where $Z_1 \subseteq Z_1'$ and $Z_2 \subseteq Z_2'$. By assumption, (Z_1', Z_2') must be a minimal 2-separation of M/e so we may suppose that $|Z_1'| = 2$. Since $|Z_1| \geq 2$ and $Z_1' \supseteq Z_1$, it follows that $Z_1' = Z_1$. By Corollary 8.2.2, Z_1' is either a circuit or a cocircuit of M/e. The former cannot occur since $Z_1' \subseteq E(\mathrm{si}(M/e))$. Thus Z_1' is a 2-element cocircuit of M/e. But this implies that Z_1' is a 2-element cocircuit of M; a contradiction to Proposition 8.2.1.

Finally, if $|E(M)| \leq 3$, then M is one of the uniform matroids in Table 8.1 and one easily checks that both $\mathrm{si}(M/e)$ and $\mathrm{co}(M \backslash e)$ are 3-connected. $\square$

For $k \geq 2$, a k-connected matroid is called *internally $(k+1)$-connected* if it has no non-minimal k-separations. Thus, restating the first part of the last result, we have that, for every element e of a 3-connected matroid M, either $M \backslash e$ or M/e is internally 3-connected. In Chapter 12, we shall discuss some structural results for internally 4-connected matroids, that is, 3-connected matroids whose only 3-separations have a triangle or a triad as one side.

Lemma 8.7.1 was proved in the case $k = 2$ by Bixby (1982) on his way to proving Lemma 8.7.3. Coullard (1985) noted that Bixby's argument also holds for arbitrary k. The type of argument used in that proof originated in Tutte's (1966b) work on matroid connectivity and is used frequently in this area.

Triangles and triads feature prominently in any discussion of 3-connected matroids. One reason for this is contained in the next result (Tutte 1966b). We leave it to the reader to deduce this result from Bixby's Lemma.

Corollary 8.7.4 *If e is an element of a 3-connected matroid M and neither $M \backslash e$ nor M/e is 3-connected, then e is in a triangle or a triad of M.* □

The following is straightforward to check (see Exercise 7 of Section 8.3).

Lemma 8.7.5 *A 3-connected matroid M contains a set that is both a triangle and a triad if and only if $M \cong U_{2,4}$.* □

The next lemma (Tutte 1966b) has been used extensively in work on 3-connected matroids and played an important role in Tutte's proof of the Wheels-and-Whirls Theorem. Our proof, a modification of Tutte's original proof, is due to Coullard (private communication). It makes frequent use of the next result, which follows easily from Lemma 8.2.4(i) and the definition of λ.

Corollary 8.7.6 *Let X and D be disjoint subsets of the ground set E of a matroid M. Suppose that $r(M \backslash D) = r(M)$. Then*

(i) $\lambda_{M \backslash D}(X) = \lambda_M(X)$ *if and only if $D \subseteq \mathrm{cl}_M(E - (X \cup D))$; and*
(ii) $\lambda_{M \backslash D}(X) = \lambda_M(X \cup D)$ *if and only if $D \subseteq \mathrm{cl}_M(X)$.* □

Lemma 8.7.7 (Tutte's Triangle Lemma) *Let M be a 3-connected matroid having at least four elements and suppose that $\{e, f, g\}$ is a triangle of M such that neither $M \backslash e$ nor $M \backslash f$ is 3-connected. Then M has a triad that contains e and exactly one of f and g.*

Proof If $|E(M)| = 4$, then $M \cong U_{2,4}$. But every single-element deletion of this matroid is 3-connected. Hence we may assume that $|E(M)| \geq 5$. It suffices to prove that M has a triad T containing e, for then, by Proposition 2.1.11, T must meet $\{f, g\}$; and $T \neq \{e, f, g\}$ by Lemma 8.7.5.

Suppose first that $M \backslash e, f$ is disconnected having (X, Y) as a 1-separation where $g \in X$. By Proposition 8.2.8, $M \backslash e$ is connected, so $\lambda_{M \backslash e}(X \cup f) = 1$. As $e \in \mathrm{cl}(X \cup f)$, Corollary 8.7.6 implies that $\lambda_M(X \cup \{e, f\}) = 1$. As M is 3-connected, we deduce that $|Y| = 1$ so the one element of Y is a coloop of $M \backslash e, f$, and $Y \cup \{e, f\}$ is the required triad of M.

We may now assume that $M \backslash e, f$ is connected. Let (X_e, Y_e) and (X_f, Y_f) be 2-separations of $M \backslash e$ and $M \backslash f$, respectively. Suppose first that (X_e, Y_e) is a minimal 2-separation of $M \backslash e$. Then, by Corollary 8.2.2(ii), $M \backslash e$ has a 2-circuit or a 2-cocircuit. Since M itself has neither a 2-circuit nor a 2-cocircuit, it follows that M has a triad containing e, as required. Thus we may assume that $\min\{|X_e|, |Y_e|\} \geq 3$. We may also suppose, without loss of generality, that $e \in Y_f$

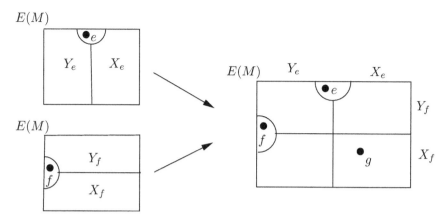

FIG. 8.14. Combining the 2-separations of $M \backslash e$ and $M \backslash f$.

and $f \in Y_e$. Then $g \notin Y_f$ otherwise $(X_f, Y_f \cup f)$ is a 2-separation of M. Hence $g \in X_f$ and, similarly, $g \in X_e$. Thus the situation is as shown in Figure 8.14. In fact, we shall establish below that $X_e \cap X_f = \{g\}$.

By Lemma 8.2.4, we have $1 = \lambda_{M \backslash e}(X_e) \geq \lambda_{M \backslash e, f}(X_e) = 1$. Therefore

$$\lambda_{M \backslash e}(X_e) = \lambda_{M \backslash e, f}(X_e) = 1. \tag{8.23}$$

Hence, by Corollary 8.7.6(i), $f \in \operatorname{cl}(Y_e - f)$. Similarly,

$$\lambda_{M \backslash f}(X_f) = \lambda_{M \backslash e, f}(X_f) = 1, \tag{8.24}$$

and so $e \in \operatorname{cl}(Y_f - e)$. Therefore, $\{e, f\} \subseteq \operatorname{cl}((Y_e - f) \cup (Y_f - e))$ so, by Corollary 8.7.6(i) again,

$$\lambda_{M \backslash e, f}(X_e \cap X_f) = \lambda_M(X_e \cap X_f). \tag{8.25}$$

Now $f \notin \operatorname{cl}(X_f)$ but $f \in \operatorname{cl}(Y_e - f)$, so $Y_e - f \not\subseteq X_f$. Hence $X_e \cup X_f \neq E(M \backslash e, f)$, so $\lambda_{M \backslash e, f}(X_e \cup X_f) \geq 1$. Using this, (8.23), (8.24), and the submodularity of λ, we get

$$1 \geq \lambda_{M \backslash e, f}(X_e) + \lambda_{M \backslash e, f}(X_f) - \lambda_{M \backslash e, f}(X_e \cup X_f) \geq \lambda_{M \backslash e, f}(X_e \cap X_f).$$

Thus, by (8.25), $\lambda_M(X_e \cap X_f) \leq 1$, so $|X_e \cap X_f| \leq 1$. But $g \in X_e \cap X_f$, so

$$X_e \cap X_f = \{g\}.$$

As $|X_e| \geq 3$ and $|X_f| \geq 2$, the last equation implies that

$$|X_e \cap Y_f| \geq 2 \quad \text{and} \quad |Y_e \cap X_f| \geq 1. \tag{8.26}$$

Now, by (8.24), $\lambda_{M \backslash e, f}(X_f) = 1$, and, by (8.23) and (8.1.3), $\lambda_{M \backslash e, f}(Y_e - f) = 1$. Thus, by the submodularity of λ, we have

$$\lambda_{M\backslash e,f}(X_f \cap Y_e) + \lambda_{M\backslash e,f}(X_f \cup (Y_e - f)) \leq 2.$$

Since $E(M\backslash e, f) - (X_f \cap Y_e) = X_e \cup (Y_f - e)$, we have

$$\lambda_{M\backslash e,f}(X_e \cup (Y_f - e)) + \lambda_{M\backslash e,f}(X_f \cup (Y_e - f)) \leq 2.$$

As $e \in \mathrm{cl}(Y_f - e)$ and $f \in \mathrm{cl}(Y_e - f)$, we deduce, by Corollary 8.7.6(ii), that

$$\lambda_{M\backslash f}(X_e \cup Y_f) + \lambda_{M\backslash e}(X_f \cup Y_e) \leq 2.$$

It follows, since $\{e, g\} \subseteq X_e \cup Y_f$ and $\{f, g\} \subseteq X_f \cup Y_e$, that

$$\lambda_M(X_e \cup Y_f \cup f) + \lambda_M(X_f \cup Y_e \cup e) \leq 2.$$

But $E(M) - (X_f \cup Y_e \cup e) = X_e \cap Y_f$. Thus, by (8.26), $\lambda_M(X_f \cup Y_e \cup e) \geq 2$, so $\lambda_M(X_e \cup Y_f \cup f) = 0$. Hence $|X_f \cap Y_e| = 0$, contradicting (8.26). □

Exercises

1. (Coullard 1985) Let e be an element of a 4-connected matroid M. Prove that, for some N in $\{M\backslash e, M/e\}$, every 3-separation (X, Y) of N has the property that $\min\{|X|, |Y|\} \leq 4$.

2. (Wong 1978) Let M be a 4-connected matroid with $|E(M)| \geq 8$. Prove that if M has an element e such that neither $M\backslash e$ nor M/e is 4-connected, then M has a circuit or a cocircuit having exactly four elements.

3. (Oxley 1981b) Let M be a *minimally 3-connected matroid*, that is, a 3-connected matroid for which no single-element deletion is 3-connected. For $|E(M)| \geq 4$ and e in $E(M)$, prove the following:

   (a) If M/e is not 3-connected, then e is in a triad.
   (b) If e is not in a triad, then M/e is minimally 3-connected.

4. (Lemos 1989) Let e and f be distinct elements of a matroid M such that $M\backslash e$ is not 3-connected. Suppose that M has triangles T and T' containing e and f, respectively, such that $|T \cap T'| = 1$ and $T \cup f$ is a cocircuit of M. Prove that e is in a triad of M.

5. (Oxley 1981a,b) Let M be a minimally 3-connected matroid with at least four elements. Prove that:

   (a) Every circuit of M meets at least two distinct triads.
   (b) M has at least $\frac{1}{2}r^*(M) + 1$ triads.

6. *(Oxley 1997) Prove that, in a 3-connected matroid M, for all pairs $\{a, b\}$ of distinct elements and for all cocircuits C^*, there is a circuit that contains $\{a, b\}$ and meets C^*. Deduce that if u, v, and w are vertices of a 3-connected graph G, then u, v, and w all lie on some common cycle of G (Dirac 1960).

8.8 Tutte's Wheels-and-Whirls Theorem

Wheels and whirls were discussed in Section 8.4. In this section, we shall prove a theorem of Tutte (1966b) that the wheels and whirls of rank at least three are the only 3-connected matroids for which no single-element deletion or contraction is 3-connected. An idea that is useful both in the proof of this theorem and, more generally, in dealing with the structure of 3-connected matroids is that of a fan, which, as Proposition 8.8.1 shows, can be thought of as a partial wheel.

In a simple, cosimple matroid M, a subset S of $E(M)$ having at least three elements is a *fan* in M if there is an ordering $(s_1, s_2, \ldots, s_n)$ of the elements of S such that, for all i in $\{1, 2, \ldots, n-2\}$,

(i) $\{s_i, s_{i+1}, s_{i+2}\}$ is a triangle or a triad; and

(ii) when $\{s_i, s_{i+1}, s_{i+2}\}$ is a triangle, $\{s_{i+1}, s_{i+2}, s_{i+3}\}$ is a triad; and, when $\{s_i, s_{i+1}, s_{i+2}\}$ is a triad, $\{s_{i+1}, s_{i+2}, s_{i+3}\}$ is a triangle.

The ordering $(s_1, s_2, \ldots, s_n)$ is called a *fan ordering* of S although we shall sometimes say simply that $(s_1, s_2, \ldots, s_n)$ is a fan. A 3-connected matroid M with at least four elements is certainly simple and cosimple. Clearly every triangle or triad in M is a fan. Moreover, for $|E(M)| \geq 5$, if T is a triangle and T^* is a triad, then $|T \cap T^*| \in \{0, 2\}$. For the graph G in Figure 8.15(a), both $(7, 6, 4, 1, 2, 3, 5, 9)$ and $(9, 12, 11, 10, 8, 7)$ are fans in $M(G)$. The reader can easily check, by considering the triangles and triads that meet these fans, that each is maximal. Clearly we can obtain other fans by removing elements from the ends of these fan orderings provided that we retain at least three elements from the original fans. Figure 8.15(b) shows a geometric representation for a rank-4 non-graphic matroid M, and (a, b, c, d, e) is a fan in M.

The occurrence of a fan S in a matroid M tells us nothing about whether or not M itself is graphic, binary, or even representable. But, by combining the next result (Oxley and Wu 2000a) with Proposition 8.8.7, we get that $M|S$ is regular unless M is a whirl with ground set S, in which case, $M|S$ is representable over all fields except $GF(2)$.

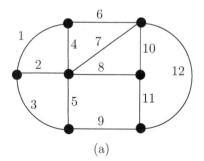
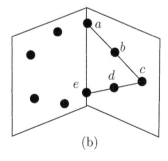

FIG. 8.15. Fans in (a) a cycle matroid; and (b) a geometric representation.

Proposition 8.8.1 *Let S be a fan in a 3-connected matroid M. Then either there is a subgraph G of a wheel such that $M|S \cong M(G)$, or $S = E(M)$ and M is a whirl.*   □

Next we show how Tutte's Triangle Lemma (8.7.7) enables us to build fans.

Lemma 8.8.2 *Let M be a 3-connected matroid with at least four elements and X be a rank-2 subset of $E(M)$ having at least three elements. Then exactly one of the following holds.*

(i)   *$M\backslash x$ is 3-connected for all x in X;*

(ii) *$|X| = 3$ and there are exactly two elements x of X for which $M\backslash x$ is 3-connected;*

(iii) *$|X| = 3$, there is a unique element x of X for which $M\backslash x$ is 3-connected, and X is contained in a 4-element fan;*

(iv) *$|X| = 3$, there is no element x of X for which $M\backslash x$ is 3-connected, and X is contained in a 5-element fan.*

Proof Suppose that neither (i) nor (ii) holds and let x_1 be an element of X such that $M\backslash x_1$ is not 3-connected. By Proposition 8.2.8, $M\backslash x_1$ is connected. Thus $M\backslash x_1$ has a 2-separation (U, V). Now $|U \cap X| \leq 1$ otherwise $x_1 \in \mathrm{cl}(U)$ and $(U \cup x_1, V)$ is a 2-separation of M. Likewise $|V \cap X| \leq 1$, so $|X| = 3$. Moreover, as (ii) does not hold, $X - x_1$ contains an element x_2 such that $M\backslash x_2$ is not 3-connected. Then, by Tutte's Triangle Lemma, M has a triad T_1^* that contains x_1 and exactly one other element of X. It follows that $X \cup T_1^*$ is a 4-element fan of M. Thus either (iii) holds, or $M\backslash x_3$ is not 3-connected where $X = \{x_1, x_2, x_3\}$. Without loss of generality, we may assume that x_2 is not in T_1^*. By Tutte's Triangle Lemma again, M has a triad T_2^* containing x_2 and exactly one other element of X. If $T_1^* - X = T_2^* - X$, then, by cocircuit elimination, we get that X contains and therefore is a cocircuit of M. Hence, by Lemma 8.7.5, $M \cong U_{2,4}$ contradicting the fact that $M\backslash x_1$ is not 3-connected. Thus $T_1^* - X \neq T_2^* - X$ and it follows easily $T_1^* \cup X \cup T_2^*$ is a 5-element fan.   □

Let M be a 3-connected matroid. Following Tutte (1966b), we call an element e of M *essential* if neither $M\backslash e$ nor M/e is 3-connected. For example, if M is the cycle matroid of the graph in Figure 8.15(a), then every element is essential except 7 and 9. Although neither $M/7$ nor $M\backslash 9$ is 3-connected, both $M\backslash 7$ and $M/9$ are 3-connected. On the other hand, the only essential elements in the matroid in Figure 8.15(b) are b, c and d. By Corollary 8.7.4, every essential element of a 3-connected matroid M is in a triangle or a triad and is therefore in a fan. If M is a wheel or whirl with at least six elements and e is an element of M, then, because e is in both a triad and a triangle, $M\backslash e$ and M/e have a 2-element cocircuit and a 2-element circuit, respectively. Therefore, by Proposition 8.2.1, neither $M\backslash e$ nor M/e is 3-connected. Thus the wheels and whirls of rank at least three have every element essential. Tutte (1966b) considered whether any other 3-connected matroids have this property having earlier (Tutte 1961a) solved the corresponding problem for graphic matroids by proving the following.

Theorem 8.8.3 *A 3-connected simple graph G has the property that, for all edges e, neither $G\backslash e$ nor G/e is both simple and 3-connected if and only if G is a wheel with at least four vertices.*

By Proposition 8.1.9, this graph result is an immediate consequence of the next theorem, a matroid generalization of it (Tutte 1966b). The full significance of this fundamental result will become clearer in Chapter 12.

Theorem 8.8.4 (Tutte's Wheels-and-Whirls Theorem) *The following are equivalent for a 3-connected matroid M having at least one element.*

(i) *For every element e of M, neither $M\backslash e$ nor M/e is 3-connected.*

(ii) *M has rank at least three and is isomorphic to a wheel or a whirl.*

The proof of this theorem uses the following lemma. Part (ii) of the lemma is from Seymour (1980b) and is also implicit in Tutte's proof; part (i) is from Oxley and Wu (2000a).

Lemma 8.8.5 *For some $n \geq 2$, let $\{x_1, x_2, \ldots, x_n\}$ and $\{y_1, y_2, \ldots, y_n\}$ be disjoint subsets X and Y of the ground set of a connected matroid M. Suppose that $\{x_i, y_i, x_{i+1}\}$ is a triangle for all i in $\{1, 2, \ldots, n-1\}$ and that $\{y_j, x_{j+1}, y_{j+1}\}$ is a triad for all j in $\{1, 2, \ldots, n\}$, where all subscripts are interpreted modulo n.*

(i) *If M is 3-connected, then $\{x_n, y_n, x_1\}$ is a triangle.*

(ii) *If $\{x_n, y_n, x_1\}$ is a triangle, then $E(M) = X \cup Y$ and M is isomorphic to $M(\mathcal{W}_n)$ or $\mathcal{W}^n$ having X as its set of spokes and Y as its rim.*

Proof Clearly $X \subseteq \mathrm{cl}^*(Y)$ and $Y - y_n \subseteq \mathrm{cl}(X)$. Thus

$$r^*(X \cup (Y - y_n)) \leq r^*(X \cup Y) = r^*(Y) \leq n \qquad (8.27)$$

and

$$r(X \cup (Y - y_n)) = r(X) \leq n. \qquad (8.28)$$

On adding (8.27) and (8.28) and using the fact that $|X \cup (Y - y_n)| = 2n - 1$, we get that $\lambda(X \cup (Y - y_n)) \leq 1$, that is,

$$r(X \cup (Y - y_n)) + r(E(M) - (X \cup (Y - y_n))) - r(M) \leq 1. \qquad (8.29)$$

If M is 3-connected, then we deduce from (8.29) that $|E(M) - (X \cup (Y - y_n))| = 1$, so $X \cup Y = E(M)$ and equality holds in (8.29) and hence in (8.27) and (8.28). In particular, $r^*(M) = n$. By Proposition 2.1.9,

$$\begin{aligned} r(\{x_n, y_n, x_1\}) &= r^*(E - \{x_n, y_n, x_1\}) + |\{x_n, y_n, x_1\}| - r^*(M) \\ &\leq r^*(\{y_1, y_2, \ldots, y_{n-1}\}) + 3 - n \\ &\leq n - 1 + 3 - n = 2. \end{aligned}$$

Thus $\{x_n, y_n, x_1\}$ is a triangle, and (i) holds.

Now, to prove (ii), assume that $\{x_n, y_n, x_1\}$ is a triangle. Then $y_n \in \mathrm{cl}(X)$, so (8.28) can be replaced by

$$r(X \cup Y) = r(X) \leq n \qquad (8.30)$$

Adding (8.27) and (8.30), we get $\lambda(X \cup Y) \leq 0$. Hence $\lambda(X \cup Y) = 0$ and equality holds in (8.30), so $r(M) = n$.

Next we show that M is a wheel or a whirl having X as its set of spokes and Y as its rim. To this end, let $\mathcal{D}$ be the set of subsets D of $X \cup Y$ for which $|D \cap X| = 2$ such that if $D \cap X = \{x_j, x_k\}$, then $D \cap Y$ is $\{y_j, y_{j+1}, \ldots, y_{k-1}\}$ or $\{y_k, y_{k+1}, \ldots, y_{j-1}\}$ where, again, subscripts are to be read modulo n.

As a first step towards finding $\mathcal{C}(M)$, we now show that every circuit C of M contains a member of $\mathcal{D} \cup \{Y\}$. By orthogonality, for all i in $\{1, 2, \ldots, n\}$,

$$|C \cap \{y_i, x_{i+1}, y_{i+1}\}| \neq 1. \qquad (8.31)$$

If $C \cap X = \emptyset$, then C contains y_j for some j. By repeatedly using (8.31), we obtain that $C \supseteq Y$. We now suppose that $C \cap X \neq \emptyset$ and that $C \not\supseteq Y$. Let x_k be an element of C. Then, by (8.31), y_{k-1} or y_k is in C. By symmetry, we may assume the latter. As $C \not\supseteq Y$, there is a non-negative integer m such that $\{y_k, y_{k+1}, \ldots, y_{k+m}\} \subseteq C$ but $y_{k+m+1} \notin C$. Then, by (8.31) again, $x_{k+m+1} \in C$. Hence C contains $\{x_k, y_k, y_{k+1}, \ldots, y_{k+m}, x_{k+m+1}\}$, a member of $\mathcal{D}$, and so every member of $\mathcal{C}(M)$ does indeed contain a member of $\mathcal{D} \cup \{Y\}$.

Next we show that every member of $\mathcal{D}$ is a circuit of M. To do this, it suffices, by symmetry, to show that $\{x_1, y_1, y_2, \ldots, y_t, x_{t+1}\}$ is a circuit for all t in $\{1, 2, \ldots, n-1\}$. We shall do this by induction on t. By assumption, $\{x_1, y_1, x_2\}$ is a circuit of M. Assume that $\{x_1, y_1, y_2, \ldots, y_t, x_{t+1}\}$ is a circuit for some $t < n-1$. Then, as $\{x_{t+1}, y_{t+1}, x_{t+2}\}$ is also a circuit, elimination implies that $\{x_1, y_1, y_2, \ldots, y_{t+1}, x_{t+2}\}$ contains a circuit C'. From above, C' must contain a member of $\mathcal{D} \cup \{Y\}$. Therefore $C' = \{x_1, y_1, y_2, \ldots, y_{t+1}, x_{t+2}\}$ and we conclude, by induction, that every member of $\mathcal{D}$ is a circuit of M.

If Y is a circuit of M, it follows that $\mathcal{C}(M) = \mathcal{D} \cup \{Y\}$. Then $M \cong M(\mathcal{W}_n)$ with X being the set of spokes. If Y is not a circuit of M, then Y is an independent set with $|Y| = r(M)$, so Y is a basis of M. Moreover, for all x_i in X and all proper subsets Y' of Y, the set $Y' \cup x_i$ contains no member of $\mathcal{D}$. Hence $Y \cup x_i \in \mathcal{C}(M)$ and so $\mathcal{C}(M) = \mathcal{D} \cup \{Y \cup x_i : 1 \leq i \leq n\}$. In this case, $M \cong \mathcal{W}^n$. $\qquad\square$

The Wheels-and-Whirls Theorem will follow easily from the next result (Oxley and Wu 2000a).

Lemma 8.8.6 *Let M be a 3-connected matroid with at least four elements. Let S be a maximal fan in M. Then exactly one of the following holds:*

(i) *M is a wheel or a whirl;*

(ii) *S is a triangle or triad that contains at least two non-essential elements;*

(iii) *$|S| > 3$ and S contains exactly two non-essential elements. Moreover, these two elements are the first and last elements in every fan ordering of S.*

When (iii) occurs, the two non-essential elements of S are the fan *ends*. In Figure 8.15(a), 7 and 9 are the ends of both the maximal fans $(7, 6, 4, 1, 2, 3, 5, 9)$ and $(9, 12, 11, 10, 8, 7)$, and both $M \backslash 7$ and $M/9$ are 3-connected.

Proof of Lemma 8.8.6 If $|E(M)| = 4$, then $M \cong U_{2,4} \cong \mathcal{W}^2$. Thus we may assume that $|E(M)| \geq 5$. Let $|S| = n$ and suppose first that $n = 3$. Then S is a triangle or a triad. If S contains at least two essential elements, then, by Tutte's Triangle Lemma (8.7.7) or its dual, S meets a triad or a triangle of M in exactly two elements, so S is not a maximal fan. Hence if $n = 3$, then S contains at least two non-essential elements. We may now assume that $n \geq 4$. Let $(s_1, s_2, \ldots, s_n)$ be a fan ordering of S. Each of $s_2, s_3, \ldots, s_{n-1}$ is in both a triangle and a triad, so each is essential. By replacing M by M^* if necessary, we may assume that $\{s_1, s_2, s_3\}$ is a triangle.

Suppose that s_1 is essential. Then, by Tutte's Triangle Lemma, M has a triad T^* that contains s_1 and exactly one of s_2 and s_3. Let t be the element of T^* that is not in $\{s_1, s_2, s_3\}$. Suppose first that $s_2 \in T^*$. If $t \notin \{s_3, s_4, \ldots, s_n\}$, then $S \cup t$ is a fan having $(t, s_1, s_2, \ldots, s_n)$ as a fan ordering. This contradicts the maximality of S. Thus $t \in \{s_3, s_4, \ldots, s_n\}$. Let m be the greatest odd number not exceeding n, so $m \in \{n-1, n\}$. By orthogonality, t is in none of the triangles $\{s_3, s_4, s_5\}, \{s_5, s_6, s_7\}, \ldots, \{s_{m-2}, s_{m-1}, s_m\}$. Hence n is even and $t = s_n$. Thus, by Lemma 8.8.5, M is a wheel or a whirl.

We may now assume that $s_2 \notin T^*$, so $s_3 \in T^*$. Then either $n = 4$; or $n \geq 5$ and, by orthogonality, $t \in \{s_4, s_5\}$. In the first case, we interchange the labels on s_2 and s_3 and thereby return to the case when $s_2 \in T^*$. Thus we may assume that $n \geq 5$ and t is s_4 or s_5. If $t = s_4$, then $\{s_1, s_3, s_4\}$ and $\{s_2, s_3, s_4\}$ are both triads. By circuit elimination, $\{s_1, s_2, s_3\}$ is a triad, which, by Lemma 8.7.5, contradicts the fact that this set is a triangle. Thus $t = s_5$. Now let $X = \{s_1, s_2, s_3, s_4, s_5\}$. Then X contains at least two triangles and at least two triads, so $r(X) + r^*(X) - |X| \leq 1$. Hence $|E(M) - X| \leq 1$ and so $|E(M)| \leq 6$. Now $r(M) \geq 3$, otherwise M has no triads. By duality, $r^*(M) \geq 3$, so $|E(M)| = 6$ and $r(M) = 3$. Letting $E(M) - X = \{s_6\}$, we deduce, since $r(\{s_1, s_2, s_3\}) = 2$, that $\{s_4, s_5, s_6\}$ is a triad of M. Likewise, as $\{s_1, s_3, s_5\}$ and $\{s_2, s_3, s_4\}$ are triads, $\{s_2, s_4, s_6\}$ and $\{s_1, s_5, s_6\}$ are triangles. It follows by the dual of Lemma 8.8.5 that M is isomorphic to a rank-3 wheel or whirl.

We conclude that if s_1 is essential, then M is a wheel or a whirl. By symmetry, M is a wheel or a whirl if s_n is essential. Hence we may assume that both s_1 and s_n are non-essential and it follows that (iii) holds. $\square$

Proof of Theorem 8.8.4 We noted just before Theorem 8.8.3 that (ii) implies (i). Now assume that every element of the 3-connected matroid M is essential. Then, by Lemma 8.8.6, M is a wheel or a whirl. If $M \cong \mathcal{W}^2$, then no element of M is essential. Hence $r(M) \geq 3$ and (i) holds. $\square$

In Example 5.1.7, we gave an explicit representation for a wheel that is valid over all fields. We show next how Lemma 8.8.5 can be used with a natural modification of that representation to show that every whirl is representable

$$
\begin{array}{c}
\begin{array}{ccccccc} y_1 & y_2 & y_3 & y_4 & \cdots & y_{r-1} & y_r \end{array}\\
\begin{array}{c} x_1 \\ x_2 \\ x_3 \\ x_4 \\ \\ x_{r-1} \\ x_r \end{array}
\left[\begin{array}{ccccccc}
1 & 0 & 0 & 0 & \cdots & 0 & \alpha \\
1 & 1 & 0 & 0 & \cdots & 0 & 0 \\
0 & 1 & 1 & 0 & \cdots & 0 & 0 \\
0 & 0 & 1 & 1 & \cdots & 0 & 0 \\
\vdots & \vdots & \vdots & \vdots & \ddots & \vdots & \vdots \\
0 & 0 & 0 & 0 & \cdots & 1 & 0 \\
0 & 0 & 0 & 0 & \cdots & 1 & 1
\end{array}\right]
\end{array}
$$

Fig. 8.16. D_α.

over all fields with more than two elements. Since every whirl has $\mathcal{W}^2$, that is $U_{2,4}$, as a minor, it is non-binary.

Proposition 8.8.7 *For all $r \geq 2$ and all fields $\mathbb{F}$, if D_α is the matrix in Figure 8.16, then $[I_r|D_\alpha]$ is an $\mathbb{F}$-representation for $M(\mathcal{W}_r)$ if $\alpha = (-1)^r$ and for $\mathcal{W}^r$ if $\alpha \notin \{0, (-1)^r\}$.*

Proof Fix a field $\mathbb{F}$ and let M_α be the matroid that is represented by $[I_r|D_\alpha]$ over $\mathbb{F}$. Then M_α^* is represented by $[-D_\alpha^T|I_r]$. Evidently $\{x_i, y_i, x_{i+1}\}$ is a triangle and $\{y_i, x_{i+1}, y_{i+1}\}$ is a triad of M_α for all i in $\{1, 2, \ldots, n\}$. Thus, by Lemma 8.8.5, M_α is a wheel or whirl having $\{y_1, y_2, \ldots, y_r\}$ as its rim. By expanding along the first row of D_α, we get that $\det D_\alpha = 1 + (-1)^{r+1}\alpha$. Hence $\{y_1, y_2, \ldots, y_r\}$ is a circuit of M_α if and only if $\alpha = (-1)^r$. The proposition follows immediately. □

In a 3-connected matroid, even though an essential element can be in more than one maximal fan, when this occurs, the essential elements of these maximal fans coincide. More precisely, Oxley and Wu (2000a) proved the following result. The exercises contain several results that can be combined to give a proof.

Theorem 8.8.8 *Let M be a 3-connected matroid. Let e be an essential element of M. Then e is in a maximal fan S in M. Moreover, S is unique unless*

(i) *every maximal fan containing e consists of a triangle and every two such triangles meet in $\{e\}$;*

(ii) *every maximal fan containing e consists of a triad and every two such triads meet in $\{e\}$; or*

(iii) *e is in exactly three maximal fans each of which has exactly five elements, the union X of these three fans has exactly six elements, and the restriction or contraction of M to X is isomorphic to $M(K_4)$.* □

By Lemma 8.8.6, wheels and whirls are the only 3-connected matroids with at most one non-essential element. We noted above that the cycle matroid of the graph in Figure 8.15(a), which consists of two maximal fans pieced together, has exactly two non-essential elements. All 3-connected matroids with exactly two non-essential elements and all 3-connected graphic matroids with exactly three non-essential elements were determined by Oxley and Wu (2000b, 2004).

When analysing the structure of 3-connected matroids, it is useful to identify the maximal fans since the behaviour of the elements within these fans is relatively predictable. Indeed, Oxley and Wu (2000a) showed that a 3-connected matroid M with a maximal fan having at least four elements can be obtained by sticking together a wheel and a 3-connected minor of M. This result underlies the proof of Proposition 8.8.1. The operation that is used here to do the sticking together generalizes parallel connection and is discussed in Section 11.4.

Exercises

1. Let M be a 3-connected matroid and X be a subset of $E(M)$ such that $M|X \cong U_{r,2r}$ for some $r \geq 2$. Prove that if $x \in X$, then $M\backslash x$ is 3-connected.

2. Let $(s_1, s_2, s_3, s_4, s_5)$ be a fan in a 3-connected matroid M. Prove that M has at most one element s_6 such that $(s_1, s_2, s_3, s_4, s_5, s_6)$ is a fan.

3. Let $(s_1, s_2, \ldots, s_n)$ be a maximal fan in a 3-connected matroid M. Suppose that $\{s_1, s_2, s_3\}$ is a triangle and $n \geq 5$. Prove that $M/s_2\backslash s_3$ is 3-connected unless M has an element e such that $M|\{s_1, s_2, s_3, s_4, s_5, e\} \cong M(K_4)$.

4. Let e_1, e_2, e_3, e_4, e_5 be distinct elements of a 3-connected matroid M where $M \not\cong M(\mathcal{W}_3)$. If M has $\{e_2, e_3, e_4\}$ as a triad and $\{e_1, e_2, e_3\}$ and $\{e_3, e_4, e_5\}$ as triangles, prove that it has no other triangles or triads containing e_3.

5. Let $(s_1, s_2, s_3, s_4, s_5, s_6)$ be a fan in a 3-connected matroid M. Suppose that s_1 is non-essential. Prove that $\{s_1, s_2, s_3\}$ and $\{s_2, s_3, s_4\}$ are the only triangles or triads of M containing s_2.

6. Let $(s_1, s_2, \ldots, s_n)$ and $(t_1, t_2, \ldots, t_n)$ be different fan orderings of a fan S in a 3-connected matroid M where $S \neq E(M)$. Prove that
   (a) if $n \geq 5$, then $(t_1, t_2, \ldots, t_n) = (s_n, s_{n-1}, \ldots, s_1)$;
   (b) if $n = 4$, then (t_1, t_2, t_3, t_4) is one of $(s_4, s_3, s_2, s_1), (s_1, s_3, s_2, s_4)$, or (s_4, s_2, s_3, s_1).

7. (a) Modify the graph in Figure 8.15(a) to produce an infinite family of 3-connected graphic matroids with exactly two non-essential elements.
   (b) Determine all 3-connected graphic matroids with exactly two non-essential elements.

8. *Prove Theorem 8.8.8.

9. *Let M be an excluded minor for the class of $\mathbb{F}$-representable matroids where $|\mathbb{F}| \neq 2$. Prove that M has no fans with four or more elements.

10. *(Wu 1998b) In a 3-connected matroid M, prove that:
    (a) If a triad T^* contains an essential element, then $M\backslash T^*$ is connected.
    (b) If e is an essential element of M, then e is in a triangle T for which M/T is connected, or e is in a triad T^* for which $M\backslash T^*$ is connected.

11. *(Geelen, Gerards, and Whittle 2006a) Let M be a 3-connected matroid with $|E(M)| \geq 4$ such that no element is in both a triangle and a triad. Prove that M has elements e and f such that, for some N in $\{M, M^*\}$, both $N\backslash e$ and $N\backslash f$ are 3-connected, and $N\backslash e, f$ is internally 3-connected.

9

BINARY MATROIDS

In earlier chapters, we have seen several examples of graph-theoretic results that have matroid analogues or generalizations. Numerous other examples will appear later in the book. When a result for graphs does not generalize to all matroids, there are two natural classes of matroids for which the result may still hold: the class of regular matroids and the larger class of binary matroids. Regular matroids will be looked at in more detail in Chapters 10 and 13. In this chapter, we shall discuss the properties of binary matroids. In particular, we prove numerous characterizations of such matroids. We shall also look at a number of special properties of binary matroids that are not possessed by matroids in general.

9.1 Characterizations

Binary matroids have been characterized in many ways. We showed in Theorem 6.5.4 that they are precisely the matroids having no $U_{2,4}$-minor. Seven other characterizations are stated in the first theorem of this section and several more will be proved later in the section. Apart from the excluded-minor characterization, the descriptions of binary matroids in the next two theorems are of two basic types: those that concern the cardinality of circuit–cocircuit intersections and those that concern symmetric differences of circuits. If X and Y are sets, then their symmetric difference, $X \triangle Y$, is the set $(X-Y) \cup (Y-X)$. One easily checks that the operation of symmetric difference is both commutative and associative. Indeed, for sets $X_1, X_2, \ldots, X_n$, their *symmetric difference*, $X_1 \triangle X_2 \triangle \cdots \triangle X_n$, consists of those elements that are in an odd number of the sets $X_1, X_2, \ldots, X_n$.

The proof of the first theorem will use the following easy result.

Lemma 9.1.1 *If e is an element of an independent set I of a matroid M, then M has a cocircuit C^* such that $C^* \cap I = \{e\}$.* □

In following theorem, a disjoint union of sets may consist of the disjoint union of the empty collection of sets and so may be empty. Recall that $C(e, B)$ is the fundamental circuit of the element e with respect to the basis B.

Theorem 9.1.2 *The following statements are equivalent for a matroid M:*

(i) *M is binary.*
(ii) *If C is a circuit and C^* is a cocircuit, then $|C \cap C^*|$ is even.*
(iii) *If C_1 and C_2 are distinct circuits, then $C_1 \triangle C_2$ contains a circuit.*
(iv) *If C_1 and C_2 are circuits, then $C_1 \triangle C_2$ is a disjoint union of circuits.*
(v) *The symmetric difference of any set of circuits is either empty or contains a circuit.*

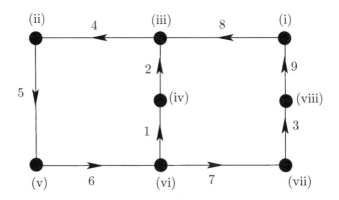

FIG. 9.1. Each arc corresponds to an implication that will be proved.

(vi) *The symmetric difference of any set of circuits is a disjoint union of circuits.*

(vii) *If B is a basis and C is a circuit, then $C = \triangle_{e \in C - B} C(e, B)$.*

(viii) *M has a basis B such that if C is a circuit, then $C = \triangle_{e \in C - B} C(e, B)$.*

Proof The digraph in Figure 9.1 summarizes the proof. The implications will be proved in the order indicated by the numbering on the arcs. The first three are immediate.

Now suppose that M satisfies (iii) but not (ii). Let C be a circuit and C^* be a cocircuit for which $|C \cap C^*|$ is odd so that $|C \cap C^*|$ is a minimum among such pairs (C, C^*). Then, by Proposition 2.1.11, C and C^* do not have exactly one common element, so we can find three distinct elements x, y, and z in their intersection. By the dual of Proposition 4.2.6, M has a circuit that meets C^* in some subset of $(C \cap C^*) - y$ containing x. Choose such a circuit C_1 so that $C \cup C_1$ is minimal.

Now $x \in C \cap C_1$ and $y \in C - C_1$, so there is a circuit C_2 such that $y \in C_2 \subseteq (C \cup C_1) - x$. Since $y \in C \cap C_2$ and $x \in C - C_2$, there is a circuit C_3 such that $x \in C_3 \subseteq (C \cup C_2) - y$. Clearly

$$C \cup C_3 \subseteq C \cup C_2 \subseteq C \cup C_1. \tag{9.1}$$

As x is in C_3 but y is not, it follows by (9.1) and the choice of C_1 that equality must hold throughout (9.1). Thus $C_2 - C = C_3 - C$, so $C_2 \triangle C_3 \subseteq C$. Since $x \in C_3 - C_2$, the circuits C_2 and C_3 are distinct. Thus, by (iii), $C_2 \triangle C_3$ contains a circuit. Hence

$$C_2 \triangle C_3 = C.$$

Therefore $C_2 \cap C^*$ and $C_3 \cap C^*$ partition $C \cap C^*$. Moreover, as $y \in C_2 \cap C^*$ and $x \in C_3 \cap C^*$, both $|C_2 \cap C^*|$ and $|C_3 \cap C^*|$ are less than $|C \cap C^*|$. It follows by the choice of (C, C^*) that both $|C_2 \cap C^*|$ and $|C_3 \cap C^*|$ are even. Hence $|C \cap C^*|$ is even; a contradiction. We conclude that (iii) implies (ii).

Now assume that (ii) holds but (v) does not. Then M has a set of distinct circuits $C_1, C_2, \ldots C_k$ such that $C_1 \triangle C_2 \triangle \cdots \triangle C_k$ is non-empty and independent. Therefore, by Lemma 9.1.1, there is a cocircuit C^* that meets $C_1 \triangle C_2 \triangle \cdots \triangle C_k$ in a single element. Thus $1 = |C^* \cap (C_1 \triangle C_2 \triangle \cdots \triangle C_k)| = |(C^* \cap C_1) \triangle (C^* \cap C_2) \triangle \cdots \triangle (C^* \cap C_k)|$. But, by (ii), $|C^* \cap C_i|$ is even for all i. Hence $|(C^* \cap C_1) \triangle (C^* \cap C_2) \triangle \cdots \triangle (C^* \cap C_k)|$ is even. It follows from this contradiction that (ii) implies (v).

Next assume that (v) holds but (vi) does not. Let $\{C_1, C_2, \ldots, C_k\}$ be a set of circuits for which $C_1 \triangle C_2 \triangle \cdots \triangle C_k$ is not a disjoint union of circuits so that, among such sets, $|C_1 \triangle C_2 \triangle \cdots \triangle C_k|$ is a minimum. By (v), $C_1 \triangle C_2 \triangle \cdots \triangle C_k$ contains a circuit D_1, and clearly $|C_1 \triangle C_2 \triangle \cdots \triangle C_k \triangle D_1| < |C_1 \triangle C_2 \triangle \cdots \triangle C_k|$. Thus $C_1 \triangle C_2 \triangle \cdots \triangle C_k \triangle D_1$ is a disjoint union of circuits; hence so too is $C_1 \triangle C_2 \triangle \cdots \triangle C_k$. This contradiction completes the proof that (v) implies (vi).

To show that (vi) implies (vii), assume that (vi) holds and let C be a circuit and B be a basis of M. Evidently $C - B = (\triangle_{e \in C-B} C(e, B)) - B$. If $C \neq \triangle_{e \in C-B} C(e, B)$, then $C \triangle (\triangle_{e \in C-B} C(e, B))$ is non-empty and so, by (vi), is dependent. But this is a contradiction since $C \triangle (\triangle_{e \in C-B} C(e, B)) \subseteq B$. Thus $C = \triangle_{e \in C-B} C(e, B)$ and so (vi) implies (vii).

Next we show that (i) implies (iii). Thus assume that (i) holds and let C_1 and C_2 be distinct circuits of M. Let $\psi : E(M) \to V(r, 2)$ be a coordinatization of M where $r = r(M)$. As C_1 and C_2 are both circuits, $\sum_{e \in C_1} \psi(e) = \underline{0} = \sum_{e \in C_2} \psi(e)$. Thus $\sum_{e \in C_1} \psi(e) + \sum_{e \in C_2} \psi(e) = \underline{0}$. If $e \in C_1 \cap C_2$, then $\psi(e)$ appears exactly twice on the left-hand side of the last equation. Thus $\sum_{e \in C_1 \triangle C_2} \psi(e) = \underline{0}$, so $C_1 \triangle C_2$ contains a circuit of M. Hence (i) implies (iii).

Finally, we shall show that (viii) implies (i). The proof will make use of the implications established above. In particular, we shall use the fact that, since a binary matroid satisfies (iii), it satisfies (ii)–(vii). Assume that M satisfies (viii), that is, M has a basis B such that if C is a circuit of M, then $C = \triangle_{e \in C-B} C(e, B)$. Let X be the B-fundamental-circuit incidence matrix of M and let $r(M) = r$. Then, viewed over $GF(2)$, the matrix $[I_r|X]$ represents a binary matroid. We shall show that this matroid is equal to M. First we note that B is a basis of $M[I_r|X]$ and that if $e \in E(M) - B$, then $C_M(e, B)$ is a circuit of $M[I_r|X]$. Hence we may drop the subscript on $C_M(e, B)$ without creating ambiguity.

Let C be a circuit of M. Then, by (viii), $C = \triangle_{e \in C-B} C(e, B)$. As $C(e, B)$ is a circuit of the binary matroid $M[I_r|X]$, it follows, by (vi), that $\triangle_{e \in C-B} C(e, B)$ is a disjoint union of circuits of $M[I_r|X]$. Thus, by Proposition 7.3.6, $M[I_r|X]$ is a quotient of M. But, since M and $M[I_r|X]$ both have rank r, it follows, by Corollary 7.3.4, that they are equal, that is, M is binary. We conclude that (viii) implies (i) and this completes the proof of the theorem. $\square$

The proof just given that (viii) implies (i) is somewhat less elementary than the other parts of the proof, relying as it does on results on quotients. Since it is this implication that completes the proof that all of (ii)–(viii) are equivalent to (i), the reader may prefer a more basic proof. We now present such a proof.

Alternative proof that (viii) implies (i) The proof proceeds as before to the point where it is shown that every circuit of M is a disjoint union of circuits of $M[I_r|X]$. Rather than using this to deduce a result about quotients, we deduce instead that every circuit of M is dependent in $M[I_r|X]$ and hence that every independent set in $M[I_r|X]$ is independent in M. To complete the proof that $M[I_r|X] = M$, and hence that M is binary, we now show that every circuit of $M[I_r|X]$ is a circuit of M. The result then follows from the elementary Lemma 2.1.22. Let D be a circuit of $M[I_r|X]$ and choose e in $D - B$. Then $D - e$ is contained in a basis B_1 of $M[I_r|X]$ such that $B_1 \subseteq (B \cup D) - e$. Moreover, D is the unique circuit of $M[I_r|X]$ that is contained in $B_1 \cup e$. Now B_1 is independent in $M[I_r|X]$, so it is independent in M. Since $|B_1| = r = r(M)$, it follows that B_1 is a basis of M. Now, from above, $C_M(e, B_1)$ is a disjoint union of circuits of $M[I_r|X]$. But $C_M(e, B_1) \subseteq B_1 \cup e$, and D is the only circuit of $M[I_r|X]$ contained in $B_1 \cup e$. Therefore $C_M(e, B_1) = D$ and so D is a circuit of M. □

The fact that (viii) characterizes binary matroids was proved by Las Vergnas (1980). The other characterizations are older and can be found in the work of Whitney (1935), Rado (1957), Lehman (1964), Tutte (1965a), and Minty (1966).

Next we use the excluded-minor characterization of binary matroids (Theorem 6.5.4) to establish that a certain natural weakening of 9.1.2(ii) also characterizes binary matroids (Seymour 1976).

Theorem 9.1.3 *The following statements are equivalent for a matroid M:*

(i) *M is binary.*

(ii) *If C is a circuit and C^* is a cocircuit, then $|C \cap C^*| \neq 3$.*

(iii) *M has no minor isomorphic to $U_{2,4}$.*

Proof By Theorem 6.5.4, (i) and (iii) are equivalent. Moreover, by the last theorem, if M is binary and C and C^* are a circuit and cocircuit respectively, then $|C \cap C^*|$ is even. Hence (i) implies (ii). To complete the proof of the theorem, we shall show that (ii) implies (iii) by proving the contrapositive.

Assume that M has a $U_{2,4}$-minor N and let Z be a 3-element subset of $E(N)$. Then Z is both a circuit and a cocircuit of N. It follows that M has a circuit and a cocircuit whose intersection is Z. Hence (ii) fails. We conclude that (ii) implies (iii). □

The lattice form of the Scum Theorem is obtained by combining Theorem 3.3.1 with Corollary 3.3.9. The next corollary uses these results together with the last theorem to characterize binary matroids in terms of their lattices of flats.

Corollary 9.1.4 *A rank-r matroid M is binary if and only if every rank-$(r-2)$ flat of M is contained in at most three hyperplanes.* □

Another variant of the Scum Theorem, namely Lemma 3.3.2, can be used to prove the next result (Oxley 1988), a straightforward extension of an interesting theorem of Fournier (1981) that characterizes binary matroids by a circuit double-elimination property. Fournier's result is stated here as a corollary of the next proposition, it being an immediate consequence of that proposition and Theorem 9.1.3.

Proposition 9.1.5 *Let M be a matroid and n be an integer exceeding one. Then M has no minor isomorphic to $U_{n,n+2}$ if and only if, whenever $C_1, C_2, \ldots, C_n$ are distinct circuits of M and e and f are distinct elements of $C_1 \cap C_2 \cap \cdots \cap C_n$, there is a circuit of M that is contained in $(C_1 \cup C_2 \cup \cdots \cup C_n) - \{e, f\}$.* □

Corollary 9.1.6 *A matroid M is binary if and only if, whenever C_1 and C_2 are distinct circuits of M and e and f are elements of $C_1 \cap C_2$, there is a circuit contained in $(C_1 \cup C_2) - \{e, f\}$.* □

The list of characterizations of binary matroids given above is long but not exhaustive. It will be supplemented in the following exercises and in Section 9.4.

Exercises

1. Let X_1, X_2, and X_3 be subsets of a set E. Show that:
   (a) $(E - X_1) \triangle (E - X_2) = X_1 \triangle X_2$.
   (b) $(E - X_1) \triangle (E - X_2) \triangle (E - X_3) = E - (X_1 \triangle X_2 \triangle X_3)$.
2. Prove that a matroid is binary if and only if, for every two distinct hyperplanes H_1 and H_2, there is a hyperplane that contains $H_1 \cap H_2$, is contained in $H_1 \cup H_2$, and is distinct from H_1 and H_2.
3. Prove that a geometric lattice $\mathcal{L}$ is the lattice of flats of a binary matroid if and only if every interval of height two in $\mathcal{L}$ contains at most three elements of height one.
4. Find all matroids for which every circuit–cocircuit intersection has odd cardinality.
5. Prove that a non-empty connected binary matroid either has a circuit of even cardinality, or is isomorphic to $U_{1,1}$ or $U_{n-1,n}$ for some odd n.
6. (Tutte 1965a) Let C be a circuit of a matroid M and x and y be elements of $\mathrm{cl}_M(C)$. If both $M|(C \cup x)$ and $M|(C \cup y)$ are binary, prove that $M|(C \cup \{x, y\})$ is binary.
7. (a) (White 1971) Prove that a matroid is binary if and only if $C_1 \triangle C_2$ is a circuit for every two distinct intersecting circuits C_1 and C_2 such that $r(C_1) + r(C_2) = r(C_1 \cup C_2) + r(C_1 \cap C_2)$.
   (b) State the corresponding characterization of binary matroids in terms of hyperplanes.
8. (Seymour 1981b) If X is a 3-element subset of a matroid M, prove that M has a $U_{2,4}$-minor whose ground set contains X if and only if M has a circuit and a cocircuit whose intersection is X.
9. Prove Proposition 9.1.5.

10. (Oxley 1984a) Let X be a 4-element set in a matroid M. Show that:
    (a) X is the intersection of a circuit and a cocircuit in M if and only if M has a minor isomorphic to one of $M(K_4), \mathcal{W}^3, Q_6, P_6, U_{3,6}$, or R_6 (see Figures 8.7 and 8.11) in which X is a circuit-cocircuit intersection.
    (b) A binary matroid has a 4-element circuit–cocircuit intersection if and only if it has an $M(K_4)$-minor.
    Now suppose M has a k-element set that is the intersection of a circuit and a cocircuit. Show that:
    (c) If $k \geq 3$, then M has a t-element set that is the intersection of a circuit and a cocircuit for some t with $\lceil k/2 \rceil \leq t \leq k - 1$.
    (d) *If $k \geq 4$, then M has a 4-element circuit–cocircuit intersection.
    (e) °If $k \geq 4$, then M has a $(k - 2)$-element set that is the intersection of a circuit and a cocircuit.
11. (Kung in Oxley and Whittle 1991) Prove that a simple binary matroid M is uniquely determined by its set of non-spanning circuits unless M is a free matroid or a circuit.
12. *(Fournier 1974) Prove that a matroid is binary if and only if, whenever C_1, C_2, and C_3 are distinct circuits such that $C_1 \cap C_2 \cap C_3 \neq \emptyset$, there are distinct elements i, j, and k of $\{1,2,3\}$ such that $C_j - C_i \neq C_k - C_i$.
13. *(Greene and White in White 1971) Prove that a matroid is binary if and only if, for all bases B_1 and B_2 and all elements f of B_2, the number of elements e of B_1 such that both $(B_1 - e) \cup f$ and $(B_2 - f) \cup e$ are bases is odd. (Hint: Show that this condition on bases is equivalent to 9.1.2(ii).)

9.2 Circuit and cocircuit spaces

In this section, we shall look at two vector spaces which are commonly associated with a binary matroid. In addition, we shall extend these ideas to a more general class of matroids and touch briefly on Tutte's theory of chain-groups. To begin, consider the following.

Example 9.2.1 The matrix $[I_3|D]$ in Figure 9.2(b) is a binary representation of $M(G)$ where G is the graph shown in Figure 9.2(a). Row spaces of matrices were considered briefly in Section 2.2. Recall that the row space $\mathcal{R}(A)$ of an $m \times n$ matrix A over a field $\mathbb{F}$ is the subspace of $V(n, \mathbb{F})$ that is spanned by the rows

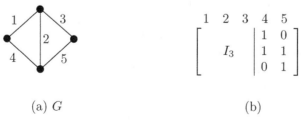

(a) G (b)

FIG. 9.2. The matrix in (b) represents $M(G)$ over $GF(2)$.

of A. Thus, the members of the row space $\mathcal{R}[I_3|D]$ of $[I_3|D]$ are the rows of the following matrix.

	1	2	3	4	5
Row 1	1	0	0	1	0
Row 2	0	1	0	1	1
Row 3	0	0	1	0	1
Row 1 + Row 2	1	1	0	0	1
Row 1 + Row 3	1	0	1	1	1
Row 2 + Row 3	0	1	1	1	0
Row 1 + Row 2 + Row 3	1	1	1	0	0
Row 1 + Row 1	0	0	0	0	0

Viewed as incidence vectors, the rows of this matrix correspond to the sets $\{1,4\},\{2,4,5\},\{3,5\},\{1,2,5\},\{1,3,4,5\},\{2,3,4\},\{1,2,3\}$, and $\emptyset$. It is straightforward to check that these sets consist of all possible disjoint unions of cocircuits of $M(G)$. As noted in Section 2.2, $\mathcal{R}[I_3|D]$ is the binary linear code with generator matrix $[I_3|D]$ and parity-check matrix $[-D^T|I_2]$. This link with coding theory will be considered further in the exercises. □

This example illustrates a general phenomenon for binary matroids which will be described in the next result. Let M be a binary matroid on the set $\{1,2,\ldots,n\}$. The *circuit space* and *cocircuit space* of M are the subspaces of $V(n,2)$ that are generated by the incidence vectors of the circuits and cocircuits, respectively, of M.

Proposition 9.2.2 *Let A be a binary representation of a rank-r binary matroid M. Then the cocircuit space of M equals the row space of A. Moreover, this space has dimension r and is the orthogonal subspace of the circuit space of M.*

Proof Let $|E(M)| = n$. If A is a zero matrix, then $M \cong U_{0,n}$ and the result follows easily. Now suppose that A is not a zero matrix. By reordering the elements of M, if necessary, we may assume that the first r columns of A are linearly independent. Then, by a sequence of elementary row operations and deletions of zero rows, A can be transformed into a matrix in the form $[I_r|D]$. Moreover, the row spaces of A and $[I_r|D]$ coincide. Thus, as the rows of $[I_r|D]$ are clearly linearly independent, the row space of A has dimension equal to $r(M)$.

Let the last $n-r$ columns of $[I_r|D]$ be labelled by the set B^*. Evidently B^* is a cobasis of M. Moreover, as noted in Section 6.4, the rows of $[I_r|D]$ are the incidence vectors of the fundamental cocircuits of the elements of $E(M) - B^*$ with respect to B^*. Thus, every member of $\mathcal{R}[I_r|D]$ is in the cocircuit space of M. Furthermore, by Theorem 9.1.2 and duality, the incidence vector of every cocircuit of M is in $\mathcal{R}[I_r|D]$. We conclude that the row space of $[I_r|D]$ coincides with the cocircuit space of M. By duality, if $r < n$, then $\mathcal{R}[-D^T|I_{n-r}]$ coincides with the circuit space of M. Hence, by Proposition 2.2.23, if $r < n$, then the cocircuit space is the orthogonal subspace of the circuit space. But one easily checks that this is also true if $r = n$, so the proposition is proved. □

The following result is an easy consequence of the last proof.

Corollary 9.2.3 *Let B be a basis of an n-element rank-r binary matroid M and X be the B-fundamental-circuit incidence matrix of M. Suppose that $1 \leq r \leq n-1$. Then the row spaces of $[I_r|X]$ and $[X^T|I_{n-r}]$ are the cocircuit and circuit spaces, respectively, of M.* □

Another consequence of the last proposition is that one can derive a binary matroid from a family $\mathcal{A}$ of subsets of a finite set E in the following natural way. Let $\mathcal{A}'$ be the set of all possible symmetric differences of members of $\mathcal{A}$ and let $\mathcal{C}$ be the set of minimal non-empty members of $\mathcal{A}'$. Then it follows immediately from Proposition 9.2.2 that $\mathcal{C}$ is the set of circuits of a binary matroid on E.

Although binary matroids are the primary focus of this chapter, we show next that many of the above ideas extend naturally to a larger class of matroids. Let $\mathbb{F}$ be an arbitrary field and $(v_1, v_2, \ldots, v_n)$ be a member $\underline{v}$ of $V(n, \mathbb{F})$. The *support* of $\underline{v}$ is $\{i : v_i \neq 0\}$. An immediate consequence of Proposition 9.2.2 is that if A is a binary representation of a matroid M, then the cocircuits of M are precisely the minimal non-empty sets that are supports of members of $\mathcal{R}(A)$. The next result captures the essence of Tutte's (1965a) work on chain-groups, although the latter is set in a somewhat more general context (see Exercise 4).

Proposition 9.2.4 *Let A be an $m \times n$ matrix over a field $\mathbb{F}$ and $M = M[A]$. Then the set of cocircuits of M coincides with the set of minimal non-empty supports of vectors from the row space of A.*

Proof As in the proof of Proposition 9.2.2, we may assume that A is non-zero and that it is of the form $[I_r|D]$ where $r = r(M)$. Let the columns of $[I_r|D]$ be labelled, in order, by $1, 2, \ldots, n$. If $r = n$, then $M \cong U_{n,n}$ and the result follows easily. Thus suppose that $r < n$. Then, by Theorem 2.2.8, M^* is represented by the matrix $[-D^T|I_{n-r}]$ where it too has columns labelled $1, 2, \ldots, n$.

We want to show that the circuits of M^* coincide with the minimal non-empty supports of members of $\mathcal{R}[I_r|D]$. The main difficulty in the proof of this is notational. Let $\underline{d}(i)$ denote row i of D, let $\underline{e}(j)$ denote row j of I_{n-r}, and let $\underline{e}'(k)$ denote row k of I_r. Then $(\underline{e}'(i), \underline{d}(i))$ is the ith row of $[I_r|D]$ and

$$[-D^T|I_{n-r}] = [-\underline{d}(1)^T \cdots -\underline{d}(r)^T \,|\, \underline{e}(1)^T \cdots \underline{e}(n-r)^T].$$

Now suppose that $\{i_1, i_2, \ldots, i_s, r+j_1, r+j_2, \ldots, r+j_t\}$ is a circuit of M^* where $1 \leq i_1 < i_2 < \cdots < i_s \leq r$ and $1 \leq j_1 < j_2 < \cdots < j_t \leq n-r$. Then, for some non-zero elements $\alpha_1, \alpha_2, \ldots, \alpha_s, \beta_1, \beta_2, \ldots, \beta_t$ of $\mathbb{F}$,

$$-\alpha_1\underline{d}(i_1)^T - \cdots - \alpha_s\underline{d}(i_s)^T + \beta_1\underline{e}(j_1)^T + \cdots + \beta_t\underline{e}(j_t)^T = \underline{0}. \qquad (9.2)$$

Clearly the sum $\alpha_1(\underline{e}'(i_1), \underline{d}(i_1)) + \alpha_2(\underline{e}'(i_2), \underline{d}(i_2)) + \cdots + \alpha_s(\underline{e}'(i_s), \underline{d}(i_s))$ is a member of the row space of $[I_r|D]$. Moreover, with $\underline{0}_k$ denoting the zero vector in $V(k, \mathbb{F})$, we can rewrite this sum as

$$\alpha_1(\underline{e}'(i_1), \underline{0}_{n-r}) + \cdots + \alpha_s(\underline{e}'(i_s), \underline{0}_{n-r}) + \alpha_1(\underline{0}_r, \underline{d}(i_1)) + \cdots + \alpha_s(\underline{0}_r, \underline{d}(i_s)).$$

By (9.2), this equals

$$\alpha_1(\underline{e}'(i_1),\underline{0}_{n-r}) + \cdots + \alpha_s(\underline{e}'(i_s),\underline{0}_{n-r}) + \beta_1(\underline{0}_r,\underline{e}(j_1)) + \cdots + \beta_t(\underline{0}_r,\underline{e}(j_t)).$$

Evidently the support of this row vector is $\{i_1, i_2, \ldots, i_s, r+j_1, r+j_2, \ldots, r+j_t\}$.

By Lemma 2.1.22, to complete the proof of the Proposition 9.2.4, it suffices to show that the support of every non-zero member of $\mathcal{R}[I_r|D]$ is dependent in M^*. The proof here essentially amounts to reversing the argument given for the first part and we omit the details. □

It is tempting to ask if one can say anything about the collection of all supports of vectors in the row space of A. In Example 9.2.1, where A represents a matroid over $GF(2)$, these supports coincide with all disjoint unions of cocircuits of M. But some unions of cocircuits, such as $\{1,2,3,4\}$ and $\{2,3,4,5\}$, do not occur as supports. The matrix A also represents M over $\mathbb{R}$ and, when viewed over this field, its row space changes. Indeed, the set of supports of vectors from the row space over $\mathbb{R}$ coincides with all unions of cocircuits of M. This occurs in general as the next result shows.

Corollary 9.2.5 *Let A be a matrix over an infinite field $\mathbb{F}$ and let $M = M[A]$. Then the set of flats of M coincides with the set of complements of supports of vectors from the row space of A.*

Proof The complement of a union of cocircuits is an intersection of hyperplanes and so is a flat. Moreover, every flat arises in this way. Thus it suffices to prove that the set of non-empty supports coincides with the set of unions of minimal non-empty supports. Suppose X is the support of some non-zero member of $\mathcal{R}(A)$. Then X contains a set X_1 that is a minimal non-empty support. Take x_1 in X_1. Then there is clearly a vector in $\mathcal{R}(A)$ whose support X' contains $X - X_1$ and is contained in $X - x_1$. By repeating this process using X' in place of X, we deduce that X can be written as a union of minimal non-empty supports. Conversely, as $\mathbb{F}$ is infinite, it is straightforward to argue by induction on n that the union of n minimal supports is a support, and we omit the details. □

Exercises

1. (a) State the extension of Proposition 6.3.12 for binary matroids that arises by recognizing the non-singular diagonal matrices over $GF(2)$.
   (b) (Bondy and Welsh 1971) Let $[I_r|A]$ be a matrix over $GF(2)$. Prove that $M[I_r|A]$ is identically self-dual if and only if A is square and $AA^T = I_r$.

2. For $r \geq 2$, consider an $r \times (2^r - 1)$ matrix A_r whose columns are all of the non-zero vectors in $V(r,2)$. The vectors in the orthogonal subspace of $\mathcal{R}(A_r)$ form the binary *Hamming code* H_r, which arises frequently in coding theory. Note that H_r is only determined up to a permutation of the coordinates. Show that:

(a) H_3 equals the row space of the following matrix over $GF(2)$.

$$\left[\quad I_4 \quad \left| \begin{array}{ccc} 0 & 1 & 1 \\ 1 & 0 & 1 \\ 1 & 1 & 0 \\ 1 & 1 & 1 \end{array} \right. \right]$$

(b) H_r has dimension $2^r - r - 1$.

(c) There is a partition of $V(2^r - 1, 2)$ into m classes where m is the number of vectors in H_r and each class contains a unique member $\underline{v}$ of H_r together with all members of $V(2^r - 1, 2)$ that differ from $\underline{v}$ in exactly one coordinate.

3. (a) Prove that if M is a non-empty binary matroid with no coloops, then there is a member of the circuit space of M whose number of non-zero coordinates exceeds $\frac{1}{2}|E(M)|$.

(b) Show that if M is the cycle matroid of a graph G and the cocircuit space of M contains $E(G)$, then G is a bipartite graph.

(c) Show that if G is a graph with no loops, then G has a bipartite subgraph H such that $|E(H)| > \frac{1}{2}|E(G)|$.

4. (Tutte 1965a) Let R be an integral domain and E be a finite set. A *chain* on E over R is a mapping of E into R. Let N be a *chain-group* on E over R, that is, N is a set of chains on E over R that is closed under the operations of addition and multiplication by an element of R. The *support* of a chain f is $\{e \in E : f(e) \neq 0\}$. Show that:

(a) The minimal non-empty supports of members of N form the circuits of a matroid $M\{N\}$ on E.

(b) The minimal sets meeting the supports of all non-zero chains in N form the bases of a matroid on E.

(c) The matroids in (i) and (ii) are duals of each other.

(d) *If N^* is the set of chains g on E over R such that $\sum_{e \in E} g(e)f(e) = 0$ for all f in N, then N^* is a chain-group and $M\{N^*\} = (M\{N\})^*$.

(e) If $S \subseteq E$ and $N|S$ denotes the set of restrictions to S of chains in N, then $N|S$ is a chain-group and $(M\{N|S\})^* = M|S$.

(f) If $S \subseteq E$ and $N.S$ denotes the set of restrictions to S of chains in N whose supports are contained in S, then $N.S$ is a chain-group and $(M\{N.S\})^* = M.S$.

9.3 The operation of 3-sum

We saw in Chapter 7 that the operation of 2-sum extends from graphs to matroids. Moreover, in Chapter 8, we showed that a 2-connected matroid is 3-connected if and only if it cannot be decomposed as a 2-sum. The graph operation of 2-sum is a special case of the *clique-sum* operation for graphs (Robertson and Seymour 1984). In this section, we show how 3-sum for graphs can be extended to binary matroids. Consideration of the issues that arise when trying to extend this operation to non-binary matroids will be delayed until Section 11.4.

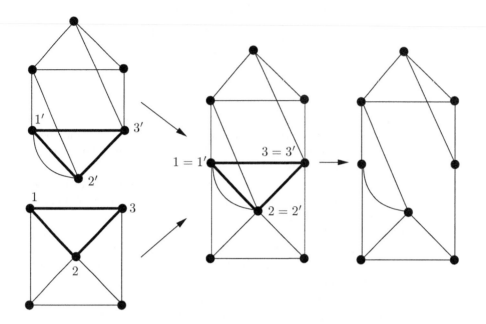

FIG. 9.3. Forming a 3-sum of two graphs.

The clique-sum of two graphs is obtained by sticking the graphs together along a common complete subgraph and then deleting all identified edges. An example of this operation is shown in Figure 9.3. This operation is formally described as follows. If G_1 and G_2 are graphs each having a K_n-subgraph for some positive integer n, then to form an n-*sum* of G_1 and G_2, one first pairs the vertices of the chosen K_n-subgraph of G_1 with distinct vertices of the chosen K_n-subgraph of G_2. The paired vertices are then identified, as are the corresponding pairs of edges. Finally, all identified edges are deleted.

The following operation for binary matroids was defined by Seymour (1980b). As in Section 9.1, a disjoint union of circuits may be empty.

Lemma 9.3.1 *Let M_1 and M_2 be binary matroids and $E = E(M_1) \bigtriangleup E(M_2)$. Then there is a matroid $M_1 \bigtriangleup M_2$ with ground set E whose set of circuits consists of the minimal non-empty subsets of E of the form $X_1 \bigtriangleup X_2$ where X_i is a disjoint union of circuits of M_i. Furthermore, if A is a matrix over $GF(2)$ whose columns are indexed by the elements of $E(M_1) \cup E(M_2)$ and whose rows consist of the incidence vectors of all the circuits of M_1 and all the circuits of M_2, then*

$$M_1 \bigtriangleup M_2 = (M[A])^* \backslash (E(M_1) \cap E(M_2)).$$

Proof Let $M = M[A]$. Then, by Proposition 9.2.2, the row space of A is the circuit space of M^*. Now if $\underline{v} \in \mathcal{R}[A]$, then we can write $\underline{v} = \underline{v}_1 + \underline{v}_2$ where $\underline{v}_i$ is the mod-2 sum of the incidence vectors of some set of circuits of M_i. Hence

$\underline{v}_i$ is the incidence vector of some disjoint union of circuits of M_i. Thus the circuits of M^* are the minimal non-empty sets of the form $X_1 \triangle X_2$ where X_i is a disjoint union of circuits of M_i. The collection of such sets that are subsets of of $E(M_1) \triangle E(M_2)$ coincides with the set of circuits of $M^* \backslash (E(M_1) \cap E(M_2))$, so $M_1 \triangle M_2 = M^* \backslash (E(M_1) \cap E(M_2))$ and $M_1 \triangle M_2$ is certainly a matroid. □

Next we observe that the operation defined in the last lemma generalizes both direct sum and 2-sum. We omit the elementary proof.

Lemma 9.3.2 *Let M_1 and M_2 be binary matroids.*

(i) *If $E(M_1) \cap E(M_2) = \emptyset$, then $M_1 \triangle M_2 = M_1 \oplus M_2$.*

(ii) *If $E(M_1) \cap E(M_2) = \{p\}$ and neither M_1 nor M_2 has p as a loop or a coloop, then $M_1 \triangle M_2 = M_1 \oplus_2 M_2$.* □

Now suppose that the ground sets of binary matroids M_1 and M_2 meet in a set T that is a triangle of both. When both $|E(M_1)|$ and $|E(M_2)|$ exceed six and neither M_1 nor M_2 has a cocircuit contained in T, we call $M_1 \triangle M_2$ the *3-sum*, $M_1 \oplus_3 M_2$, of M_1 and M_2. The reader may wonder why the technical conditions in the last sentence have been included in the definition of matroid 3-sums. Two key properties of 3-sums are captured in Propositions 9.3.4 and 9.3.5 below. These technical conditions are needed to prove these propositions. The 3-sum operation features prominently in Seymour's Decomposition Theorem for regular matroids, the proof of which is given in Chapter 13.

The next lemma shows that the circuits in a 3-sum of two binary matroids can be more simply described than in Lemma 9.3.1. This lemma quickly implies a link between the operations of 3-sum for binary matroids and for graphs. Specifically, let G_1 and G_2 be graphs whose edge sets meet in a set T that is a 3-cycle of both but a bond of neither. When both $|E(G_1)|$ and $|E(G_2)|$ exceed six, if G is the 3-sum of the graphs G_1 and G_2 with respect to the triangle T, then $M(G)$ is the binary matroid $M(G_1) \oplus_3 M(G_2)$.

Lemma 9.3.3 *Let M_1 and M_2 be binary matroids such that $E(M_1) \cap E(M_2) = T$, where T is a triangle of both M_1 and M_2. Then $\mathcal{C}(M_1 \triangle M_2)$ is the union of $\mathcal{C}(M_1 \backslash T)$, $\mathcal{C}(M_2 \backslash T)$, and the collection of minimal sets of the form $C_1 \triangle C_2$ where C_i is a circuit of M_i such that $C_1 \cap T = C_2 \cap T$ and the last set has exactly one element.*

Proof Certainly every set in $\mathcal{C}(M_1 \backslash T) \cup \mathcal{C}(M_2 \backslash T)$ is a circuit of $M_1 \triangle M_2$. Now let C be a circuit of $M_1 \triangle M_2$ that meets both $E(M_1) - T$ and $E(M_2) - T$. By Lemma 9.3.1, $C = X_1 \triangle X_2$ where X_i is a non-empty disjoint union of some set $\mathcal{X}_i$ of circuits of M_i. We show first that C can be written as the symmetric difference of a circuit of M_1 and a circuit of M_2.

For each i, choose $C_i \in \mathcal{X}_i$. If $C_1 \cap T = C_2 \cap T$, then $C_1 \triangle C_2 \subseteq X_1 \triangle X_2$. But $C_1 \triangle C_2$ is dependent in $M_1 \triangle M_2$, while $X_1 \triangle X_2$ is a circuit. Hence $C_1 \triangle C_2 = X_1 \triangle X_2 = C$. We may now assume that if $C_1 \in \mathcal{X}_1$ and $C_2 \in \mathcal{X}_2$, then $C_1 \cap T \neq C_2 \cap T$.

Now $X_1 \cap T = X_2 \cap T$. If X_1 contains T, then $X_1 \triangle T$ is a proper subset of $E(M_1)\backslash T$ and contains a circuit of $M_1\backslash T$ and hence of $M_1 \triangle M_2$. This circuit is properly contained in C; a contradiction. Thus X_1 contains at most two elements of T. From the conclusion to the previous paragraph, we deduce that $|X_1 \cap T| = 2$ and, without loss of generality, that $|\mathcal{X}_1| = 1$ and $|\mathcal{X}_2| = 2$. Let $\mathcal{X}_1 = \{C_1\}$ and $\mathcal{X}_2 = \{C_{2,1}, C_{2,2}\}$. If $T = \{a_1, a_2, a_3\}$, we may assume that $C_1 \cap T = \{a_1, a_2\}$ and $C_{2,i} \cap T = \{a_i\}$. Then $C_1 \triangle T$ is a circuit D_1 of M_1 meeting T in $\{a_3\}$, and $C_{2,1} \triangle C_{2,2} \triangle T$ contains a circuit D_2 of M_2 meeting T in $\{a_3\}$. Then $D_1 \triangle D_2 \subseteq X_1 \triangle X_2$, so equality holds.

We now know that every circuit of $M_1 \triangle M_2$ that meets $E(M_1) - T$ and $E(M_2) - T$ can be written in the form $C_1 \triangle C_2$, where C_i is a circuit of M_i and $C_1 \cap T = C_2 \cap T$. This intersection is not equal to T or $\emptyset$. If it has exactly two elements, then $C_i \triangle T$ is a circuit of M_i and $C_1 \triangle C_2 = (C_1 \triangle T) \triangle (C_2 \triangle T)$. Since $|(C_1 \triangle T) \cap T| = 1$, the lemma holds.   □

A fundamental feature of the matroid operations of direct sum and 2-sum is their link with the presence of 1- and 2-separations: M can be written as a direct sum of two of its proper minors if and only if M has an exact 1-separation; and M can be written as a 2-sum of two of its proper minors if and only if M has an exact 2-separation. For 3-sums, the situation is similar yet more complicated. Firstly, we are confined to binary matroids. Secondly, although being expressible as a 3-sum is not equivalent to having an exact 3-separation, it is equivalent to having an exact 3-separation in which each side has at least four elements.

Proposition 9.3.4 *Let (X_1, X_2) be an exact 3-separation of a binary matroid M with $|X_1|, |X_2| \geq 4$, and let Z be a 3-element set that is disjoint from $E(M)$. Then there is a binary matroid N with ground set $E(M) \cup Z$ such that $N|E(M) = M$, the set Z is a circuit of N, and $r(X_i \cup Z) = r(X_i)$ for each i. Moreover, $M = N_1 \oplus_3 N_2$ where $N_i = N|(X_i \cup Z)$ for each i, and N is unique up to a permutation on the labels of the elements of Z. Conversely, the 3-sum, $M_1 \oplus_3 M_2$, of binary matroids M_1 and M_2 has $(E(M_1) - E(M_2), E(M_2) - E(M_1))$ as an exact 3-separation, $|E(M_1) - E(M_2)|$ and $|E(M_2) - E(M_1)|$ exceed three, and*

$$r(M_1 \oplus_3 M_2) = r(M_1) + r(M_2) - 2.$$

Proof The argument here is due to P. Seymour (private communication). Let $r(M) = r$ and take an $r \times n$ matrix A representing M. For all e in $E(M)$, let $\psi_A(e)$ be the column of A labelled by e. As noted in Section 6.1, ψ_A is a rank-preserving map from $E(M)$ into $V(r, 2)$. As $r(X_1) + r(X_2) = r + 2$, we have

$$
\begin{aligned}
r + 2 &= \dim\langle \psi_A(X_1)\rangle + \dim\langle \psi_A(X_2)\rangle \\
&= \dim\langle \psi_A(X_1) \cup \psi_A(X_2)\rangle + \dim\langle \psi_A(X_1) \cap \psi_A(X_2)\rangle \\
&= r + \dim\langle \psi_A(X_1) \cap \psi_A(X_2)\rangle.
\end{aligned}
$$

Hence $\langle \psi_A(X_1) \cap \psi_A(X_2)\rangle$ is a 2-dimensional subspace of $V(r, 2)$. Adjoin new columns z_1, z_2, and z_3 to A corresponding to the three non-zero vectors in this subspace. Denote the resulting matrix by A^+, let $Z = \{z_1, z_2, z_3\}$, and let $N = M[A^+]$. Let N_1 and N_2 be $N|(X_1 \cup Z)$ and $N|(X_2 \cup Z)$. Then, for each i, the set Z is a circuit of N_i and, since $r(X_i \cup Z) = r(X_i)$, this circuit does not contain a cocircuit of N_i. One now easily checks that $M = N_1 \oplus_3 N_2$. To see that N is unique once the labels on the set Z have been specified, note that, since M is binary, Proposition 6.4.1 implies that M is uniquely $GF(2)$-representable. Thus all choices for the matrix A are equivalent. Since $\langle \psi_A(X_1) \cap \psi_A(X_2) \rangle$ has dimension 2 and we require $r(X_i \cup Z) = r(X_i)$ for each i, it follows that the vectors labelled by Z are uniquely determined by A. Clearly we can permute the labels on these vectors. But, otherwise, the matroid N is unique.

Conversely, suppose that M is the 3-sum of M_1 and M_2. Writing E_i for $E(M_i)$, we have $E_1 \cap E_2 = T$ where T is a triangle of both M_1 and M_2 that contains a cocircuit of neither matroid. Moreover, by Lemma 9.3.1, if D is a matrix over $GF(2)$ whose columns are indexed by the elements of $E_1 \cup E_2$ and whose rows consist of the incidence vectors of all the circuits of M_1 and all the circuits of M_2, then $M = (M[D])^* \backslash T$.

Let $T = \{e, f, g\}$. Then $\{e, f\}$ extends to a basis B_1 of M_1 and to a basis B_2 of M_2. Consider $B_1 \cup B_2$. It certainly spans $M[D]$. Suppose it contains a circuit C of $(M[D])^*$. Then C meets both $B_2 - B_1$ and $B_1 - B_2$. By Lemma 9.3.3, $C = C_1 \triangle C_2$ where C_i is a circuit of M_i with $|C_i \cap T| = 1$. Now C_1 and $C_1 \triangle T$ are dependent sets of M_1, one of which is contained in the independent set B_1. This contradiction implies that $B_1 \cup B_2$ is a basis of $(M[D])^*$. Thus $r((M[D])^*) = r(M_1) + r(M_2) - 2$. As T does not contain a cocircuit of M_1, it does not contain a cocircuit of $(M[D])^*$. Hence

$$r(M_1 \oplus_3 M_2) = r((M[D])^* \backslash T) = r((M[D])^*) = r(M_1) + r(M_2) - 2$$

and $r(M_1 \oplus_3 M_2) = r(E_1 - E_2) + r(E_2 - E_1) - 2$. The definition of $M_1 \oplus_3 M_2$ ensures that each of $|E_1 - E_2|$ and $|E_2 - E_1|$ exceeds three. Thus $(E_1 - E_2, E_2 - E_1)$ is an exact 3-separation of M. $\qquad \square$

There is a third way in which the behaviour of 3-sums differs from that of 1- and 2-sums: we need to impose an additional connectivity condition to guarantee that the parts of a 3-sum are isomorphic to minors of the whole.

Proposition 9.3.5 *If a 3-connected matroid M is the 3-sum of binary matroids M_1 and M_2, then M has minors that are isomorphic to each of M_1 and M_2.*

The proof below of this proposition was privately communicated to the author by J. Geelen. It uses the following preliminary lemma involving $\kappa_M(X, Y)$ which, from Section 8.5, is defined for disjoint subsets X and Y of the ground set E of a matroid M to be $\min\{\lambda_M(S) : X \subseteq S \subseteq E - Y\}$.

Lemma 9.3.6 *Let C and X be disjoint sets in a matroid M such that C is a circuit and $\kappa_M(C, X) = 2$. Then there are elements $a, b,$ and c of C and a minor N of M that has $X \cup \{a, b, c\}$ as its ground set and $\{a, b, c\}$ as a circuit such that $\kappa_N(\{a, b, c\}, X) = 2$.*

Proof Since $\kappa_M(C, X) = 2$, the circuit C has at least three elements. We argue by induction on $|E(M) - X|$ noting that, since $E(M) - X$ contains C, this number is at least 3; the result is immediate if it equals 3. Suppose first that $E(M) - (X \cup C)$ is non-empty and let e be in this set. Suppose $e \in \text{cl}_M(C)$ and let $\kappa_{M\backslash e}(C, X) = \lambda_{M\backslash e}(S)$ for some S with $C \subseteq S \subseteq E - e - X$. Then $e \in \text{cl}(S)$, so $\lambda_M(S \cup e) = \lambda_{M\backslash e}(S)$. Hence $\kappa_M(C, X) \leq \kappa_{M\backslash e}(C, X)$ so, by Lemma 8.5.5, $\kappa_M(C, X) = \kappa_{M\backslash e}(C, X)$ and the result follows by induction. Thus we may assume that $e \notin \text{cl}_M(C)$. Then C is a circuit of both $M\backslash e$ and M/e. By Theorem 8.5.7, M has a minor M' with ground set $X \cup C$ such that $\kappa_{M'}(C, X) = 2$ and $M'|C = M|C$. As M' is a minor of $M\backslash e$ or M/e, it follows, by Lemma 8.5.5 that $\kappa_{M\backslash e}(C, X) = 2$ or $\kappa_{M/e}(C, X) = 2$, and again the result follows by induction.

We may now assume that $E(M) = C \cup X$. Then $\kappa_M(C, X) = \lambda_M(X) = 2$. As $r(C) + r(X) - r(M) = 2$, the circuit C has exactly three elements if and only if X spans M. Hence we may assume that C contains an element f that is not $\text{cl}(X)$. Then $\kappa_{M/f}(C - f, X) = \lambda_{M/f}(X) = \lambda_M(X)$ by Corollary 8.2.6(iii). Hence $\kappa_{M/f}(C - f, X) = 2$ and the result follows by induction.   □

Proof of Proposition 9.3.5 Let $E(M_i) = E_i$ and $E_1 \cap E_2 = T$. For $X_i = E_i - T$, Proposition 9.3.4 implies that (X_1, X_2) is an exact 3-separation of M. It suffices to prove that M has a minor isomorphic to M_1. By the definition of a 3-sum, $|X_i| \geq 4$. Suppose X_2 is independent in M. Then

$$2 = \lambda_M(X_2) = r_M(X_2) + r_{M^*}(X_2) - |X_2| = r_{M^*}(X_2).$$

Thus, as M is binary and 3-connected, $|X_2| \leq 3$; a contradiction. We conclude that X_2 is dependent in M and so it contains a circuit C.

Let $P = (M[A])^*$ where A is a matrix over $GF(2)$ whose columns are indexed by the elements of $E_1 \cup E_2$ and whose rows are the incidence vectors of all the circuits of M_1 and all the circuits of M_2. It will be convenient here to operate inside the matroid P. By Lemma 9.3.1, $P\backslash T = M_1 \oplus_3 M_2 = M$. Moreover, by Proposition 9.3.4,

$$r(M) = r(P\backslash T) = r(P) = r(M_1) + r(M_2) - 2 = r(X_1) + r(X_2) - 2.$$

Now suppose that $\kappa_{M_2}(T, C) = \kappa_{M_2}(Y_1, Y_2)$ where (Y_1, Y_2) is a partition of E_2 with $T \subseteq Y_1 \subseteq E_2 - C$. Then, as $T \subseteq \text{cl}(X_1)$ and $T \subseteq Y_1 \subseteq E_2$, we have, by Lemma 8.2.3, that $2 \leq \sqcap(X_1, Y_1) \leq \sqcap(X_1, E_2) = 2$. Hence

$$r(X_1) + r(Y_1) - 2 = r(X_1 \cup Y_1).$$

Thus

$$\begin{aligned}
\kappa_{M_2}(T, C) &= \kappa_{M_2}(Y_1, Y_2) \\
&= r(Y_1) + r(Y_2) - r(M_2) \\
&= r(M_1) + r(Y_1) - 2 + r(Y_2) - r(M_2) - r(M_1) + 2 \\
&= (r(X_1) + r(Y_1) - 2) + r(Y_2) - (r(M_2) + r(M_1) - 2)
\end{aligned}$$

Hence

$$\kappa_{M_2}(T, C) = r(X_1 \cup Y_1) + r(Y_2) - r(M)$$
$$= r(X_1 \cup (Y_1 - T)) + r(Y_2) - r(M)$$
$$= \kappa_M(X_1 \cup (Y_1 - T), Y_2) \geq 2,$$

where the last inequality holds since M is 3-connected. But $\kappa_{M_2}(T, C) \leq r(T) + r(X_2) - r(M_2) = 2$, so $\kappa_{M_2}(T, C) = 2$. By Lemma 9.3.6, C has a subset $\{a, b, c\}$ and M_2 has a minor N with ground set $T \cup \{a, b, c\}$ that has $\{a, b, c\}$ as a circuit such that $\kappa_N(T, \{a, b, c\}) = 2$. Thus $r_N(T) + r_N(\{a, b, c\}) - r(N) = 2$. Hence $r_N(T) = r(N)$. Moreover, $2 \geq r_N(T) = r(N) \geq r_N(\{a, b, c\}) = 2$ so N is obtained from T by adding a new element in parallel to each existing element.

As $N = M_2 \backslash W / Z$ for some subsets W and Z of X_2, and $M_2 = P \backslash X_1$, we have $N = P \backslash X_1 \backslash W / Z$. We shall show that $M \backslash W / Z \cong M_1$. First note that

$$r_{M_2}(T) = 2 = r_N(T) = r_{M_2/Z}(T) = r_{M_2}(T \cup Z) - r_{M_2}(Z).$$

Thus $r_{M_2}(T \cup Z) = r_{M_2}(T) + r_{M_2}(Z)$. Hence M_2 has no circuits meeting both T and Z. Therefore $M_1 \oplus_3 M_2$ has no circuits meeting X_1 and Z, so $r_P(X_1 \cup Z) = r_P(X_1) + r_P(Z)$. But $T \subseteq \text{cl}(X_1)$, so

$$r_P(E_1 \cup Z) = r_P(X_1 \cup T \cup Z) = r_P(X_1 \cup T) + r_P(Z) = r_P(E_1) + r_P(Z).$$

Thus $P \backslash W \backslash \{a, b, c\} = P|(E_1 \cup Z) = (P|E_1) \oplus (P|Z) = M_1 \oplus (P|Z)$. Hence $P \backslash W \backslash \{a, b, c\}$ has Z as a separator, so $P \backslash W \backslash \{a, b, c\} / Z = P \backslash W \backslash \{a, b, c\} \backslash Z = P|E_1 = M_1$. But $P \backslash W / Z \backslash X_1 = N$, so, in $P \backslash W / Z$, each element of $\{a, b, c\}$ is parallel to a different element of T. Thus $P \backslash W / Z \backslash T \cong P \backslash W / Z \backslash \{a, b, c\}$, that is, $M \backslash W / Z \cong M_1$. □

Exercises

1. Suppose that $M_1 \cong F_7 \cong M_2$ and $M_3 \cong M(K_4)$. For each i in $\{2, 3\}$, let $E(M_1) \cap E(M_i)$ be a triangle of both M_1 and M_i. Show that:
    (a) $M_1 \triangle M_3 \cong F_7^*$; and
    (b) $M_1 \oplus_3 M_2 \cong AG(3, 2)$.
2. (Seymour 1980b) Let M_1 and M_2 be binary matroids. Show that:
    (a) if $M = M_1 \triangle M_2$, then $M^* = M_1^* \triangle M_2^*$;
    (b) if $M = M_1 \oplus_3 M_2$, then $M^* \neq M_1^* \oplus_3 M_2^*$.
3. Let M be a 3-connected binary matroid and T be a triangle of M. Suppose that $M/T = M_1 \oplus M_2$ where $\min\{|E(M_1)|, |E(M_2)|\} \geq 4$. Prove that:
    (a) M can be written as the 3-sum of two matroids each of which is isomorphic to a proper minor of M.
    (b) $M \backslash e$ is 3-connected for all e in T.
4. *Let M be the 3-sum of two regular matroids. Prove that M is regular.

9.4   Other special properties

In the first section of this chapter, we proved several equivalent characterizations of binary matroids. In this section, we note several other special properties of such matroids. We also show that, although there is no polynomial-time algorithm to test whether a matroid is binary, there is such an algorithm to test whether a matroid is graphic.

We begin with an analogue for binary matroids of a well-known graph-theoretic result. Recall that a graph is *Eulerian* if it contains a closed walk that traverses each edge exactly once. One of the first theorems one learns in graph theory is that a non-empty connected graph is Eulerian if and only if all its vertices have even degree. To see how to extend this result to matroids, we observe that a non-empty connected graph is Eulerian if and only if its edge set can be partitioned into cycles. Moreover, it is not difficult to show that all the vertices in a graph G have even degree if and only if all the bonds in G have even cardinality (Exercise 1). Using these ideas, Welsh (1969c) generalized the above characterization of Eulerian graphs to binary matroids. His result is the equivalence of the first two parts in the next proposition. The equivalence of the second and third parts was shown independently by Brylawski (1972) and Heron (1972a). Exercise 7 of Section 6.2 sought a direct proof of the equivalence of (i) and (iii). Another proof of the equivalence of all four parts and of several other related conditions can be found in Brylawski (1982a).

Proposition 9.4.1 *The following statements are equivalent for an n-element binary matroid M:*

(i) $E(M)$ *can be partitioned into circuits.*

(ii)   *Every cocircuit of M has even cardinality.*

(iii) M^* *is a binary affine matroid.*

(iv)   *The n-tuple of all ones is in the circuit space of M.*

Before proving this result, we note that (i) and (ii) need not be equivalent if M is non-binary. For example, the ground set of the non-Fano matroid F_7^- (see Figure 1.15(a)) can be partitioned into a 3-element and a 4-element circuit, but F_7^- has a 5-element cocircuit. Hence (i) does not imply (ii). To see that (ii) does not imply (i), consider, for example, $U_{2,5}$.

Proof of Proposition 9.4.1   Since it is clear that (i) is equivalent to (iv), it suffices to show that (iii) implies (ii), that (ii) implies (iv), and that (iv) implies (iii). We leave the first two of these as exercises and prove the third. Suppose that (iv) holds. Then M^* has non-zero rank, r say. Let $[I_r|D]$ be a binary representation for M^*. Then, as the rows of $[I_r|D]$ span the circuit space of M, the n-tuple of all ones is a sum of rows of $[I_r|D]$. From considering the first r coordinates, we see that every row of $[I_r|D]$ is involved in this sum. Thus if $(x_1, x_2, \ldots, x_r)^T$ is a column of $[I_r|D]$, then $x_1 + x_2 + \cdots + x_r = 1$. Hence the hyperplane $\{(x_1, x_2, \ldots, x_r) : x_1 + x_2 + \cdots + x_r = 0, \quad x_i \in \{0,1\} \text{ for all } i\}$ of

$V(r, 2)$ avoids $E(M^*)$. Thus M^* is a binary affine matroid, that is, (iii) holds. Hence (iv) implies (iii). $\qquad \square$

An easy consequence of the last result is the following well-known graph-theoretic result.

Corollary 9.4.2 *Let G be a connected non-empty plane graph. Then G is Eulerian if and only if G^* is bipartite.* $\qquad \square$

Next we note another property of graphs that extends to binary matroids but not to arbitrary matroids. Let C be a cycle of size exceeding one in a graph G and let e be an edge of G that is in $\mathrm{cl}_{M(G)}(C) - C$ but is not a loop. Then it is easy to see that e is a diagonal of C. Thus there is a partition (X_1, X_2) of C such that both $X_1 \cup e$ and $X_2 \cup e$ are cycles. To see that non-binary matroids need not have this property, consider the following.

Example 9.4.3 Let k be an integer exceeding one and suppose that $M \cong U_{k,k+2}$. Let $E(M) = \{1, 2, \ldots, k+2\}$ and $C = \{1, 2, \ldots, k+1\}$. Then $k+2$ is in $\mathrm{cl}(C) - C$. Evidently there is no partition (X_1, X_2) of C such that $X_1 \cup \{k+2\}$ and $X_2 \cup \{k+2\}$ are circuits of M. Note, however, that if we let $X_i = \{i\}$ for all i in $\{1, 2, \ldots, k+1\}$, then $(X_1, X_2, \ldots, X_{k+1})$ is a partition of C and $(C - X_i) \cup \{k+2\}$ is a circuit of M for all i. $\qquad \square$

This example provides the clue as to how the original graph property can be generalized not only to binary matroids but also to a somewhat larger class of matroids. The proof of the following result is left as an exercise.

Proposition 9.4.4 *Let C be a circuit of a matroid M such that $|C| \geq 2$ and let e be a non-loop element of $\mathrm{cl}(C) - C$. Then, for some integer n exceeding one, there is a partition of C into non-empty subsets $X_1, X_2, \ldots, X_n$ such that each of the sets $(C - X_1) \cup e, (C - X_2) \cup e, \ldots, (C - X_n) \cup e$ is a circuit of M, and M has no other circuits that contain e and are contained in $C \cup e$. Moreover, if M has no minor isomorphic to $U_{k,k+2}$, then $n \leq k$.* $\qquad \square$

On combining the last result with Corollary 6.5.3, we obtain the following result about the behaviour of circuits in $GF(q)$-representable matroids.

Corollary 9.4.5 *Let C be a circuit of a $GF(q)$-representable matroid M. If $e \in E(M) - C$, then there are at most q distinct circuits that contain e and are contained in $C \cup e$.* $\qquad \square$

We know by Proposition 3.1.10 that, for an element e of a matroid M, the circuits of M/e are the minimal non-empty sets of the form $C - e$ where C is a circuit of M. In particular, if $e \in C$ and $|C| \geq 2$, then $C - e$ is a circuit of M/e. The last proposition enables us to determine the effect of contracting e on those circuits of M that do not contain e. The straightforward proof of the next result is left as an exercise.

Corollary 9.4.6 *Let C be a circuit of a matroid M and e be an element of $E(M) - C$. Then either C is a circuit of M/e, or, for some integer n exceeding* one, there is a partition of C into non-empty subsets $X_1, X_2, \ldots, X_n$ such that each of the sets $C - X_1, C - X_2, \ldots, C - X_n$ is a circuit of M/e, and M/e has no other circuits contained in C. Moreover, if M has no minor isomorphic to $U_{k,k+2}$, then $n \leq k$.   □

This result specializes to binary matroids as follows.

Corollary 9.4.7 *Let C be a circuit of a binary matroid M and e be an element of $E(M) - C$. Then, in M/e, either C is a circuit, or C is a disjoint union of two circuits. In both cases, M/e has no other circuits contained in C.*   □

Next we show how Proposition 9.4.4 can be extended to give a characterization of the matroids having no $U_{k,k+2}$-minor.

Proposition 9.4.8 *Let k be an integer exceeding one. Then a matroid M has no minor isomorphic to $U_{k,k+2}$ if and only if, for every circuit C and every subset S of $E(M) - C$ that is a series class of $M|(C \cup S)$, there are at most k distinct circuits of M that contain S and are contained in $C \cup S$.*

Before proving this result, we note that, on combining it with Theorem 9.1.3, we get yet another characterization of binary matroids. This result is essentially just a restatement of an observation of Bixby (1976) (see Exercise 13).

Corollary 9.4.9 *A matroid M is binary if and only if, for every circuit C and every subset S of $E(M) - C$ that is a series class of $M|(C \cup S)$, there are at most two distinct circuits of $M|(C \cup S)$ that contain S.*   □

Proof of Proposition 9.4.8 Assume that M has no minor isomorphic to $U_{k,k+2}$. Let C be a circuit of M and S be a subset of $E(M) - C$ that is a series class of $M|(C \cup S)$. Choose e in S and contract $S - e$ from $M|(C \cup S)$ to obtain the matroid N. This matroid has C as a circuit because S is a series class of $M|(C \cup S)$. Moreover, if $D \subseteq C$, then $D \cup e$ is a circuit of N containing e if and only if $D \cup S$ is a circuit of M containing S. It follows, without difficulty, from Proposition 9.4.4 that $M|(C \cup S)$ has at most k distinct circuits that contain S. We leave the proof of the converse to the reader.   □

For the next special property of binary matroids to be noted, we look again at circuits comparing the number of these sets with the number of bases. For an arbitrary matroid M, we shall denote these numbers by $d(M)$ and $b(M)$, respectively. Welsh (1976, p. 287) suggests that 'on average a matroid has more bases than circuits' and poses the problem of determining the values of r and n such that if $\mathcal{N}$ is the set of all non-isomorphic n-element matroids of rank r, then $\sum_{M \in \mathcal{N}} b(M) \geq \sum_{M \in \mathcal{N}} d(M)$. We shall not address this problem specifically. Instead, we shall compare $b(M)$ and $d(M)$ for certain fixed matroids M and, in particular, for simple binary matroids. The following result of Oxley (1983) extends a result of Quirk and Seymour (in Welsh 1976). The proof is omitted.

Proposition 9.4.10 *If M is a simple binary matroid, then $b(M) \geq 2d(M)$. Moreover, equality holds here if and only if M is isomorphic to the direct sum of the Fano matroid and a free matroid.*   □

The last result can be used to prove another bound which sharpens the last bound unless M is the direct sum of a free matroid and a matroid of rank less than six (Oxley 1983).

Proposition 9.4.11 *For a rank-r simple binary matroid M having no coloops,*

$$b(M) > \tfrac{6}{19}(r+1)d(M).$$ $\square$

The constant $\tfrac{6}{19}$ in the last inequality does not seem to be best-possible. Indeed, the referee of Oxley's paper proposed the following conjecture.

Conjecture 9.4.12 *For a rank-r simple binary matroid M having no coloops,*

$$b(M) \geq \tfrac{1}{2}(r+1)d(M).$$ $\square$

For simple *non-binary* matroids M, the quotient $b(M)/d(M)$ can become arbitrarily close to zero (see Exercise 5(b)). Intuitively, the contrast between binary and non-binary matroids here reflects the fact that circuits behave more predictably in binary matroids than they do in matroids in general.

We opened this section by proving a generalization to arbitrary binary matroids of the well-known characterization of Eulerian graphs. In that generalization, the role of a vertex was assumed by a cocircuit. Although cocircuits are not the exact counterparts of vertices, one can often obtain a matroid analogue of a graph result by letting cocircuits take the place of vertices. To exemplify this, consider the following graph result of Kaugars (in Harary 1969, p. 31).

Proposition 9.4.13 *If a simple graph G is a block having minimum degree at least three, then G has a vertex v such that $G - v$ is also a block.* $\square$

Now recall that the notions of simplicity for graphs and matroids coincide and that a graph is a block if and only if its cycle matroid is connected. Then, letting cocircuits take the role of vertices, we have the following potential matroid analogue of Proposition 9.4.13: If M is a simple connected matroid in which every cocircuit has at least three elements, then M has a cocircuit C^* such that $M \backslash C^*$ is connected. To see that this assertion does not hold for all matroids, one needs only to consider $U_{3,5}$. However, Seymour (in Oxley 1978b) showed that the assertion is true for binary matroids. The proof is left to the reader.

Proposition 9.4.14 *Let M be a simple connected binary matroid for which every cocircuit has at least three elements. Then M has a cocircuit C^* such that $M \backslash C^*$ is connected.* $\square$

In general, we call a cocircuit C^* in a connected matroid M *non-separating* if $M \backslash C^*$ is connected. In a 3-connected simple graph G, a cocircuit C^* of $M(G)$ is non-separating if and only if C^* is the set of edges incident with some vertex of G. This fact motivated Kelmans (1980, 1981) to define a *vertex* in a connected matroid M to be a non-separating cocircuit of M. Later in this section, we shall see how we can determine whether or not a matroid is graphic by considering the matroids obtained from it by deleting a cocircuit.

The next theorem collects together several attractive properties of non-separating cocircuits in 3-connected binary matroids. The first three of these were conjectured by J. Edmonds and proved by Bixby and Cunningham (1979); the first two were also proved independently by Kelmans (1981). The last was conjectured by H. Wu and proved by Lemos (2009).

Theorem 9.4.15 *In a 3-connected binary matroid M with $|E(M)| \geq 4$:*

(i) *Every element belongs to at least two non-separating cocircuits.*

(ii) *M is graphic if and only if every element belongs to at most two non-separating cocircuits.*

(iii) *Every cocircuit of M is a symmetric difference of non-separating cocircuits.*

(iv) *M is graphic if and only if every element avoids exactly $r(M) - 1$ non-separating cocircuits.* $\square$

In Chapter 6, as an indication of the computational complexity of determining whether a matroid is $\mathbb{F}$-representable, we remarked that Seymour (1981a) has shown that this problem is difficult even when $\mathbb{F} = GF(2)$. To close this chapter, we shall prove this result and note a contrasting result for graphic matroids.

We know that matroids can be specified in many ways. Moreover, certain types of matroids can be represented by some succinct structure such as a graph or a matrix over a field. However, as Robinson and Welsh (1980) note, in general, the size of the input for a matroid problem on an n-element set is $O(2^n)$. For this reason, when discussing the computational complexity of matroid properties, it is usual to assume that the computer being used has a special subroutine or *oracle* for quickly deciding whether a given set is independent, or whether it is a basis or a circuit, or even for deciding its rank. We shall assume here that all matroids M are presented by means of an independence oracle which will determine, in unit time, whether or not any given subset of $E(M)$ is independent. Formally, an *independence oracle* is a function Ω defined for all subsets X of $E(M)$ by

$$\Omega(X) = \begin{cases} \text{YES} & \text{if } X \in \mathcal{I}(M); \\ \text{NO} & \text{if } X \notin \mathcal{I}(M). \end{cases}$$

To determine whether a matroid on a given set E has a certain property, one must apply the independence oracle to some sequence of subsets of E. Each application of the independence oracle to a particular subset of E is called a *probe*. Evidently, since we know we have a matroid, each probe gives more information than simply whether the probed set X is independent. For example, if $\Omega(X) = \text{YES}$, then we deduce that $\Omega(Y) = \text{YES}$ for all subsets Y of X. Similarly, if $\Omega(X) = \text{NO}$, we know that $\Omega(Z) = \text{NO}$ for all subsets Z of E that contain X.

Example 9.4.16 Suppose we wish to determine whether a matroid on the set $\{1, 2, 3, 4\}$ is binary. How many probes of the independence oracle are required? We know that $U_{2,4}$ is the only non-binary matroid on a 4-element set. But, there are six different labelled matroids on the set $\{1, 2, 3, 4\}$ that are isomorphic to

$$
\begin{array}{c}
\begin{array}{cccccccc}
x_1 & x_2 & \cdots & x_r & y_1 & y_2 & y_3 & \cdots & y_r
\end{array}\\
\left[
\begin{array}{cccc|cccccc}
 & & & & 0 & 1 & 1 & \cdots & 1\\
 & & & & 1 & 0 & 1 & \cdots & 1\\
 & I_r & & & 1 & 1 & 0 & \cdots & 1\\
 & & & & \vdots & \vdots & \vdots & \ddots & \vdots\\
 & & & & 1 & 1 & 1 & \cdots & 0
\end{array}
\right]
\end{array}
$$

FIG. 9.4. A binary representation for the matroid N_r.

$M(\mathcal{W}_2)$. To distinguish between $U_{2,4}$ and each of these matroids, one must probe all 2-element subsets of $\{1,2,3,4\}$. Hence, at least six probes are needed. We leave it to the reader to determine exactly how many probes are required. ☐

The last example illustrates the main idea in the proof of Seymour's result (1981a) that there is no polynomial-time algorithm to test if a matroid presented by an independence oracle is binary. As we see in Exercises 8 and 9, this result generalizes to all other fields.

Proposition 9.4.17 *There is no function f in $\mathbb{Z}[x]$ such that, for all positive integers n and every n-element matroid M, at most $f(n)$ probes of an independence oracle are required to determine whether M is binary.*

Proof This is based on the following:

Example 9.4.18 Let r be an integer exceeding two and N_r be the vector matroid of the matrix over $GF(2)$ shown in Figure 9.4. Then $N_3 \cong M(K_4)$ and $N_4 \cong AG(3,2)$. Moreover, it is straightforward to check that the set of circuits of N_r consists of all sets of the form $\{x_i, y_i, x_j, y_j\}$ for $1 \le i < j \le r$, along with the collection $\mathcal{D}$ of sets of the form $\{d_1, d_2, \ldots, d_r\}$ where each d_i is in $\{x_i, y_i\}$, and $|\{d_1, d_2, \ldots, d_r\} \cap \{y_1, y_2, \ldots, y_r\}|$ is odd. Indeed, N_r is a tipless r-spike. Clearly each member of $\mathcal{D}$ is also a hyperplane of N_r. Moreover, $|\mathcal{D}|$ is 2^{r-1}, the number of subsets of $\{y_1, y_2, \ldots, y_r\}$ with an odd number of elements. ☐

If $D \in \mathcal{D}$, let $N_r(D)$ denote the matroid obtained from N_r by relaxing the circuit–hyperplane D. It is straightforward to show that $N_r(D)$ is non-binary (Exercise 7). Now, in order to distinguish N_r from each of the matroids $N_r(D)$, one must probe each of the 2^{r-1} sets of $\mathcal{D}$. Hence the number of probes required is at least $2^{(|E(N_r)|/2)-1}$ and the proposition follows. ☐

In contrast to the last result, Seymour (1981a) also proved that there *is* a polynomial-time algorithm to determine whether a matroid presented by an independence oracle is graphic. Before discussing this, we look briefly at some other matroid oracles. The basis and circuit oracles are defined analogously to the independence oracle and provide YES or NO in answer to the questions "Is X a basis?" and "Is X a circuit?" For a matroid M, a *rank oracle* R will, in unit time, provide the rank of any given subset of $E(M)$. Formally, R is a function defined on all subsets X of $E(M)$ such that $R(X)$ gives the value of $r_M(X)$.

We show next that the independence and rank oracles are *polynomially equivalent* in the sense that, for every subset X of an n-element matroid M, the number of probes of each oracle needed to determine the other oracle's value on X is bounded above by a polynomial in n. Given a rank oracle R, a set X is independent if and only if $R(X) = |X|$. On the other hand, suppose we have an independence oracle Ω for M. Then, for each subset X of $E(M)$, we can determine its rank by at most n probes of Ω (Robinson and Welsh 1980). To see this, let $X = \{x_1, x_2, \ldots, x_k\}$, and consider the following recursive procedure. Let $X_0 = \emptyset$. For each i in $\{1, 2, \ldots, k\}$, let

$$X_i = \begin{cases} X_{i-1} \cup x_i & \text{if } \Omega(X_{i-1} \cup x_i) = \text{YES}; \\ X_{i-1} & \text{otherwise.} \end{cases} \tag{9.3}$$

One easily checks that X_k is a basis of X, so $r(X) = |X_k|$. Hence the independence and rank oracles are indeed polynomially equivalent.

Robinson and Welsh (1980) showed that, by contrast with the above, the independence and basis oracles are not polynomially equivalent. To see this, suppose that we have a given input matroid M on an n-element set E and we wish to determine if M has a loop. This can clearly be done using at most n probes of the independence oracle, but what if we are using a basis oracle? Consider the following example. For each $\lfloor n/2 \rfloor$-element subset A of E, let a_A be an element of $E - A$. Define $M_1(A)$ and $M_2(A)$ to be the matroids on E having $\{A \cup x : x \in E - A\}$ and $\{A \cup x : x \in E - (A \cup a_A)\}$ as their sets of bases. One easily checks that $M_1(A)$ and $M_2(A)$ are indeed matroids and that $M_2(A)$ has a_A as its unique loop, whereas $M_1(A)$ has no loops. For each i in $\{1, 2\}$, there are clearly $\binom{n}{\lfloor n/2 \rfloor}$ matroids on E of the form $M_i(A)$. We leave it to the reader to use these ideas to complete the argument that no polynomially bounded number of probes of the basis oracle will determine whether or not M has a loop.

Further discussion of matroid oracles and of the complexity of certain matroid properties with respect to such oracles can be found in the papers of Hausmann and Korte (1978a, 1978b, 1981), Robinson and Welsh (1980), Seymour and Walton (1981), Jensen and Korte (1982), Truemper (1982b), and Mayhew (2008).

By Proposition 9.4.17, there is no polynomial-time algorithm to test whether a matroid is binary. To see that there is such an algorithm to test whether a matroid is graphic, we first consider the problem of testing whether a binary matroid is graphic (Tutte 1960). For this, we again consider deleting cocircuits from matroids. First recall that, for a subset X of the ground set E of a matroid M, we write $M.X$ for $M/(E - X)$. Similarly, in a graph G, for a subset X of $E(G)$, we denote by $G.X$ the graph $G/(E(G) - X)$. Now let C^* be a cocircuit of the matroid M. A *bridge* of C^* is the ground set of a component of $M \backslash C^*$. Two distinct bridges Y_1 and Y_2 of C^* *overlap* if there are no parallel classes Z_1 and Z_2 of $M.(C^* \cup Y_1)$ and $M.(C^* \cup Y_2)$, respectively, such that $C^* \subseteq Z_1 \cup Z_2$.

Example 9.4.19 Let M be the cycle matroid of the graph G in Figure 9.5(a) and let C^* be the cocircuit $\{5, 6, 7, 8\}$. The bridges of $M \backslash C^*$ are $\{1, 2, 4\}, \{3\}$,

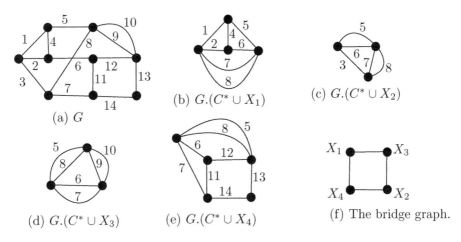

FIG. 9.5. Finding the bridge graph of $\{5,6,7,8\}$ in $M(G)$.

$\{9,10\}$, and $\{11,12,13,14\}$. Call these X_1, X_2, X_3, and X_4. All of the graphs $G.(C^* \cup X_i)$ for i in $\{1,2,3,4\}$ are shown in Figure 9.5(b)–(e). The only two pairs of non-overlapping bridges are $\{X_1, X_2\}$ and $\{X_3, X_4\}$. The graph shown in Figure 9.5(f) is the bridge graph of C^*. In general, the *bridge graph* of a cocircuit C^* of a matroid M is the simple graph whose vertices are the bridges of C^*, with two vertices being adjacent if and only if the corresponding bridges overlap. This graph has also been called the *avoidance graph* of C^* (Bixby and Cunningham 1979, Mighton 2008, Wagner 2010).

The next result identifies one situation in which we can quickly decide that a binary matroid is graphic. Exercise 3 of Section 5.1 outlines the easy proof.

Lemma 9.4.20 *Let B be a basis of a binary matroid M. If each fundamental circuit with respect to B has at most three elements, then M is graphic.* □

By combining the next result with the last lemma, Tutte (1960) gave a simple recursive algorithm for testing whether a binary matroid is graphic.

Theorem 9.4.21 *Let C^* be a cocircuit of a binary matroid M. Then M is graphic if and only if the bridge graph of C^* is bipartite, and $M.(C^* \cup X)$ is graphic for every bridge X of C^*.*

Proof We prove that if M is graphic, then the bridge graph of every cocircuit is bipartite. Our argument follows Geelen and Cunningham (private communication). Let C^* be a cocircuit of $M(G)$ for some graph G. Then we may assume that G is connected. Now C^* is a bond of G, so $G \backslash C^*$ has exactly two components, say G_1 and G_2. Clearly each bridge of C^* is contained in $E(G_1)$ or $E(G_2)$. To prove that the bridge graph is bipartite, it suffices to show that two distinct bridges X and Y contained in $E(G_1)$ do not overlap. Now X and Y are the edge sets of blocks of G_1 and any two such blocks share at most one vertex. Thus

$G_1.(X \cup Y)$ has exactly two blocks, G_{1X} and G_{1Y}, with edge sets X and Y. Let $H = G.(C^* \cup X \cup Y)$. In this graph, every edge of C^* is incident with the vertex v that results from contracting all the edges of $E(G_2)$. Moreover, in G, every edge of C^* has a unique end in $V(G_1)$. Thus, from considering $G_1.(X \cup Y)$, we see that, in H, each edge of C^* has v as one end and has its other end in $V(G_{1X})$ or $V(G_{1Y})$. Let C_X^* be the set of edges of C^* with an end in $V(G_{1X})$, and C_Y^* be the set of edges of C^* with an end in $V(G_{1Y})$. Now $H/Y = G.(C^* \cup X)$ and, in this graph, all the edges of C_X^* are parallel. Similarly, all the edges of C_Y^* are parallel in H/X. Since $C_X^* \cup C_Y^* \supseteq C^*$, we deduce that X and Y do not overlap.

The proof that the two specified conditions are sufficient to guarantee that M is graphic is more difficult and is omitted. □

Although we now have a polynomial-time algorithm to test if a binary matroid is graphic, we know by Proposition 9.4.17 that there is no such algorithm to test if a matroid is binary. Nevertheless, Seymour (1981a) was able to combine Tutte's algorithm with the following result to give a polynomial-time algorithm to test whether an arbitrary matroid, specified by an independence oracle, is graphic. The proof of the next lemma is left for the reader (see Exercise 12).

Lemma 9.4.22 *Let G be a graph with edge set E, and M be a matroid on E. Then $M = M(G)$ if and only if $r(M) = r(M(G))$ and, for each vertex v of G, every bond of G in which all the edges meet v is a cocircuit of M.* □

Proposition 9.4.23 *There is a polynomial-time algorithm to test whether a matroid specified by an independence oracle is graphic.*

Proof Let M be specified by an independence oracle, and $e_1, e_2, \ldots, e_n$ be an ordering of $E(M)$. We can find a basis B of M by using the recursive procedure in (9.3). Next we find the fundamental circuit with respect to B of each element e of $E(M) - B$ by recalling from Exercise 5 of Section 1.2 that, for f in $E(M)$, the set $(B \cup e) - f$ is a basis of M if and only if $f \in C(e, B)$. Form the B-fundamental-circuit incidence matrix A of M, and let M_B be the matroid for which $[I_r|A]$ is a binary representation, where $r = r(M)$. We now use Tutte's algorithm to determine whether M_B is graphic. Suppose that it is not, but that M is graphic. Then M is binary, so $M = M_B$, and M_B is graphic; a contradiction. Thus if M_B is not graphic, then M is not graphic. We may now assume that M_B is graphic. In that case, we find a graph G such that $M(G) = M_B$. The fact that this can be done in polynomial time was shown by Bixby and Cunningham (1980). We now observe that M is graphic if and only if $M = M(G)$, for M is certainly graphic if $M = M(G)$; and if M is graphic, then $M = M_B$ and so M is graphic since M_B is. Finally, we can determine in polynomial time whether or not $M = M(G)$ by applying Lemma 9.4.22. We leave the details here to the reader. □

Fournier (1974) gave a necessary and sufficient condition for an arbitrary matroid to be graphic in terms of special sets of three cocircuits. When C_1^*, C_2^*, and C_3^* are distinct cocircuits of a matroid M, we say that C_1^* *separates* C_2^* *from* C_2^* if no component of $M \backslash C_1^*$ contains $(C_2^* \cup C_3^*) - C_1^*$, that is, no bridge of $M \backslash C_1^*$ contains $(C_2^* \cup C_3^*) - C_1^*$. For instance, in Example 9.4.19, $\{5, 6, 7, 8\}$ separates $\{1, 4, 5\}$ from $\{5, 8, 9, 10\}$, but does not separate $\{5, 6, 8, 11, 14\}$ from $\{5, 8, 12, 13\}$. A collection (C_1^*, C_2^*, C_3^*) of three distinct cocircuits in a matroid is a *Fournier triple* if $C_1^* \cap C_2^* \cap C_3^*$ is non-empty and none of C_1^*, C_2^*, and C_3^* separates the other two.

Proposition 9.4.24 *A matroid is graphic if and only if, for every three distinct cocircuits having a common element, one of the cocircuits separates the other two.*

Proof We shall show that no graphic matroid has a Fournier triple. Again our proof is due to Geelen and Cunningham (private communication). Assume the contrary and let G be a minor-minimal graph for which $M(G)$ has a Fournier triple. Then G is connected. Let (C_1^*, C_2^*, C_3^*) be a Fournier triple in $M(G)$, suppose $e \in C_1^* \cap C_2^* \cap C_3^*$, and let a and b be the ends of e. For each i in $\{1, 2, 3\}$, let A_i and B_i be the components of $G \backslash C_i^*$ where $a \in V(A_i)$ and $b \in V(B_i)$. Let $X = C_1^* \cup C_2^* \cup C_3^*$. Since none of C_1^*, C_2^*, and C_3^* separates the other two, $X - C_i^*$ is contained in either A_i or B_i. By reordering (C_1^*, C_2^*, C_3^*) and possibly interchanging a and b, we may assume that $X - C_i^* \subseteq A_i^*$ for each i in $\{1, 2\}$. As C_1^* and C_2^* are distinct cocircuits, there is an edge f in $C_2^* - C_1^*$. Then f has an end v in $V(B_2)$, and B_2 is a connected graph containing no edges of X. Thus, in $G \backslash C_1^*$, all the edges of $B_2 \cup f$ are in B_1, or all such edges are in A_1. Now $f \in X - C_1^*$ and $X - C_1^* \subseteq A_1$. Thus $E(B_2) \subseteq E(A_1)$, so b is a vertex of A_1; a contradiction. We conclude that if M is graphic, then it has no Fournier triple. A proof of the converse may be found in Fournier's original paper. $\square$

Mighton (2008) combined the results of Tutte and Fournier to give a new characterization of when a binary matroid is graphic. Its advantage over the earlier results is that it only needs to be checked for fundamental cocircuits with respect to some fixed basis where we recall from Exercise 10 of Section 2.1 that if B is a basis of a matroid M and $e \in B$, the unique cocircuit of M contained in $(E(M) - B) \cup e$ is called the fundamental cocircuit of e with respect to B.

Theorem 9.4.25 *Let B be a fixed basis of a binary matroid M. Then M is graphic if and only if the bridge graph of every B-fundamental cocircuit is bipartite, and, for every three B-fundamental cocircuits that share a common element, there is one that separates the other two.*

Proof This follows Wagner (2010). In the proofs of Theorem 9.4.21 and Proposition 9.4.24, we showed that if M is graphic, then it satisfies the specified conditions. Now suppose that M satisfies these conditions. We may assume that M is connected, for it suffices to prove that M is graphic in this case. If every B-fundamental circuit has at most three elements, then, by Lemma 9.4.20, M is graphic. Thus we may suppose that M has some B-fundamental circuit, say $C(e, B)$, that contains at least three elements of B. Let C_1^*, C_2^*, and C_3^* be the B-fundamental cocircuits with respect to three distinct elements of $C(e, B) - e$. By orthogonality, $e \in C_1^* \cap C_2^* \cap C_3^*$. Thus if C_1^*, C_2^*, and C_3^* are all non-separating, then (C_1^*, C_2^*, C_3^*) is a Fournier triple; a contradiction. Hence we may assume that C_1^*, say, is separating. Under the hypotheses governing M, it follows by Theorem 9.4.21 that M is graphic if and only if $M.(C_1^* \cup X)$ is graphic for all bridges X of C_1^*. The proof is now completed inductively following the outline in Exercise 13. We leave it to the reader to furnish the details.   □

Theorem 9.4.25 gives an elegant polynomial-time algorithm to test whether a binary matroid is graphic. Geelen and Cunningham (private communication) have noted that one can obtain a similar algorithm for testing whether a matroid is 3-connected from the following theorem, which, although it is new, is a straightforward consequence of results of Bixby and Cunningham (1979).

Theorem 9.4.26 *A simple connected matroid with a fixed basis B is 3-connected if and only if the bridge graph of each of its B-fundamental cocircuits is connected.*   □

Exercises

1. Show that every vertex of a graph has even degree if and only if every bond has even cardinality.
2. (a) In Proposition 9.4.1, prove that (ii) implies (iv).
   (b) Let M be a binary matroid. Prove that $E(M)$ can be partitioned into circuits if and only if there is a basis of the cocircuit space all of whose members have even support.
3. Give a direct proof of 9.4.7 that does not use 9.4.6.
4. (Seymour 1986a) Prove that a binary, 3-connected, vertically 4-connected matroid is internally 4-connected.
5. Let $c_{r+1}(M)$ denote the number of $(r+1)$-element circuits of a rank-r matroid M. Show that:
   (a) If M is simple and binary, then $b(M) \geq (r+1)c_{r+1}(M)$.
   (b) If $M = PG(r-1, q)$ and $r \geq 2$, then $(q-1)^{r-1}b(M) = (r+1)c_{r+1}(M)$.
6. (Robinson and Welsh 1980) Show that none of the basis, circuit, or rank oracles is polynomially equivalent to a different one of these oracles.
7. (a) Show that none of the matroids $N_r(D)$ in the proof of Proposition 9.4.17 is binary.
   (b) For fixed r, how many non-isomorphic matroids have the form $N_r(D)$?
8. Let $q = p^k$ where p is prime, and let $N_{r,q}$ be the vector matroid of the matrix in Example 9.4.18 viewed over $GF(q)$. Then $N_{r,2} = N_r$. Show that:
   (a) $N_{r,q}$ is a tipless r-spike;
   (b) if $r \geq 5$, then the non-spanning circuits of $N_{r,q}$ with at least five elements are all the sets of the form $\{d_1, d_2, \ldots, d_r\}$ where $d_i \in \{x_i, y_i\}$ for all i and $|\{d_1, d_2, \ldots, d_r\} \cap \{y_1, y_2, \ldots, y_r\}| \equiv 1 \pmod{p}$;
   (c) Proposition 9.4.17 generalizes to $GF(q)$-representable matroids.
9. Use the matroids $M_r(k)$ from Lemma 6.5.18 and the proof of this lemma to show that Proposition 9.4.17 generalizes to all infinite fields.

10. (a) Let M be a matroid such that, for every circuit C and every element e not in C, there are at most two distinct circuits that contain e and are contained in $C \cup e$. Show that M need not be binary.

   (b) Assume that both a matroid M and its dual have the property that, for every circuit C and every series class S disjoint from C, there are at most three distinct circuits that contain S and are contained in $C \cup S$. Show that M need not be ternary.

11. (Lehman 1964) Let e be an element of a connected binary matroid M and let $\mathcal{C}_e$ be the set of circuits containing e. Show that the circuits of M not containing e are precisely the minimal non-empty sets of the form $C_1 \triangle C_2$ where C_1 and C_2 are in $\mathcal{C}_e$.

12. (Seymour 1981a) Let G be a graph with edge set E and M be a matroid with ground set E. Prove that $M = M(G)$ if and only if $r(M) \leq r(M(G))$ and, for every vertex v of G, the set of non-loop edges of G meeting v can be written as a union of cocircuits of M.

13. (Wagner 2010) Let B be a basis of a matroid M, let C^* be a B-fundamental cocircuit, and let X be a bridge of C^*. Prove that:

   (a) The set $B \cap (C^* \cup X)$ is a basis B' of $M.(C^* \cup X)$.

   (b) If $M.(C^* \cup X)$ has a Fournier triple of B'-fundamental cocircuits, then M has a Fournier triple of B-fundamental cocircuits.

   (c) For a B-fundamental cocircuit D^* of M that is also a cocircuit of $M.(C^* \cup X)$, if M is connected and the bridge graph of D^* in M is bipartite, then so is the bridge graph of D^* in $M.(C^* \cup X)$.

14. (a) (Bixby 1976) Prove that a matroid is binary if and only if it has no series minor isomorphic to any of the matroids $U_{r,r+2}$ for $r \geq 2$.

   (b) State the corresponding excluded-parallel minor characterization.

   (c) Use (a) and duality to give an alternative proof of Corollary 9.4.9.

15. *(Jensen and Korte 1982) Let M be a non-uniform matroid such that M and M^* are paving. Prove that there is no function f in $\mathbb{Z}[x]$ such that, for all positive integers n, one can determine whether an n-element matroid has an M-minor by using at most $f(n)$ probes of an independence oracle.

16. *(Wu 1998b)) Let M be a 3-connected binary matroid with at least four elements. Prove that:

   (a) If $M \backslash e$ is not 3-connected for all elements e of some circuit C of M, then M has least two non-separating triads meeting C.

   (b) If M is minimally 3-connected, then M has at least $\frac{1}{2}r^*(M) + 1$ non-separating triads.

17. *(Mayhew, Royle, and Whittle 2009a) Let M and N be 3-connected binary matroids with $|E(M)| \geq 7$. Let e be an element of M such that $M \backslash e$ has a 2-separation (X_1, X_2) where $r_M(X_1), r_M(X_2) \geq 3$. Prove that if $M \backslash e$ has an N-minor, then so does M/e.

10

EXCLUDED-MINOR THEOREMS

Many of the classes of matroids that we have met in this book are minor-closed, that is, every minor of a member of the class is also in the class. Every minor-closed class of matroids can be characterized by a list of *excluded minors*, that is, matroids that are not in the class themselves but have all their proper minors in the class. In general, the problem of explicitly determining all excluded minors for a given class can be very challenging. Indeed, we recall from 6.5.11 that, for all $q \geq 5$, it is even difficult to resolve Rota's Conjecture that the set of excluded minors for $GF(q)$-representability is finite. We proved in Section 6.5 that the class of binary matroids has a unique excluded minor, namely $U_{2,4}$. In that section, we also stated the excluded-minor characterizations for the classes of ternary and quaternary, $GF(3)$- and $GF(4)$-representable, matroids. In this chapter, we shall prove the ternary-matroid result in Section 10.2. Our proof is obtained by modifying the proof we give in Section 10.1 of Tutte's (1958) excluded-minor characterization of the class of regular matroids. We recall from Theorem 6.6.3 that a matroid is regular if and only if it is both ternary and binary. In Section 10.3, the focus switches to graphic matroids and we prove Tutte's (1959) excluded-minor characterization of the class of graphic matroids. The theorems proved in this chapter are among the most celebrated results in matroid theory.

10.1 The characterization of regular matroids

A regular matroid was defined in Chapter 1 to be a matroid that can be represented over $\mathbb{R}$ by a totally unimodular matrix, the latter being a matrix for which the determinant of every square submatrix is in $\{0, 1, -1\}$. Totally unimodular matrices are of fundamental importance in combinatorial optimization because, as Hoffman and Kruskal (1956) showed, there is a basic link between total unimodularity and integer linear programming, the details of which can be found in Schrijver's book (1986, Chapter 19). As we have seen, a totally unimodular matrix represents exactly the same matroid irrespective of the field over which we view the matrix. Indeed, by Theorem 6.6.3, the matroids that are representable over every field are precisely the regular matroids. Another important reason for studying the class of regular matroids is that this class occupies an intermediate position between the important classes of graphic and binary matroids. There are a number of results for graphs which, although they cannot be generalized to the class of all binary matroids, can be extended to the class of regular matroids.

The main purpose of this section is to present Gerards' (1989) proof of Tutte's (1958) excluded-minor characterization of the class of regular matroids. Tutte's original proof was long and difficult. An alternative characterization of regular matroids, this a constructive one due to Seymour (1980b), will be given in Chapter 13. Gerards' proof is the most elementary one known of Tutte's theorem and puts one of the most celebrated results of matroid theory within relatively easy reach. We begin by restating the theorem for ease of reference.

Theorem 10.1.1 *A matroid is regular if and only if it has no minor isomorphic to $U_{2,4}$, F_7, or F_7^*.*

This theorem is actually a combination of the excluded-minor characterization of binary matroids (Theorem 6.5.4) and the following result.

Theorem 10.1.2 *A binary matroid is regular if and only if it has no minor isomorphic to F_7 or F_7^*.*

To see this, we note that it is an immediate consequence of the elementary Proposition 6.4.5 that every regular matroid is binary. In order to prove Theorem 10.1.2, we shall reformulate it in matrix terms. First recall from Section 6.4 that if D is a matrix, then $D^\#$ denotes the matrix that is obtained from D by replacing every non-zero entry by a one. Let Y be a $(0,1)$-*matrix*, that is, a matrix for which every entry is in $\{0,1\}$. A *signing* of Y is a real matrix Z having every entry in $\{0,1,-1\}$ such that $Z^\# = Y$. Thus every signing of Y can be obtained from Y by changing some of its entries from 1 to -1.

Lemma 10.1.3 *Let $[I_r|X]$ be a $GF(2)$-representation of a binary matroid M. The following are equivalent.*

(i) *M is regular.*

(ii) *X has a totally unimodular signing.*

(iii) *$[I_r|X]$ has a totally unimodular signing.*

Proof Suppose X has a signing, Z, that is totally unimodular. Then $[I_r|Z]$ is a totally unimodular signing of $[I_r|X]$. Thus, by Proposition 6.4.5, $M[I_r|Z] = M[I_r|X] = M$, so M is regular. Hence (ii) implies (iii), and (iii) implies (i). To complete the proof, let M be regular. Then, by Lemma 2.2.21, there is a totally unimodular matrix $[I_r|Z]$ representing M such that the column labels of $[I_r|Z]$ and $[I_r|X]$ appear in the same order. By Proposition 6.4.1, $X^\# = Z^\#$. Thus Z is a totally unimodular signing of X, and (i) implies (ii). $\square$

It follows without difficulty from this lemma that Theorem 10.1.2 is equivalent to the next result where X_F is the matrix

$$\begin{bmatrix} 1 & 1 & 1 & 0 \\ 1 & 1 & 0 & 1 \\ 1 & 0 & 1 & 1 \end{bmatrix}.$$

Thus $[I_3|X_F]$ and $[I_4|X_F^T]$ are $GF(2)$-representations of F_7 and F_7^*.

Theorem 10.1.4 *The following are equivalent for a $(0,1)$-matrix $[I_r|X]$.*

(i) *X has no totally unimodular signing.*

(ii) *When viewed over $GF(2)$, the matrix $[I_r|X]$ can be transformed into $[I_3|X_F]$ or $[I_4|X_F^T]$ by a sequence of the following operations: deleting rows or columns; permuting rows or columns; and pivoting.*

The proof of this theorem, which occupies the rest of this section, will require a number of preliminaries. One of the main tools used in the proof will be the simple bipartite graph $G(X)$ that is associated with a $(0,1)$-matrix X. This was introduced in Section 6.4. We begin with a result for such a graph.

Lemma 10.1.5 *Let G be a simple, connected, bipartite graph such that, whenever two distinct vertices from the same vertex class are deleted, a disconnected graph results. Then G is a path or a cycle.*

Proof Assume that G is neither a path nor a cycle. Then G has a vertex of degree at least three and so has a spanning tree T with a vertex of degree at least three. Thus T has at least three degree-1 vertices and so has two such vertices, u and v, that are in the same vertex class of the bipartite graph G. Since deleting u and v from G produces a connected graph, we have a contradiction. □

The next two lemmas note some basic properties of totally unimodular matrices.

Lemma 10.1.6 *Let D be an $n \times n$ matrix $[d_{ij}]$ with entries in $\{0,1,-1\}$. If $G(D^\#)$ is a cycle, then D is totally unimodular if and only if the number of negative entries in D is congruent to n modulo two.*

Proof As $G(D^\#)$ is a cycle, by permuting rows and permuting columns, D can be transformed into a matrix D_1 such that

$$D_1^\# = \begin{bmatrix} 1 & 0 & 0 & \cdots & 0 & 0 & 1 \\ 1 & 1 & 0 & \cdots & 0 & 0 & 0 \\ 0 & 1 & 1 & \cdots & 0 & 0 & 0 \\ 0 & 0 & 1 & \cdots & 0 & 0 & 0 \\ \vdots & \vdots & \vdots & \ddots & \vdots & \vdots & \vdots \\ 0 & 0 & 0 & \cdots & 1 & 1 & 0 \\ 0 & 0 & 0 & \cdots & 0 & 1 & 1 \end{bmatrix}. \tag{10.1}$$

By 2.2.16, these permutations can only affect the determinant of a square submatrix of D by changing its sign. Therefore we may assume that $D = D_1$. Now deleting vertices from a cycle leaves a forest. Hence if D' is a proper square submatrix of D, then D' has a row or column with at most one non-zero entry. Using this observation iteratively, we deduce that such a submatrix D' has determinant in $\{0,1,-1\}$. Thus D is totally unimodular if and only if $\det D \in \{0,1,-1\}$. By scaling, in turn, column 1, row 2, column 2, row 3, ..., row n, and column n, we can transform D into the matrix D_2 where

$$D_2 = \begin{bmatrix} 1 & 0 & 0 & \cdots & 0 & 0 & d_{1n} \\ 1 & 1 & 0 & \cdots & 0 & 0 & 0 \\ 0 & 1 & 1 & \cdots & 0 & 0 & 0 \\ 0 & 0 & 1 & \cdots & 0 & 0 & 0 \\ \vdots & \vdots & \vdots & \ddots & \vdots & \vdots & \vdots \\ 0 & 0 & 0 & \cdots & 1 & 1 & 0 \\ 0 & 0 & 0 & \cdots & 0 & 1 & 1 \end{bmatrix}$$

and $d_{1n} \in \{1, -1\}$. Moreover, $|\det D_2| = |\det D|$, and D_2 and D have the same number of negative entries modulo two. By expanding along the first row of D_2, we get $\det D_2 = 1 + (-1)^{n+1}d_{1n}$. Hence $|\det D| = |1 + (-1)^{n+1}d_{1n}|$. Thus $\det D \in \{0, 1, -1\}$ if and only if $d_{1n} = (-1)^n$, that is, if and only if the number of negative entries of D is congruent to n modulo two. □

The next lemma is due to Camion (1963). The proof that we shall give is Seymour's (in Gerards 1989).

Lemma 10.1.7 *Let D_1 and D_2 be totally unimodular matrices. If $D_1^{\#} = D_2^{\#}$, then D_2 can be obtained from D_1 by multiplying some rows and columns by -1.*

Proof Clearly $G(D_1^{\#}) = G(D_2^{\#})$. We shall call an edge of this bipartite graph *even* if the corresponding entries in D_1 and D_2 are the same; all other edges are called *odd*. We shall show first that every cycle C of $G(D_1^{\#})$ has an even number of odd edges. The proof of this is by induction on the number n of diagonals of C, the core of the proof being contained in the case $n = 0$, which we now present. Let X_1 and X_2 be the submatrices of D_1 and D_2, respectively, that correspond to C, where we assume that C has no diagonals. Then $G(X_1^{\#}) = G(X_2^{\#}) = C$. An edge of C is odd if and only if exactly one of the corresponding entries of X_1 and X_2 is equal to -1. Thus the set of odd edges of C is the symmetric difference of the set of edges of $G(X_1^{\#})$ for which the corresponding entry of X_1

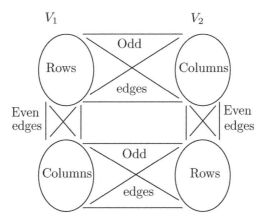

FIG. 10.1. The positions of the even and odd edges in $G(D_1^{\#})$.

is -1 and the set of edges of $G(X_1^\#)$ for which the corresponding entry of X_2 is -1. But, by the preceding lemma, the cardinalities of the last two sets are equal modulo two. Thus C has an even number of odd edges; that is, a cycle in $G(D_1^\#)$ without diagonals has an even number of odd edges. We leave it to the reader to complete the straightforward induction argument that every cycle in $G(D_1^\#)$ has an even number of odd edges.

Next we define a new partition (V_1, V_2) of the vertex set of $G(D_1^\#)$. This part of the argument mimics the argument that is used to show that a graph is bipartite if all its cycles have even length. We first choose a set W of vertices of $G(D_1^\#)$ containing exactly one vertex from each component of $G(D_1^\#)$. For each vertex u of $G(D_1^\#)$, let S_u be the set of shortest paths from u to the unique vertex of W that is in the same component of $G(D_1^\#)$ as u. If P_1 and P_2 are two members of S_u, then the symmetric difference of their edge sets is a disjoint union of cycles and so contains an even number of odd edges. Thus both P_1 and P_2 have an even number of odd edges, or both have an odd number of odd edges. Hence either every member of S_u contains an even number of odd edges, in which case we put u in V_1; or every member of S_u contains an odd number of odd edges and we put u in V_2. Certainly (V_1, V_2) is a partition of the vertex set of $G(D_1^\#)$. Moreover, it is straightforward to show that the odd edges of $G(D_1^\#)$ are precisely those edges that join vertices in V_1 to vertices in V_2 (see Figure 10.1).

If we now multiply by -1 the rows and columns of D_1 that correspond to vertices in V_2, then, when the resulting matrix D_1' is compared with D_2, we find that every edge is even. Hence $D_1' = D_2$. $\qquad\square$

$$\left[\begin{array}{c|c|c} I_r & \begin{array}{c|c} \alpha & \underline{x}^T \\ \hline \underline{y} & Z \end{array} \end{array}\right] \qquad \left[\begin{array}{c|c|c|c} & \begin{array}{c|c} \alpha^{-1} & 0 \\ \hline -\alpha^{-1}\underline{y} & I_{r-1} \end{array} & \begin{array}{c} 1 \\ \hline 0 \end{array} & \begin{array}{c} \alpha^{-1}\underline{x}^T \\ \hline Z - \alpha^{-1}\underline{y}\,\underline{x}^T \end{array} \end{array}\right]$$

$$A_1 \qquad\qquad\qquad\qquad\qquad\qquad A_2$$

$$\left[\begin{array}{c|c} I_r & \begin{array}{c|c} \alpha^{-1} & \alpha^{-1}\underline{x}^T \\ \hline -\alpha^{-1}\underline{y} & Z - \alpha^{-1}\underline{y}\,\underline{x}^T \end{array} \end{array}\right]$$

$$A_3$$

FIG. 10.2. A pivot on the entry α transforms A_1 into A_3.

To complete the preliminaries needed to prove Theorem 10.1.4, we now note some properties of the pivoting operation. This operation was introduced in Section 2.2 and modified in Section 6.6 to include the natural column interchange as part of the operation. Let the matrix A_1 be as shown in Figure 10.2 where Z is a matrix, α is a non-zero scalar, and $\underline{x}$ and $\underline{y}$ are column vectors. Recall that a pivot on α seeks to transform column $r + 1$ of A_1 into the first standard basis vector. Once this is achieved by elementary row operations, columns 1 and $r + 1$ of the resulting matrix A_2 are interchanged. Thus the effect of the whole operation is to transform A_1 into A_3 (see Figure 10.2). Moreover, if we pivot on the entry α^{-1} of A_3, then it is easy to show that we obtain A_1.

Lemma 10.1.8 (i) *If $G(A_1^{\#})$ is connected, then so is $G(A_3^{\#})$.*

(ii) *If $\left[\begin{array}{c|c}\alpha & \underline{x}^T \\ \hline \underline{y} & Z\end{array}\right]$ is square, then its determinant is $\alpha \det(Z - \alpha^{-1}\underline{y}\,\underline{x}^T)$.*

Proof Suppose that $G(A_3^{\#})$ is disconnected. Then, by permuting rows and permuting columns, A_3 can be transformed into a block diagonal matrix D_3. As A_1 can be obtained from A_3 by pivoting on the entry α^{-1} of the latter, a matrix that is a row and column permutation of A_1 can be obtained from D_3 by pivoting. Since a pivot in D_3 clearly produces another matrix in block diagonal form, we conclude that $G(A_1^{\#})$ is disconnected. Thus (i) holds.

Part (ii) follows from 2.2.15 and 2.2.17 by noting which row operations must be applied to obtain A_2 from A_1 in Figure 10.2. ◻

In the rest of the proof of Theorem 10.1.4, when we consider pivoting, it will be convenient to suppress the identity matrices in A_1 and A_3. Hence, when we pivot on the entry α of A_1' in Figure 10.3, we get the matrix A_3'.

$$\left[\begin{array}{c|c}\alpha & \underline{x}^T \\ \hline \underline{y} & Z\end{array}\right] \qquad \left[\begin{array}{c|c}\alpha^{-1} & \alpha^{-1}\underline{x}^T \\ \hline -\alpha^{-1}\underline{y} & Z - \alpha^{-1}\underline{y}\,\underline{x}^T\end{array}\right]$$

$$A_1' \qquad\qquad\qquad A_3'$$

FIG. 10.3. Pivoting on the entry α of A_1' gives A_3'.

Proof of Theorem 10.1.4 Suppose first that X has a totally unimodular signing and assume that $[I_3|X_F]$ or $[I_4|X_F^T]$ can be obtained from $[I_r|X]$ by a sequence of the operations specified in (ii). Then it follows by 2.2.16 and Lemma 2.2.20 that $[I_3|X_F]$ or $[I_4|X_F^T]$ has a totally unimodular signing. Hence $[I_3|X_F]$ has a totally unimodular signing. This implies that F_7 is $\mathbb{R}$-representable, which contradicts Proposition 6.4.8. Alternatively, it is straightforward to show, using scaling, that X_F has no totally unimodular signing. We conclude that if (i) fails, then so does (ii).

Now suppose that X has no totally unimodular signing, but that every proper submatrix of X does have such a signing. Then one easily checks that $G(X)$ is connected but is not a path or a cycle; otherwise X has a totally unimodular signing. It follows by Lemma 10.1.5 that, by permuting columns, X or X^T can be transformed into a matrix of the form $[\underline{x}|\underline{y}|Z]$ where $\underline{x}$ and $\underline{y}$ are column vectors and $G(Z)$ is connected. By assumption, both $[\underline{x}|Z]$ and $[\underline{y}|Z]$ have totally unimodular signings. Moreover, these signings can be chosen so that Z is signed the same way in both cases. To see this, suppose that $[\underline{x}_1|Z_1]$ and $[\underline{y}_2|Z_2]$ are totally unimodular signings of $[\underline{x}|Z]$ and $[\underline{y}|Z]$, respectively. Then, by Lemma 10.1.7, Z_1 can be obtained from Z_2 by multiplying some rows and columns by -1. Performing the same row and column scalings in $[\underline{y}_2|Z_2]$, we transform this matrix into $[\underline{y}_1|Z_1]$. Thus some matrix Y that is obtained from X or X^T by permuting columns has a signing $[\underline{x}_1|\underline{y}_1|Z_1]$ such that

(a) $G(Z_1^\#)$ is connected; and
(b) both $[\underline{x}_1|Z_1]$ and $[\underline{y}_1|Z_1]$ are totally unimodular.

The rest of the proof will show that we can transform Y into X_F by a sequence of pivots, row and column deletions, and row and column permutations.

By assumption, $[\underline{x}_1|\underline{y}_1|Z_1]$ is not totally unimodular. By Lemma 10.1.8(i) and Lemma 2.2.20, if we pivot in $[\underline{x}_1|\underline{y}_1|Z_1]$ on an entry of Z_1, then we maintain a matrix of the same form as $[\underline{x}_1|\underline{y}_1|Z_1]$ for which (a) and (b) both hold. We remark that we do this pivot over $\mathbb{R}$ so as to maintain a signing. By Lemma 2.2.20 again, if the resulting matrix is viewed modulo two, then we get exactly the same matrix as we would by viewing $[\underline{x}_1|\underline{y}_1|Z_1]$ over $GF(2)$ and doing the pivot over $GF(2)$.

Now let $\mathcal{Z}$ be the set of matrices that can be obtained from $[\underline{x}_1|\underline{y}_1|Z_1]$ by a sequence of pivots in Z_1. Let $[\underline{x}_3|\underline{y}_3|Z_3]$ be a member of $\mathcal{Z}$ having a square submatrix W whose determinant is not in $\{0, 1, -1\}$ such that every square matrix smaller than W that is a submatrix of a member of $\mathcal{Z}$ has determinant in $\{0, 1, -1\}$.

Lemma 10.1.9 *The matrix W is a submatrix of $[\underline{x}_3|\underline{y}_3]$ and can be obtained from $\begin{bmatrix} 1 & 1 \\ 1 & -1 \end{bmatrix}$ by multiplying some rows or columns of $[\underline{x}_3|\underline{y}_3|Z_3]$ by -1.*

Proof Clearly W meets both the columns $\underline{x}_3$ and $\underline{y}_3$. If W also meets Z_3, then we pivot in $[\underline{x}_3|\underline{y}_3|Z_3]$ on an entry of Z_3 that is also in W. Let $[\underline{x}_4|\underline{y}_4|Z_4]$ be the resulting matrix. Using Lemma 10.1.8(ii) and the fact that every non-zero entry of Z_3 is in $\{1, -1\}$, we deduce that $[\underline{x}_4|\underline{y}_4|Z_4]$ has a square submatrix W' such that $|\det W'| = |\det W|$ but W' is smaller than W. This contradiction to the choice of W implies that W itself is a submatrix of $[\underline{x}_3|\underline{y}_3]$. Thus W is a 2×2 matrix and the lemma follows without difficulty. $\square$

By this lemma, we may assume that $\begin{bmatrix} 1 & 1 \\ 1 & -1 \end{bmatrix}$ occurs as a submatrix of $[\underline{x}_3|\underline{y}_3]$. By permuting rows, we may also suppose that this submatrix occupies rows 1 and 2 of $[\underline{x}_3|\underline{y}_3]$. Let $Y_3 = [\underline{x}_3|\underline{y}_3|Z_3]$. Now $G(Z_3^\#)$ is connected and so has a shortest path P linking the vertices corresponding to rows 1 and 2. If P has length two, then Y_3 has $\begin{bmatrix} 1 & 1 & a \\ 1 & -1 & b \end{bmatrix}$ as a submatrix for some a and b in $\{1, -1\}$. But if $a = b$, then $|\det \begin{bmatrix} 1 & a \\ -1 & b \end{bmatrix}| = 2$; and, if $a = -b$, then $|\det \begin{bmatrix} 1 & a \\ 1 & b \end{bmatrix}| = 2$. In each case, we obtain a contradiction to the fact that both $[\underline{y}_3|Z_3]$ and $[\underline{x}_3|Z_3]$ are totally unimodular. Thus P must have length exceeding two. Therefore, by permuting rows of Y_3 and columns of Z_3, we can obtain, for some $k \geq 3$, a $k \times (k+1)$ matrix of the form shown in Figure 10.4, where each starred entry is in $\{1, -1\}$ and the entries in the submatrix D are unspecified. Next we note that each of the starred entries may be assumed to be equal to 1. To achieve this, we scale, in turn, column 3, row 3, column 4, ..., column k, and row k. This makes all the starred entries equal to 1 except for the two in the last column. By scaling this column and then row 2, we can make these last two starred entries equal to 1,

$$\begin{bmatrix}
\begin{array}{cc|cccccccc}
1 & 1 & * & 0 & 0 & \cdots & 0 & 0 & 0 \\
1 & -1 & 0 & 0 & 0 & \cdots & 0 & 0 & * \\
\hline
 & & * & * & 0 & \cdots & 0 & 0 & 0 \\
 & & 0 & * & * & \cdots & 0 & 0 & 0 \\
D & & 0 & 0 & * & \cdots & 0 & 0 & 0 \\
 & & \vdots & \vdots & \vdots & \ddots & \vdots & \vdots & \vdots \\
 & & 0 & 0 & 0 & \cdots & * & * & 0 \\
 & & 0 & 0 & 0 & \cdots & 0 & * & *
\end{array}
\end{bmatrix}$$

FIG. 10.4. Permuting rows and columns gives a submatrix of Y_3 like this.

although we may need to interchange columns 1 and 2 to preserve the matrix $\begin{bmatrix} 1 & 1 \\ 1 & -1 \end{bmatrix}$ in the top left corner.

Now, in the matrix in Figure 10.4, we pivot on the entry in row 3 and column 4 and then delete that row and column to obtain the $(k-1) \times k$ matrix

$$\begin{bmatrix}
\begin{array}{cc|cccccccc}
1 & 1 & 1 & 0 & 0 & \cdots & 0 & 0 & 0 \\
1 & -1 & 0 & 0 & 0 & \cdots & 0 & 0 & 1 \\
\hline
 & & -1 & 1 & 0 & \cdots & 0 & 0 & 0 \\
 & & 0 & 1 & 1 & \cdots & 0 & 0 & 0 \\
D' & & 0 & 0 & 1 & \cdots & 0 & 0 & 0 \\
 & & \vdots & \vdots & \vdots & \ddots & \vdots & \vdots & \vdots \\
 & & 0 & 0 & 0 & \cdots & 1 & 1 & 0 \\
 & & 0 & 0 & 0 & \cdots & 0 & 1 & 1
\end{array}
\end{bmatrix}.$$

Arguing as above, we see that, by multiplying some rows and columns of this matrix by -1 and then interchanging columns 1 and 2 if necessary, we can get a $(k-1) \times k$ matrix of the form shown in Figure 10.4. Repeating this process, we eventually obtain a matrix of this form having three rows. This matrix is

$$\begin{bmatrix}
1 & 1 & 1 & 0 \\
1 & -1 & 0 & 1 \\
c & d & 1 & 1
\end{bmatrix} \tag{10.2}$$

where c and d are in $\{0, 1, -1\}$. Moreover, deleting either of the first two columns of this matrix gives a totally unimodular matrix. The submatrices $\begin{bmatrix} 1 & 1 \\ d & 1 \end{bmatrix}$ and $\begin{bmatrix} -1 & 1 \\ d & 1 \end{bmatrix}$ imply that $d \neq -1$ and $d \neq 1$. Thus $d = 0$. The submatrix obtained by deleting column 2 has determinant equal to $c - 2$. Therefore $c = 1$. We conclude that, when viewed over $GF(2)$, the matrix in (10.2) is X_F. This completes the proof that (i) implies (ii) thereby finishing the proof of Theorem 10.1.4. $\square$

Exercises

1. Prove that a real matrix $[I_r|D]$ is totally unimodular if every $r \times r$ submatrix of $[I_r|D]$ has determinant in $\{0, 1, -1\}$.

2. Let A be a $GF(2)$-representation of a matroid M. Give an example for which M is regular but A has no totally unimodular signing.
3. Show that pivoting on the entry α^{-1} of the matrix A_3 in Figure 10.2 gives the matrix A_1.
4. Let A_{10} be the $GF(2)$-representation for R_{10} given after Corollary 6.6.8. Show that R_{10} is regular by giving a totally unimodular signing of A_{10}.
5. *(Geelen 1999) Let $\{x, y\}$ be a coindependent set of elements of a matroid M such that $M\backslash x$ and $M\backslash y$ are binary and $M\backslash x, y$ is connected. Prove that M is binary or M has an F_7^--minor.

10.2 The characterization of ternary matroids

Lovász and Schrijver (in Gerards 1989) observed that the proof above of the excluded-minor characterization of regular matroids can be modified to prove the excluded-minor characterization of ternary matroids. In this section, we follow this approach, drawing heavily on Truemper (1992b), who presented the details of such a proof. Thus the purpose of this section is to prove the following.

Theorem 10.2.1 *A matroid is ternary if and only if it has no minor isomorphic to $U_{2,5}$, $U_{3,5}$, F_7, or F_7^*.*

Because of the similarity of the proof to that of Theorem 10.1.1 given in the last section, we sometimes limit our discussion to a description of the changes required to the earlier proof, leaving completion of the details to the reader.

Proof of Theorem 10.2.1 We showed in Corollary 6.5.3 and Proposition 6.5.5 that $\{U_{2,5}, U_{3,5}, F_7, F_7^*\}$ is contained in the set of excluded minors for $GF(3)$-representability.

Let N be a minor-minimal matroid that is not ternary, let $E(N) = E$, and assume that N has no minor isomorphic to $U_{2,5}$, $U_{3,5}$, F_7, or F_7^*. Then both the rank and corank of N exceed two. Let $[I_r|X]$ be a partial representation for N.

Lemma 10.2.2 *If $G(X)$ is a path or a cycle, then N is ternary.*

Proof If $G(X)$ is a cycle, then N has triangles and triads satisfying the hypothesis of Lemma 8.8.5. By that result, N is a wheel or a whirl, and so, by Proposition 8.8.7, N is ternary. If $G(X)$ is a path, then, by transposing X if necessary, we may assume there is a matrix X' obtained by adding a column e to X so that $G(X')$ is a cycle. Now $[I_r|X']$ is a partial representation of a matroid N' that is obtained by freely adding e on a line spanned by two elements of N. Thus, as above, N' is a wheel or a whirl, so N', and hence N, is ternary. □

The next two lemmas are the appropriate modifications of Lemmas 10.1.6 and 10.1.7. A *ternary signing* of a $(0,1)$-matrix Y is a $(0,\pm1)$-matrix Z over $GF(3)$ such that $Z^\# = Y$.

Lemma 10.2.3 *Let D be a $(0,\pm1)$-matrix. If $G(D^\#)$ is a cycle, then $\det D$ is 0 over $GF(3)$ if and only if the sum over $\mathbb{R}$ of the entries in D is 0 modulo 4.*

Proof We may assume that $D^{\#}$ is the matrix $D_1^{\#}$ from (10.1). Every time we multiply a row or column of D by -1, we multiply its $GF(3)$-determinant by -1 and we do not affect the mod-4 sum of the entries. We deduce that D can be assumed to have all of its non-zero entries equal to 1 except for the top-right entry whose value is d for some d in $\{1, -1\}$. Thus, if D has n rows, then, by expanding the determinant along the first row, we deduce that $\det D = 1 + (-1)^{n+1}d$. Clearly the sum over $\mathbb{R}$ of the entries in D is $d + 2n - 1$. By separating the cases when n is even and when n is odd, it is straightforward to obtain the lemma. □

Lemma 10.2.4 *Let D_1 and D_2 be $(0, \pm 1)$-matrices. If $D_1^{\#} = D_2^{\#}$ and $[I_r|D_1]$ and $[I_r|D_2]$ represent the same matroid over $GF(3)$, then D_2 can be obtained from D_1 by multiplying some rows and columns by -1.*

Proof We follow the argument in the proof of Lemma 10.1.7 verbatim to the point where we consider the numbers of entries in X_1 and X_2 whose value is -1. Call these numbers k_1 and k_2, and let X_1 and X_2 be $m \times m$ matrices. Then, for each i, the sum of the entries in X_i is $2(m - k_i)$. Now suppose that $\det X_1$ is 0 over $GF(3)$. Then $\det X_2$ is also 0 because $M[I_r|D_1] = M[I_r|D_2]$. By Lemma 10.2.3, both $2(m - k_1)$ and $2(m - k_2)$ are congruent to 0 modulo 4, so $k_1 \equiv k_2 \pmod 2$. On the other hand, if $\det X_1$ is non-zero over $GF(3)$, then $\det X_2$ is also non-zero and, by Lemma 10.2.3 again, neither $2(m - k_1)$ nor $2(m - k_2)$ is congruent to 0 modulo 4. Thus, both are congruent to 2 modulo 4 and again $k_1 \equiv k_2 \pmod 2$. We conclude that, in each of the two cases, k_1 and k_2 are equal modulo two and the argument can now be completed as before. □

Now N is not ternary but all its proper minors are. Thus $G(X)$ is connected but, by Lemma 10.2.2, it is not a path or a cycle. It follows by Lemmas 10.1.5 and 10.2.4 that there is a matrix Y that can be obtained from X or X^T, say X, by permuting columns such that Y has a ternary signing $[\underline{x}_1|\underline{y}_1|Z_1]$ where the columns $\underline{x}_1$ and $\underline{y}_1$ correspond to elements a and b of N, and

(a) $G(Z_1^{\#})$ is connected; and

(b) $[I_r|\underline{x}_1|Z_1]$ and $[I_r|\underline{y}_1|Z_1]$ represent $N\backslash b$ and $N\backslash a$, respectively, over $GF(3)$.

In the event that Y is obtained from X^T instead of X, we replace N by its dual in (b) and throughout the rest of the argument.

Let M be the ternary matroid that is represented over $GF(3)$ by $[I_r|\underline{x}_1|\underline{y}_1|Z_1]$. Then $M\backslash a = N\backslash a$ and $M\backslash b = N\backslash b$. By Lemma 10.1.8(i), if we pivot over $GF(3)$ in $[\underline{x}_1|\underline{y}_1|Z_1]$ on an entry of Z_1, we get a matrix of the same form as $[\underline{x}_1|\underline{y}_1|Z_1]$ for which (a) holds. Moreover, (b) still holds since $M\backslash a = N\backslash a$ and $M\backslash b = N\backslash b$.

Now let $\mathcal{Z}$ be the set of matrices that can be obtained from $[\underline{x}_1|\underline{y}_1|Z_1]$ by performing a sequence of pivots in Z_1. For each square submatrix Y of $[\underline{x}_1|\underline{y}_1|Z_1]$, there is a *corresponding* r-element subset B_Y of $E(N)$ consisting of those elements that label the columns of Y together with those elements that label the rows of $[\underline{x}_1|\underline{y}_1|Z_1]$ that do not meet Y. We call such a submatrix Y *violating* for N if B_Y is a basis in exactly one of N and M. Let $[\underline{x}_3|\underline{y}_3|Z_3]$ be a member of

$\mathcal{Z}$ having a violating submatrix W such that no square matrix smaller than W that is a submatrix of a member of $\mathcal{Z}$ is violating.

Lemma 10.2.5 *The matrix W is a submatrix of $[\underline{x}_3|\underline{y}_3]$ that can be obtained from $\left[\begin{smallmatrix} 1 & 1 \\ 1 & \gamma \end{smallmatrix}\right]$, for some γ in $\{1, -1\}$, by multiplying some rows or columns of $[\underline{x}_3|\underline{y}_3|Z_3]$ by -1.*

Proof Clearly W meets both $\underline{x}_3$ and $\underline{y}_3$. Suppose that W also meets Z_3. Then we pivot in $[\underline{x}_3|\underline{y}_3|Z_3]$ on an entry of Z_3 that is also in W. Let $[\underline{x}_4|\underline{y}_4|Z_4]$ be the resulting matrix. By Lemma 10.1.8(ii), $[\underline{x}_4|\underline{y}_4|Z_4]$ has a submatrix W' that is smaller than W such that $\det W' = 0$ if and only if $\det W = 0$, and $B_{W'} = B_W$. Since W is violating, so too is W'; a contradiction to the choice of W. Hence W is a 2×2 matrix.

It remains to show that W has no zero entries. Assume the contrary. Let $S = B_W - \{a, b\}$. Then M/S is represented over $GF(3)$ by $[I_2|W]$, and $\{a, b\}$ is a basis of exactly one of M/S and N/S. As $[I_2|W^{\#}]$ is a partial representation of both of the last two matroids, W has no zero columns. Thus we may suppose that the column of $W^{\#}$ corresponding to a is $(1, 0)^T$. Then $\{a, u\}$ is a circuit of both M/S and N/S, where u labels the first column of I_2. Thus, for each Q in $\{M/S, N/S\}$, the set $\{a, b\}$ is independent in Q if and only if $\{u, b\}$ is independent in $Q \backslash a$. As $M/S \backslash a = N/S \backslash a$, the latter holds if and only if $\{u, b\}$ is independent in both $M/S \backslash a$ and $N/S \backslash a$. Hence either $\{a, b\}$ is independent in both M/S and N/S, or in neither M/S nor N/S; a contradiction. $\square$

The next lemma treats a situation that does not arise in the regular case.

Lemma 10.2.6 *Let D be a 2×3 matrix such that $[I_2|D]$ represents a ternary matroid M_T, and $D^{\#}$ has all its entries equal to 1. Let N_T be a matroid that is represented by $[I_2|D]$ except that one 2×2 submatrix of D is violating. Then $N_T \cong U_{2,5}$.*

Proof Let e and f label the columns of the 2×2 matrix that is violating for N_T, and let g label the third column of D. Let $\mathcal{D}_M$ and $\mathcal{D}_N$ consist of those 2-subsets of $\{e, f, g\}$ that are circuits in M_T and N_T, respectively. Then $\mathcal{D}_M = \mathcal{D}_N - \{\{e, f\}\}$ or $\mathcal{D}_N = \mathcal{D}_M - \{\{e, f\}\}$. Clearly $|\mathcal{D}_M|, |\mathcal{D}_N| \in \{0, 1, 3\}$. Thus $\{|\mathcal{D}_M|, |\mathcal{D}_N|\} = \{0, 1\}$, so $M_T \cong U_{2,5}$ or $N_T \cong U_{2,5}$. As M_T is ternary, the latter holds. $\square$

As in the last section, we may assume that W is $\left[\begin{smallmatrix} 1 & 1 \\ 1 & \gamma \end{smallmatrix}\right]$ and occupies rows 1 and 2 of $[\underline{x}_3|\underline{y}_3]$. Let $Y_3 = [\underline{x}_3|\underline{y}_3|Z_3]$. Every submatrix of Y_3 that has W as a submatrix corresponds to a minor $M \backslash U/V$ of M where $V \subseteq B_W - \{a, b\}$ and U labels some set of columns of Z_3. Clearly V is independent in both M and N. Thus, for $L \in \{M, N\}$, the set $B_W - V$ is independent in $L \backslash U/V$ if and only if B_W is independent in L. Hence W is violating for $N \backslash U/V$.

Let P be a shortest path linking the vertices corresponding to rows 1 and 2 in the connected graph $G(Z_3^{\#})$. Suppose first that P has length two. Then Y_3 has a submatrix A' given by

$$
\begin{array}{ccc}
a & b & y \\
\end{array}
$$
$$
\begin{bmatrix}
1 & 1 & \pm 1 \\
1 & \gamma & \pm 1
\end{bmatrix},
$$

and A' corresponds to a minor M' of M. As above, with $M' = M\backslash U/V$ and $N' = N\backslash U/V$, we have $M'\backslash a = N'\backslash a$ and $M'\backslash b = N'\backslash b$. Hence W is the only 2×2 submatrix of A' that is violating for N'. Thus, by Lemma 10.2.6, $N' \cong U_{2,5}$; a contradiction.

We may now assume that P has length exceeding two. By permuting rows of Y_3 and columns of Z_3 and scaling rows and columns, we can obtain a $k \times (k+1)$ matrix A' of the following form for some $k \geq 3$ and some ζ in $\{1, -1\}$:

$$
\left[
\begin{array}{cc|ccccccc}
1 & 1 & 1 & 0 & 0 & \cdots & 0 & 0 & 0 \\
1 & \gamma & 0 & 0 & 0 & \cdots & 0 & 0 & 1 \\
\hline
 & & 1 & 1 & 0 & \cdots & 0 & 0 & 0 \\
 & & 0 & 1 & 1 & \cdots & 0 & 0 & 0 \\
D & & 0 & 0 & 1 & \cdots & 0 & 0 & 0 \\
 & & \vdots & \vdots & \vdots & \ddots & \vdots & \vdots & \vdots \\
 & & 0 & 0 & 0 & \cdots & 1 & 1 & 0 \\
 & & 0 & 0 & 0 & \cdots & 0 & 1 & \zeta
\end{array}
\right].
$$

As before, A' corresponds to a minor M' of M where $M' = M\backslash U/V$, and W is violating for N' where $N' = N\backslash U/V$. Let B'_W be the subset of $E(N')$ corresponding to W. Then $B'_W = B_W - V$.

Now, as in the regular case, we reduce to a smaller matrix of the same form by pivoting, this time over $GF(3)$. Let the rows and columns of A' be labelled by $s_1, s_2, \ldots, s_k$ and $a, b, t_1, t_2, \ldots, t_{k-1}$. The first pivot we do is on the (s_3, t_2)-entry of A', the entry in row 3 and column 4. After this pivot, we delete row 3 and column 4. The resulting matrix corresponds to $M'/t_2\backslash s_3$. By multiplying some rows and columns of it by -1, we get a matrix A'' that has the same form as A' but is smaller. We want to know that W remains violating for $N'/t_2\backslash s_3$. We observe that $\{s_3, s_4, t_2\}$ is a circuit of both M' and N', and $\{s_3, s_4\} \subseteq B'_W$. Thus, for each L' in $\{M', N'\}$, the following statements are equivalent:

(i) B'_W is a basis of L';

(ii) $(B'_W - s_3) \cup t_2$ is a basis of L';

(iii) $B'_W - s_3$ is a basis of $L'/t_2\backslash s_3$.

It follows that W is, indeed, violating for $N'/t_2\backslash s_3$.

By repeating this process, we eventually obtain the following matrix A:

$$
\begin{array}{c}
\\
e \\
f \\
g
\end{array}
\begin{array}{cccc}
a & b & c & d \\
\end{array}
$$
$$
\begin{array}{c}
\\
e \\
f \\
g
\end{array}
\begin{bmatrix}
1 & 1 & 1 & 0 \\
1 & \gamma & 0 & 1 \\
\alpha & \beta & \delta & \varepsilon
\end{bmatrix},
$$

where $\alpha, \beta, \gamma, \delta$, and ε are in $\{0, 1, -1\}$ but none of γ, δ, or ε is zero. For $V' = \{t_2, t_3, \ldots, t_{k-2}\}$ and $U' = \{s_3, s_4, \ldots, s_{k-1}\}$, this matrix corresponds to $M'/V'\backslash U'$, and W is violating for $N'/V'\backslash U'$. The subset of $E(N'/V'\backslash U')$ corresponding to W is $\{a, b, g\}$ and it is a basis for exactly one of $M'/V'\backslash U'$ and $N'/V'\backslash U'$. Moreover, every violating submatrix of A meets both of the columns a and b.

As the first case, suppose that both α and β are in $\{1, -1\}$. The submatrix indexed by $\{e, g\}$ and $\{a, b\}$ is not violating, otherwise, by applying Lemma 10.2.6 to the submatrix indexed by $\{e, g\}$ and $\{a, b, c\}$, we get the contradiction that N has a $U_{2,5}$-minor. Similarly, by arguing on the submatrix indexed by $\{f, g\}$ and $\{a, b, d\}$, we deduce that the submatrix indexed by $\{f, g\}$ and $\{a, b\}$ is not violating. We conclude that, in the submatrix indexed by $\{a, b\}$ and $\{e, f, g\}$, exactly one of the 2×2 submatrices is violating. It now follows using the dual of Lemma 10.2.6 that N has a $U_{3,5}$-minor; a contradiction.

Suppose next that $\alpha = \beta = 0$. If we pivot on the (e, c)-entry of A, then we return to the first case because the resulting matrix retains W as a violating submatrix with $\{a, b, g\}$ as its corresponding set.

By symmetry and rescaling, we may now assume that $\alpha = 0$ and $\beta = 1$. If $\delta = -1$ or $\gamma = -\varepsilon$, then, by pivoting on the (e, c)-entry or the (f, d)-entry, respectively, we return to the first case. Thus, we may assume that $\delta = 1$ and $\gamma = \varepsilon$. If $\gamma = 1$, then, by pivoting on the (e, c)-entry and then on the (f, d)-entry, we again reduce to the first case. Thus, we may assume that $\gamma = -1$. Hence the matrix A is

$$
\begin{array}{c}
\\
e \\
f \\
g
\end{array}
\begin{array}{cccc}
a & b & c & d \\
\left[\begin{array}{cccc}
1 & 1 & 1 & 0 \\
1 & -1 & 0 & 1 \\
0 & 1 & 1 & -1
\end{array}\right].
\end{array}
$$

It is straightforward to check that the minor M'' of M that is represented by A is F_7^- where the line of F_7 that has been relaxed is $\{a, b, g\}$. Because $M\backslash a = N\backslash a$ and $M\backslash b = N\backslash b$, we deduce that the minor N'' of N corresponding to A can differ from M'' only by dependencies involving $\{a, b\}$. As W is a violating submatrix, $\{a, b\}$ is dependent in N'/g. Hence $\{a, b, g\}$ is a line of N'', so N'', and hence N, is isomorphic to F_7. $\qquad\square$

Theorem 10.2.1 was announced in 1971 by Ralph Reid but he never published his proof. The first published proofs are due to Bixby (1979) and Seymour (1979). Alternative proofs have since been given by Truemper (1982a), Kahn (1984), and Kahn and Seymour (1988). The last paper contains the shortest of those proofs, and it was presented in the first edition of this book. That proof has been replaced here since it is based on exploiting some extremal connectivity results and so is quite different from our proof of the excluded-minor theorem for regular matroids.

Exercises

1. Show that the 8-element rank-4 matroid J for which a geometric representation is shown in Figure 10.4 is ternary and self-dual.

2. (Kahn and Seymour 1988) Let M_1 and M_2 be distinct connected matroids on the same set E, and a and b be distinct elements of E such that

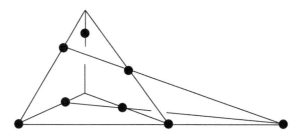

FIG. 10.5. The 8-element rank-4 matroid J.

$M_1 \backslash a = M_2 \backslash a$ and $M_1 \backslash b = M_2 \backslash b$. If $M_1 \backslash a, b$ is connected and $\{a, b\}$ is not a cocircuit of M_1 or M_2, prove that M_1 or M_2 is non-ternary.

3. *(Seymour 1979) Let e and f be distinct elements of a matroid M such that the following hold:

    (a) for any subset X of $E(M) - \{e, f\}$, the set $X \cup e$ is a circuit if and only if $X \cup f$ is;

    (b) e and f are not both loops and are not both coloops, and $\{e, f\}$ is not a circuit or a cocircuit; and

    (c) $M \backslash e, f$ is connected.

    Show that M has a $U_{2,5}$-minor.

4. *(Oxley and Whittle 1998) Let M be a matroid that is obtained by relaxing a circuit-hyperplane H in a ternary matroid N. Prove that M is ternary if and only if N is isomorphic to the cycle matroid of a graph that is obtained from a cycle with edge set H by adjoining an extra vertex v and then adding some non-zero number of edges from v to the vertices of H.

5. *(Bixby 1979) Prove that a matroid is ternary if and only if it has no parallel minor isomorphic to any of F_7, F_7^*, $U_{3,5}$, or $U_{2,n}$ for $n \geq 5$.

10.3 The characterization of graphic matroids

The main purpose of this section is to prove the following excluded-minor characterization of the class of graphic matroids (Tutte 1959), which was stated earlier as Theorem 6.6.7(i).

Theorem 10.3.1 *A matroid is graphic if and only if it has no minor isomorphic to any of the matroids $U_{2,4}$, F_7, F_7^*, $M^*(K_5)$, and $M^*(K_{3,3})$.*

This theorem is actually a combination of three results: the excluded-minor characterization of binary matroids; the assertion that every graphic matroid is binary (Proposition 5.1.2); and the following result, also due to Tutte (1965a).

Theorem 10.3.2 *A binary matroid is graphic if and only if it has no minor isomorphic to F_7, F_7^*, $M^*(K_5)$, or $M^*(K_{3,3})$.*

Alternatively, one can deduce Theorem 10.3.1 by combining three other results: the excluded-minor characterization of regular matroids (Theorem 10.1.1);

the assertion that every graphic matroid is regular (Proposition 5.1.5); and the following theorem of Tutte (1959).

Theorem 10.3.3 *A regular matroid is graphic if and only if it has no minor isomorphic to* $M^*(K_5)$ *or* $M^*(K_{3,3})$.

The proof of Theorem 10.3.2 that will be given here is due to Seymour (1980c). The form of the proof that appears below is the combination of two versions that were privately communicated to the author, one by R. Hall, D. Mayhew, and G. Whittle, and the other by J. Geelen. Although this proof is shorter and conceptually simpler than Tutte's original proof, it is still quite long occupying, after a few short remarks, all of the rest of this section. The proof makes extensive use of Whitney's 2-Isomorphism Theorem (5.3.1). Another proof of Theorem 10.3.2 due to Wagner (1985) appeared in the first edition of this book. Yet another proof, this one based on signed graphs, has been given by Gerards (1995).

We observe here that, by dualizing Theorems 10.3.1, 10.3.2, and 10.3.3, one can deduce excluded-minor characterizations of the class of cographic matroids within the classes of matroids, binary matroids, and regular matroids, respectively. The first of these was stated earlier as (ii) of Theorem 6.6.7. The other two are now stated for reference.

Corollary 10.3.4 *A binary matroid is cographic if and only if it has no minor isomorphic to* F_7, F_7^*, $M(K_5)$, *or* $M(K_{3,3})$. □

Corollary 10.3.5 *A regular matroid is cographic if and only if it has no minor isomorphic to* $M(K_5)$ *or* $M(K_{3,3})$. □

If M is a binary matroid that is an excluded minor for the class of graphic matroids and $e \in E(M)$, then $M \backslash e$ is graphic. This observation motivates the following concept, which was introduced by Seymour (1980b). A *graft* is a pair (G, γ) where G is a graph and γ is a subset of $V(G)$. The *incidence matrix* $A_{(G,\gamma)}$ of (G, γ) is the matrix that is obtained from the mod-2 vertex-edge incidence matrix of G by adjoining a new column e_γ corresponding to γ. Specifically, e_γ is the incidence vector of the set γ, that is, e_γ has a 1 in each row corresponding

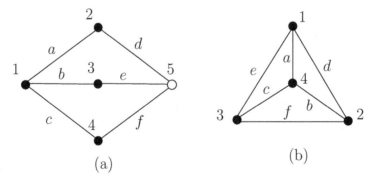

(a)

(b)

FIG. 10.6. Grafts corresponding to F_7^* and F_7.

to a vertex in γ and has a 0 in every other row. The matroid $M(G, \gamma)$ associated with the graft (G, γ) is the vector matroid $M[A_{(G,\gamma)}]$ where $A_{(G,\gamma)}$ is viewed as a matrix over $GF(2)$. Thus the *graft matroid*, $M(G, \gamma)$, has ground set $E(G) \cup e_\gamma$. If $|\gamma|$ is odd, then, since every column of $A_{(G,\gamma)}$ except e_γ has an even number of ones, $M(G, \gamma)$ has no circuits containing e_γ, so e_γ is a coloop and $M(G, \gamma)$ is graphic. For this reason, Seymour and some other authors require, as part of the definition of a graft, that $|\gamma|$ be even. We shall not impose this restriction here although it will take effect in many of our uses of grafts.

It is convenient to view a graft (G, γ) as a graph G and a collection γ of *coloured vertices*. Those vertices not in γ are *uncoloured*. In figures, the coloured and uncoloured vertices will be indicated by solid and hollow circles, respectively.

Example 10.3.6 Let (G_1, γ_1) be the graft shown in Figure 10.6(a). Then $G_1 \cong K_{2,3}$ and $\gamma_1 = V(G_1) - 5$. The matrix $A_{(G_1, \gamma_1)}$ is

$$
\begin{array}{c}
\\ 1 \\ 2 \\ 3 \\ 4 \\ 5
\end{array}
\begin{array}{c}
\begin{array}{ccccccc}
a & b & c & d & e & f & e_{\gamma_1}
\end{array} \\
\left[
\begin{array}{ccccccc}
1 & 1 & 1 & 0 & 0 & 0 & 1 \\
1 & 0 & 0 & 1 & 0 & 0 & 1 \\
0 & 1 & 0 & 0 & 1 & 0 & 1 \\
0 & 0 & 1 & 0 & 0 & 1 & 1 \\
0 & 0 & 0 & 1 & 1 & 1 & 0
\end{array}
\right]
\end{array}.
$$

To recognize $M(G_1, \gamma_1)$, which is $M[A_{(G_1,\gamma_1)}]$, note that, since the rows of $A_{(G_1,\gamma_1)}$ sum to the zero vector, we can obtain another representation for $M(G_1, \gamma_1)$ by deleting the last row of $A_{(G_1,\gamma_1)}$. Once this is done, it is straightforward to see by, for example, pivoting on the first entry of column c, that $M(G_1, \gamma_1) \cong F_7^*$.

Next let (G_2, γ_2) be the graft shown in Figure 10.6(b). Then $G_2 \cong K_4$ and $\gamma_2 = V(G_2)$. The matrix $A_{(G_2, \gamma_2)}$ is

$$
\begin{array}{c}
\\ 1 \\ 2 \\ 3 \\ 4
\end{array}
\begin{array}{c}
\begin{array}{ccccccc}
a & b & c & d & e & f & e_{\gamma_2}
\end{array} \\
\left[
\begin{array}{ccccccc}
1 & 0 & 0 & 1 & 1 & 0 & 1 \\
0 & 1 & 0 & 1 & 0 & 1 & 1 \\
0 & 0 & 1 & 0 & 1 & 1 & 1 \\
1 & 1 & 1 & 0 & 0 & 0 & 1
\end{array}
\right]
\end{array}.
$$

On deleting the last row of $A_{(G_2,\gamma_2)}$, we see that $M(G_2, \gamma_2) \cong F_7$. □

The proof of Theorem 10.3.2 will use the fact that the bond matroids of $K_{3,3}$ and K_5 also arise as grafts. The next example verifies this.

Example 10.3.7 Let (G_3, γ_3) be the graft shown in Figure 10.7(a). Then G_3 is isomorphic to the 4-spoked wheel and γ_3 consists of the vertices of the rim. After deleting the row labelled 5 from $A_{(G_3,\gamma_3)}$, we get the matrix

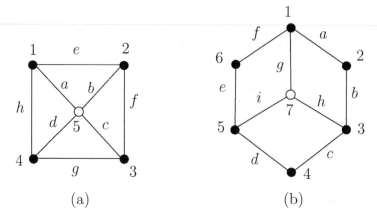

FIG. 10.7. Grafts corresponding to $M^*(K_{3,3})$ and $M^*(K_5)$.

$$
\begin{array}{c}
\quad\; a \;\; b \;\; c \;\; d \;\; e \;\; f \;\; g \;\; h \;\; e_{\gamma_3} \\
\begin{array}{c} 1 \\ 2 \\ 3 \\ 4 \end{array}
\left[
\begin{array}{ccccccccc}
1 & 0 & 0 & 0 & 1 & 0 & 0 & 1 & 1 \\
0 & 1 & 0 & 0 & 1 & 1 & 0 & 0 & 1 \\
0 & 0 & 1 & 0 & 0 & 1 & 1 & 0 & 1 \\
0 & 0 & 0 & 1 & 0 & 0 & 1 & 1 & 1
\end{array}
\right].
\end{array}
$$

Thus the dual of $M(G_3, \gamma_3)$ is represented by

$$
\begin{array}{c}
\quad\; a \;\; b \;\; c \;\; d \;\; e \;\; f \;\; g \;\; h \;\; e_{\gamma_3} \\
\left[
\begin{array}{ccccccccc}
1 & 1 & 0 & 0 & 1 & 0 & 0 & 0 & 0 \\
0 & 1 & 1 & 0 & 0 & 1 & 0 & 0 & 0 \\
0 & 0 & 1 & 1 & 0 & 0 & 1 & 0 & 0 \\
1 & 0 & 0 & 1 & 0 & 0 & 0 & 1 & 0 \\
1 & 1 & 1 & 1 & 0 & 0 & 0 & 0 & 1
\end{array}
\right].
\end{array}
$$

The last matrix represents $M(K_{3,3})$ where $K_{3,3}$ is labelled as in Figure 10.8. We conclude that $M(G_3, \gamma_3) \cong M^*(K_{3,3})$.

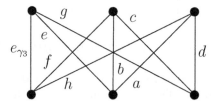

FIG. 10.8. $K_{3,3}$.

Next we assert that, for the graft, (G_4, γ_4) in Figure 10.7(b),

$$M(G_4, \gamma_4) \cong M^*(K_5).$$

To see this, observe that, after deleting the row labelled by 7 from $A_{(G_4,\gamma_4)}$, we get the matrix

$$
\begin{array}{c c}
 & \begin{array}{c c c c c c c c c c} a & b & c & d & e & f & g & h & i & e_{\gamma_4} \end{array} \\
\begin{array}{c} 1 \\ 2 \\ 3 \\ 4 \\ 5 \\ 6 \end{array} &
\left[\begin{array}{c c c c c c c c c c}
1 & 0 & 0 & 0 & 0 & 1 & 1 & 0 & 0 & 1 \\
1 & 1 & 0 & 0 & 0 & 0 & 0 & 0 & 0 & 1 \\
0 & 1 & 1 & 0 & 0 & 0 & 0 & 1 & 0 & 1 \\
0 & 0 & 1 & 1 & 0 & 0 & 0 & 0 & 0 & 1 \\
0 & 0 & 0 & 1 & 1 & 0 & 0 & 0 & 1 & 1 \\
0 & 0 & 0 & 0 & 1 & 1 & 0 & 0 & 0 & 1
\end{array} \right].
\end{array}
$$

By replacing rows 1, 3, and 5 by row 1 + row 6, row 3 + row 2, and row 5 + row 4 and then reordering the columns, we get

$$
\begin{array}{c c}
 & \begin{array}{c c c c c c c c c c} g & b & h & d & i & f & a & c & e & e_{\gamma_4} \end{array} \\
\begin{array}{c} 1 \\ 2 \\ 3 \\ 4 \\ 5 \\ 6 \end{array} &
\left[\begin{array}{c c c c c c c c c c}
1 & 0 & 0 & 0 & 0 & 0 & 1 & 0 & 1 & 0 \\
0 & 1 & 0 & 0 & 0 & 0 & 1 & 0 & 0 & 1 \\
0 & 0 & 1 & 0 & 0 & 0 & 1 & 1 & 0 & 0 \\
0 & 0 & 0 & 1 & 0 & 0 & 0 & 1 & 0 & 1 \\
0 & 0 & 0 & 0 & 1 & 0 & 0 & 1 & 1 & 0 \\
0 & 0 & 0 & 0 & 0 & 1 & 0 & 0 & 1 & 1
\end{array} \right].
\end{array}
$$

The last matrix has the form $[I_6|D]$, and the columns of $[D^T|I_4]$ consist of all binary vectors with four coordinates that have exactly one or exactly two ones. Hence $M[D^T|I_4] \cong M(K_5)$ and so $M(G_4, \gamma_4) \cong M^*(K_5)$. □

Next we show how grafts arise naturally in proving Theorem 10.3.2.

Lemma 10.3.8 *Let M be a binary excluded minor for the class of graphic matroids. If $f \in E(M)$, then there is a graft (G, γ) with $|\gamma| \geq 4$ such that G has no isolated vertices, and M can be obtained from $M(G, \gamma)$ by relabelling e_γ as f.*

Proof Clearly $M\backslash f = M(G)$ for some loopless graph G without isolated vertices, and $r(M) = r(M(G))$. Take the mod-2 vertex-edge incidence matrix A of G and let B be a basis for $M(G)$. The fundamental circuit of f with respect to B in M uniquely determines the column corresponding to f that we can adjoin to A to get a binary representation for M. Label this column by f, and let γ be the support of f, that is, the subset of $V(G)$ corresponding to those rows in which the entry in column f is 1. If $|\gamma|$ is odd, we obtain the contradiction that f is a coloop of M since the sum of all the original rows of A is the zero vector. Hence $|\gamma|$ is even. Moreover, $|\gamma| \geq 4$ otherwise M is graphic. □

Now let (G, γ) be a graft. If $x \in E(G)$, to obtain the *deletion* $(G, \gamma)\backslash x$ and the *contraction* $(G, \gamma)/x$ of x from (G, γ), we delete or contract x from G leaving the set of coloured vertices unchanged except when x is contracted and has distinct ends u and v. In the exceptional case, $(G, \gamma)/x = (G/x, \gamma')$ where the vertex w that results from identifying u and v is coloured if and only if exactly one of u and v is. Equivalently, $A_{(G/x,\gamma')}$ is obtained from $A_{(G,\gamma)}$ by deleting column x and replacing rows u and v by a single row equal to their sum modulo 2. The *minors* of (G, γ) are those grafts that can be produced by a sequence of single-edge deletions and contractions. It is not difficult to check that, for disjoint subsets X and Y of $E(G)$, the graft $(G, \gamma)\backslash X/Y$ is well-defined. We show next that performing these minor operations on grafts corresponds to performing their namesakes on the matroids of the grafts.

Lemma 10.3.9 *Let (G, γ) be a graft and suppose that $x \in E(G)$. Then*

$$M((G, \gamma)\backslash x) = M(G, \gamma)\backslash x \text{ and } M((G, \gamma)/x) = M(G, \gamma)/x.$$

Proof By definition, $(G, \gamma)\backslash x = (G\backslash x, \gamma)$. Hence $M((G, \gamma)\backslash x) = M(G\backslash x, \gamma) = M(G, \gamma)\backslash x$. For contraction, when x is a loop, the result follows from the result for deletion, so assume x has distinct ends u and v. Then $(G, \gamma)/x = (G/x, \gamma')$ where $A_{(G/x,\gamma')}$ is obtained from $A_{(G,\gamma)}$ by deleting column x and replacing rows u and v by their mod-2 sum. Now $M(G, \gamma)/x = M[A_{(G,\gamma)}]/x$. In $A_{(G,\gamma)}$, a pivot on the entry in column x and row u replaces row v by the mod-2 sum of rows u and v. The resulting matrix has a unique non-zero entry in column x and this entry is in row u. Deleting column x and row u from this matrix, we get a matrix A' that represents $M[A_{(G,\gamma)}]/x$. But $A' = A_{(G/x,\gamma')}$, so $M(G, \gamma)/x = M[A_{(G,\gamma)}]/x = M[A_{(G/x,\gamma')}] = M(G/x, \gamma') = M((G, \gamma)/x)$, as required. $\square$

The proof of Theorem 10.3.2 will use the following lemma, which is of independent interest.

Lemma 10.3.10 *Let M be a 3-connected matroid having at least two elements. Then M has at least two elements x such that $\mathrm{si}(M/x)$ is 3-connected.*

Proof Let M be a matroid for which the lemma does not hold such that $|E(M)|$ is as small as possible. Then one easily checks that M is not a wheel or a whirl and that $|E(M)| \geq 5$. Thus, by Lemma 8.8.6, M has at least two non-essential elements. Let e be such an element and suppose that $M\backslash e$ is 3-connected. Then the choice of M implies that $M\backslash e$ has at least two elements x for which $\mathrm{si}(M\backslash e/x)$ is 3-connected. But, for each such element x, it follows by Proposition 8.2.7 that $\mathrm{si}(M/x)$ is 3-connected. Hence the lemma holds for M; a contradiction. We deduce that, for each non-essential element e of M, the deletion $M\backslash e$ is not 3-connected, so the contraction M/e is 3-connected. But there are at least two non-essential elements, so we again obtain the contradiction that the lemma holds for M. $\square$

The last proof is due to H. Wu (private communication). He also proved (Wu 1998a) that if the matroid M in the last lemma has at least three elements, it has at least three elements x for which $\mathrm{si}(M/x)$ is 3-connected (Exercise 7). The existence of at least one such element was first established by Cunningham (1981) and Seymour (1980b).

Recall that a matroid is internally 3-connected precisely when it is 2-connected and has no non-minimal 2-separations. Thus every 2-separation in such a matroid corresponds to a 2-circuit or a 2-cocircuit. A *series couple* in a matroid N is a series class of N containing exactly two elements. We say that N is *3-connected up to series couples* if N is 2-connected having at least six elements and, whenever (X, Y) is a 2-separation of N, either X or Y is a series couple of N. A *parallel couple* in a matroid is a parallel class with exactly two elements, and a matroid is *3-connected up to parallel couples* if its dual is 3-connected up to series couples. Evidently a matroid that is 3-connected up to series couples or is 3-connected up to parallel couples has no non-minimal 2-separations.

Lemma 10.3.11 *Let N be a 3-connected binary matroid having at least four elements. If* $\mathrm{co}(N\backslash x)$ *is 3-connected, then* $N\backslash x$ *is 3-connected up to series couples.*

Proof If Z is a series class of $N\backslash x$, then $Z \cup x$ has rank 2 in N^*. But N^* is simple and binary, so $|Z \cup x| \leq 3$. Hence every non-trivial series class of $N\backslash x$ is a series couple. □

Lemma 10.3.12 *For a graph G, let $M(G)$ be 3-connected up to series couples. If $\{a, b\}$ is a series couple of $M(G)$, then a and b have a common end in G.*

Proof Suppose a and b do not share a common vertex in G. Let v_a and v_b be ends of a and b, respectively, that are in different components, G_a and G_b, of $G\backslash\{a, b\}$. Since $M(G)$ has no series classes of size exceeding two, each of v_a and v_b has degree exceeding two. Thus $(E(G_a) \cup a, E(G_b) \cup b)$ is a non-minimal 2-separation in $M(G)$; a contradiction. □

Now let G be a graph without isolated vertices such that $M(G)$ is 3-connected up to series couples. By Whitney's 2-Isomorphism Theorem, if H is a graph without isolated vertices and $M(H) = M(G)$, then H can be obtained from G by a sequence of moves each consisting of interchanging the labels on the edges of a series couple. We call such a move a *series switch*. If $\{a, b\}$ is a series couple in $M(G)$, then, by Lemma 10.3.12, a and b share a common vertex v. We shall call v the *tip* of $\{a, b\}$. A *foot* of $\{a, b\}$ is an end of a or b that is different from v. Now let G' be obtained from G by a series switch on $\{a, b\}$. The next lemma considers the effect of this series switch on $M(G, \gamma)$ determining, in particular, a graft (G', γ') such that $M(G', \gamma') = M(G, \gamma)$. As we shall see, if v is uncoloured in G, then γ does not change, while if v is coloured in G, the sets of coloured and uncoloured feet of $\{a, b\}$ are swapped. We shall say that the graft (G', γ') has been obtained by a *series switch on* (G, γ). Recall that the symmetric difference $X \triangle Y$ of sets X and Y is the set $(X - Y) \cup (Y - X)$.

Lemma 10.3.13 *Let (G, γ) be a graft where G has no isolated vertices and $M(G)$ is 3-connected up to series couples. In $M(G)$, let $\{a, b\}$ be a series couple with tip v and feet v_a and v_b incident with a and b, respectively. Let G' be obtained from G by a series switch on $\{a, b\}$.*

(i) *If v is coloured, then $M(G, \gamma) = M(G', \gamma')$ where $\gamma' = \gamma \triangle \{v_a, v_b\}$.*

(ii) *If v is uncoloured, then $M(G,\gamma) = M(G',\gamma')$ where $\gamma' = \gamma$.*

Proof For (ii), let $\gamma' = \gamma$ and suppose v is uncoloured. Then, in both $A_{(G,\gamma)}$ and $A_{(G',\gamma')}$, the only non-zero entries in row v are in columns a and b. Thus $\{a,b\}$ is a series couple in both $M(G,\gamma)$ and $M(G',\gamma')$, and it follows that $M(G,\gamma) = M(G',\gamma')$.

For (i), assume that v is coloured and that $\gamma' = \gamma \triangle \{v_a, v_b\}$. We shall relabel by e the columns e_γ and $e_{\gamma'}$ of $A_{(G,\gamma)}$ and $A_{(G',\gamma')}$ corresponding to γ and γ'. By the definition of (G',γ'), the submatrices of $A_{(G,\gamma)}$ and $A_{(G',\gamma')}$ corresponding to the set $\{v, v_a, v_b\}$ of vertices and the set $\{a, b, e\}$ of elements are

$$
\begin{array}{c}
\begin{array}{ccc} a & b & e \end{array} \\
\begin{array}{c} v \\ v_a \\ v_b \end{array}
\begin{bmatrix} 1 & 1 & 1 \\ 1 & 0 & x_a \\ 0 & 1 & x_b \end{bmatrix}
\end{array}
\quad \text{and} \quad
\begin{array}{c}
\begin{array}{ccc} a & b & e \end{array} \\
\begin{array}{c} v \\ v_a \\ v_b \end{array}
\begin{bmatrix} 1 & 1 & 1 \\ 0 & 1 & x_a + 1 \\ 1 & 0 & x_b + 1 \end{bmatrix},
\end{array}
$$

for some x_a and x_b in $\{0,1\}$, where the sums are calculated modulo 2. Moreover, in both $A_{(G,\gamma)}$ and $A_{(G',\gamma')}$, all the non-zero entries in row v and all the non-zero entries in columns a and b are contained in these two submatrices. Thus every circuit of $M(G,\gamma)$ or $M(G',\gamma')$ containing e must contain exactly one of a and b. The sums of columns a and e are the same in $A_{(G,\gamma)}$ and $A_{(G',\gamma')}$, as are the sums of columns b and e. Hence, for each x in $\{a,b\}$, the matroids $M(G,\gamma)$ and $M(G',\gamma')$ have the same sets of circuits containing $\{x, e\}$. Since $M(G) = M(G')$, we deduce that $M(G,\gamma) = M(G',\gamma')$, so (i) holds. □

The following proposition will be in constant use in deciding whether the matroid of a graft is graphic.

Proposition 10.3.14 *Let (G,γ) be a graft where G has no isolated vertices. Let $M(G)$ be 3-connected up to series couples and e_γ be neither a loop nor a coloop of $M(G,\gamma)$. Then $M(G,\gamma)$ is graphic if and only if (G,γ) can be transformed, by a sequence of series switches, to a graft (H,η) such that $|\eta| = 2$.*

Proof If (G,γ) can be transformed as described, then $M(G,\gamma) = M(H,\eta)$. But $M(H,\eta)$ is graphic since $|\eta| = 2$. Hence so is $M(G,\gamma)$.

Now suppose that $M(G,\gamma)$ is graphic and relabel the element e_γ of $M(G,\gamma)$ as e. We have $M(G,\gamma) = M(G')$ for some graph G' without isolated vertices, and $M(G) = M(G')\backslash e = M(G'\backslash e)$. As noted above, Whitney's 2-Isomorphism Theorem implies that $G'\backslash e$ can be obtained from G by a sequence of series switches. If we perform these series switches on (G,γ), we get $(G'\backslash e, \gamma')$ for some γ' and, by Lemma 10.3.13, $M(G,\gamma) = M(G'\backslash e, \gamma')$. Hence $M(G'\backslash e, \gamma') = M(G')$. It remains to show that $|\gamma'| = 2$. To see this, take a cycle C of G' containing e. This cycle exists as e is not a coloop of $M(G')$. Then C is a circuit of $M(G'\backslash e, \gamma')$, that is, of $M[A_{(G'\backslash e, \gamma')}]$. Now $C - e$ is the edge set of a path in $G'\backslash e$. Therefore the mod-2 sum of the columns of $A_{(G'\backslash e, \gamma')}$ corresponding to the edges of this path must be column e of $A_{(G'\backslash e, \gamma')}$. Moreover, this mod-2 sum has exactly two ones otherwise $C = \{e\}$ and e is a loop of $M(G,\gamma)$. Thus $|\gamma'|$ is indeed 2. □

The next lemma was privately communicated to the author by J. Geelen. It will be used not only in the proof of Theorem 10.3.2 but also in the proof of Seymour's Decomposition Theorem for regular matroids (13.1.1).

Lemma 10.3.15 *Let M be a 3-connected binary non-graphic matroid having an element e such that $M \backslash e$ is graphic and 3-connected up to series couples but $M \backslash e$ is not 3-connected. Then*

(i) *M is isomorphic to F_7^* or $M^*(K_5)$;*

(ii) *$M^* \cong M_1 \oplus_3 M(K_5)$ for some cographic matroid M_1; or*

(iii) *M has a triad that contains e and some element f for which M/f is non-graphic.*

Proof Since M is binary and 3-connected, $|E(M)| \notin \{4, 5\}$ and if $|E(M)| = 6$, then, by the Wheels-and-Whirls Theorem (8.8.4), $M \cong M(K_4)$. As M is not graphic, it follows that $|E(M)| \geq 7$. If $|E(M)| = 7$, then $r(M) = 3$ or $r^*(M) = 3$. Hence $M \cong F_7$ or $M \cong F_7^*$. But every single-element deletion of F_7 is isomorphic to $M(K_4)$ and so is 3-connected. Thus if $|E(M)| = 7$, then $M \cong F_7^*$. We may now assume that $|E(M)| \geq 8$.

We can represent M as a graft $M(G, \gamma)$ where G has no isolated vertices, $M(G) = M \backslash e$, and $e = e_\gamma$. Since M is non-graphic, $|\gamma|$ is even. As $M \backslash e$ is 3-connected up to series couples, Lemma 10.3.12 implies that G is a subdivision of a graph H such that $M(H)$ is 3-connected, and G has no two adjacent vertices of degree 2. Next we prove the following.

Lemma 10.3.16 *Every degree-2 vertex of G is coloured.*

Proof Let v be a degree-2 vertex of G incident with edges a and b. Then $\{a, b\}$ is a cocircuit of $M \backslash e$. As M is 3-connected, it follows that $\{a, b, e\}$ is a cocircuit of M. Assume that v labels the first row of $A_{(G, \gamma)}$. Clearly those columns of $A_{(G, \gamma)}$ whose first entry is 0 form a flat of (G, γ). Since $\{a, b, e\}$ is a cocircuit of $M(G, \gamma)$, it follows that e has a 1 as its first entry, that is, $v \in \gamma$. □

The proof of Lemma 10.3.15 will be completed by considering the various possibilities for the number of degree-2 vertices in G.

Lemma 10.3.17 *If G has more than three degree-2 vertices, then M has an element f that is in a triad with e such that M/f is non-graphic.*

Proof Suppose G has at least four degree-2 vertices. Then $M \backslash e$ has at least four series couples. Take a foot u of one such series couple $\{a, b\}$ and do a switch if necessary to produce a graft (G', γ') in which u is uncoloured. We may assume that a meets u. Now $M(G'/a)$ is 3-connected up to series couples and this matroid has at least three series couples. Since the tip of each of these couples is coloured, every graft obtained from $(G', \gamma')/a$ by a sequence of series switches has at least three coloured vertices. Proposition 10.3.14 implies that $M((G', \gamma')/a)$ is non-graphic. As $\{a, b, e\}$ is a triad of M, the lemma holds. □

Lemma 10.3.18 *$M \backslash e$ has no two series couples with the same set of feet.*

Proof Assume the contrary, letting X be the union of two series couples having the same sets of feet. Then X is a circuit, so $r_{M\backslash e}(X) = 3$. Also $r^*_{M\backslash e}(X) = 2$. Hence $r_{M\backslash e}(X) + r^*_{M\backslash e}(X) - |X| = 3 + 2 - 4 = 1$. But $|E(M\backslash e)| \geq 7$. Thus $M\backslash e$ has a non-minimal 2-separation; a contradiction. $\square$

Lemma 10.3.19 *If G has a vertex w that is the foot of exactly one series couple, $\{a, b\}$, in $M\backslash e$, then $\{a, b, e\}$ is a triad of M, and M/a or M/b is non-graphic.*

Proof Certainly $\{a, b, e\}$ is a triad of M. From among the grafts that can be obtained from (G, γ) by a sequence of series switches, let (G', γ') be one for which the number of coloured vertices is a minimum. We may assume that w is uncoloured otherwise we perform a series switch on $\{a, b\}$ and this will not alter the number of coloured vertices. By Proposition 10.3.14, since $M(G', \gamma')$ is non-graphic, $|\gamma'| \geq 4$. Without loss of generality, we may assume that a meets w in G'. Then $(G', \gamma')/a$ has $|\gamma'|$ coloured vertices including the vertex w' that results from identifying the ends, v and w, of a. Evidently w' is not the foot of a series couple in $(G', \gamma')/a$. Assume $M(G', \gamma')/a$ is graphic. Then Proposition 10.3.14 implies that we can transform $(G', \gamma')/a$, by series switches in G'/a, into a graft (G'', γ'') with $|\gamma''| = 2$. But each of these series switches in G'/a can be performed in G' before the edge a is contracted. By the choice of (G', γ'), this sequence of series switches will not reduce the number of coloured vertices below $|\gamma'|$ and will not alter the fact that v is coloured and w is not. Moreover, the contraction of a, after performing these series switches, gives (G'', γ'') but does not reduce the number of coloured vertices below $|\gamma'|$. Since $|\gamma'| \geq 4$ but $|\gamma''| = 2$, we have a contradiction. Therefore $M(G', \gamma')/a$, which equals M/a, is non-graphic. $\square$

We shall now complete the proof of Lemma 10.3.15. First note that we may assume that the following holds, otherwise part (iii) of the lemma holds.

10.3.20 *If an edge f meets a degree-2 vertex of G, then M/f is graphic.*

By the last three lemmas, we may also assume that G has exactly three degree-2 vertices and these vertices have as their neighbours $\{u, v\}$, $\{v, w\}$, and $\{w, u\}$ for some vertices u, v, and w. By series switches, we may suppose that either all of u, v, and w are coloured, or none are coloured. In the latter case, by contracting a matching consisting of one edge from each series couple, we get a graft (H, η) in which H is 3-connected and there are at least three coloured vertices. By Proposition 10.3.14, $M(H, \eta)$ is not graphic contradicting 10.3.20.

We may now assume that all of u, v, and w are coloured. Next we show that the set of coloured vertices of G consists of $\{u, v, w\}$ and the three degree-2 vertices. Suppose that G has some other coloured vertex, x say. By a series switch, we may assume that exactly one of u, v, and w, say u, is coloured. From each series couple with u as a foot, we contract the edge not incident with u to get a graft (G', γ') in which all of u, v and w are coloured. By performing a series switch on the one remaining series couple in (G', γ'), we reduce the number of coloured vertices by two but no further reductions in this number are possible. Counting the coloured vertices in the resulting graft (G'', γ''), we have

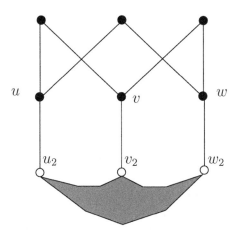

FIG. 10.9. The graft (G', γ').

one in $\{u, v, w\}$, the one remaining degree-2 vertex, and x. Hence $|\gamma''| \geq 3$ so, by Proposition 10.3.14, $M(G'', \gamma'')$ is non-graphic, again contradicting 10.3.20.

We now know that (G, γ) has exactly six coloured vertices consisting of the three degree-2 vertices and their neighbours, u, v, and w. Let X be the set of six edges of G that meet degree-2 vertices. Since $M \backslash e$ is 3-connected up to series couples, there are no edges with both ends in $\{u, v, w\}$. If $V(G) - \gamma$ contains a single vertex, then G consists of a 6-cycle, all of whose vertices are coloured, along with one uncoloured vertex that is adjacent to three mutually non-consecutive vertices of the 6-cycle. In that case, by Example 10.3.7, we have $M(G, \gamma) \cong M^*(K_5)$ and the lemma holds.

It remains to consider when $|V(G) - \gamma| \geq 2$. We shall show in this case that M^* is the 3-sum of $M(K_5)$ and a cographic matroid. Let $Y = E(M) - (X \cup e)$. Since $|V(G) - \gamma| \geq 2$, we have $|Y| \geq 5$. As the edges of X form a 6-cycle and the only coloured vertices of (G, γ) are the vertices of this 6-cycle, $e \in \mathrm{cl}_M(X)$. Hence $r(X \cup e) \leq 5$. The triads of M corresponding to the degree-2 vertices of G imply that $r^*(X \cup e) \leq 4$. Thus $\lambda_M(X \cup e) = r(X \cup e) + r^*(X \cup e) - |X \cup e| \leq 5 + 4 - 7 = 2$. But M is 3-connected, so $(X \cup e, Y)$ is an exact 3-separation of M and hence of M^*. By Proposition 9.3.4, there is a binary extension N of M^* by a 3-element set Z such that Z is a circuit of N with $r_N(X \cup e \cup Z) = r_N(X)$ and $r_N(Y \cup Z) = r_N(Y)$. Moreover, N is unique up to the labelling of Z, and M^* is the 3-sum of $N|(X \cup e \cup Z)$ and $N|(Y \cup Z)$.

Next we construct the matroid N. Let G_2 be the graph that is obtained from G by deleting the degree-2 vertices and relabelling u, v, and w as u_2, v_2, and w_2. Let G_1 be $G[X]$. Viewing G_1 and G_2 as disjoint graphs, form G' from these graphs by adding edges uu_2, vv_2, and ww_2. Then contracting these three added edges from G' gives G. Let (G', γ') be the graft whose coloured vertices are the vertices of G_1 in G' (see Figure 10.9), and let $M' = M(G', \gamma')$. Since G_2 is connected, $\{uu_2, vv_2, ww_2\}$ is a cocircuit of $M(G')$. As $\gamma = V(G_1)$, it follows that $e \in \mathrm{cl}_{M'}(X)$. Thus $\{uu_2, vv_2, ww_2\}$ is a cocircuit of M' and hence is a triangle of its dual. Call this triangle Z. Clearly both $\mathrm{cl}^*_{M'}(X \cup e)$ and $\mathrm{cl}^*_{M'}(Y)$ contain Z. Thus $(M')^*$ is the required extension N of M^*, and M^* is the 3-sum of $(M')^*|(X \cup e \cup Z)$ and $(M')^*|(Y \cup Z)$. The last two matroids are $(M'/Y)^*$ and $(M'/(X \cup e))^*$. Now $M' = M(G', \gamma')$, so $M'/Y = M'/E(G_2)$ which, by Example 10.3.7 again, is isomorphic to $M^*(K_5)$. On the other hand, since $e \in \mathrm{cl}_{M'}(X)$, we have $M'/(X \cup e) = M'/X\backslash e = (M'\backslash e)/X = M(G')/X$, so $M'/(X \cup e)$ is graphic. We conclude that M^* is indeed isomorphic to the 3-sum of $M(K_5)$ and a cographic matroid. $\square$

We now prove the excluded-minor characterization of graphic matroids.

Proof of Theorem 10.3.2 As every minor of a graphic matroid is graphic, to show that a graphic matroid has no minor isomorphic to F_7, F_7^*, $M^*(K_5)$, or $M^*(K_{3,3})$, it suffices to show that none of these four matroids is graphic. Proposition 2.3.3 noted that the last two matroids are non-graphic. One can deduce that neither F_7 nor F_7^* is graphic from the fact that neither is regular. Alternatively, one can show this directly without difficulty (Exercise 1).

To prove the converse, suppose that M is a non-graphic binary matroid for which every proper minor is graphic. If M is not 3-connected, then, by Theorem 8.3.1, it is the direct sum or 2-sum of two proper minors of M. As these minors must be graphic, Corollary 7.1.26 implies that M is too; a contradiction. We deduce that M is 3-connected. By the dual of Lemma 10.3.10, M has an element e such that $\mathrm{co}(M\backslash e)$ is 3-connected. Thus, by Lemma 10.3.11, $M\backslash e$ is 3-connected up to series couples. By Lemma 10.3.8, there is a graft (G, γ) such that $M = M(G, \gamma)$ and $M\backslash e = M(G)$. By Lemma 10.3.15, if $M\backslash e$ is not 3-connected, then M is isomorphic to F_7^* or $M^*(K_5)$, or $M^* \cong M' \oplus_3 M(K_5)$ for some matroid M'. In the last case, by Proposition 9.3.5, M^* has $M(K_5)$ as a proper minor; a contradiction. Thus we may assume that $M\backslash e$ is 3-connected, that is, $M(G)$ is 3-connected. As M is not graphic, Proposition 10.3.14 implies that $|\gamma| \geq 4$.

Lemma 10.3.21 *Let f be an edge uv of G. If G/f is 3-connected, then $|\gamma| = 4$ and $u, v \in \gamma$.*

Proof The matroid M/f is graphic and equals $M(G, \gamma)/f$. By Lemma 10.3.9, the last matroid equals $M(G/f, \gamma')$ where $(G, \gamma)/f = (G/f, \gamma')$ and $|\gamma'| = |\gamma|$ unless both u and v are in γ. In the exceptional case, $|\gamma'| = |\gamma| - 2$. The graph G/f is 3-connected and loopless but may have parallel edges. Let H be obtained from G/f by deleting all but one edge from each parallel class. Then $M(H)$ is 3-connected and $M(H, \gamma')$ is graphic so, by Proposition 10.3.14, $|\gamma'| = 2$. Thus $|\gamma| \leq 4$. But $|\gamma| \geq 4$, so $|\gamma| = 4$ and $u, v \in \gamma$. $\square$

Lemma 10.3.22 *The set γ of coloured vertices has exactly four members.*

Proof By Lemma 10.3.10, G has an edge g such that G/g is 3-connected. The assertion follows by Lemma 10.3.21. $\square$

Lemma 10.3.23 *Let f be an edge uv of G. If $u \notin \gamma$, then v is a degree-3 vertex, $v \in \gamma$, and v has two neighbours in γ.*

Proof As $u \notin \gamma$, by Lemma 10.3.21, G/f is not 3-connected. Thus, by Bixby's Lemma (8.7.3), $\mathrm{co}(M(G\backslash f))$ is 3-connected so, by Lemma 10.3.11, $M(G\backslash f)$ is 3-connected up to series couples. By Lemma 10.3.12, if $\{a, b\}$ is a series couple of $M(G\backslash f)$, then a and b meet at a degree-2 vertex of $G\backslash f$. As $M(G)$ is 3-connected, we deduce that $M(G\backslash f)$ has at most two series couples. Let (G', γ') be the graft that is obtained from (G, γ) by deleting f and contracting one edge from each series couple in $M(G\backslash f)$. Since G' is 3-connected and $M(G', \gamma')$ is graphic having no coloops, by Proposition 10.3.14, $|\gamma'| = 2$ unless $e_{\gamma'}$ is a loop, in which case, $|\gamma'| = 0$. Because $u \notin \gamma$, contracting an edge of $G\backslash f$ meeting u will not change the number of coloured vertices. Since $G\backslash f$ has four coloured vertices but G' has at most two, v must have degree 2 in $G\backslash f$. Moreover, the contraction of either of the edges of $G\backslash f$ meeting v must reduce the number of coloured vertices. Hence v and its two neighbours in $G\backslash f$ belong to γ. □

Lemma 10.3.24 *The graft (G, γ) is isomorphic either to a 3-spoked wheel in which every vertex is coloured or to a 4-spoked wheel in which every vertex other than the hub is coloured.*

Proof As $|\gamma| = 4$, the graph G has at least four vertices. If it has exactly four vertices, then, since G is 3-connected, it is isomorphic to K_4 or, equivalently, to the 3-spoked wheel, so the lemma holds. We may now assume that G has at least one uncoloured vertex. By the last lemma, for each such vertex u, each neighbour v of u is in γ, has degree 3, and is adjacent to two other vertices in γ.

Suppose every coloured vertex has some uncoloured neighbour. Then every coloured vertex has exactly two coloured neighbours. Hence the subgraph of G induced by the coloured vertices has every vertex of degree 2 and so is a 4-cycle. Moreover, every coloured vertex has degree 3 in G and so, as G is 3-connected, there is only one uncoloured vertex and G is a 4-spoked wheel in which every vertex other than the hub is coloured.

We may now assume that some coloured vertex v is adjacent only to coloured vertices. Since G is 3-connected, v has degree 3. Let x, y, and z be the neighbours of v and let u be an uncoloured vertex. By the last lemma, the neighbours of u are x, y, and z. Moreover, each of x, y, and z has degree 3 and has all its neighbours in $\{u, v, x, y, z\}$. Thus G is a 5-vertex graph in which every vertex has degree 3; a contradiction. □

If (G, γ) is either a 3-spoked wheel in which every vertex is coloured or a 4-spoked wheel in which every vertex except the hub is coloured, then, by Examples 10.3.6 and 10.3.7, $M(G, \gamma)$ is isomorphic to F_7 or $M^*(K_{3,3})$. We conclude that Theorem 10.3.2 holds. □

As noted in Section 5.2, we can combine Theorem 10.3.2 with Theorem 5.2.2 to quickly obtain Wagner's (1937a) variant of Kuratowski's Theorem, which was stated earlier as Theorem 5.2.5 and is restated below for convenience.

Corollary 10.3.25 *A graph is planar if and only if it has no minor isomorphic to K_5 or $K_{3,3}$.* □

Exercises

1. Prove directly that neither F_7 nor F_7^* is graphic by showing that, in each case, there is no graph G with $M(G)$ isomorphic to the specified matroid.
2. If γ is the set of degree-2 vertices of $K_{2,4}$, prove that $M(K_{2,4}, \gamma)$ is isomorphic to the dual of the rank-4 binary spike (with tip).
3. (Oxley 1979) Use the fact that $M(K_5)$ and $M^*(K_{3,3})$ are maximal regular (Exercise 3, Section 6.6) to prove that a regular matroid is cographic if and only if it has no parallel minor isomorphic to $M(K_5)$ and no series minor isomorphic to $M(K_{3,3})$.
4. Let M be a cyclically 3-connected non-empty matroid. Prove that M has a series class S such that $M\backslash S$ is cyclically 3-connected.
5. *(Graver 1966, Welsh 1969b) Prove that a matroid M is graphic if and only if M is binary and its cocircuit space has a basis B such that every element of $E(M)$ is in at most two members of B.
6. *(Asano, Nishizeki, and Seymour 1984) Let $\{e, f, g\}$ be a circuit of a 3-connected non-graphic matroid M. Prove that M has a minor using $\{e, f, g\}$ that is isomorphic to $U_{2,4}$, F_7, or $M^*(K_{3,3})$.
7. *(Wu 1998a) Let M be a 3-connected matroid with at least three elements. Prove that:
   (a) M has at least three elements x for which si(M/x) is 3-connected.
   (b) M has exactly three such elements if and only if $M \cong M^*(K_{3,k}''')$ for some $k \geq 2$ where $K_{3,k}'''$ is obtained from $K_{3,k}$ by adding a 3-cycle on the vertices of its three-member vertex class.

10.4 Further properties of regular and graphic matroids

In this section we survey some additional properties of regular and graphic matroids. Most of the proofs will be omitted. We begin with some extensions of Theorems 10.1.2 and 10.3.3 due to Bixby (1976, 1977). Recall from Chapter 5 that a matroid N is a parallel minor of a matroid M if N can be obtained from M by a sequence of contractions and parallel deletions, where a parallel deletion is the deletion of an element from a non-trivial parallel class. Dually, N is a series minor of M if N can be obtained from M by deletions and series contractions.

Theorem 10.4.1 *The following are equivalent for a binary matroid M:*

(i) *M is regular.*
(ii) *M has no parallel minor isomorphic to F_7 or F_7^*.*
(iii) *M has no series minor isomorphic to F_7 or F_7^*.*

Proof This is left to the reader (see Exercise 5). □

The next result extends Propositions 5.4.3 and 5.4.4. It is taken from a paper of Bixby (1977) that was the first to point out the special role that R_{10} plays

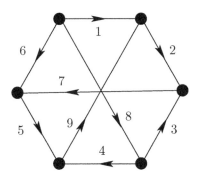

FIG. 10.10. An orientation of $K_{3,3}$.

within the class of regular matroids. The full realization of the importance of R_{10} within this class emerged with Seymour's Decomposition Theorem (13.1.1). The simple graphs $K'_{3,3}$, $K''_{3,3}$, and $K'''_{3,3}$ are obtained from $K_{3,3}$ by adding one, two, and three edges, respectively, within the same vertex class (see Figure 5.14).

Theorem 10.4.2 *The following are equivalent for a regular matroid M:*

(i) *M is graphic.*

(ii) *M has no parallel minor isomorphic to $M^*(K_5)$ or $M^*(K_{3,3})$.*

(iii) *M has no series minor isomorphic to $R_{10}, M^*(K_5), M^*(K_{3,3}), M^*(K'_{3,3}),$
$M^*(K''_{3,3}),$ or $M^*(K'''_{3,3}).$* □

Corollary 10.4.3 *The following are equivalent for a regular matroid M:*

(i) *M is cographic.*

(ii) *M has no series minor isomorphic to $M(K_5)$ or $M(K_{3,3})$.*

(iii) *M has no parallel minor isomorphic to $R_{10}, M(K_5), M(K_{3,3}), M(K'_{3,3}),$
$M(K''_{3,3}),$ or $M(K'''_{3,3}).$* □

Next we consider a property of graphs that extends to regular matroids but not to all binary matroids. As we shall see, this property can actually be used to give another characterization of the class of regular matroids. Let M be a matroid on the set $\{1, 2, \ldots, n\}$ such that $\mathcal{C}(M) = \{C_1, C_2, \ldots, C_m\}$. The *circuit incidence matrix* $A(\mathcal{C})$ of M is the $m \times n$ matrix $[a_{ij}]$ in which a_{ij} is 1 or 0 depending on whether j is or is not in C_i. The *cocircuit incidence matrix* $A(\mathcal{C}^*)$ of M is the circuit incidence matrix of M^*. Evidently each of $A(\mathcal{C})$ and $A(\mathcal{C}^*)$ is only determined to within a permutation of its rows. Following White (1987c), we call M *signable* if there are signings $A'(\mathcal{C})$ and $A'(\mathcal{C}^*)$ of $A(\mathcal{C})$ and $A(\mathcal{C}^*)$ such that, over $\mathbb{R}$, every row of $A'(\mathcal{C})$ is orthogonal to every row of $A'(\mathcal{C}^*)$.

Given a graph G, an orientation of G can be used in a natural way to give signings of the circuit and cocircuit incidence matrices of $M(G)$. For instance, let $G \cong K_{3,3}$ and suppose G is oriented as in Figure 10.10. Let C and C^*

be the circuit $\{1, 2, 3, 4, 5, 6\}$ and the cocircuit $\{1, 4, 7, 8, 9\}$ of $M(G)$. The corresponding rows, $\underline{v}_C$ and $\underline{v}_{C^*}$, of $A(\mathcal{C})$ and $A(\mathcal{C}^*)$ are $(1, 1, 1, 1, 1, 1, 0, 0, 0)$ and $(1, 0, 0, 1, 0, 0, 1, 1, 1)$. Now arbitrarily assign orientations to C and C^* and change the signs of those entries of $\underline{v}_C$ and $\underline{v}_{C^*}$ for which these directions disagree with the directions on the corresponding edges of G. For example, if C is oriented clockwise and C^* from left to right, the signed rows corresponding to C and C^* are $(1, 1, -1, 1, -1, -1, 0, 0, 0)$ and $(1, 0, 0, -1, 0, 0, -1, 1, 1)$, respectively. By repeating this process for all cycles and bonds of G, we obtain signings $A'(\mathcal{C})$ and $A'(\mathcal{C}^*)$ of $A(\mathcal{C})$ and $A(\mathcal{C}^*)$. Moreover, it is not difficult to show that, for all graphs, these signings have the property that the rows of $A'(\mathcal{C})$ are orthogonal to those of $A'(\mathcal{C}^*)$ (Exercise 7). Hence every graphic matroid is signable. As the dual of a signable matroid is clearly signable, every cographic matroid is also signable. The following result (Minty 1966) identifies the class of signable matroids. Our proof follows White (1987c).

Proposition 10.4.4 *A matroid is signable if and only if it is regular.*

Proof Let M be a signable matroid and $A'(\mathcal{C})$ and $A'(\mathcal{C}^*)$ be signings of the circuit and cocircuit incidence matrices of M such that, over $\mathbb{R}$, the rows of $A'(\mathcal{C})$ are orthogonal to those of $A'(\mathcal{C}^*)$. Clearly this orthogonality condition implies that a circuit and cocircuit of M cannot meet in an odd number of elements. Therefore, by Theorem 9.1.2, M is binary. Moreover, $A(\mathcal{C}^*)$ is a $GF(2)$-representation of M. To see this, let B be a basis of M and consider the submatrix A_1 of $A(\mathcal{C}^*)$ whose rows are the incidence vectors of the $(E(M) - B)$-fundamental cocircuits of M. Then some column permutation of A_1 has the form $[I_r | X]$ where, by Proposition 6.4.2, X is the B-fundamental-circuit incidence matrix of M. Thus A_1 is a $GF(2)$-representation of M. Since every row of $A(\mathcal{C}^*)$ is in the cocircuit space of M and is therefore a linear combination of rows of A_1, we conclude that $A(\mathcal{C}^*)$ is indeed a $GF(2)$-representation of M.

Next we show that M is equal to the matroid M_1 for which $A'(\mathcal{C}^*)$ is an $\mathbb{R}$-representation. If C is the circuit $\{i_1, i_2, \ldots, i_m\}$ of M, let the row of $A'(\mathcal{C})$ corresponding to C be $(\alpha_1, \alpha_2, \ldots, \alpha_n)$. Then α_j is 1 or -1 if j is in $\{i_1, i_2, \ldots, i_m\}$, and α_j is 0 otherwise. Since $(\alpha_1, \alpha_2, \ldots, \alpha_n)$ is orthogonal to every row of $A'(\mathcal{C}^*)$, if we sum, over all j in $\{i_1, i_2, \ldots, i_m\}$, the product of α_j and the jth column of $A'(\mathcal{C}^*)$, then we get the zero vector. Thus C is dependent in M_1, that is, every circuit of M is dependent in M_1. But, since $A(\mathcal{C}^*)$ is a $GF(2)$-representation for M, it follows easily that every circuit of M_1 is dependent in M. We conclude, by Lemma 2.1.22, that $\mathcal{C}(M) = \mathcal{C}(M_1)$ and therefore $M = M_1$.

As M is both binary and $\mathbb{R}$-representable, it follows, by Theorem 6.6.3, that M is regular. Hence if M is signable, then it is regular. We leave the proof of the converse of this to the reader (Exercise 8). $\square$

Signable matroids were introduced by Minty (1966) in an attempt to develop a theory of orientation for matroids. Indeed, Minty used the word 'orientable' rather than 'signable' to describe such matroids. However, Rockafellar (1969) suggested that, in view of Minty's result that all signable matroids are regular, 'a much broader theory of orientation ought to be possible'. Such a theory was introduced and developed by Bland and Las Vergnas (1978) and Folkman and Lawrence (1978). This theory provides a natural context for the generalization of many results concerning directed graphs, convex polyhedra, and linear programs. Indeed, this theory is rich enough to fill an entire book (Björner, Las Vergnas, Sturmfels, White, and Ziegler 1993). We shall not go into the details of this theory here. Instead, we shall be content to indicate how it extends the notion of signability discussed above.

Let M be a matroid. Clearly M is signable if and only if there are signings $A'(\mathcal{C})$ and $A'(\mathcal{C}^*)$ of the circuit and cocircuit incidence matrices of M such that, for all circuits C and all cocircuits C^*,

$$|\{e \in E(M) : \underline{v}'_C(e) = \underline{v}'_{C^*}(e) \neq 0\}| = |\{e \in E(M) : \underline{v}'_C(e) = -\underline{v}'_{C^*}(e) \neq 0\}|$$

where, for example, $\underline{v}'_C(e)$ is the entry of $A'(\mathcal{C})$ corresponding to the circuit C and the element e. The matroid M is *orientable* if there are signings $A'(\mathcal{C})$ and $A'(\mathcal{C}^*)$ of $A(\mathcal{C})$ and $A(\mathcal{C}^*)$ such that, whenever a circuit C and a cocircuit C^* have non-empty intersection, both $\{e \in E(M) : \underline{v}'_C(e) = \underline{v}'_{C^*}(e) \neq 0\}$ and $\{e \in E(M) : \underline{v}'_C(e) = -\underline{v}'_{C^*}(e) \neq 0\}$ are non-empty. Evidently every signable matroid is orientable. Indeed, Bland and Las Vergnas (1978) proved that the classes of signable matroids and binary orientable matroids coincide. The next corollary follows immediately by combining that result with the last proposition.

Corollary 10.4.5 *The following statements are equivalent for a matroid M:*

(i) M *is signable.*
(ii) M *is regular.*
(iii) M *is binary and orientable.* □

Transversal matroids share with regular matroids the property of being representable over fields of all characteristics. But exactly how do these two classes of matroids overlap? To conclude this chapter, we answer this and a related question. Unlike the class of regular matroids, the class of transversal matroids is not closed under contraction (see Example 3.2.9). However, as Bondy (1972) showed, this class is closed under the more restrictive operation of series contraction.

Proposition 10.4.6 *A matroid is transversal if and only if all of its series minors are transversal.*

Proof An outline of a proof due to Piff (1972) is given in Exercise 9. □

Bondy used the last result as a step towards characterizing those matroids that are both graphic and transversal (see also Las Vergnas 1970). Subsequently, de Sousa and Welsh (1972) showed that such matroids are precisely the matroids that are both binary and transversal. Clearly these matroids are also the matroids that are both regular and transversal. For $k \geq 3$, we denote by C_k^2 the graph that is obtained from a k-edge cycle by replacing each edge by two edges in parallel. Thus, for example, C_3^2 is isomorphic to the graph G_2 in Figure 1.20.

Theorem 10.4.7 *The following statements are equivalent for a matroid M:*

(i) M *is graphic and transversal.*
(ii) M *is regular and transversal.*
(iii) M *is binary and transversal.*
(iv) M *has no series minor isomorphic to any of the matroids $U_{2,4}$, $M(K_4)$, and $M(C_k^2)$ for $k \geq 3$.*   □

The smallest minor-closed class of matroids that contains the class of regular transversal matroids is the class of regular gammoids, where we recall from Chapter 3 that a gammoid is a minor of a transversal matroid. The next result, which characterizes the class of regular gammoids, follows immediately from a result of Brylawski (1971, 1975b) and Ingleton (1971b) which identified the class of binary gammoids.

Theorem 10.4.8 *The following statements are equivalent for a matroid M:*

(i) M *is a graphic gammoid.*
(ii) M *is a regular gammoid.*
(iii) M *is a binary gammoid.*
(iv) M *has no minor isomorphic to $U_{2,4}$ or $M(K_4)$.*
(v) M *is a direct sum of series–parallel networks.*   □

In the last result, we have followed the common practice of using the term 'series–parallel network' both for the graph of such a network and for the cycle matroid of this graph. For some authors, such as Seymour (1995), a series–parallel network is any matroid, connected or not, that has no $U_{2,4}$- or $M(K_4)$-minor.

The equivalence of (iii) and (iv) in Theorems 10.4.7 and 10.4.8 was generalized by Oxley (1986b, 1987c) in results that characterize the classes of ternary transversal matroids and ternary gammoids. One may wonder about similar characterizations of the class of all transversal matroids and the class of all gammoids. These problems look very difficult. Ingleton (1977) noted that there are infinitely many minor-minimal non-gammoids and infinitely many series-minor-minimal non-transversal matroids. Moreover, he described as 'probably futile' the task of completely characterizing the class of gammoids by excluded minors.

Exercises

1. Update the Venn diagram in Figure 6.26 by indicating where the classes of transversal matroids and gammoids fit.
2. (Bondy 1972, Las Vergnas 1970) Let G be a graph. Use Theorem 10.4.7 to prove that $M(G)$ is transversal if and only if no subgraph of G is a subdivision of K_4 or C_k^2 for any $k \geq 3$.
3. Characterize those graphs whose cycle matroids are strict gammoids.
4. Prove the equivalence of (iv) and (v) in Theorem 10.4.8.
5. (a) Show that a binary matroid has an F_7-minor if and only if it has F_7 as a parallel minor.

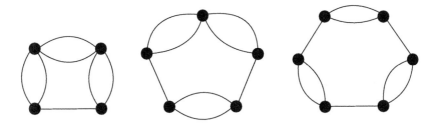

FIG. 10.11. Three graphs whose cycle matroids are not bicircular.

(b) Show that a binary matroid with no F_7-minor has an F_7^*-minor if and only if it has F_7^* as a parallel minor.

(c) Use (a) and (b) to prove Theorem 10.4.1.

(d) Prove that a binary matroid is regular if and only if it has no parallel minor isomorphic to F_7 and no series minor isomorphic to F_7^*.

6. Prove that the following are equivalent for a binary matroid M:

(a) M is graphic.

(b) M has no series minor isomorphic to $F_7, F_7^*, R_{10}, M^*(K_5), M^*(K_{3,3})$, $M^*(K_{3,3}'), M^*(K_{3,3}'')$, or $M^*(K_{3,3}''')$.

(c) M has no parallel minor isomorphic to $M^*(K_5), M^*(K_{3,3}), F_7$, or F_7^*.

7. Use the signings of the circuit and cocircuit matrices described before Proposition 10.4.4 to prove that every graphic matroid is signable.

8. Let $[I_r|A]$ be a totally unimodular matrix.

(a) Use the pivoting operation (without the natural column interchange) to show that, for each cocircuit C^* of $M[I_r|A]$, the row space of $[I_r|A]$ contains a signing of the incidence vector of C^*.

(b) Deduce that if a matroid is regular, then it is signable.

9. Let M be the strict gammoid $L(G, B)$ and suppose that x and y are parallel in M. Show that $M \backslash x$ is a strict gammoid and hence prove Proposition 10.4.6. [Hint: There are two similar cases. If $|\{x, y\} \cap B| = 0$, then G has a vertex whose removal separates $\{x, y\}$ from B. Modify G to obtain a new graph G' for which $L(G, B) = L(G', B)$ and $L(G' - x, B) = M \backslash x$.]

10. *(Lindström 1984b) Prove that $U_{1,2}$ is the only connected regular matroid that is identically self-dual.

11. *(J. Edmonds in Bondy 1972) Prove that a transversal matroid is binary if and only if it has a presentation whose associated bipartite graph is a forest.

12. *(Matthews 1977) Let G be a loopless graph. Prove that $M(G)$ is bicircular if and only if G has no subgraph that is a subdivision of K_4, C_3^2, or one of the three graphs in Figure 10.11.

13. *(Tutte 1965a) Prove that a connected binary matroid M of rank at least three has an F_7-minor or a cocircuit C^* such that $M \backslash C^*$ is disconnected.

11

SUBMODULAR FUNCTIONS AND MATROID UNION

A number of matroid constructions were discussed in Chapter 7. In this chapter, we shall consider several more such constructions. A common thread here is the association of these constructions to submodular functions where, for a set E, a function f from 2^E into $\mathbb{R}$ is *submodular* if $f(X)+f(Y) \geq f(X \cup Y)+f(X \cap Y)$ for all subsets X and Y of E. Such a function is *increasing* if $f(X) \leq f(Y)$ whenever $X \subseteq Y \subseteq E$. In Section 11.1, we shall prove a result of Edmonds and Rota (1966) that associates a matroid on E to every increasing submodular function on 2^E. Section 11.2 considers several applications of submodular functions, one of which is in the proof of Hall's Marriage Theorem, a result that specifies precisely when a family of sets has a transversal. In Section 11.3, we shall discuss the operation of matroid union and present several of its attractive applications. In Section 11.4, we consider the problem of sticking two matroids together across a common restriction. Much of the discussion there will be focused on a generalization of the operation of parallel connection. The last operation is used in Section 11.5 to define a matroid generalization of the graph operation of Δ-Y exchange.

Our treatment of the topics in Sections 11.1–11.3 is similar to that found in Chapters 7 and 8 of Welsh (1976) and a more detailed exposition of some of these topics can be found there. We note here that several authors, including Crapo and Rota (1970), refer to submodular functions as *semimodular functions*.

11.1 Deriving matroids from submodular functions

For a matroid M on a set E, the rank function is an increasing submodular function on 2^E, the set of subsets of E. Moreover, $\mathcal{C}(M) = \{C \subseteq E: C$ is minimal and non-empty such that $r(C) < |C|\}$. The next result (Edmonds and Rota 1966) generalizes this observation.

Proposition 11.1.1 *Let f be an increasing submodular function from 2^E into $\mathbb{Z}$. Let $\mathcal{C}(f) = \{C \subseteq E: C$ is minimal and non-empty such that $f(C) < |C|\}$. Then $\mathcal{C}(f)$ is the collection of circuits of a matroid on E.*

Proof Evidently $\mathcal{C}(f)$ satisfies (C1) and (C2). Now let C_1 and C_2 be distinct members of $\mathcal{C}(f)$ with e in $C_1 \cap C_2$. Clearly $|C_i - e| \leq f(C_i - e)$ for each i. Since f is increasing and $C_i \in \mathcal{C}(f)$, it follows that $|C_i - e| \leq f(C_i) < |C_i|$. Thus

$$f(C_i) = |C_i| - 1. \tag{11.1}$$

To complete the proof that $\mathcal{C}(f)$ satisfies (C3), we need only show that

$$f((C_1 \cup C_2) - e) < |(C_1 \cup C_2) - e|, \tag{11.2}$$

for it follows from this that $(C_1 \cup C_2) - e$ contains a member of $\mathcal{C}(f)$. Since f is increasing and submodular,

$$f((C_1 \cup C_2) - e) \le f(C_1 \cup C_2) \le f(C_1) + f(C_2) - f(C_1 \cap C_2).$$

Moreover, as $C_1 \cap C_2 \subsetneq C_1$, we have $f(C_1 \cap C_2) \ge |C_1 \cap C_2|$. Thus, by (11.1),

$$
\begin{aligned}
f((C_1 \cup C_2) - e) &\le |C_1| - 1 + |C_2| - 1 - f(C_1 \cap C_2) \\
&\le |C_1| - 1 + |C_2| - 1 - |C_1 \cap C_2| \\
&= |C_1 \cup C_2| - 2 = |(C_1 \cup C_2) - e| - 1.
\end{aligned}
$$

Hence (11.2) holds and the proposition is proved. □

Let $M(f)$ denote the matroid on E that has $\mathcal{C}(f)$ as its set of circuits. We shall say that $M(f)$ is *induced* by f. Next we examine such matroids more closely. We begin with a straightforward consequence of the preceding proposition.

Corollary 11.1.2 *A subset I of E is independent in $M(f)$ if and only if $|I'| \le f(I')$ for all non-empty subsets I' of I.* □

Example 11.1.3 Let G be a graph and E be its edge set. For all subsets X of E, denote $V(G[X])$ by $V(X)$ and let

$$f_{-1}(X) = |V(X)| - 1.$$

Certainly f_{-1} is integer-valued and increasing. Moreover, if $X, Y \subseteq E$, then

$$
\begin{aligned}
f_{-1}(X) + f_{-1}(Y) &= |V(X)| - 1 + |V(Y)| - 1 \\
&= |V(X) \cup V(Y)| - 1 + |V(X) \cap V(Y)| - 1.
\end{aligned}
$$

But $V(X) \cup V(Y) = V(X \cup Y)$ and $V(X) \cap V(Y) \supseteq V(X \cap Y)$. Hence

$$f_{-1}(X) + f_{-1}(Y) \ge f_{-1}(X \cup Y) + f_{-1}(X \cap Y).$$

Thus f_{-1} is submodular. Therefore f_{-1} induces a matroid on E, and it is not difficult to check that this matroid is precisely $M(G)$. □

Next consider the function f_0 defined, for all subsets X of E, by

$$f_0(X) = |V(X)| = f_{-1}(X) + 1.$$

It is an immediate consequence of the next result, the proof of which is left as an exercise, that f_0 is integer-valued, increasing, and submodular.

Lemma 11.1.4 *Let f be an integer-valued, increasing, submodular function on 2^E and, for all $X \subseteq E$, let $h(X) = f(X) + 1$. Then h is also integer-valued, increasing, and submodular.* □

In order to characterize the circuits of the matroid $M(f_0)$ induced on $E(G)$ by f_0, we shall use the following elementary graph-theoretic result.

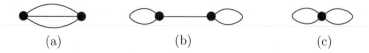

(a)   (b)   (c)

FIG. 11.1. The subdivisions of these graphs are the circuits of $B(G)$.

Lemma 11.1.5 *A connected graph H has at least two cycles if and only if* $|V(H)| \leq |E(H)| - 1$. $\qquad\qquad\qquad\qquad\qquad\qquad\qquad\qquad\qquad\qquad\qquad\qquad\qquad$ □

As we show in the next proof, $M(f_0)$ coincides with the bicircular matroid $B(G)$ of G, which was introduced in Section 6.10. Recall from Figure 6.37 that subdivisions of the graphs in (a), (b), and (c) of Figure 11.1 are called Θ-graphs, loose handcuffs, and tight handcuffs, respectively. The edge sets of all such subgraphs of G are the circuits of $B(G)$ while the ground set of $B(G)$ is $E(G)$.

Proposition 11.1.6 *For all graphs G, the matroids $M(f_0)$ and $B(G)$ are equal.*

Proof We show that the circuits of $M(f_0)$ and $B(G)$ coincide. Suppose first that C is a circuit of $B(G)$. Then $G[C]$ is a Θ-graph or a loose or tight handcuff, so it has at least two cycles. Thus, by the definition of f_0 and Lemma 11.1.5, $f_0(C) = |V(C)| < |C|$. Hence C is dependent in $M(f_0)$.

Now let C be a circuit in $M(f_0)$. Then C is a minimal non-empty set such that $|V(C)| < |C|$. Assume that C has a proper non-empty subset X such that $G[X]$ is a component of $G[C]$. Then

$$|V(C)| = |V(X)| + |V(C - X)| \geq |X| + |C - X| = |C|.$$

This contradiction implies that $G[C]$ is connected. Hence, by Lemma 11.1.5, $G(C)$ has at least two cycles and so, by Lemma 6.10.4, $G[C]$ has a subgraph that is a Θ-graph or a loose or tight handcuff. Thus C contains a circuit of $B(G)$.

We now know that a circuit in either of $M(f_0)$ and $B(G)$ is dependent in the other. Thus, by Lemma 2.1.22, $M(f_0) = B(G)$. $\qquad\qquad\qquad\qquad\qquad\qquad\qquad$ □

Some of the attractive properties of bicircular matroids will be noted in the exercises. As examples, observe that $B(K_4) \cong U_{4,6}$, while if G consists of two vertices joined by n parallel edges for some $n \geq 2$, then $B(G) \cong U_{2,n}$.

The increasing submodular functions considered so far have been allowed to take arbitrary integer values on the empty set. If we now insist that $f(\emptyset) = 0$ for such a function f, then we can specify the rank function of $M(f)$ in terms of f (Edmonds and Rota 1966). We shall call such a function f a *polymatroid* on 2^E. These functions have also been called *polymatroid functions*, *integer polymatroids*, and *integer polymatroid functions*. Detailed discussion of polymatroids and their properties, including their link to convex polyhedra, can be found in Welsh (1976, Chapter 18), Lovász and Plummer (1986, Chapter 11), and Schrijver (2003, Chapter 44).

Proposition 11.1.7 *Let f be a polymatroid on 2^E. For $X \subseteq E$, its rank $r_f(X)$ in $M(f)$ is given by*

$$r_f(X) = \min\{f(Y) + |X - Y| : Y \subseteq X\}.$$

Proof Let $a(X) = \min\{f(Y) + |X - Y| : Y \subseteq X\}$ for all subsets X of E. We need to show that, for all such X,

$$a(X) = r_f(X). \tag{11.3}$$

We leave it to the reader to show this when $X \in \mathcal{I}(M(f))$ and to show that a is increasing. Now suppose that $X \notin \mathcal{I}(M(f))$ and let B be a basis of X. Then, as (11.3) holds for all members of $\mathcal{I}(M(f))$ and a is increasing,

$$r_f(X) = |B| = r_f(B) = a(B) \le a(X).$$

To complete the proof of (11.3), it remains to show that

$$a(X) \le |B| = r_f(X). \tag{11.4}$$

Let $X - B = \{e_1, e_2, \ldots, e_m\}$ and suppose $i \in \{1, 2, \ldots, m\}$. Then $B \cup e_i$ is not in $\mathcal{I}(M(f))$. As B is in $\mathcal{I}(M(f))$, it has a subset I_i such that $f(I_i \dot\cup e_i) < |I_i \dot\cup e_i|$. Evidently $I_i \in \mathcal{I}(M(f))$, so $f(I_i) \ge |I_i|$ and therefore

$$f(I_i \cup e_i) = f(I_i) = |I_i|. \tag{11.5}$$

We now argue, by induction on j, that, for all j in $\{1, 2, \ldots, m\}$,

$$f(\bigcup_{i=1}^{j} I_i \cup \{e_1, e_2, \ldots, e_j\}) \le |\bigcup_{i=1}^{j} I_i|. \tag{11.6}$$

By (11.5), this is true for $j = 1$. Assume it true for $j \le k$ and let $j = k + 1$. Then, as f is submodular,

$$
\begin{aligned}
f(\bigcup_{i=1}^{k+1} I_i \cup \{e_1, e_2, \ldots, e_{k+1}\}) &\le f(\bigcup_{i=1}^{k} I_i \cup \{e_1, e_2, \ldots, e_k\}) \\
&\quad + f(I_{k+1} \cup e_{k+1}) - f((\bigcup_{i=1}^{k} I_i) \cap I_{k+1}) \\
&\le |\bigcup_{i=1}^{k} I_i| + |I_{k+1}| - |(\bigcup_{i=1}^{k} I_i) \cap I_{k+1}| \tag{11.7}
\end{aligned}
$$

where the last step follows by the induction hypothesis and the fact that the set $(\bigcup_{i=1}^{k} I_i) \cap I_{k+1}$ is in $\mathcal{I}(M(f))$. From (11.7), we deduce that (11.6) holds for $j = k + 1$. It follows by induction that (11.6) holds for $j = m$, that is,

$$f(\bigcup_{i=1}^{m} I_i \cup (X - B)) \le |\bigcup_{i=1}^{m} I_i|.$$

Therefore

$$f(\bigcup_{i=1}^{m} I_i \cup (X - B)) + |B - (\bigcup_{i=1}^{m} I_i)| \le |B|.$$

Thus $a(X) \le |B|$, so (11.4) holds and the proposition is proved. $\square$

A consequence of the last result is that the function a, being the rank function of $M(f)$, also induces $M(f)$, that is, $M(a) = M(f)$. The problem of characterizing those matroids that are induced by a unique polymatroid was solved independently by Duke (1981) and Dawson (1983).

Next we shall describe a useful geometric interpretation of polymatroids and of the matroids they induce (Helgason 1974, McDiarmid 1975c, Lovász 1977). The key to this description is the following.

Example 11.1.8 Figure 11.2 shows a geometric representation of a rank-4 matroid M with ground set $\{0, 1, \ldots, 9\}$. Now let $e_1 = \{2\}$, $e_2 = \{2, 4, 5\}$, $e_3 = \{6, 7, 8, 9, 0\}$, $e_4 = \{6, 0\}$, and $e_5 = \{1, 2, 3, 4, 5, 6\}$. Let $E = \{e_1, e_2, e_3, e_4, e_5\}$. Then each member of E is a flat of M. Define $f : 2^E \rightarrow \mathbb{Z}$ by $f(X) = r_M(\cup_{e_i \in X} e_i)$. Hence $f(\{e_3\}) = r_M(\{6, 7, 8, 9, 0\}) = 3$, while $f(\{e_1, e_2, e_4\}) = r_M(\{2, 4, 5, 6, 0\}) = 4$. Because r_M is a matroid rank function, it is not difficult to see that f is an increasing submodular function for which $f(\emptyset) = 0$, that is, f is a polymatroid on 2^E.

Generalizing this example, let M be a matroid on a set S and let E be a set, each member of which labels some flat of M, where the same flat can receive several different labels. For all $X \subseteq E$, define $f(X)$ to be the rank in M of the union of those sets labelled by members of X. Then it is routine to check that f is a polymatroid on 2^E. Indeed, the same conclusion holds more generally when the members of E label arbitrary subsets of S. $\square$

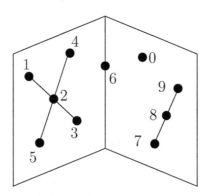

FIG. 11.2. A rank-4 matroid M.

The feature of this example which makes it particularly noteworthy is that every polymatroid is representable in this way as a collection of flats of some matroid. Formally:

Theorem 11.1.9 *Let f be a function defined on the set of subsets of a set E. Then f is a polymatroid on 2^E if and only if, for some matroid M, there is a function φ from E into the set of flats of M such that $f(X) = r_M(\bigcup_{x \in X} \varphi(x))$ for all $X \subseteq E$.*

The key idea in the proof of this theorem is that of freely adding a point on an element of a polymatroid, an operation which we shall see is effectively the same as that of freely adding a point on a flat of a matroid. Suppose f is a polymatroid on 2^E. If $x \in E$ and $x' \notin E$, then we can extend the domain of definition of f to include all subsets of $E \cup x'$ by letting

$$f(X \cup x') = \begin{cases} f(X), & \text{if } f(X \cup x) = f(X); \\ f(X) + 1, & \text{if } f(X \cup x) > f(X). \end{cases}$$

It is not difficult to check that this extended function is a polymatroid on the set of subsets of $E \cup x'$; we say that this new polymatroid has been obtained from f by *freely adding* x' to x. If we repeat this construction by freely adding a new element y' to some member y of E, then one can show that the order in which these two extensions are performed is irrelevant. In particular, by freely adding one element to each member of E, we obtain a well-defined polymatroid f' on the set of subsets $E \dot\cup E'$ where E and E' are in one-to-one correspondence, and the restriction of f' to the set of subsets of E is the original polymatroid f on 2^E. We leave it to the reader to prove the following result of Edmonds (1970).

Proposition 11.1.10 *Let f be a polymatroid on 2^E. Extend this to a polymatroid on the set of subsets of $E \dot\cup E'$ by adding one element freely on each member of E. Then the restriction of this extended function to the set of subsets of E' is the rank function of a matroid and this matroid is isomorphic to $M(f)$.* $\square$

This proposition is used in the proof of Theorem 11.1.9, which we now sketch.

Proof of Theorem 11.1.9 It suffices to show that, for a polymatroid f on 2^E, there is a matroid M having the specified properties. For each element y of E, replace y by a set A_y consisting of $f(y)$ new elements. Let A be $\bigcup_{y \in E} A_y$ and k be the function defined on all subsets X of A by $k(X) = f(\{y : A_y \cap X \neq \emptyset\})$. Then it is not difficult to check that k is a polymatroid on 2^A. Moreover, Proposition 11.1.10 implies that if we freely add a set A' of elements, one to each member of A, we obtain a matroid M on A'. For all y in E, let F_y be the closure in M of those elements that were added on members of A_y. Then one easily shows that $f(X) = r_M(\bigcup_{x \in X} F_x)$ for all $X \subseteq E$. The theorem follows by taking $\varphi(y) = F_y$ for all y in E. $\square$

We conclude this section with a generalization of Proposition 11.1.7 that was proved by McDiarmid (1973) (see also Welsh 1976, pp. 116–117). Its proof is outlined in Exercise 9. Let $\mathcal{E}$ be a collection of subsets of a set E such that $\mathcal{E}$ is a lattice under set inclusion. In such a case, we call $\mathcal{E}$ a *lattice of subsets* of E. A function t from $\mathcal{E}$ into $\mathbb{R}$ is *submodular on $\mathcal{E}$* if $t(X)+t(Y) \geq t(X \vee Y)+t(X \wedge Y)$ for all X and Y in $\mathcal{E}$.

Proposition 11.1.11 *Let $\mathcal{E}$ be a lattice of subsets of a set E such that $\mathcal{E}$ is closed under intersection and contains $\emptyset$ and E. Suppose that f is a non-negative, integer-valued, submodular function on $\mathcal{E}$ for which $f(\emptyset) = 0$. Let $\mathcal{I}(\mathcal{E}, f) = \{X \subseteq E : f(X) \geq |X \cap T| \text{ for all } T \text{ in } \mathcal{E}\}$. Then $\mathcal{I}(\mathcal{E}, f)$ is the collection of independent sets of a matroid on E. The rank of a subset X of E in this matroid is $\min\{f(T) + |X - T| : T \in \mathcal{E}\}$.* $\square$

Exercises

1. Let $E = \{1, 2, 3\}$ and f be the function defined for all X in 2^E by

$$f(X) = \begin{cases} 1.5, & \text{if } X \text{ is } \{1,2\} \text{ or } \{1,3\}; \\ 2, & \text{if } X = E; \\ |X|, & \text{otherwise.} \end{cases}$$

    (a) Show that f is increasing and submodular.
    (b) Find $\mathcal{C}(f)$.
    (c) Deduce that 11.1.1 may fail for functions whose range is $\mathbb{Q}$.

2. Let G be an undirected graph having vertex set V. Assign to every edge e of E a positive integral weight. For all $X \subseteq V$, let $f_1(X)$ be obtained by summing the weights of all those edges with at least one end in X, and let $f_2(X)$ be the corresponding sum over all edges with exactly one end in X. Show that both f_1 and f_2 are submodular.

3. Let f be a function defined on the set of subsets of a set E. Prove that f is submodular if and only if $f(X \cup y) + f(X \cup z) \geq f(X) + f(X \cup \{y, z\})$ for all subsets X of E and all distinct elements y and z of $E - X$.

4. For the non-Fano matroid labelled as in Figure 1.15(a), consider the poly-matroid f defined on the set of subsets of $\{\{1, 2, 3\}, \{1, 4, 7\}, \{1, 5, 6\}\}$ as in Example 11.1.8. Show that f is isomorphic to the corresponding poly-matroid defined on the set of subsets of $\{\{1, 5, 6\}, \{2, 4\}, \{4, 6\}\}$.

5. (Dawson 1983) Let f be a polymatroid on 2^E and let r be the rank function of $M(f)$. A subset X of E is *f-balanced* if $f(X) = r(X)$. Show that:
    (a) every circuit of $M(f)$ is f-balanced;
    (b) a union of f-balanced sets is f-balanced;
    (c) a set is f-balanced if and only if its closure is f-balanced;
    (d) if $X \cup Y \in \mathcal{I}(M(f))$ and X and Y are f-balanced, then $X \cap Y$ is f-balanced.

6. (Dawson 1983) Let f be a polymatroid on 2^E, and let cl denote the closure operator of $M(f)$. Show that $f(\text{cl}(X)) = f(X)$ for all $X \subseteq E$.

7. (Matthews 1977) Let G be a connected graph with more than one edge and assume that G is not a cycle. Prove that $B(G)$ is a connected matroid if and only if G has no vertices of degree one.

8. (Oxley, Prendergast, and Row 1982) Let f be an arbitrary integer-valued function defined on a set E. If $X \subseteq E$, define $a(X)$ to be $\max\{f(x) : x \in X\}$ when X is non-empty, and $\min\{f(x) : x \in E\}$ otherwise. Show that:

   (a) a is submodular and increasing, and hence $M(a)$ is a matroid.
   (b) The class of matroids $M(a)$ arising in this way coincides with the class of transversal matroids having a nested presentation (Section 1.6, Exercise 13).

9. Let E, $\mathcal{E}$, f, and $\mathcal{I}(\mathcal{E}, f)$ be as defined in 11.1.11 and let a be the function defined on 2^E by $a(X) = \min\{f(T) + |X - T| : T \in \mathcal{E}\}$.

   (a) Show that, for all subsets X, Y, Z, and W of E,

$$|X - Z| + |Y - W| \geq |(X \cup Y) - (Z \cup W)| + |(X \cap Y) - (Z \cap W)|.$$

   (b) Use (a) to show that a is submodular.
   (c) Now show that a is increasing and finish the proof of 11.1.11.

11.2 The theorems of Hall and Rado

Let $\mathcal{A}$ be a family $(A_j : j \in J)$ of subsets of a set S. When does $M[\mathcal{A}]$ have rank $|J|$? Equivalently, under what circumstances does $M[\mathcal{A}]$ have a transversal? This problem is often referred to as the *marriage problem* for, historically, it was originally formulated as follows. Let J be the set of unmarried men in a primitive village and, for each j in J, let A_j be the set of unmarried women in the village each of whom would be an acceptable wife for man j. The problem of finding necessary and sufficient conditions under which acceptable marriages can be arranged for every man in J is precisely the problem of finding necessary and sufficient conditions for $(A_j : j \in J)$ to have a transversal. We begin this section by noting P. Hall's (1935) solution to the marriage problem. Then we present a generalization of Hall's Theorem due to Rado (1942). Both this and Hall's result will be obtained here as corollaries of a still more general result due to Welsh (1971a). Several applications of these results will also be noted in this section including a formula for the rank function of an arbitrary transversal matroid.

If $K \subseteq J$, we shall denote $\bigcup_{j \in K} A_j$ by $A(K)$. Evidently if $(A_j : j \in J)$ has a transversal, then $|A(K)| \geq |K|$ for all $K \subseteq J$. The next result asserts that the converse of this is also true.

Theorem 11.2.1 (Hall's Marriage Theorem) *A family $(A_j : j \in J)$ of subsets of a set S has a transversal if and only if, for all $K \subseteq J$,*

$$|A(K)| \geq |K|.$$

Theorem 11.2.2 (Rado's Theorem) *Let $(A_j : j \in J)$ be a family of subsets of a set S and let M be a matroid on S having rank function r. Then $(A_j : j \in J)$ has a transversal that is independent in M if and only if, for all $K \subseteq J$,*

$$r(A(K)) \geq |K|.$$

In the last result, just as in Hall's Theorem, an obvious set of necessary conditions is also sufficient. This phenomenon is quite common in transversal theory. If we let the matroid M in Rado's Theorem be the free matroid on S, then we immediately obtain the Marriage Theorem. The next result was obtained by Welsh (1971a) by extracting the essential features from a proof of Hall's Theorem due to Rado (1967). A *system of representatives* for a family $(A_j : 1 \leq j \leq n)$ of subsets of a set S is a sequence $(e_1, e_2, \ldots, e_n)$ such that $e_j \in A_j$ for all j in $\{1, 2, \ldots, n\}$. Such a sequence differs from a transversal in that its members need not be distinct. The subset of E corresponding to this sequence will be denoted by $\{e_1, e_2, \ldots, e_n\}$.

Theorem 11.2.3 *Let $\mathcal{A}$ be a family $(A_j : j \in J)$ of non-empty subsets of a set S and let f be a non-negative, integer-valued, increasing, submodular function on 2^S. Then $\mathcal{A}$ has a system of representatives $(e_j : j \in J)$ such that*

$$f(\{e_j : j \in K\}) \geq |K| \text{ for all } K \subseteq J \tag{11.8}$$

if and only if

$$f(A(K)) \geq |K| \text{ for all } K \subseteq J. \tag{11.9}$$

Proof Let $J = \{1, 2, \ldots, n\}$. If $(e_j : j \in J)$ is a system of representatives for $\mathcal{A}$ satisfying (11.8), then $f(A(K)) \geq f(\{e_j : j \in K\}) \geq |K|$ for all subsets K of J, so (11.9) holds.

Now assume that (11.9) holds. If, for each j in J, the set A_j contains a unique element a_j, then $(a_j : j \in J)$ is the required system of representatives for $\mathcal{A}$. Thus we may assume, without loss of generality, that $|A_1| \geq 2$.

Lemma 11.2.4 *For some element x of A_1, the family $(A_1 - x, A_2, \ldots, A_n)$ satisfies (11.9).*

Proof Assume that no such element x exists. Then, for distinct elements x_1 and x_2 of A_1, there are non-empty subsets K_1 and K_2 of $\{2, 3, \ldots, n\}$ such that

$$f((A_1 - x_1) \cup A(K_1)) < |K_1| + 1,$$

and

$$f((A_1 - x_2) \cup A(K_2)) < |K_2| + 1.$$

As f is submodular and increasing,

$$
\begin{aligned}
|K_1| + |K_2| &\geq f((A_1 - x_1) \cup A(K_1)) + f((A_1 - x_2) \cup A(K_2)) \\
&\geq f((A_1 - x_1) \cup A(K_1) \cup (A_1 - x_2) \cup A(K_2)) \\
&\quad + f([(A_1 - x_1) \cup A(K_1)] \cap [(A_1 - x_2) \cup A(K_2)]) \\
&\geq f(A_1 \cup A(K_1 \cup K_2)) + f(A(K_1) \cap A(K_2)).
\end{aligned}
$$

Thus, as $A(K_1) \cap A(K_2) \supseteq A(K_1 \cap K_2)$,

$$|K_1| + |K_2| \geq f(A_1 \cup A(K_1 \cup K_2)) + f(A(K_1 \cap K_2)). \qquad (11.10)$$

But $\mathcal{A}$ satisfies (11.9) and $1 \notin K_1 \cup K_2$, so

$$f(A_1 \cup A(K_1 \cup K_2)) \geq 1 + |K_1 \cup K_2|$$

and

$$f(A(K_1 \cap K_2)) \geq |K_1 \cap K_2|.$$

Hence, by (11.10),

$$|K_1| + |K_2| \geq 1 + |K_1 \cup K_2| + |K_1 \cap K_2| = 1 + |K_1| + |K_2|.$$

This contradiction completes the proof of the lemma. $\qquad\square$

By this lemma, we may successively delete elements from A_1, maintaining a family of sets satisfying (11.9), until we arrive at a singleton set. We repeat this process for $A_2, A_3, \ldots, A_n$ reducing each to a singleton set while maintaining a family of sets that satisfies (11.9). But we have already noted that the required system of representatives exists when all the sets in the family are singletons. Hence the theorem is proved. $\qquad\square$

Clearly Rado's Theorem follows from the last result by taking f to be the rank function r of M. The following extension of Rado's Theorem is due to Perfect (in Mirsky 1971).

Corollary 11.2.5 *Let $\mathcal{A}$ be a family $(A_j : j \in J)$ of subsets of a set S and let M be a matroid on S having rank function r. Let d be a non-negative integer not exceeding $|J|$. Then $\mathcal{A}$ has a partial transversal of size $|J| - d$ that is independent in M if and only if, for all subsets K of J,*

$$r(A(K)) \geq |K| - d.$$

Proof Let $f(X) = r(X) + d$ for all $X \subseteq S$. The result now follows from Theorem 11.2.3 and we omit the straightforward details of the argument. $\qquad\square$

This corollary enables us to find the rank function of any transversal matroid.

Proposition 11.2.6 *Let $\mathcal{A}$ be a family $(A_j : j \in J)$ of subsets of a set S. The rank in $M[\mathcal{A}]$ of a subset X of S is given by*

$$r_{M[\mathcal{A}]}(X) = \min\{|A(K) \cap X| - |K| + |J| : K \subseteq J\}.$$

Proof Consider the family $\mathcal{A}_X = (A_j \cap X : j \in J)$. We shall apply the last result to this family of sets taking M to be the free matroid on S. Thus the rank function of M is just the cardinality function, so $r_{M[\mathcal{A}]}(X) = \max\{|Y| : Y$ is a partial transversal of $\mathcal{A}_X\}$. By Corollary 11.2.5, for $0 \leq d \leq |J|$, the family $\mathcal{A}_X$

S

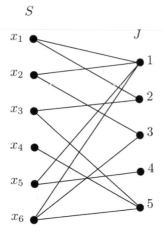

FIG. 11.3. The bipartite graph $\Delta[\mathcal{A}]$ associated with $\mathcal{A}$.

has a partial transversal of size $|J| - d$ if and only if $|A(K) \cap X| \geq |K| - d$ for all $K \subseteq J$. Thus

$$
\begin{aligned}
r_{M[\mathcal{A}]}(X) &= \max\{|J| - d : |A(K) \cap X| \geq |K| - d \text{ for all } K \subseteq J\} \\
&= \max\{|J| - d : |A(K) \cap X| - |K| + |J| \geq |J| - d \text{ for all } K \subseteq J\}.
\end{aligned}
$$

Writing t for $|J| - d$, we have

$$
r_{M[\mathcal{A}]}(X) = \max\{t : |A(K) \cap X| - |K| + |J| \geq t \text{ for all } K \subseteq J\}.
$$

But the largest integer t for which $|A(K) \cap X| - |K| + |J| \geq t$ for all $K \subseteq J$ equals the smallest value of $|A(K) \cap X| - |K| + |J|$ over all $K \subseteq J$; that is, $r_{M[\mathcal{A}]}(X) = \min\{|A(K) \cap X| - |K| + |J| : K \subseteq J\}$.   □

As an immediate consequence of this, we have the following.

Corollary 11.2.7 *Let $\mathcal{A}$ be a family $(A_j : j \in J)$ of subsets of a set S. A subset X of S is independent in $M[\mathcal{A}]$ if and only if, for all $K \subseteq J$,*

$$
|A(K) \cap X| - |K| + |J| \geq |X|.
$$
            □

We can obtain a different characterization of the independent sets in $M[\mathcal{A}]$ by reference to the bipartite graph $\Delta[\mathcal{A}]$ associated with $\mathcal{A}$ where we recall that $\Delta[\mathcal{A}]$ was introduced in Section 1.6. A subset X of S is independent in $M[\mathcal{A}]$ if and only if there is a matching W in $\Delta[\mathcal{A}]$ such that every vertex in X is incident with an edge of W. This can be expressed in terms of a new family of sets as follows. For all $Y \subseteq X$, let $N(Y)$ be those members of J that are adjacent in $\Delta[\mathcal{A}]$ to some member of Y. Let $\mathcal{N}_X = (N(x) : x \in X)$. Then X is independent in $M[\mathcal{A}]$ if and only if $\mathcal{N}_X$ has a transversal.

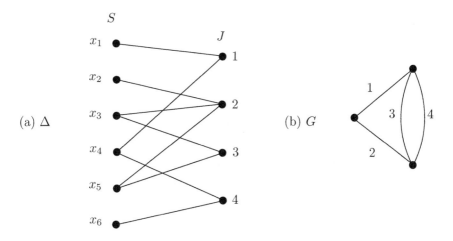

FIG. 11.4. The matroid induced from $M(G)$ by Δ is $U_{2,6}$.

Example 11.2.8 Let $\mathcal{A}$ be the family of subsets of $\{x_1, x_2, \ldots, x_6\}$ whose members are $A_1 = \{x_1, x_2, x_5, x_6\}$, $A_2 = \{x_1, x_3\}$, $A_3 = \{x_2, x_6\}$, $A_4 = \{x_5\}$, and $A_5 = \{x_3, x_4, x_6\}$. Then $\Delta[\mathcal{A}]$ is as shown in Figure 11.3. As $\{x_21, x_32, x_45, x_63\}$ is a matching in $\Delta[\mathcal{A}]$, the set $\{x_2, x_3, x_4, x_6\}$ is independent in $M[\mathcal{A}]$. Equivalently, $\mathcal{N}_X$ has a transversal $(1, 2, 5, 3)$ where

$$\mathcal{N}_X = (N(x_2), N(x_3), N(x_4), N(x_6)) = (\{1, 3\}, \{2, 5\}, \{5\}, \{1, 3, 5\}). \qquad \square$$

Using Hall's Theorem, we can now specify when X is independent in terms of the family $\mathcal{N}_X$. The straightforward proof is left to the reader.

Proposition 11.2.9 *Let $\mathcal{A}$ be a family $(A_j : j \in J)$ of subsets of a set S. For a subset X of S, let $\mathcal{N}_X = (N(x) : x \in X)$ where $N(x) = \{j : x \in A_j\}$. Then X is independent in $M[\mathcal{A}]$ if and only if*

$$|N(Y)| \geq |Y| \text{ for all } Y \subseteq X. \qquad \square$$

By combining this result with Proposition 11.1.6, one can show that bicircular matroids arise very naturally as transversal matroids (Matthews 1977).

Corollary 11.2.10 *Let G be a graph and, for each v in $V(G)$, let A_v be the set of edges incident with v. Then $B(G)$ is equal to the transversal matroid of the family $(A_v : v \in V(G))$.* $\qquad \square$

Our final application of Rado's Theorem will use the theory of submodular functions to define a class of matroids that contains the class of transversal matroids. Let Δ be a bipartite graph with vertex classes S and J and suppose that M is a matroid on J. Let $\Delta(\mathcal{I}) = \{X \subseteq S : X \text{ can be matched in } \Delta \text{ to a member of } \mathcal{I}(M)\}$.

Example 11.2.11 Let Δ be the bipartite graph in Figure 11.4(a) and let the matroid M on J be the cycle matroid of the graph G in Figure 11.4(b). Then it is not difficult to check that $\Delta(\mathcal{I}) = \{X \subseteq S : |X| \leq 2\}$. □

In the last example, $\Delta(\mathcal{I})$ is the set of independent sets of a matroid on S, namely $U_{2,6}$. Moreover, if $\mathcal{I}(M) = 2^J$, that is, if M is the free matroid on J, then $\Delta(\mathcal{I})$ is again the set of independent sets of a matroid on S, this matroid being the transversal matroid on S associated with Δ. These observations are special cases of the following basic result (Perfect 1969).

Theorem 11.2.12 *Let Δ be a bipartite graph with vertex classes S and J and let M be a matroid on J. Suppose that $\Delta(\mathcal{I})$ is the set of subsets X of S that are matched in Δ to a member of $\mathcal{I}(M)$. Then $\Delta(\mathcal{I})$ is the set of independent sets of a matroid $\Delta(M)$ on S.*

The matroid $\Delta(M)$ is said to be *induced from M by Δ*. We shall prove Theorem 11.2.12 by showing that $\Delta(M)$ is the matroid induced on S by a submodular function. This proof follows Welsh (1976). As before, for all $X \subseteq S$, we let $N(X)$ be the set of vertices in J that are adjacent in Δ to some member of X. Define $f : 2^S \to \mathbb{Z}^+ \cup \{0\}$ by

$$f(X) = r_M(N(X)).$$

Lemma 11.2.13 *The function f is non-negative, integer-valued, increasing, and submodular. A subset X of S is independent in $M(f)$ if and only if $X \in \Delta(\mathcal{I})$. Hence $\Delta(M)$ is the matroid $M(f)$.*

Proof Evidently f is non-negative and integer-valued. Moreover, it is straightforward to check that f is increasing and submodular. Now, by Corollary 11.1.2, X is independent in $M(f)$ if and only if, for all subsets Y of X,

$$r_M(N(Y)) \geq |Y|.$$

But, by Rado's Theorem, this occurs if and only if the family $(N(e) : e \in X)$ has a transversal that is independent in M. Since the latter occurs precisely when Δ has a matching of X onto a member of $\mathcal{I}(M)$, the lemma follows and hence so does Theorem 11.2.12. □

On combining the last lemma with Proposition 11.1.7, we obtain the rank function of $\Delta(M)$.

Corollary 11.2.14 *In $\Delta(M)$, the rank of a subset X of S is equal to*

$$\min\{r_M(N(Y)) + |X - Y| : Y \subseteq X\}.$$
□

A straightforward consequence of Theorem 11.2.12 is that any function σ on the ground set of a matroid can be used to induce a matroid on the range of σ (Nash-Williams 1966).

Corollary 11.2.15 *Let M be a matroid on a set J and suppose that σ is a function from J into a set S. Then there is a matroid $\sigma(M)$ on S whose set of independent sets is $\{\sigma(I) : I \in \mathcal{I}(M)\}$.* □

In general, an induced matroid $\Delta(M)$ retains very few of the properties of the original matroid M. It need not have the same rank as M, although clearly $r(\Delta(M)) \leq r(M)$. In Example 11.2.11, M is graphic and cographic and so is regular, binary, and ternary. But $\Delta(M) \cong U_{2,6}$ so it has none of these five properties. Also M and its dual are isomorphic, but $\Delta(M)$ and its dual do not even have the same rank. Although, $\mathbb{F}$-representability of M does not force $\mathbb{F}$-representability of $\Delta(M)$, Piff and Welsh (1970) were able to prove the following.

Proposition 11.2.16 *Let Δ be a bipartite graph with vertex classes S and J and let M be a matroid on J. Then there is an integer $n(M)$ such that if M is representable over a field $\mathbb{F}$ that has at least $n(M)$ elements, then the induced matroid $\Delta(M)$ is $\mathbb{F}$-representable.*

Before proving this result, we note two immediate consequences of it, which were stated earlier without proof. The first implies that a transversal matroid is representable over all sufficiently large finite fields and over all infinite fields.

Corollary 11.2.17 *Let M be a transversal matroid and $\mathbb{F}$ be a field. Then there is an integer $n(M)$ such that M is representable over every extension field of $\mathbb{F}$ having at least $n(M)$ elements.* □

Corollary 11.2.18 *A transversal matroid is representable over fields of every characteristic.* □

On combining Corollary 11.2.17 with Proposition 6.7.10 and Exercise 5 of Section 6.7, we obtain the following.

Corollary 11.2.19 *A transversal matroid is algebraic over all fields.* □

The next lemma contains the core of the proof of Proposition 11.2.16.

Lemma 11.2.20 *Let M be a matroid on a set J and let $\sigma : J \to S$ be a surjection. Then there is an integer $n(M)$ such that if M is representable over a field $\mathbb{F}$ that has at least $n(M)$ elements, the matroid $\sigma(M)$ is also $\mathbb{F}$-representable.*

Proof The surjection σ will be called *simple* if $|J| = |S|+1$. As every surjection is the composition of at most $|J|-1$ simple surjections, it will suffice to prove the proposition in the case that σ is simple. Thus assume this and let $\psi : J \to V(r, \mathbb{F})$ be an $\mathbb{F}$-coordinatization of M.

Let $J = \{j_1, j_2, \ldots, j_m\}$ and $S = \{s_2, s_3, \ldots, s_m\}$, and suppose $\sigma(j_i) = s_i$ for all i in $\{2, 3, \ldots, m\}$, while $\sigma(j_1) = s_2$. Now define $\psi_1 : S - s_2 \to V(r, \mathbb{F})$ by $\psi_1(s_i) = \psi(j_i)$ for all i in $\{3, 4, \ldots, m\}$. We shall show that there are elements c_1 and c_2 of $\mathbb{F}$ such that if $\psi_1(s_2) = c_1\psi(j_1) + c_2\psi(j_2)$, then the extended map $\psi_1 : S \to V(r, \mathbb{F})$ is an $\mathbb{F}$-coordinatization of $\sigma(M)$. This is certainly true if $\{j_1\}$, $\{j_2\}$, or $\{j_1, j_2\}$ is a circuit of M. Hence we may assume that $\{j_1, j_2\}$ is independent in M. Thus $\{s_2\}$ is independent in $\sigma(M)$.

Now let $\mathcal{B}'$ be the set of bases of $\sigma(M)/s_2$ and suppose $B \in \mathcal{B}'$. Then $B \cup s_2$ is independent in $\sigma(M)$, so $B \cup s_2 = \sigma(I)$ for some I in $\mathcal{I}(M)$. Therefore $\langle \psi_1(B) \rangle$ is a proper subspace of $\langle \psi(I) \rangle$. Hence $\langle \psi(j_1), \psi(j_2) \rangle \cap \langle \psi_1(B) \rangle$ is a 0- or 1-dimensional subspace of the 2-dimensional space $\langle \psi(j_1), \psi(j_2) \rangle$. Now if $\mathbb{F} = GF(q)$, then $V(2, \mathbb{F})$ has $q + 1$ subspaces of dimension one, while if $\mathbb{F}$ is infinite, $V(2, \mathbb{F})$ has infinitely many such subspaces. Thus if $|\mathbb{F}| \geq |\mathcal{B}'|$, then $\langle \psi(j_1), \psi(j_2) \rangle$ contains a 1-dimensional subspace that is not in $\bigcup \{ \langle \psi_1(B) \rangle : B \in \mathcal{B}' \}$. Let $c_1 \psi(j_1) + c_2 \psi(j_2)$ be a non-zero member of such a subspace and define $\psi_1(s_2) = c_1 \psi(j_1) + c_2 \psi(j_2)$. Now there are only finitely many non-isomorphic matroids M on the fixed set J and only finitely many simple surjections σ with domain J. Therefore the maximum value of $|\mathcal{B}'|$ over all such M and all such σ exists and is finite. We take this maximum to be the required integer $n(M)$.

To complete the proof, we need to show that ψ_1 is an $\mathbb{F}$-coordinatization for M, that is, if $X \subseteq S$, then X is independent in $\sigma(M)$ if and only if $\psi_1(X)$ is linearly independent in $V(r, \mathbb{F})$. This is easily seen to be true if $s_2 \notin X$, so assume that $s_2 \in X$. Suppose that X is independent in $\sigma(M)$. Then X is contained in a basis $B \,\dot{\cup}\, s_2$ of $\sigma(M)$. From above, as $B \subseteq S - s_2$, the set $\psi_1(B)$ is linearly independent. Moreover, the choice of $\psi_1(s_2)$ guarantees that $\psi_1(B \cup s_2)$ is linearly independent. Thus $\psi_1(X)$ is linearly independent. Now suppose X is a dependent set in $\sigma(M)$ containing s_2. Then it is not difficult to show that $\psi_1(X)$ is linearly dependent in $V(r, \mathbb{F})$ and this finishes the proof of the lemma. $\qquad \square$

Proof of Proposition 11.2.16 We shall assume that Δ has no isolated vertices since the general case follows easily from this case. Now let Δ_1 be the simple bipartite graph with vertex classes S and $E(\Delta)$ in which a member v of S is joined to a member e of $E(\Delta)$ precisely when e is incident with v in Δ. Let Δ_2 be defined similarly with vertex classes $E(\Delta)$ and J. An example of this construction is shown in Figure 11.5.

Now, clearly, $\mathrm{si}(\Delta_2(M)) \cong \mathrm{si}(M)$. Thus if M is $\mathbb{F}$-representable, so too is

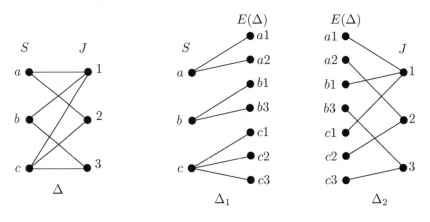

Fig. 11.5. $\Delta(M) = \Delta_1(\Delta_2(M))$.

$\Delta_2(M)$. Moreover, $\Delta(M) = \Delta_1(\Delta_2(M))$ and clearly Δ_1 corresponds to a surjection. The proposition now follows immediately from Lemma 11.2.20. □

We shall show next that both the construction of an induced matroid from a bipartite graph and the construction of a strict gammoid given in Section 2.4 are special cases of an operation that induces a matroid from an arbitrary directed graph. Lemma 2.4.3, which provided the key step in the proof that the dual of a transversal matroid is a strict gammoid, is also of fundamental importance in the proof that this more general construction does indeed produce a matroid.

Let G be a directed graph having vertex set V, and let M be a matroid on V. Let $L(G, M)$ be the collection of subsets of V that can be linked in G to an independent set of M. Thus, for example, if $B_0 \subseteq V$ and M is the direct sum of a free matroid on B_0 and a rank-0 matroid on $V - B_0$, then $L(G, M)$ is precisely the set of independent sets of a matroid, namely the strict gammoid $L(G, B_0)$. This observation is a special case of the following result, a proof of which is outlined in Exercise 10.

Proposition 11.2.21 *If G is a directed graph having vertex set V, and $L(G, M)$ is the set of subsets of V that are linked to an independent set in M, then $L(G, M)$ is the set of independent sets of a matroid on V.* □

By analogy with strict gammoids, we use $L(G, M)$ to denote both the matroid obtained here and its collection of independent sets.

Example 11.2.22 Let G be the directed graph shown in Figure 11.6(a) and M be the matroid on $\{1, 2, \ldots, 6\}$ for which a geometric representation is shown in Figure 11.6(b). Then it is not difficult to check that Figure 11.6(c) is a geometric representation for $L(G, M)$. For example, $\{3, 4, 6\}$ is independent in $L(G, M)$ but not in M as $\{3, 4, 6\}$ is linked to the independent set $\{3, 4, 1\}$ in G. □

We call $L(G, M)$ the matroid *induced from M by G*. Its rank function was determined by McDiarmid (1975a) using the theory of submodular functions. Although G has thus far been required to be a directed graph, we can easily extend the definition of $L(G, M)$ to include the possibility that G is undirected.

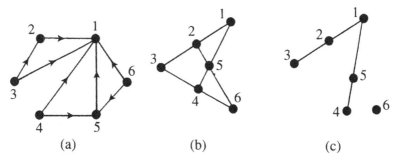

FIG. 11.6. (a) G. (b) M. (c) The induced matroid $L(G, M)$.

In this case, one replaces each non-loop edge of G by two oppositely directed edges, deletes each loop from G, and then finds the matroid induced from M by the resulting directed graph.

Next we note how two of the matroid constructions discussed in Chapter 7 can be achieved by this operation of inducing by a graph (Mason 1977). We begin with the principal extension $M +_F e$ of M in which the element e is freely added to the flat F. In the following result, a straightforward consequence of Proposition 11.2.21, the graph G is as illustrated in Figure 11.7(a).

Corollary 11.2.23 *Let F be a flat in a matroid M and let G be the graph with vertex set $E(M) \dot\cup e$ and edge set $\{ef : f \in F\}$. Then $M +_F e$ is the matroid $L(G, M \oplus U_{0,1})$, where $M \oplus U_{0,1}$ is formed from M by adjoining e as a loop.* □

The free coextension of M by the element e can also be obtained as an induced matroid. Let M' be an isomorphic copy of M for which the ground set E' is disjoint from the ground set E of M. For each x in E, let x' denote the corresponding element of E'. Now form the bipartite graph H having vertex set $E \cup E' \cup \{e\}$ and edge set $\{xx' : x \in E\} \cup \{xe : x \in E\}$ (see Figure 11.7(b)). Let M_0 be the matroid obtained from M' by adjoining e as a coloop and adjoining the elements of E as loops. We leave it to the reader to check the following.

Proposition 11.2.24 *The restriction of $L(H, M_0)$ to $E \cup e$ is the free coextension of M.* □

In the next section, we give another important example of a matroid operation that can be achieved by using this technique of inducing by a graph. Next we describe the geometric significance of the operation of inducing a matroid from

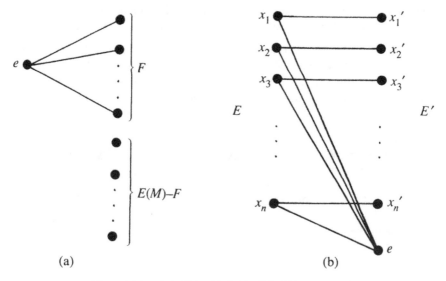

(a)   (b)

FIG. 11.7. (a) G in 11.2.23. (b) H in 11.2.24.

a bipartite graph Δ. Suppose that Δ has vertex classes S and J, and let M be a matroid on J. Then it is not difficult to show that, by adding edges to Δ, we can find a bipartite graph Γ with $\Gamma(M) = \Delta(M)$ such that, for every x in S, the set $N_\Gamma(x)$ of neighbours of x in Γ forms a flat in M (see Exercise 9). Using this fact, one can prove the following extension of Corollary 11.2.23.

Proposition 11.2.25 *Let Δ be a bipartite graph with vertex classes S and J and let M be a matroid on J. Let M^+ be the matroid that is obtained from M by adjoining all of the elements of S as loops. Then the matroids obtained by restricting $L(\Delta, M^+)$ to J and S are M and $\Delta(M)$, respectively.* $\square$

The matroid $L(\Delta, M^+)$ is constructed from M by a sequence of single-element extensions with each element x of S being freely added to the flat $\mathrm{cl}_M(N_\Gamma(x))$ of M. Since transversal matroids are precisely the matroids that are induced by bipartite graphs from free matroids, this construction can be used to give a geometric description of transversal matroids (Ingleton 1971b, Brylawski 1975d). Before giving the details of this description, we comment briefly on terminology. A matroid of the form $L(\Delta, M^+)$ where M is free is called a *principal transversal matroid* or a *fundamental transversal matroid*. We know that, when M is free, the restriction of $L(\Delta, M^+)$ to S is transversal but it may not be immediately apparent that $L(\Delta, M^+)$ itself is transversal. To see this, we modify Δ as follows. Add a set J' of $|J|$ new vertices to Δ joining each element of J to exactly one of the new vertices. Let $\{jj' : j \in J\}$ be the resulting set of new edges. Let Δ' be the bipartite graph with vertex classes $S \cup J'$ and J and with edge set $E(\Delta) \cup \{jj' : j \in J\}$. As M is free, the matroid $\Delta'(M)$ is certainly transversal and is easily seen to be isomorphic to $L(\Delta, M^+)$. Hence $L(\Delta, M^+)$ is indeed transversal. We can extract some more information from the construction of Δ'. The matroid $L(\Delta, M^+)$ has J as a distinguished basis and S as the corresponding cobasis. Using Δ', it is not difficult to show that the transversal matroid $L(\Delta, M^+)$ has, as a presentation, the family of fundamental cocircuits with respect to the cobasis S. Conversely, every transversal matroid having such a presentation is principal transversal (Exercise 13). We shall show in Proposition 11.2.28 that the class of principal transversal matroids is closed under duality.

If $|J| = m$, a free matroid on J can be represented geometrically by m affinely independent points in $\mathbb{R}^{m-1}$. So, for instance, when $m = 3$ and $m = 4$, the elements of J are the vertices of a triangle and a tetrahedron respectively. In general, such a collection of affinely independent points is called a *simplex*. A *vertex* of the simplex is any of the points of J; a *face* of the simplex is any flat of the affine matroid on J.

In view of the above discussion, a consequence of Proposition 11.2.25 is that a rank-r matroid M is transversal if and only if there is a simplex in $\mathbb{R}^{r-1}$ such that every element e of M is freely added to a face F_e of the simplex so that M has no dependencies except those that are forced by the faces F_e. Clearly the last statement needs to be made more precise. This is done in the next proposition where, for ease of statement, attention is confined to loopless non-empty matroids. Little generality is lost by this restriction since the addition or deletion of loops does not alter whether or not a matroid is transversal.

Proposition 11.2.26 *For $r \geq 1$, a loopless rank-r matroid M is transversal if and only if there are disjoint finite subsets J and S of $\mathbb{R}^{r-1}$ such that*

(i) $|J| = r$ *and the elements of J form a simplex;*

(ii) M *is isomorphic to the affine matroid N on S; and*

(iii) *every element e of S is freely added to a face F_e of the simplex J so that, for every flat X of N containing e, the closure of X in the affine matroid on $S \cup J$ contains F_e.*   □

We observe that, in the last result, the affine matroid on $S \cup J$ is a principal transversal matroid and every loopless non-empty principal transversal matroid P arises in this way. Of course, we can also apply the last proposition directly to P. To see P occurring as the affine matroid N in (ii), we simply add an element in parallel to each element of the distinguished basis of P and take the simplex J to be the set of these added elements. The last two propositions are illustrated in the following.

Example 11.2.27 Let Δ be the bipartite graph shown in Figure 11.8(a) where $S = \{a, b, \ldots, h\}$ and $J = \{1, 2, 3, 4\}$. Let M_J be the free matroid on J and M_J^+ be the matroid obtained from M_J by adjoining the elements of S as loops. A geometric representation for the principal transversal matroid $L(\Delta, M_J^+)$ is shown in Figure 11.8(b) where we note that the element c of this matroid is free on the flat spanned by $\{1, 2, 3\}$. The restriction of $L(\Delta, M_J^+)$ to J is a free matroid, in this case, $U_{4,4}$. The fundamental cocircuits of $L(\Delta, M_J^+)$ with respect to the cobasis S are precisely the complements of the faces of the tetrahedron shown, namely, $\{1, a, b, c\}, \{2, c, d, e\}, \{3, c, d, f, g\}$, and $\{4, e, f, g, h\}$. Moreover, the transversal matroid having these four sets as a presentation is $L(\Delta, M_J^+)$. The restriction of $L(\Delta, M_J^+)$ to S equals the matroid $\Delta(M_J)$. A geometric representation for the last matroid is shown in Figure 11.8(c). Of course, $\Delta(M_J)$ is precisely the transversal matroid on S that is associated with the bipartite graph Δ.   □

For many matroids M, Proposition 11.2.26 provides a quick way to show that M is not transversal. For instance, $M(K_4)$ is not transversal. If it were, then, since each of its six elements lies on the intersection of two 3-point lines, each such element must be parallel to a vertex of the associated simplex. But this simplex has only three vertices and $M(K_4)$ has no non-trivial parallel classes.

To conclude this section, we briefly consider some classes of transversal matroids whose duals are also transversal. Las Vergnas (1970) proved that every principal transversal matroid is cotransversal.

Proposition 11.2.28 *The dual of a principal transversal matroid is principal transversal.*

Proof Let M be a principal transversal matroid. Then there is a bipartite graph Δ with vertex classes S and J such that $M = L(\Delta, N)$ where N is the

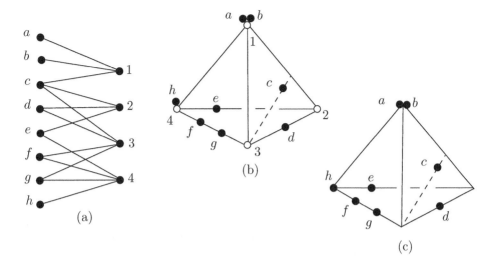

FIG. 11.8. (a) Δ. (b) $L(\Delta, M_J^+)$. (c) $L(\Delta, M_J^+)|S = \Delta(M_J)$.

direct sum of a rank-0 matroid on S with a free matroid on J. Now a subset B of $S \cup J$ is a basis of M if and only if, for some subsets $\{s_1, s_2, \ldots, s_k\}$ and $\{j_1, j_2, \ldots, j_k\}$ of S and J, respectively, $B = \{s_1, s_2, \ldots, s_k\} \cup (J - \{j_1, j_2, \ldots, j_k\})$ where $\{s_1 j_1, s_2 j_2, \ldots, s_k j_k\}$ is a matching in Δ. Now N^* is the direct sum of a rank-0 matroid on J and a free matroid on S, and it is not difficult to see, by comparing their sets of bases, that $M^* = L(\Delta, N^*)$. □

Transversal matroids that are also cotransversal are sometimes called *bi-transversal* (see, for example, Bonin and de Mier 2008). Some special classes of such matroids have received a lot of recent research attention. The next such class we consider is obtained as follows. In a Cartesian plane, consider a lattice path that begins at $(0,0)$ and ends at (m, r) and consists of $m + r$ unit-length steps each of which is north or east, that is, parallel to the y- or the x-axis. We can describe such a path by specifying precisely which of the $m + r$ steps are north. Let P and Q be two such paths whose ith north steps occur as steps l_i and u_i, respectively, of their paths where $l_i \leq u_i$. Then $l_1 < l_2 < \cdots < l_r$ and $u_1 < u_2 < \cdots < u_r$. Let $[l_i, u_i]$ denote the set of integers j with $l_i \leq j \leq u_i$. Then the transversal matroid with ground set $\{1, 2, \ldots, r+m\}$ having the family $([l_i, u_i] : 1 \leq i \leq r)$ as a presentation is a *lattice path matroid*. Observe that the bases of such a matroid are precisely the sets $\{b_1, b_2, \ldots, b_r\}$ of integers with $b_1 < b_2 < \cdots b_r$ such that $l_i \leq b_i \leq u_i$ for all i. As an example, consider the two lattice paths P and Q from $(0,0)$ to $(7,4)$ that form the upper and lower boundaries of the diagram in Figure 11.9. We have numbered the eleven steps of each of these paths. Then $(l_1, l_2, l_3, l_4) = (1, 2, 5, 7)$ and $(u_1, u_2, u_3, u_4) = (4, 7, 10, 11)$. The associated lattice path matroid $M[P, Q]$ is the transversal matroid having as a presentation $([1, 4], [2, 7], [5, 10], [7, 11])$. Such matroids seem to have first

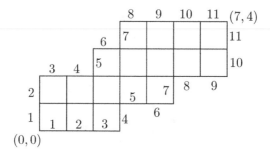

FIG. 11.9. Two lattice paths from $(0,0)$ to $(7,4)$.

appeared in the literature, albeit briefly, in the work of Stanley (1977). They were introduced independently, given their current name, and studied in detail by Bonin, de Mier, and Noy (2003) and Bonin and de Mier (2006).

In our example above, instead of considering the subintervals of $[1,11]$ determined by north steps, we can consider those determined by east steps. This gives us the family $([1,3],[2,4],[3,6],[5,8],[6,9],[8,10],[9,11])$. The transversal matroid having this family of intervals as a presentation is the dual of $M[P,Q]$, so this dual is also a lattice path matroid. This procedure works in general to show that the class of lattice path matroids is closed under taking duals. The class of lattice path matroids is also closed under taking minors and has several other attractive properties.

Another class of transversal matroids that is closed under taking duals and minors is the class of *nested matroids*. Such a matroid is a transversal matroid having a presentation $(A_1, A_2, \ldots, A_r)$ such that $A_1 \subseteq A_2 \subseteq \cdots \subseteq A_r$ (see Exercise 13 of Section 1.6). We leave it as an exercise to show that every nested matroid is a lattice path matroid. For $n \geq 2$, let M_n be the rank-n paving matroid on a $2n$-element set whose only non-spanning circuits are two disjoint n-element sets. Thus M_n is $T^{n-2}(U_{n-1,n} \oplus U_{n-1,n})$, the matroid obtained by truncating the direct sum of two n-element circuits $n-2$ times. Oxley, Prendergast, and Row (1982) showed that the set of excluded minors for the class of nested matroids is $\{M_n : n \geq 2\}$. More recently, Bonin (2009b) found the set of excluded minors for the class of lattice path matroids. Bonin and Giménez (2007) studied multi-path matroids, a generalization of lattice path matroids that also form a class of transversal matroids that is closed under taking minors and duals. The excluded minors for the last class have yet to be determined.

Nested matroids have appeared frequently in the literature under a variety of different names. The name used here follows Bonin and de Mier (2008) and seems most evocative. A history of the many appearances of nested matroids is given by Bonin and de Mier (2006, pp. 711–712). These matroids seem to have first occurred in the work of Crapo (1965) and are called *generalized Catalan matroids* by Bonin, de Mier, and Noy (2003) because they generalize a class of matroids that Ardila (2003) calls Catalan matroids. For a positive integer n, the *rank-n Catalan matroid* is the transversal matroid with ground set $\{1, 2, \ldots, 2n\}$ having the family $([1, 2i - 1] : 1 \leq i \leq n)$ of intervals as a presentation. Its name derives from the fact that its number of bases is the Catalan number $\frac{1}{n+1}\binom{2n}{n}$. Ardila uses the term *shifted matroid* for what we have called a 'nested matroid'. Bonin and de Mier (2006, Theorem 4.2) proved that the smallest minor-closed class of matroids containing all Catalan matroids is the class of nested matroids. Moreover, Bonin, de Mier, and Noy (2003, Theorem 3.14) proved that nested matroids are precisely the matroids that can be obtained from the empty matroid by iterating the operations of adding a coloop and taking a free extension.

Exercises

1. (Ore 1955) Show that a family $(A_j : j \in J)$ of subsets of a set S has a partial transversal of size $|J| - d$ if and only if $|A(K)| \geq |K| - d$ for all subsets K of J.

2. Let $(A_j : j \in J)$ be a family of subsets of a set S and let $\mathcal{I}$ be the set of subsets K of J for which $(A_j : j \in K)$ has a transversal. Prove that $\mathcal{I}$ is the set of independent sets of a matroid on J.

3. Prove Proposition 11.2.9 and Corollary 11.2.10.

4. Find $L(G, M)$ if M is the matroid in Figure 11.6(b) and G is the undirected graph underlying the directed graph in Figure 11.6(a).

5. Let Δ be a bipartite graph with vertex classes S and J and M be a matroid on J. Show that:

   (a) the characteristic set of M is a subset of the characteristic set of $\Delta(M)$, and these two sets need not be equal;

   (b) if M is transversal, then $\Delta(M)$ need not be transversal.

6. (Matthews 1977) Prove that bicircular matroids are precisely the loopless transversal matroids that have a presentation in which the intersection of any three members is empty.

7. Let $\mathcal{A}$ be a family $(A_j : j \in J)$ of subsets of a set S and m be a positive integer. Use Theorem 11.2.3 to show that:

   (a) $\mathcal{A}$ has m disjoint transversals if and only if $|A(K)| \geq m|K|$ for all $K \subseteq J$;

   (b) $\mathcal{A}$ has a system of representatives in which no element occurs more than m times if and only if $m|A(K)| \geq |K|$ for all $K \subseteq J$.

8. Let M be a rank-r transversal matroid.

   (a) Show that M has at most r non-trivial parallel classes.

   (b) Extend (a) by determining the maximum number of rank-k flats in M each of which has a spanning circuit.

   (c) Show that, for all $k \geq 3$, exactly one of the two non-isomorphic parallel extensions of $\mathcal{W}^k$ is transversal.

9. Let Δ be a bipartite graph with vertex classes S and J. Let M be a matroid on J and x be in S.

(a) Suppose that $y \in \mathrm{cl}_M(N_\Delta(x)) - N_\Delta(x)$ and let Δ_1 be obtained from Δ by adding the edge xy. Show that $\Delta_1(M) = \Delta(M)$.

(b) Deduce that there is a bipartite graph Δ' that can be obtained by adding edges to Δ such that $\Delta'(M) = \Delta(M)$ and, for every x in S, the set of neighbours of x in Δ' forms a flat in M.

10 Let M be a matroid on the vertex set V of a directed graph G and let $\widehat{G}$ be the bipartite graph corresponding to G (see Section 2.4). Let $\widehat{M}$ be an isomorphic copy of M on the set $\widehat{V}$. Prove Proposition 11.2.21 by showing, using Lemma 2.4.3, that the maximal members of $L(G, M)$ coincide with the cobases of the matroid induced by $\widehat{G}$ from the dual of $\widehat{M}$.

11. For a graph G, let Δ be the bipartite graph with vertex classes $E(G)$ and $V(G)$ and edge set $\{ev : e \in E(G), v \in V(G), e \text{ is incident with } v \text{ in } G\}$. Let M be the free matroid on $V(G)$.

(a) Show that $\Delta(M)$ is the bicircular matroid $B(G)$.

(b) Use this to show that (i)–(iii) of 11.2.26 give a geometric characterization of bicircular matroids if one adds to (iii) the requirement that, for all e in S, the face F_e of the simplex to which e is freely added has rank one or two. (Compare this with Exercise 4 of Section 6.10.)

12. (Brualdi 1969) Use Hall's Theorem to prove that $\mathcal{B}$ is the collection of bases of a matroid on E if and only if $\mathcal{B}$ satisfies (B1) and the following *bijective basis exchange axiom*:

(B2)" *If B_1 and B_2 are in $\mathcal{B}$, then there is a bijection $\alpha : B_1 \to B_2$ such that $(B_1 - b) \cup \alpha(b) \in \mathcal{B}$ for all b in B_1.*

13. Prove that a transversal matroid M is a principal transversal matroid if and only if M has a cobasis B^* such that the family $(C^*(e, B^*) : e \in E(M) - B^*)$ is a presentation for M.

14. (Bonin 2009b) Show that neither the class of principal transversal matroids nor the class of lattice path matroids is a subset of the other.

15. Show that the classes of a lattice path matroids and nested matroids are both closed under taking minors and duals, and that the first class but not the second is closed under taking direct sums.

16. (Bonin and de Mier 2006) Prove that a matroid M is a lattice path matroid if and only if M is transversal and has a presentation $(A_1, A_2, \ldots, A_r)$ such that each A_i is an interval in a linear order on $E(M)$ and none of these intervals is a subset of another.

17. (Sims 1977, Bonin and de Mier 2008) Consider the set $\mathcal{Z}$ of cyclic flats of a matroid. From Exercise 13 of Section 3.3, $\mathcal{Z}$ is a lattice under set inclusion. Show that:

(a) Every finite lattice is isomorphic to the lattice of cyclic flats of some bitransversal matroid.

(b) A matroid is nested if and only if the lattice on $\mathcal{Z}$ is a chain.

18. (Hall 1935) Let $(A_j : j \in J)$ and $(D_j : j \in J)$ be families of subsets of a set S. Prove that there is a sequence $(x_j : j \in J)$ that is a system of representatives for both of these families if and only if $|\{j \in J : D_j \cap A(K) \neq \emptyset\}| \geq |K|$ for all $K \subseteq J$.
19. (Mendelsohn and Dulmage 1958) Let $(A_j : j \in J)$ be a family of subsets of a set S. For $J_1 \subseteq J$ and $S_1 \subseteq S$, prove the following are equivalent.
    (a) $(A_j : j \in J_1)$ has a transversal and S_1 is a partial transversal of $(A_j : j \in J)$.
    (b) For some J_2 with $J_1 \subseteq J_2 \subseteq J$, there is a set S_2 such that $S_1 \subseteq S_2 \subseteq S$ and S_2 is a transversal of $(A_j : j \in J_2)$.

11.3 Matroid union and its applications

In this section, we consider a way of joining matroids which, unlike the previous such operations we have considered, is defined irrespective of the number of elements common to the matroids. This important operation was introduced by Nash-Williams (1966). We begin by defining it for two matroids on a common ground set, although eventually we shall extend the definition to include any finite set of matroids having arbitrary ground sets.

Theorem 11.3.1 *Let M_1 and M_2 be matroids on a set E having rank functions r_1 and r_2, respectively. Let*

$$\mathcal{I} = \{I_1 \cup I_2 : I_1 \in \mathcal{I}(M_1), I_2 \in \mathcal{I}(M_2)\}.$$

Then $\mathcal{I}$ is the set of independent sets of a matroid $M_1 \vee M_2$ on E. Moreover, the rank in $M_1 \vee M_2$ of a subset X of E is

$$\min\{r_1(Y) + r_2(Y) + |X - Y| : Y \subseteq X\}.$$

Proof Construct the bipartite graph Δ as follows. Let ϕ_1 and ϕ_2 be bijections from E onto disjoint sets E_1 and E_2. Then each ϕ_i induces an isomorphic copy of M_i on E_i. Hence, on $E_1 \dot\cup E_2$, we have a matroid isomorphic to $M_1 \oplus M_2$. Let Δ have E as one vertex class and $E_1 \cup E_2$ as the other. Join each element e of E to $\phi_1(e)$ and $\phi_2(e)$ and to no other vertices in $E_1 \cup E_2$ (see Figure 11.10).

Now consider the matroid N on E induced from $M_1 \oplus M_2$ by Δ. Since $M_1 \oplus M_2$ has rank function $r_1 + r_2$, Corollary 11.2.14 implies that the rank function of N is as specified in the theorem. Moreover, one easily checks that $\mathcal{I}(N) = \{I_1 \cup I_2 : I_1 \in \mathcal{I}(M_1), I_2 \in \mathcal{I}(M_2)\}$. □

The matroid $M_1 \vee M_2$ is called the *union* of M_1 and M_2. Thus, for example, if M_1 and M_2 are as shown in Figure 11.11, then $M_1 \vee M_2 \cong U_{6,7}$. Notice that both $M_1 \vee M_1$ and $M_2 \vee M_2$ are also isomorphic to $U_{6,7}$.

It is straightforward to extend 11.3.1 to show that if $(M_1, M_2, \ldots, M_n)$ is an arbitrary finite family of matroids on a common ground set E, then there is a matroid $M_1 \vee M_2 \vee \cdots \vee M_n$ on E such that

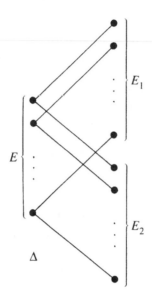

FIG. 11.10. $M_1 \vee M_2$ is induced from $M_1 \oplus M_2$ by Δ.

11.3.2 $\mathcal{I}(M_1 \vee M_2 \vee \cdots \vee M_n) = \{I_1 \cup I_2 \cup \cdots \cup I_n : I_i \in \mathcal{I}(M_i) \text{ for } 1 \le i \le n\}$.
Moreover, if M_i has rank function r_i, then the rank of X in $\bigvee_{i=1}^{n} M_i$ is

11.3.3 $\min\{\sum_{i=1}^{n} r_i(Y) + |X - Y| : Y \subseteq X\}$.

Example 11.3.4 If $m_1 + m_2 + \cdots + m_k = m$, then

$$U_{m_1,n} \vee U_{m_2,n} \vee \cdots \vee U_{m_k,n} = \begin{cases} U_{m,n} & \text{if } m < n, \\ U_{n,n} & \text{if } m \ge n. \end{cases}$$

Hence a union of uniform matroids is uniform. However, a union of graphic matroids need not even be binary. For instance, if G_1 and G_2 are as shown in Figure 11.12, then $M(G_1) \vee M(G_2) \cong U_{2,4}$. □

So far the operation of matroid union has only been defined for matroids that share a common ground set. We shall show next that the definition can be

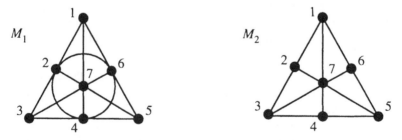

FIG. 11.11. Each of $M_1 \vee M_2$, $M_1 \vee M_1$, and $M_2 \vee M_2$ is $U_{6,7}$.

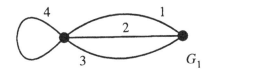
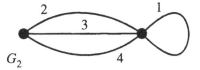

FIG. 11.12. $M(G_1) \vee M(G_2)$ is non-binary.

extended in a natural way to cover the case when this condition does not hold. Let $M_1, M_2, \ldots, M_n$ be matroids having ground sets $E_1, E_2, \ldots, E_n$, respectively, and let $E = E_1 \cup E_2 \cup \cdots \cup E_n$. For all i in $\{1, 2, \ldots, n\}$, define M_i^+ to be the matroid on E that is formed by adjoining a set of loops to M_i, these loops being labelled by the elements of $E - E_i$. As $M_1^+, M_2^+, \ldots, M_n^+$ have the same ground set, $M_1^+ \vee M_2^+ \vee \cdots \vee M_n^+$ is well-defined, and we take $M_1 \vee M_2 \vee \cdots \vee M_n$ to be equal to this matroid. Thus $E(M_1 \vee M_2 \vee \cdots \vee M_n) = E$ and $\mathcal{I}(M_1 \vee M_2 \vee \cdots \vee M_n)$ is as specified in 11.3.2. Moreover, the rank of a subset X of E in $\bigvee_{i=1}^n M_i$ is

11.3.5 $\min\{|X - Y| + \sum_{i=1}^n r_i(Y \cap E_i) : Y \subseteq X\}$

where r_i is the rank function of M_i.

The operation of matroid union has now been defined for any finite collection of matroids. Two important special cases of this operation that we have already met are direct sum and series connection.

Proposition 11.3.6 *Let M_1 and M_2 be matroids.*

(i)   *If $E(M_1) \cap E(M_2) = \emptyset$, then $M_1 \vee M_2 = M_1 \oplus M_2$.*
(ii) *If $E(M_1) \cap E(M_2) = \{p\}$, then $M_1 \vee M_2 = S(M_1, M_2)$.*

Proof   This is left as an exercise. □

Transversal matroids have the following attractive characterization in terms of the matroid union operation. We omit the straightforward proof of this result.

Proposition 11.3.7 *A matroid is transversal if and only if it is a union of rank-1 matroids.* □

As an immediate consequence of this proposition we have the following.

Corollary 11.3.8 *A union of transversal matroids is transversal. In particular, a series connection of transversal matroids is transversal.* □

Evidently every matroid M can be written as a union of itself with a rank-0 matroid. We call M *reducible* if it can be written as the union of two matroids neither of which is equal to M; otherwise M is *irreducible*. Welsh (1971b) raised the following problem that has so far defied solution.

Problem 11.3.9 *Characterize all irreducible matroids.* □

To get some feeling for the sorts of difficulty involved with this problem, the reader is urged to check that $M(K_4)$ is irreducible, although P_6, Q_6, $U_{3,6}$, and

$\mathcal{W}^3$ (see Figure 2.1) are all reducible. Currently the best partial result towards 11.3.9 is the following result of Cunningham (1979), which generalizes earlier results of his (1977) and Lovász and Recski (1973). For a short proof of this result, the reader is referred to Duke (1987).

Proposition 11.3.10 *A binary matroid M is irreducible if and only if M and all of its single-element deletions are connected.*   □

There are a number of other unsolved problems associated with decomposing matroids as unions of other matroids. These problems and the progress towards their solutions were surveyed by Recski (1985).

The remainder of this section discusses various applications of the operation of matroid union. We begin by proving a packing result and a covering result due to Edmonds (1965a,b). The original proofs of these results preceded the discovery of the operation of matroid union and were difficult. The easy proofs given below (Harary and Welsh 1969) exemplify the power of the union operation. In each of the next five results, k will denote an arbitrary positive integer.

Theorem 11.3.11 *A matroid M has k disjoint bases if and only if, for every subset X of E(M),*

$$kr(X) + |E(M) - X| \geq kr(M).$$

Theorem 11.3.12 *A matroid M has k independent sets whose union is E(M) if and only if, for every subset X of E(M),*

$$kr(X) \geq |X|.$$

Proof of Theorem 11.3.11 The matroid M has k disjoint bases if and only if $\bigvee_{i=1}^{k} M$ has rank at least $kr(M)$. By 11.3.3, the latter occurs if and only if $kr(X) + |E(M) - X| \geq kr(M)$ for all $X \subseteq E(M)$.   □

Proof of Theorem 11.3.12 The matroid M has k independent sets whose union is $E(M)$ if and only if M has k bases whose union is $E(M)$. But the latter occurs if and only if $\bigvee_{i=1}^{k} M$ has rank at least $|E(M)|$. The result follows by again using 11.3.3.   □

In conjunction with his proof of Theorem 11.3.12, Edmonds (1965a) described an algorithm that, for fixed k and a given matroid M, will either produce a partition of $E(M)$ into k independent sets or establish that no such partition exists. This algorithm, known as the *matroid partitioning algorithm*, proceeds in polynomial time provided that one has an independence oracle or, equivalently, one can decide in polynomial time whether or not any given set is independent.

An immediate consequence of Theorem 11.3.12 is a characterization of when a finite subset of a vector space can be partitioned into k linearly independent sets. This result was originally proved by Horn (1955) using an intricate algebraic argument. Rado (1966) pointed out that this argument generalizes to matroids in the natural way. In particular, a vector that is a linear combination of some set X of vectors corresponds to a matroid element that is in the closure of X.

Part of the motivation for Theorems 11.3.11 and 11.3.12 derived from the fact that the corresponding results for graphs had been proved just a few years earlier. These two graph results are derived here as consequences of Theorems 11.3.11 and 11.3.12. Both results were originally proved by Tutte (1961b), with the first also being independently derived by Nash-Williams (1961).

If π is a partition of the vertex set $V(G)$ of a graph G, then $E_\pi(G)$ will denote the set of edges of G that join vertices in different classes of π. In addition, $|\pi|$ will denote the number of classes in π.

Proposition 11.3.13 *A connected graph G has k edge-disjoint spanning trees if and only if, for every partition π of $V(G)$,*

$$|E_\pi(G)| \geq k(|\pi| - 1).$$

Proof Suppose that the specified condition holds for all partitions π of $V(G)$. By Theorem 11.3.11, to show that $M(G)$ has k disjoint bases, we need to show that, for all subsets X of $E(G)$,

$$kr(X) + |E(G) - X| \geq kr(M(G)). \tag{11.11}$$

For such a subset X, let G_X be the subgraph of G having vertex set $V(G)$ and edge set X. Let π be the partition of $V(G)$ in which the classes are the vertex sets of connected components of G_X. Then, as $E(G) - X \supseteq E_\pi(G)$,

$$|E(G) - X| \geq k(|\pi| - 1). \tag{11.12}$$

On combining (11.12) with (11.11), we see that (11.11) holds provided that

$$|\pi| - 1 \geq r(M(G)) - r(X). \tag{11.13}$$

But $r(M(G)) = |V(G)| - 1$ and $r(X) = |V(G_X)| - \omega(G_X) = |V(G)| - |\pi|$, so (11.13) holds with equality and therefore G has k edge-disjoint spanning trees. The straightforward proof of the converse is left as an exercise. $\square$

Proposition 11.3.14 *Let G be a graph. Then $E(G)$ can be partitioned into k disjoint forests if and only if, for all subsets X of $V(G)$,*

$$|E(G[X])| \leq k(|X| - 1).$$

Proof This follows by applying Theorem 11.3.12 to $M(G)$. $\square$

Next we shall show how to use the operation of matroid union to determine necessary and sufficient conditions for two matroids to have a common k-element independent set (Edmonds 1970).

Theorem 11.3.15 *Let M_1 and M_2 be matroids with rank functions r_1 and r_2 and a common ground set E. Then there is a k-element subset of E that is independent in both M_1 and M_2 if and only if, for all subsets X of E,*

$$r_1(X) + r_2(E - X) \geq k.$$

Proof By Proposition 2.1.9, $r_2(E - X) = |E - X| + r_2^*(X) - r_2^*(M_2)$. Thus $r_1(X) + r_2(E - X) \geq k$ for all $X \subseteq E$ if and only if, for all such X,

$$r_1(X) + r_2^*(X) + |E - X| \geq k + r_2^*(M_2). \tag{11.14}$$

By Theorem 11.3.1, inequality (11.14) holds for all $X \subseteq E$ if and only if

$$r(M_1 \vee M_2^*) \geq k + r_2^*(M_2). \tag{11.15}$$

To complete the proof, we shall show that (11.15) holds if and only if M_1 and M_2 have a common k-element independent set. If the latter holds and I is such a set, then I is independent in M_1, so $E - I$ is spanning in M_2^*. Thus $E - I$ contains a basis B_2^* of M_2^* and so

$$r(M_1 \vee M_2^*) \geq |I \,\dot\cup\, B_2^*| = k + r_2(M_2^*),$$

that is, (11.15) holds. Conversely, if (11.15) holds and D_2^* is a basis of M_2^*, then $M_1 \vee M_2^*$ has a basis B containing D_2^*. Now $B = I_1 \dot\cup D_2^*$ where $I_1 \in \mathcal{I}(M_1)$, so

$$|I_1| + r_2(M_2^*) = |I_1 \,\dot\cup\, D_2^*| = |B| = r(M_1 \vee M_2^*) \geq k + r_2(M_2^*).$$

Hence $|I_1| \geq k$. Moreover, $I_1 \subseteq E - D_2^*$, so $I_1 \in \mathcal{I}(M_2)$. Thus I_1 has a k-element subset that is independent in both M_1 and M_2, and the theorem follows. $\square$

As an immediate consequence of the last theorem, we have the following result that specifies the size of a largest common independent set in two matroids.

Corollary 11.3.16 *Let M_1 and M_2 be matroids with rank functions r_1 and r_2 and a common ground set E. Then*

$$\max\{|I| : I \in \mathcal{I}(M_1) \cap \mathcal{I}(M_2)\} = \min\{r_1(T) + r_2(E - T) : T \subseteq E\}. \quad \square$$

The problem of finding a maximum-sized set that is independent in both M_1 and M_2 can be solved in polynomial time assuming that one has polynomial-time subroutines for testing independence of sets in M_1 and M_2. For descriptions of this *matroid intersection algorithm* and of algorithms that solve the corresponding weighted version of this problem, the reader is referred to Lawler (1976), Edmonds (1979), Faigle (1987), and Cook, Cunningham, Pulleyblank, and Schrijver (1998). The last book, along with Bondy and Murty (2008) and Garey and Johnson (1979), is recommended to the reader who is unfamiliar with computational complexity and, in particular, the theory of NP-completeness.

The basic idea in the matroid intersection algorithm is similar to that used in the algorithm for finding a maximum-sized matching in a bipartite graph: construction of an augmenting path. By contrast with the situation for two matroids, the problem of finding a maximum-size independent set in three matroids is NP-complete as a special case of it is the problem of deciding whether a directed graph has a Hamiltonian path (see Exercise 6). The last problem is easily shown to be polynomial-time reducible to the problem of deciding whether a graph has a Hamiltonian cycle, one of the best-known NP-complete problems.

Theorem 11.3.15 will now be used to prove a result of Perfect (1968), which has as a corollary Ford and Fulkerson's theorem (1958) characterizing when two families of sets have a common transversal.

Proposition 11.3.17 *Let $(A_j : j \in J)$ and $(D_j : j \in J')$ be two families, $\mathcal{A}$ and $\mathcal{D}$, of subsets of a set E. Then there is a k-element subset of E that is a partial transversal of both $\mathcal{A}$ and $\mathcal{D}$ if and only if, for all $K \subseteq J$ and all $K' \subseteq J'$,*

$$|A(K) \cap D(K')| \geq |K| + |K'| - |J| - |J'| + k.$$

Proof Evidently $\mathcal{A}$ and $\mathcal{D}$ have a common k-element partial transversal if and only if $M[\mathcal{A}]$ and $M[\mathcal{D}]$ have a common k-element independent set. By Theorem 11.3.15, this occurs if and only if, for all $X \subseteq E$,

$$r_1(X) + r_2(E - X) \geq k, \tag{11.16}$$

where r_1 and r_2 are the rank functions of $M[\mathcal{A}]$ and $M[\mathcal{D}]$. By Corollary 11.2.7, inequality (11.16) holds for all $X \subseteq E$ if and only if, for all such X,

$$\min\{|A(K) \cap X| - |K| + |J| : K \subseteq J\}$$
$$+ \min\{|D(K') \cap (E - X)| - |K'| + |J'| : K' \subseteq J'\} \geq k.$$

In turn, this holds for all $X \subseteq E$ if and only if

$$|A(K) \cap X| + |D(K') \cap (E - X)| \geq |K| + |K'| - |J| - |J'| + k \tag{11.17}$$

for all $X \subseteq E$, all $K \subseteq J$, and all $K' \subseteq J'$. The left-hand side of (11.17) equals

$$|(A(K) \cap X) \cup (D(K') \cap (E - X))| + |A(K) \cap X \cap D(K') \cap (E - X)|,$$

which equals $|(A(K) \cap X) \cup (D(K') \cap (E - X))|$. Thus $M[\mathcal{A}]$ and $M[\mathcal{D}]$ have a common k-element independent set if and only if, for all $X \subseteq E$, all $K \subseteq J$, and all $K' \subseteq J'$,

$$|(A(K) \cap X) \cup (D(K') \cap (E - X))| \geq |K| + |K'| - |J| - |J'| + k. \tag{11.18}$$

Letting $X = D(K')$, it follows that if $M[\mathcal{A}]$ and $M[\mathcal{D}]$ have a common k-element independent set, then, for all $K \subseteq J$ and all $K' \subseteq J'$,

$$|A(K) \cap D(K')| \geq |K| + |K'| - |J| - |J'| + k. \qquad (11.19)$$

Finally, if (11.19) holds for all K and K', then (11.18) holds for all X, K, and K' since

$$\begin{aligned} A(K) \cap D(K') &= (A(K) \cap D(K') \cap X) \cup (A(K) \cap D(K') \cap (E - X)) \\ &\subseteq (A(K) \cap X) \cup (D(K') \cap (E - X)). \end{aligned} \qquad \square$$

Corollary 11.3.18 *Two families $(A_j : j \in J)$ and $(D_j : j \in J)$ of subsets of a set E have a common transversal if and only if, for all subsets K and K' of J,*

$$|A(K) \cap D(K')| \geq |K| + |K'| - |J|. \qquad \square$$

A number of other results can be derived using this operation of matroid union. Some of these will be considered in the exercises. For a detailed general account of results in transversal theory such as 11.3.17, 11.3.18, and the theorems of Hall and Rado, we refer the reader to Mirsky's book (1971).

Exercises

1. Let M_1 and M_2 be matroids on a set E and suppose $T \subseteq E$. Show that $(M_1 \vee M_2)|T = (M_1|T) \vee (M_2|T)$.

2. For matroids M_1 and M_2 on a set E, let $M_1 \wedge M_2$ be $(M_1^* \vee M_2^*)^*$. Show that $\mathcal{S}(M_1 \wedge M_2) = \{X_1 \cap X_2 : X_1 \in \mathcal{S}(M_1), X_2 \in \mathcal{S}(M_2)\}$ where $\mathcal{S}(N)$ is the set of spanning sets of N.

3. Let f_1 and f_2 be non-negative, integer-valued, increasing, submodular functions on 2^E. Show that:
   (a) (Pym and Perfect 1970) $M(f_1 + f_2) = M(f_1) \vee M(f_2)$;
   (b) (McDiarmid in Welsh 1976) if f_1 and f_2 are not increasing, then (a) may fail when $|E| = 2$.

4. Let T be a subset of the ground set of a matroid M. Show that:
   (a) if $M|T$ is irreducible and $r(T) = r(M)$, then M is irreducible;
   (b) the converse of (a) does not hold;
   (c) the assumption that $r(T) = r(M)$ is needed in (a);
   (d) F_7 is irreducible.

5. In a matroid M having ground set E, prove that

$$\max\{|X| - 2r(X) : X \subseteq E\} = \min\{|B_1^* \cap B_2^*| : B_1^*, B_2^* \in \mathcal{B}^*(M)\}.$$

6. (Held and Karp 1970) Let D be a directed graph with vertex set $\{1, 2, \ldots, n\}$ and G be the underlying graph. For each i in $V(D)$, let T_i and H_i be the sets of arcs of D having i as the tail or head, respectively. Let $\mathcal{T} = (T_i : 1 \leq i \leq n)$ and $\mathcal{H} = (H_i : 1 \leq i \leq n)$. Prove that D has a Hamiltonian path, a directed path using all the vertices of D, if and only if the three matroids $M[\mathcal{T}], M[\mathcal{H}]$, and $M(G)$ have a common basis.

7. (Welsh 1976) Prove that two families $(A_j : j \in J)$ and $(D_j : j \in J)$ of subsets of a set E have a common transversal containing a subset X of E if and only if, for all subsets K and K' of J,

$$|(A(K) \cap D(K')) - X| \quad + \quad |A(K) \cap X| + |D(K') \cap X|$$
$$\geq \quad |K| + |K'| + |X| - |J|.$$

[Hint: Let M be the matroid on E whose bases are the bases of $M[(D_j : j \in J)]$ containing X. Then the required common transversal exists if and only if $(A_j : j \in J)$ has a transversal that is a basis of M.]

8. (Pym and Perfect 1970) If M_1 and M_2 are matroids on a set E and $X \subseteq E$, show that the rank of X in $M_1 \vee M_2$ is $\max\{r_1(Y) + r_2(X - Y) : Y \subseteq X\}$.

9. (Piff and Welsh 1970) Let $M_1, M_2, \ldots, M_m$ be $\mathbb{F}$-representable matroids on a set E. Prove that there is an integer n such that $\vee_{i=1}^m M_i$ is representable over every extension of $\mathbb{F}$ with at least n elements.

10. (Brualdi and Scrimger 1968, Brualdi 1969) A matroid M is *base-orderable* if, given any two bases B_1 and B_2, there is a bijection $\alpha : B_1 \to B_2$ such that, for all e in B_1, both $(B_1 - e) \cup \alpha(e)$ and $(B_2 - \alpha(e)) \cup e$ are bases; M is *strongly base-orderable* if the bijection α above can be chosen so that, for all subsets X of B_1, both $(B_1 - X) \cup \alpha(X)$ and $(B_2 - \alpha(X)) \cup X$ are bases. Prove that the classes of base-orderable and strongly base-orderable matroids are closed under each of the following operations:
    (a) restriction;
    (b) the taking of duals;
    (c) the taking of minors;
    (d) union;
    (e) truncation.

11. Show that:
    (a) All gammoids are strongly base-orderable.
    (b) $M(K_4)$ is not base-orderable.
    (c) P_8 is base-orderable but not strongly base-orderable.
    (d) The matroid J in Figure 10.5 is not base-orderable.
    (e) P_7 is strongly base-orderable but is not a gammoid.

12. (McDiarmid 1975b) Let X and Y be independent sets in a matroid M and let (X_1, X_2) be a partition of X. By considering $(M/X_1) \vee (M/X_2)$, prove that there is a partition (Y_1, Y_2) of Y such that both $X_1 \cap Y_1$ and $X_2 \cap Y_2$ are empty, and both $X_1 \cup Y_1$ and $X_2 \cup Y_2$ are independent in M.

13. *(Brualdi 1970) Let Δ be a bipartite graph with vertex classes S_1 and S_2, and k be a non-negative integer. For each i in $\{1, 2\}$, let M_i be a matroid on S_i having rank function r_i. Show that the following are equivalent:
    (a) there is a k-element subset of M_1 that can be matched into an independent set of M_2;
    (b) $r_1(S_1 - X) + r_2(N(X)) \geq k$ for all $X \subseteq S_1$.

11.4　Amalgams and the generalized parallel connection

Let M_1 and M_2 be matroids having a common restriction N. In such circumstances, one often wants to stick these matroids together across N. We have already seen with incompatible extensions (Example 7.2.4) that this may not be possible. In this section, we investigate some circumstances under which it is possible, concentrating in particular on the operation of generalized parallel connection. The properties of this operation are discussed in the last part of the section beginning with Proposition 11.4.13. The first part of the section is based on some unpublished results of A. W. Ingleton and on an unpublished exposition of these results due to J. H. Mason.

Suppose that the matroids M_1 and M_2 have ground sets E_1 and E_2, rank functions r_1 and r_2, and closure operators cl_1 and cl_2. Let $E_1 \cup E_2 = E$. We are assuming that $M_1|T = M_2|T = N$ where $E_1 \cap E_2 = T$. The rank function of this common restriction of M_1 and M_2 will be denoted by r. If M is a matroid on E such that $M|E_1 = M_1$ and $M|E_2 = M_2$, then M is called an *amalgam* of M_1 and M_2. We call M_0 the *free amalgam* of M_1 and M_2 if M_0 is an amalgam of M_1 and M_2 that is freer than all other amalgams, that is, for any other amalgam M, every independent set in M is independent in M_0.

We have seen already that there may be no amalgam of M_1 and M_2. Next we show that, even when there is an amalgam, there may be no free amalgam.

Example 11.4.1 Consider the three rank-4 matroids for which geometric representations are shown in Figure 11.13. It is clear that the Vámos matroid, V_8, is an amalgam of M_1 and M_2. Another amalgam of these two matroids is the vector matroid of the matrix A over $\mathbb{R}$ where

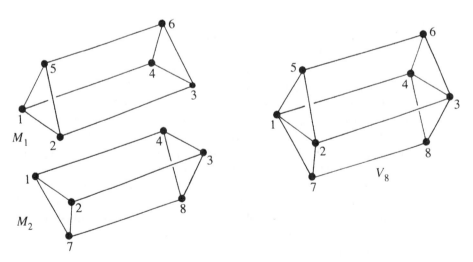

FIG. 11.13. V_8 is an amalgam of M_1 and M_2.

$$A = \begin{array}{c} \begin{array}{cccccccc} 1 & 2 & 4 & 5 & 7 & 3 & 6 & 8 \end{array} \\ \left[\begin{array}{ccccc|ccc} & & & & & 1 & 1 & 1 \\ & & & & & 1 & 0 & 0 \\ & & I_5 & & & 1 & 1 & 1 \\ & & & & & 0 & 1 & 0 \\ & & & & & 0 & 0 & 1 \end{array} \right] \end{array}.$$

Assume that the free amalgam M_0 of M_1 and M_2 exists. Then, as $\{1, 2, 4, 5, 7\}$ is independent in $M[A]$, it is independent in M_0, so $r(M_0) \geq 5$. Moreover, as $\{5, 6, 7, 8\}$ is independent in V_8, it is independent in M_0. Therefore $\{5, 6, 7, 8\}$ is contained in a 5-element independent set I in M_0. By symmetry, we may assume that I is $\{5, 6, 7, 8, 1\}$. But $\{5, 6, 1, 4\}$ is a circuit of M_1 and hence of M_0, and $\{7, 8, 1, 4\}$ is a circuit of M_2 and hence of M_0. Thus, eliminating the element 4, we get that $\{5, 6, 7, 8, 1\}$ is dependent in M_0. This contradiction shows that there is no free amalgam of M_1 and M_2. □

We observe next that if M is an arbitrary amalgam of M_1 and M_2, then, by submodularity of the rank function, for all $X \subseteq E$,

$$r_M(X) \leq \eta(X)$$

where

$$\eta(X) = r_1(X \cap E_1) + r_2(X \cap E_2) - r(X \cap T). \tag{11.20}$$

Now let

$$\zeta(X) = \min\{\eta(Y) : Y \supseteq X\}. \tag{11.21}$$

Then, for all $X \subseteq E$,

$$\zeta(X) \geq r_M(X). \tag{11.22}$$

Throughout the rest of this section, where ambiguity does not arise, we shall often drop the subscripts on the various rank functions involved.

Proposition 11.4.2 *If the function ζ defined in (11.21) is submodular, then it is the rank function of a matroid on E. Moreover, this matroid is the free amalgam of M_1 and M_2.*

Proof By hypothesis, ζ satisfies (R3) and it is not difficult to check that ζ satisfies (R1) and (R2). Thus ζ is the rank function of a matroid M on E. For $X \subseteq E_1$ and $Y \supseteq X$, by (11.20),

$$\eta(Y) = r_1(Y \cap E_1) + r_2(Y \cap E_2) - r(Y \cap T) \geq r_1(Y \cap E_1) \geq r_1(X).$$

Hence $\zeta(X) \geq r_1(X)$. But $\eta(X) = r_1(X)$, so $\zeta(X) = r_1(X)$ and $M|E_1 = M_1$. Similarly, $M|E_2 = M_2$. Thus M is indeed an amalgam of M_1 and M_2. Moreover, by (11.22) and Proposition 7.3.11, M is the free amalgam of M_1 and M_2. □

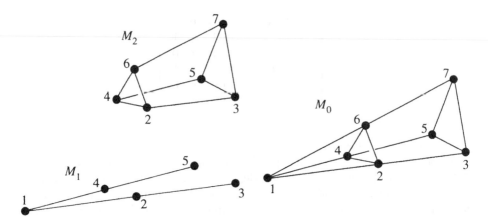

FIG. 11.14. M_0 is the free amalgam of M_1 and M_2 but is not a proper amalgam.

When ζ is submodular, the matroid on E that has ζ as its rank function is called the *proper amalgam* of M_1 and M_2. The last result established that every proper amalgam is a free amalgam. The next example shows that some free amalgams are not proper. To verify that this example has the stated properties, we shall use the following result, the proof of which is left to the reader.

Proposition 11.4.3 *A given matroid M is the proper amalgam of $M|E_1$ and $M|E_2$ if and only if, for every flat F of M,*

$$r(F) = \zeta(F) = r(F \cap E_1) + r(F \cap E_2) - r(F \cap T). \qquad \square$$

Example 11.4.4 In Figure 11.14, the matroid M_0 is clearly an amalgam of M_1 and M_2. In fact, it is the free amalgam. We recall here from Example 1.5.5 that 1, 6, and 7 must be collinear in order for M_0 to be a matroid. To see that M_0 is not the proper amalgam of M_1 and M_2, it suffices to note that the rank equation in Proposition 11.4.3 fails when F is the flat $\{1, 6, 7\}$ of M_0. $\qquad \square$

The remainder of our discussion of general amalgams will be concerned with finding conditions to guarantee the existence of the proper amalgam of M_1 and M_2. We remark here that, while Proposition 11.4.3 gives necessary and sufficient conditions for a *given* matroid to be this proper amalgam, it says nothing about when the proper amalgam exists.

Let $\mathcal{L}(M_1, M_2)$ denote the set of subsets X of E such that $X \cap E_1$ and $X \cap E_2$ are flats of M_1 and M_2, respectively. Clearly every flat of every amalgam of M_1 and M_2 is in $\mathcal{L}(M_1, M_2)$. Moreover, $\mathcal{L}(M_1, M_2)$ is a lattice of subsets of E in which $X \wedge Y = X \cap Y$. The next result shows that ζ is uniquely determined by the values taken by η on members of $\mathcal{L}(M_1, M_2)$.

Proposition 11.4.5 *For all $X \subseteq E$,*

$$\zeta(X) = \min\{\eta(Y) : Y \in \mathcal{L}(M_1, M_2) \text{ and } Y \supseteq X\}.$$

This result follows without difficulty from the following.

Lemma 11.4.6 *For $Y \subseteq E$, let Z be the least member of $\mathcal{L}(M_1, M_2)$ containing Y. Then $\eta(Z) \leq \eta(Y)$.*

Proof For all $X \subseteq E$, let $\phi_1(X) = \mathrm{cl}_1(X \cap E_1) \cup (X \cap E_2)$ and $\phi_2(X) = (X \cap E_1) \cup \mathrm{cl}_2(X \cap E_2)$. Then

$$\phi_1(Y) \cap E_2 = (\phi_1(Y) \cap T) \cup (Y \cap E_2)$$

and

$$Y \cap T \subseteq (\phi_1(Y) \cap T) \cap (Y \cap E_2).$$

Using these facts and the submodularity of r_2, we get that

$$r_2(\phi_1(Y) \cap E_2) + r_2(Y \cap T) \leq r_2(\phi_1(Y) \cap T) + r_2(Y \cap E_2),$$

so

$$r_2(\phi_1(Y) \cap E_2) - r_2(\phi_1(Y) \cap T) \leq r_2(Y \cap E_2) - r_2(Y \cap T).$$

But also $r_1(\phi_1(Y) \cap E_1) = r_1(Y \cap E_1)$. Hence

$$\eta(\phi_1(Y)) \leq r_1(Y \cap E_1) + r_2(Y \cap E_2) - r_2(Y \cap T) = \eta(Y).$$

Now applying the above argument with the subscripts 1 and 2 interchanged and Y replaced by $\phi_1(Y)$, we get that

$$\eta(\phi_2(\phi_1(Y))) \leq \eta(\phi_1(Y)).$$

Thus, by applying the operations ϕ_1 and ϕ_2 alternately, we get, after a finite number of steps, a set Z_0 such that $\eta(Z_0) \leq \eta(Y)$ and $\phi_1(Z_0) = Z_0 = \phi_2(Z_0)$. It is not difficult to check that $Z_0 \in \mathcal{L}(M_1, M_2)$ and that, in fact, $Z_0 = Z$. The lemma follows. □

The next proposition is Ingleton's main result on when the proper amalgam of two matroids is guaranteed to exist.

Proposition 11.4.7 *Suppose that η is submodular on $\mathcal{L}(M_1, M_2)$. Then the proper amalgam of M_1 and M_2 exists.*

Proof Suppose $X_1, X_2 \subseteq E$. By Proposition 11.4.5, for each i in $\{1, 2\}$, there is a member Y_i of $\mathcal{L}(M_1, M_2)$ such that $X_i \subseteq Y_i$ and $\zeta(X_i) = \eta(Y_i)$. Evidently $X_1 \cap X_2 \subseteq Y_1 \wedge Y_2$ and $X_1 \cup X_2 \subseteq Y_1 \vee Y_2$. Therefore

$$\begin{aligned} \zeta(X_1 \cap X_2) + \zeta(X_1 \cup X_2) &\leq \eta(Y_1 \wedge Y_2) + \eta(Y_1 \vee Y_2) \\ &\leq \eta(Y_1) + \eta(Y_2) = \zeta(X_1) + \zeta(X_2). \end{aligned}$$

Thus ζ is submodular on 2^E and so, by Proposition 11.4.2 and the remarks following it, the proper amalgam of M_1 and M_2 exists. □

Next we seek conditions that will ensure that η is submodular on $\mathcal{L}(M_1, M_2)$. If M is an arbitrary matroid and X and Y are subsets of $E(M)$, we define the *modular defect* $\delta(X, Y)$ of (X, Y) to be $r(X)+r(Y)-r(X\cup Y)-r(X\cap Y)$. A subset Z of $E(M)$ is said to be *fully embedded* in M if $\delta(F_1\cap Z, F_2\cap Z) \le \delta(F_1, F_2)$ for all flats F_1 and F_2 of M. Thus, in Example 11.4.4, if $Z = \{2, 3, 4, 5\}$, then Z is not fully embedded in M_0. To see this, let $F_1 = \{1, 4, 5, 6, 7\}$ and $F_2 = \{1, 2, 3, 6, 7\}$. Then $\delta(F_1, F_2) = 0$. But $\delta(F_1 \cap Z, F_2 \cap Z) = 1$.

It is not difficult to check that loops and parallel elements have no effect on whether or not a set is fully embedded.

Lemma 11.4.8 *Let M be a matroid and suppose that $Z \subseteq Z' \subseteq E(M)$. Assume that every element of $Z' - Z$ is either parallel to some element of Z or is a loop. Then Z' is fully embedded in M if and only if Z is fully embedded in M.* $\square$

The main examples of proper amalgams will be derived from the following result. Recall that $T = E_1 \cap E_2$.

Proposition 11.4.9 *If T is fully embedded in M_1, then η is submodular on $\mathcal{L}(M_1, M_2)$ and hence the proper amalgam of M_1 and M_2 exists.*

Proof Suppose X and Y are in $\mathcal{L}(M_1, M_2)$. Then, as T is fully embedded in M_1, it follows that $\delta(X\cap T, Y\cap T) \le \delta(X\cap E_1, Y\cap E_1)$. Now, by Lemma 11.4.6 and the fact that $X \wedge Y = X \cap Y$, we have that

$$\eta(X) + \eta(Y) - \eta(X \wedge Y) - \eta(X \vee Y) \ge \eta(X) + \eta(Y) - \eta(X \cap Y) - \eta(X \cup Y).$$

On expanding the right-hand side of the last line using (11.20) and then rearranging terms, we obtain

$$[\delta(X \cap E_1, Y \cap E_1) - \delta(X \cap T, Y \cap T)] + [r_2(X \cap E_2) + r_2(Y \cap E_2) - r_2((X \cup Y) \cap E_2) - r_2((X \cap Y) \cap E_2)].$$

Each quantity enclosed in square brackets is non-negative, hence η is indeed submodular on $\mathcal{L}(M_1, M_2)$. It follows by Proposition 11.4.7 that the proper amalgam of M_1 and M_2 exists. $\square$

The next result is our main application of Proposition 11.4.9.

Theorem 11.4.10 *If either of the following conditions holds, then T is fully embedded in M_1 and so the proper amalgam of M_1 and M_2 exists:*

(i) *T is a modular flat of M_1.*
(ii) *$M_1|T$ is a modular matroid.*

Proof Part (ii) is immediate from the definition. To prove (i), we shall use the following lemma which holds for an arbitrary matroid having rank function r. The proof is left to the reader.

Lemma 11.4.11 *Suppose that F and T are flats of a matroid and T is modular. If $F \cap T \subseteq W \subseteq T$, then*

$$r(F \cup W) + r(F \cap W) = r(F) + r(W). \qquad \square$$

Now suppose that T is a modular flat of M_1. To complete the proof of Theorem 11.4.10, we shall show that T is fully embedded in M_1. Let X and Y be flats of M_1. Then, it is not difficult to show using Lemma 8.2.3 that

$$r(X \cup Y) - r([X \cap (Y \cup T)] \cup [Y \cap (X \cup T)])$$
$$\leq r(X) - r(X \cap (Y \cup T)) + r(Y) - r(Y \cap (X \cup T)).$$

Moreover, by Lemma 11.4.11,

$$r(X \cap (Y \cup T)) = r(X \cap Y) + r(X \cap T) - r(X \cap Y \cap T)$$

and

$$r(Y \cap (X \cup T)) = r(X \cap Y) + r(Y \cap T) - r(X \cap Y \cap T).$$

Also $[X \cap (Y \cup T)] \cup [Y \cap (X \cup T)] = (X \cap Y) \cup [(X \cup Y) \cap T]$ and, by submodularity,

$$r((X \cap Y) \cup [(X \cup Y) \cap T]) \leq r(X \cap Y) + r((X \cup Y) \cap T) - r(X \cap Y \cap T).$$

On combining these observations, it is not difficult to show that $\delta(X, Y) \geq \delta(X \cap T, Y \cap T)$. Hence T is fully embedded in M_1. $\qquad \square$

On combining the last theorem with Lemma 11.4.8, we get the following.

Corollary 11.4.12 *Suppose that $\mathrm{cl}_1(T)$ is a modular flat of M_1 and every nonloop element of $\mathrm{cl}_1(T) - T$ is parallel to some element of T. Then T is fully embedded in M_1, so the proper amalgam of M_1 and M_2 exists.* $\qquad \square$

Now recall that, for a matroid M, the simple matroid associated with M is denoted by $\mathrm{si}(M)$. We shall denote the matroid $\mathrm{si}(M_1|T)$ by $\mathrm{si}(T)$. In this notation, the hypothesis of the last corollary is that $\mathrm{si}(T)$ is a modular flat of $\mathrm{si}(M_1)$. When this condition holds, we call the proper amalgam of M_1 and M_2 the *generalized parallel connection* of M_1 and M_2 *across* T. This matroid will be denoted by $P_N(M_1, M_2)$ or $P_T(M_1, M_2)$, where we recall that $N = M_1|T = M_2|T$. Brylawski (1975b) studied the generalized parallel connection in detail when both M_1 and M_2 are simple. The extension of his work to the general case is straightforward. Note that, by Proposition 11.4.3, for every flat F of $P_N(M_1, M_2)$, we have $\zeta(F) = r(F)$ and

$$r(F) = r(F \cap E_1) + r(F \cap E_2) - r(F \cap T). \qquad (11.23)$$

Proposition 11.4.13 *The flats of $P_N(M_1, M_2)$ are precisely the members of $\mathcal{L}(M_1, M_2)$, that is, F is a flat of $P_N(M_1, M_2)$ if and only if $F \cap E_1$ is a flat of M_1 and $F \cap E_2$ is a flat of M_2.*

Proof As $P_N(M_1, M_2)$ is an amalgam of M_1 and M_2, every flat of it is certainly in $\mathcal{L}(M_1, M_2)$. Now suppose X is in $\mathcal{L}(M_1, M_2)$ and let Z be its closure $\mathrm{cl}_P(X)$ in $P_N(M_1, M_2)$. Then Z is also in $\mathcal{L}(M_1, M_2)$ and, since ζ is the rank function of $P_N(M_1, M_2)$, we have $\zeta(X) = \zeta(Z)$. For i in $\{1, 2\}$, let $X_i = X \cap E_i$ and $Z_i = Z \cap E_i$. By (11.23), (11.20), and the fact that $Z \cap T = Z_1 \cap T$, we have

$$\zeta(Z) = r(Z_1) + r(Z_2) - r(Z_1 \cap T) = \eta(Z).$$

But $\zeta(X) \leq \eta(X)$. Hence $\eta(Z) \leq \eta(X)$, that is,

$$r(Z_1) + r(Z_2) - r(Z_1 \cap T) \leq r(X_1) + r(X_2) - r(X_1 \cap T). \tag{11.24}$$

As $\mathrm{si}(T)$ is a modular flat of $\mathrm{si}(M_1)$, it follows easily that

$$r(X_1) + r(T) = r(X_1 \cap T) + r(X_1 \cup T)$$

and

$$r(Z_1 \cap T) + r(Z_1 \cup T) = r(Z_1) + r(T).$$

Adding the last two equations and eliminating $r(T)$, we get

$$r(X_1) - r(X_1 \cap T) = r(Z_1) - r(Z_1 \cap T) + [r(X_1 \cup T) - r(Z_1 \cup T)].$$

Therefore, as $X_1 \cup T \subseteq Z_1 \cup T$,

$$r(X_1) - r(X_1 \cap T) \leq r(Z_1) - r(Z_1 \cap T). \tag{11.25}$$

But we also know, since $X_2 \subseteq Z_2$, that

$$r(X_2) \leq r(Z_2). \tag{11.26}$$

If we now combine (11.25) and (11.26) with (11.24), we deduce that equality holds in all three of those statements. In particular, as X_2 and Z_2 are flats of M_2 of the same rank and $X_2 \subseteq Z_2$, we have $X_2 = Z_2$. Thus

$$X_1 \cap T = X \cap T = X_2 \cap T = Z_2 \cap T = Z \cap T = Z_1 \cap T.$$

Hence, since (11.25) holds with equality, $r(X_1) = r(Z_1)$, so $X_1 = Z_1$. We conclude that $X = Z$, that is, X is a flat of $P_N(M_1, M_2)$. $\qquad\square$

As an alternative to proceeding through the derivation of $P_N(M_1, M_2)$ as a particular proper amalgam, one can define this matroid directly as follows. Let M_1 and M_2 be matroids with ground sets E_1 and E_2 such that $E_1 \cap E_2 = T$ and $M_1|T = M_2|T = N$. When $\mathrm{si}(M_1|T)$ is a modular flat of $\mathrm{si}(M_1)$, the generalized parallel connection, $P_N(M_1, M_2)$, is the matroid on $E_1 \cup E_2$ whose flats are those subsets X of $E_1 \cup E_2$ such that $X \cap E_1$ is a flat of M_1, and $X \cap E_2$ is a flat of M_2. The fact that this construction does indeed produce a matroid is proved directly by Brylawski (1975b, pp. 19–20) in the case that both M_1 and M_2 are simple. The extension of this argument to the general case is straightforward.

The generalized parallel connection is a very natural construction and has numerous attractive properties. The 11-point matroid in Figure 1.14(b) was obtained by sticking together two copies of F_7 along a 3-point line. We are now able to identify this matroid as $P_N(F_7, F_7)$ where $N \cong U_{2,3}$. When $N \cong U_{1,1}$, the matroid $P_N(M_1, M_2)$ is just the parallel connection $P(M_1, M_2)$. Indeed, the parallel connection of M_1 and M_2 with respect to the basepoint p can always be thought of as a generalized parallel connection provided $M_1|\{p\} = M_2|\{p\}$. The last condition fails precisely when p is a loop in exactly one of M_1 and M_2.

The proofs of the following properties of the generalized parallel connection are not difficult and are left to the reader. Recall that $\text{si}(T) = \text{si}(M_1|T)$.

Proposition 11.4.14 *The generalized parallel connection has the following properties:*

(i) $P_N(M_1, M_2)|E_1 = M_1$ *and* $P_N(M_1, M_2)|E_2 = M_2$.

(ii)   *If* $\text{si}(T)$ *is a modular flat in* $\text{si}(M_2)$ *as well as in* $\text{si}(M_1)$, *then*
$$P_N(M_2, M_1) = P_N(M_1, M_2).$$

(iii) *The ground set of* $\text{si}(M_2)$ *is a modular flat of the simple matroid associated with* $P_N(M_1, M_2)$.

(iv)   *If* $e \in E_1 - T$, *then* $P_N(M_1, M_2)\backslash e = P_N(M_1\backslash e, M_2)$.

(v)   *If* $e \in E_1 - \text{cl}_1(T)$, *then* $P_N(M_1, M_2)/e = P_N(M_1/e, M_2)$.

(vi)   *If* $e \in E_2 - T$, *then* $P_N(M_1, M_2)\backslash e = P_N(M_1, M_2\backslash e)$.

(vii) *If* $e \in E_2 - \text{cl}_2(T)$, *then* $P_N(M_1, M_2)/e = P_N(M_1, M_2/e)$.

(viii) *If* $e \in T$, *then* $P_N(M_1, M_2)/e = P_{N/e}(M_1/e, M_2/e)$.

(ix) $P_N(M_1, M_2)/T = (M_1/T) \oplus (M_2/T)$.   □

In the preceding proposition, we did not identify $P_N(M_1, M_2)\backslash e$ when $e \in T$; nor did we identify $P_N(M_1, M_2)/e$ when $e \in \text{cl}_1(T) - T$ or $e \in \text{cl}_2(T) - T$. In the first case, $\text{si}(T\backslash e)$ need not be modular in $\text{si}(M_1\backslash e)$; in the third case, N remains a restriction of M_1 but not necessarily of M_2/e. In the second case, e is a loop or is parallel to some element of T, so we can determine $P_N(M_1, M_2)/e$ from (iv) or (viii), respectively.

We saw in Theorem 7.1.16(ii) that if p is an element of a connected matroid M and if M/p is disconnected, then M can be written as a parallel connection. The next result generalizes this.

Proposition 11.4.15 *Let* M *be a matroid on a set* E *and suppose that, for some subset* T *of* E, *the matroid* $M/T = M_1 \oplus M_2$. *If* $\text{si}(T)$ *is a modular flat of the simple matroid associated with* $M\backslash E(M_2)$, *then*

$$M = P_{M|T}(M\backslash E(M_2), M\backslash E(M_1)).$$

Proof A proof for simple matroids can be found in Brylawski (1975b). As an alternative, one can argue using the modular short circuit axiom (6.9.9). Although the argument is non-trivial, the details are omitted.   □

We show next that the operation of 3-sum for binary matroids is easily described in terms of generalized parallel connection.

Proposition 11.4.16 *Let M_1 and M_2 be binary matroids and $E(M_1) \cap E(M_2) = T$. Suppose that T is a 3-circuit of both M_1 and M_2, that $|E(M_1)|$ and $|E(M_2)|$ exceed six, and that T does not contain a cocircuit of M_1 or M_2. Then*

$$M_1 \oplus_3 M_2 = P_T(M_1, M_2) \backslash T.$$

Proof By Corollary 6.9.6, si(T) is a modular flat of both si(M_1) and si(M_2), so $P_T(M_1, M_2)$ is well-defined. Let $E_i = E(M_i)$ and $X_i = E_i - T$. Since T does not contain a cocircuit of M_i, the set X_i spans E_i and so

$$r(P_T(M_1, M_2) \backslash T) = r(P_T(M_1, M_2)) = r(M_1) + r(M_2) - 2.$$

By Proposition 9.3.4, $r(M_1 \oplus_3 M_2) = r(M_1) + r(M_2) - 2$. Hence $M_1 \oplus_3 M_2$ and $P_T(M_1, M_2) \backslash T$ are matroids of the same rank having the same ground set.

We shall now show that every flat F of $P_T(M_1, M_2) \backslash T$ is a flat of $M_1 \oplus_3 M_2$. This will imply, by Proposition 7.3.6, that $P_T(M_1, M_2) \backslash T$ is a quotient of $M_1 \oplus_3 M_2$; and, since these matroids have the same rank, Corollary 7.3.4 then implies that they are equal. By definition, $F \cap X_i$ is a flat of $M_i \backslash T$. Suppose that F is not a flat of $M_1 \oplus_3 M_2$. Then there is a circuit C of the latter and an element e not in F such that $e \in C \subseteq F \cup e$. Evidently C meets both X_1 and X_2. By Lemma 9.3.3, $C = C_1 \triangle C_2$ where C_i is a circuit of M_i, and T contains an element t such that $C_1 \cap T = \{t\} = C_2 \cap T$. Hence $C \cap T = \emptyset$. Now, without loss of generality, we may assume that $e \in X_1$. In $P_T(M_1, M_2)$, the element t is in the closure of $C_2 - t$, and e is in the closure of $C_1 - e$. Thus e is in the closure of $(C_2 - t) \cup (C_1 - e - t)$, that is, of $C - e$. As $C - e \subseteq F$ but $e \notin F$, this contradicts the fact that F is a flat of $P_T(M_1, M_2) \backslash T$. $\square$

We remark that an alternative way to prove the last proposition is to show that, subject to the hypotheses there, $P_T(M_1, M_2)$ equals $(M[A])^*$ where A is the matrix over $GF(2)$ whose columns are indexed by the elements of $E(M_1) \cup E(M_2)$ and whose rows consist of the incidence vectors of all the circuits of M_1 and all the circuits of M_2. To see this, one uses Proposition 11.4.15 and the fact that contracting T from $(M[A])^*$ gives a matroid with $E(M_1) - T$ and $E(M_2) - T$ as separators. Proposition 11.4.16 then follows by Lemma 9.3.1.

To close this section, we consider how various classes of matroids behave under generalized parallel connection. We noted in Example 7.1.27 that the class of transversal matroids is not closed under parallel connection. Hence this class is certainly not closed under generalized parallel connection; nor is the class of cographic matroids (Exercise 6). Moreover, these are not the only well-known classes that are not closed under generalized parallel connection.

Example 11.4.17 Consider T_8, the vector matroid of the matrix $[I_4 | J_4 - I_4]$ over $GF(3)$. This matroid is $\mathbb{F}$-representable if and only if $\mathbb{F}$ has characteristic three (Exercise 4 of Section 6.5). If the columns of $[I_4 | J_4 - I_4]$ are labelled, in order, $1, 2, \ldots, 8$, then $T_8 = M | \{1, 2, \ldots, 8\}$ where M is the 12-element

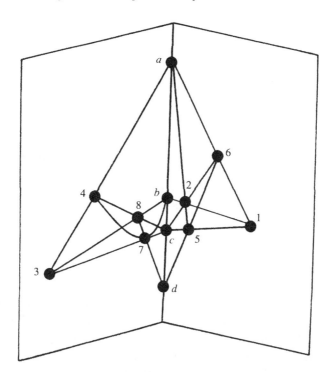

FIG. 11.15. A generalized parallel connection across a 4-point line.

matroid for which a geometric representation is shown in Figure 11.15. Let $M_1 = M \backslash \{1, 2, 5, 6\}$, $M_2 = M \backslash \{3, 4, 7, 8\}$, and $N = M | \{a, b, c, d\}$. Then, by Corollary 6.9.6, $\{a, b, c, d\}$ is a modular flat of both M_1 and M_2, and it is not difficult to check that $M = P_N(M_1, M_2)$. It is also straightforward to show that M_1 is $\mathbb{F}$-representable unless $\mathbb{F}$ has characteristic two and that $M_2 \cong M_1$. Thus if $\mathbb{F}$ has characteristic other than two or three, then both M_1 and M_2 are $\mathbb{F}$-representable yet $P_N(M_1, M_2)$, since it has T_8 as a restriction, is not. □

In contrast to the above, we have the following.

Proposition 11.4.18 *The classes of graphic, regular, binary, and ternary matroids are all closed under the operation of generalized parallel connection.* □

For a proof of this result and for a discussion of various other properties of the generalized parallel connection, we refer the reader to the work of Brylawski (1975b, 1986a). Moreover, further results on amalgams in general can be found in Bachem and Kern (1988), Nešetřil, Poljak, and Turzík (1981, 1985), and Poljak and Turzík (1982, 1984).

Exercises

1. (Poljak and Turzík 1984) Let M_1 and M_2 be matroids on sets E_1 and E_2, respectively, and let M be an amalgam of these two matroids. Show that:

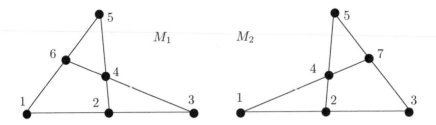

FIG. 11.16. These matroids have both F_7 and F_7^- as amalgams.

    (a) if $X \subseteq E_1 \cup E_2$, then $M|X$ is an amalgam of $M_1|X$ and $M_2|X$;

    (b) if $Y \subseteq E_1 \cap E_2$, then M/Y is an amalgam of M_1/Y and M_2/Y.

2. Let M_1 and M_2 be the matroids for which geometric representations are shown in Figure 11.16. Show that M_1 and M_2 have an amalgam isomorphic to F_7 but that their free amalgam is isomorphic to F_7^-.

3. (Poljak and Turzík 1984) Let M_1 and M_2 be $\mathbb{F}$-representable matroids on the sets E_1 and E_2, respectively. Show that if the matroids $M_1|(E_1 \cap E_2)$ and $M_2|(E_1 \cap E_2)$ are equal and are uniquely $\mathbb{F}$-representable, then there is an $\mathbb{F}$-representable matroid that is an amalgam of M_1 and M_2.

4. Let M_1 and M_2 be matroids with rank functions r_1 and r_2 and let $T = E(M_1) \cap E(M_2)$. Prove that if T is independent in both M_1 and M_2, then there is a matroid M on $E(M_1) \cup E(M_2)$ such that every flat F of M has rank $r_1(F \cap E(M_1)) + r_2(F \cap E(M_2)) - |F \cap T|$.

5. Let $M = P_N(M_1, M_2)$ where $N = M_1|T = M_2|T$. Let $\mathrm{cl}_1, \mathrm{cl}_2$, and cl_M denote the closure operators of M_1, M_2, and M, respectively. If $X \subseteq E(M)$ and $X_i = \mathrm{cl}_i(X \cap E_i) \cup X$, show that:

    (a) $\mathrm{cl}_M(X) = \mathrm{cl}_1(X_2 \cap E_1) \cup \mathrm{cl}_2(X_1 \cap E_2)$; and

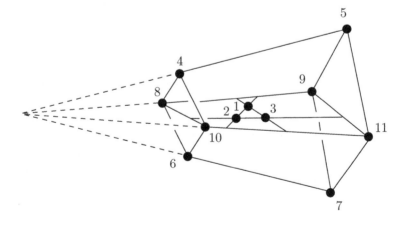

FIG. 11.17. The converse of Proposition 11.4.7 fails.

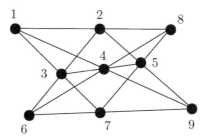

FIG. 11.18. The Pappus matroid.

(b) $r(X) = r(X_2 \cap E_1) + r(X_1 \cap E_2) - r(T \cap [X_1 \cup X_2])$.

6. (Brylawski 1975b) Show that the generalized parallel connection of two copies of $M(K_5 \backslash e)$ across an $M(K_4)$ has an $M(K_{3,3})$-minor. Deduce that the class of cographic matroids is not closed under the operation of generalized parallel connection.

7. (Mason) Let M be the matroid for which Figure 11.16 is a geometric representation. Let $E_1 = \{1, 2, \ldots, 9\}$ and $E_2 = \{1, 2, \ldots, 7\} \cup \{10, 11\}$. Show that the converse of Proposition 11.4.7 fails by showing that M is the proper amalgam of $M|E_1$ and $M|E_2$, but η is not submodular on $\mathcal{L}(M|E_1, M|E_2)$.

8. (Ingleton) Let M be the Pappus matroid labelled as in Figure 11.18. Let $E_1 = \{1, 2, \ldots, 8\}$ and $E_2 = \{1, 2, \ldots, 7, 9\}$. Show that the converse of Proposition 11.4.9 fails by showing that M is the proper amalgam of $M|E_1$ and $M|E_2$, and η is submodular on $\mathcal{L}(M|E_1, M|E_2)$, but $\{1, 2, \ldots, 7\}$ is not fully embedded in M_1 or M_2.

9. (T. J. Reid, private communication) Let $P = P_N(M_1, M_2)$ where M_1 and M_2 are n-connected matroids each having at least $2n - 2$ elements for some $n \geq 2$. Prove that P is n-connected if and only if $|E(N)| \geq n - 1$.

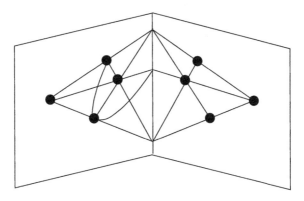

FIG. 11.19. The matroid F_8.

10. Let $F_8 = P_N(F_7, F_7^-)\backslash E(N)$ where $N \cong U_{2,3}$ (see Figure 11.19 or Exercise 5 of Section 6.4). Show that:
   (a) F_8 is self-dual but is not identically self-dual.
   (b) F_8 can be obtained from the 3-sum of two copies of F_7 by relaxing two intersecting circuit–hyperplanes.
   (c) F_8 is not a (minimal) excluded minor for $\mathbb{F}$-representability for any field $\mathbb{F}$.
   (d) F_8 is non-algebraic even though there are fields, for example $GF(2)$, over which both F_7 and F_7^- are algebraic.

11.5   Generalizations of delta-wye exchange

The operations of Δ-Y (delta-wye) exchange and Y-Δ (wye-delta) exchange are of basic importance in graph theory. In this section, we define matroid generalizations of these operations and we note some attractive properties of these matroid operations, particularly in relation to representability. First we define the graph operations. Let $\{a, b, c\}$ be the edge set of a triangle T in a graph G. Adjoin a new vertex v to G and add new edges joining v to the vertices of T. Finally, delete the edges of T. The resulting graph is said to have been obtained from G by a Δ-Y *exchange*. This operation is illustrated in Figure 11.20. There we begin with the graph $K_5 - e$ obtained by deleting a single edge from K_5. By performing a Δ-Y exchange on the specified triangle, we obtain $K_{3,3}$.

The reverse operation of Δ-Y exchange is called Y-Δ *exchange*. Formally, in a graph H, we begin with a degree-3 vertex u whose adjacent vertices x, y, and z are distinct. We add three new edges, one joining each of the pairs $\{x, y\}, \{y, z\}$, and $\{x, z\}$. The operation is completed by deleting u. Observe that if we perform a Y-Δ exchange on $K_{2,3}$, the resulting graph is K_4. Other examples of Δ-Y and Y-Δ exchanges in graphs are considered in the exercises.

Another way to view a Δ-Y exchange on a graph G is to consider it as being the 3-sum of K_4 and G. This immediately suggests a matroid generalization of the operation. Since a triangle T is a modular flat in $M(K_4)$, if T is a triangle of another matroid M and $E(M(K_4)) \cap E(M) = T$, then the generalized parallel connection of $M(K_4)$ and M across T is well-defined. The matroid

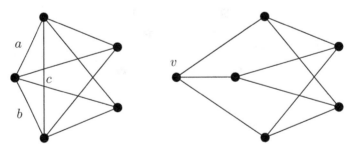

FIG. 11.20.   A Δ-Y exchange on $K_5 - e$ gives $K_{3,3}$.

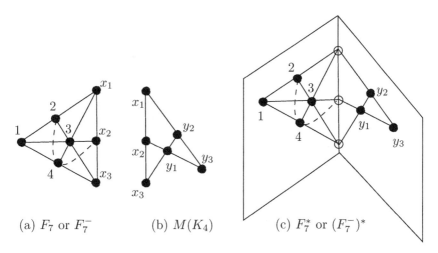

(a) F_7 or F_7^- (b) $M(K_4)$ (c) F_7^* or $(F_7^-)^*$

FIG. 11.21. A Δ-Y exchange on F_7 or F_7^- gives its dual.

$P_T(M(K_4), M)\backslash T$ we get by deleting T from this generalized parallel connection is said to have been obtained from M by a Δ-Y *exchange on* T. Figure 11.21 illustrates the fact that F_7^* and $(F_7^-)^*$ can be obtained from F_7 and F_7^-, respectively, by a Δ-Y exchange on $\{x_1, x_2, x_3\}$. The dashed line is included in F_7 and F_7^* but not in F_7^- or $(F_7^-)^*$.

A Y-Δ exchange is defined using duality. Suppose T is a triad of a matroid M and, as before, T is a triangle of $M(K_4)$ and $E(M(K_4)) \cap E(M) = T$. As T is a triangle of M^*, the matroid $[P_T(M(K_4), M^*)\backslash T]^*$ is defined. It is said to have been obtained from M by a Y-Δ *exchange on* T. We show next that Δ-Y exchange and Y-Δ exchange are special cases of two attractive matroid operations whose definitions again rely on generalized parallel connection.

To motivate our generalization of Δ-Y exchange, consider the following construction. In $PG(2, \mathbb{R})$, take a basis $\{y_1, y_2, y_3\}$ and a line L that is freely placed relative to this basis. By modularity, for each i in $\{1, 2, 3\}$, the hyperplane of $PG(2, \mathbb{R})$ that is spanned by $\{y_1, y_2, y_3\} - \{y_i\}$ meets L. Let x_i be the point of intersection. We shall denote by Θ_3 the restriction of $PG(2, \mathbb{R})$ to $\{y_1, y_2, y_3, x_1, x_2, x_3\}$. As illustrated in Figure 11.21(b), Θ_3 is isomorphic to $M(K_4)$ and has $\{x_1, x_2, x_3\}$ as a modular line. To generalize this construction, suppose that $n \geq 3$ and let $\{y_1, y_2, \ldots, y_n\}$ be a basis Y of $PG(n-1, \mathbb{R})$, and L be a line that is freely placed relative to Y, that is, L is not contained in any subspace spanned by a proper subset of Y. As before, the modularity of subspaces in a vector space (6.3) implies that, for each i, the hyperplane of $PG(n-1, \mathbb{R})$ that is spanned by $\{y_1, y_2, \ldots, y_n\} - \{y_i\}$ meets L, and we let x_i be the point of intersection. Let $X = \{x_1, x_2, \ldots, x_n\}$ and let Θ_n denote the restriction of $PG(n-1, \mathbb{R})$ to $Y \cup X$. By construction, each hyperplane of Θ_n meets the line X, so, by Corollary 6.9.3, this line is modular. It follows that if M is a matroid having an n-point line as a restriction and the points of this line are labelled by the elements of X, then the matroid $P_X(\Theta_n, M)$ is well-defined. Hence so too is $P_X(\Theta_n, M)\backslash X$. Indeed, this is how we shall define our generalization of Δ-Y exchange. This operation was introduced by Oxley, Vertigan, and Semple (2000) and the treatment here will follow that paper.

The matroid Θ_n was described above. We now give a more formal definition of this matroid by listing its set of bases. We leave it to the reader to check that these two definitions are consistent by showing that the bases of the matroid Θ_n already defined are as specified in the next result. Let n be an integer exceeding one and let X and Y be disjoint n-element sets $\{x_1, x_2, \ldots, x_n\}$ and $\{y_1, y_2, \ldots, y_n\}$.

Proposition 11.5.1 *There is a matroid Θ_n having ground set $X \cup Y$ whose set $\mathcal{B}$ of bases consists of all of the following sets:*

(i) $\{Y\}$;

(ii) $\{(Y - y_i) \cup x_j : \{i, j\} \subseteq \{1, 2, \ldots, n\} \text{ and } i \neq j\}$; and

(iii) $\{(Y - Y') \cup X' : |Y'| = |X'| = 2 \text{ and } X' \subseteq X \text{ and } Y' \subseteq Y\}$.

Proof Clearly $\mathcal{B}$ is non-empty and all its members have exactly n elements. Suppose that B_1 and B_2 are in $\mathcal{B}$ and $e \in B_1 - B_2$. We need to show that $B_2 - B_1$ contains an element f such that $(B_1 - e) \cup f \in \mathcal{B}$. If $|B_1 - B_2| = 1$, then this is certainly true. Thus we may assume that $|B_1 - B_2| \geq 2$.

The members of $\mathcal{B}$ are of three types and we shall consider first the type of B_1 and then the type of B_2. For each case, we shall specify an element f of $B_2 - B_1$ such that $(B_1 - e) \cup f \in \mathcal{B}$. Suppose first that $B_1 = Y$. Since $|B_1 - B_2| \geq 2$, we have $B_2 = (Y - Y') \cup X'$ for some 2-element subsets X' and Y' of X and Y. Then $e = y_i$ for some i and we take f in $X' - x_i$.

Next assume that $B_1 = (Y - y_i) \cup x_j$ for some distinct i and j. Then $B_2 \neq Y$. If $B_2 = (Y - y_s) \cup x_t$ for some distinct s and t, then, as $|B_1 - B_2| \geq 2$, we must have that $j \neq t$ and $i \neq s$. Thus $e = x_j$ or $e = y_s$. In the first case, we take $f = y_i$; in the second, we take $f = x_t$. Now suppose that $B_2 = (Y - Y') \cup X'$ for some 2-element subsets X' and Y' of X and Y. If $e = x_j$, then we take f in $X' - \{x_i, x_j\}$. If $e = y_u$, then we take f in $X' - x_j$.

Finally, assume that $B_1 = (Y - Y') \cup X'$ where $Y' = \{y_s, y_t\}$ and $X' = \{x_i, x_j\}$. If $B_2 = Y$, then we may assume that $e = x_i$ and we take f in Y' such that $\{f, y_j\} \neq Y'$. Next suppose that $B_2 = (Y - y_u) \cup x_v$ where u and v are distinct. If $e = x_i$, then we take $f = x_v$ unless $v = j$; in the exceptional case, $y_j \in B_2$ and we take $f = y_j$ if $y_j \notin B_1$ and take f in $\{y_s, y_t\} - y_u$ otherwise. If $e = y_w$, then we take f in $\{y_s, y_t\} - \{y_u\}$.

It remains to consider when $B_2 = (Y - Y'') \cup X''$ for some 2-element subsets X'' and Y'' of X and Y. In that case, if $e = x_i$, then we choose f in $X'' - X'$ unless $X'' = X'$. In the exceptional case, as $|B_1 - B_2| \geq 2$, we must have that $Y' \cap Y'' = \emptyset$. Then we take f in Y' such that $\{f, y_j\} \neq Y'$. If, instead, $e = y_w$, then $e \in Y'' - Y'$ and we can choose f in $Y' - Y''$. We conclude that $\mathcal{B}$ is indeed the set of bases of a matroid on $X \cup Y$. $\square$

Clearly $\Theta_2 \cong U_{1,2} \oplus U_{1,2}$ and, as noted earlier, $\Theta_3 \cong M(K_4)$. The next two results are easy consequences of the definition of Θ_n in terms of bases.

Corollary 11.5.2 *Let $n \geq 2$.*

(i) $\Theta_n \cong \Theta_n^*$ *under the map that interchanges x_i with y_i for all i.*

(ii) *If σ is a permutation of $\{1, 2, \ldots, n\}$, then the map that, for all i takes x_i and y_i to $x_{\sigma(i)}$ and $y_{\sigma(i)}$, respectively, is an automorphism of Θ_n.*

(iii) *If $n \geq 3$, then $\Theta_n/y_i \backslash x_i \cong \Theta_{n-1}$ for all i.*

(iv) *For all i, the matroid $\Theta_n/x_i \backslash y_i \cong U_{1,n-1} \oplus U_{n-2,n-1}$ where the components have ground sets $X - x_i$ and $Y - y_i$, respectively.* $\square$

Corollary 11.5.3 *For all $n \geq 2$, the circuits of Θ_n consist of the following:*

(i) *all 3-element subsets of X;*

(ii) *all sets of the form $(Y - y_i) \cup x_i$ where $1 \leq i \leq n$; and*

(iii) *all sets of the form $(Y - y_u) \cup \{x_s, x_t\}$ where $s, t,$ and u are distinct elements of $\{1, 2, \ldots, n\}$.* $\square$

By the last corollary, Θ_n is certainly connected.

Proposition 11.5.4 *For all $n \geq 3$, the matroid Θ_n is 3-connected.*

Proof We argue by induction on n noting that the lemma holds for $n = 3$ as $\Theta_3 \cong M(K_4)$. Assume that $n \geq 4$ and that the lemma holds for smaller values of n. By Corollary 11.5.2(iii), $\Theta_n/y_i \backslash x_i \cong \Theta_{n-1}$. By the induction assumption, Θ_n/y_i is a single-element extension of a 3-connected matroid. By the last lemma, Θ_n/y_i is simple and has the same rank as Θ_{n-1}. Thus, by Proposition 8.2.7, Θ_n/y_i is 3-connected and, by applying the dual of that proposition, we deduce that Θ_n is 3-connected. $\square$

To ensure that we can use Θ_n in a generalized parallel connection, we require the following result.

Lemma 11.5.5 *For all $n \geq 2$, the set X is a rank-2 modular flat of Θ_n.*

Proof It is clear from Corollary 11.5.3 that X is a rank-2 flat. From that corollary and Corollary 11.5.2(i), we can find all the hyperplanes of Θ_n and then show that each of them meets X. Thus, by Corollary 6.9.3, X is modular. $\square$

Now suppose that $n \geq 2$ and let M be a matroid such that $M|X \cong U_{2,n}$ where $X = \{x_1, x_2, \ldots, x_n\}$. By Lemma 11.5.5, X is a modular line of Θ_n. Thus the generalized parallel connection $P_X(\Theta_n, M)$ of Θ_n and M across X exists. Moreover, by the symmetry of Θ_n noted in Corollary 11.5.2, $P_X(\Theta_n, M)$ is uniquely determined to within relabelling. Hence so too is $P_X(\Theta_n, M) \backslash X$. If $n = 2$, then $P_X(\Theta_n, M) \backslash X \cong M$ since the former is obtained from M by adding an element in parallel with each of the elements of X and then deleting the elements of X. If $n = 3$, then, because $\Theta_3 \cong M(K_4)$, the matroid $P_X(\Theta_3, M) \backslash X$ is exactly the matroid that is obtained from M by performing a Δ-Y exchange on X. While such a Δ-Y exchange is defined as long as X is a triangle of M, the set Y is a triad in $P_X(\Theta_3, M)\backslash X$ only if X is coindependent in M. We omit the straightforward proof of the following extension of this observation.

Lemma 11.5.6 *For all $n \geq 2$, the restriction of $(P_X(\Theta_n, M)\backslash X)^*$ to Y is isomorphic to $U_{2,n}$ if and only if X is coindependent in M.* □

We now augment the assumption that M is an arbitrary matroid for which $M|X \cong U_{2,n}$ with the requirement that X is coindependent in M. We define $\Delta_X(M)$ to be $P_X(\Theta_n, M)\backslash X$ and call this operation a Δ_X-*exchange*, a *segment–cosegment exchange on X*, or, since $|X| = n$, a Δ_n-*exchange* or a *segment–cosegment exchange of size n*. Thus, for example, $U_{4,6}$ can be obtained from $U_{2,6}$ by a segment–cosegment exchange of size 4. As we shall see in Proposition 11.5.11, the additional requirement that X is coindependent in M will ensure that the inverse of segment–cosegment exchange coincides with the dual operation. In defining this dual operation, we mimic the definition of $Y - \Delta$ exchange in terms of $\Delta - Y$ exchange or, indeed, the definition of contraction in terms of deletion. Let M be a matroid for which M^* has a $U_{2,n}$-restriction on the set X. If X is independent in M, then $\nabla_X(M)$ is defined to be $(\Delta_X(M^*))^*$, that is, $[P_X(\Theta_k, M^*)\backslash X]^*$. This operation is called a ∇_X-*exchange*, a *cosegment–segment exchange on X*, a ∇_n-*exchange*, or a *cosegment–segment exchange of size n*.

Lemma 11.5.7 *If $|X| = n$, then*

$$r(\Delta_X(M)) = r(M) + n - 2.$$

Proof From (11.23),

$$r(P_X(\Theta_n, M)) = r(\Theta_n) + r(M) - r(X).$$

Since X is coindependent in M, it is coindependent in $P_X(\Theta_n, M)$. Therefore $r(P_X(\Theta_n, M)) = r(\Delta_X(M)) = n + r(M) - 2$. □

Next we describe the bases of $\Delta_X(M)$ in terms of the bases of M. Recall that $E(\Theta_n) - X = Y$, and that Y is a basis of Θ_n.

Proposition 11.5.8 *A subset D of $E(\Delta_X(M))$ is a basis of $\Delta_X(M)$ if and only if D satisfies one of the following:*

(i) *D contains Y, and $D - Y$ is a basis of M/X;*
(ii) *$D \cap Y = Y - y_i$ for some i in $\{1, 2, \ldots, n\}$, and $D - (Y - y_i)$ is a basis of $M/x_i \backslash (X - x_i)$; or*
(iii) *$D \cap Y = Y - \{y_i, y_j\}$ for some distinct elements i and j of $\{1, 2, \ldots, n\}$, and $D - (Y - \{y_i, y_j\})$ is a basis of $M\backslash X$.*

Proof By Lemma 11.5.7, $r(\Delta_X(M)) = r(M) + n - 2$, so every basis of $\Delta_X(M)$ contains at least $n - 2$ elements of Y. First assume that D contains Y. Then D is a basis of $\Delta_X(M)$ if and only if $D - Y$ is a basis of $\Delta_X(M)/Y$. To complete the proof of (i), it remains to show that $\Delta_X(M)/Y = M/X$. Since Y spans Θ_n, the matroid $P_X(\Theta_n, M)/Y$ has X as a set of loops. Thus $P_X(\Theta_n, M)/Y\backslash X = P_X(\Theta_n, M)/Y/X$. Now $P_X(\Theta_n, M)/Y\backslash X = P_X(\Theta_n, M)\backslash X/Y = \Delta_X(M)/Y$. Also $P_X(\Theta_n, M)/Y/X = P_X(\Theta_n, M)/X/Y$. By Proposition 11.4.14(x), the last matroid is $[(\Theta_n/X)\oplus(M/X)]/Y$, which equals M/X. Hence $\Delta_X(M)/Y = M/X$ as required.

Now assume that D contains exactly $n-1$ elements of Y. Then $D \cap Y = Y - y_i$ for some i in $\{1, 2, \ldots, n\}$. Such a set D is a basis for $\Delta_X(M)$ if and only if $D - (Y - y_i)$ is a basis for $\Delta_X(M)/(Y - y_i)$ avoiding y_i, that is, a basis of $\Delta_X(M)/(Y - y_i)\backslash y_i$. Thus to complete the proof of (ii), we need to show that $\Delta_X(M)/(Y - y_i)\backslash y_i = M/x_i\backslash(X - x_i)$. Now $\Delta_X(M)/(Y - y_i)\backslash y_i = P_X(\Theta_n, M)\backslash X/(Y - y_i)\backslash y_i = P_X(\Theta_n, M)/(Y - y_i)\backslash X\backslash y_i$. By Corollary 11.5.3, x_i is a loop of $P_X(\Theta_n, M)/(Y - y_i)$. Therefore

$$\Delta_X(M)/(Y - y_i)\backslash y_i = P_X(\Theta_n, M)/(Y - y_i)/x_i\backslash(X - x_i)\backslash y_i.$$

By Proposition 11.4.14(iv) and (viii),

$$P_X(\Theta_n, M)/x_i\backslash y_i = P_{(M|X)/x_i}(\Theta_n/x_i\backslash y_i, M/x_i).$$

By Corollary 11.5.2(iv), $\Theta_n/x_i\backslash y_i$ has $X - x_i$ and $Y - y_i$ as components. Thus $P_X(\Theta_n, M)/x_i\backslash y_i/(Y - y_i)\backslash(X - x_i) = M/x_i\backslash(X - x_i)$ and $\Delta_X(M)/(Y - y_i)\backslash y_i = M/x_i\backslash(X - x_i)$ as required.

Lastly, assume that D contains exactly $n-2$ elements of Y. Then $D \cap Y = Y - \{y_i, y_j\}$ for some distinct elements i and j of $\{1, 2, \ldots, n\}$. Such a set D is a basis of $\Delta_X(M)$ if and only if $D - (Y - \{y_i, y_j\})$ is a basis of the matroid $\Delta_X(M)/(Y - \{y_i, y_j\})\backslash\{y_i, y_j\}$. From Corollary 11.5.3, we deduce that $\Theta_n/(Y - \{y_i, y_j\})$ can be obtained from $\Theta_n|X$ by placing y_i and y_j in parallel with x_j and x_i, respectively. Therefore, by Proposition 11.4.14(iv),

$$P_X(\Theta_n, M)/(Y - \{y_i, y_j\}) = P_X(\Theta_n/(Y - \{y_i, y_j\}), M),$$

so $\Delta_X(M)/(Y - \{y_i, y_j\})\backslash\{y_i, y_j\} = M\backslash X$. Thus D is a basis of $\Delta_X(M)$ meeting Y in $Y - \{y_i, y_j\}$ if and only if $D - (Y - \{y_i, y_j\})$ is a basis of $M\backslash X$. □

A natural way to preserve the ground set of M when forming $\Delta_X(M)$ is to relabel each y_i in the latter by x_i. For the rest of this section, we adopt this convention to preserve the ground set of a matroid under both Δ_n- and ∇_n-exchanges. The elementary proof of the next lemma is omitted.

Lemma 11.5.9 (i) *If $\Delta_X(M)$ is defined, then $\Delta_X(M)\backslash X = M\backslash X$ and $\Delta_X(M)/X = M/X$. Moreover, $\Delta_X(M)\backslash x_i/(X - x_i) = M/x_i\backslash(X - x_i)$ for all x_i in X.*

(ii) *If $\nabla_X(M)$ is defined, then $\nabla_X(M)\backslash X = M\backslash X$ and $\nabla_X(M)/X = M/X$. Moreover, $\nabla_X(M)/x_i\backslash(X - x_i) = M\backslash x_i/(X - x_i)$ for all x_i in X.* □

Restating Lemma 11.5.8 under the convention that M and $\Delta_X(M)$ have the same ground sets, we get the following result.

Corollary 11.5.10 *Let $\Delta_X(M)$ be the matroid with ground set $E(M)$ that is obtained from M by a Δ_X-exchange. A subset of $E(M)$ is a basis of $\Delta_X(M)$ if and only if it is a member of one of the following sets:*

(i) $\{X \cup Y' : Y'$ *is a basis of* $M/X\}$;

(ii) $\{(X - x_i) \cup Y'' : 1 \le i \le n$ *and* Y'' *is a basis of* $M/x_i \backslash (X - x_i)\}$; *or*

(iii) $\{(X - \{x_i, x_j\}) \cup Y''' : 1 \le i < j \le n$ *and* Y''' *is a basis of* $M\backslash X\}$. $\qquad\square$

The next result asserts that the operations of Δ_n-exchange and ∇_n-exchange are duals of each other. The proof of (i) follows by comparing the lists of bases of M and $\nabla_X(\Delta_X(M))$. Part (ii) is the dual of (i).

Proposition 11.5.11 *Let M be a matroid.*

(i) *Let X be a coindependent set in M such that every 3-element subset of X is a triangle. Then $\nabla_X(\Delta_X(M))$ is well-defined and*

$$\nabla_X(\Delta_X(M)) = M.$$

(ii) *Let X be an independent set in M such that every 3-element subset of X is a triad. Then $\Delta_X(\nabla_X(M))$ is well-defined and*

$$\Delta_X(\nabla_X(M)) = M. \qquad\square$$

We noted earlier that the duals of the Fano and non-Fano matroids can be obtained from the Fano and non-Fano matroids themselves by a Δ-Y exchange. This fact hints at the following result of Akkari and Oxley (1993).

Theorem 11.5.12 *For all fields $\mathbb{F}$ with at least three elements, the set of excluded minors for $\mathbb{F}$-representability is closed under the operations of Δ-Y and Y-Δ exchange.* $\qquad\square$

This theorem was generalized by Oxley, Semple, and Vertigan (2000) who proved the next theorem, the proof of which relies on the following two lemmas. The proof of the second of these lemmas is outlined in Exercise 5.

Lemma 11.5.13 *Let $n \ge 2$ and let M be a matroid such that $M|X \cong U_{2,n}$. If M and Θ_n are both representable over a field $\mathbb{F}$, then $P_X(\Theta_n, M)$ is also representable over $\mathbb{F}$.* $\qquad\square$

Lemma 11.5.14 *Let $n \ge 2$. The matroid Θ_n is representable over a field $\mathbb{F}$ if and only if $|\mathbb{F}| \ge n - 1$.* $\qquad\square$

Theorem 11.5.15 *For all $n \ge 2$ and all fields $\mathbb{F}$, if a matroid M is an excluded minor for $\mathbb{F}$-representability, then so is every matroid that is obtained from M by a Δ_n- or ∇_n-exchange.* $\qquad\square$

By considering the matroids that can be obtained from $U_{2,q+2}$ by a sequence of segment–cosegment and cosegment–segment exchanges, we can get the following exponential lower bound on the number of excluded minors for representability over $GF(q)$. We leave it to the reader to complete the details of this argument.

$$
\begin{array}{c}
\begin{array}{ccccccc}
y_1 & y_2 & y_3 & x_4 & x_5 & & x_n
\end{array}\\
\begin{array}{c}
x_1\\x_2\\y_3\\y_4\\y_5\\\vdots\\y_n
\end{array}
\left[
\begin{array}{ccccccc}
0 & 1 & 1 & 1 & 1 & \cdots & 1\\
-1 & 0 & 1 & \alpha_1 & \alpha_2 & \cdots & \alpha_{n-3}\\
1 & 1 & 0 & 0 & 0 & \cdots & 0\\
1 & \alpha_1 & 0 & 0 & 0 & \cdots & 0\\
1 & \alpha_2 & 0 & 0 & 0 & \cdots & 0\\
\vdots & \vdots & \vdots & \vdots & \vdots & \ddots & \vdots\\
1 & \alpha_{n-3} & 0 & 0 & 0 & \cdots & 0
\end{array}
\right].
\end{array}
$$

FIG. 11.22. The matrix D_n.

The bound here is not intended to be sharp but simply to show that the number of excluded minors for representability over $GF(q)$ is at least exponential in q. Rota's Conjecture (6.5.11) is that this number is finite for all q.

Theorem 11.5.16 *For all prime powers q, the set of excluded minors for the class of $GF(q)$-representable matroids has at least 2^{q-4} distinct members.* □

In the definition of a segment–cosegment exchange on a set X of M, we insisted that X must be coindependent in M. As we have seen, this requirement ensures that a cosegment–segment exchange can be performed on $\Delta_X(M)$ to recover M. This requirement is also essential in Theorem 11.5.15 and explains why Theorem 11.5.12 has an exception for $GF(2)$. Indeed, if M is an excluded minor for representability over a field $\mathbb{F}$ but X is not coindependent in M, then it is possible for $P_X(\Theta_n, M)\backslash X$ to be $\mathbb{F}$-representable. For example, if $|X| = q + 1 \geq 3$, then $P_X(\Theta_{q+1}, U_{2,q+2})\backslash X \cong U_{q+1,q+2}$. However, although $U_{2,q+2}$ is an excluded minor for the class of $GF(q)$-representable matroids, $U_{q+1,q+2}$ is regular and so is certainly $GF(q)$-representable.

Exercises

1. Show that the graph of the octahedron can be obtained from the graph of the cube (see Figure 14.3) by performing Y-Δ exchanges on two diagonally opposite vertices of the latter.

2. Show that the *Petersen family*, that is, the class of graphs that can be obtained from the Petersen graph by a sequence of Y-Δ and Δ-Y exchanges, has exactly seven non-isomorphic members, including K_6.

3. Prove that if $n \geq 2$, then $\Delta_n(U_{2,n+2}) = U_{n,n+2}$.

4. For $n \geq 3$ and a field $\mathbb{F}$, let $\alpha_1, \alpha_2, \ldots, \ldots, \alpha_{n-3}$ be distinct elements of $\mathbb{F} - \{0,1\}$. Let D_n be the matrix shown in Figure 11.22. Show that:

   (a) Every non-zero subdeterminant of D_n can be written as a product of positive or negative powers of differences of distinct elements of $\{0, 1, \alpha_1, \alpha_2, \ldots, \alpha_{n-3}\}$.

   (b) If $|\mathbb{F}| \geq n - 1$, then Θ_n is represented over $\mathbb{F}$ by $[I_n | D_n]$.

   (c) If Θ_n is $\mathbb{F}$-representable, then $|\mathbb{F}| \geq n - 1$.

5. (Oxley, Semple, and Vertigan 2000) Suppose $\Delta_X(M)$ is defined. Prove that:
   (a) If $x \in X$ and $|X| = n \geq 3$, then $\Delta_{X-x}(M\backslash x)$ is also defined and $\Delta_X(M)/x = \Delta_{X-x}(M\backslash x)$.
   (b) If $y \in E(M) - X$ and X is coindependent in $M\backslash y$, then $\Delta_X(M)\backslash y$ is defined and $\Delta_X(M)\backslash y = \Delta_X(M\backslash y)$.
   (c) If $z \in E(M) - \mathrm{cl}(X)$, then $\Delta_X(M/z)$ is defined and $\Delta_X(M)/z = \Delta_X(M/z)$.

6. Let M be a connected matroid and suppose $\Delta_n(M)$ is defined. Show that:
   (a) $\Delta_n(M)$ is connected.
   (b) If M is $\mathbb{F}$-representable, then $\Delta_n(M)$ is $\mathbb{F}$-representable.
   (c) $\Delta_n(M)$ need not be 3-connected even if M is 3-connected and $n \geq 3$.
   (d) (Oxley, Semple, and Vertigan 2000) If $m \geq 4$, every matroid that can be obtained from $U_{2,m}$ by a sequence of segment–cosegment and cosegment–segment exchanges is 3-connected.
   (e) If $n \geq 3$ and M is 3-connected, then $\mathrm{co}(\Delta_n(M))$ is 3-connected.

7. A 2-connected graph G is ΔY-*reducible* if G can be reduced to a 2-cycle by a sequence of operations each consisting of a Δ-Y or Y-Δ exchange, deletion of one edge from a pair of parallel edges, or contraction of one edge from a pair of edges meeting at a degree-2 vertex. Show that:
   (a) $K_{3,3}$ and K_5 are ΔY-reducible.
   (b) Each of the graphs in the Petersen family is a minor-minimal graph that it is both 2-connected and not ΔY-reducible.
   (c) (Archdeacon, Colbourn, Gitler, and Provan 2000) The graph obtained from $K_{5,5}$ by deleting a 5-edge (perfect) matching is a minor-minimal graph that is both 2-connected and not ΔY-reducible.

8. The $n \times n$ *grid* is the graph with vertex set $\{(i,j) : i,j \in \{1,2,\ldots,n\}\}$ such that (i,j) and (i',j') are adjacent if and only if $|i - i'| + |j - j'| = 1$. Show that:
   (a) Every planar graph is a minor of some grid graph.
   (b) Every grid graph is ΔY-reducible.
   (c) (Epifanov 1966; Truemper 1989) Every 2-connected planar graph is ΔY-reducible.

9. *Prove Theorem 11.5.16.

10. °Find the collection of minor-minimal graphs that are both 2-connected and not ΔY-reducible. (For recent work on this, see Yu (2004, 2006).)

12

THE SPLITTER THEOREM

In this chapter, we shall state and prove Seymour's Splitter Theorem. This result, which is a generalization of Tutte's Wheels-and-Whirls Theorem (8.8.4), is a very powerful general tool for deriving matroid structure results. The Wheels-and-Whirls Theorem determines when we can find some element in a 3-connected matroid M to delete or contract in order to preserve 3-connectedness. The Splitter Theorem considers when such an element removal is possible that will not only preserve 3-connectedness but will also maintain the presence of an isomorphic copy of some specified minor of M. We shall illustrate the power of the Splitter Theorem by noting a variety of consequences of it. In addition, we shall discuss some extensions and generalizations of the theorem. Surveys of the role played by the Splitter Theorem in an analysis of matroid structure have been given by Seymour (1995) and Oxley (1996), but those papers contain no detailed proofs.

12.1 The theorem and its proof

Let $\mathcal{N}$ be a class of matroids that is closed under minors and under isomorphism. A member N of $\mathcal{N}$ is a *splitter* for $\mathcal{N}$ if, whenever M is a member of $\mathcal{N}$ having an N-minor, either $M \cong N$, or M has a 1- or 2-separation. Thus N is a splitter for $\mathcal{N}$ if and only if $\mathcal{N}$ has no 3-connected member having a proper N-minor.

Although, in general, a splitter need not be 3-connected, most interest here will be focused on 3-connected splitters. This is because, by Corollary 8.3.4, a matroid that is not 3-connected can be decomposed into smaller matroids by using the operations of direct sum and 2-sum. Thus a 3-connected splitter for a class $\mathcal{N}$ is, in a very natural sense, a basic building block for $\mathcal{N}$.

An important precursor of the Splitter Theorem is the following result of K. Wagner (1960). We shall prove this result after we have stated the Splitter Theorem to indicate how the latter can be applied. We denote by H_8 the 8-vertex graph shown in Figure 12.1. We shall call this graph the *Wagner graph*.

Proposition 12.1.1 *Let G be a simple 3-connected graph having no K_5-minor. Then either G has no H_8-minor or $G \cong H_8$.*

Observe that, although this result is stated for graphs, Proposition 8.1.9 implies that it could equally well have been stated in terms of graphic matroids. In that context, the result asserts that $M(H_8)$ is a splitter for the class of graphic matroids with no $M(K_5)$-minor.

The next result, which is known as the Splitter Theorem, is the main theorem of this chapter (Seymour 1980b).

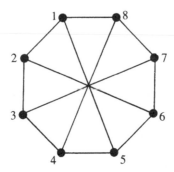

FIG. 12.1. The Wagner graph H_8.

Theorem 12.1.2 *Let N be a non-empty, connected, simple, cosimple minor of a 3-connected matroid M. Suppose that if N is a wheel, then M has no larger wheel as a minor, while if N is a whirl, M has no larger whirl as a minor. Then either $M = N$, or M has a connected, simple, cosimple minor M_1 such that some single-element deletion or some single-element contraction of M_1 is isomorphic to N. Moreover, if N is 3-connected, so too is M_1.*

The Splitter Theorem was also proved by J. J.-M. Tan (1981) in his Ph.D. thesis. The theorem is of sufficient importance that it warrants restatement. Indeed, we shall give three reformulations of the theorem in the case that N is 3-connected. The second and third of these will be given in Section 12.2. The first is as follows.

Corollary 12.1.3 *Let $\mathcal{N}$ be a class of matroids that is closed under minors and under isomorphism. Let N be a 3-connected member of $\mathcal{N}$ having at least four elements such that if N is a wheel, it is the largest wheel in $\mathcal{N}$, while if N is a whirl, it is the largest whirl in $\mathcal{N}$. Suppose there is no 3-connected member of $\mathcal{N}$ that has N as a minor and has one more element than N. Then N is a splitter for $\mathcal{N}$, that is, no 3-connected member of $\mathcal{N}$ has N as a proper minor.* □

On specializing the last result to graphs, we obtain the following result that was also derived by Negami (1982), independently of Seymour.

Corollary 12.1.4 *Let $\mathcal{G}$ be a class of graphs that is closed under minors and under isomorphism. Let H be a simple 3-connected graph such that if it is a wheel, it is the largest wheel in $\mathcal{G}$. Suppose that there is no simple 3-connected graph in $\mathcal{G}$ having a single-edge deletion or a single-edge contraction isomorphic to H. Then, for every simple 3-connected graph G in $\mathcal{G}$, either G has no minor isomorphic to H, or $G \cong H$.* □

Before proving the Splitter Theorem, we use the last corollary to prove Wagner's result.

Proof of Proposition 12.1.1 Let $\mathcal{G}$ be the class of graphs with no minor isomorphic to K_5, and let $H = H_8$. Now suppose that G is a graph having a non-loop edge e such that $G/e = H$. As every vertex of H has degree three, G has a vertex of degree at most two and so is not 3-connected. Hence $\mathcal{G}$ does not contain a simple 3-connected graph having a single-edge contraction isomorphic to H. Suppose next that G is a simple 3-connected graph in $\mathcal{G}$ for which some single-edge deletion is isomorphic to H. Then, by symmetry, we may suppose that G is obtained from H by adding one of the edges 13 and 14 (see Figure 12.2). In each case, it is straightforward to show that $G \notin \mathcal{G}$: in the first case, we obtain a graph isomorphic to K_5 by contracting the edges 26, 78, and 45; in the second case, we obtain such a graph by contracting 23, 56, and 78. Proposition 12.1.1 now follows immediately from Corollary 12.1.4. □

Most of the remainder of this section will be devoted to proving the Splitter Theorem. The proof given here is due to C. R. Coullard and L. L. Gardner (private communication). It uses Lemma 8.8.5, which was also used in the proof of the Wheels-and-Whirls Theorem. Recall that the fact that a circuit and a cocircuit cannot meet in exactly one element is referred to as 'orthogonality'.

Proof of Theorem 12.1.2 Observe that it suffices to prove the theorem when N is connected, simple, and cosimple, for it follows from Proposition 8.2.7 that if N is 3-connected, then so is M_1. Now suppose the theorem is false and that, among all pairs of matroids for which the theorem fails, (N, M) is chosen so that $|E(M) - E(N)|$ is as small as possible. Certainly $|E(M) - E(N)| > 0$. As N is non-empty, simple, and cosimple, $|E(N)| \geq 4$. Define the sets X and Y by

$$X = \{x \in E(M) : M \backslash x \text{ has an } N\text{-minor}\}$$

and

$$Y = \{y \in E(M) : M/y \text{ has an } N\text{-minor}\}.$$

Clearly either

$$\text{(i)} \quad X \cap Y = \emptyset; \quad \text{or} \quad \text{(ii)} \quad X \cap Y \neq \emptyset.$$

Assume that (i) holds. In this case, we shall establish the contradiction that the hypothesis concerning wheels and whirls is violated. Evidently X or Y is

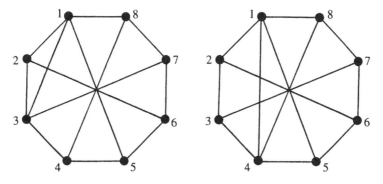

FIG. 12.2. The two simple graphs with H_8 as a single-edge deletion.

non-empty. We shall assume the former noting that if X is empty, then Y is not and we may apply the argument that follows to the pair (N^*, M^*), which is also a minimal counterexample to the theorem.

Lemma 12.1.5 *Let x be an element of X. Then M has connected minors N_1 and N_2 and elements y, y', and x' such that*

(a) $y \in Y - E(N_1);$ $y' \in Y \cap E(N_1);$ *and* $x' \in X \cap E(N_1);$
(b) $N \cong N_1 = N_2/y \backslash x$ *and* $r(N_1) = r(N_2) - 1;$ *and*
(c) $\{x, y, y'\}$ *and* $\{x, y, x'\}$ *are a triad and a triangle, respectively, of N_2.*

Proof As $x \in X$, the matroid $M \backslash x$ has an N-minor N_1, say. Let N_3 be the minor of M for which $N_3 \backslash x = N_1$. As the theorem fails for the pair (N, M), we must have that N_3 is disconnected, non-simple, or non-cosimple. Now, since $X \cap Y$ is empty, x is not a loop or a coloop of N_3. Thus, by Proposition 8.2.7, N_3 is connected. Since N is cosimple, so is N_3. Hence N_3 is non-simple and so x is in a 2-circuit with some element x' of N_1. Clearly $x' \in X$.

As M is 3-connected, it does not have $\{x, x'\}$ as a circuit. Thus there is an element y of $E(M) - E(N_3)$ and a minor N_2 of M such that $N_3 = N_2/y$ and $\{x, y, x'\}$ is a triangle of N_2. Evidently $y \in Y$. The dual of Proposition 8.2.7 implies that, since N_3 is connected, so is N_2. Consider $N_2 \backslash x$. As $(N_2 \backslash x)/y = N_1$, we must have that $N_2 \backslash x$ is disconnected, non-simple, or non-cosimple. Again using the fact that $X \cap Y = \emptyset$ and the dual of Proposition 8.2.7, we deduce that y is in a 2-cocircuit, $\{y, y'\}$ say, of $N_2 \backslash x$. Clearly $y' \in Y$, so $y' \neq x'$. As $\{x, y, x'\}$ is a circuit of N_2, orthogonality implies that $\{y, y'\}$ is not a cocircuit of N_2. Hence $\{x, y, y'\}$ is a triad of N_2 and so $r(N_1) = r(N_2) - 1$.   □

We now look more closely at N_2.

Lemma 12.1.6 *The matroid N_2 is simple and cosimple.*

Proof As N_2 is connected, we need only show that N_2 has no 2-circuits and no 2-cocircuits. Assume that $\{a, b\}$ is a 2-circuit of N_2. Then $\{a, b\} \subseteq X$, so $\{a, b\} \cap Y = \emptyset$. Now $N_2/y \backslash x$ is isomorphic to N and so is simple. Therefore $\{a, b\}$ does not contain a circuit of $N_2/y \backslash x$. Thus $x \in \{a, b\}$. As $\{y, y'\} \subseteq Y$, it follows that the circuit $\{a, b\}$ and the cocircuit $\{x, y, y'\}$ of N_2 meet in a single element; a contradiction. Thus N_2 has no 2-circuits. A dual argument establishes that N_2 has no 2-cocircuits.   □

The next lemma completes the proof that case (i) cannot occur.

Lemma 12.1.7 *Both N and N_2 are wheels or both are whirls.*

Proof Let $X_2 = \{e \in E(N_2) : N_2 \backslash e$ has an N-minor$\}$ and $Y_2 = \{e \in E(N_2) : N_2/e$ has an N-minor$\}$. Then $X_2 \subseteq X$ and $Y_2 \subseteq Y$, so $X_2 \cap Y_2 = \emptyset$. Since $N_2/y \backslash x \cong N$, we have that $x \in X_2$ and $y \in Y_2$. Moreover, $x' \in X_2$ and $y' \in Y_2$ since $N_2/y \backslash x \cong N_2/y \backslash x'$, and $N_2 \backslash x/y \cong N_2 \backslash x/y'$. Now let a be an arbitrary element of X_2. Then $N_2 \backslash a$ has an N-minor. Thus, since $r(N_2) - r(N) = 1$ and $|E(N_2) - E(N)| = 2$, the matroid $N_2 \backslash a$ has an element b that is not a loop

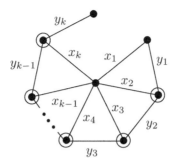

FIG. 12.3. The circled vertices correspond to known triads of N_2.

or a coloop such that $N_2 \backslash a/b \cong N$. Hence $N_2 \backslash a$ is connected. By the choice of (N, M), it follows that $N_2 \backslash a$ is not cosimple. Thus $N_2 \backslash a$ has a 2-cocircuit, $\{c, d\}$ say. Clearly $\{c, d\} \subseteq Y_2$. Moreover, as N_2 is cosimple, $\{a, c, d\}$ is a triad of N_2. Thus every element of X_2 is in a triad with two elements of Y_2. A dual argument shows that every element of Y_2 is in a triangle with two elements of X_2.

Now let k be the largest integer for which there are subsets $\{x_1, x_2, \ldots, x_k\}$ of X_2 and $\{y_1, y_2, \ldots, y_k\}$ of Y_2 such that $(x_1, y_1, x_2, y_2, \ldots, x_k, y_k)$ is a fan F in N_2 having $\{x_1, y_1, x_2\}$ as a triangle. By taking $(x', y, x, y') = (x_1, y_1, x_2, y_2)$, we see that $k \geq 2$. Although N_2 need not be graphic, we follow Tutte (1966b) and keep track of the triangles and triads in N_2 by using the graph in Figure 12.3. Note that in that diagram a circled vertex corresponds to a *known* triad.

Consider the element y_k. Since it is in Y_2, it is in a triangle T with two members of X_2. As T meets the triad $\{y_{k-1}, x_k, y_k\}$, it must contain x_k. Therefore T is $\{x_k, y_k, x_{k+1}\}$ for some x_{k+1} in X_2. Clearly $x_{k+1} \neq x_k$. By orthogonality between T and the triads in F, we deduce that $x_{k+1} \in X_2 - \{x_2, x_3, \ldots, x_k\}$.

We show next that $x_{k+1} = x_1$. Assume the contrary. As $x_{k+1} \in X_2$, there is a triad T^* that contains x_{k+1} and two members of Y_2. Since T^* meets the triangle $\{x_k, y_k, x_{k+1}\}$, we must have that $y_k \in T^*$. Hence T^* is $\{y_k, x_{k+1}, y_{k+1}\}$ for some y_{k+1} in Y_2. Clearly $y_{k+1} \neq y_k$. Moreover, the orthogonality between T^* and the triangles in the fan $(x_1, y_1, x_2, y_2, \ldots, x_k, y_k, x_{k+1})$ implies that y_{k+1} is not in $\{y_1, y_2, \ldots, y_k\}$. Thus $(x_1, y_1, x_2, y_2, \ldots, x_{k+1}, y_{k+1})$ is a fan that contradicts the choice of k. Thus we do indeed have that $x_{k+1} = x_1$ and Figure 12.3 can be updated as in Figure 12.4.

This diagram suggests that $\{y_k, x_1, y_1\}$ is a triad of N_2 and we can show this by arguing as follows: x_1 is in a triad with two elements of Y_2, and the triangles $\{x_k, y_k, x_1\}$ and $\{x_1, y_1, x_2\}$ imply that these two elements must be y_k and y_1. It now follows immediately from Lemma 8.8.5 and the fact that $N \cong N_2/y \backslash x$ that both N_2 and N are wheels or both are whirls. □

As we now know that case (i) cannot occur, we may assume that case (ii) occurs; that is, we may suppose that, for some element e of M, both $M \backslash e$ and M/e have an N-minor. The choice of (N, M) implies that neither $M \backslash e$ nor M/e

is 3-connected. By Bixby's Lemma (8.7.3), $M\backslash e$ or M/e has no non-minimal 2-separations. We shall assume the latter, noting that if the former holds, we may apply the argument that follows to $(M\backslash e)^*$, which equals M^*/e. By Bixby's Lemma again, $\mathrm{si}(M/e)$ is 3-connected. Moreover, by the minimality of $|E(M) - E(N)|$, we must have that $\mathrm{si}(M/e) \cong N$ and $M/e \not\cong N$. Thus N is 3-connected.

As $M/e \neq \mathrm{si}(M/e)$, there is a triangle in M containing e. Let $\{T_1, T_2, \ldots, T_k\}$ be the set of such triangles where $T_i = \{a_i, b_i, e\}$ for all i. If $(T_i \cap T_j) - e \neq \emptyset$ for some distinct i and j in $\{1, 2, \ldots, k\}$, then M/e has a parallel class of size at least three. As $|E(M/e)| \geq |E(N)| + 2 \geq 6$, we deduce that M/e has a non-minimal 2-separation; a contradiction. Thus the $2k$ elements $a_1, b_1, a_2, b_2, \ldots, a_k, b_k$ are distinct. Moreover, all these elements are in X. Hence, by the choice of (N, M), the matroid $M\backslash e$ and all $M\backslash a_i$ and $M\backslash b_i$ fail to be 3-connected. Therefore, by Tutte's Triangle Lemma (8.7.7), each of a_i and b_i is in a triad with e. Consider such a triad containing $\{a_1, e\}$. By orthogonality, it must meet the disjoint sets $T_2 - e, T_3 - e, \ldots, T_k - e$. Therefore $k \leq 2$. Furthermore, since $|E(M)| > 4$, the triangle $\{a_1, e, b_1\}$ is not a triad of M.

Suppose that $k = 1$. Then $M/e\backslash a_1 \cong N$ so $|E(M) - E(N)| = 2$. From above, M has triads $\{a_1, e, a\}$ and $\{b_1, e, b\}$ for some a and b not in $\{a_1, b_1\}$. Thus $M\backslash e$ has at least three elements that are in 2-cocircuits. But $|E(M) - E(N)| = 2$ and $M\backslash e$ has an N-minor. Hence N cannot be cosimple; a contradiction. We conclude that $k \neq 1$.

We may now suppose that $k = 2$. Then $|E(M) - E(N)| = 3$. Moreover, by relabelling a_2 and b_2 if necessary, we may assume that $\{a_1, e, a_2\}$ and $\{b_1, e, c\}$ are triads of M for some c in $\{a_2, b_2\}$. If $c = a_2$, then, by cocircuit elimination, we obtain the contradiction that $\{a_1, e, b_1\}$ is a triad of M. Thus $c \neq a_2$, so $\{b_1, e, b_2\}$ is a triad of M. Now let $Z = \{a_1, b_1, a_2, b_2, e\}$. Then $\{a_1, a_2, e\}$ and $\{a_1, b_1, e\}$ span Z in M and M^*, respectively. Thus

$$\lambda_M(Z) = r(Z) + r^*(Z) - |Z| \leq 3 + 3 - 5 = 1.$$

Since M is 3-connected, it follows that $|E(M) - Z| \leq 1$, so $|E(M)| \leq 6$. As $|E(M) - E(N)| \geq 3$ and $|E(N)| \geq 4$, this is a contradiction that completes the proof of the Splitter Theorem. $\square$

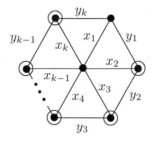

FIG. 12.4. The sequence of triangles and triads closes up.

In the last section, we shall discuss several variations on the Splitter Theorem. To close this section, we observe certain ways in which this theorem cannot be extended. First, it does not remain valid if, instead of requiring that M be 3-connected, we insist only that it be connected, simple, and cosimple.

Example 12.1.8 Let $N = U_{2,4}$ and $M = U_{2,4} \oplus_2 U_{2,4}$. Then both N and M are connected, simple, and cosimple. Yet M has no connected, simple, cosimple minor having a single-element deletion or a single-element contraction that is isomorphic to N. □

The requirement in the Splitter Theorem that N be non-empty, simple, and cosimple implies that $|E(N)| \geq 4$. But not every 3-connected matroid has four or more elements and one may ask, in the case that N is 3-connected, whether the theorem remains valid for $|E(N)| \leq 3$. It does not, as one can see by letting $N = U_{2,3}$ and $M = M(K_5)$.

Finally, we draw attention to the fact that the matroid M_1 in the Splitter Theorem is not required to have a single-element deletion or contraction *equal* to N but only to have such a minor *isomorphic* to N. Indeed, it is not difficult to show (Exercise 7) that the theorem is no longer valid if M_1 is subject to the stronger requirement. In one of the variants of the Splitter Theorem to be discussed in Section 12.3, we shall insist on obtaining a 3-connected minor of M that has, as a proper minor, the matroid N itself rather than just an isomorphic copy of N.

Exercises

1. Show that the Wagner graph, H_8, in Figure 12.1 is isomorphic to the graph in Figure 12.5, the *(cubic) Möbius ladder with four rungs*.

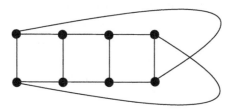

FIG. 12.5. Another drawing of the Wagner graph H_8.

2. Let $\mathcal{N}$ be a class of matroids that is closed under isomorphism, minors, and duality. Show that if M is a splitter for $\mathcal{N}$, then so is M^*.

3. Let $\mathcal{N}_1$ and $\mathcal{N}_2$ be classes of matroids, each of which is closed under isomorphism and minors, and let $\mathcal{N}_1$ be a subclass of $\mathcal{N}_2$. Suppose that $M \in \mathcal{N}_1$. Show that if M is a splitter for $\mathcal{N}_2$, then it is a splitter for $\mathcal{N}_1$, but that the converse of this fails.

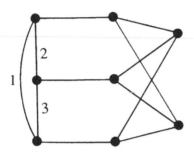

FIG. 12.6. The graph G in Exercise 7.

4. (Welsh 1982) Show that each of the following classes of matroids has no splitters: $GF(q)$-representable matroids; graphic matroids; and cographic matroids.

5. Give an example of a class $\mathcal{N}$ of matroids and a disconnected member N of $\mathcal{N}$ such that $\mathcal{N}$ is closed under isomorphism and under minors, and N is a splitter for $\mathcal{N}$.

6. State and prove the analogue of Corollary 12.1.3 when N is connected, simple, and cosimple.

7. For the graph G in Figure 12.6, let $M = M(G)$ and $N = M(G)/\{1,2,3\}$. Show that M has no 3-connected minor M_1 having an element e such that $M_1\backslash e$ or M_1/e equals N, but that M does have such a minor M_1 and an element e for which $M_1\backslash e \cong N$.

8. (Oxley 1989a) Show that, for each of the graphs G in Figure 12.7, $M(G)$ is a splitter for the class of graphic matroids having no $M(\mathcal{W}_5)$-minor.

9. (Oxley 1990a) Show that the free extension of $AG(3,2)$ is a splitter for the class of matroids with no minor isomorphic to $\mathcal{W}^3$ or P_6.

10. (Oxley 1987c) Show that the matroid in Figure 10.5 is a splitter for the class of matroids with no minor isomorphic to $M(K_4)$, $U_{2,5}$, or $U_{3,5}$.

11. (Lemos, private communication) Let M be a connected matroid and X be a subset of $E(M)$ having at least four elements. Suppose that, for each e in X, there is a triangle T and a triad T^* such that $T \cup T^* \subseteq X$ and $e \in T \cap T^*$. Prove that M is isomorphic to a wheel or a whirl.

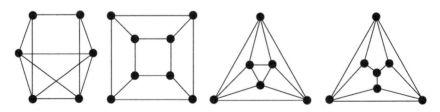

FIG. 12.7. Splitters for the graphic matroids with no $M(\mathcal{W}_5)$-minor.

12.2 Applications of the Splitter Theorem

This section contains a number of examples of structure theorems for various classes of matroids and graphs. All of these results can be derived from the Splitter Theorem. We begin by stating a second reformulation of the theorem in the case that N is 3-connected. The reader will find it instructive to compare this result with Corollary 4.3.8, its analogue for connected matroids.

Corollary 12.2.1 *Let M and N be 3-connected matroids such that N is a minor of M with at least four elements and if N is a wheel, then M has no larger wheel as a minor, while if N is a whirl, then M has no larger whirl as a minor. Then there is a sequence $M_0, M_1, \ldots, M_n$ of 3-connected matroids with $M_0 \cong N$ and $M_n = M$ such that M_i is a single-element deletion or a single-element contraction of M_{i+1} for all i in $\{0, 1, \ldots, n-1\}$.* □

We now use the last result to give another proof of the Wheels-and-Whirls Theorem (8.8.4), which is restated below for convenient reference. But first a reminder of a basic fact concerning wheels and whirls which will be used repeatedly throughout this chapter: the smallest whirl, $\mathcal{W}^2$, is isomorphic to $U_{2,4}$, while the smallest 3-connected wheel, $M(\mathcal{W}_3)$, is isomorphic to $M(K_4)$.

Corollary 12.2.2 *The following statements are equivalent for a non-empty 3-connected matroid M.*

(i) *For every element e of M, neither $M \backslash e$ nor M/e is 3-connected.*

(ii) *M has rank at least three and is isomorphic to a wheel or a whirl.*

Proof It was noted prior to Theorem 8.8.3 that (ii) implies (i). Now suppose that (i) holds and assume that M has a whirl or a 3-connected wheel as a minor. Take such a minor N of largest rank. If $N = M$, then, since $M \not\cong \mathcal{W}^2$, (ii) holds. Thus we may assume that $N \neq M$. Then, by Corollary 12.2.1, M has an element e such that $M \backslash e$ or M/e is 3-connected and has an N-minor. As this contradicts (i), we may suppose that no minor of M is a whirl or a 3-connected wheel.

If $1 \leq |E(M)| \leq 5$, then we deduce from Table 8.1 in Section 8.4 that M has an element e for which $M \backslash e$ or M/e is 3-connected. Thus we may assume that $|E(M)| \geq 6$. Then, as M is 3-connected, $r(M) \geq 2$. But if $r(M) = 2$, then $M \cong U_{2,n}$ for some $n \geq 6$ and again (i) fails. Thus $r(M) \geq 3$ and, similarly, $r^*(M) \geq 3$.

Now choose x in $E(M)$. Then, by Bixby's Lemma (8.7.3), $\text{si}(M/x)$ or $\text{co}(M \backslash x)$ is 3-connected. By switching to the dual if necessary, we may assume the former. If $|E(\text{si}(M/x))| \geq 4$, then applying Corollary 12.2.1 taking N equal to $\text{si}(M/x)$, we obtain a contradiction to (i). Hence $|E(\text{si}(M/x))| \leq 3$, and, since $r(\text{si}(M/x)) \geq 2$, it follows that $r(\text{si}(M/x)) = 2$. Thus $r(M) = 3$. Now M has no $U_{2,4}$-minor and so is binary. Hence M is a simple binary matroid of rank three having at least six elements. Thus M has $PG(2,2) \backslash p$ as a minor for some point p. But $PG(2,2) \backslash p \cong M(\mathcal{W}_3)$ and this contradiction completes the proof. □

For the next two applications of the Splitter Theorem, we shall need to recall Tutte's result (10.1.2) that a binary matroid is regular if and only if it has no minor isomorphic to F_7 or F_7^*. The first application (Seymour 1980b) asserts that F_7 is a splitter for the class of binary matroids having no F_7^*-minor.

Proposition 12.2.3 *If M is a 3-connected binary matroid, then M has no F_7^*-minor if and only if either M is regular or $M \cong F_7$.*

Proof If M is regular or is isomorphic to F_7, then, by Tutte's result just noted, M has no F_7^*-minor. For the converse, let $\mathcal{N}$ be the class of binary matroids having no F_7^*-minor and let $N = F_7$. Then, by Tutte's result again, it suffices to show that if N_1 is a 3-connected member of $\mathcal{N}$, then either N_1 has no N-minor, or $N_1 \cong N$. By Corollary 12.1.3, this will follow if we can show that $\mathcal{N}$ contains no 3-connected members that are single-element extensions or coextensions of N. Since $N \cong PG(2,2)$, it has no 3-connected binary single-element extensions.

Now consider the binary single-element coextensions of F_7. Two such coextensions are the vector matroids of the following matrices over $GF(2)$:

$$
\left[\begin{array}{c|cccc}
 & 0 & 1 & 1 & 1 \\
I_4 & 1 & 0 & 1 & 1 \\
 & 1 & 1 & 0 & 1 \\
 & 1 & 1 & 1 & 0
\end{array} \right]
\quad \text{and} \quad
\left[\begin{array}{c|cccc}
 & 0 & 1 & 1 & 1 \\
I_4 & 1 & 0 & 1 & 1 \\
 & 1 & 1 & 0 & 1 \\
 & 1 & 1 & 1 & 1
\end{array} \right].
$$

Clearly both these matroids are isomorphic to their duals. The first is $AG(3,2)$; the second will be denoted by S_8. Seymour (1985a) proved the following.

Lemma 12.2.4 *The matroids $AG(3,2)$ and S_8 are the only 3-connected binary single-element coextensions of F_7. Dually, $AG(3,2)$ and S_8 are the only 3-connected binary single-element extensions of F_7^*.*

Proof Evidently it suffices to prove the second statement. Now F_7^* is represented by the matrix A over $GF(2)$ where

$$
A = \left[\begin{array}{c|ccc}
 & \begin{matrix} 1 & 2 & 3 & 4 \end{matrix} & \begin{matrix} 5 & 6 & 7 \end{matrix} \\
 & & 0 & 1 & 1 \\
 & I_4 & 1 & 0 & 1 \\
 & & 1 & 1 & 0 \\
 & & 1 & 1 & 1
\end{array} \right].
$$

Suppose that M is a 3-connected binary single-element extension of F_7^*. Then, as binary matroids are uniquely $GF(2)$-representable, M can be represented by

$$
\left[\begin{array}{c|cccc}
 & \begin{matrix} 5 & 6 & 7 & 8 \end{matrix} \\
 & 0 & 1 & 1 & x_1 \\
I_4 & 1 & 0 & 1 & x_2 \\
 & 1 & 1 & 0 & x_3 \\
 & 1 & 1 & 1 & x_4
\end{array} \right]
$$

where each of x_1, x_2, x_3, and x_4 is in $\{0, 1\}$. We shall denote this matrix by $A + 8$. As M is 3-connected and therefore simple, at least two of x_1, x_2, x_3, and

x_4 are non-zero. By symmetry, we may assume that $(x_1, x_2, x_3, x_4)^T$ is one of $(1,1,1,1)^T$, $(1,1,1,0)^T$, $(1,1,0,0)^T$, and $(1,0,0,1)^T$. Label these columns by e_1, e_2, e_3, and e_4, respectively.

Evidently $M[A + e_1] = S_8$ and $M[A + e_2] \cong AG(3, 2)$. Moreover, as we shall show next, $M[A + e_3] \cong S_8 \cong M[A + e_4]$. The matrices $A + e_3$ and $A + e_4$ are

$$
\left[\begin{array}{c|cccc} & 0 & 1 & 1 & 1 \\ & \boxed{1} & 0 & 1 & 1 \\ I_4 & 1 & 1 & 0 & 0 \\ & 1 & 1 & 1 & 0 \end{array} \right]
\quad \text{and} \quad
\left[\begin{array}{c|cccc} & 0 & 1 & 1 & 1 \\ & 1 & 0 & 1 & 0 \\ I_4 & 1 & 1 & 0 & 0 \\ & \boxed{1} & 1 & 1 & 1 \end{array} \right] .
$$

On pivoting on the boxed entries in each case, we obtain the matrices

$$
\left[\begin{array}{c|cccc} & 0 & 1 & 1 & 1 \\ & 1 & 0 & 1 & 1 \\ I_4 & 1 & 1 & 1 & 1 \\ & 1 & 1 & 0 & 1 \end{array} \right]
\quad \text{and} \quad
\left[\begin{array}{c|cccc} & 0 & 1 & 1 & 1 \\ & 1 & 1 & 0 & 1 \\ I_4 & 1 & 0 & 1 & 1 \\ & 1 & 1 & 1 & 1 \end{array} \right] ,
$$

where we recall from Section 6.6 that a pivot includes the natural column interchange. Clearly the vector matroids of the last two matrices are isomorphic to S_8 and so the lemma is proved. □

To complete the proof of Proposition 12.2.3, it suffices to note that, by the last lemma, neither $AG(3, 2)$ nor S_8 is in $\mathcal{N}$. □

Now we recall from Corollary 8.3.4 that a matroid that is not 3-connected can be built up from some of its 3-connected minors by a sequence of direct sums and 2-sums. Hence, as immediate consequences of the last proposition and duality, we have the following structural results (Seymour 1981c).

Corollary 12.2.5 *Every binary matroid with no F_7^*-minor can be obtained from regular matroids and copies of F_7 by a sequence of direct sums and 2-sums.* □

Corollary 12.2.6 *Every binary matroid with no F_7-minor can be obtained from regular matroids and copies of F_7^* by a sequence of direct sums and 2-sums.* □

Both these corollaries are special cases of the following result (Seymour 1981c), a third reformulation of the Splitter Theorem in the case when N is 3-connected.

Corollary 12.2.7 *Let $\mathcal{N}$ be a class of matroids that is closed under isomorphism and under minors and let N be a 3-connected member of $\mathcal{N}$ having at least four elements such that no 3-connected member of $\mathcal{N}$ with an N-minor has exactly $|E(N)| + 1$ elements. Suppose that if N is a wheel, then $\mathcal{N}$ contains no larger wheels, while if N is a whirl, $\mathcal{N}$ contains no larger whirls. Then every matroid in $\mathcal{N}$ can be obtained by a sequence of direct sums and 2-sums from copies of N and members of $\mathcal{N}$ having no N-minor.* □

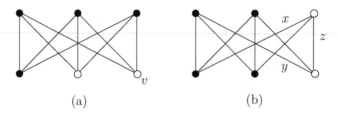

FIG. 12.8. The two grafts $(K_{3,3}, \gamma)$ with $|\gamma| = 4$.

Recall from the end of Section 6.6 that R_{10} is the vector matroid of the matrix A_{10} over $GF(2)$, where

$$
A_{10} = \begin{array}{c}
\begin{array}{cccccccccc}
1 & 2 & 3 & 4 & 5 & 6 & 7 & 8 & 9 & 10
\end{array} \\
\left[\begin{array}{ccccc|ccccc}
 & & & & & 1 & 1 & 0 & 0 & 1 \\
 & & & & & 1 & 1 & 1 & 0 & 0 \\
 & & I_5 & & & 0 & 1 & 1 & 1 & 0 \\
 & & & & & 0 & 0 & 1 & 1 & 1 \\
 & & & & & 1 & 0 & 0 & 1 & 1
\end{array}\right]
\end{array}.
$$

Seymour (1980b) proved the Splitter Theorem on his way to establishing his Decomposition Theorem for regular matroids (13.1.1). The latter theorem, which will be proved in Chapter 13, asserts that every regular matroid can be built using direct sums, 2-sums, and 3-sums from graphic matroids, cographic matroids, and copies of R_{10}. An important step in the proof of this theorem is to show that R_{10} is a splitter for the class of regular matroids. Our proof is due to Geelen (private communication).

Proposition 12.2.8 *If M is a 3-connected regular matroid, then either M has no R_{10}-minor or $M \cong R_{10}$.*

We showed in Lemma 6.6.9 that every single-element deletion of R_{10} is isomorphic to $M(K_{3,3})$. Hence R_{10} arises as a graft matroid, $M(K_{3,3}, \gamma)$, where we recall that graft matroids were defined just after Corollary 10.3.5. The next lemma shows that, in this graft, every vertex is coloured, that is, every vertex is in γ. This lemma will be used not only in the proof of Proposition 12.2.8 but also in our proof of Seymour's Decomposition Theorem (13.1.1). Recall, from Example 10.3.6, that the graft matroid $M(K_4, V(K_4))$ is isomorphic to F_7. On the other hand, F_7^* is the matroid of the graft that is formed from $K_{2,3}$ by colouring all of the vertices except for one of degree 3.

Lemma 12.2.9 *Let $(K_{3,3}, \gamma)$ be a graft such that $M(K_{3,3}, \gamma)$ is regular. Then*

(i) *$|\gamma|$ is odd and $M(K_{3,3}, \gamma)$ is graphic; or*
(ii) *$|\gamma| \in \{0, 2\}$ and $M(K_{3,3}, \gamma)$ is graphic; or*
(iii) *$|\gamma| = 6$ and $M(K_{3,3}, \gamma) \cong R_{10}$.*

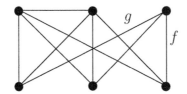

FIG. 12.9. The graft (G, γ'').

Proof If $|\gamma|$ is odd, then, as we showed when introducing grafts in Section 10.3, e_γ is a coloop of $M(K_{3,3}, \gamma)$, and so $M(K_{3,3}, \gamma)$ is graphic. If $|\gamma| \in \{0, 2\}$, then $A_{(K_{3,3}, \gamma)}$ is the mod-2 vertex-edge incidence matrix of some graph, so $M(K_{3,3}, \gamma)$ is again graphic. Now suppose that $|\gamma| = 4$. Up to symmetry, there are two possibilities for $(K_{3,3}, \gamma)$. These are shown in Figure 12.8. Let $(K_{3,3}, \gamma_a)$ be the graft in (a). Then, by contracting one edge incident with v and deleting the other two, we get a copy of $K_{2,3}$ in which every vertex is coloured except for one of degree three. As noted above, the matroid of this graft is F_7^*. Thus, by Lemma 10.3.9, $M(K_{3,3}, \gamma_a)$ has F_7^* as a minor, so $M(K_{3,3}, \gamma_a)$ is not regular.

Let $(K_{3,3}, \gamma_b)$ be as shown in Figure 12.8(b). Then, by deleting z and contracting x and y, we get a graft of the form $(K_4, V(K_4))$. As noted above, $M(K_4, V(K_4)) \cong F_7$, so $M(K_{3,3}, \gamma_b)$ has F_7 as a minor and hence is not regular.

It remains to consider the case when $|\gamma| = 6$. We deduce that this case is the only one that could produce a regular, non-graphic single-element extension of $M(K_{3,3})$. From Lemma 6.6.9 and Corollary 6.6.10, R_{10} is such an extension of $M(K_{3,3})$. Hence, when $|\gamma| = 6$, we must have $M(K_{3,3}, \gamma) \cong R_{10}$. $\square$

Let G be a graph and γ_1 and γ_2 be subsets of $V(G)$. Now adjoin the incidence vectors of γ_1 and γ_2 to the mod-2 vertex–edge incidence matrix of G. We define $M(G, \gamma_1, \gamma_2)$ to be the binary matroid that is represented by this augmented matrix, letting e_{γ_1} and e_{γ_2} label the columns corresponding to γ_1 and γ_2. Thus $M(G, \gamma_1)$ and $M(G, \gamma_2)$ can be obtained from $M(G, \gamma_1, \gamma_2)$ by deleting e_{γ_2} and e_{γ_1}, respectively.

We can now verify that R_{10} is a splitter for the class of regular matroids.

Proof of Proposition 12.2.8 Suppose R_{10} is not a splitter. Then, as it is isomorphic to its dual, there is a regular 3-connected single-element extension M of R_{10} by an element z. By Lemma 12.2.9, $R_{10} \cong M(K_{3,3}, \gamma)$ where $|\gamma| = 6$. Hence $M \cong M(K_{3,3}, \gamma_1, \gamma_2)$ where $|\gamma_1| = 6$ and e_{γ_2} corresponds to z. Clearly $M \backslash e_{\gamma_1} \cong M(K_{3,3}, \gamma_2)$. By Lemma 12.2.9 again, since M is simple, $|\gamma_2| \neq 6$, so $|\gamma_2| = 2$. Thus, without loss of generality, $M = M(G, \gamma'')$, where (G, γ'') is as shown in Figure 12.9. On contracting the edges f and g, and simplifying the resulting graft, we get the graft $(K_4, V(K_4))$, which corresponds to F_7. Hence, by Lemma 10.3.9, M is non-regular; a contradiction. $\square$

The terms 'extension' and 'coextension' have so far been used just for matroids. In our next application of the Splitter Theorem, these terms will also be used for graphs. In particular, if G is a graph and $T \subseteq E(G)$, then G will be called an *extension* of $G \backslash T$ and a *coextension* of G/T. We now look more closely at single-edge coextensions of graphs. Suppose that the graph H has been obtained from the graph G by contracting the edge e. Since contracting a loop in a graph has the same effect as deleting the loop, we shall assume that e has distinct ends, u and v, in G. Then, in H, the vertices u and v are identified as a new vertex w, and we shall say that G has been obtained from H by *splitting the vertex w* into the vertices u and v (Tutte 1984). Clearly the graph obtained by splitting a vertex need not be unique. For example, in Figure 12.10, both G_1 and G_2 can be obtained from K_5 by splitting w. We remark here, in case the use of similar terms causes confusion, that the notions of a 'splitter' for a class of matroids and 'splitting' a vertex in a graph are quite distinct.

The routine proof of the next result (Tutte 1961a) is left to the reader.

Lemma 12.2.10 *Let H be a simple 3-connected graph and suppose that the graph G is a single-edge coextension of H. Then G is simple and 3-connected if and only if G can be obtained from H by splitting a vertex of degree at least four into two vertices of degree at least three.* □

We show next how the Splitter Theorem can be used to prove a result of Hall (1943) that strengthens Kuratowski's Theorem in the case of 3-connected graphs. Recall that, when G is a graph, $\widetilde{G}$ denotes its associated simple graph.

Proposition 12.2.11 *If G is a 3-connected graph, then G has no $K_{3,3}$-minor if and only if either G is planar or its associated simple graph is K_5.*

Proof If $\widetilde{G} \cong K_5$, then G certainly has no $K_{3,3}$-minor. Moreover, by Proposition 5.2.7, if G is planar, it has no $K_{3,3}$-minor. To prove the converse, let $\mathcal{G}$ be the class of graphs having no $K_{3,3}$-minor and let $H = K_5$. Evidently H has no simple 3-connected single-element extensions. Moreover, by symmetry and Lemma 12.2.10, a simple 3-connected single-element coextension of K_5 is isomorphic to the graph G_1 in Figure 12.10. But $G_1 \backslash \{ab, cd\} \cong K_{3,3}$. We conclude by Corollary 12.1.4 that if G is a 3-connected member of $\mathcal{G}$, then either $\widetilde{G}$ has no K_5-minor or $\widetilde{G} \cong K_5$. Thus, by Proposition 5.2.7 again, either $\widetilde{G}$ is planar or

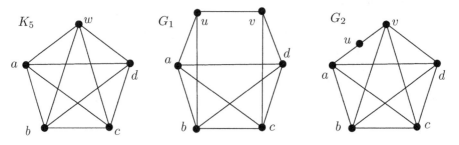

FIG. 12.10. Both G_1 and G_2 are obtained from K_5 by splitting w.

$\widetilde{G} \cong K_5$. But, if $\widetilde{G}$ is planar, so is G. Hence we have, as required, that either G is planar or $\widetilde{G} \cong K_5$. ☐

Leaving graphs for the moment, we return to matroids in general focusing again on Corollary 12.2.1 and its applications. In order to be able to use this result, we need not only a 3-connected matroid M but also some 3-connected minor N of M. The next result shows that one potential choice for N will always be a wheel or a whirl.

Proposition 12.2.12 *Every 3-connected matroid M having at least four elements has a minor isomorphic to $U_{2,4}$ or $M(K_4)$.*

Proof If M is non-binary, it certainly has a $U_{2,4}$-minor. Thus suppose that M is binary. Then M has no whirl as a minor. If M has a wheel of rank at least three as a minor, it has an $M(K_4)$-minor. Hence we may assume that M has no such wheel as a minor. Now $|E(M)| \geq 4$. But, from Table 8.1 in Section 8.4, since M is binary, $|E(M)| \notin \{4,5\}$. Therefore $|E(M)| \geq 6$. As M has no 3-connected wheel or whirl as a minor, by repeated application of the Wheels-and-Whirls Theorem (12.2.2), we can construct a sequence $N_0, N_1, \ldots, N_k$ of 3-connected matroids with $N_0 = M$ and $|E(N_k)| = 6$ such that N_i is a single-element deletion or a single-element contraction of N_{i-1} for all i in $\{1, 2, \ldots, k\}$. As N_k is binary and 3-connected, it is simple and cosimple and cannot have rank or corank less than three. Thus $r(N_k) = 3 = r^*(N_k)$. Every rank-3 simple binary matroid on six elements can be obtained from $PG(2,2)$ by deleting a single element and so is isomorphic to $M(K_4)$. Thus $N_k \cong M(K_4)$; a contradiction. ☐

We now note two corollaries of the last result. The first is immediate. The second appeared earlier as part of Theorem 10.4.8.

Corollary 12.2.13 *Every 3-connected binary matroid having at least four elements has a minor isomorphic to $M(K_4)$.* ☐

Corollary 12.2.14 *A non-empty connected matroid M has no minor isomorphic to $U_{2,4}$ or $M(K_4)$ if and only if M is isomorphic to $M(G)$ for some series-parallel network G.* ☐

By Proposition 12.2.12, every 3-connected matroid M with four or more elements has a minor isomorphic to $U_{2,4}$ or $M(K_4)$. We now use the Splitter Theorem to obtain some more explicit information about the minors of such a matroid M. Suppose that M has a $U_{2,4}$-minor but has no $\mathcal{W}^3$-minor. Then if $|E(M)| \geq 5$, the Splitter Theorem implies that M has a $U_{2,5}$- or a $U_{3,5}$-minor. The next result (Oxley 1989b) is obtained by extending this idea. The matroids Q_6 and P_6 have occurred many times earlier (see, for example, Figure 8.11).

Proposition 12.2.15 *The following are equivalent for a 3-connected matroid having rank and corank at least three:*

(i) *M has a $U_{2,5}$-minor.*

(ii) *M has a $U_{3,5}$-minor.*

(iii) *M has a minor isomorphic to P_6, Q_6, or $U_{3,6}$.*

Before outlining the proof of this result, we note the following interesting corollary that comes from combining it with Theorem 10.2.1.

Corollary 12.2.16 *A 3-connected matroid having rank at least three is ternary if and only if it has no minor isomorphic to $U_{3,5}$, F_7, or F_7^*.*   □

The proof of Proposition 12.2.15 will use the following.

Lemma 12.2.17 *Let N be a 3-connected matroid having an element e such that $N/e \cong U_{2,k}$ for some $k \geq 5$. Then N has a minor isomorphic to one of P_6, Q_6, and $U_{3,6}$.*

Proof We argue by induction on k. If $k = 5$, then N is a 3-connected coextension of $U_{2,5}$ and it is not difficult to check that N is isomorphic to one of P_6, Q_6, and $U_{3,6}$. Thus the lemma is true for $k = 5$. Now assume it true for $k = n$ where $n \geq 5$ and let $k = n + 1$. Suppose that f is an element of $E(N/e)$. Then $N\backslash f/e \cong U_{2,n}$ and the induction assumption implies that, provided $N\backslash f$ is 3-connected, $N\backslash f$, and hence N, has a minor isomorphic to one of P_6, Q_6, and $U_{3,6}$. Thus we may assume that, for all elements f of $E(N) - e$, the matroid $N\backslash f$ is not 3-connected. But both N and $N\backslash f/e$ are 3-connected and so it follows, by Proposition 8.2.7, that N has a triad containing $\{e, f\}$. Hence every series class of $N\backslash e$ is non-trivial. As $r(N\backslash e) = 3$, it follows without difficulty that $|E(N\backslash e)| \leq 4$. This contradiction completes the proof of the lemma.   □

Proof of Proposition 12.2.15 This is not difficult using the last lemma and the Splitter Theorem, and we leave the details to the reader.   □

As a consequence of Theorem 6.5.4, Corollary 12.2.1, and Proposition 12.2.15, we have the following result (Oxley 1987a), whose proof is left to the reader.

Corollary 12.2.18 *A 3-connected non-binary matroid whose rank and corank exceed two has a minor isomorphic to one of $\mathcal{W}^3$, P_6, Q_6, and $U_{3,6}$.*   □

The next result is obtained by combining this with Corollary 12.2.13.

Corollary 12.2.19 *A 3-connected matroid whose rank and corank exceed two has a minor isomorphic to one of $M(\mathcal{W}_3)$, $\mathcal{W}^3$, Q_6, P_6, and $U_{3,6}$.*   □

The last corollary (Walton 1981; Oxley 1984a) explains why the five specified 6-element matroids arise so frequently: any reasonably sized 3-connected matroid must have one of these five as a minor.

In view of the fundamental role played by wheels and whirls within the class of 3-connected matroids, it is natural to consider what can be said about the structure of a minor-closed class of matroids which avoids some small wheel or some small whirl. Previously we have sought, for a given minor-closed class, a complete list of the excluded minors for the class. We are now approaching this situation from the other end: first we specify the list of excluded minors and then we seek properties of the minor-closed class determined by this list. When these excluded minors are wheels or whirls, we already have characterizations of some of the classes of matroids that arise. By Theorem 9.1.3, if the smallest whirl $\mathcal{W}^2$ is the unique excluded minor, then the resulting class of matroids coincides with the class of binary matroids. If the excluded minors are the smallest whirl $\mathcal{W}^2$ and the smallest 3-connected wheel $M(\mathcal{W}_3)$, then, by Corollary 12.2.14, the only non-empty matroids we get are direct sums of series–parallel networks. Next we shall describe the structure of the class of matroids that have no minor isomorphic to $\mathcal{W}^2$ or $M(\mathcal{W}_4)$. Since every member of this class that is not 3-connected can be constructed from 3-connected members of the class by direct sums and 2-sums, it suffices to specify the 3-connected members of the class. In general, for a set $\{M_1, M_2, \ldots\}$ of matroids, $EX(M_1, M_2, \ldots)$ will denote the class of matroids having no minor isomorphic to any of $M_1, M_2, \ldots$.

To find the 3-connected members of $EX(\mathcal{W}^2, M(\mathcal{W}_4))$, we use the following strategy. First we note that all 3-connected matroids with fewer than four elements are trivially in the class. Next we let M be a 3-connected member of the class having four or more elements. By Proposition 12.2.12, M has an $M(K_4)$-minor. Moreover, since M has no $M(\mathcal{W}_4)$-minor, it has no minor isomorphic to $M(\mathcal{W}_r)$ for any $r \geq 4$. Thus, by Corollary 12.2.1, there is a sequence $M_0, M_1, \ldots, M_n$ of 3-connected matroids with $M_0 \cong M(K_4)$ and $M_n = M$ such that M_i is a single-element extension or a single-element coextension of M_{i-1} for each i in $\{1, 2, \ldots, n\}$. Clearly each of $M_0, M_1, \ldots, M_n$ is binary. The unique 3-connected binary extension of $M(K_4)$ is F_7; by duality, the unique 3-connected binary coextension is F_7^*. Thus M_1 is F_7 or F_7^*. It now follows, by Lemma 12.2.4, that M_2 is $AG(3,2)$ or S_8. The fact that each of $\mathcal{W}^2$ and $M(\mathcal{W}_4)$ is self-dual means that $EX(\mathcal{W}^2, M(\mathcal{W}_4))$ is closed under duality. Moreover, as both $AG(3,2)$ and S_8 are self-dual, either M_3 or M_3^* is a binary 3-connected extension of S_8 or $AG(3,2)$. To determine the possible such extensions, we take matrices representing S_8 and $AG(3,2)$ over $GF(2)$ and consider what columns can be added to these matrices so as to avoid creating an $M(\mathcal{W}_4)$-minor. We are relying here on the unique representability of binary matroids, which follows from Proposition 6.4.1. Continuing to analyse the sequence $M_0, M_1, \ldots, M_n$ in this way, we see a pattern emerging and, from this, we can formulate and then prove the structure theorem (12.2.21) stated below (Oxley 1987b).

Let r be an integer exceeding two and Z_r be the vector matroid over $GF(2)$ of the matrix shown in Figure 12.11. Evidently $Z_3 \cong F_7$, while $Z_4 \backslash t \cong AG(3,2)$; and $Z_4 \backslash y_i \cong S_8$ for all i. Moreover, the matroid Z_r is from an important and familiar class of matroids.

Proposition 12.2.20 *The matroid Z_r is the unique binary r-spike with tip t.*

Proof By Proposition 1.5.18, Z_r is a binary spike with tip t. Now let M be an arbitrary binary spike with tip t and legs $L_1, L_2, \ldots, L_r$ where $L_i - t = \{x_i, y_i\}$. Then, from the definition of the circuits of M in Proposition 1.5.17, we deduce that $\{t, x_1, x_2, \ldots, x_{r-1}\}$ is a basis B_t of M and that $B_t \cup x_r$ or $B_t \cup y_r$ is a spanning circuit of M. By relabelling if necessary, we may assume that $B_t \cup x_r$

$$
\begin{array}{ccccccccc}
x_1 & x_2 & \cdots & x_r & y_1 & y_2 & y_3 & \cdots & y_r & t
\end{array}
$$

$$
\left[\begin{array}{cccc|ccccc|c}
 & & & & 0 & 1 & 1 & \cdots & 1 & 1 \\
 & & & & 1 & 0 & 1 & \cdots & 1 & 1 \\
 & & I_r & & 1 & 1 & 0 & \cdots & 1 & 1 \\
 & & & & \vdots & \vdots & \vdots & \ddots & \vdots & \vdots \\
 & & & & 1 & 1 & 1 & \cdots & 0 & 1
\end{array}\right]
$$

FIG. 12.11. A binary representation for Z_r.

is a circuit of M. Then $\{x_1, x_2, \ldots, x_r\}$ is a basis B of M. Moreover, if D is the B-fundamental-circuit incidence matrix of M with respect to B, then $[I_r|D]$ coincides with the representation above for Z_r. Hence $M \cong Z_r$.   □

Theorem 12.2.21 *Let M be a binary matroid. Then M is 3-connected and has no $M(\mathcal{W}_4)$-minor if and only if*

(i) $M \cong Z_r$, Z_r^*, $Z_r \backslash y_r$, *or* $Z_r \backslash t$ *for some* $r \geq 3$; *or*

(ii) $M \cong U_{0,0}$, $U_{0,1}$, $U_{1,1}$, $U_{1,2}$, $U_{1,3}$, *or* $U_{2,3}$.   □

Theoretically, the technique used above of building up, an element at a time, from a wheel or a whirl could also be applied to find the structure of $EX(\mathcal{W}^2, M(\mathcal{W}_r))$ for any $r \geq 5$. But, even when $r = 5$, the number of possibilities to be considered is large and has so far defied analysis. Note, however, that the smaller class of regular matroids with no $M(\mathcal{W}_5)$-minor has been characterized (Oxley 1989a).

We now look at $EX(M(\mathcal{W}_3))$, the class of matroids obtained by excluding the smallest 3-connected wheel as a minor. By Theorem 10.4.8, this class includes the class of gammoids and hence includes the class of transversal matroids. On the other hand, it contains no 3-connected binary matroids with more than three elements. Corollary 12.2.14 characterizes the connected binary members of $EX(M(\mathcal{W}_3))$. The next theorem (Oxley 1987c) determines the ternary members of $EX(M(\mathcal{W}_3))$. As in the last theorem, only the 3-connected such matroids need be specified. We note that, since the class of ternary matroids is $EX(U_{2,5}, U_{3,5}, F_7, F_7^*)$, and both F_7 and F_7^* have an $M(\mathcal{W}_3)$-minor, we have $EX(U_{2,5}, U_{3,5}, F_7, F_7^*) \cap EX(M(\mathcal{W}_3)) = EX(U_{2,5}, U_{3,5}, M(\mathcal{W}_3))$.

Two special matroids appear in the next result, namely the splitters for $EX(U_{2,5}, U_{3,5}, M(\mathcal{W}_3))$. One of these matroids is J, the 8-element rank-4 matroid for which a geometric representation is shown in Figure 10.5. The second such matroid is the vector matroid over $GF(3)$ of the matrix D_{12} shown in Figure 12.12. The last matroid is actually very well known in a slightly different context which we shall now describe. A *Steiner system* $S(t, k, v)$ is a pair $(S, \mathcal{D})$ where S is a v-element set and $\mathcal{D}$ is a collection of k-element subsets of S called *blocks* such that every t-element subset of S is contained in exactly one block. The blocks of such a Steiner system form a t-partition of S and so, by Proposi-

$$
\begin{bmatrix}
 & & & & & 0 & 1 & 1 & 1 & 1 & 1 \\
 & & & & & 1 & 0 & 1 & -1 & -1 & 1 \\
 & I_6 & & & & 1 & 1 & 0 & 1 & -1 & -1 \\
 & & & & & 1 & -1 & 1 & 0 & 1 & -1 \\
 & & & & & 1 & -1 & -1 & 1 & 0 & 1 \\
 & & & & & 1 & 1 & -1 & -1 & 1 & 0
\end{bmatrix}
$$

FIG. 12.12. D_{12}.

tion 2.1.24, there is a paving matroid of rank $t+1$ on S having $\mathcal{D}$ as its set of hyperplanes.

Now let E be the set of elements of $M[D_{12}]$ and $\mathcal{H}$ be its set of hyperplanes. Then the pair $(E, \mathcal{H})$ is an $S(5, 6, 12)$ (Coxeter 1958). Although, in general, there may be more than one Steiner system with a given set of parameters, t, k, and v, the Steiner system $S(5, 6, 12)$ is unique (Witt 1940). Moreover, this Steiner system has numerous attractive properties (see, for example, Cameron 1980). For example, its automorphism group is the Mathieu group M_{12} and the complement of every block is a block. From these observations, we deduce that both the set of circuits and the set of cocircuits of $M[D_{12}]$ equal $\mathcal{H}$, and so $M[D_{12}]$ is identically self-dual. Moreover, since M_{12} is a 5-transitive group, if $(e_1, e_2, \ldots, e_5)$ and $(f_1, f_2, \ldots, f_5)$ are ordered 5-tuples of distinct elements of $M[D_{12}]$, there is an automorphism of $M[D_{12}]$ that, for all i, maps e_i to f_i.

We are now ready to characterize $EX(U_{2,5}, U_{3,5}, M(K_4))$. In this result, we denote by $S(5, 6, 12)$ the paving matroid associated with this Steiner system.

Theorem 12.2.22 *A matroid M is 3-connected, ternary and has no $M(K_4)$-minor if and only if M is isomorphic to J, to $\mathcal{W}^r$ for some $r \geq 2$, or to a 3-connected minor of $S(5, 6, 12)$.* $\square$

Extensions of this result to determine, say, the quaternary matroids with no $M(K_4)$-minor again run into the difficulty of large numbers of cases. However, the same technique has been used to characterize the quaternary matroids with no $M(K_4)$- or $\mathcal{W}^3$-minor (Oxley 1987c). We leave to the exercises consideration of some further results of this type. In addition, Theorem 14.10.25 states a result of Mayhew, Royle, and Whittle (2009a) that describes all internally 4-connected binary matroids with no $M(K_{3,3})$-minor.

Exercises

1. Let M be a binary matroid having no $M(K_4)$-minor. Show that M can be obtained by a sequence of direct sums and 2-sums starting with matroids on at most three elements.
2. Show that, for all prime powers q, there is at least one Steiner system $S(2, q + 1, q^2 + q + 1)$ and at least one $S(2, q, q^2)$.
3. Show, without using Theorem 12.2.21, that $M(\mathcal{W}_4)$, $AG(3, 2)$, and S_8 are the only 8-element binary 3-connected matroids.

4. Show that the hyperplanes of $AG(3,2)$ are the blocks of an $S(3,4,8)$ on $E(AG(3,2))$.

5. To a copy of $K_{2,n}$ with vertex classes V_1 and V_2 where $|V_1| = 2$ and $|V_2| = n \geq 2$, add an edge joining the vertices in V_1 to get the graph $K'_{2,n}$. Let γ_1 and γ_2 be the two distinct subsets of $V_1 \cup V_2$ that contain V_2 and have an even number of elements. Prove that:
    (a) for each i, the graft matroid $M(K'_{2,n}, \gamma_i)$ is isomorphic to $Z_{n+1} \backslash x_1$;
    (b) $Z_{n+1} \cong M(K'_{2,n}, \gamma_1, \gamma_2)$, where the latter is defined after 12.2.9.

6. Let M be a 3-connected regular single-element extension of R_{10} and assume M is represented over $GF(2)$ by a matrix of the form $A_{10} + e$. Give an alternative proof of Proposition 12.2.8 by verifying the following.
    (a) By the symmetry of A_{10}, we may assume that e labels one of the vectors: $(1, 0, 1, x_4, x_5)^T$ for some x_4 and x_5 in $\{0, 1\}$; $(1, 1, 0, 0, 0)^T$; or $(1, 1, 1, 1, 1)^T$.
    (b) In the first of these three cases, $M/\{4, 5\}\backslash\{9, 10\} \cong F_7$; in the second, $M/\{4, 8\}\backslash\{2, 7\} \cong F_7$; and, in the third, $M/\{1, 3\}\backslash\{7, 10\} \cong F_7$.
    (c) R_{10} is a splitter for the class of regular matroids.

7. Show that a non-ternary 3-connected matroid with rank and corank exceeding two has a minor isomorphic to F_7, F_7^*, P_6, Q_6, or $U_{3,6}$.

8. Characterize the ternary matroids with no $\mathcal{W}^3$-minor.

9. Let $(S, \mathcal{D})$ be a Steiner system $S(t, k, v)$. If $x \in S$, let $\mathcal{D}_x = \{H - x : x \in H \in \mathcal{D}\}$. Show that:
    (a) $(S - x, \mathcal{D}_x)$ is an $S(t - 1, k - 1, v - 1)$. This Steiner system is said to be *derived* from $(S, \mathcal{D})$.
    (b) $\mathcal{D}_x$ is the set of hyperplanes of the matroid M/x where M is the matroid on S that has $\mathcal{D}$ as its set of hyperplanes.

10. (Acketa 1988) Show that the only 3-connected binary paving matroids are $U_{0,0}$, $U_{0,1}$, $U_{1,1}$, $U_{1,2}$, $U_{1,3}$, $U_{2,3}$, $M(K_4)$, F_7, F_7^*, and $AG(3,2)$.

11. (Oxley 1984a) Prove that a 3-connected matroid has a 4-element subset that is the intersection of a circuit and a cocircuit if and only if both its rank and its corank exceed two.

12. (Oxley 1988) Consider the following circuit elimination property: whenever C_1, C_2, and C_3 are circuits, none of which is contained in the union of the others, and e_1, e_2, and e_3 are distinct elements of $C_1 \cap C_2 \cap C_3$, there is a circuit contained in $(C_1 \cup C_2 \cup C_3) - \{e_1, e_2, e_3\}$. Show that:
    (a) A matroid has this property if and only if it has no minor isomorphic to any of $U_{3,6}$, $\mathcal{W}^3$, P_6, or Q_6.
    (b) A 3-connected matroid whose rank and corank exceed two has the property if and only if it is binary.

13. (Seymour 1981c) Show that:
    (a) Every binary matroid with no $AG(3,2)$- or S_8-minor can be obtained from regular matroids and copies of F_7 and F_7^* by a sequence of direct sums and 2-sums.

(b) If G_1 is as shown in Figure 12.10, then every binary matroid with no F_7- or $M(G_1)$-minor can be obtained from copies of $M(K_5)$ and from binary matroids with no F_7- or $M(K_5)$-minor by a sequence of direct sums and 2-sums.

14. A graph is *outerplanar* if it is isomorphic to a plane graph in which every vertex is on the boundary of the infinite face. Show that:

    (a) (Chartrand and Harary 1967) The following statements are equivalent for a graph G:

        (i) G is outerplanar.

        (ii) G has no subgraph that is a subdivision of K_4 or $K_{2,3}$.

        (iii) G has no minor isomorphic to K_4 or $K_{2,3}$.

    (b) (Seymour 1995) G has no $K_{2,3}$-minor if and only if every block of G is either outerplanar or has fewer than five vertices.

15. (Wagner 1960) Let G be a 3-connected simple graph with no $(K_5\backslash e)$-minor. Show that G is a wheel, or G is isomorphic to $K_{3,3}$ or $(K_5\backslash e)^*$.

16. (Oxley 1989b)

    (a) *Let M be a 3-connected matroid having a $U_{2,5}$-minor. Show that either M is uniform or M has a minor isomorphic to P_6 or Q_6.

    (b) Deduce the following extension of 12.2.18: A non-binary 3-connected matroid whose rank and corank exceed two is either uniform or has a minor isomorphic to $\mathcal{W}^3$, P_6, or Q_6.

    (c) Show the following are equivalent for a 3-connected matroid M:

        (i) $M \in EX(M(K_4), P_6, Q_6) - EX(M(K_4), U_{2,5}, U_{3,5})$.

        (ii) M is a non-binary member of $EX(\mathcal{W}^3, P_6, Q_6)$ having rank and corank exceeding two.

        (iii) M is uniform having rank and corank exceeding two.

17. (a) *(Seymour 1980b) Show that $M(K_5)$ is a splitter for the class of regular matroids with no minor isomorphic to $M(K_{3,3})$.

    (b) (Walton and Welsh 1980) Deduce that $M(K_5)$ is a splitter for each of the classes $EX(U_{2,4}, F_7, M(K_{3,3}))$ and $EX(U_{2,4}, F_7^*, M(K_{3,3}))$.

18. *(Oxley 1991) Show that the 3-connected ternary paving matroids are precisely the 3-connected minors of $PG(2,3)$, $S(5,6,12)$, R_8, and T_8 where the last two matroids were considered in Exercise 9 of Section 2.2.

19. *(Oxley, Semple, and Vertigan 2000) Prove that the following statements are equivalent for a matroid M.

    (i) M is 3-connected member of $EX(M(\mathcal{W}_3), \mathcal{W}^3, Q_6, U_{3,6})$;

    (ii) M is isomorphic to a member of $\{U_{0,0}, U_{0,1}, U_{1,1}, U_{1,2}, U_{1,3}, U_{2,3}\}$, or M can be obtained from $U_{2,m}$ for some $m \geq 4$ by a sequence of segment–cosegment or cosegment–segment exchanges.

12.3 Variations on the Splitter Theorem

The Splitter Theorem has been extended in several directions, some of which we shall discuss in this section. Most of the results here will not be proved. We begin by noting that the restriction on excluding wheels and whirls can be weakened so that instead of applying to all such matroids, it applies only to the smallest 3-connected wheels and whirls (Coullard 1985). The result here is stated in the case when N is 3-connected since the more general case, when N is connected, simple, and cosimple, is covered by the Splitter Theorem.

Proposition 12.3.1 *Let N be a 3-connected proper minor of a 3-connected matroid M such that $|E(N)| \geq 4$ and M is not a wheel or a whirl. Suppose that if $N \cong \mathcal{W}^2$, then M has no $\mathcal{W}^3$-minor, while if $N \cong M(\mathcal{W}_3)$, then M has no $M(\mathcal{W}_4)$-minor. Then M has a 3-connected minor M_1 and an element e such that $M_1 \backslash e$ or M_1/e is isomorphic to N.*

Proof This can be found in Coullard and Oxley (1992). $\qquad\qquad\square$

We noted at the end of Section 12.1 that the Splitter Theorem cannot be strengthened to require that the minor M_1 of M be a single-element extension or coextension of N itself. Indeed, if M and N are as given in Exercise 7 of Section 12.1, then, for every 3-connected minor M_2 of M having N as a minor, $|E(M_2) - E(N)| \geq 3$. This example prompts the question as to, in general, how large the gap can be between N and a next largest 3-connected minor of M having N as a minor. Truemper (1984) answered this question by showing that this gap has size at most three. This result was strengthened slightly by Bixby and Coullard (1986) who proved the following result.

Theorem 12.3.2 *Let N be a 3-connected proper minor of a 3-connected matroid M. Then M has a 3-connected minor M_1 and an element e such that N is a cosimple matroid associated with $M_1 \backslash e$ or a simple matroid associated with M_1/e, and $|E(M_1) - E(N)| \leq 3$.* $\qquad\qquad\square$

In fact, Bixby and Coullard proved a somewhat stronger result than this since they analysed more closely what happens when $|E(M_1) - E(N)|$ takes the values 2 and 3. For an explicit statement of their result, the reader is referred to Coullard and Oxley (1992). Alternatively, this result can be deduced from Theorem 12.3.6, which will be stated later in the section.

In the case when N has no circuits or cocircuits of size less than four, Bixby and Coullard (1986) improved the bound in the last theorem by one.

Proposition 12.3.3 *Let N be a 3-connected minor of a 3-connected matroid M and suppose that N has no circuits or cocircuits with fewer than four elements. Then there is a sequence $M_0, M_1, \ldots, M_n$ of 3-connected matroids with $M_0 = N$ and $M_n = M$ such that, for all i in $\{1, 2, \ldots, n\}$, there is an element e_i of M_i for which M_{i-1} is a cosimple matroid associated with $M_i \backslash e_i$ or a simple matroid associated with M_i/e_i, and $|E(M_i) - E(M_{i-1})| \leq 2$.* $\qquad\qquad\square$

Next we consider a variant of Theorem 12.3.2 in which we seek a 3-connected minor M_1 of M that not only has an N-minor but also contains some nominated element e of $E(M) - E(N)$. We leave it to the reader to show (Exercise 8) using Proposition 4.3.7 that there is such a matroid M_1 with $|E(M_1)| - |E(N)| \leq 1$.

Note that this allows the N-minor of M_1 to be an isomorphic copy of N. If we require this N-minor to be N itself, then we have the following result, also due to Bixby and Coullard (1987).

Proposition 12.3.4 *Let N be a 3-connected minor of a 3-connected matroid M. Suppose that $|E(N)| \geq 4$ and $e \in E(M) - E(N)$. Then M has a 3-connected minor M_1 with N as a minor such that $e \in E(M_1)$ and $|E(M_1) - E(N)| \leq 4.\square$*

Returning to the situation of seeking a 3-connected single-element extension or coextension of some isomorphic copy N_1 of N, we now add the requirement that N_1 uses some nominated element e of N in the same way that e is used in N. To make this precise, we introduce the following concept. Let M_1 and M_2 be matroids and e be an element that is in both $E(M_1)$ and $E(M_2)$. Then M_1 and M_2 are *e-isomorphic* if there is an isomorphism between M_1 and M_2 under which e is fixed. The following result is due to Tseng and Truemper (1986).

Theorem 12.3.5 *Let N be a 3-connected proper minor of a 3-connected matroid M. Suppose that $|E(N)| \geq 4$ and e is an element of N. Then M has a 3-connected minor M_1 having a minor N_1 that is e-isomorphic to N such that either*

(i) $|E(M_1) - E(N_1)| = 1$; *or*
(ii) $N_1 \cong M(\mathcal{W}_n)$ *for some* $n \geq 3$ *and* $M_1 \cong M(\mathcal{W}_{n+1})$; *or*
(iii) $N_1 \cong \mathcal{W}^n$ *for some* $n \geq 2$ *and* $M_1 \cong \mathcal{W}^{n+1}$. $\qquad\square$

An extension of the last result in which the restriction on $|E(N)|$ is dropped will be noted in Exercise 9. Proposition 12.3.4 is actually a simplified version of another result of Bixby and Coullard (1987), which we state next. The detailed information provided by this theorem makes it useful in proving a number of results for 3-connected matroids. We shall say that a matroid M *uses* an element e or a set Z if $e \in E(M)$ or $Z \subseteq E(M)$.

Theorem 12.3.6 *Let N be a 3-connected minor of a 3-connected matroid M with $|E(N)| \geq 4$. Suppose that $e \in E(M) - E(N)$ and M has no 3-connected proper minor that both uses e and has N as a minor. Then, for some (N_1, M_1) in $\{(N, M), (N^*, M^*)\}$, one of the following holds where $|E(M) - E(N)| = n$:*

(i) $n = 1$ *and* $N_1 = M_1 \backslash e$;
(ii) $n = 2$ *and* $N_1 = M_1 \backslash e/f$ *for some element f; and N_1 has an element x such that $\{e, f, x\}$ is a triangle of M_1;*
(iii) $n = 3$ *and* $N_1 = M_1 \backslash e, g/f$ *for some elements f and g; and N_1 has an element x such that M_1 has $\{e, f, x\}$ is a triangle and $\{f, g, x\}$ as a triad; moreover, $M_1 \backslash e$ is 3-connected;*
(iv) $n = 3$ *and M_1 has a triad $\{e, f, g\}$ such that $N_1 = M_1 \backslash e, f, g$; moreover, N_1 has distinct elements x and y such that $\{e, g, x\}$ and $\{e, f, y\}$ are triangles of M_1; or*
(v) $n = 4$ *and* $N_1 = M_1 \backslash e, g/f, h$ *for some elements f, g, and h; and N_1 has an element x such that $\{e, f, x\}$ and $\{g, h, x\}$ are triangles of M_1, and $\{f, g, x\}$ is a triad of M_1; moreover, $M_1 \backslash e$ and $M_1 \backslash e/f$ are 3-connected.* $\qquad\square$

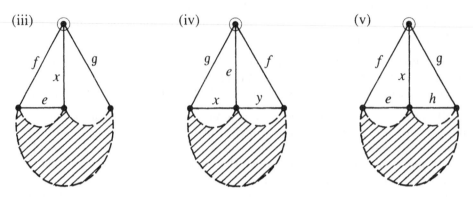

FIG. 12.13. A graphic depiction of 12.3.6(iii)–(v).

Although the last result applies to all matroids and not just graphic ones, Bixby and Coullard use graphs to depict what happens in (iii)–(v) (see Figure 12.13). Each circled vertex corresponds to a known triad in the matroid; all cycles shown are indeed circuits of the matroid; and the shaded part of the diagram corresponds to the rest of the matroid. Observe that, in (iii), (iv), and (v), we have fans in M_1 containing e, specifically $(e, f, x, g), (x, g, e, f, y)$, and (e, f, x, g, h), respectively.

The applications of the last theorem that we shall describe next are concerned with relating certain minors in a matroid to particular elements of the matroid. Interest in results of this type began with some results for non-binary matroids, the first of which was proved by Bixby (1974).

Proposition 12.3.7 *Let M be a 2-connected matroid having a $U_{2,4}$-minor and suppose that $e \in E(M)$. Then M has a $U_{2,4}$-minor using e.*

Proof Let N be a $U_{2,4}$-minor of M and suppose that $e \notin E(N)$. Then, by Corollary 4.3.8, M has a 2-connected minor M_1 such that $M_1 \backslash e$ or M_1/e is isomorphic to N. As N is self-dual, we may assume the former. But then either M_1 is a 5-point line, or M_1 is a parallel extension by e of a 4-point line. In both cases, it follows easily that M has a $U_{2,4}$-minor using e.   □

It is not difficult to see that without the requirement that M is 2-connected, the last result may fail. However, Seymour (1981b) showed that if one imposes a stronger condition on the connectivity of M, then one can strengthen the conclusion as follows.

Proposition 12.3.8 *Let M be a 3-connected matroid having a $U_{2,4}$-minor and let e and f be distinct elements of M. Then M has a $U_{2,4}$-minor using $\{e, f\}$.*□

This proposition can be verified straightforwardly using Theorem 12.3.6. Kahn (1988) used this result to prove that 3-connected quaternary matroids are uniquely $GF(4)$-representable (Proposition 14.6.3).

The last two results prompt several questions including whether the analogue of these results holds for 4-connected matroids, and whether $\{U_{2,4}\}$ is the only set of matroids with the properties noted there. To formalize these questions, we introduce the following definition (Seymour 1985a). Let t be a positive integer. A class $\mathcal{N}$ of matroids is *t-rounded* if every member of $\mathcal{N}$ is $(t+1)$-connected and the following condition holds: If M is a $(t+1)$-connected matroid having an $\mathcal{N}$-minor and X is a subset of $E(M)$ with at most t elements, then M has an $\mathcal{N}$-minor using X. We caution the reader that this t-rounded property of a class of matroids is completely unrelated to the property defined earlier of a matroid M being 'round', the latter meaning that every cocircuit of M is spanning.

Restated using the terminology above, Propositions 12.3.7 and 12.3.8 assert that $\{U_{2,4}\}$ is both 1- and 2-rounded. In general, the task of checking whether a given class of matroids is t-rounded is potentially infinite. However, Seymour (1977a, 1985a) showed that in the two most frequently studied cases, when t is 1 or 2, this task is finite.

Theorem 12.3.9 *Let t be 1 or 2 and $\mathcal{N}$ be a collection of $(t+1)$-connected matroids. Then $\mathcal{N}$ is t-rounded if and only if the following condition holds: If M is a $(t+1)$-connected matroid having an $\mathcal{N}$-minor N such that $|E(M) - E(N)| = 1$, and X is a subset of $E(M)$ with at most t elements, then M has an $\mathcal{N}$-minor using X.*

Proof For $t = 1$, the proof is similar to that of Proposition 12.3.7. We leave the details here to the reader, as we do for the case when $t = 2$, a proof of which can be derived from Theorem 12.3.6. $\qquad\square$

Using this theorem, it is not difficult to show that each of the following collections of matroids is 2-rounded: $\{U_{2,4}, M(K_4)\}$, $\{U_{2,4}, F_7, F_7^*, S_8\}$ (Seymour 1985a); $\{U_{2,4}, M(\mathcal{W}_4)\}$ (Oxley and Reid 1990); $\{U_{3,6}, P_6, Q_6, \mathcal{W}^3\}$, and $\{\mathcal{W}^3, P_6, Q_6\}$ (Oxley 1984a, 1989b). Since every 2-rounded set is also 1-rounded (Exercise 2), all of these collections are 1-rounded as are, for example, $\{U_{2,4}, F_7, F_7^*\}$, $\{U_{2,5}, U_{3,5}, F_7, F_7^*\}$, and $\{U_{2,4}, F_7, F_7^*, M^*(K_5), M^*(K_{3,3}), M^*(K_{3,3}')\}$ (Seymour 1977a). In the last of these sets, $K_{3,3}'$ denotes the graph that is obtained from $K_{3,3}$ by adding an edge joining two non-adjacent vertices.

We began this discussion of roundedness with some questions arising from Propositions 12.3.7 and 12.3.8. We now answer these questions. Firstly, suppose that N is a matroid having at least four elements. Then $\{N\}$ is 2-rounded if and only if $N \cong U_{2,4}$, while $\{N\}$ is 1-rounded if and only if $N \cong U_{2,4}$, Q_6, or $M(\mathcal{W}_2)$ (Oxley 1984b). Secondly, we observe that Kahn (1985) and Coullard (1986) independently showed that $\{U_{2,4}\}$ is not 3-rounded thereby disproving a conjecture of Seymour (1981b). In light of their result and Proposition 12.3.8, it is natural to try to characterize when a set of three elements in a 4-connected non-binary matroid is not in a $U_{2,4}$-minor. This problem is still open although a partial answer to it is provided by the following result for non-binary 3-connected matroids (Oxley 1987a). A proof of this result can be obtained by using Theorem 12.3.6 and Proposition 12.3.8.

Proposition 12.3.10 *Let* $\{x, y, z\}$ *be contained in a 3-connected matroid* M *that has a* $U_{2,4}$-*minor. If* M *has no* $U_{2,4}$-*minor using* $\{x, y, z\}$, *then* M *has a* $\mathcal{W}^3$-*minor that uses* $\{x, y, z\}$ *as its rim or as its spokes.* □

Throughout most of this chapter, we have been dealing with a pair of 3-connected matroids M and N such that M has an N-minor. Our concern has been to remove elements from M to maintain both 3-connectedness and the presence of an N-minor. This has meant that we cannot guarantee whether these element removals are deletions or contractions. Whittle (1999) considered what can be said when we want to maintain an N-minor but we insist on doing contractions. One obvious impediment to being able to maintain 3-connectedness under some single-element contraction is if every element of M is in a triangle. The following straightforward consequence of the Splitter Theorem ensures that, even in this case, we can still find an element e for which M/e is vertically 3-connected and has an N-minor.

Lemma 12.3.11 *Let* N *be a 3-connected minor of a 3-connected matroid* M *with* $|E(N)| \geq 4$. *If* $r(M) > r(N)$, *then* M *has an element* e *such that* $\mathrm{si}(M/e)$ *is 3-connected having an* N-*minor.*

Proof If M is a wheel or a whirl, then so is N and the result holds if we take e to be a rim element of M. We may now suppose that M is neither a wheel nor a whirl. If N is a wheel or a whirl, we shall assume that N is a largest wheel minor or a largest whirl minor of M for it suffices to prove the result in this case.

By Corollary 12.2.1, there is a sequence $M_0, M_1, \ldots, M_n$ of 3-connected matroids such that $M = M_0$ and $M_n \cong N$ where each M_j is obtained by deleting or contracting an element e_j from M_{j-1}. As $r(M) > r(N)$, at least one e_j must be contracted. Let k be the least integer i such that $M_i = M_{i-1}/e_i$. Then $M_k = M \backslash e_1, e_2, \ldots, e_{k-1}/e_k$. We argue by induction on k to show that $\mathrm{si}(M/e_k)$ is 3-connected. This is immediate if $k = 1$. Assume that $k \geq 2$ and that the result holds for smaller values of k. Then, by the induction assumption, $\mathrm{si}(M \backslash e_1/e_k)$ is 3-connected. Now $\mathrm{si}(M \backslash e_1/e_k) \cong M \backslash e_1/e_k \backslash X$ for some subset X of $\{e_2, e_3, \ldots, e_{k-1}\}$. As $r(M \backslash e_1/e_k \backslash X) = r(M/e_k)$ and M is 3-connected, e_1 is neither a loop nor a coloop of $M/e_k \backslash X$. As $M/e_k \backslash X \backslash e_1$ is 3-connected, by Proposition 8.2.7, either $M/e_k \backslash X$ is 3-connected, or e_1 is in a 2-circuit of $M/e_k \backslash X$. In the first case, $\mathrm{si}(M/e_k) \cong M/e_k \backslash X$ while, in the second, $\mathrm{si}(M/e_k) \cong M/e_k \backslash (X \cup e_1)$. In each case, $\mathrm{si}(M/e_k)$ is 3-connected. We conclude that the lemma holds. □

Whittle (1999) needed an extension of the last lemma to prove a result on the behaviour of inequivalent representations over a field, which will be discussed in Section 14.8. In particular, he proved the following result.

Lemma 12.3.12 *Let* N *be a 3-connected minor of a 3-connected matroid* M *with* $|E(N)| \geq 4$ *and* $r(M) \geq r(N) + 2$. *Then* M *has distinct elements* x *and* y *such that each of* $\mathrm{si}(M/x)$ *and* $\mathrm{si}(M/y)$ *is 3-connected having an* N-*minor.*

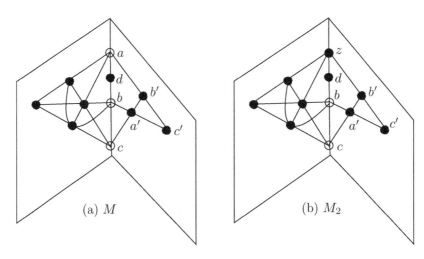

FIG. 12.14. (a) The matroid M in Example 12.3.14. (b) M_2.

Moreover, if $r(M) \geq r(N) + 3$, then x and y can be found satisfying these conditions so that, in addition, si$(M/x, y)$ is 3-connected having an N-minor. □

By duality, we immediately obtain the following consequence of the last result.

Corollary 12.3.13 *Let N be a 3-connected minor of a 3-connected matroid M with $|E(N)| \geq 4$. If $|E(M) - E(N)| \geq 5$, then M has distinct elements x and y such that either each of si(M/x), si(M/y), and si$(M/x, y)$ is 3-connected having an N-minor, or each of co$(M\backslash x)$, co$(M\backslash y)$, and co$(M\backslash x, y)$ is 3-connected having an N-minor.* □

We should like to reduce the lower bound on $|E(M) - E(N)|$ in the last corollary. The case when $|E(M) - E(N)| = 3$ creates problems, but when these arise, a very specific structure occurs.

Example 12.3.14 We construct a matroid M as follows. Begin with a copy of F_7 having $\{a, b, c\}$ as a line and freely add a point d on this line to get the matroid M_1. Let M be obtained from M_1 by performing a Δ-Y exchange on $\{a, b, c\}$. Then M is the matroid on the eight solid points in Figure 12.14(a). Let $N = U_{2,5}$. It is easily checked that we cannot find two elements x and y of M such that either each of si(M/x), si(M/y), and si$(M/x, y)$ is 3-connected having an N-minor; or each of co$(M\backslash x)$, co$(M\backslash y)$, and co$(M\backslash x, y)$ is 3-connected having an N-minor. However, suppose we allow ourselves to modify M by performing a Y-Δ exchange, in this case returning us to the matroid M_1. Then M_1 is 3-connected and each of co$(M_1\backslash a)$, co$(M_1\backslash b)$, and co$(M_1\backslash a, b)$ is 3-connected having an N-minor. □

The situation that arises in the last example is representative of the only type of problem that can occur with finding elements x and y for which the conclusion of Corollary 12.3.13 holds under a weaker hypothesis on $|E(M) - E(N)|$. Indeed, Whittle (1999) proved the following.

Theorem 12.3.15 *Let N be a 3-connected minor of a 3-connected matroid M with $|E(N)| \geq 4$. If $r(M) - r(N) \geq 2$ or $r(M^*) - r(N^*) \geq 2$, then at least one of the following holds:*

(i) *M has distinct elements x and y such that each of $\mathrm{si}(M/x)$, $\mathrm{si}(M/y)$, and $\mathrm{si}(M/x, y)$ is 3-connected having an N-minor.*

(ii) *M has distinct elements x and y such that each of $\mathrm{co}(M \backslash x)$, $\mathrm{co}(M \backslash y)$, and $\mathrm{co}(M \backslash x, y)$ is 3-connected having an N-minor.*

(iii) *$|E(M) - E(N)| = 3$ and $r(M) = r(N) + 1$, and M has a triangle T such that if M_1 is obtained from M by a Δ-Y exchange on T, then M_1 is 3-connected having distinct elements x and y for which each of $\mathrm{si}(M_1/x)$, $\mathrm{si}(M_1/y)$, and $\mathrm{si}(M_1/x, y)$ is 3-connected and has an N-minor.*

(iv) *$|E(M) - E(N)| = 3$ and $r(M) = r(N) + 2$, and M has a triad T^* such that if M_1 is obtained from M by a Y-Δ exchange on T^*, then M_1 is 3-connected having distinct elements x and y for which each of $\mathrm{co}(M_1 \backslash x)$, $\mathrm{co}(M_1 \backslash y)$, and $\mathrm{co}(M_1 \backslash x, y)$ is 3-connected and has an N-minor.* $\square$

In the last theorem, if (i) holds, then both M/x and M/y are guaranteed to be connected, but $M/x, y$ need not be since it may have loops. Whittle (1999) was interested in when we can find elements x and y for which we can, for example, add to (i) the requirement that $M/x, y$ is 3-connected. The same four outcomes arise as in the last theorem but one needs to resort to Δ-Y or Y-Δ exchanges in a few more situations. There is one further complication that can arise. Consider the matroid M_2 in Figure 12.14(b) that is obtained by modifying the matroid M in Example 12.3.14 by adding the element z where a had been. When we had M, we were able to perform a Y-Δ exchange on the triad $\{a', b', c'\}$ to get a 3-connected matroid for which (ii) holds. If we perform the same Y-Δ exchange in M_1, the resulting matroid is not 3-connected as it has a single pair of parallel elements, namely $\{a, z\}$. To proceed in this case, we first delete one of these parallel elements, which, in general, can be chosen to be in the exchange triangle. In the dual case, we do a single series contraction if needed where, again, the element to be contracted can be chosen from the exchange triad. The following is Whittle's main result.

Theorem 12.3.16 *Let N be a 3-connected minor of a 3-connected matroid M with $|E(N)| \geq 4$. If $r(M) - r(N) \geq 2$ or $r(M^*) - r(N^*) \geq 2$, then at least one of the following holds:*

(i) *There is a pair of distinct elements x and y of M such that $M/x, y$ is connected and each of $\mathrm{si}(M/x)$, $\mathrm{si}(M/y)$, and $\mathrm{si}(M/x, y)$ is 3-connected having an N-minor.*

(ii) *There is a pair of distinct elements x and y of M such that $M \backslash x, y$ is connected and each of $\mathrm{co}(M \backslash x)$, $\mathrm{co}(M \backslash y)$, and $\mathrm{co}(M \backslash x, y)$ is 3-connected having an N-minor.*

(iii) $|E(M)-E(N)| \in \{3,4\}$, *and it is possible to perform a single Δ-Y exchange on M followed by at most one series contraction to obtain a 3-connected matroid for which (i) holds.*

(iv) $|E(M)-E(N)| \in \{3,4\}$, *and it is possible to perform a single Y-Δ exchange on M followed by at most one parallel deletion to obtain a 3-connected matroid for which (ii) holds.* □

To conclude this section, we follow Seymour (1978, 1995) in sketching how the type of structural results considered above can be used to prove a matroid extension of the edge form of Menger's Theorem. The statement of Menger's Theorem given in Chapter 8 is sometimes called the vertex form of the theorem. The companion edge form (Menger 1927) asserts that if u and v are distinct vertices in a graph G, then the minimum number of edges whose removal from G leaves u and v in different components is equal to the maximum number of edge-disjoint paths joining u and v. Now suppose that we add a new edge e to G joining u and v and let the resulting graph be $G + e$. Then, for $M = M(G + e)$, the edge form of Menger's Theorem asserts that M has the following property.

12.3.17 *The minimum size of a cocircuit of M containing e is one more than the maximum size of a set $\mathcal{P}$ of circuits in M such that each member of $\mathcal{P}$ contains e, and any two distinct members of $\mathcal{P}$ meet in $\{e\}$.*

This formulation of Menger's Theorem suggests the problem of determining all pairs (M, e) such that e is a non-loop element of a matroid M for which 12.3.17 holds. One difficulty with this problem is that the class of matroids satisfying 12.3.17 is not minor-closed. For example, although 12.3.17 holds for $(\mathcal{W}^3, e)$ when e is a spoke of $\mathcal{W}^3$, it fails for $(\mathcal{W}^2, e)$ for every choice of an element e of $\mathcal{W}^2$.

This difficulty suggests that we should seek another reformulation of Menger's Theorem, one which will yield a matroid condition that is preserved under the taking of minors. We shall now describe such a reformulation due to Seymour (1977b). This was obtained be generalizing the statement of the max-flow min-cut theorem, the latter being essentially a digraph version of Menger's Theorem.

Let e be an element of an arbitrary matroid M and assign a non-negative integer capacity $c(f)$ to each element f of $E(M) - e$. Assume that

12.3.18 *M has a multiset $\{C_1, C_2, \ldots, C_k\}$ of circuits each containing e such that every element f of $E(M) - e$ is in at most $c(f)$ members of this multiset.*

Then it is not difficult to show (Exercise 6(a)) that the following does not hold.

12.3.19 *M has a cocircuit C^* containing e such that $\sum_{f \in C^*-e} c(f) < k$.*

The matroid M has the *integer max-flow min-cut* ($\mathbb{Z}^+$-*MFMC*) *property* with respect to the element e if, for all positive integers k and all non-negative integer-valued functions c on $E(M) - e$, exactly one of 12.3.18 and 12.3.19 holds.

It is straightforward to show (see Exercises 5(a) and 6(b)) that the edge form of Menger's Theorem is equivalent to the assertion that if e is an element of a graphic matroid M, then M has the $\mathbb{Z}^+$-MFMC property with respect to e. We note that e can even be a loop here since, in that case, 12.3.19 fails while 12.3.18 holds. If M is an arbitrary matroid having the $\mathbb{Z}^+$-MFMC property with respect to an element e, and N is a minor of M using e, then we leave it to the reader to check (Exercise 6(c)) that N has the $\mathbb{Z}^+$-MFMC property with respect to e. Thus we have obtained a reformulation of Menger's Theorem of the type we were seeking. We are now faced with the problem of determining precisely when a matroid has the $\mathbb{Z}^+$-MFMC property with respect to a specified element. This problem, which is very difficult, was solved by Seymour (1977b).

Theorem 12.3.20 *A matroid M has the $\mathbb{Z}^+$-MFMC property with respect to an element e if and only if M has no $U_{2,4}$- or F_7^*-minor using e.*   □

As an alternative to his original 'long and cumbersome' proof of this theorem, Seymour (1995) outlined a proof which relies mainly on the type of structural results that we have been considering in this section. In particular, this proof uses Proposition 12.3.7; Gallai (1959) and Minty's (1966) result that every regular matroid has the $\mathbb{Z}^+$-MFMC property with respect to each of its elements; and the following structural theorem of Tseng and Truemper (1986). Recall that a matroid M is internally 4-connected if M is 3-connected and every 3-separation of it has a triangle or a triad on one side.

Theorem 12.3.21 *Let e be an element of an internally 4-connected binary matroid M. Then exactly one of the following holds.*

(i)   *There is an F_7^*-minor of M using e.*

(ii)   *M is regular.*

(iii) *$M \cong F_7$.*   □

The original proof of this theorem used Theorem 12.3.5. Bixby and Rajan (1989) gave a shorter proof that relies on some further structural results of Truemper (1986). Seymour (1978, 1981c) discussed numerous results and problems related to Theorem 12.3.20 and proved several other interesting structural results. Both of the last two papers are rich in open conjectures.

On combining Theorem 12.3.20 with Corollary 12.2.5, we immediately obtain the following.

Corollary 12.3.22 *A matroid M has the $\mathbb{Z}^+$-MFMC property with respect to every element if and only if M can be obtained from regular matroids and copies of F_7 by a sequence of direct sums and 2-sums.*   □

Exercises

1. Find all non-binary 3-connected matroids M such that, for some element e, either $M \backslash e$ or M/e is isomorphic to $M(\mathcal{W}_3)$.
2. Prove that if a class of matroids is 2-rounded, then it is also 1-rounded.
3. Show that both $\{Q_6\}$ and $\{M(\mathcal{W}_2)\}$ are 1-rounded but not 2-rounded.

4. (Oxley and Reid 1990) Show that $\{U_{2,4}, M(\mathcal{W}_n)\}$ is 2-rounded if and only if $n \in \{3, 4\}$.

5. (a) Show that if a matroid M has the $\mathbb{Z}^+$-MFMC property with respect to an element e, then e is a loop or 12.3.17 holds.
   (b) Deduce that 12.3.17 holds if M is a regular matroid and e is a non-loop element of M.
   (c) Show that the converse of (a) fails.

6. (a) By considering the set of pairs (f, C_i) such that f is an element of M that is in C_i, prove that if 12.3.18 holds, then 12.3.19 fails.
   (b) Let e be a non-loop edge of a graph G. Let $c : E(G) - e \to \mathbb{Z}^+ \cup \{0\}$ and k be a positive integer such that 12.3.19 does not hold for $M = M(G)$. Use the following technique to show that M satisfies 12.3.18: for each edge f in $E(G) - e$ with $c(f) > 0$, replace f by $c(f)$ edges in parallel; then delete all those edges f with $c(f) = 0$; finally use the fact that 12.3.17 holds in the cycle matroid of the resulting graph.
   (c) Prove that if M is an arbitrary matroid having the $\mathbb{Z}^+$-MFMC property with respect to an element e, and N is a minor of M using e, then N also has the $\mathbb{Z}^+$-MFMC property with respect to e.

7. (Whittle 1999) Let e be an element of a 3-connected matroid M. Prove that $\mathrm{si}(M/e)$ is not 3-connected if and only if M has a 3-separation (X, Y) such that $\min\{r(X), r(Y)\} \geq 3$ and $e \in \mathrm{cl}(X) \cap \mathrm{cl}(Y)$.

8. (Whittle 1999) Let C^* be an independent triad of a matroid M. Prove that if $\mathrm{si}(M \backslash C^*)$ is 3-connected, then so is $\mathrm{si}(M)$.

9. (Tan 1981; Truemper 1984) Let N be a 3-connected minor of a 3-connected matroid M with $|E(N)| \geq 4$. Show that there is a sequence $N_0, N_1, \ldots, N_n$ of 3-connected matroids, each a proper minor of its successor, with $N_0 \cong N$ and $N_n = M$, such that $|E(N_{i+1}) - E(N_i)| \leq 2$ for all i in $\{0, 1, \ldots, n-1\}$ and equality holds only if N_{i+1} and N_i are both wheels or are both whirls.

10. Let N be a 3-connected minor of a 3-connected matroid M and suppose that $e \in E(M) - E(N)$. Show that M has a 3-connected minor M_1 such that M_1 has an N-minor, $e \in E(M_1)$, and $|E(M_1)| - |E(N)| \leq 1$.

11. (Geelen, private communication) Let M be a 3-connected matroid. Prove that if M has a 3-element rank-2 modular flat, then M is binary.

12. (Coullard in Oxley and Row 1989) Let N be a 3-connected proper minor of a 3-connected matroid M and suppose that $e \in E(N)$. Prove that, provided M is neither a wheel nor a whirl of rank at least three, there is an element f of M such that some member M_0 of $\{M \backslash f, M/f\}$ is 3-connected and has a minor N_0 that is e-isomorphic to N.

13. (a) (Oxley 1987a) Prove that $U_{2,4}$ is the only 3-connected non-binary matroid M with an element e such that $M \backslash e$ and M/e are binary.
    (b) Let M be a 3-connected binary matroid with $|E(M)| \geq 4$. Prove that if M has an element e such that neither $M \backslash e$ nor M/e has an $M(K_4)$-minor, then $M \cong M(K_4)$.

14. (Whittle 1999) Let M be a 3-connected matroid and let x and p be elements of $E(M)$ such that both si(M/x) and si$(M/x,p)$ are 3-connected but si(M/p) is not. Prove that $r(M) \geq 4$ and M has a rank-3 cocircuit C^* containing x such that $p \in \mathrm{cl}(C^*) \cap (E(M) - C^*)$.

15. (Whittle 1999) Let C^* be a rank-3 cocircuit of a 3-connected matroid M. Suppose that an element x of C^* has the property that $\mathrm{cl}_M(C^*) - x$ contains a triangle of M/x. Prove that si(M/x) is 3-connected.

16. (Whittle 1999) Let M be a 3-connected matroid with a triad $\{x, a, b\}$ and a triangle $\{a, b, p\}$. Let N be a 3-connected matroid and assume that si(M/x) and si$(M/x,p)$ are both 3-connected with N-minors and that $M\backslash p$ is not 3-connected. Prove that there is an element y of $E(M) - x$ such that both si(M/y) and si$(M/x,y)$ are 3-connected with N-minors.

17. *(Reid 1988) Prove that there is no matroid M such that $\{M\}$ is 3-rounded.

18. (Coullard and Oxley 1992) Let n be an integer exceeding two. Consider the following assertion: Every 3-connected matroid M with a $\mathcal{W}^2$- and an $M(\mathcal{W}_n)$-minor has a 3-connected non-binary minor N and an element e such that $N\backslash e$ or N/e is isomorphic to $M(\mathcal{W}_n)$.
    (a) *Prove the assertion is false for all $n \geq 5$.
    (b) *Prove the assertion is true for $n = 3$.
    (c) °Is the assertion true or false for $n = 4$?
    (d) Find an 8-element 3-connected non-ternary matroid M having an $M(\mathcal{W}_3)$-minor such that M has no proper minor that is 3-connected, non-ternary, and has an $M(\mathcal{W}_3)$-minor.

19. *(Oxley 1987a) Let x, y, and z be distinct elements of a 3-connected matroid M for which $|E(M| \geq 4$. Prove that:
    (a) If M is non-binary, then either M has a $U_{2,4}$-minor using $\{x, y, z\}$, or M has a $\mathcal{W}^3$-minor in which $\{x, y, z\}$ is the rim or the set of spokes.
    (b) If M is binary, then M has an $M(K_4)$-minor using $\{x, y, z\}$.

20. *(Reid 1991) Let $\{e, f, g\}$ be a circuit of a 3-connected binary matroid M and N be a 3-connected minor of M that uses e. Prove that M has a 3-connected minor M_1 using $\{e, f, g\}$ such that M_1 has an N-minor and $|E(M_1)| \leq |E(N)| + 2$.

21. *(Reid 1991) Let T be a triangle of a 3-connected matroid M. Prove that:
    (a) If M is binary and $|E(M)| \geq 8$, then M has a minor using T that is isomorphic to S_8, $M(\mathcal{W}_4)$, or $M(K_5\backslash e)$.
    (b) If M is regular and has an $M(K_5\backslash e)$-minor, then M has such a minor using T.

13

SEYMOUR'S DECOMPOSITION THEOREM

The excluded-minor theorem (10.1.1) provides one characterization of the class of regular matroids. In this chapter, we shall present a different characterization of this class: a decomposition theorem for members of the class, which was proved by Seymour (1980b). This theorem has a number of important consequences one of which is that it leads to a polynomial-time algorithm to test whether a given real matrix is totally unimodular.

The presentation given in this chapter follows J. Geelen (private communication). This chapter is closely linked to the proof of Tutte's excluded-minor characterization given in Chapter 10. In particular, the power of the rather technical Lemma 10.3.15 is revealed as it plays an important role in the main proof.

13.1 Overview

In this section, we state Seymour's Decomposition Theorem and outline the main results used in the proof. The details of the proofs of these results will be given in later sections. The reader is cautioned that the arguments in this chapter are the most difficult in the book.

We proved in Proposition 5.1.5 and Corollary 5.1.6 that every graphic matroid and every cographic matroid is regular. Seymour's theorem addresses the question of what else is in the class of regular matroids apart from graphic and cographic matroids. Evidently direct sums and 2-sums of regular matroids are also regular. Thus, for example, $M(K_5) \oplus M^*(K_5)$ and $M(K_5) \oplus_2 M^*(K_{3,3})$ are regular, but neither is graphic or cographic. However, these examples are not 3-connected. A particularly important example of a 3-connected regular matroid that is neither graphic nor cographic is R_{10}, which was introduced in Section 6.6 as the vector matroid of the matrix A_{10} over $GF(2)$ where

$$A_{10} = \left[\begin{array}{c|ccccc} & 1 & 1 & 0 & 0 & 1 \\ & 1 & 1 & 1 & 0 & 0 \\ I_5 & 0 & 1 & 1 & 1 & 0 \\ & 0 & 0 & 1 & 1 & 1 \\ & 1 & 0 & 0 & 1 & 1 \end{array} \right].$$

Alternatively, as we showed in Lemma 12.2.9, R_{10} arises as a graft matroid from $K_{3,3}$ by taking every vertex to be coloured.

In Proposition 12.2.8, we proved that the only 3-connected regular matroid having an R_{10}-minor is R_{10} itself. By combining this fact with Corollary 8.3.4, we get that every regular matroid can be obtained from copies of R_{10} and from regular matroids with no R_{10}-minors by a sequence of direct sums and 2-sums. The building blocks used in Seymour's Decomposition Theorem for the class of regular matroids are graphic matroids, cographic matroids, and copies of R_{10}. The operations that are used to stick these building blocks together are direct sums, 2-sums, and 3-sums. The following is the main result of the chapter.

Theorem 13.1.1 (Seymour's Decomposition Theorem) *Every regular matroid M can be constructed by using direct sums, 2-sums, and 3-sums starting with matroids each of which is either graphic, cographic, or isomorphic to R_{10}, and each of which is isomorphic to a minor of M.*

In view of the remarks above, it is natural to consider the 3-connected regular matroids that are neither graphic nor cographic and have no R_{10}-minor. The first big step in the proof of Theorem 13.1.1 is to establish that every such matroid must have an R_{12}-minor, where R_{12} is the self-dual matroid that is represented over $GF(2)$ by the matrix A_{12} where

$$A_{12} = \left[\begin{array}{c|cccccc} & 1 & 1 & 1 & 0 & 0 & 0 \\ & 1 & 1 & 0 & 1 & 0 & 0 \\ & 1 & 0 & 0 & 0 & 1 & 0 \\ I_6 & 0 & 1 & 0 & 0 & 0 & 1 \\ & 0 & 0 & 1 & 0 & 1 & 1 \\ & 0 & 0 & 0 & 1 & 1 & 1 \end{array} \right].$$

As we shall soon see, R_{12} is isomorphic to a 3-sum of $M^*(K_{3,3})$ and $M(K_5 \backslash e)$.

Theorem 13.1.2 *Let M be a 3-connected regular matroid. Then M is graphic or cographic, or M has a minor isomorphic to R_{10} or R_{12}.*

This theorem, which will be proved in the next section, motivates an examination of the regular matroids that have an R_{12}-minor. In Section 13.4, we prove the following theorem for such matroids.

Theorem 13.1.3 *Every regular matroid with an R_{12}-minor has an exact 3-separation (X_1, X_2) in which both $|X_1|$ and $|X_2|$ exceed five.*

In order to prove Theorem 13.1.3, we develop the theory of blocking sequences in Section 13.3. Such a sequence is a certificate that a k-separation of a minor N of a matroid M does not extend to a k-separation of M.

Once we have proved Theorems 13.1.2 and 13.1.3, we will use the fact that an exact 3-separation gives rise to a 3-sum (Proposition 9.3.4) to complete the proof of Theorem 13.1.1. The details are given in Section 13.4 after the proof of Theorem 13.4.3.

Next we show how R_{12} arises as a graft. Let H_{11} be the graph shown in Figure 13.1. If we add the edges $\{3,4\}, \{3,5\}$, and $\{4,5\}$ to H_{11}, we see that H_{11} is a 3-sum of $K_5 \backslash e$ and $\mathcal{W}_4$, with the vertices of the rim of the 4-wheel being, in order, $3, 5, 7, 6$, and the missing edge from K_5 being $\{1,2\}$. From Example 10.3.7, if the graft $(\mathcal{W}_4, \gamma)$ has γ equal to the vertex set of the rim, then $(\mathcal{W}_4, \gamma) \cong M^*(K_{3,3})$. Thus $M(H_{11}, \{3,5,6,7\})$ is a 3-sum of $M(K_5 \backslash e)$ and $M^*(K_{3,3})$.

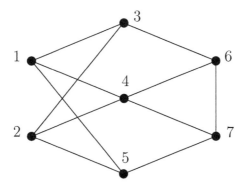

FIG. 13.1. The graph H_{11}.

Lemma 13.1.4 $M(H_{11}, \{3, 5, 6, 7\}) \cong R_{12}$.

Proof Label the edges of H_{11} as in Figure 13.2 and the columns of A_{12} as:

$$A_{12} = \begin{array}{c c c c c c c c c c c c}
 & a & b & c & d & e & f & g & h & i & j & k & l \\
\end{array}
\left[\begin{array}{cccccc|cccccc}
 & & & & & & 1 & 1 & 1 & 0 & 0 & 0 \\
 & & & & & & 1 & 1 & 0 & 1 & 0 & 0 \\
 & & I_6 & & & & 1 & 0 & 0 & 0 & 1 & 0 \\
 & & & & & & 0 & 1 & 0 & 0 & 0 & 1 \\
 & & & & & & 0 & 0 & 1 & 0 & 1 & 1 \\
 & & & & & & 0 & 0 & 0 & 1 & 1 & 1 \\
\end{array} \right].$$

Then one easily checks, by comparing fundamental circuits with respect to the basis $\{a, b, c, d, e, f\}$, that $M(H_{11}) = M[A_{12}] \backslash i$. Since $\{a, e, e_\gamma\}$ and $\{a, e, i\}$ are triangles of $M(H_{11}, \{3, 5, 6, 7\})$ and $M[A_{12}]$, respectively, we conclude that $M(H_{11}, \{3, 5, 6, 7\}) \cong M[A_{12}] \cong R_{12}$. □

When we work with R_{12} in the rest of the chapter, we shall identify it with the graft $M(H_{11}, \gamma)$ where $\gamma = \{3, 5, 6, 7\}$ (see Figure 13.3). Equivalently, R_{12}

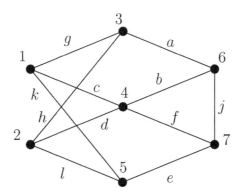

FIG. 13.2. $M(H_{11}) \cong R_{12} \backslash i$.

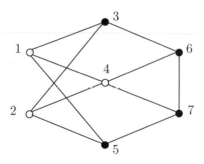

FIG. 13.3. The matroid of the graft (H_{11}, γ) is R_{12}.

can be obtained from $M[A_{12}]$ by relabelling i as e_γ. Let X be the set of edges of the subgraph of H_{11} induced by $\{1, 2, 3, 4, 5\}$ and let $Y_1 = E(H_{11}) - X$. Clearly (X, Y_1) is a 3-separation of $M(H_{11})$. Since $e_\gamma \in \text{cl}_{R_{12}}(Y_1)$, we deduce that $(X, Y_1 \cup e_\gamma)$ is a 3-separation of R_{12}. Let $Y = Y_1 \cup e_\gamma$. Note that X is the disjoint union of the only two triads of R_{12}, and Y is the disjoint union of the only two triangles of R_{12}. It is not difficult to check, using Lemma 8.1.7, that (X, Y_1) is the only 3-separation of $M(H_{11})$ into a 6-element and a 5-element set. It follows that (X, Y) is the only 3-separation of R_{12} into two 6-element sets. This property of R_{12} will be important in the proofs of Theorems 13.1.2 and 13.1.3. Clearly every isomorphism from R_{12} to R_{12}^* interchanges the members of X with the members of Y. Moreover, as the next lemma shows, R_{12} has two different types of elements, those in X and those in Y. The *orbits* of the automorphism group of a matroid M are the maximal subsets Z of $E(M)$ such that if z_1 and z_2 are in Z, there is an automorphism of M mapping z_1 to z_2.

Lemma 13.1.5 *The orbits of the automorphism group of R_{12} are X and Y.*

Proof When R_{12} is viewed as a 3-sum of $M(K_5 \backslash e)$ and $M^*(K_{3,3})$ across a triangle Δ, this triangle is the only triangle of $K_5 \backslash e$ in which all vertices have degree four, and it is an arbitrary triangle of $M^*(K_{3,3})$. Moreover, $X = E(M(K_5 \backslash e)) - \Delta$ and $Y = E(M^*(K_{3,3})) - \Delta$. The symmetry of each of $M(K_5 \backslash e)$ and $M^*(K_{3,3})$ implies that if $Z \in \{X, Y\}$ and $z_1, z_2 \in Z$, then R_{12} has an automorphism that maps z_1 to z_2. As X is the union of the only two triads of R_{12}, no automorphism maps an element of X to an element of Y.   □

The next result is obtained by combining the last two lemmas.

Corollary 13.1.6 *The matroid R_{12} is a minor-minimal matroid that is neither graphic nor cographic.*

Proof Clearly $M(H_{11}, \{3, 5, 6, 7\})/c, l \backslash k \cong M^*(K_{3,3})$. Thus, by Lemma 13.1.4 and the fact that R_{12} is self-dual, we deduce that R_{12} is neither graphic nor cographic. By definition, the triangle $\{a, e, e_\gamma\}$ is contained in Y. Now $R_{12} \backslash e_\gamma = M(H_{11})$, which is graphic; and R_{12}/a has $\{e, e_\gamma\}$ as a circuit, so R_{12}/a is graphic since $R_{12}/a \backslash e_\gamma$ is graphic. By Lemma 13.1.5, $R_{12} \backslash y$ and R_{12}/y are graphic for all y in Y. By duality, $R_{12} \backslash x$ and R_{12}/x are cographic for all x in X.   □

The converse of Seymour's Decomposition Theorem, namely that taking the direct sum, 2-sum, or 3-sum of two regular matroids produces a regular matroid, comes from combining Proposition 4.2.11, Corollary 7.1.26, and the next result.

Lemma 13.1.7 *The 3-sum of two regular matroids is regular.*

Proof Let M_1 and M_2 be regular matroids with ground sets E_1 and E_2. Suppose that $|E_1|, |E_2| \geq 7$ and $E_1 \cap E_2 = T = \{a, b, c\}$ where T is a triangle of both M_1 and M_2 and contains a cocircuit of neither matroid. By Proposition 11.4.16, $M_1 \oplus_3 M_2 = P_T(M_1, M_2) \backslash T$. We shall show that $P_T(M_1, M_2)$ is regular.

By Proposition 9.3.4, $(E_1 - E_2, E_2 - E_1)$ is an exact 3-separation of $M_1 \oplus_3 M_2$ and

$$r(M_1 \oplus_3 M_2) = r(M_1) + r(M_2) - 2 = r(P_T(M_1, M_2)).$$

For each i in $\{1, 2\}$, let $B_i \dot{\cup} \{a, b\}$ be a basis of M_i containing $\{a, b\}$ and let $|B_i| = r_i$. Then $B_1 \cup B_2 \cup \{a, b\}$ is a basis for $P_T(M_1, M_2)$. Thus if

$$
\begin{array}{ccc}
a & b & c
\end{array}
$$
$$
C = \begin{bmatrix} 1 & 0 & 1 \\ 0 & 1 & 1 \end{bmatrix},
$$

then $P_T(M_1, M_2)$ has a binary representation of the form

$$
\begin{array}{cccccc}
E_1 - B_1 - T & B_1 & T & B_2 & E_2 - B_2 - T
\end{array}
$$
$$
\begin{bmatrix}
A_1 & I_{r_1} & 0 & 0 & 0 \\
D_1 & 0 & C & 0 & D_2 \\
0 & 0 & 0 & I_{r_2} & A_2
\end{bmatrix}.
$$

Because M_1 and M_2 are regular, it follows, by Lemma 10.1.3, that there are signings C_1' and C_2' of C and signings $A_1', D_1', A_2',$ and D_2' of $A_1, D_1, A_2,$ and D_2 such that

$$
\begin{bmatrix} A_1' & I_{r_1} & 0 \\ D_1' & 0 & C_1' \end{bmatrix} \quad \text{and} \quad \begin{bmatrix} C_2' & 0 & D_2' \\ 0 & I_{r_2} & A_2' \end{bmatrix}
$$

are totally unimodular. By using row and column scalings, we can ensure that $C_1' = C_2' = C$. Moreover, every non-zero column of D_1' or D_2' is a scalar multiple of one of the columns of C.

To show that $P_T(M_1, M_2)$ is regular, it suffices to show that

$$
\begin{bmatrix}
A_1' & 0 & 0 \\
D_1' & C & D_2' \\
0 & 0 & A_2'
\end{bmatrix}
$$

is totally unimodular. Let A be a square submatrix of the last matrix. We shall show that $\det A \in \{0, 1, -1\}$. Evidently A has the form

$$\begin{bmatrix} A_1'' & 0 & 0 \\ D_1'' & C'' & D_2'' \\ 0 & 0 & A_2'' \end{bmatrix}$$

where $A_1'', D_1'', A_2'', D_2''$, and C'' are, possibly empty, submatrices of A_1', D_1', A_2', D_2', and C, respectively. If A_1'' is empty, then A has a zero column; or A has two columns that are scalar multiples of each other; or A can be obtained from a submatrix of

$$\begin{bmatrix} C & D_2'' \\ 0 & A_2'' \end{bmatrix}$$

by scaling some columns. In each case, $\det A \in \{0, 1, -1\}$.

We may now assume that A_1'' is non-empty. If A_1'' is zero, then $\det A = 0$. If A_1'' is non-zero, then we can pick a non-zero entry of it and pivot in A on this entry. This produces a matrix A' having the same form as A with a column, say t, that is a standard basis vector and with $|\det A'| = |\det A|$. By removing column t from A' and the row determined by its unique non-zero entry, we get a submatrix of A' whose determinant is $\pm \det A$. This submatrix has the same form as A but the size of its A_1''-part has been reduced. We can repeat this process until we eliminate the A_1''-part of the matrix. Once this happens, as shown above, we get a matrix with determinant in $\{0, 1, -1\}$. Hence $\det A \in \{0, 1, -1\}$ and the lemma follows. $\square$

Exercises

1. Let $\{a_1, a_2, a_3\}$ be a triangle of $M^*(K_{3,3})$. For each i, add an element b_i in parallel to a_i. Then do Δ-Y exchanges on both $\{a_1, a_2, a_3\}$ and $\{b_1, b_2, b_3\}$. Show the resulting matroid is isomorphic to R_{12}.

2. Show that $M(\mathcal{W}_5)$ has exactly two non-isomorphic graphic 3-connected coextensions and that $M(H_{11})$ is one of these.

3. (Seymour 1980b) Suppose that $M[A_{12} + e]$ is a simple regular matroid. Show that the column vector labelled by e is one of $(1, 1, 0, 0, 0, 0)^T$, $(1, 1, 0, 0, 1, 1)^T$, $(0, 0, 0, 0, 1, 1)^T$, or $(0, 0, 1, 1, 0, 0)^T$, and that, while the first three cases give isomorphic matroids, the fourth does not.

4. Give an alternative proof of Lemma 13.1.7 by showing that the 3-sum of two regular matroids is binary having no minor isomorphic to F_7 or F_7^*.

5. *Show that R_{10} is the unique splitter for the class of regular matroids.

13.2   Graphic, cographic, or a special minor

The purpose of this section is to prove Theorem 13.1.2, namely that a 3-connected matroid is either graphic, cographic, or has one of R_{10} or R_{12} as a minor. The second lemma below is the tool we use for detecting the presence of an R_{12}-minor. To prove that lemma, we first establish a preliminary result. Recall that we use the term *orthogonality* to refer to the fact that a circuit and a cocircuit of a matroid cannot share exactly one common element (Proposition 2.1.11); and disjoint sets X and Y in a matroid are *skew* if $r(X \cup Y) = r(X) + r(Y)$.

Lemma 13.2.1 *Let M be a 3-connected binary matroid having disjoint triangles T_1 and T_2 such that $T_1 \cup T_2$ is 3-separating. Then $r(M) \geq r(T_1 \cup T_2) = 4$ and $|E(M)| \geq 8$. Moreover, for every 2-element subset X_1 of T_1, there is a unique 2-element subset X_2 of T_2 such that $X_1 \cup X_2$ is a cocircuit.*

Proof Since M is binary and 3-connected and the triangles T_1 and T_2 are disjoint, $r(M) \geq r(T_1 \cup T_2) = 4$. As $M|(T_1 \cup T_2)$ is disconnected, it follows that $|E(M)| \geq |T_1 \cup T_2| + 2 = 8$. Since $\lambda(T_1 \cup T_2) \leq 2$, we deduce that $\lambda(T_1 \cup T_2) = 2$, that is, $2 = r(T_1 \cup T_2) + r^*(T_1 \cup T_2) - |T_1 \cup T_2|$. Hence $r^*(T_1 \cup T_2) = 4$ so $T_1 \cup T_2$ contains at least two cocircuits of M. Furthermore, by Theorem 9.1.2, since M is binary and cosimple, each such cocircuit must contain exactly two elements of each of T_1 and T_2. By Theorem 9.1.2 again, the symmetric difference of two such cocircuits contains a cocircuit, and the lemma follows. $\square$

Lemma 13.2.2 *Let M be a 3-connected binary matroid. Let T_1 and T_2 be disjoint triangles, and T_1^* and T_2^* be disjoint triads of M. If $T_1 \cup T_2$ and $T_1^* \cup T_2^*$ are both 3-separating, then either $M \cong M(\mathcal{W}_4)$, or M has an R_{12}-minor.*

Proof By the last lemma and duality, $r(M) \geq 4$ and $r^*(M) \geq 4$. Moreover, we can label T_1 and T_2 as $\{1,2,3\}$ and $\{4,5,6\}$ so that each of $\{1,2,4,5\}$, $\{1,3,4,6\}$, and $\{2,3,5,6\}$ are cocircuits.

Next we show:

13.2.3 *If $T_1^* \cup T_2^*$ meets $T_1 \cup T_2$, then M is a rank-4 wheel.*

Without loss of generality, we may assume that T_1^* meets T_1 in $\{1,2\}$. Let $T_1^* - T_1 = \{0\}$ and $T_2^* = \{7,8,9\}$. Note that the members of each of $\{0,1,\ldots,6\}$ and $\{0,1,2,7,8,9\}$ are distinct, but these two sets may overlap. By the dual of Lemma 13.2.1, we may assume that $\{1,2,7,8\}, \{1,0,7,9\}$, and $\{2,0,8,9\}$ are circuits of M. Since $\{1,2,7,8\}$ and $\{1,2,3\}$ are circuits, $\{3,7,8\}$ is also a circuit. The orthogonality of this circuit and the cocircuits $\{1,3,4,6\}$ and $\{2,3,5,6\}$ gives that

$$|\{7,8\} \cap \{4,6\}| = 1 = |\{7,8\} \cap \{5,6\}|. \tag{13.1}$$

Similarly, as $\{1,2,4,5\}$ and $\{1,2,0\}$ are cocircuits, so is $\{4,5,0\}$. Orthogonality with the circuits $\{1,0,7,9\}$ and $\{2,0,8,9\}$ implies that

$$|\{4,5\} \cap \{7,9\}| = 1 = |\{4,5\} \cap \{8,9\}|. \tag{13.2}$$

Because $\{1,2,7,8\}$ is a circuit but the triangles $\{1,2,3\}$ and $\{4,5,6\}$ are skew, $\{7,8\} \neq \{4,5\}$. Thus, by (13.1), $6 \in \{7,8\}$ and $\{4,5\} \cap \{7,8\} = \emptyset$. Hence, by (13.2), $9 \in \{4,5\}$. Since $\{3,7,8\}$ is a circuit, its elements are distinct, so $|\{0,1,2,3,4,5,7,8\}| = 8$. Thus, if $Z = \{0,1,\ldots,9\}$, then $|Z| = 8$. Since $6 \in \{7,8\}$ and $9 \in \{4,5\}$, the circuits $\{1,2,7,8\}$ and $\{1,0,7,9\}$ imply that $T_1 \cup T_2$ spans Z. Similarly, $T_1^* \cup T_2^*$ spans Z in M^*. Thus $\lambda(Z) = 0$, so $|E(M) - Z| = 0$. Hence M is a 3-connected binary matroid of rank 4 having 8 elements. Since M has two skew triangles, it follows, by, for example, Theorem 12.2.21, that $M \cong M(\mathcal{W}_4)$. Hence 13.2.3 holds.

We may now assume that $T_1 \cup T_2$ and $T_1^* \cup T_2^*$ are disjoint. Then, by Corollary 8.5.7, as $\lambda_M(T_1 \cup T_2) = 2$ and M is 3-connected, M has a minor N on $T_1 \cup T_2 \cup T_1^* \cup T_2^*$ with $\lambda_N(T_1 \cup T_2) = 2$ such that $M|(T_1 \cup T_2) = N|(T_1 \cup T_2)$ and $M|(T_1^* \cup T_2^*) = N|(T_1^* \cup T_2^*)$. Thus $r_N(T_1 \cup T_2) = 4$, so $r_N^*(T_1 \cup T_2) = 4$.

Recall that T_1 and T_2 are $\{1,2,3\}$ and $\{4,5,6\}$, respectively. Each of the cocircuits $\{1,2,4,5\}, \{1,3,4,6\}$, and $\{2,3,5,6\}$ of M is a disjoint union of cocircuits of N. As $r_N^*(T_1 \cup T_2) = 4$, it follows that all of $\{1,2,4,5\}, \{1,3,4,6\}$, and $\{2,3,5,6\}$ are cocircuits of N, otherwise one of these cocircuits of M is the union of two disjoint 2-cocircuits of N, and $r_N^*(T_1 \cup T_2) \leq 3$.

If $T_1^* = \{a,b,c\}$ and $T_2^* = \{d,e,f\}$, then we may assume that M has $\{a,b,d,e\}, \{a,c,d,f\}$, and $\{b,c,e,f\}$ as circuits. Thus every element of $T_1^* \cup T_2^*$ is in some circuit of N and so is not a coloop of N. Because T_1^* and T_2^* are disjoint unions of cocircuits of N, we deduce that T_1^* and T_2^* are triads of N. Thus

$$r_N(T_1 \cup T_2) \leq r(N) - 2. \tag{13.3}$$

Moreover, the circuits contained in $T_1^* \cup T_2^*$ imply that

$$r_N(T_1^* \cup T_2^*) \leq 4. \tag{13.4}$$

Hence, as $\lambda_N(T_1 \cup T_2) = 2$, we have

$$\begin{aligned} r(N) + 2 &= r_N(T_1 \cup T_2) + r_N(T_1^* \cup T_2^*) \\ &\leq r(N) - 2 + 4 = r(N) + 2. \end{aligned} \tag{13.5}$$

Since equality must hold throughout (13.5), it also holds in (13.3) and (13.4). As $r_N(T_1 \cup T_2) = 4$, we have $r(N) = 6$. Moreover, as $r_N(T_1^* \cup T_2^*) = 4$ while T_1^* and T_2^* are triads of N, we deduce, by arguing as in the last paragraph, that $\{a,b,d,e\}, \{a,c,d,f\}$, and $\{b,c,e,f\}$ are circuits of N.

The triangles, T_1 and T_2, and the triads, T_1^* and T_2^*, imply that N is represented over $GF(2)$ by the matrix $[I_6|A]$ where A is

$$
\begin{array}{c|cccccc}
 & 3 & 6 & b & c & e & f \\
\hline
1 & 1 & 0 & z_{13} & z_{14} & z_{15} & z_{16} \\
2 & 1 & 0 & z_{23} & z_{24} & z_{25} & z_{26} \\
4 & 0 & 1 & z_{33} & z_{34} & z_{35} & z_{36} \\
5 & 0 & 1 & z_{43} & z_{44} & z_{45} & z_{46} \\
a & 0 & 0 & 1 & 1 & 0 & 0 \\
d & 0 & 0 & 0 & 0 & 1 & 1 \\
\end{array}.
$$

Now $N/a,d$ has $\{b,e\}$ and $\{c,f\}$ as circuits, while $N\backslash 3,6$ has $\{1,4\}$ and $\{2,5\}$ as cocircuits. Hence $z_{i3} = z_{i5}$ and $z_{i4} = z_{i6}$ for all i in $\{1,2,3,4\}$; and $z_{1j} = z_{3j}$ and $z_{2j} = z_{4j}$ for all j in $\{3,4,5,6\}$. Thus

$$\begin{bmatrix} z_{13} & z_{14} \\ z_{23} & z_{24} \end{bmatrix} = \begin{bmatrix} z_{15} & z_{16} \\ z_{25} & z_{26} \end{bmatrix} = \begin{bmatrix} z_{33} & z_{34} \\ z_{43} & z_{44} \end{bmatrix} = \begin{bmatrix} z_{35} & z_{36} \\ z_{45} & z_{46} \end{bmatrix} = Z', \text{say}.$$

The columns of Z' are distinct and non-zero, otherwise $\{a,b,c\}$ contains a 2-circuit of N; a contradiction. Likewise, the rows of Z' are distinct and non-zero.

Up to symmetry, this leaves only two possibilities for Z', namely I_2 and $\left[\begin{smallmatrix} 1 & 0 \\ 1 & 1 \end{smallmatrix}\right]$. In the first case, if we pivot on the (a,c)-entry of Z and then on the (d,f)-entry of the result, we obtain the matrix corresponding to the second case. Thus, up to isomorphism, there is a single possibility for N. But, as noted at the end of Section 10.1, R_{12} satisfies the hypotheses of the lemma. Hence $N \cong R_{12}$. □

Before beginning the proof of Theorem 13.1.2, we present two more lemmas. The first is a straightforward connectivity result. The second shows that $M(K_5)$ and $M^*(K_{3,3})$ are maximal regular matroids (see Exercise 3 of Section 6.6).

Lemma 13.2.4 *Let $z_1, z_2, \ldots, z_k, y$ be distinct elements of a matroid M such that M has at least four elements, $M/z_1, z_2, \ldots, z_k \backslash y$ is 3-connected, and $M/z_1, z_2, \ldots, z_i$ is 3-connected for all i in $\{0, 1, \ldots, k\}$. If $k = 1$ or M is binary, then $M \backslash y$ is internally 3-connected.*

Proof If $|E(M)| = 4$, then $M \cong U_{2,4}$, so $M \backslash y$ is internally 3-connected. Hence we may assume that $|E(M)| \geq 5$. Suppose that $k = 1$. Certainly $M \backslash y$ is 2-connected. Suppose $M \backslash y$ has a non-minimal 2-separation (X_1, X_2). We may assume that $z_1 \in X_1$. Then, by Corollary 8.2.5,

$$\lambda_{M/z_1 \backslash y}(X_1 - z_1) \leq \lambda_{M \backslash y}(X_1) = 1. \tag{13.6}$$

But $M/z_1 \backslash y$ is 3-connected; a contradiction. Hence the lemma holds if $k = 1$.

To establish the result when M is binary, we argue by induction on k. We just showed that it holds when $k = 1$. Assume it holds when $k < n$ and let $k = n \geq 2$. Then, by applying the induction assumption to M/z_1, we deduce that $M/z_1 \backslash y$ is internally 3-connected. Assume that (X_1, X_2) is a non-minimal 2-separation of $M \backslash y$ with z_1 in X_1. Then, by (13.6), $(X_1 - z_1, X_2)$ is a 2-separation of $M/z_1 \backslash y$, which must be minimal. Hence $|X_1| = 3$. If X_1 is dependent in M, then the 3-connected matroid M/z_1 has a 2-circuit, contradicting the fact that $|E(M/z_1)| \geq 4$. Thus X_1 is independent in M and, as $\lambda_M(X_1 \cup y) > \lambda_{M \backslash y}(X_1) = 1$, we deduce that $X_1 \cup y$ is independent and $\lambda_M(X_1 \cup y) = 2$. As $2 = r(X_1 \cup y) + r^*(X_1 \cup y) - |X_1 \cup y|$, we have $r^*(X_1 \cup y) = 2$. Since $|X_1 \cup y| = 4$ and M^* is simple, M^* has a $U_{2,4}$-minor, contradicting the fact that M is binary. It follows, by induction, that $M \backslash y$ is internally 3-connected. □

Lemma 13.2.5 *Let M be a 3-connected matroid having an element e such that $M \backslash e$ is isomorphic to $M(K_5)$ or $M^*(K_{3,3})$. Then M is not regular.*

Proof Assume that M is regular. Then M is binary. Suppose $M \backslash e \cong M(K_5)$. Then, by representing $M(K_5)$ by all non-zero binary vectors having four coordinates and at most two ones, we deduce that the vector e has at least three ones. We can now easily check that M has F_7 as a minor; a contradiction.

Now suppose that $M \backslash e \cong M^*(K_{3,3})$. Since every vertex of $K_{3,3}$ has degree 3, the simple matroid M cannot be cographic. Thus, as $r(M) = 4$, by Corollary 10.3.5, M has $M(K_5)$ as a minor. We deduce that $M^*(K_{3,3})$ is a restriction of $M(K_5)$ contradicting Theorem 10.3.2. □

We are now ready to prove the main result of this section.

Proof of Theorem 13.1.2   Assume that M is regular, 3-connected, and neither graphic nor cographic, that M has no R_{12}-minor, and that there is no 3-connected proper minor of M that is neither graphic nor cographic. We shall show that $M \cong R_{10}$. The following lemma will be particularly useful.

Lemma 13.2.6 *For some M' in $\{M, M^*\}$ and some e in $E(M')$,*

(i)   *$M' \backslash e$ is internally 3-connected; and*
(ii)   *$M' \backslash e$ has an $M(K_{3,3})$-minor; and*
(iii) *M' has no two disjoint triangles T_1 and T_2 such that $e \in T_1 \cup T_2$ and $T_1 \cup T_2$
       is 3-separating.*

Proof By Lemma 13.2.2 and duality, we may assume that M does not have a pair, $\{T_1, T_2\}$, of disjoint triangles such that $T_1 \cup T_2$ is 3-separating. Because M is regular but not cographic, Corollary 10.3.5 implies that M has an N-minor for some N in $\{M(K_5), M(K_{3,3})\}$. By the Splitter Theorem (12.1.2), M has an element f such that some M_1 in $\{M \backslash f, M/f\}$ is 3-connected having an N-minor. By assumption, M_1 is graphic. By Proposition 12.2.11, M_1 has an $M(K_{3,3})$-minor or is isomorphic to $M(K_5)$. Clearly the lemma holds if $M_1 = M \backslash f$ and $M \backslash f$ has an $M(K_{3,3})$-minor. If $M_1 = M/f$ and M/f has an $M(K_{3,3})$-minor, then, by the Splitter Theorem again, there is a sequence of elements $f_1, f_2, \ldots, f_k$ of M such that $M/f/f_1, f_2, \ldots, f_i$ is 3-connected having an $M(K_{3,3})$-minor for all i in $\{0, 1, \ldots, k\}$. For a maximal such sequence, either $M/f, f_1, f_2, \ldots, f_k \cong M(K_{3,3})$, which contradicts Lemma 13.2.5, or there is an element y such that $M/f, f_1, f_2, \ldots, f_k \backslash y$ is 3-connected having an $M(K_{3,3})$-minor. In the latter case, by Lemma 13.2.4, $M \backslash y$ is internally 3-connected and the lemma holds.

We may now assume that $M_1 \cong M(K_5)$. Then, by Lemma 13.2.5, $M_1 \neq M \backslash f$, so $M_1 = M/f$. Since M is not graphic, Corollary 10.3.5 implies that it has $M^*(K_5)$ or $M^*(K_{3,3})$ as a minor. But $r(M) = 5$, so the first possibility is excluded. Thus $M \backslash g/h \cong M^*(K_{3,3})$ for some elements g and h. By Lemma 13.2.5, M/h is not 3-connected. Hence $\{g, h\}$ is in a triangle of M, so $\{g, h\} \cap \{f\} = \emptyset$. Now $M^*(K_{3,3})$ and $M(K_5)$ have unique single-element contractions, and it is straightforward to check that $M \backslash g/h/f$ cannot be a minor of both $M^*(K_{3,3})$ and $M(K_5)$. We conclude that $M_1 \not\cong M(K_5)$, so the lemma holds.   □

The reader may have observed that (iii) of the last lemma holds without the restriction that $e \in T_1 \cup T_2$. However, this restriction will be important in what follows. From among all M' in $\{M, M^*\}$ and all e in $E(M')$ satisfying 13.2.6(i)–(iii), choose (M', e) such that $|E(\mathrm{co}(M' \backslash e))|$ is maximum. By replacing M by M^* if necessary, we may assume that $M' = M$. Then, by Lemma 10.3.11, $M \backslash e$ is 3-connected up to series couples, that is, up to 2-element series classes.

Lemma 13.2.7 *$M \backslash e$ is 3-connected.*

Proof Suppose not. If $M \backslash e$ is non-graphic, then $\mathrm{co}(M \backslash e)$ is 3-connected and neither graphic nor cographic; a contradiction to the choice of M. Thus $M \backslash e$

is graphic. Since M is regular but not cographic, Lemma 10.3.15 implies that either

(a) $M^* \cong M'' \oplus_3 M(K_5)$ for some matroid M''; or
(b) M has a triad $\{e, f, g\}$ such that M/f is not graphic.

Assume that (b) holds. Then $\mathrm{co}(M \backslash e) = \mathrm{co}(M \backslash e/f)$. This matroid is 3-connected and non-cographic. The choice of M implies that $\mathrm{co}(M \backslash e/f)$ is graphic. But M/f is non-graphic, so M/f has no 2-circuit containing e. Now M/f is neither graphic nor cographic, so the choice of M implies that $\mathrm{si}(M/f)$ is not 3-connected. Thus M/f has a non-minimal 2-separation (X, Y) with $e \in X$. As $M \backslash e/f$ is 3-connected up to series couples and $(X - e, Y)$ is a 2-separation of $M \backslash e/f$, it follows that $X - e$ is a series couple of $M \backslash e/f$, so $r_{M \backslash e/f}(X - e) = 2$. Moreover, as M is 3-connected, X is a triad of M and hence of M/f. Since

$$1 = \lambda_{M/f}(X) = r_{M/f}(X) + r^*_{M/f}(X) - |X|,$$

we deduce that $r_{M/f}(X) = 2 = r_{M/f}(X - e)$. As M/f is binary, Theorem 9.1.2 implies that every circuit-cocircuit intersection has even cardinality, so X is not a triangle of M/f. But $e \in \mathrm{cl}_{M/f}(X - e)$, so e is in a 2-circuit of M/f; a contradiction. We conclude that (a) holds.

Let M'' and $M(K_5)$ have $\{a, b, c\}$ as their common triangle and let f be the edge $v_1 v_2$ of K_5 that has no vertices in common with any of $a, b,$ or c. Let $X'' = E(M'') - \{a, b, c\}$. We assert that, with $(M', e) = (M^*, f)$, (i)–(iii) of 13.2.6 hold.

To see that (i) holds, we prove a stronger assertion, that $M^* \backslash f$ is 3-connected. First observe that if $M^* \backslash f$ has a 2-cocircuit C^*, then $C^* \cup f$ is a triad of M^*. By orthogonality, C^* must contain an element of each of the three triangles of $M(K_5) \backslash \{a, b, c\}$ containing f, so $|C^*| \geq 3$. This contradiction implies that $M^* \backslash f$ is cosimple. Now suppose that $M^* \backslash f$ has a 2-separation (Z_1, Z_2). For each i in $\{1, 2\}$, let Y_i be the set of edges of $K_5 \backslash f$ meeting v_i. Then Y_i and $Y_i \cup f$ are cocircuits of $M^* \backslash f$ and M^*, respectively. Since Z_1 or Z_2 contains at least two of the elements of Y_i, and $M^* \backslash f$ is cosimple, we may assume, by Proposition 8.2.14, that all three elements of Y_i are in the same Z_j. In particular, we may suppose that $Y_1 \subseteq Z_1$. By Corollary 8.2.6, Z_1 does not span f in M^*. Therefore $Y_2 \subseteq Z_2$. Now $|X''| \geq 4$, so $|Z_1 \cap X''| \geq 2$ or $|Z_2 \cap X''| \geq 2$. Without loss of generality, assume the latter. As $r(Z_1 \cup Y_2) = r(Z_1) + 1$ and $r(Z_2 - Y_2) \leq r(Z_2) - 1$, it follows that $(Z_1 \cup Y_2, Z_2 - Y_2)$ is a 2-separation of $M^* \backslash f$ since $Z_2 - Y_2 = Z_2 \cap X''$. As f is in the closure of $Z_1 \cup Y_2$, we deduce that M^* has a 2-separation; a contradiction. We conclude that $M^* \backslash f$ is 3-connected, so (i) holds.

For (ii), we show first that M'' has an $M(K_4)$-minor using $\{a, b, c\}$. Although this follows immediately from Exercise 12.3.14(b), for completeness we prove it. Let C^* be a cocircuit of M'' avoiding $\{a, b, c\}$. Then $\kappa_{M''}(C^*, \{a, b, c\}) = 2$ for it is certainly at most 2 as $\{a, b, c\}$ is a triangle; and if it is less than 2, then so is $\kappa_{M^*}(C^*, E(M(K_5)) - \{a, b, c\})$, contradicting the fact that M^* is 3-connected. We may now apply the dual of Lemma 9.3.6. This gives that C^*

contains elements x, y, and z so that $(M'')^*$ has a minor N_1 with ground set $\{a, b, c, x, y, z\}$ such that $\{x, y, z\}$ is a circuit of N_1 and $\lambda_{N_1}(\{a, b, c\}) = 2$. Thus $r(N_1) = r_{N_1}(\{a, b, c\})$. Since $\{a, b, c\}$ is a cocircuit of $(M'')^*$, this set is a disjoint union of cocircuits of N_1. As N_1 has $\{a, b, c\}$ as a spanning set and $\{x, y, z\}$ as a triangle, it follows that $\{a, b, c\}$ is a triad of N_1 and $N_1 \cong M(K_4)$. Thus N_1^* is an $M(K_4)$-minor of M'' having $\{a, b, c\}$ as a triangle. We now see that $M^* \backslash f$ has $P_{\{a,b,c\}}(M(K_4), M(K_5)\backslash f))\backslash\{a, b, c\}$ as a minor. This matroid is isomorphic to $M(K_{3,3})$. Hence (ii) holds.

To see (iii), assume that M^* has a pair, $\{T_1, T_2\}$, of disjoint triangles such that $f \in T_1 \cup T_2$ and $T_1 \cup T_2$ is 3-separating. Let $T_1 = \{f, g, h\}$. By orthogonality with the cocircuits $Y_1 \cup f$ and $Y_2 \cup f$ of M^*, we deduce that $\{g, h\}$ meets Y_1 and Y_2, so T_1 is a triangle of $K_5\backslash\{a, b, c\}$. By Lemma 13.2.1, T_2 can be labelled $\{u, v, w\}$ such that each of $\{f, g, u, v\}$, $\{f, h, u, w\}$, and $\{g, h, u, w\}$ is a cocircuit of M^*. Hence $M^* \backslash T_1$ has $\{u, v, w\}$ as a circuit and a series class. Thus $M^* \backslash T_1$ has $\{u, v, w\}$ as a component. Since the edges of $E(K_5) - \{a, b, c\} - \{f, g, h\}$ form a 4-circuit D, this circuit is contained in $E(M^*) - (T_1 \cup T_2)$. But f is in the closure of D, so we contradict orthogonality with $\{f, g, u, v\}$ say. We conclude that (iii) holds.

We now know that (i)–(iii) hold when $(M', e) = (M^*, f)$. But $M^* \backslash f$ is 3-connected, so $|E(\mathrm{co}(M^* \backslash f))| = |E(M^* \backslash f)| = |E(M)| - 1$. This contradicts the choice of (M, e) as $|E(\mathrm{co}(M \backslash e))| \leq |E(M)| - 2$ since $M \backslash e$ is not 3-connected. We conclude that Lemma 13.2.7 holds.   □

By Lemmas 13.2.6 and 13.2.7, $M \backslash e$ is 3-connected having an $M(K_{3,3})$-minor. Thus $M \backslash e$ is certainly not cographic. By the choice of M, it follows that $M \backslash e$ is graphic. Hence there is a graft (G, γ) with $|\gamma|$ even such that $M = M(G, \gamma)$ and $M \backslash e = M(G)$. Observe that G is 3-connected and simple and, since every vertex of $K_{3,3}$ has degree 3, the graph G contains a subgraph H that is isomorphic to a subdivision of $K_{3,3}$. Moreover, since M is not graphic,

$$|\gamma| \geq 4.$$

We note here that the next result considers a 3-connected graph. This graph is not required to be simple. By contrast, we recall that a 3-connected matroid with at least four elements must be simple.

Lemma 13.2.8 *Let f be an edge $\{u, v\}$ of G and let G/f be 3-connected having a $K_{3,3}$-minor. Then $|\gamma| = 4$ and $u, v \in \gamma$.*

Proof Suppose that $|\gamma| > 4$ or that u or v is not in γ. Then $(G, \gamma)/f = (G/f, \gamma')$ where $|\gamma'| \geq 4$. Thus, by Lemma 10.3.14, since $\mathrm{si}(M(G/f))$ is 3-connected, $M(G/f, \gamma')$ is not graphic; that is, M/f is not graphic.

Now $\mathrm{si}(M(G/f))$ is 3-connected and $M(G/f) = M \backslash e/f$, so $\mathrm{si}(M/f)$ must also be 3-connected. Since the last matroid has an $M(K_{3,3})$-minor, it is not cographic. But it is also not graphic, so $\mathrm{si}(M/f)$ contradicts the choice of M. Thus the lemma holds.   □

Lemma 13.2.9 *Let f be an edge $\{u, v\}$ of G. If $M(G\backslash f)$ is 3-connected up to series couples and has an $M(K_{3,3})$-minor, then f has an end w that has degree 3 in G such that w and its two neighbours in $G\backslash f$ belong to γ.*

Proof By Lemma 10.3.12, each series couple in $M(G\backslash f)$ corresponds to a degree-2 vertex in $G\backslash f$. Assume $G\backslash f$ has no degree-2 vertices. Then $M(G\backslash f)$ is 3-connected. As $M(G\backslash f) = M\backslash e\backslash f$, it follows that $M\backslash f$ is 3-connected. Then the choice of M implies that $M\backslash f$ is graphic and so, as $M\backslash f = M(G\backslash f, \gamma)$, it follows, by Lemma 10.3.14, that $|\gamma| = 2$; a contradiction. Hence $G\backslash f$ has a degree-2 vertex, so at least one end of f has degree 3 in G.

Assume that the lemma fails. Then, for each end w of f that has degree 3 in G, there is an incident edge h_w different from f so that at most one end of h_w is in γ. Let (H, η) be obtained from $(G\backslash f, \gamma)$ by contracting the edges in $\{h_u, h_v\}$ that exist. Clearly $|\eta| = |\gamma| \geq 4$ and H is 3-connected. Thus, by Lemma 10.3.14, $M(H, \eta)$ is not graphic. Hence $M(H, \eta)$ is 3-connected and neither graphic nor cographic; a contradiction to the choice of M. $\qquad\square$

Lemma 13.2.10 *If $|V(G)| = 6$, then $M \cong R_{10}$.*

Proof By Lemmas 13.2.9 and 12.2.9, $G \cong K_{3,3}$ and $M(G, \gamma) \cong R_{10}$. $\qquad\square$

By the last lemma, we may now assume that

$$|V(G)| > 6.$$

Lemma 13.2.11 *There are exactly four coloured vertices in G, that is, $|\gamma| = 4$.*

Proof Assume that $|\gamma| \geq 6$. Then, by Lemma 13.2.8, G has no edge f such that G/f is 3-connected having a $K_{3,3}$-minor. Hence, by the Splitter Theorem (12.1.2), G has an edge f such that $G\backslash f$ is 3-connected having a $K_{3,3}$-minor. Thus, by Lemma 13.2.9, $G\backslash f$ has a degree-2 vertex; a contradiction. $\qquad\square$

Lemma 13.2.12 *There is no degree-3 vertex w in G such that w and all of its neighbours are in γ.*

Proof Suppose that G does have such a vertex w and let its neighbours be w_1, w_2, and w_3. Since G is 3-connected and $|V(G)| > 6$, there is a vertex w_4 that is not in $\{w, w_1, w_2, w_3\}$ and three internally disjoint paths, P_1, P_2, and P_3, in G from w_4 to w. Delete the ends of these paths, letting P_1', P_2', and P_3' be the resulting paths. Since $G - w - w_4$ is connected, there is a minimum-length path Q in this graph whose ends are in different members of $\{P_1', P_2', P_3'\}$. By using Q and the paths $P_1 - w, P_2 - w$, and $P_3 - w$, we get that $G - w$ has as a minor a triangle with vertex set $\{w_1, w_2, w_3\}$. Thus the graft (G, γ) has $(K_4, V(K_4))$ as a minor. Hence, by Example 10.3.6, M has F_7 as a minor; a contradiction. $\qquad\square$

Lemma 13.2.13 *Let f be an edge $\{u, v\}$ of G. If $M(G\backslash f)$ is 3-connected up to series couples and has an $M(K_{3,3})$-minor, then exactly one of u and v is in γ.*

Proof Suppose that both u and v are in γ. Then, by Lemma 13.2.9, some w in $\{u,v\}$ has degree 3 in G, and w and its three neighbours w_1, w_2, and w_3 are in γ. This contradicts Lemma 13.2.12. $\square$

We know that G has a subgraph that is a subdivision of $K_{3,3}$. The next result implies that G is obtained from such a subdivision by adding edges.

Lemma 13.2.14 *There is no vertex v of G such that $G - v$ has a $K_{3,3}$-minor.*

Proof Suppose otherwise. Then, for every edge f incident with v, both $M(G\backslash f)$ and $M(G/f)$ have an $M(K_{3,3})$-minor. Moreover, by Bixby's Lemma (8.7.3), one of $M(G\backslash f)$ and $M(G/f)$ is internally 3-connected.

Next we show that
$$v \in \gamma.$$
Assume not. Then, by Lemma 13.2.8, there is no edge f incident with v such that $M(G/f)$ is internally 3-connected. Thus, for every edge f meeting v, the matroid $M(G\backslash f)$ is internally 3-connected having an $M(K_{3,3})$-minor. Hence, by Lemma 13.2.9, if $f = \{v, v_f\}$, then v_f has degree 3 in G, and v_f and its neighbours in $G\backslash f$ belong to γ. Thus the set $N(v)$ of neighbours of v in G is a subset of γ. Since $|\gamma| = 4$, at most one vertex of $V(G) - (N(v) \cup v)$ has a neighbour in $N(v)$. Since $G - v$ is 2-connected having at least six vertices, this is a contradiction. We conclude that v is indeed in γ.

Let f be an edge $\{v, u\}$. Since $v \in \gamma$, by Bixby's Lemma (8.7.3), either

(a) $M(G\backslash f)$ is internally 3-connected; or

(b) $M(G/f)$ is internally 3-connected.

Lemmas 13.2.13 and 13.2.8 imply that $u \notin \gamma$ if (a) holds, while $u \in \gamma$ if (b) holds. Thus outcomes (a) and (b) are mutually exclusive. Moreover, if (b) holds for each f, then, as $|\gamma| = 4$, it follows that v has exactly three neighbours, all of which are in γ; a contradiction to Lemma 13.2.12.

We may now assume that (a) holds for some edge f. Then, by Lemma 13.2.9, v has degree 3 and its neighbours in $G\backslash f$ belong to γ. As $M(G/f)$ is not internally 3-connected, G has a 3-vertex cut that contains v, u, and some third vertex, w. Also $G - v$ has a $K_{3,3}$-minor, G_0. As v has degree three, $G - \{v, u, w\}$ has exactly two components G_1 and G_2. Let $G'_i = G[V(G_i) \cup \{v, u, w\}]$. Then G'_1 or G'_2 contains at least eight edges of G_0 and we may assume the former. Subject to all this, we also assume that the vertex w is chosen so that $|V(G'_2)|$ is maximized. In $G\backslash f$, let v_i be the neighbour of v that is in G'_i. As $G - v$ is 2-connected, there are paths P_u and P_w from v_2 to u and w, respectively, such that these paths have only the vertex v_2 in common.

Let v_3 be the vertex in $\gamma - \{v, v_1, v_2\}$. Suppose that $v_3 \in V(G'_2)$. We now show that we can choose P_u and P_w so that v_3 is in one of them. In G, there are (v_3, v_2)-, (v_3, u)-, and (v_3, w)-paths, Q_1, Q_2, and Q_3, that all share the vertex v_3 but are otherwise disjoint. Beginning at v_3, let y_i be the first vertex of Q_i that is also in $V(P_u) \cup V(P_w)$. For some z in $\{u, w\}$, the path P_z contains at least two of y_1, y_2, and y_3. We now replace P_z by the path formed by that part of

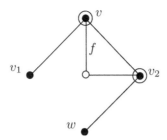

FIG. 13.4. A subgraph of the graft (G, γ).

the original P_z from v_2 to the first time one meets a member y_j of $\{y_1, y_2, y_3\}$; the path Q_j from y_j to v_3; the path Q_k from v_3 to y_k where y_k is also in P_z and $k \neq j$; and the original P_z from y_k to z. This new path contains v_3. Thus, when $v_3 \in V(G_2')$, we will choose P_u and P_w so that v_3 is in one of them. When $v_3 \notin V(G_2')$, we retain the original choices of P_u and P_w.

We now delete from G the edges in $E(G_2') - (E(G_1') \cup E(P_u) \cup E(P_w) \cup \{v_2, v\})$. Our choice of P_u and P_w ensures that v_3 does not become an isolated vertex as a result of these deletions. Next we contract all but one edge from each of P_u and P_w ensuring that we never contract an edge joining two vertices in γ. Thus if (H, η) is the corresponding minor of (G, γ), then $|\eta| = |\gamma| = 4$.

We show next that H is 3-connected. Assume it is not, and let $g = \{v_2, w\}$. Using Menger's Theorem (8.5.1), it is straightforward to check that H/g is 3-connected, although each of the edges $\{u, v_2\}$ and $\{v, v_2\}$ may be in a 2-cycle. Clearly H is simple. It follows by the dual of Lemma 8.2.7 that, as H is not 3-connected, g is in a series couple of $M(H)$. Then, by Lemma 10.3.12, w has degree 2 in H. Let w' be the neighbour of w other than v_2. Then $\{u, v, w'\}$ is a vertex cut of G, and w' contradicts the choice of w. Thus H is indeed 3-connected.

As $|\eta| = 4$, it now follows by Lemma 10.3.14 that $M(H, \eta)$ is not graphic. But H has a $K_{3,3}$-minor, so $M(H, \eta)$ is not cographic. The minimality of $M(G, \gamma)$ implies that $(H, \eta) = (G, \gamma)$. Thus, from above, G/g is 3-connected. Moreover, as $G - v$ has a $K_{3,3}$-minor, so do $(G - v)/g$ and G/g. Hence, by Lemma 13.2.8, $w \in \gamma$. A subgraph of (G, γ) is shown in Figure 13.4. The circled vertices have degree 3 in G. Now $M(G, \gamma)$ has $\{e, \{v_1, v\}, \{v_2, w\}\}$ and $\{\{v, u\}, \{u, v_2\}, \{v, v_2\}\}$ as triangles, T_1 and T_2, so $r(T_1 \cup T_2) \leq 4$. Moreover, the triads $\{\{v, v_1\}, \{v, u\}, \{v, v_2\}\}$ and $\{\{v_2, v\}, \{v_2, u\}, \{v_2, w\}\}$ of $M \backslash e$ imply that $r^*(T_1 \cup T_2) \leq 4$, so $T_1 \cup T_2$ is 3-separating in M. But this contradicts the fact that (M, e) satisfies 13.2.6(iii) with $M = M'$. Thus Lemma 13.2.14 holds. $\square$

We now know that G is obtained from a subdivision of $K_{3,3}$ by adding edges. Next we derive various properties of an arbitrary such subdivision, H. A *series path* in a graph J is a path P in J of length at least two that is maximal with the property that $V(P)$ meets $V(J[E(J) - P])$ in the set of end-vertices of P.

Lemma 13.2.15 *If P is a series path in H, then no edge of $E(G) - E(H)$ has both ends in $V(P)$.*

Proof As G is simple, if such an edge f exists, there is an internal vertex v of the subpath of P that joins the ends of f. Then $G - v$ has a $K_{3,3}$-minor, contradicting Lemma 13.2.14. □

Lemma 13.2.16 *Let f be an edge of $E(G){-}E(H)$. Then $M(G\backslash f)$ is 3-connected up to series couples. Moreover, f has exactly one end, v_f, in γ; the vertex v_f has degree 3 in G; and both of the neighbours of v_f in $G\backslash f$ are in γ.*

Proof Let (X, Y) be a 2-separation of $M(G\backslash f)$. As $r(M(G\backslash f)) = r(M(H))$,

$$
\begin{aligned}
1 &= r(X) + r(Y) - r(M(G\backslash f)) \\
&\geq r(X \cap E(H)) + r(Y \cap E(H)) - r(M(H)).
\end{aligned}
$$

Since $M(G\backslash f)$ is simple, $r(X) \geq 2$, so

$$
r(Y \cap E(H)) \leq r(Y) \leq r(M(G\backslash f)) - 1 = r(M(H)) - 1.
$$

As $M(H)$ is 2-connected and so has no coloops, $|X \cap E(H)| \geq 2$. By symmetry, $|Y \cap E(H)| \geq 2$, so $(X \cap E(H), Y \cap E(H))$ is a 2-separation of $M(H)$. Hence $X \cap E(H)$ or $Y \cap E(H)$, say the former, is a subset of a series path in H. Moreover, $r(X) + r(Y) - r(M(G\backslash f)) = r(X \cap E(H)) + r(Y \cap E(H)) - r(M(H))$, so $r(X \cap E(H)) = r(X)$. The last lemma implies that $X \cap E(H) = X$. As $1 = r_{M(G\backslash f)}(X) + r^*_{M(G\backslash f)}(X) - |X|$ and X is independent in $M(G\backslash f)$, we deduce that X is a subset of a series class in $M(G\backslash f)$. By Lemma 10.3.11, it follows that $M(G\backslash f)$ is 3-connected up to series couples. The rest of the lemma follows immediately by combining Lemmas 13.2.9 and 13.2.13. □

Since $|V(H)| = |V(G)| \geq 7$, there is a degree-2 vertex in H. This vertex is incident with some edge f of $E(G) - E(H)$. Consider v_f. It has degree 2 in H, so it lies in a series path P in H. Moreover, v_f and its two neighbours, u_1 and u_2, on P are in γ. Since $|\gamma| = 4$, there is exactly one more vertex, say t, in γ.

Lemma 13.2.17 *For every edge g of $E(G) - E(H)$, the vertex v_g is in $V(P)$.*

Proof Suppose some v_g is not in $V(P)$. Then $v_g = t$. The two neighbours of v_g in H are in γ and so equal u_1 and u_2 as $\{v_g, v_f\}$ is not an edge. Thus H has two series paths joining u_1 and u_2, contradicting the fact that H is a subdivision of $K_{3,3}$. □

Lemma 13.2.18 *Every internal vertex of P is in γ and has degree 3 in G.*

Proof Take an internal vertex w of P. It meets an edge g of $E(G) - E(H)$ and, by Lemma 13.2.15, g has exactly one end in $V(P)$. Thus, by Lemma 13.2.17, $w = v_g$ so, by Lemma 13.2.16, $w \in \gamma$ and w has degree 3. □

As $|\gamma| = 4$ and every neighbour in P of an internal vertex of P is in γ, either
(a) $|V(P)| = 4$ and $|E(G) - E(H)| = 2$; or
(b) $|V(P)| = 3$ and $|E(G) - E(H)| = 1$.

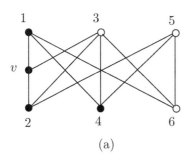 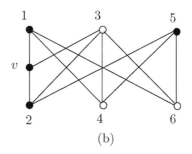

Fig. 13.5. The two possibilities for (G, γ) when $|V(G)| = 7$.

Lemma 13.2.19 *Every degree-2 vertex of H is in $V(P)$.*

Proof Assume that there is a degree-2 vertex w of H that is not in $V(P)$. There is an edge g of $E(G) - E(H)$ meeting w but $w \neq v_g$, so $w \notin \gamma$. Let e_1 and e_2 be the edges of H incident with w where $e_i = \{w, u_i\}$. Since H does not have two series paths joining vertices of P, we may assume that

$$u_1 \notin V(P).$$

As $w \notin \gamma$, Lemma 13.2.8 implies that G/e_1 is not 3-connected. Let u'_1 be the vertex of G/e_1 that results when u_1 and w are identified. Then $\{u'_1, y\}$ is a vertex cut in G/e_1 for some vertex y. Thus there is a partition $(\{u'_1, y\}, X, Y)$ of $V(G/e_1)$ into non-empty sets such that no vertex in X is adjacent to any vertex in Y. As H/e_1 is a subdivision of $K_{3,3}$, we may assume that all the vertices in $X \cup \{u'_1, y\}$ are in the same series path P' of H/e_1. Let x be the vertex of X that is adjacent to u'_1 in P', and h be the edge of P' joining x and u'_1. Then x meets an edge k of $E(G) - E(H)$. By Lemma 13.2.15, k does not meet another vertex of the series path of H containing x. Since $\{u'_1, y\}$ is a vertex cut in G/e_1, we deduce that in G, we must have $\{h, k\} = \{\{x, w\}, \{x, u_1\}\}$. If $h = \{x, w\}$, then $k = \{x, u_1\}$, so both ends of k are in the same series path of H; a contradiction to Lemma 13.2.15. Hence $h = \{x, u_1\}$ and k meets w. Thus $x = v_k$, so, by Lemma 13.2.17, $x \in V(P)$. Hence $u_1 \in V(P)$; a contradiction. □

By Lemma 13.2.19, we may assume that H is obtained from a copy of $K_{3,3}$ with vertex classes $\{1, 3, 5\}$ and $\{2, 4, 6\}$ by replacing the edge $\{1, 2\}$ by a path P of length two or three. Moreover, each degree-2 vertex in H has degree 3 in G, and each vertex of P is in γ.

Lemma 13.2.20 $|V(G)| \neq 7$.

Proof Assume that $|V(G)| = 7$. Then the edge $\{1, 2\}$ is subdivided by inserting a single vertex, say v. By symmetry, we may assume that $\{v, 3\} \in E(G)$. Then Lemma 13.2.16 implies that $3 \notin \gamma$. By symmetry, we may assume that 4 or 5 is in γ (see Figure 13.5). In the first case, delete the edges $\{4, 5\}, \{2, 3\}$, and $\{1, 4\}$ from (G, γ) and contract the edges $\{3, 4\}$ and $\{5, 6\}$. By Example 10.3.6, the resulting graft, whose underlying graph is $K_{2,3}$, corresponds to F_7^*; a contradiction.

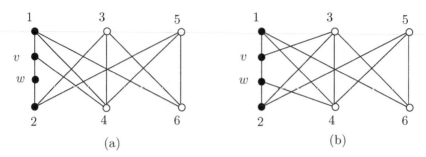

FIG. 13.6. (a) When $\{v, 4\} \in E(G)$. (b) (G, γ) when $|V(G)| = 8$.

We may now assume $5 \in \gamma$. Then M has $\{\{v, 2\}, \{v, 3\}, \{2, 3\}\}$ and $\{\{v, 1\}, \{2, 5\}, e\}$ as disjoint triangles, T_1 and T_2. Also $\{\{v, 1\}, \{v, 2\}, \{v, 3\}\}$ and $\{\{v, 2\}, \{2, 3\}, \{2, 5\}\}$ are triads of $M \backslash e$ contained in $T_1 \cup T_2$. Thus $r(T_1 \cup T_2) + r^*(T_1 \cup T_2) - |T_1 \cup T_2| \leq 2$, so $T_1 \cup T_2$ is 3-separating in M; a contradiction. □

We now know that $|V(G)| = |V(H)| = 8$. Thus the edge $\{1, 2\}$ is subdivided by inserting adjacent vertices, v and w, that are also adjacent to 1 and 2, respectively. Moreover, both v and w have degree 3 in G.

Suppose that $\{v, 4\}$ is an edge of G (see Figure 13.6(a)). Then the graph obtained from H by deleting the edge $\{1, 4\}$ and adding the edge $\{v, 4\}$ is another subdivision of $K_{3,3}$. This new graph is an alternative choice for H. But this choice does not have all of its degree-2 vertices in a single series path; a contradiction to Lemma 13.2.19. Hence $\{v, 4\} \notin E(G)$ and, by symmetry, none of $\{v, 6\}, \{w, 3\}$, and $\{w, 5\}$ is in $E(G)$. Therefore $\{v, 3\}$ or $\{v, 5\}$ is in $E(G)$, and $\{w, 4\}$ or $\{w, 6\}$ is in $E(G)$. By symmetry, we may assume that $\{v, 3\}$ and $\{w, 4\}$ are edges of G (see Figure 13.6(b)). Then, contracting the edge $\{2, 3\}$ in (G, γ) and deleting the vertices 5 and 6 and the edge $\{w, 2\}$, we get a graft that, by Example 10.3.6, corresponds to F_7^*. This contradiction completes the proof of Theorem 13.1.2. □

Exercises

1. Prove that a 3-connected regular matroid has no minor isomorphic to R_{12} if and only if it is graphic, cographic, or isomorphic to R_{10}.

2. Let T_1 and T_2 be triangles of a 3-connected binary matroid M and suppose that $T_1 \cap T_2 = \{e\}$ and that $T_1 \cup T_2$ is 3-separating. Show that $T_1 \cup T_2$ is a fan of M, or $T_1 \triangle T_2$ is a circuit and a cocircuit of M.

3. Use the fact that binary single-element extensions of $M(K_5)$ and $M^*(K_{3,3})$ can be written in the form $M(K_5, \gamma)$ and $M(\mathcal{W}_4, \gamma_1, \gamma_2)$, respectively, to give an alternative proof of Lemma 13.2.5.

4. Let N be a restriction of a 3-connected matroid M with $r(N) = r(M)$. Suppose that $co(N)$ is 3-connected. Let f be an element of $E(M) - E(N)$ that is not in the closure in M of any series class of N. Prove that:
   (a) $co(M \backslash f)$ is 3-connected; and
   (b) if M is binary, then $M \backslash f$ is internally 3-connected.

13.3 Blocking sequences

A 3-connected matroid M has no 2-separations. Thus if N is a minor of M, a
2-separation of N does not extend to a 2-separation of M. A blocking sequence
is a certificate verifying this fact. In this section, we formally define blocking
sequences, give examples of them, and discuss some of their properties. Blocking
sequences were used by Bouchet, Cunningham, and Geelen (1998) in the study
of delta-matroids and later by Geelen, Gerards, and Kapoor (2000) in their
proof of the excluded-minor characterization of quaternary matroids. Our main
application of them will be in the next section in the proof of Theorem 13.1.3. The
treatment given here follows J. Geelen (private communication) and uses different
notation from the treatments of Geelen, Gerards, and Kapoor (2000) and of
Geelen, Hliněný, and Whittle (2005). It should be noted that Seymour's (1980b)
original proof of Theorem 13.1.3 used very similar ideas to those presented here.

For a positive integer k, let (X, Y) be an exact k-separation of a matroid
N, and let M be a matroid having N as a minor. We say that (X, Y) *induces*
a k-separation of M if M has a k-separation (X', Y') such that $X' \supseteq X$ and
$Y' \supseteq Y$. If (X, Y) does not induce a k-separation of M, we call M a *bridge* for
the k-separation (X, Y) and also say that M *bridges* (X, Y).

In this section, we rely heavily on the properties of the connectivity function,
λ, presented in Section 8.2, most notably Lemma 8.2.4 and its two corollaries, and
Proposition 8.2.14. In particular, we repeatedly use the fact that λ is monotone
under taking minors. Thus if (W, Z) is a partition of the ground set of a matroid
N with $|W|, |Z| \geq k$, and (W, Z) is not a k-separation of N, then, for every
matroid M having N as a minor, (W, Z) does not induce a k-separation of M.

Example 13.3.1 Let M be the 10-element rank-4 matroid for which a geometric representation is shown in Figure 13.7(a), and let $N = M/4\backslash\{8, 9, 0\}$ (see
Figure 13.7(b)). Evidently $(\{1, 2, 5\}, \{3, 6, 7\})$ is a 2-separation of N. But this
2-separation does not induce a 2-separation of M. In fact, $(\{1, 2, 5\}, \{3, 6, 7\})$
does not even induce a 2-separation of $M\backslash 0$. To witness this fact, let $M_0 = M\backslash 0$

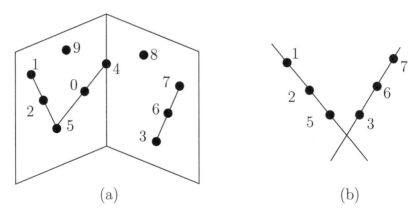

(a)   (b)

FIG. 13.7. (a) M. (b) $M/4\backslash\{8, 9, 0\}$.

and observe the following:

(i) $(\{1,2,5\},\{3,6,7\} \cup 9)$ is not a 2-separation of $M_0/4\backslash 8$;

(ii) $(\{1,2,5\} \cup 8, \{3,6,7\})$ is not a 2-separation of $M_0/4\backslash 9$;

(iii) $(\{1,2,5\} \cup 9, \{3,6,7\} \cup 4)$ is not a 2-separation of $M_0\backslash 8$; and

(iv) $(\{1,2,5\} \cup 4, \{3,6,7\} \cup 8)$ is not a 2-separation of $M_0\backslash 9$.

Suppose that M_0 has a 2-separation (X',Y') induced by $(\{1,2,5\},\{3,6,7\})$. Then, using Corollary 8.2.5, (i) and (ii) imply, respectively, that $9 \in X'$ and $8 \in Y'$. But (iii) now implies that $4 \notin Y'$, while (iv) implies that $4 \notin X'$. Hence (X',Y') does not exist, so M_0 is a bridge for the 2-separation $(\{1,2,5\},\{3,6,7\})$ of N. Thus M is also a bridge for $(\{1,2,5\},\{3,6,7\})$. We call $(9,4,8)$ a blocking sequence for $(\{1,2,5\},\{3,6,7\})$. It is a minimal sequence demonstrating the failure of the 2-separation $(\{1,2,5\},\{3,6,7\})$ of N to induce a 2-separation of M because $(\{1,2,5\}\cup 9\cup 4, \{3,6,7\})$, $(\{1,2,5\}\cup 9, \{3,6,7\}\cup 8)$, and $(\{1,2,5\},\{3,6,7\}\cup 4 \cup 8)$ are 2-separations of $M_0\backslash 8$, $M_0/4$, and $M_0\backslash 9$, respectively.   □

Before formally defining blocking sequences in general, we introduce some notation. Let M be a matroid and D and C be disjoint subsets of $E(M)$ such that D is coindependent and C is independent. Let $N = M\backslash D/C$. For any subset Z of $E(M)$, let $M[D,C;Z] = M\backslash(D-Z)/(C-Z)$. Thus $M[D,C;Z]$ is a minor of M with ground set $E(N)\cup Z$, and $M[D,C;Z]$ has N as a minor. In particular, $N = M[D,C;\emptyset]$. For notational convenience, we do not require Z and $E(N)$ to be disjoint, so $M[D,C;Z] = N$ for all subsets Z of $E(N)$.

Now let k be a positive integer and (X,Y) be an exact k-separation of N. A sequence $(v_1, v_2, \ldots, v_t)$ of elements of $E(M) - E(N)$ is a *blocking sequence* for the k-separation (X,Y) of N, with respect to D and C, if

(i)   **(a)** $(X, Y \cup v_1)$ is not a k-separation of $M[D,C;\{v_1\}]$;

      (b) $(X \cup v_t, Y)$ is not a k-separation of $M[D,C;\{v_t\}]$; and

      (c) for all i in $\{1,2,\ldots,t-1\}$, the pair $(X \cup v_i, Y \cup v_{i+1})$ is not a k-separation of $M[D,C;\{v_i, v_{i+1}\}]$; and

(ii)   no proper subsequence of $(v_1, v_2, \ldots, v_t)$ satisfies (i).

The following is an immediate consequence of the definition.

Lemma 13.3.2 *The sequence* $(v_1, v_2, \ldots, v_t)$ *is a blocking sequence with respect to D and C for the k-separation (X,Y) of N if and only if $(v_t, v_{t-1}, \ldots, v_1)$ is a blocking sequence for (Y,X) with respect to D and C.*   □

In order to detect blocking sequences, we define an auxiliary directed graph $G(X,Y)$ with vertex set $\{v_X, v_Y\}\cup(E(M)-E(N))$ and with set of arcs consisting of the following:

(a) (v_X, w) for all w in $E(M)-E(N)$ such that $(X, Y\cup w)$ is not a k-separation of $M[D,C;\{w\}]$;

(b) (w, v_Y) for all w in $E(M)-E(N)$ such that $(X\cup w, Y)$ is not a k-separation of $M[D,C;\{w\}]$; and

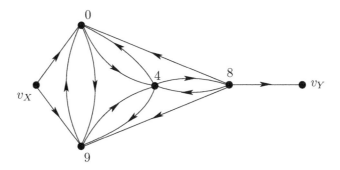

FIG. 13.8. The directed graph $G(X,Y)$ for $(X,Y) = (\{1,2,5\}, \{3,6,7\})$.

(c) (u,w) for all distinct u and w in $E(M) - E(N)$ such that $(X \cup u, Y \cup w)$ is not a k-separation of $M[D, C; \{u, w\}]$.

In a directed graph G, a *minimal directed path* from a vertex u to a vertex v is a directed path $u, v_1, v_2, \ldots, v_k, v$ for which no proper subsequence is also a directed (u, v)-path.

Example 13.3.3 In Example 13.3.1, let (X, Y) be the 2-separation $(\{1, 2, 5\}, \{3, 6, 7\})$ of N. It is not difficult to check that the corresponding directed graph $G(X, Y)$ is as shown in Figure 13.8. Thus $v_X, 9, 4, 8, v_Y$ is a minimal directed path from v_X to v_Y in $G(X, Y)$. Recall that $(9, 4, 8)$ is a blocking sequence for (X, Y). Note that $v_X, 0, 4, 8, v_Y$ is also a minimal directed path from v_X to v_Y in $G(X, Y)$. We leave it to the reader to check that $(0, 4, 8)$ is also a blocking sequence for (X, Y). The next lemma generalizes these observations. □

Lemma 13.3.4 *There is a blocking sequence for (X, Y) if and only if there is a directed path from v_X to v_Y in $G(X, Y)$.*

Proof If $v_X, v_1, v_2, \ldots, v_t, v_Y$ is a minimal directed path from v_X to v_Y in $G(X, Y)$, then $(v_1, v_2, \ldots, v_t)$ is a blocking sequence, and conversely. □

Geelen, Gerards, and Kapoor (2000) proved that a blocking sequence is indeed a certificate for verifying that an exact k-separation of a minor of M does not induce a k-separation of M.

Theorem 13.3.5 *Let D and C be disjoint subsets in a matroid M with D coindependent and C independent. Let $N = M \backslash D / C$ and (X, Y) be an exact k-separation of N for some positive integer k. Then M bridges (X, Y) if and only if there is a blocking sequence for (X, Y) in M with respect to D and C.*

The proof of this theorem will rely heavily on the following result.

Lemma 13.3.6 *For a matroid M, let $N = M \backslash D / C$ where D is coindependent and C is independent in M. Let (X, Y) and (X', Y') be partitions of $E(N)$ and*

$E(M)$ with $X \subseteq X'$ and $Y \subseteq Y'$. Then $\lambda_M(X') > \lambda_N(X)$ if and only if X' and Y' contain elements x and y such that $\lambda_{M[D,C;\{x,y\}]}(X \cup x) > \lambda_N(X)$.

Note that the elements x and y in this lemma are allowed to be in X and Y.

Proof of Lemma 13.3.6 Assume that $\lambda_M(X') = \lambda_N(X)$. If $x \in X'$ and $y \in Y'$, then, by two applications of Corollary 8.2.5, we have

$$\begin{aligned}
\lambda_M(X') &\geq \lambda_{M[D,C;\{x,y\}]}(X' - [(C \cup D) - \{x,y\}]) \\
&= \lambda_{M[D,C;\{x,y\}]}(X \cup x) \\
&\geq \lambda_N(X) = \lambda_M(X').
\end{aligned}$$

We deduce that $\lambda_{M[D,C;\{x,y\}]}(X \cup x) = \lambda_N(X)$ for all x in X' and all y in Y'.

Conversely, assume that $\lambda_M(X') > \lambda_N(X)$. Then

$$\lambda_M(Y') = \lambda_M(X') > \lambda_N(X) = \lambda_N(Y).$$

By symmetry, we may assume that $|Y'-Y| \geq |X'-X|$. As $M \neq N$, we must have $|Y'-Y| \geq 1$. Assume equality holds here and let $Y' - Y = \{y\}$. If $|X'-X| = 1$, choose x in $X' - X$; otherwise choose x in X. As $D \cup C = \{x,y\}$, we have $\lambda_{M[D,C;\{x,y\}]}(X') = \lambda_M(X') > \lambda_N(X)$, so $(X \cup x, Y \cup y)$ is not a k-separation of $M[D,C;\{x,y\}]$. Thus the result holds if $|Y'-Y| = 1$. Now assume it holds if $|Y'-Y| < n$ and let $|Y'-Y| = n \geq 2$. Choose e in $Y'-Y$. If $e \in D$ and $\lambda_{M \backslash e}(X') > \lambda_N(X)$, then, by the induction assumption, there is an element x in X' and an element y in $Y'-e$ such that $\lambda_{(M \backslash e)[D-e,C;\{x,y\}]}(X \cup x) > \lambda_N(X)$, that is, $\lambda_{M[D,C;\{x,y\}]}(X \cup x) > \lambda_N(X)$. Hence we may assume that if $e \in (Y'-Y) \cap D$, then $\lambda_{M \backslash e}(X') = \lambda_N(X)$. But $\lambda_M(X') > \lambda_N(X)$, so

$$\lambda_M(X') = \lambda_{M \backslash e}(X') + 1 = \lambda_N(X) + 1. \tag{13.1}$$

By Corollary 8.2.6(ii), the second-last equality implies that $e \in \mathrm{cl}^*_M(X')$. Thus

$$Y' \cap D = (Y' - Y) \cap D \subseteq \mathrm{cl}^*_M(X').$$

By duality, if $e \in (Y' - Y) \cap C$, we may assume that $\lambda_{M/e}(X') = \lambda_N(X)$, so

$$Y' \cap C \subseteq \mathrm{cl}_M(X').$$

By Lemma 8.2.4(ii), as $Y' \cap C$ is independent and $Y' \cap D$ is coindependent,

$$\begin{aligned}
\lambda_{M \backslash (Y' \cap D)/(Y' \cap C)}(X') &= \lambda_M(X') - |Y' \cap D| - |Y' \cap C| \\
&= \lambda_M(X') - |Y' - Y|. \tag{13.2}
\end{aligned}$$

Since $\lambda_{M \backslash e}(X') = \lambda_N(X)$ when $e \in Y' \cap D$, while $\lambda_{M/e}(X') = \lambda_N(X)$ when $e \in X' \cap D$, we have, by Corollary 8.2.5, that

$$\lambda_N(X) = \lambda_{M \backslash (Y' \cap D)/(Y' \cap C)}(X'). \tag{13.3}$$

Combining (13.2) with (13.3) and using (13.1), we get

$$\lambda_N(X) = \lambda_M(X') - |Y' - Y| = \lambda_N(X) + 1 - |Y' - Y|.$$

Thus $|Y' - Y| = 1$; a contradiction. $\qquad\square$

Proof of Theorem 13.3.5 Suppose that M does not bridge the k-separation (X, Y) in N. Then there is a k-separation (X', Y') in M with $X \subseteq X'$ and $Y \subseteq Y'$. Thus, in $G(X, Y)$, there is no arc from v_X to a vertex of $Y' - Y$, no arc from a vertex of $X' - X$ to a vertex of $Y' - Y$, and no arc from a vertex of $X' - X$ to v_Y. Thus there is no directed path from v_X to a vertex not in $v_X \cup (X' - X)$. Hence, by Lemma 13.3.4, there is no blocking sequence.

Conversely, suppose there is no blocking sequence for (X, Y) in M. Then there is no directed (v_X, v_Y)-path in $G(X, Y)$. Let X_1 be the set of vertices u of $G(X, Y)$ for which there is a directed (v_X, u)-path. Let Y_1 be those vertices of $G(X, Y)$ not in X_1. Then $v_X \in X_1$ and $v_Y \in Y_1$, and there is no arc from a vertex in X_1 to a vertex in Y_1. Let $X' = (X_1 - v_X) \cup X$ and $Y' = (Y_1 - v_Y) \cup Y$. Then the definition of $G(X, Y)$ implies that $(X \cup x, Y \cup y)$ is a k-separation of $M[D, C; \{x, y\}]$ for all x in X' and all y in Y'. Therefore, by Lemma 13.3.6, $\lambda_M(X') \leq \lambda_N(X)$. Thus, by Corollary 8.2.5, $\lambda_M(X') = \lambda_N(X)$, so (X', Y') is a k-separation of M, and M does not bridge (X, Y). $\square$

Recall that, for a matroid M, we are assuming that D and C are disjoint subsets of $E(M)$ with D coindependent and C independent, that $N = M \backslash D / C$, and that (X, Y) is an exact k-separation of N. To establish some basic properties of blocking sequences, we shall use the following consequence of Lemma 13.3.6.

Corollary 13.3.7 *Let X', Y', and Z be disjoint subsets of $E(M)$. Suppose that $\lambda_{M[D,C;X' \cup Y']}(X') = \lambda_{M[D,C;X' \cup Y' \cup z]}(X')$ for all z in Z. Then*

$$\lambda_{M[D,C;X' \cup Y' \cup Z]}(X') = \lambda_{M[D,C;X' \cup Y']}(X').$$

Proof Suppose $\lambda_{M[D,C;X' \cup Y' \cup Z]}(X') > \lambda_{M[D,C;X' \cup Y']}(X')$. By Lemma 13.3.6, $\lambda_{M[D,C;X' \cup Y' \cup \{x,y\}]}(X' \cup x) > \lambda_{M[D,C;X' \cup Y']}(X')$ for some x in X' and some y in $Y' \cup Z$. Thus $\lambda_{M[D,C;X' \cup Y' \cup y]}(X') > \lambda_{M[D,C;X' \cup Y']}(X')$. Hence $y \notin Y'$, so $y \in Z$ and we contradict the corollary's hypothesis. $\square$

For the next two lemmas, we shall assume that $(v_1, v_2, \ldots, v_t)$ is a blocking sequence, with respect to D and C, of the k-separation (X, Y) of N. Let $X_0 = X$ and $Y_{t+1} = Y$. For each i in $\{1, 2, \ldots, t\}$, let $X_i = X \cup \{v_1, v_2, \ldots, v_i\}$ and $Y_i = Y \cup \{v_i, v_{i+1}, \ldots v_t\}$.

Lemma 13.3.8 *Let i and j be integers with $0 \leq i < j - 1 \leq t$. Then (X_i, Y_j) is an exact k-separation in $M[D, C; X_i \cup Y_j]$ and $(v_{i+1}, v_{i+2}, \ldots, v_{j-1})$ is a blocking sequence for this k-separation.*

Proof For convenience, arbitrarily choose v_0 in X and v_{t+1} in Y. Suppose $g \in \{0, 1, \ldots, i\}$. By the definition of a blocking sequence, $(X \cup v_g, Y \cup v_h)$ is a k-separation of $M[D, C; \{v_g, v_h\}]$ for all h in $\{j, j + 1, \ldots, t + 1\}$. Thus, by Corollary 13.3.7, $(X \cup v_g, Y_j)$ is a k-separation of $M[D, C; v_g \cup Y_j]$. Since this holds for all g in $\{0, 1 \ldots, i\}$, we deduce, by Corollary 13.3.7 again, that (X_i, Y_j) is a k-separation in $M[D, C; X_i \cup Y_j]$.

It remains to show that $(v_{i+1}, v_{i+2}, \ldots, v_{j-1})$ is a blocking sequence for (X_i, Y_j). If $i + 1 \leq h \leq j - 2$, then $(X \cup v_h, Y \cup v_{h+1})$ is not a k-separation of

$M[D, C; \{v_h, v_{h+1}\}]$ since $(v_1, v_2, \ldots, v_t)$ is a blocking sequence for (X, Y). Thus, by Corollary 8.2.5, $(X_i \cup v_h, Y_j \cup v_{h+1})$ is not a k-separation of $M[D, C; X_i \cup Y_j \cup \{v_h, v_{h+1}\}]$. Similarly, $(X_i \cup v_{j-1}, Y_j)$ and $(X_i, Y_j \cup v_{i+1})$ are not k-separations of $M[D, C; X_i \cup Y_j \cup v_{j-1}]$ and $M[D, C; X_i \cup Y_j \cup v_{i+1}]$, respectively. We deduce that $v_{X_i}, v_{i+1}, v_{i+2}, \ldots, v_{j-1}, v_{Y_j}$ is a directed path in the directed graph associated with the k-separation (X_i, Y_j) of $M[D, C; X_i \cup Y_j]$. If it is not a minimal (v_{X_i}, v_{Y_j})-path in this directed graph, then $(X_i \cup v_h, Y_j \cup v_{h+s})$ is not a k-separation of $M[D, C; X_i \cup Y_j \cup \{v_h, v_{h+s}\}]$ for some h and s with $i \leq h \leq h + 2 \leq h + s \leq j$. As $X_i \cup v_h \subseteq X_h$ and $Y_j \cup v_{h+s} \subseteq Y_{h+s}$, we deduce by Corollary 8.2.5, that (X_h, Y_{h+s}) is not a k-separation of $M[D, C; X_h \cup Y_{h+s}]$. This contradicts what we proved in the last paragraph. □

Lemma 13.3.9 *The sequence $(v_1, v_2, \ldots, v_t)$ contains no two consecutive elements of C and no two consecutive elements of D.*

Proof Assume the contrary. Then, by Lemma 13.3.8 and duality, we may assume that $t = 2$ and that $v_1, v_2 \in C$. Let $M' = M\backslash D/(C - \{v_1, v_2\})$. Then $\lambda_{M'/\{v_1, v_2\}}(X) = \lambda_{M'/v_1}(X)$ as $(X, Y \cup v_2)$ is a k-separation of M'/v_1. Thus, by Corollary 8.2.6(iii), $v_2 \notin \mathrm{cl}_{M'/v_1}(X)$. Hence $v_2 \notin \mathrm{cl}_{M'}(X \cup v_1)$. On the other hand, $(X \cup v_1, Y \cup v_2)$ is not a k-separation of M', so $\lambda_{M'/v_2}(X \cup v_1) < \lambda_{M'}(X \cup v_1)$. Thus, by Corollary 8.2.6(iv), $v_2 \in \mathrm{cl}_{M'}(X \cup v_1)$; a contradiction. □

When we apply blocking sequences in the proof of Theorem 13.1.3, we shall be particularly interested in minor-minimal bridges. The next two results derive properties of such bridges.

Theorem 13.3.10 *Let N be a matroid with an exact k-separation (X, Y) and let M be a matroid that is a minor-minimal bridge for (X, Y). Then*

(i)   *there is a unique partition (D, C) of $E(M) - E(N)$ such that $N = M\backslash D/C$;*

(ii) $\lambda_M(X) = \lambda_M(Y) = k$ *and there is a unique ordering $(v_1, v_2, \ldots, v_t)$ of $E(M) - E(N)$ such that $(X \cup \{v_1, v_2, \ldots, v_i\}, Y \cup \{v_{i+1}, v_{i+2}, \ldots, v_t\})$ is a $(k + 1)$-separation of M for all i in $\{0, 1, \ldots, t\}$.*

Moreover, C is an independent set and D is a coindependent set of M; the sequence $(v_1, v_2, \ldots, v_t)$ is the unique blocking sequence for (X, Y) in M; and, in this sequence, no two consecutive members are in C, and no two consecutive members are in D.

Observe that this theorem means that if M is a minor-minimal bridge for an exact k-separation (X, Y) of a minor N of M, then, when we refer to a blocking sequence for (X, Y) in M, we do not need to indicate the coindependent set D and independent set C with respect to which this blocking sequence is defined. Indeed, there is exactly one blocking sequence for (X, Y) in M.

Proof of Theorem 13.3.10 We have $\kappa_N(X, Y) = k - 1$. Moreover, as M is a minor-minimal bridge for the k-separation (X, Y) of N, we have $\kappa_M(X, Y) \geq k$. Choose e in $E(M) - E(N)$. By Tutte's Linking Theorem (8.5.2) or Exercise 2 of Section 8.5, either $\kappa_{M\backslash e}(X, Y) \geq k$ or $\kappa_{M/e}(X, Y) \geq k$. Certainly N is a minor of $M\backslash e$ or M/e. But, since M is a minor-minimal bridge for (X, Y), exactly one of $M\backslash e$ and M/e has N as a minor. Hence there is a unique partition (D, C) of $E(M) - E(N)$ such that $N = M\backslash D/C$. By Lemma 3.3.2, $E(M) - E(N)$ has a partition (I, I^*) such that $N = M\backslash I^*/I$, where I is independent and I^* is coindependent. Thus $I^* = D$ and $I = C$, so D is coindependent and C is independent.

By Theorem 13.3.5, since M bridges the k-separation (X, Y) in N, there is a blocking sequence $(v_1, v_2, \ldots, v_t)$ for (X, Y) in M. Let $V = \{v_1, v_2, \ldots, v_t\}$ and let $M' = M\backslash(D - V)/(C - V) = M[D, C; V]$. Now if $Z \subseteq V$, then

$$
\begin{aligned}
M'[D \cap V, C \cap V; Z] &= M'\backslash((D \cap V) - Z)/((C \cap V) - Z) \\
&= M\backslash(D - V)/(C - V)\backslash((D \cap V) - Z)/((C \cap V) - Z) \\
&= M\backslash(D - Z)/(C - Z) = M[D, C; Z].
\end{aligned}
$$

The conditions that ensure that $(v_1, v_2, \ldots, v_t)$ is a blocking sequence for (X, Y) in M with respect to D and C also ensure that $(v_1, v_2, \ldots, v_t)$ is a blocking sequence for (X, Y) in M' with respect to $D \cap V$ and $C \cap V$. Thus, by Theorem 13.3.5, M' bridges the k-separation (X, Y) in N. The minimality of M implies that $M' = M$, that is, $E(M) = E(N) \cup V$. Using the notation introduced just before Lemma 13.3.8, we have, by that lemma, that (X_i, Y_{i+2}) is an exact k-separation of $M[D, C; X_i \cup Y_{i+2}]$ for all i in $\{0, 1, \ldots, t - 1\}$. The last matroid is obtained from M by deleting or contracting v_{i+1}. Since (X, Y) does not induce a k-separation of M, both (X_i, Y_{i+1}) and (X_{i+1}, Y_{i+2}) are exact $(k + 1)$-separations of M. Taking $i = 0$ in the first of these and $i = t - 1$ in the second, we get $\lambda_M(X) = k = \lambda_M(Y)$. Lemma 13.3.9 implies that the members of $(v_1, v_2, \ldots, v_t)$ alternate between D and C.

It remains to prove the uniqueness of the ordering $(v_1, v_2, \ldots, v_t)$. It suffices to show that if (X', Y') is a $(k + 1)$-separation of M with $X \subseteq X'$ and $Y \subseteq Y'$, then $X' = X_j$ for some j in $\{0, 1, \ldots, t\}$. Suppose not. Then, for some i in $\{1, 2, \ldots, t - 1\}$, we have v_i in Y' and v_{i+1} in X'. Now both $X' \cap X_{i+1}$ and $X' \cup X_{i+1}$ contain X and avoid Y. Therefore, as λ_M is submodular,

$$
\begin{aligned}
k + k = \lambda_M(X') + \lambda_M(X_{i+1}) &\geq \lambda_M(X' \cap X_{i+1}) + \lambda_M(X' \cup X_{i+1}) \\
&\geq \kappa_M(X, Y) + \kappa_M(X, Y) \\
&\geq k + k.
\end{aligned}
$$

Thus $\lambda_M(X' \cap X_{i+1}) = k$. Hence if we relabel $X' \cap X_{i+1}$ as X' and $E(M) - (X' \cap X_{i+1})$ as Y', we still have (X', Y') as a $(k + 1)$-separation of M with v_i in Y' and v_{i+1} in X'. Moreover, $Y' \supseteq Y_{i+2}$. By arguing as above, we get $\lambda_M(Y' \cap Y_i) = k$. Now, replacing Y' and X' by $Y' \cap Y_i$ and $E(M) - (Y' \cap Y_i)$, respectively, we maintain a $(k + 1)$-separation (X', Y') of M with v_i in Y' and v_{i+1} in X'. In addition, we have $Y' \supseteq Y_{i+2}$ and $X' \supseteq X_{i-1}$. Thus $Y' = Y_{i+2} \cup v_i$ and $X' = X_{i-1} \cup v_{i+1}$, that is, $(X_{i-1} \cup v_{i+1}, Y_{i+2} \cup v_i)$ is a $(k + 1)$-separation of M. By duality, we may assume that $v_{i+1} \in C$. Hence $v_i \in D$. By Lemma 13.3.8,

(X_{i-1}, Y_{i+2}) is an exact k-separation of $M\backslash v_i/v_{i+1}$ and (v_i, v_{i+1}) is a blocking sequence for this k-separation. Thus $(X_{i-1} \cup v_{i+1}, Y_{i+2})$ is not a k-separation of $M\backslash v_i$. But $(X_{i-1} \cup v_{i+1}, Y_{i+2} \cup v_i)$ is a $(k+1)$-separation of M. Hence

$$\lambda_{M\backslash v_i}(X_{i-1} \cup v_{i+1}) = k = \lambda_M(X_{i-1} \cup v_{i+1}).$$

Thus, by Corollary 8.2.6(i), $v_i \in \mathrm{cl}_M(Y_{i+2})$. Therefore, as (X_{i-1}, Y_{i+2}) is a k-separation of $M\backslash v_i/v_{i+1}$, it follows that $(X_{i-1}, Y_{i+2} \cup v_i)$ is a k-separation of M/v_{i+1}. This contradiction to the fact that (v_i, v_{i+1}) is a blocking sequence for (X_{i-1}, Y_{i+2}) completes the proof of the theorem. □

Theorem 13.3.11 *Let $\mathcal{M}$ be a class of matroids that is closed under isomorphism and taking minors. Let N be a 3-connected member of $\mathcal{M}$ and (X, Y) be an exact k-separation in N such that Y is closed and coclosed in N. Let M be a smallest minor-minimal bridge for (X, Y) that is in $\mathcal{M}$. Let $N = M\backslash D/C$ and let $(v_1, v_2, \ldots, v_t)$ be an ordering of $E(M) - E(N)$ such that $\lambda_M(X \cup \{v_1, v_2, \ldots, v_i\}) = k$ for all i in $\{0, 1, \ldots, t\}$. Then, for all such i, the matroid $M[D, C; \{v_1, v_2, \ldots, v_i\}]$ is 3-connected.*

Proof By Theorem 13.3.10, C and D are independent and coindependent, respectively, in M, and $(v_1, v_2, \ldots, v_t)$ is a blocking sequence for (X, Y). We show first that $M[D, C; \{v_1\}]$ is 3-connected. By duality, we may suppose that

$$v_1 \in D.$$

We shall show that $M[D, C; \{v_1\}]$ is 3-connected and then iterate to obtain the theorem. Assume that $M[D, C; \{v_1\}]$ is not 3-connected. Then, by Proposition 8.2.7, v_1 is a loop, a coloop, or a parallel element of $M[D, C; \{v_1\}]$. In the first two cases, v_1 can be moved from D to C, contradicting the uniqueness of the partition (D, C), which was established in Theorem 13.3.10. Thus v_1 is parallel to some element e of $M[D, C; \{v_1\}]$, so (X, Y) induces a k-separation of $M[D, C; \{v_1\}]$ by adjoining v_1 to whichever of X and Y contains e. Hence $t > 1$. Moreover, since $(X, Y \cup v_1)$ is not a k-separation of $M[D, C; \{v_1\}]$, we have $e \in X$.

Clearly $M[D, C; \{v_1\}]\backslash e \cong N$. Therefore, since M is a smallest minor-minimal bridge for (X, Y), the k-separation $((X - e) \cup v_1, Y)$ of $M[D, C; \{v_1\}]\backslash e$ is not bridged by $M\backslash e$. Hence $\kappa_{M\backslash e}((X - e) \cup v_1, Y) = k - 1$, so, by Lemma 8.5.5,

$$\kappa_{M\backslash e}(X - e, Y) \leq k - 1.$$

Now suppose that $|X| = k$. Then, by Corollary 8.2.2, X is independent or coindependent in N. But Y is closed and coclosed in N, so X contains both a circuit and a cocircuit. This contradiction implies that $|X| > k$.

Since Y is closed in N and $e \in X$, we have $e \notin \mathrm{cl}_N(Y)$. Thus, by Corollary 8.2.6(iii), $(X - e, Y)$ is an exact k-separation of N/e. Next we show that $(X - e, Y \cup v_i)$ is a k-separation of $M[D, C; \{v_i\}]/e$ for all i in $\{1, 2, \ldots, t\}$. This certainly holds when $i = 1$, since v_1 is a loop of $M[D, C; \{v_1\}]/e$. For $i > 1$, since $(v_1, v_2, \ldots, v_t)$ is a blocking sequence for (X, Y), we have that $(X, Y \cup v_i)$ is a k-separation of $M[D, C; \{v_i\}]$. Thus, by Lemmas 8.5.5 and 8.2.4,

$$
\begin{aligned}
k - 1 = \kappa_{N/e}(X - e, Y) &\leq \kappa_{M[D,C;\{v_i\}]/e}(X - e, Y) \\
&\leq \lambda_{M[D,C;\{v_i\}]/e}(Y \cup v_i) \\
&\leq \lambda_{M[D,C;\{v_i\}]}(Y \cup v_i) = k - 1,
\end{aligned}
$$

so $(X - e, Y \cup v_i)$ is indeed a k-separation of $M[D, C; \{v_i\}]/e$.

Suppose that $\lambda_{M/e}(X - e) > \lambda_{N/e}(X - e)$. Then, by Lemma 13.3.6, for some x in $X - e$ and some y in $Y \cup \{v_1, v_2, \ldots, v_t\}$, the partition $(X - e, Y \cup y)$ is not a k-separation of $M[D, C \cup e; \{x, y\}]$; that is, either some $(X - e, Y \cup v_i)$ is not a k-separation of $M[D, C; \{v_i\}]/e$, or $(X - e, Y)$ is not a k-separation of N/e. As neither of the last two possibilities occurs, we deduce that $\lambda_{M/e}(X - e) = \lambda_{N/e}(X - e) = k - 1$. Hence $\kappa_{M/e}(X - e, Y) \leq k - 1$.

As both $\kappa_{M \backslash e}(X - e, Y) \leq k - 1$ and $\kappa_{M/e}(X - e, Y) \leq k - 1$, Tutte's Linking Theorem (8.5.2) implies that $\kappa_M(X - e, Y) \leq k - 1$ (see Exercise 2 of Section 8.5). Then $\lambda_M(Z) \leq k - 1$ for some Z with $X - e \subseteq Z \subseteq E - Y$. But $\kappa_M(X, Y) \geq k$, so $e \notin Z$. Hence $X - e \subseteq Z \subseteq E - (Y \cup e)$, so $\kappa_M(X - e, Y \cup e) \leq k - 1$. Thus, by Lemma 8.5.5, $\lambda_N(X - e) = \kappa_N(X - e, Y \cup e) \leq k - 1$; that is, $\lambda_N(X - e) \leq \lambda_N(X)$. Hence, by Proposition 8.2.14, e is in the closure or the coclosure of Y in N. But Y is closed and coclosed in N. This contradiction completes the proof that $M[D, C; \{v_1\}]$ is 3-connected. We should like to repeat the above argument, so we need to check that the hypotheses are retained by $M[D, C; \{v_1\}]$.

By Lemma 13.3.8, $(X \cup v_1, Y)$ is an exact k-separation of $M[D, C; \{v_1\}]$ having $(v_2, v_3, \ldots, v_t)$ as a blocking sequence. Evidently M is a smallest minor-minimal bridge for the k-separation $(X \cup v_1, Y)$ in $M[D, C; \{v_1\}]$. Since $(X, Y \cup v_1)$ is not an exact k-separation of $M[D, C; \{v_1\}]$, Corollary 8.2.6 implies that v_1 is not in the closure or the coclosure of Y in $M[D, C; \{v_1\}]$. Since Y is closed in $M[D, C; \{v_1\}] \backslash v_1$, we deduce that Y is closed in $M[D, C; \{v_1\}]$. Moreover, as Y is coclosed in $M[D, C; \emptyset]$ and $v_1 \in D$, the set $Y \cup v_1$ is coclosed in $M[D, C; \{v_1\}]$. Hence so is Y. We conclude that Y is closed and coclosed in $M[D, C; \{v_1\}]$, and we can repeat the original argument to get that $M[D, C; \{v_1, v_2\}]$ is 3-connected. Continuing in this way, we get the theorem. $\square$

Although we have now developed all the material about blocking sequences that we will use to prove Seymour's Decomposition Theorem, we close this section with some additional results about minor-minimal bridges. The first of these was proved by Geelen, Hliněný, and Whittle (2005).

Proposition 13.3.12 *If (X, Y) is an exact k-separation of a $GF(q)$-representable matroid N, and M is a minor-minimal $GF(q)$-representable matroid that bridges (X, Y), then*

$$|E(M)| \leq |E(N)| + (2k + 1)(q^{k-1} - 1)(q^{k-1} - q) \ldots (q^{k-1} - q^{k-2}). \qquad \square$$

This result means that a minor-minimal bridge for a k-separation (X, Y) of N exceeds N in size by only a bounded number of elements provided we confine our attention to $GF(q)$-representable matroids. This last restriction is crucial. Without it, as we shall see in the exercises, we cannot obtain such a bound even when $k = 3$ and we restrict attention to matroids representable over some fixed infinite field. However, as the next result shows, when k is 1 or 2, we do have such bounds. The first part of this result was proved by Lemos and Oxley (1998); the second part was proved independently by Lemos and Oxley (2003) and by Geelen, Hliněný, and Whittle (2005).

Proposition 13.3.13 *Let (X, Y) be an exact k-separation of a matroid N, and M be a minor-minimal matroid that bridges (X, Y).*

(i)   *If $k = 1$, then $|E(M)| \leq |E(N)| + 2$.*
(ii) *If $k = 2$, then $|E(M)| \leq |E(N)| + 5$.*   □

Exercises

1. Let M be the matroid in Figure 13.7 and let $M_9 = M \backslash 9$ and $N_9 = M_9/4 \backslash 0$. Show that:
    (a) N_9 has $(\{3, 6, 7, 8\}, \{1, 2, 5\})$ as a 2-separation (X, Y);
    (b) M_9 is a minor-minimal bridge for (X, Y);
    (c) $\{1, 2, 5\}$ is closed and coclosed in N_9;
    (d) $M_9/4 \backslash 5 \cong N_9$ and $(\{3, 6, 7, 8\}, \{1, 2, 0\})$ is a 2-separation of $M_9/4 \backslash 5$ that is bridged by $M_9 \backslash 5$; and
    (e) M_9 is not a smallest minor-minimal bridge for (X, Y).

2. Let (X, Y) be an exact k-separation of a matroid N, and M be a minor-minimal matroid that bridges (X, Y). Show that if $e \in E(M) - E(N)$, then either $\kappa_{M \backslash e}(X, Y) < k$ and N is a minor of M/e; or $\kappa_{M/e}(X, Y) < k$ and N is a minor of $M \backslash e$.

3. Prove Corollary 13.3.7 directly without using Lemma 13.3.6.

4. (Geelen, Hliněný, and Whittle 2005) Let (X, Y) be an exact 2-separation of a minor N of a matroid M. Suppose that $\lambda_M(X) = \lambda_M(Y) = 1$ and N' is a minor of M with $E(N') = X \cup Y$ and $\lambda_{N'}(X) = 1$. Prove that $N' = N$.

5. (a) Prove Proposition 13.3.13(i).
    (b) Let N be a non-empty matroid having k connected components and M be a minor-minimal connected matroid having N as a minor. Show that $|E(M)| - |E(N)| \leq 2k - 2$.
    (c) (Lemos and Oxley 1998) Show that, for all choices of N with $r(N)$ is not in $\{0, |E(N)|\}$, there is a minor-minimal connected matroid M having N as a minor for which $|E(M)| - |E(N)| = 2k - 2$.

6. *(Geelen, Hliněný, and Whittle 2005) Let (X, Y) be an exact k-separation of a matroid N, and M be a minor-minimal matroid that bridges (X, Y). Let $N = M[D, C; \emptyset]$, let $(v_1, v_2, \ldots, v_t)$ be a blocking sequence for (X, Y), and let $M' = M[D, C; \{v_2, v_3, \ldots, v_t\}]$. Prove that if $t \geq 2k + 2$, then there

is an element i of $\{2, 3, \ldots, t-1\}$ such that $\kappa_{M'\backslash v_i}(X, Y) = k - 1$ and $\kappa_{M'/v_i}(X, Y) = k - 1$.

7. (a) Prove Proposition 13.3.13(ii). [Hint: Use Exercises 4 and 6.]
   (b) (Lemos and Oxley 2003) Give an example to show that the bound in Proposition 13.3.13(ii) is sharp.

8. *(Geelen, Hliněný, and Whittle 2005) Let $\mathbb{F}$ be an arbitrary infinite field, t be an integer, and A be the following matrix over $\mathbb{F}$:

	y_1	y_2	x_3	x_4	v_1	v_2	v_3	$\cdots$	v_{t-1}	v_t
x_1	0	0	-1	1	0	0	0	$\cdots$	0	1
x_2	0	0	-1	1	0	0	0	$\cdots$	0	0
y_3	1	1	0	1	1	1	1	$\cdots$	1	1
y_4	1	1	1	0	0	0	0	$\cdots$	0	0
u_1	1	0	1	0	α_1	0	0	$\cdots$	0	0
u_2	0	0	1	0	α_2	α_3	0	$\cdots$	0	0
u_3	0	0	1	0	0	α_4	α_5	$\cdots$	0	0
$\vdots$	$\vdots$	$\vdots$	$\vdots$	$\vdots$	$\vdots$	$\ddots$	$\ddots$	$\ddots$	$\vdots$	$\vdots$
$\vdots$	$\vdots$	$\vdots$	$\vdots$	$\vdots$	$\vdots$	$\ddots$	$\ddots$	$\ddots$	$\ddots$	$\vdots$
u_t	0	0	1	0	0	0	0	$\cdots$	α_{2t-2}	α_{2t-1}

Let $U = \{u_1, u_2, \ldots, u_t\}$ and $V = \{v_1, v_2, \ldots, v_t\}$, and let D be the submatrix of A with rows indexed by $U \cup y_3$ and columns indexed by $V \cup x_3$. Choose $\alpha_1, \alpha_2, \ldots, \alpha_{2t-1}$ in $\mathbb{F}$ so that each subdeterminant of D is non-zero unless it is identically zero as a polynomial in $\alpha_1, \alpha_2, \ldots, \alpha_{2t-1}$. Let M be the matroid that is represented over $\mathbb{F}$ by $[I_{t+4}|A]$, let $X = \{x_1, x_2, x_3, x_4\}$, let $Y = \{y_1, y_2, y_3, y_4\}$, and let $N = M\backslash V/U$. Show that:

(a) (X, Y) is an exact 3-separation in N;
(b) $(u_1, v_1, u_2, v_2, \ldots, u_t, v_t)$ is a blocking sequence for (X, Y) with respect to V and U;
(c) M is a minor-minimal bridge for (X, Y);
(d) for any integer n, there are $\mathbb{F}$-representable matroids N' and M' such that N' has an exact 3-separation (X', Y') and M' is a minor-minimal bridge for (X', Y') with $|E(M'| \geq |E(N')| + n$.

13.4 R_{12}-minors

In this section, we use the results for blocking sequences from the last section to prove Theorem 13.1.3, namely that every regular matroid with an R_{12}-minor has a 3-separation in which each side has at least six elements. We then complete the proof of Seymour's Decomposition Theorem and show how this theorem leads to a polynomial-time algorithm to test whether a real matrix is totally unimodular.

The first two lemmas investigate the regular 1- and 2-element extensions of $M(H_{11})$, where H_{11} is the graph shown in Figure 13.1.

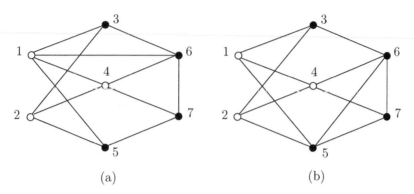

Fig. 13.9. The possibilities for (H_{12}, γ_1).

Lemma 13.4.1 *Let* (H_{11}, γ) *be a graft with* $|\gamma|$ *even. If* $M(H_{11}, \gamma)$ *is regular, then either*

(i) $|\gamma| \leq 2$ *and* $M(H_{11}, \gamma)$ *is graphic; or*
(ii) $\gamma = \{3, 5, 6, 7\}$ *and* $M(H_{11}, \gamma) \cong R_{12}$.

Proof Clearly if $|\gamma| \leq 2$, then $M(H_{11}, \gamma)$ is graphic. Hence we may assume that $|\gamma| \geq 4$. By symmetry, we may also assume that $|\gamma \cap \{3, 6\}| \leq |\gamma \cap \{5, 7\}|$. Let $(H_{11}, \gamma)/\{3, 6\} = (H', \gamma')$. Then $H' = H_{11}/\{3, 6\}$, so $H' \backslash \{4, 6\} \cong K_{3,3}$. As $M(H' \backslash \{4, 6\}, \gamma')$ is regular, it follows by Lemma 12.2.9 that either $|\gamma'| = 6$ and $M(H' \backslash \{4, 6\}, \gamma') \cong R_{10}$; or $|\gamma'| \leq 2$. In the first case, since $M(H', \gamma')$ is clearly 3-connected, we have a contradiction to Proposition 12.2.8. Thus $|\gamma'| \leq 2$. As $|\gamma| \geq 4$, we deduce that $|\gamma| = 4$, and both 3 and 6 are coloured. As $|\gamma \cap \{3, 6\}| \leq |\gamma \cap \{5, 7\}|$, both 5 and 7 are also coloured, so $\gamma = \{3, 5, 6, 7\}$. Hence, by Lemma 13.1.4, $M(H_{11}, \gamma) \cong R_{12}$. □

For a graph G and subsets γ_1 and γ_2 of $V(G)$, recall that $M(G, \gamma_1, \gamma_2)$ is the matroid represented over $GF(2)$ by the matrix obtained by adjoining the incidence vectors of γ_1 and γ_2 to the mod-2 vertex–edge incidence matrix of G.

Lemma 13.4.2 *Let* $M(H_{11}, \gamma_1, \gamma_2)$ *be a regular 3-connected matroid with* $\gamma_1 = \{3, 5, 6, 7\}$ *and* $|\gamma_2|$ *even. Then* $|\gamma_2| = 2$ *and either* $\gamma_2 \subseteq \{3, 4, 5\}$ *or* $\gamma_2 = \{1, 2\}$

Proof As $M(H_{11}, \gamma_1, \gamma_2)$ is simple, Lemma 13.4.1 implies that $|\gamma_2| = 2$ and the two vertices in γ_2 are not adjacent in H_{11}. Suppose that $\gamma_2 \not\subseteq \{3, 4, 5\}$ and $\gamma_2 \neq \{1, 2\}$. Then, by symmetry, we may assume γ_2 is $\{1, 6\}$ or $\{5, 6\}$. Thus $M(H_{11}, \gamma_1, \gamma_2)$ is $M(H_{12}, \gamma_1)$ where (H_{12}, γ_1) is one of the grafts in Figure 13.9. In each case, contracting the edges $\{2, 3\}$, $\{2, 4\}$, and $\{1, 5\}$ and simplifying, we get the graft $(K_4, V(K_4))$ corresponding to F_7; a contradiction. □

Recall that we are identifying R_{12} with $M(H_{11}, \{3, 5, 6, 7\})$. We noted at the end of Section 13.1 that R_{12} has a unique 3-separation (X, Y) with $|X| = |Y| = 6$. This 3-separation corresponds, in $M(H_{11}, \gamma)$, to taking X to be the set of edges of H_{11} induced by $\{1, 2, 3, 4, 5\}$.

Theorem 13.1.3 will follow immediately from the next result whose proof occupies most of the rest of this section.

Theorem 13.4.3 *Let (X, Y) be the unique 3-separation in R_{12} with $|X| = |Y| = 6$. If M is a 3-connected regular matroid with R_{12} as a minor, then the 3-separation (X, Y) of R_{12} is induced in M.*

Proof Assume the contrary and let M be a smallest minor-minimal regular matroid that bridges the 3-separation (X, Y) in R_{12}. We retain the labelling from above with R_{12} being identified with the graft $M(H_{11}, \{3, 5, 6, 7\})$. Observe that Y is both closed and coclosed in $M(H_{11}, \{3, 5, 6, 7\})$.

By Theorem 13.3.10,

(i)   there is a unique partition (D, C) of $E(M) - E(R_{12})$ such that $R_{12} = M \backslash D / C$; and

(ii)   there is a unique ordering $(v_1, v_2, \ldots, v_t)$ of $E(M) - E(R_{12})$ such that $(X \cup \{v_1, v_2, \ldots, v_i\}, Y \cup \{v_{i+1}, v_{i+2}, \ldots, v_t\})$ is a 4-separation of M for all i in $\{0, 1, \ldots, t\}$.

Moreover, the elements of $v_1, v_2, \ldots, v_t$ are alternately in D and C.

Lemma 13.4.4 $M[D, C; \{v_1\}] \cong M(H_{11}, \{3, 5, 6, 7\}, \{1, 2\})$ *and* t *is even.*

Proof First observe that, in $M(H_{11}, \{3, 5, 6, 7\})$, both X and Y are closed and coclosed. By Lemma 13.3.2, $(v_t, v_{t-1}, \ldots, v_1)$ is a blocking sequence for (Y, X) in M with respect to D and C. Thus, by Theorem 13.3.11, $M[D, C; \{v_t\}]$ is 3-connected. Suppose $v_t \in D$. Then, by Lemma 13.4.2, $M[D, C; \{v_t\}] = M(H_{11}, \{3, 5, 6, 7\}, \gamma_2)$ where $v_t = e_{\gamma_2}$, and either $\gamma_2 = \{1, 2\}$ or γ_2 is a 2-element subset of $\{3, 4, 5\}$. But both of the possibilities for γ_2 mean that $(X \cup v_t, Y)$ is a 3-separation of $M[D, C; \{v_t\}]$. This contradicts the fact that $(v_1, v_2, \ldots, v_t)$ is a blocking sequence for (X, Y). We conclude that $v_t \in C$.

Now $R_{12}^* \cong R_{12}$ and (X, Y) is the unique 3-separation of R_{12} with $|X| = |Y| = 6$. Moreover, X is the union of the only two triads of $M(H_{11}, \{3, 5, 6, 7\})$, and Y is the union of the only two triangles. Thus the isomorphism between $M(H_{11}, \{3, 5, 6, 7\})$ and its dual must interchange X and Y. Now $(v_1, v_2, \ldots, v_t)$ is a blocking sequence in M for the 3-separation (X, Y) of $M(H_{11}, \{3, 5, 6, 7\})$ with respect to D and C. Thus, by Lemma 13.3.2 and duality, $(v_t, v_{t-1}, \ldots, v_1)$ is a blocking sequence in M^* with respect to C and D for the 3-separation (Y, X) of $(M(H_{11}, \{3, 5, 6, 7\}))^*$. Since the isomorphism between $M(H_{11}, \{3, 5, 6, 7\})$ and its dual interchanges X and Y, while the isomorphism between M and M^* interchanges C and D, we deduce from the previous paragraph that $v_1 \in D$. Hence t is even.

By Theorem 13.3.11 $M[D, C; \{v_1\}]$ is 3-connected. Moreover, $(X, Y \cup v_1)$ is not a 3-separation of $M[D, C; \{v_1\}]$. As $v_1 \in D$, it follows, by Lemma 13.4.2, that $M[D, C; \{v_1\}] \cong M(H_{11}; \{3, 5, 6, 7\}, \{1, 2\})$.   □

Let H be the graph that is obtained by adjoining the edge $\{1, 2\}$ to H_{11} and labelling it v_1 (see Figure 13.10). Then we may identify $M[D, C; \{v_1\}]$ with $M(H, \gamma)$ where $\gamma = \{3, 5, 6, 7\}$.

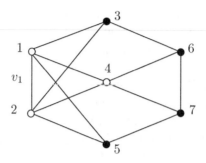

FIG. 13.10. The graft (H, γ).

Let $M' = M[D, C; \{v_1, v_2\}]$. Then $v_2 \in C$ since $v_1 \in D$. Thus $M'/v_2 = M(H, \gamma)$ and, by Theorem 13.3.10, M' is 3-connected. By (ii) in the definition of a blocking sequence, $(X, Y \cup v_2)$ must be a 3-separation of $M'\backslash v_1$. As (X, Y) is a 3-separation of $M'\backslash v_1/v_2$, we have $\lambda_{M'\backslash v_1/v_2}(X) = \lambda_{M'\backslash v_1}(X)$. Hence, by Corollary 8.2.6,

$$v_2 \notin \mathrm{cl}_{M'}(X).$$

Moreover, since $(X \cup v_1, Y \cup v_2)$ is not a 3-separation of M' but $(X, Y \cup v_2)$ is a 3-separation of $M'\backslash v_1$,

$$v_1 \notin \mathrm{cl}_{M'}(X).$$

Since $M'/v_2 = M(H, \gamma)$, we leave it to the reader to show (Exercise 4) that M' has a representation of the form

$$
V(H)
\begin{bmatrix}
\begin{array}{c|c|c}
\begin{matrix} v_2 \\ 1 \\ 0 \\ 0 \\ \vdots \\ 0 \\ 0 \end{matrix} &
\begin{matrix} E(H) \\ \underline{w}^T \\ \\ \\ A \\ \\ \\ \end{matrix} &
\begin{matrix} e_\gamma \\ \sigma \\ \\ \\ \underline{u} \\ \\ \\ \end{matrix}
\end{array}
\end{bmatrix},
$$

where A is the mod-2 vertex–edge incidence matrix of H, and the support of $\underline{u}$ is γ. In particular, the following matrix represents M' where $Y_1 = Y - e_\gamma$:

	v_2	X (x^T)						Y_1 (y^T)					v_1 (τ)	e_γ (σ)
1	0	1	1	1	0	0	0	0	0	0	0	0	1	0
2	0	0	0	0	1	1	1	0	0	0	0	0	1	0
3	0	1	0	0	1	0	0	1	0	0	0	0	0	1
4	0	0	1	0	0	1	0	0	1	1	0	0	0	0
5	0	0	0	1	0	0	1	0	0	0	1	0	0	1
6	0	0	0	0	0	0	0	1	1	0	0	1	0	1
7	0	0	0	0	0	0	0	0	0	1	1	1	0	1

We now manipulate this representation for M'. We shall refer to each row in the representation by the number of the corresponding vertex, taking the unnumbered row as the zeroth row. Evidently adding other rows to the zeroth maintains the form of the representation. Let Σ be the support of $\underline{w}^T$. Then $\Sigma \subseteq E(H)$. For $v \in V(H)$, let $\delta(v)$ be the set of edges of H that meet v. Then adding row v to the zeroth row changes Σ to $\Sigma \triangle \delta(v)$ and changes σ to $\sigma + |\{v\} \cap \gamma|$. By, if necessary, adding some of rows $3, 4$, and 5 to the zeroth row, we can ensure that the first three entries of $\underline{x}^T$ are 0. Then, by adding row 2 to the zeroth row if necessary, we can ensure that, among the last three entries of $\underline{x}^T$, there is either a single one or no ones. In the former case, we can easily check that $v_2 \in \mathrm{cl}_{M'}(X)$; a contradiction. Hence we may assume that $\underline{x} = \underline{0}$. Then, as $v_1 \notin \mathrm{cl}_{M'}(X)$, we deduce that $\tau = 1$.

By, if necessary, adding row 6 or row 7 to the zeroth row, we can also assume that the edges $\{3, 6\}$ and $\{5, 7\}$ are not in Σ. Thus $\underline{y} = (0, a, b, 0, c)^T$ for some a, b, and c in $\{0, 1\}$.

Lemma 13.4.5 *The sets Σ and Y_1 are disjoint.*

Proof From M', we first contract the edges $\{3, 6\}$ and $\{5, 7\}$ of H. Thus, in the representation, we add row 6 to row 3, and add row 7 to row 5. In the resulting matrix, we delete the last two rows and the columns corresponding to $\{3, 6\}$ and $\{5, 7\}$. The values of a, b, and c are unaltered by these operations. Thus, after deleting the column e_γ, we obtain the following matrix:

$$
\begin{array}{ccccccccccc}
v_2 & & & & & & & & & & v_1 \\
\end{array}
$$
$$
\begin{bmatrix}
1 & 0 & 0 & 0 & 0 & 0 & 0 & a & b & c & 1 \\
0 & 1 & 1 & 1 & 0 & 0 & 0 & 0 & 0 & 0 & 1 \\
0 & 0 & 0 & 0 & 1 & 1 & 1 & 0 & 0 & 0 & 1 \\
0 & 1 & 0 & 0 & 1 & 0 & 0 & 1 & 0 & 1 & 0 \\
0 & 0 & 1 & 0 & 0 & 1 & 0 & 1 & 1 & 0 & 0 \\
0 & 0 & 0 & 1 & 0 & 0 & 1 & 0 & 1 & 1 & 0
\end{bmatrix}.
$$

Assume $\Sigma \cap Y_1 \neq \emptyset$. Then, by symmetry, we may assume that $a = 1$. Now, referring to rows just by their position, if we delete the second row and the last row from the last matrix and delete the fourth, seventh, ninth, and tenth columns, we get the following matrix representing a minor of M':

$$
\begin{bmatrix}
1 & 0 & 0 & 0 & 0 & 1 & 1 \\
0 & 0 & 0 & 1 & 1 & 0 & 1 \\
0 & 1 & 0 & 1 & 0 & 1 & 0 \\
0 & 0 & 1 & 0 & 1 & 1 & 0
\end{bmatrix}.
$$

Finally, replace the third row by the sum of the second and third rows to get a matrix that clearly represents F_7^*; a contradiction. $\square$

Since $\underline{x}$ and $\underline{y}$ are both zero vectors and $\tau = 1$, we have $|\Sigma| = 1$. Thus $\sigma = 1$ as the 3-connected matroid M' has no 2-cocircuits. Hence M' is represented by

$$
\begin{array}{c}
\\
1\\
2\\
3\\
4\\
5\\
6\\
7
\end{array}
\left[
\begin{array}{cccccccccccccc}
1 & 0 & 0 & 0 & 0 & 0 & 0 & 0 & 0 & 0 & 0 & 0 & 1 & 1\\
0 & 1 & 1 & 1 & 0 & 0 & 0 & 0 & 0 & 0 & 0 & 0 & 1 & 0\\
0 & 0 & 0 & 0 & 1 & 1 & 1 & 0 & 0 & 0 & 0 & 0 & 1 & 0\\
0 & 1 & 0 & 0 & 1 & 0 & 0 & 1 & 0 & 0 & 0 & 0 & 0 & 1\\
0 & 0 & 1 & 0 & 0 & 1 & 0 & 0 & 1 & 1 & 0 & 0 & 0 & 0\\
0 & 0 & 0 & 1 & 0 & 0 & 1 & 0 & 0 & 0 & 1 & 0 & 0 & 1\\
0 & 0 & 0 & 0 & 0 & 0 & 0 & 1 & 1 & 0 & 0 & 1 & 0 & 1\\
0 & 0 & 0 & 0 & 0 & 0 & 0 & 0 & 0 & 1 & 1 & 1 & 0 & 1
\end{array}
\right].
$$

From M', we now contract the edges $\{3,6\}, \{4,6\}$, and $\{6,7\}$ of H and delete the edge $\{4,7\}$. Thus, we replace the row labelled 6 by the sum of rows 3, 4, 6, and 7, and then delete rows 3, 4, and 7 along with all zero columns. Retaining a single copy of each repeated column, we get the matrix

$$
\left[
\begin{array}{cccccccc}
1 & 0 & 0 & 0 & 0 & 0 & 1 & 1\\
0 & 1 & 1 & 0 & 0 & 0 & 1 & 0\\
0 & 0 & 0 & 1 & 1 & 0 & 1 & 0\\
0 & 0 & 1 & 0 & 1 & 1 & 0 & 1\\
0 & 1 & 0 & 1 & 0 & 1 & 0 & 1
\end{array}
\right].
$$

Replace the fourth row of this matrix by the sum of the third and fourth rows. Then delete the third-last column. Finally, since the sum of the last four rows is zero, we can delete the second row. The resulting matrix represents F_7^*. This contradiction completes the proof of Theorem 13.4.3. □

Proof of Theorem 13.1.3   Let M be a regular matroid with an R_{12}-minor. The theorem follows immediately from Theorem 13.4.3 when M is 3-connected. The straightforward proof when M is not 3-connected is left as an exercise. □

We now complete the proof of Seymour's Decomposition Theorem.

Proof of Theorem 13.1.1   First we remark that, by Propositions 4.2.19 and 7.1.21, if N is the direct sum or 2-sum of two non-empty matroids N_1 and N_2, then each of N_1 and N_2 is isomorphic to a proper minor of N.

Now let M be a minor-minimal regular matroid that cannot be constructed by direct sums and 2-sums from graphic matroids, cographic matroids, and copies of R_{10}. Then M is 3-connected and is neither graphic nor cographic. Thus, by Theorem 13.1.2, M has a minor isomorphic to R_{10} or R_{12}. In the first case, by Proposition 12.2.8, $M \cong R_{10}$; a contradiction. In the second case, by Theorem 13.4.3, M has an exact 3-separation (X_1, X_2) in which both $|X_1|$ and $|X_2|$ exceed five. Thus, by Propositions 9.3.5 and 9.3.4, M is a 3-sum of two of its proper minors, and the theorem follows without difficulty. □

The next result extracts a little bit more information about 3-connected regular matroids from the results above. We leave the proof as an exercise.

Corollary 13.4.6 *Let M be a 3-connected regular matroid. Then*

(i) *M is graphic; or*

(ii) *M is cographic; or*

(iii) $M \cong R_{10}$; *or*

(iv) *there are regular matroids M_1 and M_2 such that $E(M_1) \cap E(M_2) = T$ where T is a triangle of both M_1 and M_2, and $M = M_1 \oplus_3 M_2$; and, for each i in $\{1, 2\}$,*

    (a) *M_i is internally 3-connected and every 2-element 2-separating set of it meets T;*

    (b) *M_i is isomorphic to a minor of M; and*

    (c) *$|E(M_i) - \text{cl}_{M_i}(T)| \geq 6$ and $|E(\text{si}(M_i))| \geq 9$.* □

Theorem 13.1.1 has numerous interesting and important applications, some of which appear in a series of papers by Seymour (1980c, 1981c, 1981d, 1985b, 1986b). We shall conclude this section by looking at one such application that relates to problems in combinatorial optimization. We have already commented on the important role played by totally unimodular matrices in linear programming. In view of this, one of the most significant applications of Theorem 13.1.1 is an algorithm that tests in polynomial time whether a given real matrix is totally unimodular. Since we have put very little emphasis here on the well-developed subject of matroid algorithms, we shall be content to follow Bixby (1981) and Welsh (1982) in sketching the main ideas of this algorithm. A thorough description and justification of the algorithm has been given by Schrijver (1986). The reader who is interested in a detailed general account of the algorithmic aspects of matroid theory is referred to the books by Lawler (1976) and Recski (1989) and the survey papers of Bixby (1981) and Faigle (1987).

The main parts of the algorithm for testing whether a real matrix is totally unimodular are polynomial-time subroutines that will

(i) test whether the vector matroid M of a $(0, \pm 1)$-matrix is (a) graphic, (b) cographic, and (c) isomorphic to R_{10}; and

(ii) find 1-, 2-, and 3-separations in such a matroid.

Evidently one can easily check if $M \cong R_{10}$. Moreover, by Lemma 9.4.23, there is a polynomial-time algorithm to test whether the vector matroid M is graphic. Applying such an algorithm to M^* will test whether M is cographic. Finally, Cunningham and Edmonds (in Cunningham 1973) have given a polynomial-time algorithm for finding k-separations in a matroid for any fixed positive integer k.

We conclude this discussion with some brief remarks on Cunningham and Edmonds's algorithm. It relies on the next result which relates the problem of finding k-separations to a matroid intersection problem. As noted in Section 11.3, problems of the latter type can be solved in polynomial time using the matroid intersection algorithm.

Proposition 13.4.7 *Let M be a matroid and k be a positive integer. If X_1 and X_2 are disjoint subsets of $E(M)$ each having at least k elements, then M has a k-separation (Y_1, Y_2) with $X_1 \subseteq Y_1$ and $X_2 \subseteq Y_2$ if and only if $M/X_1 \backslash X_2$*

and $M/X_2\backslash X_1$ do not have a common t-element independent set where $t = r(M) + k - r(X_1) - r(X_2)$.

Proof Let $M_1 = M/X_1\backslash X_2$ and $M_2 = M/X_2\backslash X_1$, and let r_1 and r_2 be their rank functions. For $X \subseteq E(M) - (X_1 \cup X_2)$ and i in $\{1, 2\}$, we have $r_i(X) = r(X \cup X_i) - r(X_i)$. By Theorem 11.3.15, M_1 and M_2 do not have a common t-element independent set if and only if, for some $T \subseteq E(M) - (X_1 \cup X_2)$,

$$r_1(T) + r_2(E(M) - (X_1 \cup X_2) - T) \le t - 1,$$

that is,

$$r(T \cup X_1) + r(E(M) - (T \cup X_1)) \le t - 1 + r(X_1) + r(X_2).$$

But, as $t = r(M) + k - r(X_1) - r(X_2)$, the last inequality is equivalent to the inequality

$$r(T \cup X_1) + r(E(M) - (T \cup X_1)) \le r(M) + k - 1. \tag{13.1}$$

Now $T \cup X_1 \supseteq X_1$ and $E(M) - (T \cup X_1) \supseteq X_2$. Hence both $T \cup X_1$ and $E(M) - (T \cup X_1)$ have at least k elements. Therefore (13.1) holds if and only if $(T \cup X_1, E(M) - (T \cup X_1))$ is a k-separation of M. The proposition follows without difficulty. □

The matroid intersection algorithm finds, in polynomial time, not only a maximum-sized common independent set I of two matroids M_1 and M_2 on the same set E, but also a subset X of E that minimizes $r_1(X) + r_2(E - X)$, where r_i is the rank function of M_i. By Corollary 11.3.16, each of I and X verifies that the other has the specified property. For fixed k, by applying this algorithm to all pairs $M/X_1\backslash X_2$ and $M/X_2\backslash X_1$ for which X_1 and X_2 are disjoint k-element subsets of $E(M)$, we get a polynomial-time algorithm for finding k-separations. Combining this algorithm with those discussed earlier gives the desired polynomial-time algorithm for testing whether a real matrix is totally unimodular.

Exercises

1. Prove that a matroid is regular having no minor isomorphic to R_{10} if and only if it can be obtained from graphic matroids and cographic matroids by a sequence of direct sums, 2-sums, and 3-sums.
2. Complete the proof of Theorem 13.1.3.
3. Show that a 4-connected regular matroid is either graphic or cographic.
4. Let e be a non-loop element of a binary matroid M and X be a binary representation of M/e. Prove that there is a vector $\underline{x}$ such that M has $\begin{bmatrix} 1 & \underline{x}^T \\ \underline{0} & X \end{bmatrix}$ as a representation, where the first column is labelled by e.
5. Prove Corollary 13.4.6.

14

RESEARCH IN REPRESENTABILITY AND STRUCTURE

Ever since matroids were introduced by Whitney (1935), representable matroids have received considerable research attention. Moreover, there has been much recent activity in this area, particularly relating to matroids representable over finite fields. One focus of this work has been Rota's Conjecture (6.5.11) that, for all prime powers q, there are only finitely many excluded minors for the class of $GF(q)$-representable matroids (Rota 1971). The Graph-Minors Project of Robertson and Seymour has motivated a second focus for this work. Their 2004 result that, in every infinite set of graphs, there is one that is isomorphic to a minor of another prompts the question as to whether this theorem generalizes to matroids. As we shall see, it is easy to construct examples to show that the theorem does not extend to all matroids or even to all $\mathbb{R}$-representable matroids. Robertson and Seymour conjectured (see, for example, Geelen, Gerards, and Whittle 2006b, 2007b) that the theorem does generalize to the class of $GF(q)$-representable matroids. In this chapter, we consider recent progress on this Well-Quasi-Ordering Conjecture for matroids and on Rota's Conjecture. We also discuss some related work on matroids representable over finite fields. One constant theme throughout this chapter will be the many contrasts between the classes of $GF(q)$-representable and $\mathbb{R}$-representable matroids. Most of the chapter will only survey results and many proofs will be omitted.

14.1 The Well-Quasi-Ordering Conjecture for Matroids

K. Wagner (see, for example, Robertson and Seymour 1985) conjectured that every minor-closed class of (finite) graphs has a finite list of excluded minors. We call a set of graphs or a set of matroids an *antichain* if no member of the set is isomorphic to a minor of another member of the set. Although the extension of Wagner's conjecture to infinite graphs fails (Thomas 1988), in the twentieth of a series of difficult papers, Robertson and Seymour (2004) proved that the conjecture holds for finite graphs.

Theorem 14.1.1 (Well-Quasi-Ordering Theorem for Graphs) *There is no infinite antichain of finite graphs.* □

For those readers unfamiliar with the Graph-Minors Project, we now explain some terminology. A *quasi-order* is a pair $(\mathcal{X}, \leq)$ where $\mathcal{X}$ is a set, finite or infinite, and $\leq$ is a reflexive and transitive binary relation on $\mathcal{X}$. Thus, for example, if $\mathcal{X}$ is a set of graphs or a set of matroids and, for members Y and Z of $\mathcal{X}$, we define $Y \preccurlyeq Z$ if Y is isomorphic to a minor of Z, then $(\mathcal{X}, \preccurlyeq)$ is a quasi-order. We call $\preccurlyeq$ the *minor order* on $\mathcal{X}$. A *well-quasi-order* is a quasi-order $(\mathcal{X}, \leq)$ such

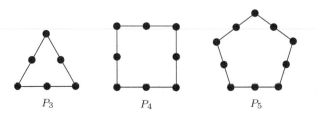

$$P_3 \qquad\qquad P_4 \qquad\qquad P_5$$

FIG. 14.1. The first three matroids in an infinite antichain.

that, in every infinite sequence $X_1, X_2, \ldots$ of elements of $\mathcal{X}$, there are integers i and j with $i < j$ such that $X_i \preccurlyeq X_j$. It is not difficult to check that, when $\mathcal{X}$ is a set of (finite) graphs or a set of matroids, the quasi-order $(\mathcal{X}, \preccurlyeq)$ defined above is a well-quasi-order if and only if $\mathcal{X}$ contains no infinite antichains.

Given the close ties between graphs and matroids, it is natural to ask whether Theorem 14.1.1 extends to matroids. But, even before that graph theorem had been proved, infinite antichains of matroids were known to exist. The following example is due to Brylawski (1986a).

Example 14.1.2 Let C_n be the graph consisting of a single n-vertex cycle. Then no member of $\{C_3, C_4, C_5, \ldots\}$ is isomorphic to a subgraph of another. We can use this set of graphs, or, indeed, any infinite set of simple graphs with the last property, to build an infinite antichain of matroids as follows: embed each C_n in the plane so that no three vertices are collinear, viewing its vertices geometrically as points of a rank-3 matroid and its edges as lines of the matroid. Then add one extra point freely on each of the specified lines to get a rank-3 matroid P_n that consists of a ring of n three-point lines (see Figure 14.1). It is not difficult to check that $\{P_3, P_4, P_5, \ldots\}$ is an infinite antichain of matroids. Moreover, one can easily show that each P_n is representable over every infinite field. □

Despite the existence of such infinite antichains of matroids, in view of Seymour's Decomposition Theorem (13.1.1), one may hope that Theorem 14.1.1 can be extended to regular matroids. This was done, but not published, by Seymour and relies on additional results from the Graph-Minors Project (Robertson and Seymour 2003b, 2009). Thus infinite antichains exist within the class of all matroids, but not within the class of regular matroids. Which other special classes of matroids contain no infinite antichains? For example, what happens for binary matroids? These are, of course, the matroids with no $U_{2,4}$-minor. The next proposition (Kahn, private communication) shows that the class of matroids with no $U_{2,5}$-minor and no $U_{3,5}$-minor does contain an infinite antichain. Recall from Proposition 12.2.20 that, for all $r \geq 3$, the tipless binary r-spike is the vector matroid of $[I_r | J_r - I_r]$ over $GF(2)$ where J_r is the $r \times r$ matrix of all ones.

Proposition 14.1.3 *For all $r \geq 2$, let N_r be the tipless binary spike of rank $2r$ and let M_r be a matroid that is obtained from N_r by relaxing a pair of complementary circuit–hyperplanes. Then no member of $\{M_r : r \geq 2\}$ is isomorphic to a minor of any other. Moreover, no M_r has a $U_{2,5}$- or a $U_{3,5}$-minor.*

To prove this proposition, we shall use the following result.

Lemma 14.1.4 *Let M' be a matroid that is obtained by relaxing a circuit–hyperplane C in a binary matroid M. Then M' is binary if and only if $M \cong U_{1,n} \oplus U_{m-1,m}$ for some positive integers m and n.*

Proof By Proposition 1.5.14, $\mathcal{C}(M') = (\mathcal{C}(M) - \{C\}) \cup \{C \cup e : e \in E(M) - C\}$. Assume that M' is binary. For distinct elements e_1 and e_2 of $E(M) - C$, the set $\{e_1, e_2\}$ is the symmetric difference of two circuits of M' and so, by Theorem 9.1.2, $\{e_1, e_2\}$ is a disjoint union of circuits of M' and hence of M. But C is a hyperplane of M, so neither e_1 nor e_2 is a loop of M. Hence M has $\{e_1, e_2\}$ as a circuit. Thus $E(M) - C$ is a parallel class of M. As $E(M) - C$ is also a circuit of M^*,

$$
\begin{aligned}
\lambda_M(E(M) - C) &= r_M(E(M) - C) + r_M^*(E(M) - C) - |E(M) - C| \\
&= 1 + (|E(M) - C| - 1) - |E(M) - C| = 0.
\end{aligned}
$$

Thus $E(M) - C$ is a component of M, and if $|E(M) - C| = n$, this component is isomorphic to $U_{1,n}$. Since C is a circuit, we deduce that $M \cong U_{1,n} \oplus U_{m-1,m}$ where $|C| = m$. We omit the straightforward proof of the converse. $\square$

Proof of Proposition 14.1.3 For all k, let X_k and Y_k be the complementary circuit–hyperplanes of N_k that are relaxed to produce M_k. Such sets certainly exist since, for example, we can take X_k to be those elements that label columns 2 through $2k + 1$ of the representation $[I_{2k} | J_{2k} - I_{2k}]$ of N_k. First we observe that every single-element deletion or contraction of M_k, and hence M_k itself, is non-binary. To see this, note that, by Proposition 3.3.5, if $e \in E(M_k)$, then $M_k \backslash e$ can be obtained from $N_k \backslash e$ by relaxing a circuit–hyperplane. Thus, by Lemma 14.1.4, $M_k \backslash e$ is non-binary and, by duality, so too is M_k / e.

By Lemma 3.3.2, every proper minor of M_j of rank and corank at least two has the form $M_j \backslash I^* / I$ for some independent set I and coindependent set I^* where both $|I|$ and $|I^*|$ are in $\{1, 2, \ldots, 2j - 2\}$. Using Proposition 3.3.5 again, it is not difficult to show that either $M_j \backslash I^* / I$ is binary, or it has both a single-element deletion and a single-element contraction that are binary. We conclude that $M_j \backslash I^* / I$ cannot be isomorphic to any M_i for $i < j$, or to $U_{2,5}$ or $U_{3,5}$. $\square$

The last result paints a rather gloomy picture. But it is straightforward to show that if $r \geq 3$, then M_r has both the Fano and non-Fano matroids as minors, so it is not representable. The examples we have seen so far still leave open the question of what happens for matroids representable over finite fields. For that important case, the following conjecture was made by Robertson and Seymour though apparently not in print (Geelen, Gerards, and Whittle 2006b, 2007b).

Conjecture 14.1.5 (Well-Quasi-Ordering Conjecture) *For all prime powers q, every infinite set of $GF(q)$-representable matroids contains two members one of which is isomorphic to a minor of the other.* $\square$

The proof of the Well-Quasi-Ordering Theorem for graphs is based on the Graph-Minors Structure Theorem (Robertson and Seymour 2003). In 2008, Geelen, Gerards, and Whittle announced they had extended the structure theorem to binary matroids and, in 2009, they announced they had used this result to prove the Well-Quasi-Ordering Conjecture when $q = 2$. The extensions of the structure theorem to the cases when q is an arbitrary prime and when q is not prime also look promising. In the last case, the structure of minor-closed classes of $GF(q)$-representable matroids is complicated by the need to account for matroids represented over subfields of $GF(q)$. For example, the class of $GF(16)$-representable matroids contains the classes of binary and quaternary matroids.

The Graph-Minors Structure Theorem (Robertson and Seymour 2003) provides a constructive characterization of the class of graphs that do not contain a given graph as a minor. In their pursuit of the Well-Quasi-Ordering Conjecture for $GF(q)$-representable matroids, Geelen, Gerards, and Whittle have taken a broadly similar approach. The details of their theorems become increasingly technical but we shall be able to state at least the initial results in this programme. A useful concept in this discussion is that of branch-width, which we shall define and discuss in the next section.

We began this section with an infinite antichain of $\mathbb{R}$-representable matroids all but the first of which has $U_{3,6}$ as a minor. Geelen (2008) proposed that, by excluding this matroid, one obtains a well-quasi-ordered class.

Conjecture 14.1.6 *The class of $\mathbb{R}$-representable matroids with no $U_{3,6}$-minor has no infinite antichains.* □

Exercises

1. Let M be a matroid. Show that there is an infinite antichain within the class of matroids having no M-minor if and only if there is an infinite antichain of matroids that includes M as a member.

2. Prove that the matroid P_r defined in Example 14.1.2 is representable over every infinite field.

3. Prove that, for the matroid M_r in Proposition 14.1.3, if $r \geq 3$, then M_r has both the Fano and non-Fano matroids as minors, so it is not representable.

4. Prove that $\{PG(2,p) : p \text{ is prime}\}$ is an infinite antichain of matroids.

5. Prove that if $\mathcal{X}$ is a set of (finite) graphs or a set of matroids, the quasi-order $(\mathcal{X}, \preccurlyeq)$ is a well-quasi-order if and only if $\mathcal{X}$ contains no infinite antichains.

6. (Kingan and Oxley 1996) Let $\mathcal{M}$ be a minor-closed class of matroids and let $\mathcal{M}_1$ be the class of matroids M such that $M|H \in \mathcal{M}$ for all hyperplanes H of M. Prove that $\mathcal{M}_1$ is a minor-closed class for which the set of excluded minors is $\{N \oplus U_{1,1} : N \text{ is an excluded minor for } \mathcal{M}\}$.

7. (Oxley, Semple, Vertigan, and Whittle 2002) For a positive integer n, take a geometric representation of $U_{3,4n}$, labelling its points by $b_1, b_2, \ldots, b_{4n}$. Now add points $a_1, a_2,$ and a_3 so that, for all i in $\{1, 2, \ldots, 2n\}$ and all j in $\{2, 3, \ldots, 2n\}$, the following sets of points are collinear: $\{a_1, b_{2i}, b_{2i-1}\}$,

$\{a_2, b_{2i}, b_{2i+1}\}$, and $\{a_3, b_j, b_{4n+2-j}\}$, where $b_{4n+1} = b_1$. Also assume that $\{a_1, a_2, a_3\}$ and $\{a_3, b_1, b_{2n+1}\}$ are collinear. Show that:

(a) The resulting configuration is a rank-3 matroid M_n.
(b) $M_1 \cong F_7$.
(c) M_n has characteristic set $\{2\}$.
(d) *The matroid M_n is an excluded minor for $\mathbb{Q}$-representability.
(e) *For all primes p, there is an infinite antichain of matroids each having characteristic set $\{p\}$.

14.2 Branch-width

The notions of tree-width and branch-width were introduced for graphs by Robertson and Seymour (1990, 1991) as part of the Graph-Minors Project. The second of these has a natural extension to matroids and the first appearance in the literature of the concept of branch-width for matroids seems to have been in the work of one of Robertson's students, Dharmatilake (1994, 1996). To try to convey some intuition for the meaning of branch-width, we observe that a vector matroid has small branch-width if and only if it can be obtained from small matroids by sticking these together in a tree-like structure across subspaces of the underlying vector space. Although the traditional definition of tree-width for graphs involves vertices, Hliněný and Whittle (2006, 2009) have given an alternative definition that extends to matroids. Moreover, generalizing a graph result of Robertson and Seymour (1991), they proved that there is an upper bound on the tree-width of every member of a class of matroids if and only if there is an upper bound on the branch-width of every member of the class (see Exercise 7). Thus branch-width and tree-width are qualitatively equivalent and we shall concentrate on the former because it is easier to work with.

In this section, we shall define branch-width for matroids and consider some of its basic properties. We also discuss the relationship of this concept to its namesake for graphs. The significance of branch-width will be made clear in the next section where we give partial results towards both Rota's Conjecture and the Well-Quasi-Ordering Conjecture.

The vertices of degree 1 of a tree are its *leaves*. A tree is *cubic* (or *ternary*) if every vertex that is not a leaf has degree three. A *branch-decomposition* of a matroid M consists of a cubic tree T with at least $|E(M)|$ leaves together with a labelling of $|E(M)|$ of the leaves by different elements of $E(M)$. Two examples are shown in Figure 14.2. In general, for technical reasons, we allow T to have some unlabelled leaves. For each edge e of such a branch-decomposition T of M, the graph $T \backslash e$ has exactly two components. Thus the set of labelled leaves of T, and hence $E(M)$, is partitioned into two subsets X and $E(M) - X$, say. We say that X and $E(M) - X$ are *displayed* by the edge e. The *width* $w(e)$ *of the edge* e is defined to be $\lambda_M(X)+1$, that is, $w(e) = r(X)+r(E(M)-X)-r(M)+1$. The *width* $w(T)$ *of the tree* T is the maximum of the widths of the edges of T. The *branch-width* $\mathrm{bw}(M)$ of M is the minimum, over all such labelled cubic trees T, of the width of T. Hence $\mathrm{bw}(M) \geq 1$ for all matroids M. This definition disagrees

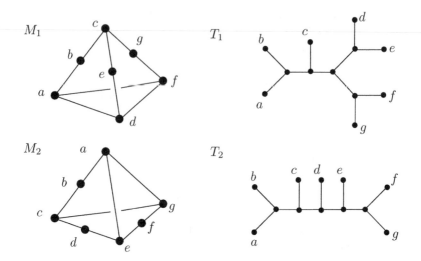

FIG. 14.2. Two matroids and width-two branch-decompositions of them.

with that given by Dharmatilake (1996) in that he defined the branch-width of matroids with at most one element to be zero.

Example 14.2.1 Consider the matroids M_1 and M_2 shown on the left-hand side of Figure 14.2. Then M_1 is the parallel connection of three 3-point lines across a common basepoint, and M_2 is the parallel connection of three 3-point lines across different basepoints. Next to each matroid M_i is a cubic tree T_i whose leaves have been labelled by the elements of the matroid, that is, T_i is a branch-decomposition of M_i. Moreover, one easily checks that $w(T_i) = 2$ for each i, so $\mathrm{bw}(M_i) \leq 2$. In fact, equality holds here since if T is a branch-decomposition of M_i, every edge that meets a labelled leaf of T has width two.

Clearly T_1 is a branch-decomposition of M_2, and T_2 is a branch-decomposition of M_1. However, each of these branch-decompositions has width three.   □

For a matroid M with at least two elements, a branch-decomposition T of M that has unlabelled leaves is quickly turned into one with the same width, but no unlabelled leaves, as follows. Consider the minimal tree induced by the labelled leaves of T. In this tree, replace each maximal path in which all internal vertices have degree two by a single edge. The resulting tree T' is once again cubic and it is easily seen that it is a branch-decomposition for M in which every leaf is labelled. Moreover, a proper non-empty subset of $E(M)$ is displayed by an edge of T' if and only if it is displayed by an edge of T.

Lemma 14.2.2 *Every cubic tree T has an edge e such that each of the two components of $T \backslash e$ contains at least one-third of the leaves of T.*

Proof Assume that T has no such edge. Then T has at least four leaves, otherwise the lemma holds for an arbitrary edge e of T. Now direct each edge f of T

away from the incident vertex that is in the component of $T\backslash f$ containing fewer than one-third of the leaves of T. Each edge of T meeting a leaf is directed away from that leaf. Choose some leaf u of T and let v be the vertex at the end of a longest directed path in T that begins at u. Then v has degree 3 and all three edges incident with v are directed towards v. But each of the three components of $T - v$ contains fewer than a third of the leaves of T; a contradiction. $\square$

The next result (Dharmatilake 1996) summarizes some elementary properties of branch-width.

Proposition 14.2.3 *Let M be a matroid. Then*

(i) $\mathrm{bw}(M^*) = \mathrm{bw}(M)$;

(ii) $\mathrm{bw}(N) \leq \mathrm{bw}(M)$ *for every minor N of M;*

(iii) *for all $e \in E(M)$, both $\mathrm{bw}(M\backslash e)$ and $\mathrm{bw}(M/e)$ are in $\{\mathrm{bw}(M)-1, \mathrm{bw}(M)\}$.*

Proof Part (i) follows immediately from the fact that $\lambda_M = \lambda_{M^*}$. Using (i), we can prove (iii) by showing that $\mathrm{bw}(M\backslash e) \in \{\mathrm{bw}(M) - 1, \mathrm{bw}(M)\}$, and (ii) will then follow from this. Let T be a branch-decomposition of M and let T_e be the branch-decomposition of $M\backslash e$ obtained from T by deleting e as a leaf label. If an edge of T determines the partition (X, Y) of $E(M)$ with e in Y, then, by Corollary 8.2.6, $\lambda_{M\backslash e}(X) \in \{\lambda_M(X) - 1, \lambda_M(X)\}$. Thus

$$w(T_e) \in \{w(T) - 1, w(T)\}. \tag{14.1}$$

As we could have chosen T with $w(T) = \mathrm{bw}(M)$, we have $\mathrm{bw}(M\backslash e) \leq \mathrm{bw}(M)$. On the other hand, if T'_e is a branch-decomposition of $M\backslash e$ with $w(T'_e) = \mathrm{bw}(M\backslash e)$, then we can take an edge x of T'_e subdivide it by inserting a new vertex v and then add another new vertex w that is adjacent only to v. By labelling w by e, we get a branch-decomposition T' of M. Moreover, by (14.1), $w(T'_e) \in \{w(T') - 1, w(T')\}$. But $w(T'_e) = \mathrm{bw}(M\backslash e)$ and $w(T') \geq \mathrm{bw}(M)$. Thus $\mathrm{bw}(M\backslash e) \geq \mathrm{bw}(M) - 1$. We conclude that (iii) holds. $\square$

For a positive integer k, let $\mathcal{B}_k$ be the class of matroids of branch-width at most k. By the last proposition, $\mathcal{B}_k$ is closed under taking duals and minors. Clearly $\mathcal{B}_1 = \{U_{0,m} \oplus U_{n,n} : m, n \geq 0\}$ since $\mathrm{bw}(M) \geq 2$ unless every element of M is a loop or a coloop. The next result is well known (see, for example, Hall, Oxley, Semple, and Whittle 2002).

Lemma 14.2.4 *For all positive integers k, the class $\mathcal{B}_k$ is closed under direct sums and 2-sums.*

Proof Let M_1 and M_2 be members of $\mathcal{B}_k$ on disjoint ground sets. Let T_1 and T_2 be branch-decompositions of M_1 and M_2, each of width at most k. We construct a branch-decomposition for $M_1 \oplus M_2$ as follows. For each i, subdivide an edge of T_i by inserting a new vertex v_i. Add a new edge x joining v_1 and v_2. Let the leaves of the resulting cubic tree inherit their labels from T_1 and T_2. This gives a branch-decomposition T for $M_1 \oplus M_2$ in which x has width 1. Moreover, it is easily checked that $w(T) = \max\{w(T_1), w(T_2)\}$, so $\mathrm{bw}(M_1 \oplus M_2) \leq k$.

Next consider the 2-sum of M_1 and M_2 with respect to the basepoints p_1 and p_2. Then p_i is neither a loop nor a coloop of M_i, so $k \geq 2$. Now identify the vertices of T_1 and T_2 labelled by p_1 and p_2, removing the label from this composite vertex v. Let f and g be the edges incident with v. Contract g to produce a branch-decomposition T' of the 2-sum of M_1 and M_2 in which f has width 2. We leave it to the reader to check that T' has width at most k. □

The following result of Geelen, Gerards, Robertson, and Whittle (2003) proves that, for all k, the set of excluded minors for $\mathcal{B}_k$ is finite.

Proposition 14.2.5 *If M is an excluded minor for the class of matroids of branch-width at most k and $k \geq 2$, then $|E(M)| \leq (6^k - 1)/5$.* □

We specified all the members of $\mathcal{B}_1$ above. It follows that $U_{1,2}$ is the unique excluded minor for $\mathcal{B}_1$. Dharmatilake (1996) characterized the members of $\mathcal{B}_2$ and, from this, the set of excluded minors for $\mathcal{B}_2$ can be determined.

Proposition 14.2.6 *The following statements are equivalent for a matroid M:*
(i) $\operatorname{bw}(M) \leq 2$;
(ii) *every component of M is a series–parallel network;*
(iii) *M has no minor isomorphic to $U_{2,4}$ or $M(K_4)$.*

Proof The equivalence of (ii) and (iii) is immediate from Corollary 12.2.14. In view of Lemma 14.2.4, it suffices to prove the equivalence of (i) and (ii) when M is a connected matroid having at least two elements. Suppose that $\operatorname{bw}(M) \leq 2$. Let N be $U_{2,4}$ or $M(K_4)$. Then N has a branch-decomposition T with exactly $|E(N)|$ leaves such that $\operatorname{bw}(N) = w(T)$. By Lemma 14.2.2, T has an edge e such that both of the sets displayed by e have at least two elements. Since N is 3-connected, it follows that e has width at least three. Thus $\operatorname{bw}(N) \geq 3$. We deduce by Proposition 14.2.3(ii) that M has no minor isomorphic to $U_{2,4}$ or $M(K_4)$. Hence M is a series–parallel network. Thus (i) implies (ii).

Finally suppose that (ii) holds. We argue by induction on $|E(M)|$ to show that $\operatorname{bw}(M) \leq 2$. If $|E(M)| = 2$, then, since M is connected, $M \cong U_{1,2}$ and $\operatorname{bw}(M) = 2$. Now suppose that (i) holds if $|E(M)| < n$ and let $|E(M)| = n \geq 3$. In the construction of the series–parallel network M, we may assume, by duality, that the last step involved adding an element p_2 in parallel to some element p_1. Thus $M \backslash p_2$ is a series–parallel network. Moreover, M is isomorphic to a 2-sum of $M \backslash p_2$ and $U_{1,3}$. By the induction assumption, $\operatorname{bw}(M \backslash p_2) \leq 2$ and one easily checks that $\operatorname{bw}(U_{1,3}) = 2$. Thus, by Lemma 14.2.4, we have $\operatorname{bw}(M) \leq 2$. It follows by induction that (ii) implies (i), so the proposition holds. □

To this point, we have defined branch-width for matroids but not for graphs. The graph definition mimics that for matroids but uses a different definition of the width of an edge e in a branch-decomposition T of a graph G. If e displays the sets X and $E(G) - X$ of edges of G, we take the width of e to be $\eta_G(X)$, the number of vertices common to the subgraphs $G[X]$ and $G[E(G) - X]$ of G induced by X and $E(G) - X$. As before, maximizing this quantity over all the edges of T gives the width of T, and the *branch-width* $\beta(G)$ of the graph G is the minimum value of this width over all such labelled cubic trees T. In order to compare $\beta(G)$ and $\text{bw}(M(G))$, we recall, from Lemma 8.1.7, that

$$\lambda_{M(G)}(X) = \eta_G(X) + \omega(G) - \omega(G[X]) - \omega(G[E(G) - X]) \qquad (14.2)$$

when G has no isolated vertices, where $\omega(H)$ is the number of components of a graph H. Hence if G is connected having at least two edges, then

$$\text{bw}(M(G)) \le \beta(G). \qquad (14.3)$$

The last inequality is strict, for example, when G is the graph that is obtained from K_2 by adding a loop at each vertex. In that case, $\text{bw}(M(G)) = 1$ and $\beta(G) = 2$. When G is disconnected, inequality (14.3) need not hold. For example, if G is the disjoint union of k copies of K_2 for some $k \ge 2$, then $\beta(G) = 0$ but $\text{bw}(M(G)) = 1$. These examples in which $\text{bw}(M(G))$ and $\beta(G)$ differ seem very specialized. Indeed, Geelen, Gerards, and Whittle (private communication, 2000) conjectured that $\text{bw}(M(G)) = \beta(G)$ unless G can be obtained from a forest by adjoining loops and this conjecture was verified independently by Hicks and McMurray (2007) and Mazoit and Thomassé (2007).

Proposition 14.2.7 *If G is a graph having a cycle with at least two edges, then*

$$\text{bw}(M(G)) = \beta(G). \qquad \square$$

Dharmatilake, Chopra, Johnson, and Robertson (in Dharmatilake 1994) showed that the class of graphs of branch-width at most three has four excluded minors (see Figure 14.3). The second graph is H_8, the four-rung cubic Möbius ladder (see also Figure 12.5). The third and fourth graphs are the *cube* and the *octahedron*. The latter is isomorphic to $K_{2,2,2}$; the cube is its geometric dual.

Proposition 14.2.8 *A graph has branch-width at most three if and only if it has no minor isomorphic to K_5, H_8, the cube, or the octahedron.* $\qquad \square$

Proposition 14.2.5 generalizes an earlier result of Hall, Oxley, Semple, and Whittle (2002) that an excluded minor for $\mathcal{B}_3$ has at most fourteen elements. Although the full list of excluded minors for $\mathcal{B}_3$ has not been explicitly determined, Hliněný (2002) was able to use his computer program MACEK to prove that

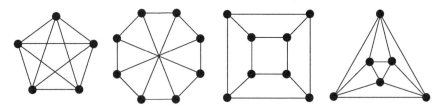

FIG. 14.3. The graphic excluded minors for $\mathcal{B}_3$.

there are exactly ten binary excluded minors for $\mathcal{B}_3$. This verified a conjecture of Dharmatilake (1994). Let N_{11} and N_{23} be the binary matroids that have the following matrices over $GF(2)$ as reduced standard representations:

$$
\begin{array}{c}
\begin{array}{cccccc} 5 & 6 & 7 & 8 & 9 & 10 \end{array} \\
\begin{array}{c} 1 \\ 2 \\ 3 \\ 4 \end{array}
\left[\begin{array}{cccccc}
1 & 0 & 0 & 1 & 1 & 1 \\
0 & 1 & 0 & 1 & 1 & 0 \\
0 & 0 & 1 & 1 & 0 & 1 \\
1 & 1 & 1 & 1 & 0 & 0
\end{array}\right]
\end{array}
\quad \text{and} \quad
\begin{array}{c}
\begin{array}{ccccc} 6 & 7 & 8 & 9 & 10 \end{array} \\
\begin{array}{c} 1 \\ 2 \\ 3 \\ 4 \\ 5 \end{array}
\left[\begin{array}{ccccc}
1 & 1 & 0 & 1 & 1 \\
1 & 1 & 1 & 0 & 1 \\
0 & 1 & 1 & 0 & 1 \\
1 & 0 & 0 & 0 & 1 \\
1 & 1 & 1 & 1 & 0
\end{array}\right]
\end{array}.
$$

Clearly N_{23} is self-dual. A geometric representation of N_{11} is shown in Figure 14.4. Observe that $N_{11}\backslash 5 \cong M^*(K_{3,3})$ and $N_{11}\backslash 8 \cong M(K_5\backslash e)$.

Proposition 14.2.9 *A binary matroid has branch-width at most three if and only if it has no minor isomorphic to* $M(K_5), M^*(K_5), M(H_8), M^*(H_8),$ $M(K_{2,2,2}), M^*(K_{2,2,2}), R_{10}, N_{11}, N_{11}^*,$ *or* N_{23}. $\quad\square$

As we shall see in the next section, restricting attention to matroids of bounded branch-width has proved very helpful in obtaining partial results towards Rota's Conjecture and the Well-Quasi-Ordering Conjecture. To conclude

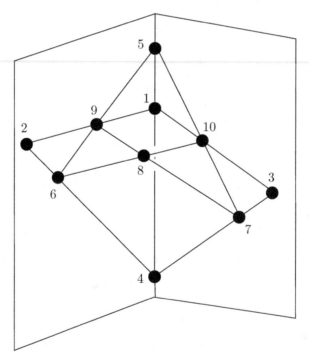

FIG. 14.4. A geometric representation for N_{11}.

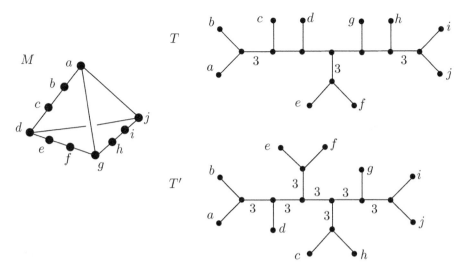

FIG. 14.5. T and T' are linked and unlinked branch-decompositions of M.

this section, we identify a specific type of branch-decomposition that assists in proving well-quasi-ordering theorems for matroids of bounded branch-width.

Example 14.2.10 The matroid M shown in Figure 14.5 is the parallel connection across different basepoints of three 4-point lines. Since M has $U_{2,4}$ as a minor, its branch-width is at least three. In fact, $\mathrm{bw}(M) = 3$, since each of T and T' is easily shown to be a width-three branch-decomposition of M. In each of T and T', the edges of width three have been labelled. All unlabelled edges have width two. The tree T' is a weaker reflection of the structure of M than T. In particular, $\{a, b, d\}$ and $\{g, i, j\}$ are displayed by edges of T'. Moreover, all the edges of T' that display partitions of $E(M)$ with $\{a, b, d\}$ in one part and $\{g, i, j\}$ in the other have width three. However, M has a 2-separation $(\{a, b, c, d\}, \{e, f, g, h, i, j\})$ with $\{a, b, d\}$ on one side and $\{g, i, j\}$ on the other. This 2-separation is not revealed by T'. □

We now formalize the ideas in the last example. Let f and g be distinct edges in a branch-decomposition T of a matroid M. Let F be the subset of $E(M)$ displayed by the component of $T\backslash f$ not containing g, and let G be the subset of $E(M)$ displayed by the component of $T\backslash g$ not containing f. Let P be the shortest path in T containing f and g. Each edge of P displays a partition (X, Y) of $E(M)$ such that $F \subseteq X$ and $G \subseteq Y$, so the widths of the edges of P are upper bounds on $\kappa_M(F, G) + 1$, where we recall that $\kappa_M(F, G) = \min\{\lambda_M(S) : F \subseteq S \subseteq E(M) - G\}$. We call f and g *linked* if the minimum width of an edge of P is $\kappa_M(F, G) + 1$. A branch-decomposition is *linked* if all of its edge pairs are linked. In Example 14.2.10, we showed that T' is not a linked branch-decomposition of M. It is not difficult to check that, by contrast, T is a linked branch-decomposition of M.

Geelen, Gerards, and Whittle (2002) proved that every matroid M has a linked branch-decomposition of width $\mathrm{bw}(M)$. This theorem is an analogue of a theorem of Thomas (1990) on linked tree-decompositions of a graph.

Theorem 14.2.11 *A matroid M of branch-width n has a linked branch-decomposition of width n.* □

The significance of the last theorem is that linked branch-decompositions are a more powerful proof tool than unlinked branch-decompositions because they provide a more faithful reflection of a matroid's structure.

Exercises

1. (Whittle, private communication) Show that if M is a rank-3 matroid, then $\mathrm{bw}(M) \leq 4$, and $\mathrm{bw}(M) \leq 3$ if and only if $E(M)$ can be written as the union of three lines. Deduce that $\mathrm{bw}(F_7) = 3 = \mathrm{bw}(F_7^-)$.

2. Show that the branch-width of a matroid is the maximum of the branch-widths of its components.

3. Find the sets of excluded minors for the classes of graphic matroids of branch-width at most 3, regular matroids of branch-width at most 3, and binary matroids of branch-width at most 3.

4. (Dharmatilake 1996) Prove that if M is an n-connected matroid and $n \geq 3$, then $\mathrm{bw}(M) \geq n$ if and only if $|E(M)| \geq 3n - 5$.

5. Recall that $\mathrm{bw}(U_{r,n}) = 1$ if $n \leq 1$. Show that if $n \geq 2$, then

$$\mathrm{bw}(U_{r,n}) = \begin{cases} r + 1 & \text{when } 3r \leq n; \\ \lceil \frac{n}{3} \rceil + 1 & \text{when } \frac{3r}{2} < n < 3r; \\ n - r + 1 & \text{when } n \leq \frac{3r}{2}. \end{cases}$$

6. (Oporowski 2002) Use Exercise 16 of Section 4.3 to prove that if M is a matroid of positive rank and largest cocircuit size c^*, then $\mathrm{bw}(M) \leq \binom{c^*+1}{2}$.

7. (Hliněný and Whittle 2006) Let M be a matroid, T be a tree, and σ be an arbitrary map from $E(M)$ into $V(T)$. For a vertex v of T, let $T_1, T_2, \ldots, T_d$ be the components of $T - v$, and let $F_i = \sigma^{-1}(V(T_i))$ for each i. The width of the vertex v is $r(M) - \sum_{i=1}^{d}[r(M) - r(E - F_i)]$ and the width of the pair (T, σ) is the maximum of its vertex widths. The minimum value of the width (T, σ) over all such pairs is the *tree-width* $\mathrm{tw}(M)$ of M. Prove that:
   (a) $\mathrm{tw}(N) \leq \mathrm{tw}(M)$ for all minors N of M; and
   (b) $\mathrm{bw}(M) - 1 \leq \mathrm{tw}(M) \leq \max\{2(\mathrm{bw}(M) - 1), 1\}$.

14.3 Rota's Conjecture and the Well-Quasi-Ordering Conjecture

In this section, we shall review the partial results that have been proved towards Rota's Conjecture and the Well-Quasi-Ordering Conjecture. As we noted in Sections 6.6 and 14.1, the former has only been proved when q is at most 4, while a proof of the latter has been announced for $q = 2$. The results in this section highlight a clear dichotomy between what is known in general and what is known when the branch-width is bounded.

Geelen and Whittle (2002) proved the following striking partial result in support of Rota's Conjecture.

Theorem 14.3.1 *For every prime power q and every positive integer k, the set of excluded minors for the class of $GF(q)$-representable matroids contains only finitely many members having branch-width k.* $\square$

For a field $\mathbb{F}$, a matroid M is *almost $\mathbb{F}$-representable* if it contains an element e such that both $M \backslash e$ and M/e are $\mathbb{F}$-representable. Evidently an excluded minor for $\mathbb{F}$-representability is almost $\mathbb{F}$-representable. Indeed, Geelen and Whittle (2002) derived Theorem 14.3.1 from the following result.

Theorem 14.3.2 *For every prime power q and every positive integer k, the class of matroids of branch-width k that are almost $GF(q)$-representable is well-quasi-ordered under the minor relation.* $\square$

Noting that an excluded-minor for a class of matroids of branch-width at most k has branch-width at most $k + 1$, Geelen and Whittle (2002) deduced the following result from the last theorem.

Corollary 14.3.3 *For every prime power q and every positive integer k, if $\mathcal{M}$ is a minor-closed class of $GF(q)$-representable matroids of branch-width at most k, then $\mathcal{M}$ has a finite set of excluded minors.*

Every $GF(q)$-representable matroid is also almost $GF(q)$-representable, so another consequence of the last theorem is the following partial result (Geelen, Gerards, and Whittle 2002) towards the Well-Quasi-Ordering Conjecture.

Theorem 14.3.4 *For every prime power q and every positive integer k, every infinite set of $GF(q)$-representable matroids of branch-width at most k contains two members such that one is isomorphic to a minor of the other.* $\square$

For an excluded minor M for $GF(q)$-representability and an element e of M, Vertigan (in Geelen and Whittle 2002) suggested considering the polymatroid g_e defined on all subsets A of $E(M) - e$ by $g_e(A) = r_{M \backslash e}(A) + r_{M/e}(A)$. Because $M \backslash e$ and M/e are both $GF(q)$-representable matroids, the polymatroid g_e is $GF(q)$-representable. In general, a polymatroid f defined on 2^E is *representable over a field $\mathbb{F}$* if there is a function ψ from E into the set of subspaces of a vector space over $\mathbb{F}$ such that $f(X) = \dim\langle \psi(x) : x \in X \rangle$ for all $X \subseteq E$. The polymatroid g_e is useful in the proof of Theorem 14.3.2.

Since both Rota's Conjecture and the Well-Quasi-Ordering Conjecture hold for matroids of bounded branch-width, we are now left with the question: What can be said about a matroid for which the branch-width is large? The next theorem, a result of Robertson and Seymour (1986), answers the corresponding question for graphs and points the way towards what holds for matroids. This theorem is usually stated in terms of tree-width. But, as noted earlier, Robertson and Seymour (1991) proved that a class of graphs has bounded tree-width if and only if it has bounded branch-width. The $n \times n$ *grid*, which appeared earlier in Exercise 8 of Section 11.5, is the graph with vertex set $\{(i, j) : i, j \in \{1, 2, \ldots, n\}\}$ such that (i, j) and (i', j') are adjacent if and only if $|i - i'| + |j - j'| = 1$.

Theorem 14.3.5 (Grid Theorem for Graphs) *For each positive integer n, there is an integer $k(n)$ such that every graph of branch-width at least $k(n)$ has a minor isomorphic to the $n \times n$ grid.* ☐

It is well known that every planar graph H is a minor of some grid. Diestel (1997) sketched the following proof of this fact. Take a planar embedding of H and fatten its vertices and edges. By superimposing a sufficiently fine plane grid, we obtain the result. In view of this, we can restate the last theorem as follows.

Theorem 14.3.6 *For every planar graph H, there is an integer $h(H)$ such that if G is a graph with branch-width at least $h(H)$, then G has an H-minor.* ☐

An unpublished handwritten manuscript of Johnson, Robertson, and Seymour conjectured that the last theorem extends to $GF(q)$-representable matroids and made significant progress towards this extension, although much remained to be done. Eventually, Geelen, Gerards, and Whittle (2007a) proved the conjecture. This theorem is the first major step on the path to describing the structure of $GF(q)$-representable matroids.

Theorem 14.3.7 (Grid Theorem for Matroids) *Let H be a planar graph and q be a prime power. Then there is an integer $n(H, q)$ such that if M is a $GF(q)$-representable matroid having branch-width at least $n(H, q)$, then M has an $M(H)$-minor.* ☐

An immediate consequence of this theorem is that a minor-closed class of $GF(q)$-representable matroids that does not contain the cycle matroids of all planar graphs must have bounded branch-width. By combining this result with Theorem 14.3.4, we obtain the following partial result towards the Matroid Well-Quasi-Ordering Conjecture (Geelen, Gerards, and Whittle 2007a).

Corollary 14.3.8 *Let q be a prime power and $\mathcal{M}$ be a minor-closed class of $GF(q)$-representable matroids that does not contain the cycle matroids of all planar graphs. Then $\mathcal{M}$ contains no infinite antichains.* ☐

Alternatively, by combining the Grid Theorem for Matroids with Theorem 14.3.1, we get the following result.

Corollary 14.3.9 *Let H be a planar graph and q be a prime power. Then there are only finitely many excluded minors for $GF(q)$-representability that do not have $M(H)$ as a minor.* ☐

In contrast to this, an excluded minor for $GF(q)$-representability cannot contain a large projective geometry (Geelen, Gerards, and Whittle 2006a).

Theorem 14.3.10 *For each prime power q, there is an integer k such that no excluded minor for $GF(q)$-representability has a $PG(k, q)$-minor.* ☐

Geelen, Gerards, and Whittle conjectured that k can be taken to be 2.

Conjecture 14.3.11 *For each prime power q, no excluded minor for the class of $GF(q)$-representable matroids has a $PG(2,q)$-minor.* $\square$

In addition to conjecturing that Theorem 14.3.6 holds, Johnson, Robertson, and Seymour (1999) proposed the following appealing conjecture for what must be found inside an arbitrary matroid of large branch-width.

Conjecture 14.3.12 *For each positive integer n, there is an integer $k(n)$ such that every matroid of branch-width at least $k(n)$ has a minor isomorphic to $U_{n,2n}$ or to the cycle matroid, the bicircular matroid, or the dual of the bicircular matroid of the $n \times n$ grid.* $\square$

If Rota's Conjecture is true, then, for all finite fields $\mathbb{F}$, there are only finitely many matroids that are minors of some excluded minor for $\mathbb{F}$-representability. In stark contrast to this, Mayhew, Newman, and Whittle (2009) proved the following result, thereby verifying a conjecture of Geelen (2008). This is a dramatic strengthening of Theorem 6.5.17, which showed that, for every infinite field $\mathbb{F}$, the set of excluded minors for $\mathbb{F}$-representability is infinite.

Theorem 14.3.13 *Let $\mathbb{F}$ be an infinite field and N be an $\mathbb{F}$-representable matroid. Then there is an excluded minor for the class of $\mathbb{F}$-representable matroids that is not representable over any field and has N as a minor.* $\square$

Mayhew, Newman, and Whittle indicate that the techniques used to prove the last result can be adapted to show that if a matroid is representable over $GF(q)$, then, for some finite extension field $\mathbb{F}$ of $GF(q)$, there is an excluded minor for $\mathbb{F}$-representability that has M as a minor.

By Proposition 14.2.6, the class $\mathcal{B}_2$ of matroids of branch-width at most two consists of direct sums of series–parallel networks. Hence, by Theorem 14.1.1 or Theorem 14.3.4, $\mathcal{B}_2$ contains no infinite antichains. To conclude this section, we use tipless spikes to show that not only does $\mathcal{B}_3$ contain an infinite antichain, but it contains such an antichain all of whose members are representable over every infinite field (Geelen, Gerards, and Whittle 2002). This contrasts sharply with Theorem 14.3.4, which shows that, for all k, the members of $\mathcal{B}_k$ that are representable over some fixed finite field are well-quasi-ordered. We begin with a straightforward lemma whose proof is left as an exercise.

Lemma 14.3.14 *Every spike has branch-width three.* $\square$

For all $r \geq 5$, let M_r be the tipless r-spike with legs $\{x_1, y_1\}, \{x_2, y_2\}, \ldots,$ $\{x_r, y_r\}$ and exactly two circuit–hyperplanes, $\{y_1, x_2, x_3, \ldots, x_r\}$ and $\{x_1, y_2, y_3, \ldots, y_r\}$.

Proposition 14.3.15 *The set $\{M_r : r \geq 5\}$ is an infinite antichain of matroids all having branch-width three and being representable over every infinite field.*

Proof It follows by the last lemma that $\mathrm{bw}(M_r) = 3$. By mimicking the argument used to prove Lemma 6.5.18, we can show that M_r is representable over every infinite field. Finally, suppose $5 \leq s < r$ and M_s is isomorphic to a minor of M_r. Then $M_s \cong M_r \backslash I^* / I$ for some independent set I and coindependent set I^*. Since $r(M_t) = r^*(M_t)$ for all t, we have $|I| = |I^*|$. For each i, the matroid $M_r / \{x_i, y_i\}$ is obtained from an $(r-1)$-element circuit by replacing every element by two elements in parallel. The last matroid has no tipless spike as a minor. Hence I does not contain any $\{x_i, y_i\}$. By duality, nor does I^*.

We show next that, for all i, the set $\{x_i, y_i\}$ is either contained in or avoids $I \cup I^*$. If not, then, for some $\{u_i, v_i\} = \{x_i, y_i\}$, we have $(I \cup I^*) \cap \{u_i, v_i\} = \{u_i\}$. By duality, we may assume that $u_i \in I$. Then $\{v_i, x_j, y_j\}$ is a circuit of M_r / u_i for all $j \neq i$. As $M_r \backslash I^* / I$ has no circuit with fewer than four elements and has v_i as an element, I^* must meet $\{x_j, y_j\}$ for all $j \neq i$. Thus $|I^*| \geq r - 1$; a contradiction. We conclude that there is a non-empty set $\{\{a_1, b_1\}, \{a_2, b_2\}, \ldots, \{a_k, b_k\}\}$ of legs of M_r such that $M_s \cong M_r \backslash \{a_1, a_2, \ldots, a_k\} / \{b_1, b_2, \ldots, b_k\}$. As $\{a_1, a_2, \ldots, a_k\}$ must meet one of the two circuit–hyperplanes of M_r, it follows that M_s has at most one circuit–hyperplane; a contradiction. $\qquad \square$

By Theorem 14.3.4, for each fixed integer k, the Well-Quasi-Ordering Conjecture holds for $GF(q)$-representable matroids of branch-width at most k. It remains, then, to consider those $GF(q)$-representable matroids with large branch-width. By the Grid Theorem for Matroids (14.3.7), such a matroid has a large grid as a minor. But such a matroid may have several parts of large branch-width separated by low-order separations. To deal with these parts, Robertson and Seymour (1991) introduced tangles for graphs, and their results have been generalized to matroids by Geelen, Gerards, and Whittle (2009a). Although these results are technical, tangles play a crucial role in matroid structure theory. In Geelen, Gerards, and Whittle (2006b, 2007b), the reader will find intuitive accounts of the way in which tangles are used in this theory.

Exercises

1. Prove that every spike, whether tipless or tipped, has branch-width three.
2. Prove that the polymatroid g_e defined after Theorem 14.3.4 is $GF(q)$-representable.
3. Show that the cycle matroid of the $n \times n$ grid has branch-width at least n.
4. For all integers k exceeding two, give an example of an infinite antichain of connected matroids all having branch-width k.
5. (Geelen and Whittle 2002) Prove that:
   (a) If N is an excluded minor for a minor-closed class of matroids each having branch-width at most k, then $\mathrm{bw}(N) \leq k + 1$.
   (b) Corollary 14.3.3 follows from (a) and Theorem 14.3.2.

14.4 Algorithmic consequences

In this section, we consider the implications of the theorems stated earlier for a number of computational problems involving graphs and matroids. We begin with a result of Robertson and Seymour (1995), which is one of the profound algorithmic consequences of the Graph-Minors Project.

Theorem 14.4.1 *For every graph H, there is a polynomial-time, indeed an $O(|V(G)|^3)$, algorithm to test if a graph G has a minor isomorphic to H.* $\square$

By combining this theorem with Theorem 14.1.1, we obtain the following.

Corollary 14.4.2 *For every minor-closed class $\mathcal{G}$ of graphs, there is a polynomial-time algorithm to test whether a graph is in $\mathcal{G}$.* $\square$

To see this, note that Theorem 14.1.1 implies that $\mathcal{G}$ has only finitely many excluded minors. For each such H, we can test in polynomial time whether a given graph G has an H-minor. An immediate problem for implementing this algorithm is that it relies on actually knowing the set of excluded minors. This is a common feature of most of the results in this section. They establish the existence of a certain type of algorithm, but it cannot be practically implemented.

Geelen, Gerards, and Whittle (2006b, 2007b) have conjectured that the last theorem extends to matroids represented over finite fields and, in 2009, they announced that they had proved this conjecture for $q = 2$. A $GF(q)$-*represented matroid* is a matroid M together with a given $GF(q)$-representation.

Conjecture 14.4.3 (Minor-Recognition Conjecture) *For every prime power q and every $GF(q)$-representable matroid N, there is a polynomial-time algorithm for testing whether a $GF(q)$-represented matroid M has a minor isomorphic to N.* $\square$

Notice here that we do not insist that we have a $GF(q)$-representation for the fixed minor N since all of its $GF(q)$-representations can be found exhaustively.

Geelen, Gerards, and Whittle (2006b, 2007b) observed that the following partial result towards this conjecture can be obtained by combining the Grid Theorem for Matroids with results of Hliněný (2006a).

Theorem 14.4.4 *For every planar graph H and prime power q, there is a polynomial-time algorithm for testing whether or not a $GF(q)$-represented matroid has an $M(H)$-minor.* $\square$

Hliněný (2003) proved Conjecture 14.4.3 for matroids of fixed branch-width.

Theorem 14.4.5 *For every prime power q, positive integer k, and $GF(q)$-representable matroid N, there is a polynomial-time algorithm that determines whether a given $GF(q)$-represented matroid M of branch-width k has an N-minor.* $\square$

In contrast to the last two theorems, Hliněný (2006b) also proved the following result. Let C_5^+ be the graph that is obtained from a cycle of length five by adding an edge joining two non-adjacent vertices of the cycle.

Proposition 14.4.6 *It is NP-hard to determine whether a given $\mathbb{Q}$-represented matroid of branch-width at most 3 contains a minor isomorphic to $M(C_5^+)$.* $\square$

The insistence on having a $GF(q)$-representation for the matroid M in Conjecture 14.4.3 raises the question as to how hard it is find such a representation. Geelen (private communication) has proposed the following conjecture noting that it holds when q is 2 or 3 but is open for all larger values of q.

Conjecture 14.4.7 *For all prime powers q, there is a polynomial-time algorithm to find a GF(q)-representation for a given GF(q)-representable matroid that is specified by an independence oracle.* □

The structure theorem for $GF(q)$-representable matroids, one of the main targets of the work of Geelen, Gerards and Whittle, will have numerous consequences. For example, Geelen (2008) believes it will help to resolve the following.

Conjecture 14.4.8 *For every prime power q, there is a polynomial-time algorithm that, given any two matrices A_1 and A_2 over GF(q) with the same set of column labels, tests whether $M[A_1] = M[A_2]$.* □

The next result (Geelen 2008) contrasts with this conjecture.

Proposition 14.4.9 *It is NP-hard to test whether $M[A_1] = M[A_2]$ for two given rational matrices A_1 and A_2, even when $M[A_1]$ is a tipless free spike.*

Proof Hliněný (2007) showed that it is NP-hard to test, for a rational matrix A, whether $M[A]$ is isomorphic to a tipless free spike. By Lemma 6.5.14, the free n-spike is represented by the rational matrix $[I_n|J_n + nI_n]$. The proposition follows without difficulty. □

How hard is it to recognize whether a matroid has bounded branch-width? Oum and Seymour (2007) proved that, for every constant k, there is a polynomial-time algorithm to determine whether a matroid M given by an independence oracle has branch-width at most k. Earlier, Hliněný (2005) had proved this for $GF(q)$-represented matroids. Oum and Seymour also showed that, when M has branch-width k, one can find, in polynomial time, a branch-decomposition of M of width k. A better algorithm for the last problem was given by Hliněný and Oum (2008) when M is $GF(q)$-represented. It uses the finite list of excluded minors for the class $\mathcal{B}_k$ of matroids of branch-width at most k.

We showed in Section 9.4 that, for every field $\mathbb{F}$, it may require exponentially many probes of a rank oracle to decide if a matroid is $\mathbb{F}$-representable. For finite fields, certifying non-representability is easier. Indeed, Geelen and Whittle (2010b) have proved that, for every prime field $GF(p)$, certifying non-representability over $GF(p)$ requires at most $O(n^2)$ rank evaluations for matroids on n elements. If Rota's Conjecture were true, then, for all prime powers q, certifying non-representability over $GF(q)$ would require only $O(1)$ rank evaluations (see Exercise 2). By contrast, Ben-David and Geelen (2010) proved that non-representability over $\mathbb{R}$ cannot be certified by using only a polynomial number of rank evaluations, thereby answering a question of Geelen (2008).

Proposition 14.4.10 *There is no function f in $\mathbb{Z}[x]$ such that, for all positive integers n and every n-element matroid M, at most $f(n)$ probes of a rank oracle will certify that M is not representable over $\mathbb{R}$.* □

The construction used to prove the last proposition also shows that imposing an upper bound on the branch-width does not ensure the existence of a polynomial-length certificate of non-representability over $\mathbb{R}$.

Exercises

1. Show that Conjecture 14.4.7 holds when q is 2 or 3.
2. (Hliněný 2007) Show that for all $r \geq 5$, an r-spike is free if and only if it has no $M(C_5^+)$-minor.
3. Show that:
   (a) For a k-element matroid N, it requires at most 2^k rank evaluations in a matroid M to certify that N is isomorphic to a minor of M.
   (b) If Rota's Conjecture holds, then certifying non-representability over $GF(q)$ requires only $O(1)$ rank evaluations.
4. Let M be a $\mathbb{Q}$-represented matroid. Show that one can determine in polynomial time whether or not M is binary.

14.5 Intertwining

By Theorem 6.6.3, the class of regular matroids is the intersection of the classes of binary and ternary matroids. Moreover, the set of excluded minors for the class of regular matroids consists of the minor-minimal matroids in $\{U_{2,4}\} \cup \{U_{2,5}, U_{3,5}, F_7, F_7^*\}$, the union of the sets of excluded minors for the classes of binary and ternary matroids. These observations are easily generalized as follows.

Lemma 14.5.1 *Let* $\mathcal{M}_1$ *and* $\mathcal{M}_2$ *be minor-closed classes of matroids having* $\mathcal{E}_1$ *and* $\mathcal{E}_2$ *as their sets of excluded minors. Then* $\mathcal{M}_1 \cap \mathcal{M}_2$ *is a minor-closed class of matroids whose set of excluded minors consists of the minor-minimal members of the set* $\mathcal{E}_1 \cup \mathcal{E}_2$. □

Having considered the intersection of two minor-closed classes $\mathcal{M}_1$ and $\mathcal{M}_2$ of matroids, it is natural to consider their union, $\mathcal{M}_1 \cup \mathcal{M}_2$, which is also easily seen to be minor-closed. Determining the set of excluded minors for $\mathcal{M}_1 \cup \mathcal{M}_2$ is, however, much more complex than finding the excluded minors for $\mathcal{M}_1 \cap \mathcal{M}_2$. We shall use the following concept. A matroid M is an *intertwine* of two matroids N_1 and N_2 if M has minors isomorphic to both N_1 and N_2, but no proper minor of M has both N_1- and N_2-minors. For example, both F_7^- and $(F_7^-)^*$ are intertwines of $M(K_4)$ and $U_{2,4}$. Moreover, by Lemma 6.6.9, R_{10} is an intertwine of $M(K_{3,3})$ and $M^*(K_{3,3})$. Evidently, if N_1 is isomorphic to a not-necessarily-proper minor of N_2, then every intertwine of N_1 and N_2 is isomorphic to N_2. In this section, we shall consider various intertwining problems and show how such problems relate to the representability problems considered above.

We begin by specifying the excluded minors for $\mathcal{M}_1 \cup \mathcal{M}_2$ in terms of intertwines. The proof of the next lemma is left as an exercise for the reader. Although it is longer than that of Lemma 14.5.1, its form is completely predictable.

Lemma 14.5.2 *Let* $\mathcal{M}_1$ *and* $\mathcal{M}_2$ *be minor-closed classes of matroids having* $\mathcal{E}_1$ *and* $\mathcal{E}_2$ *as their sets of excluded minors. Then* $\mathcal{M}_1 \cup \mathcal{M}_2$ *is a minor-closed class of matroids whose set of excluded minors consists of all minor-minimal matroids that are intertwines of some member of* $\mathcal{E}_1$ *and some member of* $\mathcal{E}_2$. □

As we saw in Chapter 13, much of the work in proving Seymour's Decomposition Theorem involves proving Theorem 13.1.2. We show next that, by combining that theorem with the last lemma, we can specify the set of excluded minors for the union of the classes of graphic and cographic matroids. The author's attention was drawn to this result by Jim Geelen (private communication), who views the result as folklore. Recall that $K'_{3,3}$ is the graph obtained from $K_{3,3}$ by adding a new edge z joining two vertices in the same vertex class of $K_{3,3}$.

Proposition 14.5.3 *The excluded minors for the classes of matroids that are graphic or cographic are*

(i) $U_{2,4}, F_7, F_7^*, R_{10}, R_{12}$;

(ii) *the four matroids that are obtained by taking the direct sum of a member of $\{M(K_5), M(K_{3,3})\}$ with a member of $\{M^*(K_5), M^*(K_{3,3})\}$; and*

(iii) *the nine matroids that are obtained by taking the 2-sum of a member of $\{M(K_5), M(K_{3,3}), M(K'_{3,3})\}$ with a member of $\{M^*(K_5), M^*(K_{3,3}), M^*(K'_{3,3})\}$ where the basepoints of these 2-sums are arbitrary in each of $M(K_5)$, $M(K_{3,3})$, $M^*(K_5)$, and $M^*(K_{3,3})$, but must equal z in each of $M(K'_{3,3})$ and $M^*(K'_{3,3})$.*

Proof Clearly $U_{2,4}$, F_7, and F_7^* are excluded minors. Let M be another excluded minor. As M has no $U_{2,4}$-, F_7-, or F_7^*-minor, Theorem 10.1.1 implies that M is regular. Assume that M is 3-connected. Then, by Theorem 13.1.2, M has an R_{10}-minor or an R_{12}-minor. By Lemmas 6.6.9 and 13.1.6, neither R_{10} nor R_{12} is graphic or cographic although every proper minor of each is graphic or cographic. Thus $M \cong R_{10}$ or R_{12}.

We may now assume that M is not 3-connected. Then M is a direct sum or 2-sum of matroids M_1 and M_2 where we may assume that M_1 is graphic but not cographic, while M_2 is cographic but not graphic. Thus M_1 has a minor N_1 in $\{M(K_5), M(K_{3,3})\}$ and M_2 has a minor N_2 in $\{M^*(K_5), M^*(K_{3,3})\}$. It follows immediately from the last lemma that if $M = M_1 \oplus M_2$, then M_1 is in $\{M(K_5), M(K_{3,3})\}$ and M_2 is in $\{M^*(K_5), M^*(K_{3,3})\}$. It remains to consider the case when $M = M_1 \oplus_2 M_2$. For each i, let p_i be the basepoint of M_i in this 2-sum. Now if there is an element e in $E(M_i) - (E(N_i) \cup p_i)$, then, by Proposition 4.3.7, $M_i \backslash e$ or M_i / e is connected and has N_i as a minor. Thus, by Proposition 7.1.21, $M \backslash e$ or M/e has both N_1- and N_2-minors, so is neither graphic nor cographic; a contradiction. We deduce that $E(M_i) = E(N_i) \cup p_i$. Thus either $M_i = N_i$, or M_i is a single-element extension or coextension of N_i. Moreover, one easily checks that no circuit or cocircuit of N_i with at most two elements contains p_i, so M_i is 3-connected. Thus, if $M_1 \neq N_1$, then M_1 is $M(K'_{3,3})$ or a 3-connected graphic single-element coextension of $M(K_5)$. In the latter case, by Proposition 12.2.11, M_1 has an $M(K_{3,3})$-minor and we may alter our choice of N_1 taking it to be this $M(K_{3,3})$-minor. This gives the contradiction that $|E(M_1) - E(N_1)| = 2$. We conclude that M_1 is $M(K_5)$, $M(K_{3,3})$, or $M(K'_{3,3})$ where, in the last case, $p_1 = z$. It is now straightforward to complete the proof of the lemma. □

The minimality condition in the definition of an intertwine means that the set of intertwines of any two matroids forms an antichain. Geelen (2008) proposed the following conjecture, noting that it would follow from the Well-Quasi-Ordering Conjecture for Matroids (14.1.5).

Conjecture 14.5.4 *For every prime power q and every two matroids N_1 and N_2, there are only finitely many non-isomorphic $GF(q)$-representable intertwines of N_1 and N_2.* □

Brylawski (1986a), Robertson (private communication), and Welsh (private communication) independently asked whether the number of intertwines of two matroids is always finite. This question was answered in the negative by Vertigan (1998) who constructed, subject to quite mild conditions on N_1 and N_2, infinitely many intertwines of N_1 and N_2. We shall describe a weaker construction due to Geelen (2008), which uses tipless spikes. Yet another appearance of spikes in a problem relating to minors may strike the reader as being more than coincidence. These repeated appearances of spikes can be explained by an attractive result of Vertigan (in Geelen 2008) that the partial order on spikes determined by the minor relation contains an isomorphic copy of the partial order on all matroids determined by the minor relation. To prove this result and to describe the construction of an infinite family of intertwines, we now give an alternative description of tipless spikes. Our presentation here follows Geelen (2008).

Lemma 14.5.5 *Let M be a tipless spike of rank at least three whose set of legs is $\{\{x_u, y_u\} : u \in U\}$. Let $\mathcal{T}$ be the set of subsets J of U such that $\{\{x_j : j \in J\} \cup \{y_k : k \in U - J\}$ is a circuit of M. Then M is uniquely determined by the pair $(U, \mathcal{T})$. Moreover,*

$$|T_1 \bigtriangleup T_2| \geq 2 \text{ for all distinct } T_1 \text{ and } T_2 \text{ in } \mathcal{T}. \tag{14.4}$$

Conversely, let U be a set with at least three elements and $\mathcal{T}$ be a set of subsets of U satisfying (14.4). Then there is a unique tipless spike with $\{\{x_u, y_u\} : u \in U\}$ as its set of legs such that the set of non-spanning circuits meeting each $\{x_u, y_u\}$ is $\{\{x_j : j \in J\} \cup \{y_k : k \in U - J\} : J \in \mathcal{T}\}$.

Proof Let M be a tipless spike whose set of legs is $\{\{x_u, y_u\} : u \in U\}$ for some set U with $|U| \geq 3$. By Proposition 1.5.17, which introduced spikes, M is uniquely determined by the collection $\mathcal{C}_3$ of non-spanning circuits that contain exactly one element of each leg. Thus M is uniquely determined by the pair $(U, \mathcal{T})$ where a subset J of U is in $\mathcal{T}$ if and only if $\{x_j : j \in J\} \cup \{y_k : k \in U - J\}$ is a circuit of M. Furthermore, by definition, no two members of $\mathcal{C}_3$ can have more than $|U| - 2$ common elements. This requirement is equivalent to condition (14.4). The proof of the converse is similar and is omitted. □

If U is a set with at least three elements and $\mathcal{T}$ is a set of subsets of U satisfying (14.4), then the unique tipless spike associated with the pair $(U, \mathcal{T})$ will be denoted by $S(U, \mathcal{T})$. Given such a pair $(U, \mathcal{T})$, suppose that $|U| \geq 4$ and

$e \in U$. To prove the next two results, we define $(U, \mathcal{T}) \backslash e$ and $(U, \mathcal{T})/e$ to be $(U - e, \mathcal{T} \backslash e)$ and $(U - e, \mathcal{T}/e)$ respectively, where $\mathcal{T} \backslash e = \{T \in \mathcal{T} : e \notin T\}$ and $\mathcal{T}/e = \{T - e : e \in T \in \mathcal{T}\}$. Clearly both $\mathcal{T} \backslash e$ and $\mathcal{T}/e$ satisfy (14.4). Moreover, it is straightforward to check that

$$S((U, \mathcal{T}) \backslash e) = S(U, \mathcal{T}) \backslash x_e / y_e \quad \text{and} \quad S((U, \mathcal{T})/e) = S(U, \mathcal{T})/x_e \backslash y_e.$$

The following theorem is due to Vertigan (in Geelen 2008).

Theorem 14.5.6 *There is a function ψ from the set of all matroids with at least three elements to the set of all tipless spikes such that if M and N are matroids with at least three elements, then N is a minor of M if and only if $\psi(N)$ is a minor of $\psi(M)$.*

Proof For a matroid M with ground set $E(M)$ and set $\mathcal{B}(M)$ of bases, let $\psi(M)$ be $S(E(M), \mathcal{B}(M))$. Since the members of $\mathcal{B}(M)$ are equicardinal, condition (14.4) holds when $\mathcal{T} = \mathcal{B}(M)$. Now suppose $|E(M)| \geq 4$ and e is an element of M. Then, using the notation defined above, we see that $(E(M), \mathcal{B}(M)) \backslash e = (E(M \backslash e), \mathcal{B}(M \backslash e))$ provided e is not a coloop of M; and $(E(M), \mathcal{B}(M))/e = (E(M/e), \mathcal{B}(M/e))$ provided e is not a loop of M.

Now let N be a minor of M. Then, by Lemma 3.3.2, N can be written in the form $M/I \backslash I^*$ where I is independent and I^* is coindependent. Thus, to prove that $\psi(N)$ is a minor of $\psi(M)$, it suffices to show this when N is $M \backslash e$ and $r(M \backslash e) = r(M)$, and when N is M/e and $r(M/e) = r(M) - 1$. In the first case,

$$\psi(M \backslash e) = S(E(M \backslash e), \mathcal{B}(M \backslash e)) = S((E(M), \mathcal{B}(M)) \backslash e)$$
$$= S(E(M), \mathcal{B}(M)) \backslash x_e / y_e = \psi(M) \backslash x_e / y_e.$$

Similarly, in the second case, $\psi(M/e) = \psi(M)/x_e \backslash y_e$.

For the converse, if $\psi(N)$ is a minor of $\psi(M)$, we need to show that N is a minor of M. We shall assume that $S(U_1, \mathcal{T}_1)$ is a minor of $S(U_2, \mathcal{T}_2)$ and that $e \in U_2 - U_1$ and show that $S(U_1, \mathcal{T}_1)$ is a minor of $S((U_2, \mathcal{T}_2) \backslash e)$ or $S((U_2, \mathcal{T}_2)/e)$. Now $S(U_1, \mathcal{T}_1)$ is a minor of either $S(U_2, \mathcal{T}_2) \backslash x_e$ or $S(U_2, \mathcal{T}_2)/x_e$. We shall prove the result in the former case, noting that a similar argument can be used to treat the latter case. As y_e is not an element of $S(U_1, \mathcal{T}_1)$, this matroid is a minor of either $S(U_2, \mathcal{T}_2) \backslash x_e / y_e$ or $S(U_2, \mathcal{T}_2) \backslash x_e, y_e$. In the second case, if f is in U_1, then $\{x_f, y_f\}$ is a cocircuit of $S(U_2, \mathcal{T}_2) \backslash x_e, y_e$. Hence $\{x_f, y_f\}$ contains a cocircuit of $S(U_1, \mathcal{T}_1)$; a contradiction. We deduce that $S(U_1, \mathcal{T}_1)$ is a minor of $S(U_2, \mathcal{T}_2) \backslash x_e / y_e$, that is, of $S((U_2, \mathcal{T}_2) \backslash e)$, so the theorem holds. $\square$

In a tipless spike of rank at least five, the legs are uniquely determined because the 4-element circuits coincide with all unions of pairs of distinct legs. This means that the last theorem can be strengthened as follows.

Corollary 14.5.7 *There is a function ψ from the set of all matroids with at least five elements to the set of all tipless spikes such that if M and N are matroids with at least five elements, then N is isomorphic to a minor of M if and only if $\psi(N)$ is isomorphic to a minor of $\psi(M)$.* $\square$

Next we give an example of Geelen (2008) to show that the number of intertwines of two tipless spikes is infinite provided the spikes satisfy some fairly weak conditions. In particular, any two distinct spikes from the infinite antichain identified in Proposition 14.3.15 satisfy these conditions.

Proposition 14.5.8 *Let N_1 and N_2 be tipless spikes of rank at least five. Suppose that each of N_1 and N_2 has every element in a non-spanning circuit that meets every leg. If neither N_1 nor N_2 is isomorphic to a minor of the other, then there are arbitrarily large tipless spikes that are intertwines of N_1 and N_2. Hence there are infinitely many intertwines of N_1 and N_2.*

Proof Let $N_1 = S(U_1, \mathcal{T}_1)$ and $N_2 = S(U_2, \mathcal{T}_2)$. We may assume that U_1 and U_2 are disjoint. Let Z be a set that is disjoint from $U_1 \cup U_2$. Now let $U = U_1 \cup U_2 \cup Z$ and $\mathcal{T} = \mathcal{T}_1 \cup \{Z \cup T_2 : T_2 \in \mathcal{T}_2\}$. Because $\mathcal{T}_1$ and $\mathcal{T}_2$ both obey (14.4), it is straightforward to check that $\mathcal{T}$ does too. Thus, by Lemma 14.5.5, $S(U, \mathcal{T})$ is a well-defined tipless spike. Moreover, $S((U, \mathcal{T}) \backslash (Z \cup U_2)) = S(U_1, \mathcal{T} \backslash (Z \cup U_2)) = S(U_1, \mathcal{T}_1)$ and $S((U, \mathcal{T}) \backslash U_1 / Z) = S(U_2, \mathcal{T} \backslash U_1 / Z) = S(U_2, \mathcal{T}_2)$. Thus $S(U, \mathcal{T})$ has an intertwine of $S(U_1, \mathcal{T}_1)$ and $S(U_2, \mathcal{T}_2)$ as a minor. The core of the proof involves showing the following.

14.5.9 *Every intertwine of $S(U_1, \mathcal{T}_1)$ and $S(U_2, \mathcal{T}_2)$ that is a minor of $S(U, \mathcal{T})$ has the form $S(U', \mathcal{T}')$ where $Z \subseteq U' \subseteq U$.*

First we observe that, by Corollary 3.2.14, to obtain one tipless spike as a minor of another, we cannot delete or contract both elements of a leg. Since the contraction of one element of a leg makes the other element of that leg into the tip of a spike of lower rank, that tip must subsequently be deleted. Using these observations and duality, we deduce that every tipless spike that is a proper minor of $S(U, \mathcal{T})$ is a minor of $S((U, \mathcal{T}) \backslash e)$ or $S((U, \mathcal{T}) / e)$.

Suppose $e \in Z$. Then $\mathcal{T} \backslash e = \mathcal{T}_1$. Thus if $f \in U_2 \cup (Z - e)$, then the tipless spike $S((U, \mathcal{T}) \backslash e)$ has no non-spanning circuit that meets every leg and contains x_f. Thus every tipless spike that is a minor of $S((U, \mathcal{T}) \backslash e)$ and has x_f as an element has no non-spanning circuit containing x_f and meeting every leg. But every element of $S(U_2, \mathcal{T}_2)$ is in a non-spanning circuit that meets every leg. Therefore if $S(U_2, \mathcal{T}_2)$ is isomorphic to a minor of $S((U, \mathcal{T}) \backslash e)$, then this minor avoids all the legs indexed by $Z \cup U_2$. Hence $S(U_2, \mathcal{T}_2)$ is isomorphic to a minor of $S(U_1, \mathcal{T}_1)$; a contradiction. We conclude that $S(U_2, \mathcal{T}_2)$ is not isomorphic to a minor of $S((U, \mathcal{T}) \backslash e)$. A similar argument establishes that $S(U_1, \mathcal{T}_1)$ is not isomorphic to a minor of $S((U, \mathcal{T}) / e)$. We conclude that 14.5.9 holds.

To see that $S(U_1, \mathcal{T}_1)$ and $S(U_2, \mathcal{T}_2)$ have infinitely many intertwines, take a finite set of such intertwines and choose a set Z so that $2|Z|$ exceeds the cardinality of the ground set of any of these intertwines. The process above finds a new intertwine with more elements than any in the original set. $\quad\square$

Bonin (2010) gave another construction for infinitely many intertwines of certain pairs of matroids N_1 and N_2 on disjoint ground sets. In particular, he proved that if, up to isomorphism, neither matroid can be obtained from the

$$\begin{bmatrix} & & \begin{array}{|cccccc} 1 & 1 & 0 & 0 & 0 & 1 \\ 1 & 0 & 0 & 0 & 1 & 1 \\ 0 & 0 & 0 & 1 & 1 & 1 \\ 0 & 0 & 1 & 1 & 1 & 0 \\ 0 & 1 & 1 & 1 & 0 & 0 \\ 1 & 1 & 1 & 0 & 0 & 0 \end{array} \end{bmatrix}$$

FIG. 14.6. A binary representation of T_{12}.

other by any combination of taking minors, free extensions, and free coextensions, then, for all sufficiently large integers k, there is a rank-k intertwine of N_1 and N_2. Bonin's result is closely related to Vertigan's original construction and, indeed, both approaches can be extended to yield the same collections of intertwines.

Next we consider the implications of the last proposition for minor-closed classes of matroids. Let N_1 and N_2 be tipless spikes satisfying the hypotheses of Lemma 14.5.8. For each i, let $\mathcal{M}_i$ be $EX(N_i)$, the class of matroids having no minor isomorphic to N_i. Then, by Lemma 14.5.2, the intertwines of N_1 and N_2 are precisely the excluded minors for the class $\mathcal{M}_1 \cup \mathcal{M}_2$. Thus although $\mathcal{M}_1$ and $\mathcal{M}_2$ have finite sets of excluded minors, their union does not. Nevertheless, there are cases where $\mathcal{M}_1$ and $\mathcal{M}_2$ have a special form that we may hope that $\mathcal{M}_1 \cup \mathcal{M}_2$ has a finite set of excluded minors. Indeed, Geelen, Gerards, and Whittle announced in 2009 that their structure theorem for binary matroids can be used to show that if $\mathcal{M}_1$ is a minor-closed class of binary matroids and $\mathcal{M}_2$ is a minor-closed class of matroids having finitely many excluded minors, then $\mathcal{M}_1 \cup \mathcal{M}_2$ has a finite set of excluded minors. Moreover, Mayhew, Oporowski, Oxley, and Whittle (2009) proposed the following strengthening of Rota's Conjecture.

Conjecture 14.5.10 *Let $\mathcal{F}$ be a finite family of finite fields. The set of excluded minors for the class of matroids that are representable over at least one field in $\mathcal{F}$ is finite.*   □

Mayhew *et al.* also proved the first case of this conjecture that does not reduce to a special case of Rota's Conjecture by establishing it when $\mathcal{F}$ is $\{GF(2), GF(3)\}$. They did this by explicitly determining the list of excluded minors for the union of the classes of binary and ternary matroids. Some obvious excluded minors for this class are $U_{2,5}, U_{3,5}$, and the direct sum and 2-sum of $U_{2,4}$ with F_7 and with F_7^*. There are, in fact, just two more excluded minors. One is the unique matroid $AG(3,2)'$ that is obtained by relaxing a circuit–hyperplane of $AG(3,2)$. The second is obtained from the matroid T_{12} (Kingan 1997) for which a binary representation is shown in Figure 14.6 and which was considered earlier in Exercise 5 of Section 5.1. It is not difficult to show that T_{12} has a transitive automorphism group and that it has exactly two circuit–hyperplanes. Moreover, these circuit–hyperplanes are disjoint. It follows that, up to isomorphism, there is a unique matroid that is obtained by relaxing one of these circuit–hyperplanes.

Theorem 14.5.11 *The excluded minors for the class of matroids that are binary or ternary are $U_{2,5}$, $U_{3,5}$, $U_{2,4} \oplus F_7$, $U_{2,4} \oplus F_7^*$, $U_{2,4} \oplus_2 F_7$, $U_{2,4} \oplus_2 F_7^*$, and the unique matroids that are obtained by relaxing a circuit–hyperplane in either $AG(3,2)$ or T_{12}.* □

The proof of the last theorem relies crucially on work of Truemper (1992a) that characterizes a class of matroids that he calls *almost regular*. Such matroids are binary but not regular; for every element e, the deletion or contraction of e is regular; and an additional technical parity condition holds.

Conjecture 14.5.10 may fail if $\mathcal{F}$ is infinite. For example, let $\mathcal{F}$ be the collection of all finite fields. By Theorem 14.3.13, there are infinitely many excluded minors for $\mathbb{R}$-representability that are not representable over any field. By Proposition 6.8.2 , every $\mathbb{R}$-representable matroid is representable over at least one finite field. Thus, the class of matroids that are representable over at least one finite field has infinitely many excluded minors.

Next we note a conjecture of Kelly and Rota (1973) that is another strengthening of Rota's Conjecture and is a natural companion to Conjecture 14.5.10.

Conjecture 14.5.12 *Let $\mathcal{F}$ be a family of finite fields. There are only finitely many excluded minors for the class of matroids that are representable over every field in $\mathcal{F}$.* □

Whittle (2005) proposed the following strengthening of the last conjecture.

Conjecture 14.5.13 *Let $\mathcal{F}$ be a family of fields at least one of which is finite. There are only finitely many excluded minors for the class of matroids that are representable over every field in $\mathcal{F}$.* □

This conjecture holds when $\mathcal{F}$ contains $GF(2)$ because, by Theorem 6.6.3, the class of matroids representable over every field in $\mathcal{F}$ is either the class of regular matroids or the class of binary matroids depending on whether $\mathcal{F}$ does or does not contain a field of characteristic other than two.

We close this section with a problem and a conjecture of Geelen (2008).

Problem 14.5.14 *For a matroid M and a uniform matroid $U_{r,n}$, are there only finitely many non-isomorphic intertwines of M and $U_{r,n}$?* □

Conjecture 14.5.15 *For every two matroids N_1 and N_2 and every positive integer k, there is an integer n such that if M is an intertwine of N_1 and N_2, and M has no $U_{k,2k}$-minor, then M has branch-width at most n.* □

If the last conjecture holds, then so does Conjecture 14.5.4 (see Exercise 4). It is straightforward to answer Problem 14.5.14 affirmatively if r or $n - r$ is at most one by bounding the cardinality of an intertwine. More significantly, Geelen, Gerards, and Whittle announced in 2009 that their structure theorem for binary matroids can be used to give an affirmative answer when $U_{r,n}$ is $U_{2,4}$.

Exercises

1. Show that the only binary intertwines of F_7 and F_7^* are $F_7 \oplus F_7^*$, $F_7 \oplus_2 F_7^*$, $AG(3,2)$, and S_8, and find at least one non-binary intertwine.

2. Prove that if Rota's Conjecture and the Well-Quasi-Ordering Conjecture for Matroids hold, then so do Conjectures 14.5.12 and 14.5.13.

3. Show that the set of excluded minors for the class of matroids that are representable over some finite field is infinite.

4. Use Theorem 14.3.4 to prove that if Conjecture 14.5.15 holds, then so does Conjecture 14.5.4.

5. *(Geelen 1999) Let $M_1, M_2, \ldots, M_k$ be connected matroids and $\mathcal{M}$ be $EX(M_1, M_2, \ldots, M_k)$. Prove that if $\mathcal{M}$ contains neither F_7^- nor $(F_7^-)^*$, then there are only finitely many excluded minors for the union of $\mathcal{M}$ and the class of binary matroids.

14.6   Inequivalent representations

One of the main issues that has arisen in attempts to settle Rota's Conjecture has been gaining control of the number of inequivalent representations of $GF(q)$-representable matroids. In this section, we begin a discussion of the progress that has been made towards dealing with this problem. We shall continue this discussion in Section 14.8.

For matrices A_1 and A_2 over a field $\mathbb{F}$, suppose that $M[A_1] = M[A_2] = M$, say. Recall from Section 6.3 that A_1 and A_2 are equivalent representations of M if A_2 can be obtained from A_1 by a sequence of operations each consisting of a row or column swap, a row or column scaling, replacement of a row by the sum of that row and another, addition or removal of a zero row, or replacement of each matrix entry by its image under an automorphism of $\mathbb{F}$. If we never use the last of these operations, then A_1 and A_2 are projectively equivalent. A matroid is uniquely $\mathbb{F}$-representable if all of its $\mathbb{F}$-representations are equivalent.

By Proposition 6.6.5, for all fields $\mathbb{F}$, every $\mathbb{F}$-representable binary matroid is uniquely $\mathbb{F}$-representable. The following result of Brylawski and Lucas (1976) is a straightforward consequence of Lemma 10.2.4.

Corollary 14.6.1 *Ternary matroids are uniquely $GF(3)$-representable.*   □

To see the relevance of this result to Rota's Conjecture, we now briefly review the proof of the excluded-minor theorem for ternary matroids. If N is an excluded minor for this class, and x and y are elements of N, then $N \backslash x$ and $N \backslash y$ are both ternary, and the unique $GF(3)$-representability of ternary matroids ensures that a ternary representation for $N \backslash x, y$ can be extended to representations for both $N \backslash x$ and $N \backslash y$. This enables us to construct a ternary matroid M that is very like N. The rest of the proof involves exploiting this similarity.

Example 14.6.2 Let P_7 and P_7' be the matroids for which geometric representations are shown in Figure 14.7. Then, for

$$A_1 = \begin{bmatrix} & & & \overset{a}{} & \overset{b}{} & \overset{c}{} & \vline & \overset{d}{1} & \overset{e}{1} & \overset{f}{1} & \overset{g}{1} \\ & I_3 & & \vline & 1 & 2 & 2 & 1 \\ & & & \vline & 0 & 1 & 2 & 1 \end{bmatrix} \quad \text{and} \quad A_2 = \begin{bmatrix} & & & \vline & \overset{d}{1} & \overset{e}{1} & \overset{f}{1} & \overset{g}{1} \\ & I_3 & & \vline & 1 & 2 & 2 & 1 \\ & & & \vline & 0 & 1 & 3 & 1 \end{bmatrix},$$

it is not difficult to check that A_1 and A_2 are $GF(5)$-representations for P_7 and P_7', respectively. Evidently the identity map on $\{a, b, c, d, e, f\}$ is an isomorphism between $P_7\backslash g$ and $P_7'\backslash g$. Consider the matrices, $A_1\backslash g$ and $A_2\backslash g$, obtained from A_1 and A_2 by deleting column g. The entries of these matrices agree in row 1 and in column e. As $A_1\backslash g \neq A_2\backslash g$, it follows by Theorem 6.4.7 that $A_1\backslash g$ and $A_2\backslash g$ are projectively inequivalent representations for $P_7\backslash g$. As $GF(5)$ has no field automorphisms, the representations $A_1\backslash g$ and $A_2\backslash g$ are inequivalent. Thus a $GF(5)$-representable matroid need not be uniquely $GF(5)$-representable. $\square$

Observe, in the last example, that $P_7\backslash g \cong U_{2,4} \oplus_2 U_{2,4}$. This example extends easily to show that, for all prime powers q exceeding three, $U_{2,4} \oplus_2 U_{2,4}$ is $GF(q)$-representable but is not uniquely $GF(q)$-representable. What hope is there then for a unique representability result for such fields? In the case $q = 4$, the matroid $U_{2,4} \oplus_2 U_{2,4}$ provides a clue. This matroid is the 2-sum of two non-binary matroids. Kahn (1988) showed that such matroids are the only connected $GF(4)$-representable matroids that are not uniquely representable over $GF(4)$.

Proposition 14.6.3 *Let M be a $GF(4)$-representable matroid. Then M is uniquely $GF(4)$-representable if and only if M cannot be written as a direct sum or a 2-sum of two non-binary matroids. In particular, if M is 3-connected, then it is uniquely $GF(4)$-representable.* $\square$

In Section 14.8, we shall give a proof of the last part of this result based on a general technique introduced by Whittle (1999). Kahn (1988) also gave examples to show that, for all $q \geq 5$, there is no positive integer $m(q)$ such that every $m(q)$-connected $GF(q)$-representable matroid is uniquely $GF(q)$-representable. In addition, Kahn made the following attractive conjecture, which has received considerable research attention.

Conjecture 14.6.4. (Kahn's Conjecture) *For each prime power q, there is an integer $n(q)$ such that no 3-connected $GF(q)$-representable matroid has more than $n(q)$ inequivalent $GF(q)$-representations.* $\square$

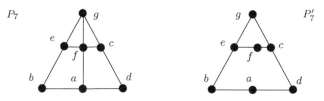

FIG. 14.7. P_7 and P_7'.

Observe that, for all a in $GF(5) - \{0,1\}$ and all b in $GF(5) - \{0,1,a\}$, the matroid $U_{2,5}$ is represented over $GF(5)$ by the matrix $\begin{bmatrix} 1 & 0 & 1 & 1 & 1 \\ 0 & 1 & 1 & a & b \end{bmatrix}$. Since there are three choices for a and two choices for b, there are six such representations. Moreover, all six are inequivalent. Oxley, Vertigan, and Whittle (1996a) verified Kahn's Conjecture for $q = 5$, showing that $U_{2,5}$ is an extremal example. Again a proof of this result is outlined in Section 14.8.

Proposition 14.6.5 *A 3-connected matroid has at most six inequivalent $GF(5)$-representations.* □

More significantly, Oxley, Vertigan, and Whittle showed that Kahn's Conjecture fails for all $q \geq 7$.

Theorem 14.6.6 *For all prime powers q exceeding five and all positive integers n, there is a 3-connected matroid having more than n inequivalent $GF(q)$-representations.* □

Two classes of examples were used to prove this theorem, free spikes and free swirls. The former were defined following Proposition 1.5.18. We now define the latter. For $r \geq 3$, begin with a rank-r whirl having spokes $s_1, s_2, \ldots, s_r$ in cyclic order. Let $\{s_1, a_1, s_2\}, \{s_2, a_2, s_3\}, \ldots, \{s_r, a_r, s_1\}$ be the non-trivial lines of this whirl. For each i in $\{1, 2, \ldots, r\}$, freely add a new point b_i on the line $\{s_i, a_i, s_{i+1}\}$ where $s_{r+1} = s_1$. Call the resulting matroid the *rank-r jointed free swirl* Ψ_r^+. By deleting $\{s_1, s_2, \ldots, s_r\}$ from Ψ_r^+, we obtain the rank-r *free swirl* Ψ_r. It is not difficult to show that a_i and b_i are clones in Ψ_r and that this matroid is isomorphic to its dual. The proof of the next result is outlined in Exercise 5.

Lemma 14.6.7 *Let q be a prime power exceeding four that is not of the form 2^p where $2^p - 1$ is prime. Then the jointed rank-r free swirl is a 3-connected $GF(q)$-representable matroid. Moreover, when $q > 5$, this matroid has at least 2^r inequivalent $GF(q)$-representations.* □

Next we deal with those prime powers that are not covered by the last lemma.

Lemma 14.6.8 *Let p be a prime and $q = p^k$ where $k \geq 2$. The rank-r free spike with tip is 3-connected and $GF(q)$-representable. Moreover, when $q \geq 8$, this matroid has at least 2^{r-1} inequivalent $GF(q)$-representations.* □

Proposition 14.6.6 was proved by combining the last two lemmas. Note that both the tipless free 3-spike and the rank-3 free swirl are isomorphic to $U_{3,6}$. Thus one way to prevent free spikes and free swirls from appearing in our matroids is to insist that they have no $U_{3,6}$-minor. Geelen, Mayhew, and Whittle (2004) proved that if one imposes this restriction, then Kahn's Conjecture holds.

Proposition 14.6.9 *Let q be a prime power and M be a 3-connected $GF(q)$-representable matroid with no $U_{3,6}$-minor. Then M has at most $(q-2)!$ inequivalent $GF(q)$-representations.* □

Excluding $U_{3,6}$ as a minor is quite a strong restriction, so a weaker condition that would exclude free spikes and free swirls was sought. Both free spikes and free swirls have numerous 3-separations and there was initially some hope that, by restricting attention to 4-connected matroids, one could recover the existence of the bound conjectured by Kahn. Indeed, Geelen and Whittle (2010b) showed that such a bound exists for all finite *prime* fields.

Theorem 14.6.10 *For all primes p, there is an integer $f(p)$ such that every 4-connected matroid has at most $f(p)$ inequivalent $GF(p)$-representations.* □

Unfortunately, as Geelen, Gerards, and Whittle (2009b) showed, the last theorem does not extend to non-prime finite fields. In fact, there is a striking dichotomy between the prime and non-prime cases.

Proposition 14.6.11 *For all non-prime finite fields $GF(q)$ with $q > 8$ and for all integers n, there is a 4-connected matroid having more than n inequivalent $GF(q)$-representations.* □

The only non-prime finite field not covered by Propositions 14.6.11 and 14.6.3 is $GF(8)$. For it, Geelen, Gerards, and Whittle (2009b) proposed the following:

Conjecture 14.6.12 *There is an integer k such that every 4-connected matroid has at most k inequivalent representations over $GF(8)$.* □

The family of examples used to prove Proposition 14.6.11 is easy to describe. For $m \leq n$, consider the complete bipartite graph $K_{m,n}$. The (m,n)-*mace* is obtained by freely adding a point to $M(K_{m,n})$. For a prime p and integer k exceeding one, let

$$m_{p,k} = \begin{cases} (k-1)(p-1)+1 & \text{when } p \geq 3; \\ k-1 & \text{when } p = 2. \end{cases}$$

Lemma 14.6.13 *Let p be a prime and k be an integer exceeding one. If $m \leq m_{p,k}$, then the (m,n)-mace is a vertically $(m+1)$-connected matroid having at least $\frac{2^n}{k-1}$ inequivalent $GF(p^k)$-representations.* □

For $n \geq 3$, since $K_{3,n}$ has no triangles, the last lemma gives that the $(3,n)$-mace is 4-connected and that this matroid verifies Proposition 14.6.11. This lemma also shows that Kahn's Conjecture cannot be redeemed by replacing '3-connected' by 'vertically k-connected' for any fixed integer k. Although the (m,n)-mace is vertically $(m+1)$-connected, it is not even 5-connected. To further limit what can be salvaged from Kahn's Conjecture in the non-prime case, Geelen, Gerards, and Whittle (2009b) show that the number of inequivalent $GF(p^k)$-representations of a $(m_{p,k}+1)$-connected matroid can be made arbitrarily large. They also conjecture that raising the connectivity by one more will prevent this phenomenon from occurring; that is, by making the connectivity large enough relative to the field size, one can limit the number of inequivalent representations.

Conjecture 14.6.14 *For every prime p and integer k exceeding one, there is an integer $h(p,k)$ such that every $(m_{p,k}+2)$-connected matroid has at most $h(p,k)$ inequivalent $GF(p^k)$-representations.* □

Given an n-vertex simple graph G, it is quite common to consider the *complement* G^c of G, this being the simple graph having the same vertex set as G such that two distinct vertices are adjacent in G^c if and only if they are non-adjacent in G. One can attempt to mimic this construction for simple matroids representable over $GF(q)$ by recalling the analogy between projective geometries and complete graphs. The existence of inequivalent representations poses the potential for problems here. But if M is a simple uniquely $GF(q)$-representable matroid, we can define its $(GF(q), k)$-*complement* to be the matroid $PG(k-1, q)\backslash T$ where $M \cong PG(k-1, q)|T$ and $k \geq r(M)$. It is not difficult to check that this complement is well-defined up to a relabelling of the elements of its ground set. In particular, $(GF(2), k)$-complements exist for all simple binary matroids. Moreover, by Corollary 14.6.1, $(GF(3), k)$-complements exist for all simple ternary matroids. We noted above that $U_{2,4} \oplus_2 U_{2,4}$ is not uniquely $GF(5)$-representable and that if T is the first six columns of the matrix A_1 in Example 14.6.2, then $PG(2,5)|T \cong U_{2,4} \oplus_2 U_{2,4}$. We leave it as an exercise to check that $PG(2,5)\backslash T$ is uniquely $GF(5)$-representable. Thus, somewhat disconcertingly, a $(GF(q), k)$-complement of a uniquely $GF(q)$-representable matroid can have inequivalent $GF(q)$-representations (Brylawski and Lucas 1976).

Exercises

1. Show that both a deletion and a contraction of a uniquely $\mathbb{F}$-representable matroid can fail to be uniquely $\mathbb{F}$-representable.

2. Prove that if M is uniquely $\mathbb{F}$-representable, then so too is M^*.

3. Give two inequivalent $GF(4)$-representations of $U_{2,4} \oplus_2 U_{2,4}$.

4. Show that the $(GF(2), 4)$-complements of $M(K_5)$ and $M^*(K_{3,3})$ are $U_{4,5}$ and $U_{2,3} \oplus U_{2,3}$, respectively.

5. (Oxley, Vertigan, and Whittle 1996a) Let $\mathbb{F}$ be a field and r be an integer exceeding two. Let Ψ_r^+ denote the rank-r jointed free swirl. Prove that:

   (a) If Ψ_r^+ is $\mathbb{F}$-representable, there are elements $\alpha_1, \alpha_2, \ldots, \alpha_{r-1}$ of $\mathbb{F} - \{0, 1\}$ and distinct non-zero elements δ_1 and δ_2 of $\mathbb{F}$ such that the following matrix A is a reduced standard representation for Ψ_r^+.

$$\begin{bmatrix} 1 & 1 & 0 & 0 & \cdots & 0 & 0 & 1 & 1 \\ 1 & \alpha_1 & 1 & 1 & \cdots & 0 & 0 & 0 & 0 \\ 0 & 0 & 1 & \alpha_2 & \cdots & 0 & 0 & 0 & 0 \\ \vdots & \vdots & \vdots & \vdots & \ddots & \vdots & \vdots & \vdots & \vdots \\ 0 & 0 & 0 & 0 & \cdots & 1 & 1 & 0 & 0 \\ 0 & 0 & 0 & 0 & \cdots & 1 & \alpha_{r-1} & \delta_1 & \delta_2 \end{bmatrix}$$

   (b) The matrix $[I_r|A]$ represents Ψ_r^+ over $\mathbb{F}$ if and only if the sets $\{\delta_1, \delta_2\}$ and $\{(-1)^r \gamma_1 \gamma_2 \cdots \gamma_{r-1} : \gamma_i \in \{1, \alpha_i\}$ for all $i\}$ are disjoint.

   (c) Ψ_r^+ is representable over $GF(q)$ for all prime powers q that are not of the form 2^p where $2^p - 1$ is prime because, in these cases, $GF(q)^\times$ has a proper multiplicative subgroup.

(d) If $q = 2^p$ where $2^p - 1$ is prime, then, for r sufficiently large, Ψ_r^+ is not $GF(q)$-representable.

(e) Lemma 14.6.7 holds.

6. Show that there are exactly three non-isomorphic 9-element rank-3 simple ternary matroids and give geometric representations for each.

7. Show that a simple $GF(4)$-representable matroid M of rank r is uniquely $GF(4)$-representable if $|E(M)| \geq \frac{1}{3}(4^{r-1} + 14)$.

8. *(Geelen, Gerards, and Whittle 2009b) Let p be a prime and k be an integer. For $n \geq p^{k+1}$, show that the (m, n)-mace is $GF(p^k)$-representable if and only if $k \geq 2$ and $m \leq (k-1)(p-1) + 1$.

14.7 Ternary matroids

In Section 6.6, we proved Tutte's theorem that a matroid M is representable over $GF(2)$ and some field of characteristic not 2 if and only if M is regular or, equivalently, if and only if M can be represented over $\mathbb{Q}$ by a matrix all of whose subdeterminants are in $\{0, 1, -1\}$. In this section, following Whittle (1995, 1997), we specify what classes of matroids arise when one considers the matroids that are representable over $GF(3)$ and some other field.

We begin with a natural analogue of totally unimodular matrices. A *dyadic matrix* is a matrix over $\mathbb{Q}$ all of whose subdeterminants are in $\{0, \pm 2^i : i \in \mathbb{Z}\}$. A matroid M is *dyadic* if there is a dyadic matrix A such that $M \cong M[A]$. Whittle (1997) proved the following attractive characterization of dyadic matroids.

Theorem 14.7.1 *The following are equivalent for a matroid M.*

(i) M *is dyadic.*

(ii) M *is representable over $GF(3)$ and $GF(5)$.*

(iii) M *is representable over $GF(p)$ for all odd primes p.*

(iv) M *is representable over $GF(3)$ and $\mathbb{Q}$.*

(v) M *is representable over $GF(3)$ and $\mathbb{R}$.*

(vi) M *is representable over $GF(3)$ and $GF(q)$ where q is an odd prime power that is congruent to 2 mod 3.* □

Whittle also showed that only a small number of classes arise when one considers the matroids that are representable over $GF(3)$ and some other field.

Theorem 14.7.2 *Let $\mathcal{F}$ be a set of fields that contains $GF(3)$ and let $\mathcal{M}$ be the class of matroids that are representable over all fields in $\mathcal{F}$. Then $\mathcal{M}$ coincides with the class of matroids that are representable over $GF(3)$ and $GF(q)$ for some q in $\{2, 3, 4, 5, 7, 8\}$.* □

This is an intriguing result. Although $\mathcal{F}$ can be an infinite set and can even contain infinite fields, the only six classes that arise do so when $\mathcal{F}$ contains at most two finite fields. One may wonder whether the last theorem extends to, for example, sets of fields containing $GF(4)$. The next proposition (Whittle 2005 and private communication) shows emphatically that it does not. A prime number of the form $2^n - 1$ is a *Mersenne prime*. It is well known that if $2^n - 1$ is prime, then so is n. Very few Mersenne primes are known and it is a famous unsolved problem to determine whether the number of them is finite or infinite There are several heuristics (see, for example, Crandall and Pomerance (2005, p. 26)) which support the existence of infinitely many.

Proposition 14.7.3 *Let q and s be prime powers exceeding three and $\mathcal{M}(q, s)$ be the set of matroids that are representable over both $GF(q)$ and $GF(s)$. For fixed q, the number of distinct classes of the form $\mathcal{M}(q, s)$ is infinite if q is not prime. This number is also infinite when q is prime provided there are infinitely many Mersenne primes.*

Proof Let $q = p^k$ where p is a prime and k is an integer exceeding one. By Lemma 14.6.8, the rank-r tipless free spike Φ_r is $GF(p^k)$-representable. It is straightforward to show that Φ_r is representable over all sufficiently large finite fields. On the other hand, Geelen, Oxley, Vertigan, and Whittle (2002, Lemma 11.6) have shown that if p_1 is a prime exceeding four, then Φ_r is not $GF(p_1)$-representable if $r \geq p_1 - 1$. Thus Φ_{p_1-1} is representable over $GF(q)$ but not over $GF(p_1)$. Moreover, Φ_{p_1-1} is representable over $GF(p')$ for all primes p' greater than or equal to some prime p_2. Thus the set of fields over which Φ_{p_1-1} is representable differs from the set of fields over which Φ_{p_2-1} is representable. Clearly one can continue this process to prove that there are infinitely many choices for $\mathcal{M}(q, s)$ when q is not prime.

Now suppose that q is prime. Then, by Lemma 14.6.7, the rank-r jointed free swirl Ψ_r^+ is $GF(q)$-representable. Again it is clear that Ψ_r^+ is representable over all sufficiently large finite fields. If $2^p - 1$ is a Mersenne prime, then it is not difficult to show (see Exercise 5 of the last section) that if r is sufficiently large, then Ψ_r^+ is not representable over $GF(2^p)$. The result now follows as before assuming there are infinitely many Mersenne primes. $\square$

From above, the classes of matroids that are representable over $GF(3)$ and $GF(q)$ coincide with the classes of regular, ternary, and dyadic matroids when q is $2, 3$, and 5, respectively. To describe the other classes of matroids that arise here, we require some more definitions. A $\sqrt[6]{1}$-*matrix* is a matrix over the field, $\mathbb{C}$, of complex numbers such that all its non-zero subdeterminants are complex sixth roots of unity. A matroid M is *sixth-root-of-unity* or is a $\sqrt[6]{1}$-*matroid* if $M \cong M[A]$ for some $\sqrt[6]{1}$-matrix A. Most of the work required to prove Theorem 14.7.1 and the other results in this section is involved with establishing the following result.

Lemma 14.7.4 *If M is a ternary 3-connected matroid that is representable over some field of characteristic other than 3, then M is either a dyadic matroid or a $\sqrt[6]{1}$-matroid.* $\square$

The next two theorems characterize the classes of matroids that are representable over $GF(3)$ and $GF(q)$ when q is 4 and when q is 7.

Theorem 14.7.5 *The following are equivalent for a matroid M.*

(i) M is a $\sqrt[6]{1}$-matroid.

(ii) M is representable over $GF(3)$ and $GF(4)$.

(iii) M is representable over $GF(3)$ and $GF(2^k)$ for some even integer k.   □

Theorem 14.7.6 *The following are equivalent for a matroid M.*

(i) M is representable over $GF(3)$ and $GF(7)$.

(ii) M is representable over $GF(3)$ and $\mathbb{C}$.

(iii) M is representable over $GF(3)$ and $GF(q)$ where q is an odd prime power that is congruent to $1 \bmod 3$.

(iv) M is a dyadic matroid, a $\sqrt[6]{1}$-matroid, or can be constructed from dyadic matroids and $\sqrt[6]{1}$-matroids by taking direct sums and 2-sums.   □

By Theorem 14.7.2, it remains to characterize the class of matroids that are representable over $GF(3)$ and $GF(8)$. This too is an interesting class. Let $\mathbb{Q}(\alpha)$ be the field that is obtained by extending $\mathbb{Q}$ by a transcendental α. A matrix over $\mathbb{Q}(\alpha)$ is *near-unimodular* if all of its non-zero subdeterminants are in $\{\pm\alpha^i(\alpha - 1)^j : i, j \in \mathbb{Z}\}$. A matroid M is *near-regular* if $M \cong M[A]$ for some near-unimodular matrix A.

Theorem 14.7.7 *The following are equivalent for a matroid M.*

(i) M is near-regular.

(ii) M is representable over $GF(3)$ and $GF(8)$.

(iii) M is representable over $GF(3)$, $GF(4)$, and $GF(5)$.

(iv) M is representable over $GF(3)$, $GF(4)$, and $\mathbb{Q}$.

(v) M is representable over all fields except possibly $GF(2)$.   □

The results above motivate the study of the classes of dyadic matroids, $\sqrt[6]{1}$-matroids, and near-regular matroids. Each of these classes has several useful properties.

Proposition 14.7.8 *Let $\mathcal{M}$ be the class of dyadic matroids, the class of $\sqrt[6]{1}$-matroids, or the class of near-regular matroids. Then $\mathcal{M}$ is closed under duality and the taking of minors, direct sums, and 2-sums.*   □

Oxley, Vertigan, and Whittle (1996b) gave conjectured lists of excluded minors for the classes of dyadic matroids, $\sqrt[6]{1}$-matroids, and near-regular matroids. The last two of these lists have now been confirmed. Using Lemma 14.5.1, we can determine the excluded minors for the class of $\sqrt[6]{1}$-matroids from Theorem 14.7.5. Indeed, Geelen, Gerards, and Kapoor (2000) derived the next theorem from their excluded-minor characterization of quaternary matroids (Theorem 6.5.9) and the corresponding result for ternary matroids (Theorem 6.5.7).

Theorem 14.7.9 *A matroid is a $\sqrt[6]{1}$-matroid if and only if it has no minor isomorphic to any of $U_{2,5}$, $U_{3,5}$, F_7, F_7^*, F_7^-, $(F_7^-)^*$, or P_8.*   □

Ternary matroids

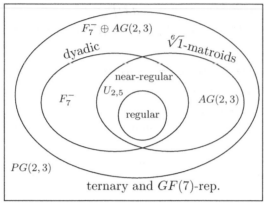

FIG. 14.8. Some special classes of ternary matroids.

Geelen announced that, by adapting the proof of the excluded-minor characterization of quaternary matroids, one could determine the complete list of excluded minors for the class of near-regular matroids. Hall, Mayhew, and Van Zwam (2009) have now done this. The ternary affine plane, $AG(2,3)$, has a transitive automorphism group, so has a unique single-element deletion, $AG(2,3)\backslash e$. Moreover, the last matroid has an automorphism mapping any three-point line to any other three-point line. Hence there is a unique matroid, $\Delta_3(AG(2,3)\backslash e)$, that is obtained by performing a Δ-Y exchange on $AG(2,3)\backslash e$. It is not difficult to check that $\Delta_3(AG(2,3)\backslash e)$ is self-dual.

Theorem 14.7.10 *A matroid is near-regular if and only if it has no minor isomorphic to any of* $U_{2,5}$, $U_{3,5}$, F_7, F_7^*, F_7^-, $(F_7^-)^*$, $AG(2,3)\backslash e$, $(AG(2,3)\backslash e)^*$, $\Delta_3(AG(2,3)\backslash e)$, *or* P_8.   □

Figure 14.8 is a Venn diagram showing the relationships between the various classes of ternary matroids considered in this section. It shows, for example, that the class of near-regular matroids contains the class of regular matroids and is the intersection of the classes of dyadic and $\sqrt[6]{1}$-matroids. Examples are given to show that each of the indicated inclusions is proper. For instance, F_7^- is dyadic but is not sixth-root-of-unity.

As pointed out by Geelen, Gerards, and Kapoor (2000), the list of excluded minors for the class of dyadic matroids conjectured by Oxley, Vertigan, and Whittle (1996b) is incomplete and should also include the matroid T_8 that is represented over $GF(3)$ by the matrix $[I_4|J_4 - I_4]$. Pendavingh (private communication, 2008) identified two other matroids that also need to be added to the list. These matroids, N_1 and N_2, have the following matrices over $GF(3)$ as reduced standard representations:

$$\begin{bmatrix} 0 & 1 & 1 & 0 & 1 \\ 1 & -1 & 0 & 0 & 1 \\ 1 & 0 & 0 & 1 & -1 \\ 0 & 0 & 1 & -1 & 1 \\ 1 & 1 & -1 & 1 & 0 \end{bmatrix} \quad \text{and} \quad \begin{bmatrix} -1 & 1 & 0 & 1 & 0 & 1 \\ 1 & 0 & 0 & 0 & 1 & 1 \\ 0 & 0 & 1 & 1 & 0 & 1 \\ 1 & 0 & 1 & 0 & 1 & 0 \\ 0 & 1 & 0 & 1 & -1 & 0 \\ 1 & 1 & 1 & 0 & 0 & 0 \end{bmatrix}.$$

Problem 14.7.11 *Determine if* $\{U_{2,5}, U_{3,5}, F_7, F_7^*, AG(2,3)\backslash e, (AG(2,3)\backslash e)^*,$ $\Delta_3(AG(2,3)\backslash e), T_8, N_1, N_2\}$ *is the complete set of excluded minors for the class of dyadic matroids.*   □

By combining the results above, we obtain the following variant on Theorem 14.7.2. We leave the proof as an exercise.

Corollary 14.7.12 *Let $\mathcal{F}$ be a set of fields that contains $\{GF(3), GF(4)\}$ and let $\mathcal{M}$ be the class of matroids that are representable over all fields in $\mathcal{F}$. Then $\mathcal{M}$ coincides with the class of regular matroids, the class of near-regular matroids, or the class of sixth-root-of-unity matroids.*   □

Using this result, together with the excluded-minor characterizations of the classes of regular, near-regular, and sixth-root-of-unity matroids, we get that Conjecture 14.5.13 holds when $\mathcal{F}$ contains $\{GF(3), GF(4)\}$.

The classes of regular, dyadic, sixth-root-of-unity, and near-regular matroids are all obtained by taking matroids representable over some field and restricting the possible values of non-zero subdeterminants. This observation prompted Semple and Whittle (1996) to introduce the following class of matroids. Let $\mathbb{F}$ be a field and G be a subgroup of the multiplicative group of $\mathbb{F}$ such that $-g \in G$ for all g in G. A $(G, \mathbb{F})$-*matroid* is a matroid that can be represented over $\mathbb{F}$ by a matrix all of whose non-zero subdeterminants are in G. Thus the class of regular matroids coincides with the class of $(\{1, -1\}, \mathbb{Q})$-matroids and with the class of $(\{1, -1\}, \mathbb{R})$-matroids. The classes of dyadic, near-regular, and sixth-root-of-unity matroids coincide with the classes of $(\{\pm 2^i : i \in \mathbb{Z}\}, \mathbb{Q})$-, $(\{\pm \alpha^i (\alpha - 1)^j : i, j \in \mathbb{Z}\}, \mathbb{Q}(\alpha))$-, and $(G_6, \mathbb{C})$-matroids, where G_6 is the group of complex sixth roots of unity. Choe, Oxley, Sokal, and Wagner (2004) extended the last observation by proving the following result.

Proposition 14.7.13 *The class of $\sqrt[6]{1}$-matroids coincides with the class of matroids that can be represented over $\mathbb{C}$ by a matrix all of whose non-zero subdeterminants have modulus one.*   □

Let τ be the *golden mean*, that is, $\frac{1+\sqrt{5}}{2}$, the positive real root of the equation $x^2 - x - 1 = 0$. We call a $(\{\pm \tau^i : i \in \mathbb{Z}\}, \mathbb{R})$-matroid a *golden-mean matroid*. Thus a golden-mean matroid is one that can be represented over $\mathbb{R}$ by a matrix all of whose non-zero determinants are ± 1 times a power of τ. Vertigan (in Whittle 2005) gave the following appealing characterization of such matroids. A proof of this theorem was published by Pendavingh and Van Zwam (2010a).

Theorem 14.7.14 *The following are equivalent for a matroid M.*

(i) *M is golden-mean.*

(ii) *M is representable over $GF(4)$ and $GF(5)$.*

(iii) *M is representable over $GF(p)$ for all primes p such that $p \equiv \pm 1 \pmod 5$ or $p = 5$, and M is representable over $GF(p^2)$ for all primes p.* □

Again consider a field $\mathbb{F}$ and a subgroup G of $\mathbb{F}^{\times}$ such that $-g \in G$ for all g in G. Then the set $G \cup \{0\}$ with the operations induced from $\mathbb{F}$ behaves much like a field except that the sum of two elements from this set may not be in $G \cup \{0\}$. This lead Semple and Whittle (1996) to give an axiomatic definition of partial fields, a class of algebraic structures that behave like fields except that the sum of two elements need not be defined. Subsequently, Vertigan (in Whittle 2005) showed that partial fields coincide with certain structures that arise in a natural way from a commutative ring with unity. A proof that these approaches are equivalent appears in Pendavingh and Van Zwam (2010a).

Let R be a commutative ring with unity, and let $U(R)$ be the set of *units* of R, that is, those elements of R that have a multiplicative inverse. A *partial field* P is a subset of R that contains $\{0, 1, -1\}$ such that $P - \{0\}$ is a multiplicative subgroup of $U(R)$. Thus multiplication is a binary operation on P but addition is only a partial binary operation because the sum $a + b$ of elements a and b of P is defined if and only if the element $a + b$ of R is also in P. Semple and Whittle (1996) extended the theory of matroids representable over fields to matroids representable over partial fields. Vertigan (in Whittle 2005) continued this study and showed that, for every set $\mathcal{F}$ of fields, there is a partial field P such that the class of matroids representable over all fields in $\mathcal{F}$ coincides with the class of matroids representable over P. Pendavingh and Van Zwam (2010a, 2010b) developed this theory further, giving new proofs of Whittle's results above and proving some interesting new results. In particular, they established that every matroid that is representable over a partial field is representable over some field.

Exercises

1. (Whittle 1997) Let A be a near-unimodular matrix and D be a matrix over $\mathbb{Q}(\alpha)$. Prove that if D is obtained from A by a sequence of pivots and row and column scalings by elements of the form $\pm \alpha^i (\alpha - 1)^j$, then D is near-unimodular.

2. (Whittle 1997) Prove that if A is a dyadic matrix representing a dyadic matroid M and $\mathbb{F}$ is a field of characteristic other than two, then the matrix obtained by interpreting the elements of A as elements of $\mathbb{F}$ is a representation of M over $\mathbb{F}$.

3. Prove that a ternary matroid is a $\sqrt[6]{1}$-matroid if and only if it has no minor isomorphic to F_7^-, $(F_7^-)^*$, or P_8.

4. Prove that every signed-graphic matroid is dyadic.

5. Prove Proposition 14.7.13.

6. (Semple and Whittle 1996) Prove that the class of matroids representable over a fixed partial field is closed under taking duals, minors, direct sums, and 2-sums.

14.8 Stabilizers

As noted in Corollary 14.6.1, ternary matroids are uniquely $GF(3)$-representable. However, in general, a ternary $GF(q)$-representable matroid M is not uniquely $GF(q)$-representable. In proving his theorems on ternary matroids from the last section, Whittle overcame this difficulty by showing that the behaviour of the inequivalent $GF(q)$-representations of such a matroid M is predictable. In particular, by Proposition 6.6.5, if M is binary, then it is uniquely $GF(q)$-representable. Thus we may assume M is non-binary, that is, M has a $U_{2,4}$-minor. Now $U_{2,4}$ itself is representable over $GF(q)$ by $\begin{bmatrix} 1 & 0 & 1 & 1 \\ 0 & 1 & 1 & a \end{bmatrix}$ for every element a of $GF(q) - \{0, 1\}$. Moreover, when q is prime, all such representations are inequivalent. Whittle (1996) showed that if N is a ternary 3-connected matroid with a $U_{2,4}$-minor, then a representation of N over a field $\mathbb{F}$ is uniquely determined by an $\mathbb{F}$-representation of an arbitrary fixed $U_{2,4}$-minor of N. Essentially, then, the behaviour of inequivalent representations of N is no more complex than the behaviour of inequivalent representations of $U_{2,4}$.

Motivated by a desire to generalize the ideas above to other classes of matroids, Whittle (1999) introduced the notion of a stabilizer, which is exemplified above by $U_{2,4}$. In this section, we define stabilizers and show that they can be used, for example, to prove several of the results from Section 14.6. The definitions and results are stated here for fields although Whittle developed the theory more generally for matroids representable over partial fields.

Let $\mathbb{F}$ be a field and A be an $\mathbb{F}$-representation of a matroid M. The process of using A to find $\mathbb{F}$-representations for minors of M is familiar (see Section 3.2). For the deletion of an element e, we simply delete the column with that label. Now suppose that e is not a loop. To construct a representation for M/e, we pivot on some non-zero entry of the column labelled by e and then delete that column and the row corresponding to the pivot entry. Although the representation we obtain in this way will depend on the choice of the pivot entry, all resulting representations of M/e are easily shown to be projectively equivalent. Now let I and I^* be an independent set and a coindependent set of M with $I \cap I^* = \emptyset$. By iterating the above process, we obtain a family of projectively equivalent representations for $M/I\backslash I^*$. Each such representation is said to be *induced* by A. Now let N be a minor of M. We say that N is an $\mathbb{F}$-*stabilizer for M* or that N *stabilizes M over* $\mathbb{F}$ if, whenever $N \cong M/I\backslash I^*$ for some independent set I and coindependent set I^*, two $\mathbb{F}$-representations A_1 and A_2 of M that induce projectively equivalent representations of $M/I\backslash I^*$ are themselves projectively equivalent. Let $\mathcal{M}$ be a minor-closed class of matroids that is also closed under duality and let $\mathbb{F}$ be a field. A matroid N in $\mathcal{M}$ *stabilizes $\mathcal{M}$* over $\mathbb{F}$ if N stabilizes over $\mathbb{F}$ every 3-connected member of $\mathcal{M}$ that has an N-minor. When no danger of ambiguity arises, we sometimes omit reference to the field $\mathbb{F}$.

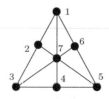

FIG. 14.9. F_7^-.

Example 14.8.1 Let $\mathbb{F}$ be a field of characteristic other than two. Let the non-Fano matroid, F_7^-, be labelled as in Figure 14.9. We shall show that $U_{2,4}$ stabilizes F_7^- over $\mathbb{F}$. Let A_1 and A_2 be $\mathbb{F}$-representations of F_7^- and suppose that $F_7^-/I\backslash I^* \cong U_{2,4}$ for some independent set I and coindependent set I^*. By symmetry, we may assume that $I = \{6\}$. Thus I^* contains one member of each of $\{3,7\}$ and $\{1,5\}$. By symmetry again, we may assume that $I^* = \{3,5\}$. Every 2×4 matrix representing $U_{2,4}$ over $\mathbb{F}$ is projectively equivalent to a matrix of the form $\left[\begin{smallmatrix} 1 & 0 & 1 & 1 \\ 0 & 1 & 1 & a \end{smallmatrix}\right]$ for some a in $\mathbb{F} - \{0,1\}$. Moreover, if B is an arbitrary basis of F_7^- containing I and avoiding I^*, then every $\mathbb{F}$-representation of F_7^- is projectively equivalent to a standard representation relative to B. Hence A_1 and A_2 are both projectively equivalent to matrices of the form $[I_4|D]$ where D is

$$
\begin{array}{c}
\begin{array}{cccc} 2 & 7 & 5 & 3 \end{array} \\
\begin{array}{c} 1 \\ 4 \\ 6 \end{array}
\left[\begin{array}{cccc}
1 & 1 & 1 & 1 \\
1 & a & 0 & a \\
b & c & d & e
\end{array}\right]
\end{array}
$$

for some a, b, c, d, and e in $\mathbb{F}$ where $a \notin \{0,1\}$. The circuits $\{1,4,7\}, \{3,4,5\}$, and $\{1,2,3\}$ imply that $c = 0$, that $d = e$, and that $ab = e$. Moreover, as $\{1,2,4,6\}$ is a circuit, $b \neq 0$. Thus D is

$$
\begin{array}{c}
\begin{array}{cccc} 2 & 7 & 5 & 3 \end{array} \\
\begin{array}{c} 1 \\ 4 \\ 6 \end{array}
\left[\begin{array}{cccc}
1 & 1 & 1 & 1 \\
1 & a & 0 & a \\
b & 0 & ab & ab
\end{array}\right].
\end{array}
$$

The circuit $\{2,5,7\}$ implies that $ba(a - 2) = 0$, so $a = 2$. Furthermore, we can scale the last row of D to make $b = 1$. We conclude that $U_{2,4}$ does indeed stabilize F_7^- over $\mathbb{F}$. Although not every $\mathbb{F}$-representation of $U_{2,4}$ extends to one for F_7^-, when a representation does extend, it does so uniquely. ☐

The next proposition (Whittle 1999) provides one of the main reasons for studying stabilizers.

Proposition 14.8.2 *Let $\mathbb{F}$ be a field and N be an $\mathbb{F}$-stabilizer for the class $\mathcal{M}$ of $\mathbb{F}$-representable matroids. If N has k inequivalent $\mathbb{F}$-representations, then every 3-connected matroid in $\mathcal{M}$ with an N-minor has at most k inequivalent*

$\mathbb{F}$-*representations. In particular, if N is uniquely representable over $\mathbb{F}$, then so is every 3-connected matroid in $\mathcal{M}$ that has an N-minor.*

Proof Let M be a 3-connected matroid in $\mathcal{M}$ having an N-minor, and let A_1 and A_2 be inequivalent $\mathbb{F}$-representations of M. Let N_0 be a fixed N-minor of M and let A_1'' and A_2'' be $\mathbb{F}$-representations of N_0 induced by A_1 and A_2. The result will hold if we can show that A_1'' and A_2'' are inequivalent. Assume they are equivalent. By the definition of a stabilizer, A_1'' and A_2'' are not projectively equivalent. Moreover, by Lemma 6.3.10, A_1'' is projectively equivalent to a matrix A_2''' that can be obtained from A_2'' by applying a field automorphism σ to all of the entries of the latter. If we apply σ to all the entries of A_2, we obtain a matrix A_2' that induces a representation of N_0 that is projectively equivalent to A_2'''. As N is a stabilizer for M, and A_1'' and A_2''' are projectively equivalent, A_1 and A_2' are projectively equivalent. But A_2' and A_2 are equivalent. Hence so are A_1 and A_2; a contradiction. We conclude that A_1'' and A_2'' are inequivalent. $\square$

The next result, whose straightforward proof is omitted, notes some basic properties of stabilizers.

Lemma 14.8.3 *Let $\mathbb{F}$ be a field, N be a 3-connected $\mathbb{F}$-representable matroid, and $\mathcal{M}$ be a minor-closed class of $\mathbb{F}$-representable matroids that contains N and is also closed under duality.*

(i) *If M is an $\mathbb{F}$-representable matroid having an N-minor, then N stabilizes M over $\mathbb{F}$ if and only if N^* stabilizes M^* over $\mathbb{F}$.*

(ii) *N stabilizes $\mathcal{M}$ over $\mathbb{F}$ if and only if N^* stabilizes $\mathcal{M}$ over $\mathbb{F}$.*

(iii) *Let N_1 and N_2 be $\mathbb{F}$-representable matroids such that N is a minor of N_1, and N_1 is a minor of N_2. Assume that N stabilizes N_1. Then N stabilizes N_2 if and only if N_1 stabilizes N_2.*

(iv) *If N is an $\mathbb{F}$-stabilizer for $\mathcal{M}$, and N' is a 3-connected matroid in $\mathcal{M}$ having an N-minor, then N' is an $\mathbb{F}$-stabilizer for $\mathcal{M}$.* $\square$

The task of checking whether a 3-connected matroid N stabilizes a class $\mathcal{M}$ is potentially infinite. Whittle's (1999) main result is that this task is finite. Indeed, it can be reduced to looking at the 3-connected members of $\mathcal{M}$ that have an N-minor and have at most two more elements than N.

Theorem 14.8.4 *Let $\mathbb{F}$ be a field, $\mathcal{M}$ be a minor-closed class of $\mathbb{F}$-representable matroids that is also closed under duality, and N be a 3-connected matroid in $\mathcal{M}$. Then N stabilizes $\mathcal{M}$ if and only if N stabilizes every 3-connected matroid M in $\mathcal{M}$ of one of the following types.*

(i) *M has an element x such that $M\backslash x = N$.*

(ii) *M has an element y such that $M/y = N$.*

(iii) *M has elements x and y such that $M\backslash x/y = N$ and both $M\backslash x$ and M/y are 3-connected.* $\square$

The last theorem has a similar form to Seymour's results that check whether a matroid is a splitter for a class (Corollary 12.1.3) and whether a class of matroids is 1- or 2-rounded (Theorem 12.3.9). But both of Seymour's results involve considering only single-element extensions and coextensions of the candidate matroids.

The next proposition presents some bounds on the number of inequivalent representations of certain classes of matroids. To prove this result, we shall use the following application of the last theorem.

Lemma 14.8.5 (i) *For all fields* $\mathbb{F}$, *the matroid* $U_{2,4}$ *is an* $\mathbb{F}$-*stabilizer for the class of* $\mathbb{F}$-*representable matroids with no* $U_{2,5}$- *or* $U_{3,5}$-*minor.*

(ii) $U_{2,4}$ *is a* $GF(4)$-*stabilizer for the class of quaternary matroids.*

(iii) $U_{2,5}$ *is a* $GF(5)$-*stabilizer for the class of* $GF(5)$-*representable matroids.*

Proof Because $U_{2,4}$ has $U_{2,5}$ as its only 3-connected single-element extension, part (i) follows immediately from Theorem 14.8.4 using duality since the required case checks are vacuous. For part (ii), we note that a $GF(4)$-representation of $U_{2,4}$ stabilizes $U_{2,5}$. Indeed, the representation of $U_{2,4}$ effectively labels four of the five one-dimensional subspaces of the vector space $V(2,4)$, so the fifth point of $U_{2,5}$ must be labelled by the one remaining one-dimensional subspace. By duality, $U_{2,4}$ stabilizes $U_{3,5}$ over $GF(4)$. To check the 3-connected quaternary matroids M for which $M\backslash x/y \cong U_{2,4}$ and $M\backslash x$ and M/y are 3-connected, we observe that $M\backslash x \cong U_{3,5}$. The only 3-connected quaternary single-element extensions of $U_{3,5}$ are $U_{3,6}$ and Q_6, where we recall that the latter is obtained by placing a point p on the intersection of two lines of $U_{3,5}$. Because of the placement of p on this intersection, $U_{3,5}$ stabilizes Q_6. Hence so does $U_{2,4}$. On the other hand, a quaternary representation for $U_{3,6}$ may be assumed to have a reduced standard representative matrix of the following form where $(a,b) \neq (c,d)$:

$$\begin{bmatrix} 1 & 1 & 1 \\ 1 & a & c \\ 1 & b & d \end{bmatrix}.$$

Moreover, a, b, c, and d are in $\{\omega, \omega+1\}$, while neither a nor d is in $\{b, c\}$. Thus $\{(a,b),(c,d)\} = \{(\omega, \omega+1),(\omega+1,\omega)\}$. The choice of (c,d) is forced once we decide on (a,b). Thus $U_{3,5}$ stabilizes $U_{3,6}$ over $GF(4)$, and (ii) follows. The proof of (iii) may be found in Oxley, Vertigan, and Whittle (1996a). $\square$

The next result exemplifies the use of stabilizers. In the first part, we give another proof of Corollary 14.6.1, while parts (iv) and (v) prove the last part of Proposition 14.6.3 and Proposition 14.6.5.

Proposition 14.8.6 (i) *Ternary matroids are uniquely* $GF(3)$-*representable.*

(ii) *If q is a prime power other than 2 or 3, then a 3-connected matroid with no* $U_{2,5}$- *or* $U_{3,5}$-*minor has at most* $q-2$ *inequivalent* $GF(q)$-*representations.*

(iii) *If q is a prime power other than 2 or 3, then a 3-connected ternary matroid has at most* $q-2$ *inequivalent* $GF(q)$-*representations.*

(iv) *Every 3-connected quaternary matroid is uniquely GF(4)-representable.*

(v) *Every 3-connected GF(5)-representable matroid has at most six inequivalent GF(5)-representations.*

Proof In each part, we use the fact that an $\mathbb{F}$-representable binary matroid is uniquely $\mathbb{F}$-representable, which was proved in Proposition 6.6.5. For (i), let M be a ternary matroid. As just noted, we may assume it is non-binary, so it has a $U_{2,4}$-minor. Certainly $U_{2,4}$ is uniquely $GF(3)$-representable. Hence, by Proposition 14.8.2, if M is 3-connected, it is also uniquely $GF(3)$-representable. If M is not 3-connected, it can be built using direct sums and 2-sums from 3-connected matroids. Part (i) follows by an easy induction argument using the fact that it holds for 3-connected matroids and that $GF(3)$ has no automorphisms.

As noted in Example 14.8.1, every $GF(q)$-representation of $U_{2,4}$ is equivalent to a matrix of the form $\left[\begin{smallmatrix} 1 & 0 & 1 & 1 \\ 0 & 1 & 1 & a \end{smallmatrix}\right]$ for some a in $GF(q) - \{0,1\}$. Thus $U_{2,4}$ has at most $q - 2$ inequivalent $GF(q)$-representations. Moreover, equality holds here if and only if $GF(q)$ has no automorphisms, that is, if and only if q is prime. Part (ii) follows from this using Proposition 14.8.2 and Lemma 14.8.5(i). Part (iii) follows immediately from part (ii).

From the last paragraph, $U_{2,4}$ is uniquely $GF(4)$-representable and part (iv) follows by Proposition 14.8.2 since, by Lemma 14.8.5(ii), $U_{2,4}$ is a $GF(4)$-stabilizer for the class of quaternary matroids.

For (v), let M be a 3-connected $GF(5)$-representable matroid. If M has no $U_{2,5}$- or $U_{3,5}$-minor, then, by (ii), M has at most three inequivalent $GF(5)$-representations. By duality, we may now assume that M has a $U_{2,5}$-minor. As shown after Conjecture 14.6.4, $U_{2,5}$ has exactly six inequivalent $GF(5)$-representations. Part (v) now follows by Lemma 14.8.5(iii) and Proposition 14.8.2. □

Geelen, Gerards, and Whittle (2007b) note that Theorem 14.6.10 seems inadequate for proving Rota's Conjecture for prime fields and that what seems to be needed is to extend the theory of stabilizers to 4-connected matroids. They also observe that $GF(5)$ is peculiar with respect to Rota's Conjecture in that the results of this section resolve all issues caused by inequivalent representations. But to be able to extend the methods used by Geelen, Gerards, and Kapoor (2000) to settle Rota's Conjecture for $GF(4)$, one encounters other problems for which it would be very useful to establish the truth of the following conjecture. Indeed, Mayhew, Whittle, and Van Zwam (2010) have proved that if this conjecture holds for $GF(5)$, then Rota's Conjecture also holds for $GF(5)$.

Conjecture 14.8.7 *Let N be a $GF(q)$-representable matroid. Then there is an integer k such that if M is a $GF(q)$-representable matroid with branch width at least k, and M has N as a minor, then there is an element e in $E(M) - E(N)$ such that both $M\backslash e$ and M/e have N as a minor.* □

Geelen, Gerards, and Whittle (2007b) hope to prove this conjecture as a corollary of their proposed structure theorem for $GF(q)$-representable matroids.

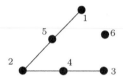

FIG. 14.10. Q_6.

Geelen (private communication) established that the truth of the last conjecture actually implies the truth of Conjecture 14.5.4, an intertwining conjecture.

Proposition 14.8.8 *If Conjecture 14.8.7 holds for a prime power q, then, for every two matroids N_1 and N_2, there are only finitely many $GF(q)$-representable intertwines of N_1 and N_2.*

Proof Assume that Conjecture 14.8.7 holds. We may also assume that N_1 and N_2 are $GF(q)$-representable. Let k_1 be the constant obtained by applying Conjecture 14.8.7 to N_1. By Theorem 14.3.4, there are only finitely many $GF(q)$-representable intertwines of N_1 and N_2 having branch-width at most $k_1 + |E(N_2)| - 1$.

Let M be an arbitrary $GF(q)$-representable intertwine of N_1 and N_2. To complete the proof, we shall show that $\mathrm{bw}(M) \le k_1 + |E(N_2)| - 1$. Assume that $\mathrm{bw}(M) \ge k_1 + |E(N_2)|$. Let N_1' and N_2' be minors of M that are isomorphic to N_1 and N_2, respectively. Now let M' be a minor of M obtained by removing the elements of $E(N_2') - E(N_1')$ in such a way that N_1' remains a minor of M'. By Proposition 14.2.3(iii), $\mathrm{bw}(M') \ge k_1$. By Conjecture 14.8.7, there is an element e of $E(M') - E(N_1')$ such that $M'\backslash e$ and M'/e both have N_1' as a minor. Now $E(N_2') \cap E(M') \subseteq E(N_1')$, so $e \in E(M) - E(N_2')$, and both $M\backslash e$ and M/e have an N_1-minor. Moreover, at least one of $M\backslash e$ and M/e has N_2' as a minor, so we have a contradiction to the fact that M is an intertwine of N_1 and N_2. □

To close this discussion of inequivalent representations, we briefly consider the notion of a fixed element in a matroid. Consider the matroid Q_6 labelled as in Figure 14.10. The only fixed element in this matroid is 2. The point 6 is freely placed in the plane while 5, for example, is freely placed on the line through 1 and 2. To make this formal, we first recall that elements e and f of a matroid M are clones if the map that interchanges e and f but acts as the identity on $E(M) - \{e, f\}$ is an automorphism of M. Thus clones are elements that are indistinguishable up to labelling. A *clonal class* of M is a maximal subset X of $E(M)$ such that every two distinct elements of X are clones. An element x of M is *fixed* in M if there is no single-element extension M_y of M by y such that x and y are clones in M_y and $\{x, y\}$ is independent in M_y. Dually, the element z of M is *cofixed* in M if it is fixed in M^*. For example, 6 is cofixed in Q_6.

Geelen, Oxley, Vertigan, and Whittle (1998) showed, by proving the following result, that every element in a 3-connected binary matroid is fixed and cofixed.

Proposition 14.8.9 *Let M be a connected binary matroid with at least two elements and let x be an element of M. If $M\backslash x$ is connected, then x is fixed in M. Dually, if M/x is connected, then x is cofixed in M.* □

Duke (1988) introduced a notion of freedom of an element of a matroid as a measure of how freely placed the element is in the matroid. He proved the following characterization of fixed elements.

Proposition 14.8.10 *Let e be a non-loop element of a matroid M. Then e is fixed in M if and only if $\mathrm{cl}(\{e\})$ is in the modular cut generated by the cyclic flats of M containing e.* □

Exemplifying this, we see that, in Q_6, the cyclic flats $\{1, 2, 5\}$ and $\{2, 3, 4\}$ form a modular pair. Thus their intersection, $\{2\}$, is in the modular cut generated by the cyclic flats containing 2.

Geometrically, it is the existence of elements with freedom that gives rise to the potential for inequivalent representations. If a matroid M is $\mathbb{F}$-representable and $x \in E(M)$, an $\mathbb{F}$-representation of $M\backslash x$ need not extend to an $\mathbb{F}$-representation for M. Next we note that, when x is fixed, if a representation does extend, it does so uniquely. The straightforward proof is omitted.

Lemma 14.8.11 *For a matroid M that is representable over a field $\mathbb{F}$, if x is fixed in M and A is an $\mathbb{F}$-representation of $M\backslash x$ that extends to a representation $[A|\underline{x}]$ of M, then $\underline{x}$ is unique up to multiplication by a non-zero scalar.* □

Geelen, Oxley, Vertigan, and Whittle (2002) call a 3-connected matroid M *totally free* if it has at least four elements and, for all x in $E(M)$, if $\mathrm{co}(M\backslash x)$ is 3-connected, then x is not fixed in M, while if $\mathrm{si}(M/x)$ is 3-connected, then x is not cofixed in M. The significance of such matroids is that they determine the number of inequivalent representations a matroid can have.

Proposition 14.8.12 *Let q be a prime power and M be a 3-connected $GF(q)$-representable matroid. If M has no totally free minors, then it is uniquely $GF(q)$-representable. Otherwise, the number of inequivalent $GF(q)$-representations of M is bounded above by the maximum, over all totally free minors of M, of the number of inequivalent $GF(q)$-representations of M.* □

A pleasing feature of totally free matroids is that they do not occur sporadically and can be found by an inductive search.

Theorem 14.8.13 *Let M be a totally free matroid with at least five elements and suppose that, for all x in $E(M)$, neither $M\backslash x$ nor M/x is totally free. Then the ground set of $E(M)$ is the union of 2-element clonal classes. Moreover, if $a \in E(M)$, then $M\backslash a$ and M/a are both 3-connected, and, if a' is the unique clone of a in M, then $M\backslash a/a'$ is totally free.* □

By Proposition 14.8.9, there are no totally free binary matroids. It is not difficult to show that $U_{2,4}$ is the only totally free ternary matroid. Using Theorem 14.8.13, Geelen, Oxley, Vertigan, and Whittle (2002) determined all totally free quaternary and $GF(5)$-representable matroids showing that free spikes and free swirls (Section 14.6) dominate these lists. Recall that P_6 is the 6-element rank-3 simple matroid with a single 3-point line and no other non-trivial lines.

Theorem 14.8.14 (i) *A quaternary matroid is totally free if and only if it is isomorphic to $U_{2,4}$, $U_{2,5}$, $U_{3,5}$, or a tipless free spike of rank at least three.*

(ii) *A $GF(5)$-representable matroid is totally free if and only if it is isomorphic to $U_{2,4}$, $U_{2,5}$, $U_{2,6}$, $U_{3,5}$, $U_{4,6}$, P_6, or a free swirl of rank at least three.* □

Exercises

1. (Geelen, Oxley, Vertigan, and Whittle 2002) Let a, b, and x be elements in a matroid M. Prove that:
   (a) If $r(\{a, b, x\}) = 3$ and x is fixed in both M/a and M/b, then x is fixed in M.
   (b) If b is fixed in M/a but not in M, then a and b are clones, or a is fixed in M.

2. (Geelen *el al.* 2002) Let $\{a, b, c\}$ be a triangle in a matroid M. Prove that:
   (a) The element a is fixed in M if and only if M has a circuit whose closure meets $\{a, b, c\}$ in $\{a\}$.
   (b) If M is totally free matroid, then a, b, and c are clones.

3. (Geelen *el al.* 2002) Let e be an element of a totally free matroid M. Prove that $M \backslash e$ or M/e is 3-connected.

4. (Whittle 1999) Let x be an element of a connected $\mathbb{F}$-representable matroid M. Assume that $M \backslash x$ is connected and that $M \backslash x$ stabilizes M. If A represents $M \backslash x$ over $\mathbb{F}$, and $[A|\underline{x}]$ and $[A|\underline{x}']$ both represent M over $\mathbb{F}$, prove that $\underline{x}$ is a scalar multiple of $\underline{x}'$.

5. (Whittle 1999) Let x and y be distinct elements of an $\mathbb{F}$-representable matroid M. Suppose that $M \backslash x, y$ is connected and stabilizes both $M \backslash x$ and $M \backslash y$. Show that $M \backslash x, y$ stabilizes M and that if A represents $M \backslash x, y$ and $[A|\underline{x}]$ and $[A|\underline{y}]$ represent $M \backslash y$ and $M \backslash x$, then $[A|\underline{x}|\underline{y}]$ represents M.

6. (Whittle 1999) For all fields $\mathbb{F}$, prove that $U_{2,5}$ is an $\mathbb{F}$-stabilizer for the class of $\mathbb{F}$-representable matroids with no $U_{2,6}$-, P_6-, $U_{4,6}$-, or $U_{3,6}$-minor.

7. (Whittle 1999) Prove that, if q is a prime power exceeding three, then a 3-connected quaternary matroid with no $U_{3,6}$-minor has at most $(q-2)(q-3)$ inequivalent $GF(q)$-representations.

8. (Geelen *el al.* 2002) Let a and b be clones in a 3-connected matroid M of rank at least three. Prove that M/a is non-binary and that if $r^*(M) \geq 3$, then $M/a \backslash b$ is non-binary.

9. (Geelen and Whittle 2010a) Let q be a prime power. Prove that:
   (a) *A rank-3 simple $GF(q)$-representable matroid with at least $q^2 + q - 1$ elements is a stabilizer for the class of $GF(q)$-representable matroids.
   (b) If M is a 3-connected $GF(q)$-representable matroid with a $PG(2, q)$-minor, then M is uniquely $GF(q)$-representable.

14.9 Unavoidable minors

It is elementary to show that every sufficiently large connected graph contains either a vertex of large degree or a long path. More formally, for every positive integer n, there is an integer $k(n)$ such that every connected graph with at least $k(n)$ edges contains a path with at least n vertices or a vertex of degree at least n. This is an example of a Ramsey-theoretic result. The form of the result is worth noting: a very large connected graph has as a minor, indeed as a subgraph, a large connected graph of one of two special types. In this section, we consider results of this type for both graphs and matroids. One such result can be derived from Theorem 4.3.15, which asserts that, in a connected matroid whose largest circuit and largest cocircuit have c and c^* elements, there are at most $\frac{1}{2}cc^*$ elements. If a matroid M has a c-element circuit, then it has a restriction isomorphic to $U_{c-1,c}$. Dually, if M has a c^*-element cocircuit, it has a contraction isomorphic to U_{1,c^*}. Hence Theorem 4.3.15 has the following immediate consequence.

Corollary 14.9.1 *For all positive integers n, a connected matroid with more than $\frac{1}{2}n^2$ elements has a minor isomorphic to either $U_{1,n}$ or $U_{n-1,n}$.* $\square$

Applied to graphs, the last result gives a result of Pou-Lin Wu (1997) that a 2-connected loopless graph with more than $n^2/2$ edges has a cycle or a bond with at least n edges. The following analogue of this result for 3-connected graphs was proved by Oporowski, Oxley, and Thomas (1993).

Theorem 14.9.2 *For every integer n exceeding two, there is an integer $k(n)$ such that every 3-connected simple graph with at least $k(n)$ edges has a minor isomorphic to $\mathcal{W}_n$ or $K_{3,n}$.* $\square$

This result was extended to 3-connected binary matroids and then to all 3-connected matroids by Ding, Oporowski, Oxley, and Vertigan (1996, 1997). Recall, from Proposition 12.2.20, that the unique binary n-spike is the matroid that is represented over $GF(2)$ by $[I_n|J_n - I_n|\mathbf{1}]$ where $\mathbf{1}$ is a column of ones.

Theorem 14.9.3 *For every integer n exceeding two, there is an integer $k(n)$ such that every 3-connected binary matroid with at least $k(n)$ elements has a minor isomorphic to $M(\mathcal{W}_n)$, $M(K_{3,n})$, $M^*(K_{3,n})$, or the binary n-spike.* $\square$

Theorem 14.9.4 *For every integer n exceeding two, there is an integer $k(n)$ such that every 3-connected matroid with at least $k(n)$ elements has a minor isomorphic to $M(\mathcal{W}_n)$, $\mathcal{W}^n$, $M(K_{3,n})$, $M^*(K_{3,n})$, $U_{2,n}$, $U_{n-2,n}$, or an n-spike.* $\square$

For matroids, this theory of unavoidable minors has not been developed beyond the 3-connected case. But, for graphs, Oporowski, Oxley, and Thomas (1993) determined the families of unavoidable minors in the 4-connected case. Although their technique could possibly be extended to prove the corresponding result for 5-connected graphs, this has not been done. Moreover, a new technique will be needed to extend the result for k-connected graphs with $k \geq 6$.

We now describe the unavoidable families of 4-connected graphs. Let $n \geq 3$. The *double wheel*, D_n, is the graph that is obtained from $\mathcal{W}_n$ by adding a new vertex and joining it to every vertex of the rim. The *quartic planar ladder*, X_n, is obtained from two vertex-disjoint cycles whose vertices, in cyclic order, are $u_1, u_2, \ldots, u_n$ and $v_1, v_2, \ldots, v_n$ by adding the set $\{u_i v_i, u_i v_{i+1} : 1 \leq i \leq n\}$ of edges, where $v_{n+1} = v_1$. The *quartic Möbius ladder*, Y_n, is obtained from a cycle whose vertices, in cyclic order, are $w_1, w_2, \ldots, w_{2n-1}$ by adding the set $\{w_i w_{i+n-1}, w_i w_{i+n} : 1 \leq i \leq n-1\} \cup \{w_n w_{2n-1}\}$ of edges. Hence $D_3 \cong K_5 - e$ and $Y_3 \cong K_5$, while both D_4 and X_3 are isomorphic to the octahedron.

Theorem 14.9.5 *For every integer n exceeding three, there is an integer $k(n)$ such that every 4-connected simple graph with at least $k(n)$ edges has a minor isomorphic to one of $D_n, X_n, Y_n,$ or $K_{4,n}$.* □

Oporowski *et al.* (1993) and Chun, Ding, Oporowski, and Vertigan (2009) proved results corresponding to Theorems 14.9.2 and 14.9.5 for unavoidable series and unavoidable parallel minors in graphs. Chun and Oxley (2009) combined these results with Seymour's Decomposition Theorem to specify the unavoidable series and unavoidable parallel minors for 3-connected regular matroids.

Exercises

1. Prove that, for every integer n exceeding two, there is an integer $k(n)$ such that every 3-connected regular matroid with at least $k(n)$ elements has a minor isomorphic to one of $M(\mathcal{W}_n)$, $M(K_{3,n})$, or $M^*(K_{3,n})$.

2. (Ding, Oporowski, Oxley, and Vertigan 1996) Let n be a positive integer and the *Ramsey number* $R(s, t)$ be the least integer m such that, whenever the edges of K_m are coloured either red or blue, there is either a red K_s-subgraph or a blue K_t-subgraph. Prove that if G is a simple connected graph with $(R(n, 2))^n$ vertices, then G has an induced subgraph isomorphic to $K_n, K_{1,n}$, or a path on n vertices.

3. (Ding *et al.* 1996) Let M be a 3-connected matroid.
   (a) *Let N be a series minor of M having no coloops. Prove that M has a 3-connected minor M_1 that has a restriction N_1 such that $r(N_1) = r(M_1)$, and N_1 is isomorphic to a matroid that can be obtained from N by a sequence of series extensions.
   (b) Let $|E(M)| \geq 2$ and C be a maximum-sized circuit of M. Prove that M has a 3-connected minor in which C is a spanning circuit.

4. *(Wu 2001) Let M be a matroid with a maximum-sized circuit C such that $|C| \geq 4$. Prove that if M is k-connected and $k \in \{2, 3\}$, then every element of M is in a circuit of size at least $\lceil |C|/2 \rceil + k - 1$.

14.10   Growth rates

Let $\mathcal{M}$ be a minor-closed class of matroids and let $h_{\mathcal{M}}(r)$ denote the maximum number of elements in a simple rank-r member of $\mathcal{M}$, where $h_{\mathcal{M}}(r) = \infty$ when this maximum does not exist. Kung (1993) conjectured that if $\mathcal{M}$ does not contain all rank-2 matroids, then the *growth rate* $h(r)$ is finite and is either linear, quadratic, or exponential. This Growth-Rate Conjecture was proved by Geelen, Kung, and Whittle (2009). In this section, we formally state their theorem and indicate some of the history of this beautiful result. One striking feature of the Growth-Rate Theorem (14.10.6) is the way in which the classes of graphic and $GF(q)$-representable matroids naturally emerge.

Much of the motivation for problems such as the Growth-Rate Conjecture derived from the existence of a substantial body of results in extremal graph theory. For example, Euler's polyhedron formula (Exercise 8, Section 5.2) implies that $|E(G)| \leq 3|V(G)| - 6$ for all simple planar graphs G with $|V(G)| \geq 3$. Moreover, the same bound holds for all simple graphs with no K_5-minor and at least three vertices (see Bollobás 1978, Corollary VII.1.13). Mader (1967) proved the following far-reaching extension of this result.

Theorem 14.10.1 *For every integer n exceeding one, if G is a simple graph with no K_n-minor, then $|E(G)| \leq (2^n - 1)|V(G)|$.*

Proof Assume that the result fails and let G be a counterexample with $|V(G)|$ minimum. Clearly $n \geq 3$. Then $|E(G)| > (2^n - 1)|V(G)| > (n - 1)|V(G)|$. Thus $\frac{2|E(G)|}{|V(G)|}$, the average vertex degree in G, exceeds $2(n - 1)$, so G certainly has a vertex v of degree at least $n - 1$. Let H be the subgraph of G induced by the set of neighbours of v. As G has no K_n-minor, H has no K_{n-1}-minor. Since $|V(H)| < |V(G)|$, we have $|E(H)| \leq (2^{n-1} - 1)|V(H)|$. Now H has a vertex w whose degree, $d_H(w)$, in H is at most the average degree in H and so is at most $2(2^{n-1} - 1)$. Let G' be the simple graph associated with G/vw. Then

$$
\begin{aligned}
|E(G')| &= |E(G)| - d_H(w) - 1 \\
&> (2^n - 1)|V(G)| - 2(2^{n-1} - 1) - 1 \\
&= (2^n - 1)(|V(G)| - 1) \\
&= (2^n - 1)|V(G')|.
\end{aligned}
$$

But $|V(G')| < |V(G)|$, so G' has a K_n-minor. Since G' is a minor of G, it follows that G has a K_n-minor; a contradiction. □

The number of edges of an arbitrary simple graph G is, of course, bounded above by $\binom{|V(G)|}{2}$, a quadratic function of $|V(G)|$. The striking feature of the last result is that, by excluding the presence of K_n as a minor, we obtain a bound on $|E(G)|$ that is linear in $|V(G)|$.

Now let M be a simple matroid. The next result shows that, when M is in three familiar minor-closed classes, $|E(M)|$ is bounded by linear, quadratic, and exponential functions of $r(M)$.

Lemma 14.10.2 *Let M be a simple rank-r matroid.*

(i) *If M is cographic and $r \geq 2$, then $|E(M)| \leq 3(r - 1)$, and equality holds if and only if $M \cong M^*(G)$ where G is a cubic, 3-edge-connected graph.*

(ii) *If M is graphic, then $|E(M)| \leq \binom{r+1}{2}$, and equality holds if and only if $M \cong M(K_{r+1})$.*

(iii) *If M is $GF(q)$-representable, then $|E(M)| \leq \frac{q^r-1}{q-1}$, and equality holds if and only if $M \cong PG(r-1, q)$.*

Proof Part (ii) is immediate, and part (iii) is proved in Corollary 6.1.7(ii). For (i), let M be cographic and suppose $r \geq 2$. As M is simple, there is a 3-edge-connected graph G such that $M^* \cong M(G)$. Since G is connected, $r(M(G)) = |V(G)| - 1$ and $r = r(M) = |E(G)| - |V(G)| + 1$. As every vertex of G has degree at least three, $3|V(G)| \leq 2|E(G)|$. Then

$$|E(M)| \leq |E(M)| + [2|E(G)| - 3|V(G)|] = 3[|E(G)| - |V(G)| + 1] - 3 = 3r - 3.$$

Equality holds in the last line if and only if G is cubic, and (iii) follows. $\square$

The first part of the last lemma is implicit in Jaeger (1979). Our proof follows Lindström (in Kung 1986b). These observations for graphic and cographic matroids prompt one to seek a similar result when M is regular. Heller (1957) proved that, in this case, both the bound on $|E(M)|$ and the matroids attaining this bound are the same as when M is graphic; that is, a simple rank-r regular matroid M has at most $\binom{r+1}{2}$ elements with equality holding if and only if $M \cong M(K_{r+1})$. This result can be proved using Seymour's Decomposition Theorem (13.1.1) along with the bounds for graphic matroids and cographic matroids from the last lemma. However, there are a number of more elementary proofs of the following stronger result of Murty (1976) (see, for example, Baclawski and White 1979 or Bixby and Cunningham 1987).

Proposition 14.10.3 *Let M be a simple rank-r binary matroid having no F_7-minor. Then $|E(M)| \leq \binom{r+1}{2}$ with this bound being attained if and only if $M \cong M(K_{r+1})$.* $\square$

A simple matroid M is *extremal* in a minor-closed class $\mathcal{M}$ of matroids if $M \in \mathcal{M}$ but $\mathcal{M}$ contains no simple single-element extension of M that has the same rank as M. In particular, if M is a maximum-sized simple rank-r member of $\mathcal{M}$, then M is certainly extremal. But the converse of this is not true. For example, by Lemma 13.2.5, $M^*(K_{3,3})$ is an extremal regular matroid. It is a 9-element rank-4 simple regular matroid. But, by Proposition 14.10.3, the unique maximum-sized simple rank-4 regular matroid is $M(K_5)$, which has ten elements.

The matroid $U_{r,n}$ shows that a simple rank-r matroid M can have arbitrarily many elements. Kung (1993, Theorem 4.3) showed that, by bounding the size of the largest rank-2 uniform matroid that can occur as a minor of M, we ensure that $|E(M)|$ is bounded by an exponential function of $r(M)$. This result extends Lemma 14.10.2(iii).

Proposition 14.10.4 *Let q and r be non-negative integers with $q > 1$. Let M be a simple rank-r matroid having no $U_{2,q+2}$-minor. Then $|E(M)| \leq \frac{q^r-1}{q-1}$. Moreover, when equality holds for $r \geq 3$, either $r \geq 4$ and $M \cong PG(r-1, q)$ for some prime power q; or $r = 3$ and M is a projective plane of order q.*

Proof To prove the bound on $|E(M)|$, we argue by induction on r. The bound certainly holds if $r \leq 1$. Now assume that $r \geq 2$ and that the bound holds for smaller values of r. Take e in $E(M)$. Since every line containing e has at most q other points, $|E(M)| \leq q|E(\mathrm{si}(M/e))| + 1$. Thus, by the induction assumption,

$$|E(M)| \leq q(\tfrac{q^{r-1}-1}{q-1}) + 1 = \tfrac{q^r-1}{q-1}.$$

We leave to the exercises the characterization of the extremal matroids for this bound (Exercise 9). □

Geelen and Nelson (2009) verified a conjecture of Kung (1993) by showing that the last proposition has the following extension.

Proposition 14.10.5 *Let k be an integer exceeding one and q be the largest prime power not exceeding k. Then there is an integer c such that if M is a simple rank-r matroid with no $U_{2,k+2}$-minor and with $r \geq c$, then $|E(M)| \leq \tfrac{q^r-1}{q-1}$. Moreover, equality holds if and only if $M \cong PG(r-1,q)$.* □

Proposition 14.10.4 gives an exponential bound on the growth rate $h_{\mathcal{M}}(r)$ of a minor-closed class $\mathcal{M}$ of matroids when the length of the longest line in $\mathcal{M}$ is bounded. Subject to this last condition, can we be more explicit about the other possibilities for the growth rate? In Lemma 14.10.2, we saw examples where $h_{\mathcal{M}}(r)$ has linear and quadratic bounds. The next theorem (Geelen, Kung, and Whittle 2009), the main result of this section, proves that there are no other possibilities. This remarkable result verifies another conjecture of Kung (1993).

Theorem 14.10.6 (Growth-Rate Theorem) *For a minor-closed class $\mathcal{M}$ of matroids, one of the following holds:*

(i)   *there is a real constant c_1 such that $|E(M)| \leq c_1 r(M)$ for all simple matroids M in $\mathcal{M}$;*

(ii) *$\mathcal{M}$ contains all graphic matroids and there is a real constant c_2 such that $|E(M)| \leq c_2(r(M))^2$ for all simple matroids M in $\mathcal{M}$;*

(iii) *there is a prime power q and a real constant c_3 such that $\mathcal{M}$ contains all $GF(q)$-representable matroids and $|E(M)| \leq c_3 q^{r(M)}$ for all simple matroids M in $\mathcal{M}$; or*

(iv) *$\mathcal{M}$ contains all simple rank-2 matroids.* □

The next three results give some interesting consequences of the Growth-Rate Theorem. The second of these is a result of Kung (1990).

Corollary 14.10.7 *Let q be a power of a prime p and let $\mathcal{M}$ be a minor-closed class of $GF(q)$-representable matroids. If $\mathcal{M}$ does not contain all $GF(p)$-representable matroids, then there is a constant c such that $|E(M)| \leq c(r(M))^2$ for all M in $\mathcal{M}$.* □

Corollary 14.10.8 *Let $\mathcal{M}$ be a minor-closed class of $\mathbb{R}$-representable matroids. If $\mathcal{M}$ does not contain all simple rank-2 matroids, then there is a constant c such that $|E(M)| \leq c(r(M))^2$ for all M in $\mathcal{M}$.* □

Corollary 14.10.9 *For fields $GF(q_1)$ and $GF(q_2)$ of different characteristics, there is a constant c such that if M is a simple rank-r matroid that is representable over both $GF(q_1)$ and $GF(q_2)$, then $|E(M)| \leq cr^2$.* □

The proof of Theorem 14.10.6 relies heavily on earlier results of Geelen and Whittle (2003) and of Geelen and Kabell (2009a). A new proof of the first of these, the linear part of the Growth-Rate Theorem (14.10.6(i)) was given by Geelen (2009), who established the following.

Theorem 14.10.10 *For all positive integers n and q, if M is a simple matroid having no $U_{2,q+2}$-minor and*

$$|E(M)| > q^{q^{3n}} r(M),$$

then M has an $M(K_n)$-minor. □

In the same paper, Geelen proved the following attractive result.

Theorem 14.10.11 *Let k, n, and q be positive integers. Then there is a positive integer ρ such that if M is a simple matroid of rank at least ρ and M has neither a $U_{2,q+2}$-minor nor an $M(K_n)$-minor, then M has a cocircuit with at most $r(M)/k$ elements.*

The next corollary is an immediate consequence of this theorem. The original proof of the linear part of the Growth-Rate Theorem actually used this corollary in its proof but Geelen's new proof of Theorem 14.10.10 does not.

Corollary 14.10.12 *For all positive integers k, n, and q, there is an integer $\alpha(k,n,q)$ such that a matroid with no $M(K_n)$-minor and no $U_{2,q+2}$-minor either has k disjoint cocircuits, or has rank at most $\alpha(k,n,q)$.* □

If we apply this corollary to cographic matroids, we get the following result.

Corollary 14.10.13 *For every positive integer k, there is an integer $\delta(k)$ such that if a graph G has $|E(G)| \geq |V(G)| + \delta(k)$, then G has k edge-disjoint cycles.*

Proof Let $\delta(k) = \alpha(k,5,2)$ and G be a graph for which $|E(G)| \geq |V(G)| + \delta(k)$ but G does not have k edge-disjoint cycles. Let $M = M^*(G)$. As M is cographic, it has no $M(K_5)$- or $U_{2,4}$-minor. Thus, by Corollary 14.10.12, $r(M) \leq \delta(k)$, so $r(M^*(G)) \leq \delta(k)$. Hence if G has $\omega(G)$ connected components, then $|E(G)| - |V(G)| + \omega(G) \leq \delta(k)$, so $|E(G)| < |V(G)| + \delta(k)$; a contradiction. □

The last corollary is due to Erdős and Pósa (1962). Let $\delta(k)$ be the least integer for which the corollary holds. Clearly $\delta(1) = 0$. For $k \geq 2$, Erdős and Pósa proved that $\delta(k) = O(k \log k)$.

To provide a flavour for the techniques used in this area, we now prove Theorem 14.10.11. We shall use three lemmas, the first of which is from Geelen, Gerards, and Whittle (2003). For each positive integer q, let $\mathcal{U}(q)$ be the class of matroids having no $U_{2,q+2}$-minor. Thus $\mathcal{U}(2)$ is the class of binary matroids, while $\mathcal{U}(1)$ consists of those matroids whose associated simple matroids are free.

Lemma 14.10.14 *Let q be a positive integer, M be a matroid in $\mathcal{U}(q)$, and D be a minimum-sized cocircuit of M. If D' is a cocircuit of $M \backslash D$, then $|D'| \geq |D|/q$.*

Proof Let $F = E(M) - (D \cup D')$. Then F is a flat of M of rank $r(M) - 2$, so M/F has rank two and has at most $q + 1$ rank-one flats. Thus, for some $q' \leq q$, there are exactly $q' + 1$ hyperplanes of M containing F. One of these is $E(M) - D$. Let the others be $H_1, H_2, \ldots, H_{q'}$. Then $\{H_1 - F, H_2 - F, \ldots, H_{q'} - F\}$ is a partition of D. As D is a cocircuit of minimum size,

$$
q'|D| \leq \sum_{i=1}^{q'} |E(M) - H_i| = \sum_{i=1}^{q'} (|E(M) - F| - |H_i - F|)
$$

$$
= q'|E(M) - F| - \sum_{i=1}^{q'} |H_i - F|.
$$

Since $E(M) - F = D \cup D'$ and $\sum_{i=1}^{q'} |H_i - F| = |D|$, we deduce that

$$
q'|D| \leq q'(|D| + |D'|) - |D|.
$$

Thus $|D| \leq q'|D'|$, so $|D'| \geq |D|/q$. $\qquad \square$

Lemma 14.10.15 *Let k and m be integers exceeding one, and t be a real number that is at least one. Let M be a matroid of rank at least $(\frac{k}{k-1})^{m-2}t$. Then either*

(i) *M has m disjoint cocircuits; or*

(ii) *M has a contraction N of rank at least t such that every cocircuit of N has rank exceeding $r(N)/k$.*

Proof We argue by induction on m. We may assume that M has a cocircuit D of rank at most $r(M)/k$ otherwise (ii) holds with $N = M$. Then $r(D) < r(M)$, so $E(M) - D$ contains a cocircuit. Hence M has two disjoint cocircuits. In particular, this means that the lemma holds for $m = 2$, so we may assume that $m \geq 3$. Now

$$
r(M/D) = r(M) - r(D) \geq r(M) - \tfrac{r(M)}{k} = \tfrac{k-1}{k}r(M) \geq (\tfrac{k}{k-1})^{m-3}t.
$$

By the induction hypothesis, either M/D has $m - 1$ disjoint cocircuits, or M/D has a contraction N of rank at least t such that every cocircuit of N has rank exceeding $r(N)/k$. In the latter case, N is a contraction of M and (ii) holds. In the former case, adding D to the specified collection of $m-1$ disjoint cocircuits of M/D gives a collection of m disjoint cocircuits of M, so (i) holds. We conclude, by induction, that the lemma holds. $\qquad \square$

Lemma 14.10.16 *There is a real-valued function $f(s, k, t, q)$ such that, for all positive integers $s, k, t,$ and q, if M is a matroid in $\mathcal{U}(q)$ having rank at least $f(s, k, t, q)$, then either M has a cocircuit of rank at most $r(M)/k$, or M has a contraction N of rank at least t such that $|E(\mathrm{si}(N))| > s(r(N) - s)$.*

Proof We shall prove the result when $k \geq 2$. To cover the case when $k = 1$, we let $f(s, 1, t, q) = f(s, 2, t, q)$ noting that our definition below ensures that $f(s, 2, t, q) > 0$, and every matroid M of positive rank has a cocircuit of rank at most $r(M)$. Let $f(1, k, t, q) = t$ and, for $s > 1$, recursively define

$$f(s, k, t, q) = k \left(\frac{k}{k-1} \right)^{skq-2} [f(s-1, k, t, q) + 1].$$

We argue by induction on s noting that the result is immediate when $s = 1$. Observe that it suffices to prove the result when M is simple. Suppose that $s > 1$ and that the result holds for smaller values of s.

Let M be a simple matroid in $\mathcal{U}(q)$ having rank at least $f(s, k, t, q)$. The lemma certainly holds if M has a cocircuit of rank at most $r(M)/k$. Thus we may assume that every cocircuit of M has rank exceeding $r(M)/k$. Let D be a cocircuit of M of minimum size. Then

$$|D| \geq r_M(D) > r(M)/k.$$

Thus, by Lemma 14.10.14, every cocircuit of $M \backslash D$ has size exceeding $\frac{r(M)}{kq}$.

We show next that we may assume that

14.10.17 *M has no minor N_0 such that $r(N_0) \geq t$ and $|E(\mathrm{si}(N_0))| > sr(N_0)$.*

Assume M has such a minor N_0. Then, by Lemma 3.3.2, $N_0 = M/I \backslash I^*$ for some independent set I and coindependent set I^* of M. Then $r(M/I) = r(N_0) \geq t$ and $|E(\mathrm{si}(M/I))| > sr(M/I)$. Thus the lemma holds with $N = M/I$.

As $f(1, k, t, q) = t$ and $s \geq 2$, it follows easily from the definition that $f(s, k, t, q) \geq t + 1$, so $r(M \backslash D) = r(M) - 1 \geq f(s, k, t, q) - 1 \geq t$. Thus, by 14.10.17, $|E(\mathrm{si}(M \backslash D))| \leq sr(M \backslash D)$. But M is simple so $|E(M \backslash D)| \leq sr(M \backslash D)$. As every cocircuit of $M \backslash D$ has more than $\frac{r(M)}{kq}$ elements, every such cocircuit has more than $\frac{r(M \backslash D)}{kq}$ elements, so $M \backslash D$ does not have skq disjoint cocircuits. In preparation for using Lemma 14.10.15, let $m = skq$.

We now construct a new matroid M_1 as follows. Take a basis B_D for $M|D$ and a subset X of $E(M) - D$ such that $B_D \cup X$ is a basis of M. Let $M_1 = M/X$. As $M \backslash D$ does not have m disjoint cocircuits, neither does $M \backslash D/X$; that is, $M_1 \backslash D$ does not have m disjoint cocircuits. Now

$$r(M_1) = r(M) - |X| = r(M) - [r(M) - r(D)] = r(D).$$

Therefore D is a spanning cocircuit of M_1. Hence

$$
\begin{aligned}
r(M_1 \backslash D) &= r(D) - 1 \\
&> r(M)/k - 1 \\
&\geq \left(\tfrac{k}{k-1} \right)^{m-2} [f(s-1, k, t, q) + 1] - 1.
\end{aligned}
$$

Thus $r(M_1 \backslash D) \geq \left(\frac{k}{k-1} \right)^{m-2} f(s-1, k, t, q)$, so, applying Lemma 14.10.15 to $M_1 \backslash D$, we get that $r(M_1 \backslash D/X_1) \geq f(s-1, k, t, q)$ for some $X_1 \subseteq E(M_1 \backslash D)$ such that every cocircuit of $M_1 \backslash D/X_1$ has rank exceeding $r(M_1 \backslash D/X_1)/k$.

Applying the induction assumption to $M_1 \backslash D/X_1$, we get that there is a subset X_2 of $E(M_1 \backslash D)$ containing X_1 such that $r(M_1 \backslash D/X_2) \geq t$ and

$$|E(\mathrm{si}(M_1 \backslash D/X_2))| > (s-1)[r(M_1 \backslash D/X_2) - (s-1)]. \qquad (14.1)$$

Now $M_1 = M/X$, so $M_1 \backslash D/X_2 = M \backslash D/(X \cup X_2)$. Let $M_2 = M/(X \cup X_2) = M_1/X_2$. Then $r(M_2) \geq r(M_1 \backslash D/X_2) \geq t$. As D is a spanning cocircuit of M_1, it is a spanning cocircuit of M_2. Using this and (14.1), we see, as $M_2 \backslash D = M_1 \backslash D/X_2$, that

$$
\begin{aligned}
|E(\mathrm{si}(M_2))| \;&\geq\; |E(\mathrm{si}(M_2 \backslash D))| + r(M_2) \\
&>\; (s-1)[r(M_2 \backslash D) - (s-1)] + r(M_2) \\
&=\; (s-1)[r(M_2) - 1 - (s-1)] + r(M_2) \\
&=\; (s-1)[r(M_2) - s] + r(M_2) \\
&>\; s(r(M_2) - s).
\end{aligned}
$$

We conclude that M_2 is a contraction of M with the desired properties, and Lemma 14.10.16 follows by induction. $\qquad\square$

Proof of Theorem 14.10.11 Let $s = q^{q^{3n}}$ and let $\rho = \lceil f(s+1, sk, (s+1)^2, q) \rceil$. Take a counterexample M to the theorem. By Theorem 14.10.10, if N is a simple minor of M, then $|E(N)| \leq sr(N)$. In particular, if D is a cocircuit of M, then $|D| \leq sr(D)$. But $|D| > \frac{r(M)}{k}$, so $r(D) \geq \frac{|D|}{s} > \frac{r(M)}{ks}$. Therefore, by Lemma 14.10.16, M has a contraction N of rank at least $(s+1)^2$ such that

$$|E(\mathrm{si}(N))| > (s+1)(r(N) - s - 1) = sr(N) + r(N) - (s+1)^2 \geq sr(N).$$

This contradicts the fact that $|E(N)| \leq sr(N)$ and so completes the proof. $\qquad\square$

The next corollary, a generalization of Theorem 14.10.1, follows immediately from Theorem 14.10.6.

Corollary 14.10.18 *For every prime power q and every graph G, there is an integer $k(q, G)$ such that if M is a simple $GF(q)$-representable matroid with no $M(G)$-minor, then $|E(M)| \leq k(q, G)r(M)$.* $\qquad\square$

Recall from Lemma 8.6.2 that a matroid is round if it has no two disjoint cocircuits. Clearly $M(K_n)$ and $U_{n,2n}$ are round for all positive integers n. Although $B(K_n)$, the bicircular matroid of K_n, need not be round, it has no three disjoint cocircuits (Exercise 1). Geelen and Kabell (2009b) proved the following extension of Corollary 14.10.12, which they hope will help to prove Conjecture 14.3.12.

Theorem 14.10.19 *There is an integer-valued function $\nu(n, k)$ such that, for all positive integers n and k, if M is a matroid of rank at least $\nu(n, k)$ having no minor isomorphic to $U_{n,2n}$ or the cycle or bicircular matroid of K_n, then M has k disjoint cocircuits.* $\qquad\square$

Although the Growth-Rate Theorem separates minor-closed classes of matroids into four categories, it gives no information at all about the actual values of the constants involved in bounding the sizes of the matroids in the classes. The next proposition is an example of one of relatively few exact results that have been proved here. A number of other such results may be found in Kung (1993). Recall that $P(M_1, M_2)$ denotes a parallel connection of the matroids M_1 and M_2, there being potentially several different possibilities for this matroid depending on the choice of the basepoints in M_1 and M_2.

Proposition 14.10.20 *Let M be a simple $GF(q)$-representable rank-r matroid having no $M(K_4)$-minor.*

(i) *If $q = 2$ and M is non-empty, then $|E(M)| \leq 2r-1$. Moreover, equality holds if and only if M is isomorphic to a member of $\mathcal{P}_r$ where $\mathcal{P}_r$ is $\{PG(r-1,2)\}$ when $r \leq 2$, and $\mathcal{P}_r$ is $\{P(M, U_{2,3}) : M \in \mathcal{P}_{r-1}\}$ when $r \geq 3$.*

(ii) *If $q = 3$, then*

$$|E(M)| \leq \begin{cases} 4r - 3 & \text{if } r \text{ is odd;} \\ 4r - 4 & \text{if } r \text{ is even.} \end{cases}$$

Moreover, equality holds if and only if M is isomorphic to a member of $\mathcal{P}_r$ where $\mathcal{P}_r$ is $\{PG(r-1,3)\}$ when $r \leq 2$; $\mathcal{P}_3 = \{AG(2,3)\}$; and $\mathcal{P}_r$ is $\{P(M, AG(2,3)) : M \in \mathcal{P}_{r-2}\}$ when $r \geq 4$.

(iii) *If $q \geq 4$, then $|E(M)| \leq (6q^3 - 1)r$.* □

The first part of the last result follows by combining Corollary 12.2.14 with a result of Dirac (1960) bounding the number of edges in a simple series–parallel network (see Exercise 9 of Section 5.4). The second part was proved by Oxley (1987c) as a consequence of Theorem 12.2.22. The third part is from Kung (1988). It was conjectured by Kung (1993) that the constant $6q^3 - 1$ in (iii) can be replaced by a linear function of q.

In the last result, we excluded $M(\mathcal{W}_3)$, the cycle matroid of the 3-wheel. In the next result, a straightforward consequence of Theorem 12.2.21, we exclude the the cycle matroid of $\mathcal{W}_4$. Recall that Z_4 is the rank-4 (tipped) binary spike.

Corollary 14.10.21 *Let M be a simple rank-r binary matroid having no minor isomorphic to $M(\mathcal{W}_4)$. Then*

$$|E(M)| \leq \begin{cases} 3r - 2 & \text{if } r \text{ is odd;} \\ 3r - 3 & \text{if } r \text{ is even.} \end{cases}$$

Moreover, equality holds if and only if M is isomorphic to a member of $\mathcal{P}_r$ where

$$\mathcal{P}_r = \begin{cases} \{PG(r-1,2)\} & \text{when } r \leq 3; \\ \{Z_4, P(F_7, U_{2,3})\} & \text{when } r = 4; \\ \{P(M, F_7) : M \in \mathcal{P}_{r-2}\} & \text{when } r > 4. \end{cases}$$
□

The exact bounds just described are for minor-closed classes with linear growth rate. Next we note that the growth rates of the special classes of ternary matroids considered in Section 14.7 are all quadratic. For $r \geq 3$, we let T_r be the rank-r matroid that is obtained by freely adding a point on a 3-point line of $M(K_{r+2})$, then contracting out that point and simplifying the resulting matroid. We also let T_1 and T_2 be $U_{1,1}$ and $U_{2,4}$. A near-unimodular representation for T_r is given in Exercise 4.

Proposition 14.10.22 *Let M be a simple rank-r matroid.*

(i)   *If M is dyadic, then $|E(M)| \leq r^2$. Moreover, equality holds if and only if M is isomorphic to the ternary Dowling geometry $Q_r(GF(3)^\times)$.*

(ii)   *If M is a $\sqrt[6]{1}$-matroid, then*

$$|E(M)| \leq \begin{cases} \binom{r+2}{2} - 2 & \text{if } r \neq 3; \\ 9 & \text{if } r = 3. \end{cases}$$

   Moreover, equality is attained in this bound if and only if $M \cong T_r$ when $r \neq 3$, or $M \cong AG(2,3)$ when $r = 3$.

(iii) *If M is near-regular, then $|E(M)| \leq \binom{r+2}{2} - 2$, and equality holds if and only if $M \cong T_r$.* □

The first part is obtained by combining Theorem 14.7.1 with results of Kung (1990) and Kung and Oxley (1988). Parts (ii) and (iii) were proved by Oxley, Vertigan, and Whittle (1998).

Although Corollary 14.10.21 gave the growth rate of the class of binary matroids with no $M(\mathcal{W}_4)$-minor, far less is known about the growth rate of the class of binary matroids with no $M(K_5)$-minor. Kung (1987) proved the following bound which, although it is not believed to be sharp, has yet to be improved.

Proposition 14.10.23 *Let M be a simple binary matroid having no $M(K_5)$-minor. Then $|E(M)| \leq 8r(M)$.* □

By contrast, Mayhew, Royle, and Whittle (2009b) determined an exact bound on the size of a binary matroid having no $M(K_{3,3})$-minor.

Theorem 14.10.24 *For $r \geq 1$, let M be a simple rank-r binary matroid having no $M(K_{3,3})$-minor. Then*

$$|E(M)| \leq \begin{cases} \frac{14}{3}r - 7 & \text{if } r \equiv 0 \mod 3; \\ \frac{14}{3}r - \frac{11}{3} & \text{if } r \equiv 1 \mod 3; \\ \frac{14}{3}r - \frac{19}{3} & \text{if } r \equiv 2 \mod 3. \end{cases}$$

Moreover, equality holds here if and only if M is isomorphic to a member of $\mathcal{P}_r$ where

$$\mathcal{P}_r = \begin{cases} \{PG(r-1,2)\} & \text{when } r \leq 4; \\ \{P(M, PG(3,2)) : M \in \mathcal{P}_{r-3}\} & \text{when } r > 4. \end{cases}$$ □

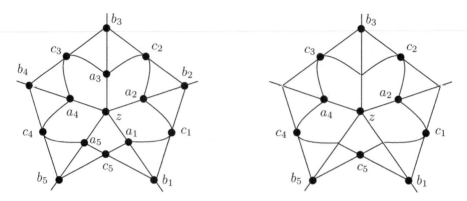

FIG. 14.11. The rank-6 triangular and triadic Möbius matroids.

The last theorem is obtained as a consequence of an impressive result of Mayhew, Royle, and Whittle (2009a) that determines all of the binary internally 4-connected matroids with no $M(K_{3,3})$-minor. This theorem is similar in form to Theorems 12.2.21 and 12.2.22 in that, when the rank is small, some sporadic examples occur but, when we raise the rank, the only matroids that occur fall into a small number of infinite families. But, whereas the main tool in the proof of these earlier results was the Splitter Theorem, Mayhew, Royle, and Whittle's theorem depends on some new connectivity results that they developed for dealing with binary internally 4-connected matroids. Their proof is also unusual in that it uses computer checks to verify whether two binary matroids are isomorphic, to test whether a binary matroid has a particular minor, and to generate all binary single-element extensions and coextensions of a given binary matroid.

The new infinite families that arise in this work are the triangular and triadic Möbius matroids, which we shall now define. Let r be an integer exceeding 2. Begin by taking the parallel connection with common basepoint z of $r-1$ three-point lines $\{z, a_1, b_1\}, \{z, a_2, b_2\}, \ldots, \{z, a_{r-1}, b_{r-1}\}$. View the resulting binary matroid as being embedded in the binary projective space $PG(r-1, 2)$ and, for all i in $\{1, 2, \ldots, r-2\}$, add the unique third point c_i of $PG(r-1, 2)$ on the line spanned by a_i and a_{i+1}. Then c_i also lies on the line spanned by b_i and b_{i+1}. Finally, add c_{r-1} as the unique third point on the line of $PG(r-1, 2)$ spanned by a_{r-1} and b_1, noting that c_{r-1} also lies on the line spanned by b_{r-1} and a_1. This gives the *rank-r triangular Möbius matroid*, which has $3r-2$ elements. Now suppose that r is even. Delete $a_1, b_2, a_3, b_4, \ldots, a_{r-1}$ from the rank-r triangular Möbius matroid to get a rank-r matroid with $2r-1$ elements and no triangles. This is the *rank-r triadic Möbius matroid*. Figure 14.11 should help the reader visualize the triangular and triadic Möbius matroids of rank 6. Each has an umbrella-like shape in rank 6 with z as the tip of the umbrella. Note that these diagrams are not actually geometric representations as such geometric representations can only accurately depict matroids of rank at most four.

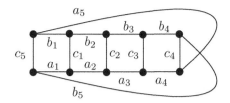 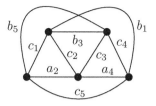

FIG. 14.12. A cubic and a quartic Möbius ladder.

If we delete the element z from each matroid in Figure 14.11, then the duals of the resulting matroids are the cycle matroids of the two graphs in Figure 14.12. In general, deleting z from the triangular and triadic Möbius matroids and dualizing gives the cycle matroids of the cubic and quartic Möbius ladder graphs having $2r - 2$ and $r - 1$ vertices, respectively. Quartic Möbius ladders were defined at the end of Section 14.9. The three smallest cubic Möbius ladders coincide with K_4, $K_{3,3}$, and the Wagner graph, H_8, from Figure 12.1. In general, for $n \geq 2$, a *cubic Möbius ladder* has vertex set $\{v_1, v_2, \ldots, v_{2n}\}$ and consists of a Hamiltonian cycle $\{v_1 v_2, v_2 v_3, \ldots, v_{2n-1} v_{2n}, v_{2n} v_1\}$ and a matching $\{v_1 v_{n+1}, v_2 v_{n+2}, \ldots, v_n v_{2n}\}$.

The following is the main result of Mayhew, Royle, and Whittle's (2009a) paper. The eighteen sporadic matroids in the third part of the theorem have at most 21 elements and have rank at most 11.

Theorem 14.10.25 *An internally 4-connected binary matroid M has no minor isomorphic to $M(K_{3,3})$ if and only if M is either*

(i) *cographic;*

(ii) *isomorphic to a triangular or triadic Möbius matroid; or*

(iii) *isomorphic to one of eighteen sporadic matroids.* □

Exercises

1. For the bicircular matroid $B(K_n)$ of K_n and for $X \subseteq E(K_n)$, show that:
   (a) X is a hyperplane of $B(K_n)$ if and only if X is a spanning tree of K_n, or there is a partition (V_1, V_2) of $V(K_n)$ into non-empty sets where $|V_1| \geq 3$ such that the edges of X consist of a complete graph on the vertices of V_1 together with a spanning tree on the vertices of V_2.
   (b) If $n \geq 4$, then $B(K_n)$ has two but not three disjoint cocircuits.
2. (a) (Oxley 1990a) Show that a 3-connected ternary matroid with no $\mathcal{W}^3$-minor is regular or is uniform of rank or corank at most two.
   (b) (Kung 1993) Deduce that if M is a simple rank-r ternary member of $EX(\mathcal{W}^3)$, then $|E(M)|$ is at most $\binom{r+1}{2}$ when $r \geq 4$ and is at most $3r - 2$ when $r \leq 3$.
3. (Geelen and Whittle 2003) Let M be a matroid and let F_1 and F_2 be flats such that $M|F_1$ and $M|F_2$ are round, $r(F_1) = r(F_2) = k$, and $r(F_1 \cup F_2) = k + 1$. Prove that if $\mathrm{cl}(F_1 \cup F_2) \neq F_1 \cup F_2$, then $M|\mathrm{cl}(F_1 \cup F_2)$ is round.

4. (Oxley, Vertigan, and Whittle 1998) Prove that, for all $r \geq 3$, the matrix

$$\begin{bmatrix} 1 & 0\ldots0 & 1\ldots1 & \alpha\ldots\alpha & 0\ldots0 \\ 0 & I_{r-1} & I_{r-1} & I_{r-1} & D_{r-1} \end{bmatrix}$$

is a near-unimodular representation for T_r, where $\mathbf{0}$ is a column of zeros and D_{r-1} is the $r \times \binom{r}{2}$ matrix whose columns consist of all r-tuples with exactly two non-zero entries, the first equal to 1 and the second -1.

5. (Bonin 1996) Let M be a rank-r simple matroid in $\mathcal{U}(q)$. Prove that if $r > 3$ and q is not a power of a prime, then $|E(M)| \leq q^{r-1} - 1$.

6. (Bonin 1996) Let M be a rank-r simple matroid in $\mathcal{U}(q)$ having no $(q+1)$-point line. Prove that $|E(M)| \leq q^{r-1}$ and that, when $r \geq 4$, equality holds if and only if q is a prime power and $M \cong AG(r-1,q)$.

7. (Kung 1993) Suppose $q \geq 2$ and M is a rank-r simple matroid. Prove that:
   (a) For $r \geq 1$, if M is in $\mathcal{U}(q)$ and H is a hyperplane of M, then $|E(M)| - |H| \leq q^{r-1}$.
   (b) *(Kung 1986b) If M is in a minor-closed class $\mathcal{M}$ and $M|X$ is an extremal member of $\mathcal{M}$, then X is a modular flat of M.
   (c) For $r \geq 3$, if M is in $\mathcal{U}(q)$ and $|E(M)| = \frac{q^r-1}{q-1}$, then either $r \geq 4$ and $M \cong PG(r-1,q)$ for some prime power q; or $r = 3$ and M is a projective plane of order q.

8. *(Kung 2008) Let M be a simple rank-r binary matroid. Give an example of such a matroid M with no $PG(3,2)$-minor having $2\binom{r}{2}+1$ elements and prove that if $|E(M)| > 15r(r+1)/4$, then M has a $PG(3,2)$-minor.

9. *(Oxley 1989a) Let M be a rank-r simple regular matroid having no $M(\mathcal{W}_5)$-minor.
   (a) Prove that
   $$|E(M)| \leq \begin{cases} 3r - 2 & \text{if } r \equiv 1 \mod 3; \\ 3r - 3 & \text{otherwise.} \end{cases}$$
   (b) Show that M attains this bound if and only if either $M = P(M_1, M_2)$ where each of M_1 and M_2 attains the bound, has rank at least two, and $r(M_1)$ or $r(M_2)$ is congruent to 1 (mod 3); or $M \cong M(K_n)$ for some n in $\{2,3,4,5\}$; or $M \cong M(K_{2,2,2})$; or $M \cong M(K'''_{3,k})$ for some $k \geq 3$ such that $k \neq 2$ (mod 3) where $K'''_{3,k}$ is shown in Figure 5.14(c).

10. (Geelen and Nelson 2009) Let q be a prime power and M be a round matroid. Prove that:
    (a) M has a U_{2,q^2+1}-minor if it has $U_{2,q+2}$ and $PG(2,q)$ as restrictions.
    (b) *There is a positive integer n such that if M has a $PG(n-1,q)$-minor but no U_{2,q^2+1}-minor, then $|E(\text{si}(M))| \leq \frac{q^{r(M)}-1}{q-1}$.
    (c) °There is a positive integer m such that if M has a $PG(m-1,q)$-minor but no U_{2,q^2+1}-minor, then M is $GF(q)$-representable.

11. °(Bonin, private communication) For a rank-r simple matroid M in $\mathcal{U}(q)$ and $2 \leq k \leq r-1$, is the number of rank-k flats of M at most $\left[\begin{smallmatrix} r \\ k \end{smallmatrix}\right]_q$?

15

UNSOLVED PROBLEMS

The coverage of matroid theory in this book has been far from encyclopaedic. Many important areas of the subject have been ignored completely or have received just a cursory treatment. However, some of the topics that are not fully covered here are treated in more detail in one of the other books on matroids that have appeared since 1985. The first two volumes of White's trilogy on matroid theory (White 1986, 1987a) have been referred to frequently above. The third volume (White 1992) surveys infinite matroids (Oxley 1992) along with a variety of applications of matroids. A related monograph (Björner, Las Vergnas, Sturmfels, White, and Ziegler 1993) is entirely devoted to orientable matroids.

Greedoids are another generalization of matroids that have received considerable attention. One of the many equivalent ways to define a *greedoid* is as a finite set E and a collection $\mathcal{I}$ of subsets of E satisfying (I1) and (I3) (Section 1.1). A detailed treatment of greedoids and their properties can be found in Björner and Ziegler (1992) and in Korte, Lovász, and Schrader (1991). Truemper's (1992b) monograph on matroid decompositions concentrates on the structural theory of matroids, while Recski (1989), in addition to developing the basic theory of matroids, devotes much attention to the applications of matroids in electrical network theory and statics. The proceedings of the 1995 Seattle meeting on matroid theory (Bonin, Oxley, and Servatius 1996) provide an indication of the state of the subject at that time; and Schrijver's (2003) three-volume work on combinatorial optimization devotes considerable attention to matroids. Two other valuable resources are Kung's (1986a) *Source book*, an annotated reprinting of eighteen matroid papers which formed the basis for the development of the first fifty years of the subject; and Welsh's (1976) book on matroid theory, which was a major influence not only on this book but also on the growth of the area as a whole and which was reprinted in 2010 by Dover.

The primary purpose of this chapter is to bring together, in one place, a number of unsolved problems. Several of these have appeared in earlier chapters and, for these, we have noted the earlier reference since the background to the problem is usually described there. For many of the problems that have not appeared earlier, some details of the background are provided. Several of the books listed above contain further details relating to a number of these problems. Much of the previous chapter is devoted to unsolved problems and conjectures; those problems are not repeated here. The reader who is looking for an update on a problem from the corresponding chapter of the first edition of this book and cannot find it discussed here or in the previous chapter should consult the author's website.

While the selection of problems listed here is partially influenced by the author's preference, it has also been dictated to a large extent by the topics covered in earlier chapters. Where possible, problems have been grouped by topic, although the reader will note that some problems actually belong in more than one section.

15.1   Representability: linear and algebraic

Problems in linear representability tend to be among the oldest and hardest in matroid theory. Many of the problems on linear representability from the corresponding section of the first edition of this book were discussed in detail in the preceding chapter. We begin with a problem (6.5.19) that has received considerable attention from projective geometers.

Problem 15.1.1 *Determine all of the finite fields $GF(q)$ over which $U_{r,n}$ is representable.*   □

Settling the following conjecture (6.5.20), the main conjecture for maximum distance separable codes, would complete the answer to Problem 15.1.1. Ball (2010) proved the conjecture when q is prime (Corollary 6.5.22).

Conjecture 15.1.2 *For all prime powers q, the matroid $U_{r,q+2}$ is an excluded minor for the class of $GF(q)$-representable matroids for all r in $\{2,3,\ldots,q\}$ when q is odd, and for all r in $\{2,3,\ldots,q\} - \{3,q-1\}$ when q is even.*   □

A class of $\mathcal{M}$ of matroids is *polynomial-time recognizable* if there is a polynomial $f(x)$ and an algorithm that determines after at most $f(n)$ rank evaluations, whether or not a fixed n-element matroid M is in $\mathcal{M}$. As noted in Section 9.4, Seymour (1981a) proved that the class of binary matroids is not polynomial-time recognizable although the class of graphic matroids is. Moreover, Truemper (1982b) showed that the class of regular matroids is polynomial-time recognizable. Mayhew, Whittle, and Van Zwam (2009) note that Geelen and Mayhew independently showed that the class of signed-graphic matroids is not polynomial-time recognizable. Since every signed-graphic matroid is dyadic (Exercise 14.7.4), it follows that the class of dyadic matroids is not polynomial-time recognizable. Whittle (2005) had conjectured that the classes of dyadic, sixth-root-of-unity, and near-regular matroids are all polynomial-time recognizable. This conjecture remains open for the last two classes and was reiterated by Mayhew, Whittle, and Van Zwam (2009) for the last class.

Conjecture 15.1.3 *The classes of sixth-root-of-unity and near-regular matroids are polynomial-time recognizable.*   □

From Section 14.7, each of the classes of dyadic, sixth-root-of-unity, and near-regular matroids coincides with a class of matroids that is representable over two different finite fields. Whittle (2005) noted that a folklore theorem for such matroids is that, for prime powers q_1 and q_2 with $q_1 \leq q_2$, if neither q_1 nor q_2 is prime, or if both q_1 and q_2 are prime exceeding four, then the class of matroids that are representable over both $GF(q_1)$ and $GF(q_2)$ is not polynomial-time recognizable. The examples that show this are based on free spikes and free swirls (see Section 14.6).

The following is a variant on Problem 14.7.11. The matroids N_1 and N_2 are ternary matroids that were identified by Pendavingh (private communication, 2008). Representations of them are given just before Problem 14.7.11.

Problem 15.1.4 *Determine what matroids, if any, need to be added to $\{U_{2,5},$ $U_{3,5}, F_7, F_7^*, AG(2,3)\backslash e, (AG(2,3)\backslash e)^*, \Delta_3(AG(2,3)\backslash e), T_8, N_1, N_2\}$ to obtain the complete set of excluded minors for the class of dyadic matroids.* □

Whittle (2005) asked whether there are analogues of Seymour's Decomposition Theorem for other classes of matroids. In particular, Mayhew, Whittle, and Van Zwam (2009) considered the following.

Problem 15.1.5 *Is there an analogue of Seymour's Decomposition Theorem for the class of near-regular matroids?* □

A plausible answer to this problem, which was conjectured by Whittle, is that the class of near-regular matroids can be built from signed-graphic matroids, their duals, and some finite set of special matroids by direct sums, 2-sums, and generalized parallel connections across modular lines. However, Mayhew, Whittle, and Van Zwam showed that this answer is incomplete and that yet another type of joining operation must be allowed.

The following question was stated earlier as Problem 6.8.15. As noted there, Sturmfels (1987) showed that a conjecture of Grünbaum (1972, Conjecture 2.14) is equivalent to the assertion that the answer to this question is negative.

Problem 15.1.6 *Is there a finite algorithm to determine whether a given matroid is $\mathbb{Q}$-representable?* □

Turning briefly to algebraic representability, we recall from 6.7.15 that the main unsolved problem in this subject is the following.

Problem 15.1.7 *Is the dual of an algebraic matroid also algebraic?* □

The intractability of this problem prompted Lindström (1988b) to propose the following question (6.7.16).

Problem 15.1.8 *If a matroid is algebraic over some field, is its dual also algebraic over a field of the same characteristic?* □

Several other related unsolved problems for algebraic matroids were discussed by Lindström (1985c, 1988b, 1993), Gordon (1988), and Lemos (1988).

15.2 Unimodal conjectures

Let $a_1, a_2, \ldots, a_n$ be a finite sequence of real numbers. This sequence is *unimodal* if $a_j \geq \min\{a_i, a_k\}$ for all i, j, and k such that $1 \leq i \leq j \leq k \leq n$. Equivalently, $(a_i : 1 \leq i \leq n)$ is unimodal if there is a member t of $\{1, 2, \ldots, n\}$ such that

$a_1 \leq a_2 \leq \cdots \leq a_t$ and $a_t \geq a_{t+1} \geq \cdots \geq a_n$. A somewhat stronger property of a sequence is that it be *logarithmically concave* or *log concave* for short, that is, $a_k^2 \geq a_{k-1}a_{k+1}$ for all k in $\{2, 3, \ldots, n-1\}$.

In this section, we look at three important sequences associated with a matroid that are conjectured to be log concave. This section has very few updates to the corresponding section in the first edition of this book, reflecting how little progress has been made on these conjectures. Let i_k denote the number of k-element independent sets of a matroid. Welsh (1971b) proposed the following.

Conjecture 15.2.1 *For all matroids M, the sequence $(i_k : 0 \leq k \leq r(M))$ is unimodal.* ☐

Mason (1972b) offered three successive strengthenings of this conjecture.

Conjecture 15.2.2 *If M is a matroid, then, for all k in $\{1, 2, \ldots, r(M)-1\}$:*

(i) $i_k^2 \geq i_{k-1}i_{k+1}$;

(ii) $i_k^2 \geq \frac{k+1}{k}i_{k-1}i_{k+1}$;

(iii) $i_k^2 \geq \left(\frac{k+1}{k}\right)\left(\frac{i_1-k+1}{i_1-k}\right)i_{k-1}i_{k+1}$. ☐

Progress on these conjectures has been very limited. Dowling (1980) proved (i) for $k \leq 7$ and proposed a further extension of it; Zhao (1985) proved (ii) for $k \leq 5$; Kahn and Neiman (2008) proved (iii) for $k \leq 5$ after Hamidoune and Salaün (1989) had earlier proved (iii) for $k \leq 3$; Seymour (1975) proved (iii) for those k such that M has no circuits of cardinality in $\{3, 4, \ldots, k-1\}$; and Mahoney (1985) proved (i) when M is the cycle matroid of an outerplanar graph.

The number of rank-k flats in a matroid M is denoted by W_k, and the numbers $W_0, W_1, \ldots, W_{r(M)}$ are called the *Whitney numbers of M of the second kind*. Rota (1971) conjectured that this sequence is unimodal.

Conjecture 15.2.3 *If M is a matroid, then*

$$W_k \geq \min\{W_{k-1}, W_{k+1}\} \text{ for all } k \text{ in } \{1, 2, \ldots, r(M)-1\}. \quad ☐$$

Clearly, it suffices to prove the last conjecture for simple matroids. Interestingly, Kung (2000) comments that 'Among the many conjectures bearing his [Rota's] name in matroid theory, the unimodality conjecture is perhaps the most intractable.' Kung proved that, in a simple matroid of rank at least five in which all the lines have the same number of points, $W_2 \leq W_3$.

In the same paper that Mason (1972b) proposed Conjecture 15.2.2, he offered the following analogous strengthenings of Conjecture 15.2.3.

Conjecture 15.2.4 *If M is a rank-r matroid, then, for all k in $\{1, 2, \ldots, r-1\}$,*

(i) $W_k^2 \geq W_{k-1}W_{k+1}$;

(ii) $W_k^2 \geq \frac{k+1}{k}W_{k-1}W_{k+1}$;

(iii) $W_k^2 \geq \left(\frac{k+1}{k}\right)\left(\frac{W_1-k+1}{W_1-k}\right)W_{k-1}W_{k+1}$. ☐

The following special case of (ii) has been called the *Points–Lines–Planes Conjecture* (Mason 1972b, Welsh 1976).

Conjecture 15.2.5 *If W_1 points in Euclidean 3-space determine W_2 lines and W_3 planes, then*

$$W_2^2 \geq \tfrac{3}{2} W_1 W_3.$$

$\square$

For W_1, W_2, and W_3 defined as in 15.2.5, Purdy (1986) proved the existence of a positive constant c for which $W_2^2 \geq c W_1 W_3$. There has also been some progress on certain special cases of 15.2.4(iii). Stonesifer (1975) proved that it holds when $k = 2$ for all graphic matroids, and his result was extended by Seymour (1982) who showed, again in the case $k = 2$, that 15.2.4(iii) holds for all matroids in which no line has five or more points. Incidentally, for Seymour (1982), the Points–Lines–Planes Conjecture is the case $k = 2$ of 15.2.4(iii). Dukes (2004/05) proved that the latter version of the Points–Lines–Planes Conjecture holds for all rank-4 simple matroids if and only if it holds for a special subcollection of such matroids. Aigner and Schoene (in Aigner 1987) have verified 15.2.4(i) for binary matroids in the case $k = 2$. Further support for 15.2.3 is provided by some striking results of Dowling and Wilson (1974, 1975). These and other results on Whitney numbers are surveyed by Aigner (1987). The last paper contains results not only for Whitney numbers of the second kind but also for Whitney numbers of the first kind, which we shall define in a moment. The latter results are also relevant here since the sequence of Whitney numbers of the first kind is also conjectured to be unimodal.

For a matroid M having ground set E, the *characteristic* (or *chromatic*) *polynomial* of M is defined by

$$p(M; z) = \sum_{X \subseteq E} (-1)^{|X|} z^{r(M)-r(X)}.$$

The *Whitney numbers of the first kind* are the coefficients of this polynomial. More precisely, they are the numbers $w_0, w_1, \ldots, w_{r(M)}$ where

$$p(M; z) = \sum_{k=0}^{r(M)} w_k z^{r(M)-k}.$$

Surveys of the properties of characteristic polynomials can be found in Welsh (1976) and Zaslavsky (1987a). In particular, we note that it can be shown that if M has a loop, then $p(M; z)$ is identically zero, whereas if M is loopless, then $w_0 = 1$ and $w_0, w_1, \ldots, w_{r(M)}$ alternate in sign. Moreover, if G is a graph, then $p(M(G); z)$ is closely related to the chromatic polynomial $P_G(z)$ of the graph G; when z is a positive integer, $P_G(z)$ is the number of ways to colour the vertices of G using members of $\{1, 2, \ldots, z\}$ as colours so that no two adjacent vertices are coloured alike. In particular, if G has $\omega(G)$ connected components, then

$$P_G(z) = z^{\omega(G)} p(M(G); z).$$

Implicit in Rota (1971) and explicit in Heron (1972b) is the following generalization of an earlier graph-theoretic conjecture of Read (1968).

Conjecture 15.2.6 *If M is a loopless matroid, then*

$$|w_k| \geq \min\{|w_{k-1}|, |w_{k+1}|\} \text{ for all } k \text{ in } \{1, 2, \ldots, r(M) - 1\}. \qquad \square$$

Welsh (1976) proposed that Read's conjecture could be extended further and that the Whitney numbers of the first kind are actually log concave.

Conjecture 15.2.7 *Let M be a matroid. Then*

$$w_k^2 \geq w_{k-1} w_{k+1} \text{ for all } k \text{ in } \{1, 2, \ldots, r(M) - 1\}. \qquad \square$$

Very recently, June Huh proved Conjecture 15.2.7 for matroids that are representable over a field of characteristic zero (Milnor numbers of projective hypersurfaces and the chromatic polynomial of graphs, http://arxiv.org/abs/1008.4749).

15.3 Critical problems

The characteristic polynomial of a matroid is one of a number of important matroid isomorphism invariants that are evaluations of a much-studied two-variable polynomial called the Tutte polynomial. A detailed discussion of this polynomial, its properties, and its applications can be found in Brylawski and Oxley (1992). We shall not attempt to list all of the many interesting unsolved problems related to the Tutte polynomial but instead we concentrate on a selection of such problems that relate to the characteristic polynomial. For details of the background to these problems, we refer the reader to Brylawski and Oxley's paper.

Let M be a loopless rank-r matroid representable over $GF(q)$ and let $\psi : E(M) \to V(r, q)$ be a $GF(q)$-coordinatization of M. The *critical exponent* $c(M; q)$ of M is defined to be the least natural number j for which $V(r, q)$ has j hyperplanes $H_1, H_2, \ldots, H_j$ such that $(\bigcap_{i=1}^{j} H_i) \cap \psi(E(M))$ is empty. If M has a loop and is representable over $GF(q)$, then $c(M; q)$ is taken to be ∞. Hence if M is loopless, then $c(M; q) = c(\mathrm{si}(M); q)$. Ostensibly, $c(M; q)$ depends upon the particular coordinatization ψ. However, a fundamental result of Crapo and Rota (1970) establishes that this is not the case. In particular, if M is $GF(q)$-representable and loopless, then

$$c(M; q) = \min\{j \in \mathbb{Z}^+ \cup \{0\} : p(M; q^j) > 0\}.$$

Clearly $c(M; q) = 1$ if and only if M is affine over $GF(q)$. More generally, for a loopless $GF(q)$-representable matroid M, the critical exponent $c(M; q)$ is the minimum number of sets $S_1, S_2, \ldots, S_c$ into which $E(M)$ can be partitioned so that $M|S_i$ is affine over $GF(q)$ for all i. It follows, since an independent set is certainly affine, that if $E(M)$ can be partitioned into k independent sets, then $c(M;q) \leq k$. Now let M be simple and $GF(q)$-representable. A direct consequence of Theorem 11.3.12 is that if $|X| \leq kr(X)$ for every subset X of $E(M)$, then $c(M;q) \leq k$. Thus the results from Section 14.10 on linear growth rates have immediate implications for critical exponents. In particular, Proposition 14.10.23, implies that $c(M;2) \leq 8$ for all loopless binary matroids with no $M(K_5)$-minor. That result of Kung (1987) is still the best-known partial result towards the following conjecture of Walton and Welsh (1980).

Conjecture 15.3.1 *If M is a loopless binary matroid having no minor isomorphic to $M(K_5)$, then $c(M;2) \leq 3$.* □

Another result on linear growth rates, Proposition 14.10.20, implies that the following conjecture of Brylawski (1975d) holds for $q \leq 3$. This conjecture is open for all $q \geq 4$.

Conjecture 15.3.2 *If M is a loopless $GF(q)$-representable matroid and M has no $M(K_4)$-minor, then $c(M;q) \leq 2$.* □

Numerous problems from various diverse branches of combinatorics can be expressed in terms of evaluations of critical exponents. Kung (1996) has given a comprehensive survey of such critical problems for matroids, and this section will do little more than draw attention to the existence of a rich body of work in this area. The critical exponent can be thought of as an analogue of the *chromatic number* $\chi(G)$ of a loopless graph G since

$$\chi(G) = \min\{j \in \mathbb{Z}^+ : p(M(G);j) > 0\}.$$

Thus, for $M = M(G)$,

$$q^{c(M;q)-1} < \chi(G) \leq q^{c(M;q)}.$$

Hence the Four-Colour Theorem (Appel and Haken 1976) is equivalent to the assertion that $c(M(G);2) \leq 2$ for all loopless planar graphs G.

Hadwiger's (1943) famous conjecture for graphs can be stated as follows.

Conjecture 15.3.3 *For all positive integers m, the unique graph with chromatic number m for which every loopless proper minor has chromatic number less than m is K_m.* □

By a result of Wagner (1937a), this conjecture is equivalent to the Four-Colour Conjecture when $m = 5$. Moreover, Robertson, Seymour, and Thomas (1993) proved that Hadwiger's Conjecture for $m = 6$ is also equivalent to the Four-Colour Conjecture. Thus Hadwiger's Conjecture holds for $m \leq 6$ but is open for all larger values of m.

Tutte's (1966c) geometrical version of the Four-Colour Problem is the following. Recall that P_{10} is the Petersen graph (see Figure 3.5).

Conjecture 15.3.4 (Tutte's Tangential 2-Block Conjecture) *The only binary matroids with critical exponent 3 for which every proper loopless minor has critical exponent less than 3 are $M(K_5)$, F_7, and $M^*(P_{10})$.* □

The name of this conjecture derives from the fact that, for a positive integer k, a $GF(q)$-representable matroid M with critical exponent exceeding k is a *tangential k-block over $GF(q)$* if every loopless non-empty proper minor of M has critical exponent at most k. Tutte (1966c) proved his conjecture for tangential 2-blocks of rank at most 6, and B. T. Datta (1976, 1981) proved that there are no binary tangential 2-blocks of rank 7 or 8. The most significant advance towards the resolution of this conjecture was made by Seymour (1981d) who proved that a binary tangential 2-block is either $M(K_5)$ or F_7, or it is the dual of a graphic matroid. By the Well-Quasi-Ordering Theorem for Graphs (14.1.1), this result implies that there are only finitely many binary tangential 2-blocks. Seymour's result also implies that Tutte's Tangential 2-Block Conjecture is equivalent to the following conjecture for cographic matroids, which is also due to Tutte (1966c) and is a variant of his celebrated 5-Flow Conjecture (Tutte 1954).

Conjecture 15.3.5 *Let M be a loopless cographic matroid for which $p(M;4) = 0$. Then M has a minor isomorphic to $M^*(P_{10})$.* □

Although this conjecture remains open in general, Robertson, Sanders, Seymour, and Thomas (in Thomas 1999) settled it in the important special case when M is the bond matroid of a cubic graph.

Walton and Welsh (1982) proved that the only tangential 1-blocks over $GF(3)$ are $M(K_4)$ and $U_{2,4}$. Initially, it was believed (Welsh 1980) that tangential k-blocks are quite scarce. In a sequence of four papers, Whittle (1987, 1988, 1989a, 1989b) presented a family of examples showing that this is not the case. Indeed, there are rank-r tangential blocks over $GF(q)$ for all r such that $k + 1 \le r \le q^k$. Whittle used the complete principal truncation and then the more general quotient operation to construct new tangential k-blocks from existing ones. He also showed how jointless Dowling geometries can be tangential k-blocks. Perhaps the simplest of his constructions is that of a q-lift. Let M be a simple rank-r $GF(q)$-representable matroid. Then there is a subset X of $PG(r,q)$ such that $PG(r,q)|X \cong M$. Evidently X spans a hyperplane H of $PG(r,q)$. Take a point e of $PG(r,q)$ that is not in H. Let M' be the restriction of $PG(r,q)$ to the union of all lines of the latter that contain e and some point of X. The matroid M' is called a *q-lift* of M, and Kung (1996) showed that $p(M';\lambda) = (\lambda - 1)q^r p(M; \frac{\lambda}{q})$. Earlier, Whittle (1989a) proved that if M is a tangential k-block over $GF(q)$, then a q-lift M' is a tangential $(k + 1)$-block over $GF(q)$.

Whittle's constructions of tangential blocks are very geometrical in nature and show that the intuition one gets from graphs about tangential blocks can be very misleading. Kung's (1996) paper includes a comprehensive treatment of known results on tangential blocks and ends with the following interesting conjecture, which is very much in the spirit of several problems that were discussed in the last chapter and are of central interest in matroid theory at present.

Conjecture 15.3.6 *For all prime powers q and all positive integers k, there are only finitely many tangential k-blocks over $GF(q)$.* □

Whittle proved that, for given q and k, only finitely many tangential k-blocks over $GF(q)$ have modular hyperplanes.

15.4 From graphs to matroids

In the entry in *Mathematical Reviews* (MR0427112 (55#148)) for Welsh's (1976) matroid theory book, Tutte wrote: 'It has been said that to get a theorem on matroids we should take a known one on graphs, rewrite it and its proof so as to make no mention of vertices, and then replace the word "graph" by "matroid".' Many important matroid problems are motivated by corresponding problems for graphs. Some of the most important of these were discussed in the last chapter.

The *Reconstruction Conjecture* of Kelly and Ulam (in Kelly 1942) is another famous problem for graphs that suggests numerous interesting matroid problems. The graph conjecture is that if G and H are graphs having at least three vertices and $\alpha : V(G) \to V(H)$ is a bijection such that $G - v \cong H - \alpha(v)$ for all v in $V(G)$, then $G \cong H$. There is a corresponding *Edge-Reconstruction Conjecture* for graphs with at least four edges. Brylawski (1974, 1975a) considered matroid analogues of these conjectures. Let M be a matroid on a set E and let N be another matroid on E such that $M \backslash e \cong N \backslash e$ for all e in E. If all such matroids N are isomorphic to M, then M is said to be *reconstructible*. The matroids $U_{n-1,n}$ and $U_{n,n}$ are not reconstructible since each has every single-element deletion isomorphic to $U_{n-1,n-1}$. Thus any matroid analogue of the Edge-Reconstruction Conjecture for graphs would need to make an exception for circuits and free matroids. Brylawski (1974) conjectured that these are the only exceptions, that is, all matroids other than circuits and free matroids are reconstructible. However, soon after, he (Brylawski 1975a) found a counterexample to this conjecture. Indeed, as noted in Section 6.1, if M^* is a projective plane of order p^2 where p is a prime exceeding two, then M is not reconstructible (Brylawski 1986a). In spite of these examples, the following question remains unanswered.

Problem 15.4.1 *Let M be a binary matroid that is neither a circuit nor a free matroid. Is M reconstructible?* □

How should one formulate a matroid analogue of the (vertex) Reconstruction Conjecture for graphs? We have seen several instances earlier in the book when the matroid analogue of a vertex has been taken to be a cocircuit. This will be done again here. Let M be a matroid and N be another matroid for which there is a bijection $\alpha : \mathcal{H}(M) \to \mathcal{H}(N)$ such that $M|H \cong N|\alpha(H)$ for all H in $\mathcal{H}(M)$, the set of hyperplanes of M. If all such matroids N are isomorphic to M, then M is said to be *hyperplane-reconstructible*. Since there is such a bijection between the sets of hyperplanes of the non-isomorphic matroids Q_6 and $U_{2,4} \oplus_2 U_{2,4}$, not all matroids are hyperplane-reconstructible (Brylawski 1974). However, the following conjecture (Brylawski 1974) is still open.

Conjecture 15.4.2 *All binary matroids are hyperplane-reconstructible.*   □

Since the vertex bonds in a loopless 3-connected graph G are exactly the cocircuits C^* of $M(G)$ for which $M(G)\backslash C^*$ is connected, Brylawski also proposed that a binary matroid of fixed rank and corank could be reconstructed from its connected hyperplanes. However, Welsh (private communication) noted that this fails since if G is a graph that is formed from a single cycle by adding one diagonal, then provided G has no 3-cycles, $M(G)$ has no connected hyperplanes.

Although a matroid need not be hyperplane-reconstructible, Brylawski (1981) showed that its characteristic polynomial, and more generally its Tutte polynomial, can be determined from the multiset of isomorphism classes of its hyperplanes. A matroid M is *minor-reconstructible* if $N \cong M$ whenever N is a matroid on $E(M)$ such that $N\backslash e \cong M\backslash e$ and $N/e \cong M/e$ for all e in $E(M)$. Brylawski (1986a) raised the following.

Problem 15.4.3 *Are all matroids minor-reconstructible?*   □

As with the corresponding problems for graphs, progress on these reconstruction problems for matroids has been limited. Some partial results were obtained by Miller (1997, 2003).

The background to the next two problems was discussed in Section 4.3. In particular, Problem 15.4.4, which appeared as Exercise 16 of that section, was settled by Neumann-Lara, Rivera-Campo, and Urrutia (1999) for graphic matroids and by McGuinness (2005) for cographic matroids.

Problem 15.4.4 *Let M be a connected matroid with at least two elements whose largest circuit has c elements. Prove that M has a collection of c not-necessarily-distinct cocircuits such that every element of M is in at least two of these cocircuits.*   □

The next problem is a variant on Exercise 7(c) of Section 4.3. It was proposed by Royle, who noted that the problem holds when $|E(M)| \leq 9$.

Problem 15.4.5 *Let M be a 3-connected matroid with at least four elements whose largest circuit has c elements and whose largest cocircuit has c^* elements. Prove that $|E(M)| < \frac{1}{2}cc^*$ unless $M \cong AG(3,2)$.*   □

We close this section with two matroid connectivity problems that are suggested by results for graphs. An n-connected graph or matroid is *minimally n-connected* if no deletion of a single edge or element, respectively, is n-connected. Halin (1969) showed that every minimally n-connected graph G has a vertex of degree n and this result was strengthened by Mader (1979) who showed that G has at least $[(n-1)|V(G)|+2n]/(2n-1)$ vertices of degree n. These graph results motivate one to ask whether a minimally n-connected matroid M with at least $2(n-1)$ elements has an n-element cocircuit. Murty (1974) and Wong (1978) answered this question affirmatively for $n=2$ (see Exercise 10 of Section 4.3) and $n=3$. Reid, Wu, and Zhou (2008) showed that, when $n=4$, the answer to the question is again affirmative with one exception, a 9-element matroid of rank

4. In addition, for all $n \geq 5$, Reid *et al.* found a minimally n-connected matroid with $2n + 1$ elements having no n-element cocircuit. They then proposed the following modification to the initial matroid question.

Conjecture 15.4.6 *For $n \geq 5$, let M be a minimally n-connected matroid with $|E(M)| \geq 2(n - 1)$. If $|E(M)| \neq 2n + 1$, then M has an n-element cocircuit.* $\square$

Mader (1974) proved that a simple k-connected graph with minimum degree at least $k + 2$ has a cycle C such that $G \backslash C$, the graph obtained from G by deleting the edges of C, is k-connected. Next we consider some attempts to obtain matroid analogues of this result. In the first edition of this book, it was asked whether, in every connected binary matroid M for which every circuit has at least three elements and every cocircuit has at least four elements, there is a circuit C such that $M \backslash C$ is connected. Lemos (in Lemos and Oxley 1999) answered this question negatively using the following cographic matroid. Let $M = M^*(G)$ where G is constructed from $K_{5,5}$ by, for every set $\{u, v, w\}$ of three vertices in the same vertex class, adding two new vertices and joining each to u, v, and w, and to nothing else. This example leaves open the following.

Problem 15.4.7 *Let M be a connected binary matroid in which $|C^*| \geq 5$ for every cocircuit C^*. Does M have a circuit C such that $M \backslash C$ is connected?* $\square$

Goddyn and Jackson (1999) answered this question affirmatively when M does not have both F_7 and F_7^* as minors, showing that M has a circuit C so that $M \backslash C$ is connected and $r(M \backslash C) = r(M)$. They also noted that if the answer to Problem 15.4.7 is negative, there is still the question of whether there is some integer t so that requiring every cocircuit of M to have at least t elements ensures an affirmative answer to the problem. Such results on what are called 'removable circuits' in 2- and 3-connected matroids were surveyed by Oxley (2001).

15.5 Enumeration

Blackburn, Crapo, and Higgs (1973) computed a catalogue of all non-isomorphic simple matroids on at most eight elements. Acketa (1984) used their list to determine, for all $n \leq 8$, all non-isomorphic n-element matroids and all non-isomorphic n-element binary matroids. More recently, Mayhew and Royle (2008) catalogued all non-isomorphic 9-element matroids incorporating this into an on-line database, which is accessible at Gordon Royle's website. Tables 15.1 and 15.2 summarize some of the data found by these three sets of authors. Royle (private communication) was also able to determine the numbers of binary matroids and simple binary matroids on a 10-element set and these appear in the first table. However, Royle wrote on his website that finding the number of all 10-element matroids 'is likely to be out of reach for some time to come.' Indeed, Mayhew and Royle (2008) have estimated that the number of 10-element rank-5 paving matroids exceeds 2.5×10^{12}. Thus each of the four blank entries in Table 15.1 has yet to be determined.

TABLE 15.1. The numbers of non-isomorphic matroids with various properties on an n-element set for $0 \leq n \leq 10$.

n	0	1	2	3	4	5	6	7	8	9	10
matroids	1	2	4	8	17	38	98	306	1724	383172	
simple matroids	1	1	1	2	4	9	26	101	950	376467	
binary matroids	1	2	4	8	16	32	68	148	342	848	2297
simple binary	1	1	1	2	3	5	10	20	42	102	276
paving matroids	1	2	4	7	12	19	34	71	468	266822	
simple paving	1	1	1	2	4	8	18	50	439	266784	

Of the six problems or conjectures from the section on enumeration in the first edition of this book, two have now been settled. In particular, we noted in Corollary 7.4.12 that, for all non-negative integers n and m,

$$f(n + m) \geq f(n)f(m),$$

where $f(n)$ is the number of non-isomorphic matroids on an n-element set. This result, settling an old conjecture of Welsh (1969a), was proved independently by Crapo and Schmitt (2005a) and Lemos (2004).

Let $f_r(n)$ denote the number of rank-r non-isomorphic matroids on an n-element set. Table 15.2 gives the value of $f_r(n)$ for all $n \leq 9$. This table supports the next two conjectures, both of which are due to Welsh (1971b and private communication).

Conjecture 15.5.1 *The sequence $(f_r(n) : 1 \leq r \leq n)$ is unimodal.* □

Conjecture 15.5.2 *The number $f_r(n)$ is a maximum when r is $\lfloor n/2 \rfloor$.* □

The number $f(n)$ of non-isomorphic matroids on an n-element set can also be viewed as the number of unlabelled n-element matroids. Let $l(n)$ be the

TABLE 15.2. The number of non-isomorphic rank-r matroids on an n-set.

n / r	0	1	2	3	4	5	6	7	8	9
0	1	1	1	1	1	1	1	1	1	1
1		1	2	3	4	5	6	7	8	9
2			1	3	7	13	23	37	58	87
3				1	4	13	38	108	325	1275
4					1	5	23	108	940	190214
5						1	6	37	325	190214
6							1	7	58	1275
7								1	8	87
8									1	9
9										1

number of labelled n-element matroids. This distinction between labelled and unlabelled matroids can be confusing. As an example, consider the unlabelled matroids $U_{0,1} \oplus U_{1,1}$ and $U_{1,2}$. If we label the ground set of each by the set $\{1,2\}$, then permuting the labels on the elements will produce an isomorphic labelled matroid in the second case but a non-isomorphic labelled matroid in the first case. The other two unlabelled 2-element matroids are $U_{0,2}$ and $U_{2,2}$ and each has a unique labelling. Thus $l(2) = 5$, whereas $f(2) = 4$. If we consult Table 1.1, we see that $l(n) = f(n)$ for $n < 2$ and it is not difficult to check that $l(3) = 16$ whereas $f(3) = 8$. Let $A_u(n)$ be the set of unlabelled n-element matroids with no non-trivial automorphisms. Mayhew, Newman, Welsh, and Whittle (2009) have conjectured that, asymptotically, almost every unlabelled n-element matroid is in $A_u(n)$.

Conjecture 15.5.3 *The limit* $\lim_{n \to \infty} |A_u(n)|/f(n)$ *exists and equals one.* □

If this conjecture is true, then it implies that the asymptotic behaviour of classes of labelled and unlabelled n-element matroids are the same. We now make this precise. Define a matroid *property* to be a class of matroids that is closed under isomorphism. Let P be a matroid property and denote by $P_u(n)$ and $P_l(n)$ the sets of unlabelled and labelled n-element matroids with property P.

Mayhew, Newman, Welsh, and Whittle (2009) proved the following.

Proposition 15.5.4 *Assume that Conjecture 15.5.3 holds. Let P be a matroid property. Then*

$$\lim_{n \to \infty} \frac{|P_u(n)|}{f(n)}$$

exists and equals L if and only if

$$\lim_{n \to \infty} \frac{|P_l(n)|}{l(n)}$$

exists and equals L. □

In what follows, we shall assume, as do Mayhew, Newman, Welsh, and Whittle (2009), that Conjecture 15.5.3 holds. The statement that *asymptotically almost every matroid has property P* means that both $\frac{|P_u(n)|}{f(n)}$ and $\frac{|P_l(n)|}{l(n)}$ tend to one as n tends to infinity. Mayhew *et al.* made seven other conjectures about the asymptotic behaviour of matroids. These interesting conjectures indicate how little is known about the properties of almost all matroids.

Conjecture 15.5.5 *Asymptotically, almost every matroid is connected.* □

Letting $l_c(n)$ be the number of connected labelled n-element matroids, Mayhew *et al.* proved, in support of this conjecture, that, for every $\varepsilon > 0$, there is an integer N such that, whenever $n \geq N$,

$$\frac{l_c(n)}{l(n)} \geq \tfrac{1}{2} - \varepsilon.$$

The next conjecture generalizes the last one.

Conjecture 15.5.6 *Let k be a fixed integer exceeding one. Asymptotically, almost every matroid is k-connected.*   □

The catalogue of Blackburn *et al.* prompted Welsh (1976) to ask whether most simple matroids are paving. In the first edition of this book, this was formalized as a question. We have refined the original problem below to account for the fact that, as Mayhew and Royle (2008) pointed out, when n is 7 or 8, fewer than half of the simple matroids are paving.

Problem 15.5.7 *Is it true that, for all non-negative integers n other than 7 and 8, more than half of the non-isomorphic simple matroids on an n-element set are paving matroids?*   □

From Table 15.1, we deduce that 69.6% of the matroids on a 9-element set are paving. Mayhew, Newman, Welsh, and Whittle (2009) offered the following.

Conjecture 15.5.8 *Asymptotically, almost every matroid is paving.*   □

They also proposed a striking conjecture about the presence of paving matroids.

Conjecture 15.5.9 *Let N be a fixed matroid such that both N and N^* are paving. Asymptotically, almost every matroid has an N-minor.*   □

Many of the special matroids we have met in this book satisfy the hypotheses of the last conjecture. One of these is the Vámos matroid, V_8 (see Figure 2.4 and Example 2.1.25). Mayhew *et al.* suggested that the following special case of the last conjecture may be 'more approachable'.

Conjecture 15.5.10 *Asymptotically, almost every matroid has a V_8-minor.* □

We proved in Proposition 2.2.26 that V_8 is not representable over any field. Thus the truth of the last conjecture would also imply the truth of the following.

Conjecture 15.5.11 *Asymptotically, almost every matroid is not representable over any field.*   □

In support of this conjecture, we note that, by combining Knuth's asymptotic lower bound on the number of non-isomorphic n-element matroids given in equation (1.1) of Section 1.1 with the generalization of Problem 12 of that section, we can show that, for each prime power q, asymptotically almost every matroid is not $GF(q)$-representable. Extending this, Mayhew *et al.* observe that a result of Ronyai, Babai, and Ganapathy (2001) implies that, for every fixed field $\mathbb{F}$, almost every matroid is not $\mathbb{F}$-representable.

Mayhew, Newman, Welsh, and Whittle's (2009) final conjecture is an asymptotic strengthening of Conjecture 15.5.2.

Conjecture 15.5.12 *Asymptotically, almost every matroid M satisfies*

$$\frac{|E(M)| - 1}{2} \leq r(M) \leq \frac{|E(M)| + 1}{2}.$$ □

Let $a(n)$ be the number of non-isomorphic binary matroids on an n-element set. The other problem to have been solved from the enumeration section of the first edition of this book is the determination of the long-term behaviour of $a(n)$. Marcel Wild (2000, 2005) proved that, asymptotically, $n!\,a(n)$ equals the total number of subspaces of $V(n,2)$.

Theorem 15.5.13 *For all sufficiently large positive integers n,*

$$(1 + 2^{-\frac{n}{2}+2\log_2 n + 0.2499})\frac{1}{n!}\sum_{j=0}^{n}\begin{bmatrix}n\\j\end{bmatrix}_2$$

$$\leq a(n) \leq (1 + 2^{-\frac{n}{2}+2\log_2 n + 0.2501})\frac{1}{n!}\sum_{j=0}^{n}\begin{bmatrix}n\\j\end{bmatrix}_2.$$

Hence $a(n) \sim \frac{1}{n!}\sum_{j=0}^{n}\begin{bmatrix}n\\j\end{bmatrix}_2$. $\qquad\square$

Another proof of the asymptotic behaviour of $a(n)$ was given by Xiang-Dong Hou (2007). Both of these proofs were obtained in the context of counting the number of inequivalent binary codes. Just as for matroid representations, there are several different notions of equivalence for linear q-ary codes. Our terminology follows Huffman and Pless (2003). Recall that a linear q-ary code of length n is a subspace of the vector space $V(n,q)$. Two such subspaces V_1 and V_2 are *equivalent* if V_2 can be obtained from V_1 by permuting the coordinates, scaling each coordinate, and applying an automorphism of $GF(q)$ coordinate-wise to every vector. If V_2 can be obtained from V_1 by using the first two but not the third of these three operations, then V_1 and V_2 are called *monomially equivalent*. It should be noted, however, that some authors use the term 'equivalent' for what we have called 'monomially equivalent'. Now let A_1 and A_2 be non-zero $r \times n$ matrices over $GF(q)$. Then the row spaces, $\mathcal{R}(A_1)$ and $\mathcal{R}(A_2)$, of these matrices are equivalent codes if and only if A_1 and A_2 are equivalent $GF(q)$-representations of some fixed matroid. Moreover, $\mathcal{R}(A_1)$ and $\mathcal{R}(A_2)$ are monomially equivalent codes if and only if A_1 and A_2 are projectively equivalent $GF(q)$-representations of some fixed matroid. Hou (2005, 2009) determined the asymptotic numbers of monomially inequivalent and inequivalent q-ary codes. The first of these verified a conjecture of R. F. Lax (2004).

Theorem 15.5.14 *Let n and k be positive integers, p be a prime, and $q = p^k$. The numbers of monomially inequivalent and inequivalent q-ary codes of length n are asymptotic to*

$$\frac{1}{n!(q-1)^{n-1}}\sum_{j=0}^{n}\begin{bmatrix}n\\j\end{bmatrix}_q \qquad and \qquad \frac{1}{n!k(q-1)^{n-1}}\sum_{j=0}^{n}\begin{bmatrix}n\\j\end{bmatrix}_q. \qquad\square$$

Because ternary matroids are uniquely $GF(3)$-representable, the last result immediately gives that the number of non-isomorphic ternary matroids on an

n-element set is asymptotic to $\frac{1}{n!2^{n-1}}\sum_{j=0}^{n}\left[{n\atop j}\right]_3$. But, since $GF(q)$-representable matroids are not uniquely $GF(q)$-representable when $q > 3$, the second part of the last theorem provides only an upper bound on the asymptotic behaviour of the number $a_q(n)$ of non-isomorphic n-element $GF(q)$-representable matroids.

Problem 15.5.15 *Let p be a prime, k be a positive integer, and $q = p^k$. For $q > 3$, what is the asymptotic behaviour of $a_q(n)$ as n tends to infinity? In particular, is $a_q(n) \sim \frac{1}{n!k(q-1)^{n-1}}\sum_{j=0}^{n}\left[{n\atop j}\right]_q$?* □

The next conjecture (9.4.12) compares the numbers, $b(M)$ and $d(M)$, of bases and circuits in a simple binary matroid M.

Conjecture 15.5.16 *In a rank-r simple binary matroid M having no coloops,*

$$b(M) \geq \tfrac{1}{2}(r+1)d(M).$$ □

15.6 Gammoids and transversal matroids

In general, gammoids seem rather difficult to handle. Ingleton and Piff (1973) showed that a matroid of rank at most three is a gammoid if and only if it is a strict gammoid. Moreover, Mason (1972a) gave the following criterion for determining when a matroid M is a strict gammoid (see also Kung 1978). Let the function α be defined recursively on the subsets X of $E(M)$ by $\alpha(X) = |X| - r(X) - \sum \alpha(F)$ where the summation is over all flats F of M that are properly contained in X. Mason proved that M is a strict gammoid if and only if $\alpha(X) \geq 0$ for all $X \subseteq E(M)$. Thus one can determine when a rank-3 matroid is a gammoid. However, the following problem seems to be open even for rank-4 matroids.

Problem 15.6.1 *Give an algorithm to test whether or not a given matroid is a gammoid.* □

The classes of binary gammoids and ternary gammoids have both been characterized by finite sets of excluded minors as has the class of quaternary matroids.

Problem 15.6.2 *Characterize the class of quaternary gammoids by excluded minors.* □

The classes of binary transversal and ternary transversal matroids have been characterized by excluded series minors (de Sousa and Welsh 1972, Oxley 1986b).

Problem 15.6.3 *Characterize the class of quaternary transversal matroids by excluded series minors.* □

In the last chapter, we looked at which classes of matroids have no infinite antichains. Whittle (private communication) proposed the following.

Conjecture 15.6.4 *For every integer q exceeding one, the class of gammoids with no $U_{2,q+2}$- or $U_{q,q+2}$-minors has no infinite antichains.* □

The next problem is another that was originally posed by Welsh (1971b).

Problem 15.6.5 *Characterize those transversal matroids whose duals are also transversal.* □

From Proposition 11.2.28, every principal transversal matroid is cotransversal, and Jensen (1978) showed that every binary transversal cotransversal matroid is principal transversal. However, Jensen also noted that $U_{2,4} \oplus_2 U_{2,4}$ is a (ternary) transversal cotransversal matroid that is not principal transversal.

15.7 Excluding a uniform matroid

As discussed in Section 14.1, Geelen, Gerards, and Whittle have been pursuing a structure theorem for $GF(q)$-representable matroids, which aims, in part, to resolve the Well-Quasi-Ordering Conjecture for Matroids (14.1.5). Geelen (2008) has indicated that he hopes that such structural results for matroids can be developed even further. In particular, he raised the following.

Problem 15.7.1 *For fixed integers r and n with $0 \le r \le n$, find a qualitative structure theorem for the class of matroids with no $U_{r,n}$-minor.* □

In pursuit of this problem, Geelen has proposed several conjectures, of which we shall note four. Each relates to precluding 'the possibility of certain pairs of matroids coexisting.' Let N_1 and N_2 be minors of a matroid M. Geelen defines N_1 and N_2 to be *k-interconnected* in M if $\kappa_M(E(N_1), E(N_2)) \ge k$, that is, for all sets X with $E(N_1) \subseteq X \subseteq E(M) - E(N_2)$, we have $r(X) + r(E(M) - X) - r(M) \ge k$.

Conjecture 15.7.2 *Let k be a positive integer and q_1 and q_2 be powers of distinct primes. Then there is an integer n such that if a matroid M has $PG(n, q_1)$ and $PG(n, q_2)$ as n-interconnected minors, then M has a $U_{k,2k}$-minor.* □

Conjecture 15.7.3 *For every positive integer k, there is an integer n such that if a matroid M has $M(K_n)$ and $M^*(K_n)$ as n-interconnected minors, then either M has a $U_{k,2k}$-minor, or M has a $PG(k, q)$-minor for some prime power q.* □

The last conjecture is also open when K_n is replaced by G_n^+, the graph that is obtained from an $n \times n$ grid by adding a new vertex that is adjacent to every vertex of the grid. The third and fourth conjectures of this type involve the bicircular matroid, $B(G)$, of a graph G and the dual, $B^*(G)$, of this matroid.

Conjecture 15.7.4 *For every positive integer k, there is an integer n such that if a matroid M has $B(G_n^+)$ and $B^*(G_n^+)$ as n-interconnected minors, then M has a $U_{k,2k}$-minor.* □

Conjecture 15.7.5 *For every positive integer k, there is an integer n such that if a matroid M has $B(G_n^+)$ and $M^*(G_n^+)$ as n-interconnected minors, then M has a $U_{k,2k}$-minor.* □

The last two conjectures are also open when G_n^+ is replaced by K_n. We end this section with a conjecture of Geelen and Whittle (private communication).

Conjecture 15.7.6 *Let $\mathcal{N}$ be an infinite antichain of matroids such that, for all non-negative integers k, there is a member of $\mathcal{N}$ of Tutte connectivity k. Then every uniform matroid is a minor of some member of $\mathcal{N}$.* □

15.8 Negative correlation

Recently, there has been considerable interest in a collection of matroid inequalities that are motivated by electrical network theory. The background to such problems is discussed by Wagner (2004/06). It follows from the work of Kirchhoff (1847) that if e and f are distinct edges of a connected graph G, and T is a spanning tree of G chosen uniformly at random, then

$$Pr(T \text{ contains } e|T \text{ contains } f) \leq Pr(T \text{ contains } e),$$

that is, the probability that T contains e given that it contains f does not exceed the probability that T contains e. In other words, the events that T contains e and that T contains f are negatively correlated.

A matroid M is called *negatively correlated* if, for a basis B of M chosen uniformly at random and for every two distinct elements e and f of $E(M)$,

$$Pr(B \text{ contains } e|B \text{ contains } f) \leq Pr(B \text{ contains } e).$$

Seymour and Welsh (1975) considered which matroids are negatively correlated and showed that the binary matroid S_8 is not. Feder and Mihail (1992) defined a matroid to be *balanced* if M and all of its minors are negatively correlated. Such matroids have an important computational property which essentially allows the number of bases to be closely approximated quickly. Details can be found in Jerrum (2006). The reader should be aware that the term 'balanced' has also been used in matroid theory with at least three other different meanings.

For a matroid M, recall that $b(M)$ denotes its number of bases. Extending this notation, for disjoint subsets U and V of $E(M)$, we write $b_U^V(M)$ for the number of bases of M that contain U but avoid V. Rewriting the definition, we see that M is balanced if and only if, for all minors N of M and all distinct elements e and f of N,

$$b_e(N)b_f(N) - b_{ef}(N)b(N) \geq 0. \tag{15.1}$$

Using the fact that $b(N) = b_e(N) + b^e(N)$, we can show that the last inequality is equivalent to the inequality

$$b_e^f(N)b_f^e(N) - b_{ef}(N)b^{ef}(N) \geq 0. \tag{15.2}$$

The definition ensures that the class of balanced matroids is minor-closed. Moreover, since $b(N) = b(N^*)$, we see that the class is also closed under duality. It is not, however, closed under taking 2-sums (Choe and Wagner 2006).

Feder and Mihail (1992) proved that every regular matroid is balanced and asserted that a binary matroid is balanced if and only if it has no S_8-minor. The first complete published proof of this result was given by Choe and Wagner (2006). Characterizing the class of all balanced matroids by excluded minors seems hard. Indeed, Welsh (private communication) has proposed the following.

Conjecture 15.8.1 *The class of balanced matroids has infinitely many excluded minors.* ☐

Some natural special classes of balanced matroids seem more approachable.

Problem 15.8.2 *Find the excluded minors for the class of ternary balanced matroids.* ☐

Problem 15.8.3 *Find the excluded minors for the class of quaternary balanced matroids.* ☐

Next we consider an extension of (15.1). For a matroid M with ground set E, let $\mathbf{z} : E \to \mathbb{R}$ be a real-valued weight function defined on E. For a subset A of E, define $\mathbf{z}(A) = \prod_{a \in A} \mathbf{z}(a)$ if A is non-empty, and let $\mathbf{z}(\emptyset) = 1$. For disjoint subsets U and V of E, set

$$b_U^V(M; \mathbf{z}) = \sum \mathbf{z}(A),$$

where the summation is over all sets A such that A is a basis of M containing U and avoiding V. Motivated by a property of linear resistive networks, Choe and Wagner (2006) define M to be *Rayleigh* if

$$b_e(N; \mathbf{z}) b_f(N; \mathbf{z}) - b_{ef}(N; \mathbf{z}) b(N; \mathbf{z}) \geq 0 \qquad (15.3)$$

for all distinct e and f in E whenever $\mathbf{z}(c) > 0$ for all c in E. In addition, Choe and Wagner call M *strongly Rayleigh* if (15.3) holds for all distinct e and f in E and all real-valued weight functions $\mathbf{z}$. The classes of Rayleigh and strongly Rayleigh matroids are closed under taking minors, duals, and 2-sums (Choe and Wagner 2006). By taking $\mathbf{z}(c) = 1$ for all c in E, we see that the class of balanced matroids contains the class of Rayleigh matroids; and the latter clearly contains the class of strongly Rayleigh matroids. Choe and Wagner gave an example of an 8-element rank-4 matroid that is balanced but not Rayleigh; and they show that the Fano matroid is Rayleigh but not strongly Rayleigh. Wagner (2005) proved that all rank-3 matroids are Rayleigh. Choe and Wagner raised a number of problems for Rayleigh matroids including the following.

Problem 15.8.4 *Find the minimal rank-4 matroids that are not Rayleigh.* ☐

Problem 15.8.5 *Find the excluded minors for the class of ternary Rayleigh matroids.* ☐

Problem 15.8.6 *Find the excluded minors for the class of quaternary Rayleigh matroids.* ☐

The introduction of Rayleigh matroids was motivated in part by the work of Choe, Oxley, Sokal, and Wagner (2004) on matroids with the half-plane property, which we shall now define. The *basis generating polynomial* of a matroid M on a set E is $\sum_{B \in \mathcal{B}(M)} (\prod_{e \in B} z_e)$. Viewing this as a polynomial in $|E|$ complex variables, we say that M has the *half-plane property* or is *HPP* if, whenever $\mathrm{Re}(z_e) > 0$ for all e in E, the basis generating polynomial is non-zero. Brändén (2007) answered two questions of Choe and Wagner by proving the following.

Theorem 15.8.7 *A matroid is strongly Rayleigh if and only if it is HPP.* □

Combining this with results above, we get that a binary matroid is HPP if and only if it is regular; and Choe, Oxley, Sokal, and Wagner (2004) proved that a ternary matroid is HPP if and only if it is a $\sqrt[6]{1}$-matroid. Thus the excluded minors for the classes of binary HPP matroids and ternary HPP matroids are known (6.6.6, 14.7.9). A natural next question is the following.

Problem 15.8.8 *Find the excluded minors for the class of quaternary HPP matroids.* □

Determining whether or not a matroid is HPP seems difficult. Choe, Oxley, Sokal, and Wagner (2004) have observed that, although this problem is decidable, the best known algorithms are exponential and seem thoroughly infeasible in practice. Choe *el al.* did succeed in showing that all matroids with at most six elements or with rank or corank at most two are HPP. Moreover, by combining their results with work of Wagner and Wei (2009), we know precisely which 7-element matroids are HPP. Wagner and Wei also showed, for example, that the Vámos matroid is HPP. Choe *el al.* (2004) proposed the following.

Conjecture 15.8.9 *All rank-3 transversal matroids are HPP.* □

They also asked whether all transversal matroids are HPP, but Choe and Wagner (2006) answered this negatively by giving an example of a rank-4 transversal matroid that is not even balanced. Our last problem about HPP matroids addresses the difficulty of recognizing such matroids (Choe *el al.* 2004, Problem 13.18). The corresponding problem for Rayleigh matroids is also open.

Problem 15.8.10 *Find a practically feasible algorithm for testing whether a matroid is HPP.* □

We now return to the correlation problems with which we began this section. As before, let $\mathbf{z} : E \to \mathbb{R}$ be a real-valued weight function defined on the ground set E of a matroid M. For disjoint subsets U and V of E, we again consider $\sum \mathbf{z}(A)$, where the summation is over all sets A of E that contain U and avoid V. Imposing the additional restriction that A is independent, we write $i_U^V(M; \mathbf{z})$ for $\sum \mathbf{z}(A)$. If, instead, we require A to be spanning, we write $s_U^V(M; \mathbf{z})$ for the resulting sum. Semple and Welsh (2008) define M to be *independence-correlated* if, for all distinct elements e and f of E, and all positive weight functions $\mathbf{z}$,

$$i_e(M; \mathbf{z})i_f(M; \mathbf{z}) - i_{ef}(M; \mathbf{z})i(M; \mathbf{z}) \geq 0. \qquad (15.4)$$

Likewise, M is *spanning-correlated* if the corresponding inequality holds with $i_U^V(M; \mathbf{z})$ replaced by $s_U^V(M; \mathbf{z})$. Semple and Welsh show that a matroid is independence-correlated if and only if its dual is spanning-correlated, and they call M *correlated* if it is both independence-correlated and spanning-correlated. They prove that the classes of correlated, independence-correlated, and spanning-correlated matroids are all closed under taking minors and direct sums (see also

Wagner 2008). Moreover, all three classes are closed under taking 2-sums (Cocks 2008 and Wagner 2008). The classes of independence-correlated and spanning-correlated matroids are both contained in the class of Rayleigh matroids. Indeed, for Wagner (2008), independence-correlated and spanning-correlated matroids are **I**-*Rayleigh* and **S**-*Rayleigh* matroids, respectively. Semple and Welsh (2008) and Wagner (2008) raised several questions about the relationships between these various classes of matroids including the following.

Problem 15.8.11 *Are the classes of Rayleigh matroids and correlated matroids equal?* ☐

An affirmative answer to the last question would prove the strongest in a sequence of conjectures of Wagner, although Wagner himself suggests this conjecture is 'almost certainly false'. His next strongest conjecture is the following.

Conjecture 15.8.12 *Every HPP matroid is correlated.* ☐

A negative answer to Problem 15.8.11 would leave open the following question of Semple and Welsh, which has not been settled even in the rank-3 case, despite the fact that Cocks (2008) has shown that every rank-3 matroid is independence-correlated.

Problem 15.8.13 *Is there a matroid that is independence-correlated but not spanning-correlated?* ☐

Semple and Welsh's work was motivated in part by the following two graph-theoretic conjectures due to J. Kahn (2000) and G. Grimmett and S. Winkler (2004), respectively. The first of these conjectures also appears as a question of P. Winkler in Pemantle (2000). For a subset X of the edge set of a graph G, let $i_X(G)$ be the number of forests of G containing X and let $j_X(G)$ be the number of connected subgraphs of G containing X.

Conjecture 15.8.14 *For a graph G and distinct edges e and f of G,*

$$i_e(G)i_f(G) - i_{ef}(G)i(G) \geq 0.$$ ☐

Conjecture 15.8.15 *For a graph G and distinct edges e and f of G,*

$$j_e(G)j_f(G) - j_{ef}(G)j(G) \geq 0.$$ ☐

Expressed probabilistically, Conjecture 15.8.14 asserts that if e and f are distinct edges of a graph G and F is a forest in G chosen uniformly at random, then $Pr(F$ contains $e|F$ contains $f) \leq Pr(F$ contains $e)$. Grimmett and Winkler (2004) proved this conjecture for all simple 8-vertex graphs and for all 9-vertex graphs with at most 18 edges. Cocks proved that the class of graphic matroids is independence-correlated if and only if Conjecture 15.8.14 holds for all graphs G; and the class of graphic matroids is spanning-correlated if and only if Conjecture 15.8.15 holds for all graphs G. Kahn and Neiman (2008) and Wagner (2008) considered even more properties of negatively correlated random variables, in particular, in relation to the unimodal conjectures considered in Section 15.2.

15.9 A miscellany

We conclude this chapter with a collection of problems that do not seem to belong in any of the earlier sections. We begin with yet another conjecture of Rota (in Huang and Rota 1994), which is known as Rota's Basis Conjecture.

Conjecture 15.9.1 *Let $B_1, B_2, \ldots, B_n$ be disjoint bases of a rank-n matroid M. Then the family $(B_1, B_2, \ldots, B_n)$ has n disjoint transversals each of which is a basis of M.* □

Equivalently, the n^2 elements of $B_1 \cup B_2 \cup \cdots \cup B_n$ can be arranged as entries of an $n \times n$ matrix so that the entries in row i are the elements of B_i, and the entries in each column form a basis of M. By the basis exchange axiom, the conjecture holds for $n = 2$, and Chan (1995) proved it for $n = 3$. Several authors have proved partial results towards the conjecture. For example, Geelen and Humphries (2006) proved the conjecture when M is a paving matroid; for an odd prime p, Drisko (1997) and Glynn (2010) proved the conjecture for n in $\{p+1, p-1\}$ when M is $\mathbb{F}$-representable and the characteristic of $\mathbb{F}$ is zero, or is not in some finite set of primes, or is p when $n = p - 1$; Aharoni and Berger (2006) proved that there is a partition of $B_1 \cup B_2 \cup \cdots \cup B_n$ into at most $2n$ sets each of which is a partial transversal of $(B_1, B_2, \ldots, B_n)$ that is independent in M; and Geelen and Webb (2007) proved two other partial results including the existence of at least $\lfloor \sqrt{n} \rfloor$ disjoint transversals of $(B_1, B_2, \ldots, B_n)$ that are bases of M. Chow (2009) proposed a conjecture for matroids that are the disjoint union of three bases and showed that Rota's Basis Conjecture is implied by this newer conjecture. However, very recently, Nicholas J. A. Harvey, Tamás Király, and Lap Chi Lau (On disjoint common bases in two matroids, 2010) have disproved Chow's conjecture. Nevertheless, Rota's Basis Conjecture remains open.

Ingleton (1971a) noted that every simple rank-3 matroid can be embedded in some projective plane and Welsh (1971b) asked whether this plane can be chosen to be finite.

Problem 15.9.2 *Let M be a simple rank-3 matroid. Is there a finite projective plane of which M is a restriction?* □

As a partial result towards this conjecture, Wanner and Ziegler (1991) proved that every simple rank-3 matroid is a restriction of a matroid in which every rank-one flat is the complement (Section 6.9) of some modular hyperplane.

Poljak and Turzík (1982) called a matroid N *sticky* if an amalgam of M_1 and M_2 exists whenever M_1 and M_2 are extensions of N whose ground sets meet in $E(N)$. It follows by Theorem 11.4.10 that every modular matroid is sticky. Poljak and Turzík conjectured that the converse of this also holds.

Conjecture 15.9.3 *Every sticky matroid is modular.* □

Bachem and Kern (1988) asserted that this conjecture is true if and only if it holds for rank-4 matroids, and Bonin (2009a) corrected an error in their proof. These three authors proved other partial results which mean that, to complete the proof of the conjecture, one needs only consider rank-4 matroids in which the following hold: each pair of planes meet in a line; and there is a set of four lines, no three coplanar, such that five of the six pairs of lines are coplanar, but the sixth pair is not. Bonin points out that the sticky matroid conjecture would follow if one could prove the rank-4 case of the following weakening of a conjecture of Kantor (1974).

Conjecture 15.9.4 *For sufficiently large r, if a rank-r matroid M has the property that every two distinct hyperplanes intersect in a rank-$(r-2)$ flat, then M is a restriction of a rank-r modular matroid.* □

The next conjecture (Exercise 10(e) of Section 9.1) was settled for regular matroids by Kingan and Lemos (2006) having earlier been proved for graphic matroids by Kingan (1999)

Conjecture 15.9.5 *If $k \geq 4$ and the matroid M has a k-element set that is the intersection of a circuit and a cocircuit, then M has a $(k-2)$-element set that is a circuit–cocircuit intersection.* □

Seymour (1986a) solved the following problem for binary matroids, thereby extending graph results of Chakravarti and Robertson (1980) and Watkins and Mesner (1967).

Problem 15.9.6 *Let e, f, and g be distinct elements of an internally 4-connected matroid M. Characterize when M has no circuit containing $\{e, f, g\}$.* □

One way in which elements e, f, and g of an internally 4-connected $GF(q)$-representable matroid M can fail to be in a circuit of M is if M is a restriction of a Dowling geometry associated with $GF(q)^\times$ whose joints include e, f, and g. When $q = 2$, Seymour's result shows that, provided $\{e, f, g\}$ is not a cocircuit, this is the only obstruction to $\{e, f, g\}$ being in a circuit. But Geelen, Gerards, and Whittle (private communication) have observed that, for arbitrary q, other structures arise. Thus, except for the class of binary matroids, Problem 15.9.6 remains open even for $GF(q)$-representable matroids.

A 3-element subset of the ground set of a matroid M is in a $U_{2,4}$-minor of M if and only if it is the intersection of a circuit and a cocircuit. Seymour (1981b) conjectured that, for every 3-element set X in a 4-connected non-binary matroid M, there is a $U_{2,4}$-minor of M containing X. Counterexamples were given by Kahn (1985) and Coullard (1986). Kahn's example is obtained from a binary matroid by relaxing a circuit–hyperplane; Coullard's example comes from a binary matroid by relaxing two complementary circuit–hyperplanes.

Problem 15.9.7 *Characterize all 4-connected non-binary matroids that have a 3-element set which is not in any $U_{2,4}$-minor.* □

Many matroid problems seek to characterize, for some property P, all those matroids M that do not have P such that, for all elements e, both $M \backslash e$ and M/e have P. A variant of such problems seeks to determine all such M in which, for all elements e, at least one of $M\backslash e$ and M/e has P. For example, Oxley (1990b) found all non-binary matroids M such that, for every element, its deletion or contraction from M is binary. Gubser (1990) determined all graphic matroids M in which, for all elements e, at least one of $M\backslash e$ and M/e is the cycle matroid of a planar graph. Moreover, Kingan and Lemos (2002) solved a problem from the first edition of this book by finding all binary matroids M such that, for all elements e, at least one of $M\backslash e$ and M/e is graphic. Kingan and Lemos also reduced to the following the problem of finding all non-regular matroids M such that $M\backslash e$ or M/e is regular for all elements e.

Problem 15.9.8 *Find all non-regular matroids M such that, for all elements e, exactly one of $M\backslash e$ and M/e is regular.* □

Seymour (private communication) posed the following variant of the last problem, which also suggests analogues for several other classes of matroids.

Problem 15.9.9 *Find all 3-connected non-regular matroids M such that M has an element e for which both $M\backslash e$ and M/e are regular.* □

The only 3-connected non-binary matroid having an element so that both the deletion and contraction of this element are binary is $U_{2,4}$ (Oxley 1987a).

Next we consider a problem relating to basis exchanges. The following was proposed by Van den Heuvel and Thomassé (2009).

Conjecture 15.9.10 *For bases $\{b_1, b_2, \ldots, b_r\}$ and $\{b_1', b_2', \ldots, b_r'\}$ of a matroid M, there is a ordering of the sequence $(b_1, b_2, \ldots, b_r, b_1', b_2', \ldots, b_r')$ such that every set of r cyclically consecutive terms in the sequence forms a basis of M.* □

This conjecture is a special case of one first proposed by Gabow (1976).

Conjecture 15.9.11 *Every two bases $\{b_1, b_2, \ldots, b_r\}$ and $\{b_1', b_2', \ldots, b_r'\}$ of a matroid M have orderings $(b_{\sigma(1)}, b_{\sigma(2)}, \ldots, b_{\sigma(r)})$ and $(b_{\sigma'(1)}', b_{\sigma'(2)}', \ldots, b_{\sigma'(r)}')$ such that, in the sequence $(b_{\sigma(1)}, b_{\sigma(2)}, \ldots, b_{\sigma(r)}, b_{\sigma'(1)}', b_{\sigma'(2)}', \ldots, b_{\sigma'(r)}')$, every set of r cyclically consecutive terms forms a basis of M.* □

Both of the last two conjectures hold if M is graphic (Cordovil and Moreira 1993; Kajitani, Ueno, and Miyano 1988; Wiedemann 2006). However, both are open in general even for representable matroids. Kajitani, Ueno, and Moreira (1988) proposed the following conjecture, which Van den Heuvel and Thomassé (2009) proved for matroids M in which $r(M)$ and $|E(M)|$ are relatively prime.

Conjecture 15.9.12 *In a matroid M, there is an ordering of $E(M)$ in which every set of $r(M)$ cyclically consecutive elements forms a basis if and only if*

$$|X|\,r(M) \le |E(M)|\,r(X) \text{ for all subsets } X \text{ of } E(M).$$ □

The next problem is related to combinatorial number theory. In an additive abelian group G, for subsets A and B of G, let $A+B = \{a+b : a \in A \text{ and } b \in B\}$. For every subset S of G, let $stab(S) = \{g \in G : \{g\} + S = S\}$. It is easily checked that *stab*(*S*) is a subgroup of *G*. A classical result of Cauchy (1813) and Davenport (1935) asserts that if *G* is a cyclic group of prime order *p*, and *A* and *B* are non-empty subsets of *G*, then $|A + B| \geq \min\{p, |A| + |B| - 1\}$. This result was generalized to arbitrary (possibly infinite) abelian groups by Kneser (1953) who showed that if *A* and *B* are finite and $H = stab(A+B)$, then $|A + B| \geq |A + H| + |B + H| - |H|$. Schrijver and Seymour (1990) proposed the following generalization of Kneser's theorem.

Conjecture 15.9.13 *Let M be a matroid with ground set E and G be an abelian group. For a function* $w : E \to G$, *let* $w(M) = \{\sum_{b\in B} w(b) : B$ *is a base of* $M\}$ *and* $H = stab(w(M))$. *Then*

$$|w(M)| \geq |H| \left(1 - r(M) + \sum_{Q\in G/H} r(w^{-1}(Q))\right). \qquad \square$$

Note that, in the last statement, G/H is the quotient group of *G* by *H*. Schrijver and Seymour proved their conjecture when *G* is cyclic of prime order. In that case, if *M* is taken to be a matroid that is obtained from $U_{2,2}$ by a sequence of parallel extensions, the result reduces to the Cauchy–Davenport Theorem. DeVos, Goddyn, and Mohar (2009) proved the conjecture for all abelian groups when *M* can be obtained from a uniform matroid by a sequence of parallel extensions. They also verified the conjecture for all matroids *M* when $|G|$ is a product of two distinct primes and when *G* is the additive group of $GF(p^n)$ where *p* is prime and *n* is a positive integer.

The following conjecture of Aharoni and Berger (2006) is a generalization of Edmonds's (1970) result (Corollary 11.3.16) determining the size of a largest common independent set of two matroids on the same ground set. By that result, the conjecture holds for $k = 2$, and Aharoni and Berger proved it for $k = 3$.

Conjecture 15.9.14 *Let* $M_1, M_2, \ldots, M_k$ *be matroids on a common ground set E having rank functions* $r_1, r_2, \ldots, r_k$. *The cardinality of a largest set that is independent in all* M_i *is at least*

$$\frac{1}{k-1} \min\{\sum_{i=1}^{k} r_i(X_i) : X_1, X_2, \ldots, X_k \text{ are disjoint sets whose union is } E\}. \square$$

Finally, we refer the reader seeking still more problems to Geelen's paper (2008) on problems relating to excluding a uniform matroid and to Kung's (1993, 1996) surveys of extremal matroid theory and of problems concerned with the critical exponent. Updates of progress on several of Kung's problems appear earlier in this chapter and in Chapter 14.

REFERENCES

Acketa, D. M. (1984). A construction of non-simple matroids on at most 8 elements. *J. Combin. Inform. System Sci.* **9**, 121–132. [15.5]

Acketa, D. M. (1988). On binary paving matroids. *Discrete Math.* **70**, 109–110. [12.2, Appendix]

Ádám, A. (1957). Über zweipolige elektrische Netze. I. *Magyar Tud. Akad. Mat. Kutató Int. Közl.* **2**, 211–218. [5.4]

Adams, W. W. and Loustaunau, P. (1994). *An introduction to Gröbner bases.* Amer. Math. Soc., Providence. [6.8]

Aharoni, R. and Berger, E. (2006). The intersection of a matroid and a simplicial complex. *Trans. Amer. Math. Soc.* **358**, 4895–4917. [15.9]

Aigner, M. (1979). *Combinatorial theory.* Springer-Verlag, Berlin. Reprinted 1997, Springer-Verlag, Berlin. [6.6]

Aigner, M. (1987). Whitney numbers. In *Combinatorial geometries* (ed. N. White), pp. 139–160. Cambridge University Press, Cambridge. [15.2]

Akkari, S. (1988). *On matroid connectivity.* Ph.D. thesis, Louisiana State University. [8.1]

Akkari, S. and Oxley, J. (1993). Some local extremal connectivity results for matroids. *Combin. Probab. Comput.* **2**, 367–384. [11.5]

Appel, K. and Haken, W. (1976). Every planar map is four-colorable. *Bull. Amer. Math. Soc.* **82**, 711–712. [15.3]

Archdeacon, D., Colbourn, C. J., Gitler, I., and Provan, J. S. (2000). Four-terminal reducibility and projective-planar wye-delta-wye-reducible graphs. *J Graph Theory* **33**, 83–93. [11.5]

Ardila, F. (2003). The Catalan matroid. *J. Combin. Theory Ser. A* **104**, 49–62. [11.2]

Artin, E. (1957). *Geometric algebra.* Interscience, New York. [6.3]

Asano, T., Nishizeki, T., and Seymour, P. D. (1984). A note on nongraphic matroids. *J. Combin. Theory Ser. B* **37**, 290–293. [10.3]

Asche, D. S. (1966). Minimal dependent sets. *J. Austral. Math. Soc.* **6**, 259–262. [1.4]

Atiyah, M. F. and MacDonald, I. G. (1969). *Introduction to commutative algebra.* Addison-Wesley, Reading. [6.8]

Bachem, A. and Kern, W. (1988). On sticky matroids. *Discrete Math.* **69**, 11–18. [11.4, 15.9]

Baclawski, K. and White, N. (1979). Higher order independence in matroids. *J. London Math. Soc. (2)* **19**, 193–202. [14.10]

Baer, R. (1952). *Linear algebra and projective geometry.* Academic Press, New York. [6.3]

Baines, R. and Vámos, P. (2003). An algorithm to compute the set of characteristics of a system of polynomial equations over the integers. *J. Symbolic Comput.* **35**, 269–279. [6.8]

Ball, S. (2010). On large subsets of a finite vector space in which every subset of basis size is a basis (submitted). [6.5, 15.1]

Basterfield, J. G. and Kelly, L. M. (1968). A characterization of sets of *n* points which determine *n* hyperplanes. *Proc. Cambridge Philos. Soc.* **64**, 585–588. [6.9]

Ben-David, S. and Geelen, J. (2010). Certifying non-representability of matroids over the reals (in preparation). [14.4]

Birkhoff, G. (1935). Abstract linear dependence in lattices. *Amer. J. Math.* **57**, 800–804. [1.0, 1.7]

Birkhoff, G. (1967). *Lattice theory.* Third edition. Amer. Math. Soc., Providence. [6.1, 6.9]

Bixby, R. E. (1972). *Composition and decomposition of matroids and related topics.* Ph.D. thesis, Cornell University. [8.3]

Bixby, R. E. (1974). ℓ-matrices and a characterization of non-binary matroids. *Discrete Math.* **8**, 139–145. [12.3]

Bixby, R. E. (1976). A strengthened form of Tutte's characterization of regular matroids. *J. Combin. Theory Ser. B* **20**, 216–221. [9.4, 10.4]

Bixby, R. E. (1977). Kuratowski's and Wagner's theorems for matroids. *J. Combin. Theory Ser. B* **22**, 31–53. [5.4, 6.6, 10.4, Appendix]

Bixby, R. E. (1979). On Reid's characterization of the ternary matroids. *J. Combin. Theory Ser. B* **26**, 174–204. [6.5, 10.2]

Bixby, R. E. (1981). Matroids and operations research. In *Advanced techniques in practice of operations research* (eds. H. J. Greenberg, F. H. Murphy, and S. H. Shaw), pp. 333–458. North-Holland, New York. [1.2, 1.8, 13.4]

Bixby, R. E. (1982). A simple theorem on 3-connectivity. *Linear Algebra Appl.* **45**, 123–126. [8.7]

Bixby, R. E. and Coullard, C. R. (1986). On chains of 3-connected matroids. *Discrete Appl. Math.* **15**, 155–166. [12.3]

Bixby, R. E. and Coullard, C. R. (1987). Finding a smallest 3-connected minor maintaining a fixed minor and a fixed element. *Combinatorica* **7**, 231–242. [12.3]

Bixby, R. E. and Cunningham, W. H. (1979). Matroids, graphs, and 3-connectivity. In *Graph theory and related topics* (eds. J. A. Bondy and U. S. R. Murty), pp. 91–103, Academic Press, New York. [9.4]

Bixby, R. E. and Cunningham, W. H. (1980). Converting linear programs to network problems. *Math. Oper. Res.* **5**, 321–357. [13.4]

Bixby, R. E. and Cunningham, W. H. (1987). Short cocircuits in binary matroids. *European J. Combin.* **8**, 213–225. [14.10]

Bixby, R. E. and Rajan, A. (1989). A short proof of the Truemper–Tseng theorem on max-flow min-cut matroids. *Linear Algebra Appl.* **114/115**, 277–292. [12.3]

Björner, A., Las Vergnas, M., Sturmfels, B., White, N., and Ziegler, G. (1993). *Oriented matroids*. Cambridge University Press, Cambridge. [10.4, 15.0]

Björner, A. and Ziegler, G. M. (1992). Introduction to greedoids. In *Matroid applications* (ed. N. White), pp. 284–357. Cambridge University Press, Cambridge. [15.0]

Blackburn, J. E., Crapo, H. H., and Higgs, D. A. (1973). A catalogue of combinatorial geometries. *Math. Comp.* **27**, 155–166, with loose microfiche supplement A12–G12. [6.4, 7.2, 15.5]

Bland, R. G. and Las Vergnas, M. (1978). Orientability of matroids. *J. Combin. Theory Ser. B* **24**, 94–123. [10.4]

Bollobás, B. (1978). *Extremal graph theory*. Academic Press, London. Reprinted 2004, Dover, Mineola. [14.10]

Bondy, J. A. (1972). Transversal matroids, base-orderable matroids, and graphs. *Quart. J. Math. Oxford Ser. (2)* **23**, 81–89. [3.2, 10.4]

Bondy, J. A. and Murty, U. S. R. (2008). *Graph theory*. Springer, New York. [Preliminaries, 1.8, 11.3]

Bondy, J. A. and Welsh, D. J. A. (1971). Some results on transversal matroids and constructions for identically self-dual matroids. *Quart. J. Math. Oxford Ser. (2)* **22**, 435–451. [2.1, 2.4, 9.2]

Bonin, J. E. (1996). Matroids with no $(q + 2)$-point-line minors. *Advances in Appl. Math.* **17**, 460–476. [14.10]

Bonin, J. E. (2009a). A note on the sticky matroid conjecture. *Ann. Comb.*, to appear. http://arxiv.org/abs/0906.2553 [15.9]

Bonin, J. E. (2009b). Lattice path matroids: the excluded minors. *J. Combin. Theory Ser. B*, to appear. http://arxiv.org/abs/0909.4342 [11.2]

Bonin, J. E. (2010). A construction of infinite sets of intertwines for pairs of matroids (submitted). http://arxiv.org/abs/1003.1120 [14.5]

Bonin, J. E. and de Mier, A. (2006). Lattice path matroids: structural properties. *European J. Combin.* **27**, 701–738. [11.2]

Bonin, J. E. and de Mier, A. (2008). The lattice of cyclic flats of a matroid. *Ann. Combin.* **12**, 155–170. [3.3, 11.2]

Bonin, J. E., de Mier, A., and Noy, M. (2003). Lattice path matroids: enumerative aspects and Tutte polynomials. *J. Combin. Theory Ser. A* **104**, 63–94. [11.2]

Bonin, J. E. and Giménez, O. (2007). Multi-path matroids. *Combin. Probab. Comput.* **16**, 193–217. [11.2]

Bonin, J. E., McNulty, J., and Reid, T. J. (1999). The matroid Ramsey number $n(6, 6)$. *Combin. Probab. Comput.* **8**, 229–235. [4.3]

Bonin, J. E., Oxley, J. G., and Servatius, B. (1996). *Matroid theory. Proc. AMS-IMS-SIAM Joint Summer Research Conference on Matroid Theory, July 2–6, 1995, University of Washington, Seattle.* Contemp. Math. **197**, Amer. Math. Soc., Providence. [15.0]

Borůvka, O. (1926). O jistém problému minimálním. *Práce Mor. Přírodověd Spol. v Brně (Acta Societ. Scient. Natur. Moravicae)* **3**, 37–58. [1.8]

Bose, R. C. (1947). Mathematical theory of the symmetrical factorial design. *Sankyhā* **8**, 107–166. [6.5]

Bouchet, A., Cunningham, W. H., and Geelen, J. F. (1998). Principally unimodular skew-symmetric matrices. *Combinatorica* **18**, 461–486. [13.3]

Brändén, P. (2007). Polynomials with the half-plane property and matroid theory. *Adv. Math.* **216**, 302–320. [15.8]

Brown, T. (1971). Deriving closure relations with exchange property. Notes (and an editorial appendix) by H. Crapo and G. Roulet. In *Möbius algebras*, pp. 51–55. University of Waterloo, Waterloo. [7.3]

Brualdi, R. A. (1969). Comments on bases in dependence structures, *Bull. Austral. Math. Soc.* **1**, 161–167. [2.1, 11.2, 11.3]

Brualdi, R. A. (1970). Admissible mappings between dependence spaces. *Proc. London Math. Soc. (3)* **21**, 296–312. [11.3]

Brualdi, R. A. and Scrimger, E. B. (1968). Exchange systems, matchings, and transversals. *J. Combin. Theory* **5**, 244–257. [11.3]

Bruck, R. H. and Ryser, H. J. (1949). The nonexistence of certain finite projective planes. *Canad. J. Math.* **1**, 88–93. [6.1]

Brylawski, T. H. (1971). A combinatorial model for series–parallel networks. *Trans. Amer. Math. Soc.* **154**, 1–22. [7.1, 10.4]

Brylawski, T. H. (1972). A decomposition for combinatorial geometries. *Trans. Amer. Math. Soc.* **171**, 235–282. [4.3, 9.4]

Brylawski, T. H. (1973). Some properties of basic families of subsets. *Discrete Math.* **6**, 333–341. [1.4]

Brylawski, T. H. (1974). Reconstructing combinatorial geometries. In *Graphs and combinatorics* (eds. R. A. Bari and F. Harary). Lecture Notes in Math. Vol. 406, pp. 226–235. Springer-Verlag, Berlin. [6.1, 15.4]

Brylawski, T. H. (1975a). On the nonreconstructibility of combinatorial geometries. *J. Combin. Theory Ser. B* **19**, 72–76. [6.1, 15.4]

Brylawski, T. H. (1975b). Modular constructions for combinatorial geometries. *Trans. Amer. Math. Soc.* **203**, 1–44. [6.9, 10.4, 11.4]

Brylawski, T. H. (1975c). A note on Tutte's unimodular representation theorem. *Proc. Amer. Math. Soc.* **52**, 499–502. [6.6]

Brylawski, T. H. (1975d). An affine representation for transversal geometries. *Studies in Appl. Math.* **54**, 143–160. [6.5, 11.2, 15.3]

Brylawski, T. H. (1981). Hyperplane reconstruction of the Tutte polynomial of a geometric lattice. *Discrete Math.* **35**, 25–38. [15.4]

Brylawski, T. H. (1982a). The Tutte polynomial. Part I: General theory. In *Matroid theory and its applications* (ed. A. Barlotti), pp. 125–275. Liguori editore, Naples. [9.4]

Brylawski, T. H. (1982b). Finite prime-field characteristic sets for planar configurations. *Linear Algebra Appl.* **46**, 155–176. [6.8]

Brylawski, T. H. (1986a). Constructions. In *Theory of matroids* (ed. N. White), pp. 127–223. Cambridge University Press, Cambridge. [1.3, 7.0, 7.3, 11.4, 14.1, 14.5, 15.4]

Brylawski, T. H. (1986b). Appendix of matroid cryptomorphisms. In *Theory of matroids* (ed. N. White), pp. 298–312. Cambridge University Press, Cambridge. [2.1]

Brylawski, T. H. and Kelly, D. (1980). *Matroids and combinatorial geometries.* Department of Mathematics, University of North Carolina, Chapel Hill. [1.5, 6.2, 6.4, 6.8]

Brylawski, T. H. and Lucas, D. (1976). Uniquely representable combinatorial geometries. In *Teorie combinatorie* (Proc. 1973 Internat. Colloq.), pp. 83–104. Accademia Nazionale dei Lincei, Rome. [6.3, 6.4, 6.6, 14.6]

Brylawski, T. H. and Oxley, J. (1992). The Tutte polynomial and its applications. In *Matroid applications* (ed. N. White), pp. 123–225. Cambridge University Press, Cambridge. [15.3]

Bush, K. A. (1952). Orthogonal arrays of index unity. *Ann. Math. Statist.* **23**, 426–434. [6.5]

Cameron, P. J. (1980). Extremal results and configuration theorems for Steiner systems. In *Topics on Steiner systems.* Ann. Discrete Math. **7**, pp. 43–63. North-Holland, Amsterdam. [12.2]

Camion, P. (1963). Caractérisation des matrices unimodulaires. *Cahiers Centre Études Rech. Opér.* **5**, 181–190. [10.1]

Casse, L. R. A. (1969). A solution to Beniamino Segre's "Problem $I_{r,q}$" for q even. *Atti Accad. Naz. Lincei Rend. Cl. Sci. Fis. Mat. Natur.* (8) **46**, 13–20. [6.5]

Cauchy, A. L. (1813). Recherches sur les nombres. *J. École Polytech.* **9**, 99–116. [15.9]

Chakravarti, K. and Robertson, N. (1980). Covering three edges with a bond in a non-separable graph. In *Combinatorics 79. Part I* (eds. M. Deza and I. G. Rosenberg) Ann. Discrete Math. **8**, p. 247. North-Holland, Amsterdam [15.9].

Chan, W. (1995). An exchange property of matroid. *Discrete Math.* **146**, 299–302. [15.9]

Chartrand, G. and Harary, F. (1967). Planar permutation graphs. *Ann. Inst. H. Poincaré Sect. B* (N. S.) **3**, 433–438. [12.2]

Cheung, A. L. C. (1974). *Compatibility of extensions of a combinatorial geometry.* Ph.D. thesis, University of Waterloo. [7.2]

Choe, Y.-B., Oxley, J. G., Sokal, A. D., and Wagner, D. G. (2004). Homogeneous multivariate polynomials with the half-plane property. *Adv. in Appl. Math.* **32**, 88–187. [14.7, 15.8]

Choe, Y.-B. and Wagner, D. G. (2006). Rayleigh matroids. *Combin. Probab. Comput.* **15**, 765–781. [15.8]

Chow, T. Y. (2009). Reduction of Rota's basis conjecture to a problem on three bases. *SIAM J. Discrete Math.* **23**, 369–371. [15.9]

Chun, C., Ding, G., Oporowski, B., and Vertigan, D. (2009). Unavoidable parallel minors of 4-connected graphs. *J. Graph Theory* **60**, 313–326. [14.9]

Chun, C. and Oxley, J. (2009). Unavoidable parallel minors and series minors of regular matroids. *European J. Combin.*, to appear. [14.9]

Cocks, C. C. (2008). Correlated matroids. *Combin. Probab. Comput.* **17**, 511–518. [15.8]

Cook, W. J., Cunningham, W. H., Pulleyblank, W. R., and Schrijver, A. (1998). *Combinatorial optimization.* Wiley, New York. [1.8, 5.1, 11.3]

Cordovil, R. and Moreira, M. L. (1993). Bases-cobases graphs and polytopes of matroids. *Combinatorica* **13**, 157–165. [15.9]

Coullard, C. R. (1985). *Minors of 3-connected matroids and adjoints of binary matroids.* Ph.D. thesis, Northwestern University. [8.7, 12.3]

Coullard, C. R. (1986). Counterexamples to conjectures on 4-connected matroids. *Combinatorica* **6**, 315–320. [12.3, 15.9]

Coullard, C. R. and Oxley, J. G. (1992). Extension of Tutte's wheels-and-whirls theorem. *J. Combin. Theory Ser. B* **56**, 130–140. [12.3]

Coxeter, H. S. M. (1958). Twelve points in PG(5,3) with 95040 self-transformations. *Proc. Roy. Soc. London Ser. A* **247**, 279–293. [12.2]

Crandall, R. and Pomerance, C. (2005). *Prime numbers: A computational perspective.* Second edition. Springer, New York. [14.7]

Crapo, H. H. (1965). Single-element extensions of matroids. *J. Res. Nat. Bur. Standards Sect. B* **69B**, 55–65. [7.2, 7.3, 11.2]

Crapo, H. H. (1967). Structure theory for geometric lattices. *Rend. Sem. Mat. Univ. Padova* **38**, 14–22. [7.3]

Crapo, H. H. and Rota, G.-C. (1970). *On the foundations of combinatorial theory: combinatorial geometries* (preliminary edition). MIT Press, Cambridge, Mass. [1.7, 3.3, 11.0, 15.3]

Crapo, H. and Schmitt, W. (2005a). The free product of matroids. *European J. Combin.* **26**, 1060–1065. [7.4, 15.5]

Crapo, H. and Schmitt, W. (2005b). A unique factorization theorem for matroids. *J. Combin. Theory Ser. A* **112**, 222–249. [7.4]

Cunningham, W. H. (1973). *A combinatorial decomposition theory.* Ph.D. thesis, University of Waterloo. [4.3, 8.3, 13.4]

Cunningham, W. H. (1977). Chords and disjoint paths in matroids. *Discrete Math.* **19**, 7–15. [11.3, 14.6]

Cunningham, W. H. (1979). Binary matroid sums. *Quart. J. Math. Oxford Ser. (2)* **30**, 271–281. [11.3]

Cunningham, W. H. (1981). On matroid connectivity. *J. Combin. Theory Ser. B* **30**, 94–99. [8.6, 10.3]

Cunningham, W. H. and Edmonds, J. (1980). A combinatorial decomposition theory. *Canad. J. Math.* **32**, 734–765. [8.3]

Datta, B. T. (1976). Nonexistence of six-dimensional tangential 2-blocks. *J. Combin. Theory Ser. B* **21**, 1–22. [15.3]

Datta, B. T. (1981). Nonexistence of seven-dimensional tangential 2-blocks. *Discrete Math.* **36**, 1–32. [15.3]

Davenport, H. (1935). On the addition of residue classes. *J. London Math. Soc.* **10**, 30–32. [15.9]

Dawson, J. E. (1980). Optimal matroid bases: an algorithm based on cocircuits. *Quart. J. Math. Oxford Ser. (2)* **31**, 65–69. [2.1]

Dawson, J. E. (1981). A simple approach to some basic results in matroid theory. *J. Math. Anal. Appl.* **84**, 555–559. [2.1]

Dawson, J. E. (1983). Balanced sets in an independence structure induced by a submodular function. *J. Math. Anal. Appl.* **95**, 214–222. [11.1]

Dembowski, P. (1968). *Finite geometries.* Springer-Verlag, New York. [6.1]

de Sousa, J. and Welsh, D. J. A. (1972). A characterisation of binary transversal matroids. *J. Math. Anal. Appl.* **40**, 55–59. [10.4, 15.6]

DeVos, M., Goddyn, L., and Mohar, B. (2009) A generalization of Kneser's addition theorem. *Adv. Math.* **220**, 1531–1548. [15.9]

Dharmatilake, J. S. (1994). *Binary matroids of branch-width* 3. Ph.D. thesis, The Ohio State University. [14.2]

Dharmatilake, J. S. (1996). A min-max theorem using matroid separations. In *Matroid theory* (eds. J. Bonin, J. Oxley, and B. Servatius). Contemp. Math. **197**, pp. 333–342, Amer. Math. Soc., Providence. [14.2]

Diestel, R. (1997). *Graph theory.* Springer-Verlag, New York. [2.3, 14.3]

Dilworth, R. P. (1944). Dependence relations in a semimodular lattice. *Duke Math. J.* **11**, 575–587. [1.7]

Ding, G., Oporowski, B., and Oxley, J. (1995). On infinite antichains of matroids. *J. Combin. Theory Ser. B* **63**, 21–40. [4.3]

Ding, G., Oporowski, B., Oxley, J., and Vertigan, D. (1996). Unavoidable minors of large 3-connected binary matroids. *J. Combin. Theory Ser. B* **66**, 334–360. [14.9]

Ding, G., Oporowski, B., Oxley, J., and Vertigan, D. (1997). Unavoidable minors of large 3-connected matroids. *J. Combin. Theory Ser. B* **71**, 244–293. [14.9]

Dirac, G. A. (1952). A property of 4-chromatic graphs and some remarks on critical graphs. *J. London Math. Soc.* **27**, 85–92. [5.4]

Dirac, G. A. (1960). In abstrakten Graphen vorhande vollständige 4-Graphen und ihre Unterteilungen. *Math. Nach.* **22**, 61–85. [5.4, 14.10]

Dirac, G. A. (1967). Minimally 2-connected graphs. *J. Reine Angew. Math.* **228**, 204–216. [4.3]

Doob, M. (1973). An interrelation between line graphs, eigenvalues, and matroids. *J. Combin. Theory Ser. B* **15**, 40–50. [6.10]

Dowling, T. A. (1973a). A class of geometric lattices based on finite groups. *J. Combin. Theory Ser. B* **14**, 61–86; erratum **15**, 211. [6.0, 6.10]

Dowling, T. A. (1973b). A *q*-analog of the partition lattice. In *A survey of combinatorial theory* (eds. J. N. Srivastava *et al.*), pp. 101–115. North-Holland, Amsterdam, 1973. [6.0, 6.10]

Dowling, T. A. (1980). On the independent set numbers of a finite matroid. In *Combinatorics 79. Part I* (eds. M. Deza and I. G. Rosenberg). Ann. Discrete Math. **8**, pp. 21–28. North-Holland, Amsterdam. [15.2]

Dowling, T. A. and Wilson, R. M. (1974). The slimmest geometric lattices. *Trans. Amer. Math. Soc.* **196**, 203–215. [15.2]

Dowling, T. A. and Wilson, R. M. (1975). Whitney number inequalities for geometric lattices. *Proc. Amer. Math. Soc.* **47**, 504–512. [15.2]

Drisko, A. A. (1997). On the number of even and odd Latin squares of order $p + 1$. *Adv. Math.* **128**, 20–35. [15.9]

Duffin, R. J. (1965). Topology of series–parallel networks. *J. Math. Anal. Appl.* **10**, 303–318. [5.4]

Duke, R. (1981). *Freedom in matroids*. Ph.D. thesis, The Open University. [11.1]

Duke, R. (1987). On binary reducibility. *European J. Combin.* **9**, 109–111. [11.3]

Duke, R. (1988). Freedom in matroids. *Ars Combin.* **26**, 191–216. [14.8]

Dukes, W. M. B. (2004/05). On the number of matroids on a finite set. *Sém. Lothar. Combin.* **51**, Art.B51g, 12 pp. (electronic). [15.2]

Edmonds, J. (1965a). Minimum partition of a matroid into independent sets. *J. Res. Nat. Bur. Standards Sect. B* **69B**, 67–72. [11.3]

Edmonds, J. (1965b). Lehman's switching game and a theorem of Tutte and Nash-Williams. *J. Res. Nat. Bur. Standards Sect. B* **69B**, 73–77. [11.3]

Edmonds, J. (1970). Submodular functions, matroids and certain polyhedra. In *Combinatorial structures and their applications* (Proc. Calgary Internat. Conf. 1969), pp. 69–87. Gordon and Breach, New York. [11.1, 11.3, 15.9]

Edmonds, J. (1979). Matroid intersection. In *Discrete optimization I* (eds. P. L. Hammer, E. L. Johnson, and B. H. Korte). Ann. Discrete Math. **4**, pp. 39–49. North-Holland, Amsterdam. [11.3]

Edmonds, J. and Fulkerson, D. R. (1965). Transversals and matroid partition. *J. Res. Nat. Bur. Standards Sect. B* **69B**, 147–153. [1.6]

Edmonds, J. and Fulkerson, D. R. (1970). Bottleneck extrema. *J. Combin. Theory* **8**, 299–306. [2.1]

Edmonds, J. and Rota, G.-C. (1966). Submodular set functions (Abstract). *Waterloo combinatorics conference*. [1.3, 11.0, 11.1]

Epifanov, G. V. (1966). Reduction of a plane graph to an edge by star-triangle transformations. *Soviet Math. Dokl.* **7**, 13–17. [11.5]

Erdős, P. and Pósa, L. (1962). On the maximal number of disjoint circuits of a graph. *Publ. Math. Debrecen* **9**, 3–12. [14.10]

Faigle, U. (1987). Matroids in combinatorial optimization. In *Combinatorial geometries* (ed. N. White), pp. 161–210. Cambridge University Press, Cambridge. [11.3, 13.4]

Feder, T. and Mihail, M. (1992). Balanced matroids. In *Proc. 24th ACM Symposium on Theory of Computing*, pp. 26–38. [15.8]

Folkman, J. and Lawrence, J. (1978) Oriented matroids. *J. Combin. Theory Ser. B* **25**, 199–236. [10.4]

Ford, L. R. and Fulkerson, D. R. (1958). Network flow and systems of representatives. *Canad. J. Math.* **10**, 78–84. [11.3]

Fournier, J.-C. (1971). Représentation sur un corps des matroïdes d'ordre ≤ 8. In *Théorie des matroïdes* (ed. C. P. Bruter). Lecture Notes in Math. Vol. 211, pp. 50–61. Springer-Verlag, Berlin. [6.4]

Fournier, J.-C. (1974). Une relation de séparation entre cocircuits d'un matroïde. *J. Combin. Theory Ser. B* **16**, 181–190. [9.1, 9.4]

Fournier, J.-C. (1981). A characterization of binary geometries by a double elimination axiom. *J. Combin. Theory Ser. B* **31**, 249–250. [9.1]

Fournier, J.-C. (1987). Binary matroids. In *Combinatorial geometries* (ed. N. White), pp. 28–39. Cambridge University Press, Cambridge. [6.9]

Gabow, H. (1976). Decomposing symmetric exchanges in matroid bases. *Math. Programming* **10**, 271–276. [15.9]

Gale, D. (1968). Optimal assignments in an ordered set: an application of matroid theory. *J. Combin. Theory* **4**, 176–180. [1.8]

Gallai, T. (1959). Über reguläre Kettengruppen. *Acta Math. Acad. Sci. Hungar.* **10**, 227–240. [12.3]

Garey, M. R. and Johnson, D. S. (1979). *Computers and intractability. A guide to the theory of NP-completeness.* Freeman, San Francisco. [11.3]

Geelen, J. F. (1999). On matroids without a non-Fano minor. *Discrete Math.* **203**, 279–285. [10.1, 14.5]

Geelen, J. (2008). Some open problems on excluding a uniform matroid. *Adv. in Appl. Math.* **41**, 628–637. [14.1, 14.3, 14.4, 14.5, 15.7, 15.9]

Geelen, J. (2009). Small cocircuits in matroids. *European J. Combin.*, to appear. [14.10]

Geelen, J. F., Gerards, A. M. H., and Kapoor, A. (2000). The excluded minors for $GF(4)$-representable matroids. *J. Combin. Theory Ser. B* **79**, 247–299. [6.4, 13.3, 14.7, 14.8]

Geelen, J. F., Gerards, A. M. H., Robertson, N., and Whittle, G. P. (2003). On the excluded minors for the matroids of branch-width k. *J. Combin. Theory Ser. B* **88**, 261–265. [14.2]

Geelen, J. F., Gerards, A. M. H., and Whittle, G. (2002). Branch-width and well-quasi-ordering in matroids and graphs. *J. Combin. Theory Ser. B* **84**, 270–290. [6.5, 8.2, 8.5, 14.2, 14.3]

Geelen, J. F., Gerards, A. M. H., and Whittle, G. (2003). Disjoint cocircuits in matroids with large rank. *J. Combin. Theory Ser. B* **87**, 270–279. [8.2, 8.6, 14.10]

Geelen, J., Gerards, B., and Whittle, G. (2006a). On Rota's conjecture and excluded minors containing large projective geometries. *J. Combin. Theory Ser. B* **96**, 405–425. [8.2, 8.8, 14.3]

Geelen, J., Gerards, B., and Whittle, G. (2006b). Towards a structure theory for matrices and matroids. In *International Congress of Mathematicians* Vol. III, pp. 827–842. Eur. Math. Soc., Zürich. [14.0, 14.1, 14.3, 14.4]

Geelen, J., Gerards, B., and Whittle, G. (2007a). Excluding a planar graph from GF(q)-representable matroids. *J. Combin. Theory Ser. B* **97**, 971–998. [8.2, 8.5, 14.3]

Geelen, J., Gerards, B., and Whittle, G. (2007b). Towards a matroid-minor structure theory. In *Combinatorics, complexity, and chance*, pp. 72–82, Oxford Lecture Ser. Math. Appl., 34, Oxford Univ. Press, Oxford. [14.0, 14.1, 14.3, 14.4, 14.8]

Geelen, J., Gerards, B., and Whittle, G. (2009a). Tangles, tree-decompositions and grids in matroids. *J. Combin. Theory Ser. B* **99**, 657–667. [14.3]

Geelen, J., Gerards, B., and Whittle, G. (2009b). On inequivalent representations of matroids over non-prime fields (submitted). [14.6]

Geelen, J., Hliněný, P., and Whittle, G. (2005). Bridging separations in matroids. *SIAM J. Discrete Math.* **18**, 638–646. [13.3]

Geelen, J. and Humphries, P. J. (2006). Rota's basis conjecture for paving matroids. *SIAM J. Discrete Math.* **20**, 1042–1045. [15.9]

Geelen, J. and Kabell, K. (2009a). Projective geometries in dense matroids. *J. Combin. Theory Ser. B* **99**, 1–8. [14.10]

Geelen, J. and Kabell, K. (2009b). The Erdős–Pósa property for matroid circuits. *J. Combin. Theory Ser. B* **99**, 407–419. [14.10]

Geelen, J., Kung, J. P. S., and Whittle, G. (2009). Growth rates of minor-closed classes of matroids. *J. Combin. Theory Ser. B* **99**, 420–427. [14.10]

Geelen, J., Mayhew, D., and Whittle, G. (2004). Inequivalent representations of matroids having no $U_{3,6}$-minor. *J. Combin. Theory Ser. B* **92**, 55–67. [14.6]

Geelen, J. and Nelson, P. (2009). The number of points in a matroid with no n-point line as a minor (submitted). [14.10]

Geelen, J., Oxley, J., Vertigan, D., and Whittle, G. (1998). Weak maps and stabilizers of classes of matroids. *Adv. in Appl. Math.* **21**, 305–341. [14.8]

Geelen, J., Oxley, J., Vertigan, D., and Whittle, G. (2002). Totally free expansions of matroids. *J. Combin. Theory Ser. B* **84**, 130–179. [6.5, 14.7, 14.8]

Geelen, J. and Webb, K. (2007). On Rota's basis conjecture. *SIAM J. Discrete Math.* **21**, 802–804. [15.9]

Geelen, J. and Whittle, G. (2002). Branch-width and Rota's conjecture. *J. Combin. Theory Ser. B* **86**, 315–330. [14.3]

Geelen, J. and Whittle, G. (2003). Cliques in dense GF(q)-representable matroids. *J. Combin. Theory Ser. B* **87**, 264–269. [14.10]

Geelen, J. and Whittle, G. (2010a). The projective plane is a stabilizer. *J. Combin. Theory Ser. B* **100**, 128–131. [14.8]

Geelen, J. and Whittle, G. (2010b). Inequivalent representations of matroids over prime fields (submitted). [14.4]

Gerards, A. M. H. (1989). A short proof of Tutte's characterization of totally unimodular matrices. *Linear Algebra Appl.* **114/115**, 207–212. [6.5, 10.1, 10.2]

Gerards, A. M. H. (1995). On Tutte's characterization of graphic matroids – a graphic proof. *J. Graph Theory* **20**, 351–359. [10.3]

Glynn, D. G. (2010). The conjectures of Alon–Tarsi and Rota in dimension prime minus one. *SIAM J. Discrete Math.* **24**, 394–399. [15.9]

Goddyn, L. A. and Jackson, B. (1999). Removable circuits in binary matroids. *Combin. Probab. Comput.* **6**, 539–545. [15.4]

Gordon, G. (1984). Matroids over F_p which are rational excluded minors. *Discrete Math.* **52**, 51–65. [6.7]

Gordon, G. (1988). Algebraic characteristic sets of matroids. *J. Combin. Theory Ser. B* **44**, 64–74. [15.1, Appendix]

Graver, J. E. (1966). *Lectures on the theory of matroids.* University of Alberta. [10.3]

Greene, C. (1970). A rank inequality for finite geometric lattices. *J. Combin. Theory* **9**, 357–364. [6.9]

Greene, C. (1971). *Lectures on combinatorial geometries* (with notes, footnotes and other comments by Daniel Kennedy). NSF advanced science seminar, Bowdoin College, Brunswick, Maine. [5.3]

Greene, C. (1973). A multiple exchange property for bases. *Proc. Amer. Math. Soc.* **39**, 45–50. [1.4]

Grimmett, G. R. and Winkler, S. N. (2004). Negative association in uniform forests and connected graphs. *Random Structures Algorithms* **24**, 444–460. [15.8]

Grünbaum, B. (1972). *Arrangements and spreads.* CBMS Regional Conf. Ser. No. 10, Amer. Math. Soc., Providence. [6.8, 15.1]

Gubser, B. (1990). *Some problems for graph minors.* Ph.D. thesis, Louisiana State University. [15.9]

Gulati, B. R. and Kounias, E. G. (1970). On bounds useful in the theory of symmetrical factorial designs. *J. Roy. Statist. Soc. Ser. B* **32**, 123–133. [6.5]

Hadwiger, H. (1943). Über eine Klassifikation der Streckencomplexe. *Vierteljsch. Naturforsch. Ges. Zurich* **88**, 133–142. [15.3]

Halin, R. (1969). A theorem on n-connected graphs. *J. Combin. Theory* **7**, 150–154. [15.4]

Hall, D. W. (1943). A note on primitive skew curves. *Bull. Amer. Math. Soc.* **49**, 935–937. [12.2]

Hall, M., Jr. (1986). *Combinatorial theory.* Second edition. Wiley, New York. [6.1]

Hall, P. (1935). On representatives of subsets. *J. London Math. Soc.* **10**, 26–30. [11.2]

Hall, R., Mayhew, D., and van Zwam, S. (2009). The excluded minors for near-regular matroids. *European J. Combin.*, to appear. [14.7]

Hall, R., Oxley, J., Semple, C., and Whittle, G. (2002) On matroids of branch-width three. *J. Combin. Theory Ser. B* **86**, 148–171. [14.2]

Hamidoune, Y. O. and Salaün, I. (1989). On the independence numbers of a matroid. *J. Combin. Theory Ser. B* **47**, 146–152. [15.2]

Harary, F. (1969). *Graph theory.* Addison-Wesley, Reading. [4.1, 9.4]

Harary, F. and Tutte, W. T. (1965). A dual form of Kuratowski's Theorem. *Canad. Math. Bull.* **8**, 17–20, 373. [5.2]

Harary, F. and Welsh, D. J. A. (1969). Matroids versus graphs. In *The many facets of graph theory*. Lecture Notes in Math. Vol. 110, pp. 155–170. Springer-Verlag, Berlin. [11.3]

Hartmanis, J. (1959). Lattice theory of generalized partitions. *Canad. J. Math.* **11**, 97–106. [2.1]

Hausmann, D. and Korte, B. (1978a). Oracle algorithms for fixed point problems – an axiomatic approach. In *Optimization and operations research*. Lecture Notes in Econom. and Math. Systems Vol. 157, pp. 137–156. Springer-Verlag, Berlin. [9.4]

Hausmann, D. and Korte, B. (1978b). Lower bounds on the worst-case complexity of some oracle algorithms. *Discrete Math.* **24**, 261–276. [9.4]

Hausmann, D. and Korte, B. (1981). Algorithmic versus axiomatic definitions of matroids. *Math. Prog. Study* **14**, 98–111. [8.6, 9.4]

Held, M. and Karp, R. M. (1970). The traveling-salesman problem and minimum spanning trees. *Operations Res.* **18**, 1138–1162. [11.3]

Helgason, T. (1974). Aspects of the theory of hypermatroids. In *Hypergraph seminar* (eds. C. Berge and D. K. Ray-Chaudhuri). Lecture Notes in Math. Vol 411, pp. 191–214. Springer-Verlag, Berlin. [11.1]

Heller, I. (1957). On linear systems with integral valued solutions. *Pacific J. Math.* **7**, 1351–1364. [14.10]

Heron, A. P. (1972a). *Some topics in matroid theory*. D.Phil. thesis, University of Oxford. [9.4]

Heron, A. P. (1972b). Matroid polynomials. In *Combinatorics* (eds. D. J. A. Welsh and D. R. Woodall), pp. 164–202. Institute of Math. and its Applications, Southend-on-Sea. [15.2]

Heron, A. P. (1973). A property of the hyperplanes of a matroid and an extension of Dilworth's theorem. *J. Math. Anal. Appl.* **42**, 119–132. [6.9]

Herstein, I. N. (1975) *Topics in algebra*. Second edition. Xerox, Lexington. [6.8]

Hicks, I. V. and McMurray, N. B. (2007). The branchwidth of graphs and their cycle matroids. *J. Combin. Theory Ser. B* **97**, 681–692. [14.2]

Higgs, D. A. (1966a). A lattice order on the set of all matroids on a set. *Canad. Math. Bull.* **9**, 684–685. [7.3]

Higgs, D. A. (1966b). Maps of geometries. *J. London Math. Soc.* **41**, 612–618. [7.3]

Higgs, D. A. (1968). Strong maps of geometries. *J. Combin. Theory* **5**, 185–191. [7.3, 7.4]

Hill, R. (1986). *A first course in coding theory*. Oxford University Press, Oxford. [2.2]

Hirschfeld, J. W. P. (1997). Complete arcs. *Discrete Math.* **174**, 177–184. [6.5]

Hlinený, P. (2002). On the excluded minors for matroids of branch-width three. *Electron. J. Combin.* **9**, Research Paper 32, 13 pp. (electronic). [14.2]

Hlinený, P. (2003). On matroid properties definable in the MSO logic. In *Mathematical foundations of computer science 2003*. Lecture Notes in Comput. Sci. Vol. 2747, pp. 470–479, Springer, Berlin. [14.4]

Hliněný, P. (2005). A parametrized algorithm for matroid branch-width. *SIAM J. Comput.* **35**, 259–277. [14.4]

Hliněný, P. (2006a). Branch-width, parse trees, and monadic second-order logic for matroids. *J. Combin. Theory Ser. B* **96**, 325–351. [14.4]

Hliněný, P. (2006b). On matroid representability and minor problems. In *Mathematical foundations of computer science 2006.* Lecture Notes in Comput. Sci. Vol. 4162, pp. 505–516, Springer, Berlin. [14.4]

Hliněný, P. (2007). Some hard problems on matroid spikes. *Theory Comput. Syst.* **41**, 551–562. [14.4]

Hliněný, P. and Oum, S. (2008). Finding branch-decompositions and rank-decompositions. *SIAM J. Comput.* **38**, 1012–1032. [14.4]

Hliněný, P. and Whittle, G. (2006). Matroid tree-width. *European J. Combin.* **27**, 1117–1128. [14.2]

Hliněný, P. and Whittle, G. (2009). Addendum to matroid tree-width. *European J. Combin.* **30**, 1036–1044. [14.2]

Hoffman, A. J. and Kruskal, J. B. (1956). Integral boundary points of convex polyhedra. In *Linear inequalities and related systems* (eds. H. W. Kuhn and A. W. Tucker) Ann. Math. Studies **38**, pp. 223–246. Princeton University Press, Princeton. [10.1]

Hoffman, A. J. and Kuhn, H. W. (1956). On systems of distinct representatives. In *Linear inequalities and related systems* (eds. H. W. Kuhn and A. W. Tucker). Ann. Math. Studies **38**, pp. 199–206. Princeton University Press, Princeton. [1.6]

Horn, A. (1955). A characterisation of unions of linearly independent sets. *J. London Math. Soc.* **30**, 494–496. [11.3]

Hou, X.-D. (2005). On the asymptotic number of non-equivalent q-ary linear codes. *J. Combin. Theory Ser. A* **112**, 337–346. [15.5]

Hou, X.-D. (2007). On the asymptotic number of inequivalent binary codes. *Finite Fields Appl.* **13**, 318–326. [15.5]

Hou, X.-D. (2009). Asymptotic numbers of non-equivalent codes in three notions of equivalence. *Linear Multilinear Algebra* **57**, 111–122. [15.5]

Huang, R. and Rota, G.-C. (1994). On the relations of various conjectures on Latin squares and straightening coefficients. *Discrete Math.* **128**, 225–236. [15.9]

Huffman, W. C. and Pless, V. (2003). *Fundamentals of error-correcting codes.* Cambridge University Press, Cambridge. [15.5]

Hughes, D. R. and Piper, F. (1973). *Projective planes.* Springer-Verlag, New York. [6.1]

Hungerford, T. W. (1974). *Algebra.* Springer-Verlag, New York. [6.3, 6.8]

Ingleton, A. W. (1971a). Representation of matroids. In *Combinatorial mathematics and its applications* (ed. D. J. A. Welsh), pp. 149–167. Academic Press, London. [6.1, 6.7, 6.8, 15.9, Appendix]

Ingleton, A. W. (1971b). A geometrical characterization of transversal independence structures. *Bull. London Math. Soc.* **3**, 47–51. [10.4, 11.2]

Ingleton, A. W. (1977). Transversal matroids and related structures. In *Higher combinatorics* (ed. M. Aigner), pp. 117–131. Reidel, Dordrecht. [10.4]

Ingleton, A. W. and Main, R. A. (1975). Non-algebraic matroids exist. *Bull. London Math. Soc.* **7**, 144–146. [6.7]

Ingleton, A. W. and Piff, M. J. (1973). Gammoids and transversal matroids. *J. Combin. Theory Ser. B* **15**, 51–68. [2.4, 15.6]

Inukai, T. and Weinberg, L. (1978). Theorems on matroid connectivity. *Discrete Math.* **22**, 311–312. [8.6]

Inukai, T. and Weinberg, L. (1981). Whitney connectivity of matroids. *SIAM J. Alg. Disc. Methods* **2**, 108–120. [8.6]

Jacobson, N. (1953). *Lectures in abstract algebra. Volume II. Linear algebra.* Van Nostrand, Princeton. [6.1]

Jacobson, N. (1974). *Basic algebra. I.* Freeman, San Francisco. [6.8]

Jaeger, F. (1979). Flows and generalized coloring theorems in graphs. *J. Combin. Theory Ser. B* **26**, 205–216. [5.2, 14.10]

Jaeger, F., Vertigan, D. L., and Welsh, D. J. A. (1990). On the computational complexity of the Jones and Tutte polynomials. *Math. Proc. Cambridge Philos. Soc.* **108**, 35–53. [7.1]

Jensen, P. M. (1978). Binary fundamental matroids. In *Algebraic methods in graph theory* (eds. L. Lovász and V. T. Sós). Colloq. Math. Soc. János Bolyai **25**, pp. 281–296. North-Holland, Amsterdam. [15.6]

Jensen, P. M. and Korte, B. (1982). Complexity of matroid property algorithms. *SIAM J. Comput.* **11**, 184–190. [9.4]

Jerrum, M. (2006). Two remarks concerning balanced matroids. *Combinatorica* **26**, 733–742. [15.8]

Johnson, T., Robertson, N., and Seymour, P. (1999) Grids in binary matroids. Conference talk by P. Seymour. *Graph Theory Workshop*, Oberwolfach, Germany. [14.3]

Kahn, J. (1982). Characteristic sets of matroids. *J. London Math. Soc. (2)* **26**, 207–217. [6.8]

Kahn, J. (1984). A geometric approach to forbidden minors for GF(3). *J. Combin. Theory Ser. A* **37**, 1–12. [10.2]

Kahn, J. (1985). A problem of P. Seymour on nonbinary matroids. *Combinatorica* **5**, 319–323. [3.3, 8.4, 12.3, 15.9]

Kahn, J. (1988). On the uniqueness of matroid representations over GF(4). *Bull. London Math. Soc.* **20**, 5–10. [14.6, Appendix]

Kahn, J. (2000). A normal law for matchings. *Combinatorica* **20**, 339–391. [15.8]

Kahn, J. and Kung, J. P. S. (1982). Varieties of combinatorial geometries. *Trans. Amer. Math. Soc.* **271**, 485–491. [6.10]

Kahn, J. and Neiman, M. (2008). Negative correlation and log-concavity. *Random Structures Algorithms*, to appear. [15.2, 15.8]

Kahn, J. and Seymour, P. (1988). On forbidden minors for GF(3). *Proc. Amer. Math. Soc.* **102**, 437–440. [10.2]

Kajitani, Y., Ueno, S., and Miyano, H. (1988). Ordering of the elements of a matroid such that its consecutive w elements are independent. *Discrete Math.* **72**, 187–194. [15.9]

Kantor, W. (1974). Dimension and embedding theorems for geometric lattices. *J. Combin. Theory Ser. A* **17**, 173–195. [15.9]

Kelly, D. and Rota, G.-C. (1973). Some problems in combinatorial geometry. In *A survey of combinatorial geometry*, pp. 309–312. North-Holland, Amsterdam. [14.5]

Kelly, P. J. (1942). *On isometric transformations*. Ph.D. thesis, University of Wisconsin. [15.4]

Kelmans, A. K. (1980). Concept of a vertex in a matroid and 3-connected graphs. *J. Graph Theory* **4**, 13–19. [9.4]

Kelmans, A. K. (1981). The concept of a vertex in a matroid, the nonseparating cycles of a graph and a new criterion for graph planarity. In *Algebraic methods in graph theory* (eds. L. Lovász and V. T. Sós). Colloq. Math. Soc. János Bolyai **25**, pp. 345–388. North-Holland, Amsterdam. [9.4]

Kingan, S. R. (1997). A generalization of a graph result of D. W. Hall. *Discrete Math.* **173**, 129–135. [5.1, 14.5, Appendix]

Kingan, S. R. (1999). Intersections of circuits and cocircuits in binary matroids. *Discrete Math.* **195**, 157–165. [15.9]

Kingan, S. R. and Lemos, M. (2002). Almost-graphic matroids. *Adv. in Appl. Math.* **28**, 438–477. [15.9]

Kingan, S. R. and Lemos, M. (2006). On the circuit-cocircuit intersection conjecture. *Graphs Combin.* **22**, 471–480. [15.9]

Kingan, S. R. and Oxley, J. G. (1996). On the matroids in which all hyperplanes are binary. *Discrete Math.* **160**, 265–271. [14.1]

Kirchhoff, G. (1847). Über die Auflösung der Gleichungen, auf welche man bei der Untersuchungen der linearen Vertheilung galvanischer Ströme geführt wird. *Ann. Phys. Chem.* **72**, 497–508. [15.8]

Kneser, M. (1953). Abschätzung der asymptotischen Dichte von Summenmengen. *Math. Z.* **58**, 459–484. [15.9]

Knuth, D. E. (1974). The asymptotic number of geometries. *J. Combin. Theory Ser. A* **17**, 398–401. [1.1]

Korte, B., Lovász, L., and Schrader, R. (1991). *Greedoids*. Springer-Verlag, Berlin. [15.0]

Krogdahl, S. (1977). The dependence graph for bases in matroids. *Discrete Math.* **19**, 47–59. [4.3]

Kruskal, J. B. (1956). On the shortest spanning tree of a graph and the traveling salesman problem. *Proc. Amer. Math. Soc.* **7**, 48–50. [1.8]

Kung, J. P. S. (1977). The core extraction algorithm for combinatorial geometries. *Discrete Math.* **19**, 167–175. [7.3]

Kung, J. P. S. (1978). The alpha function of a matroid – I. Transversal matroids. *Studies in Appl. Math.* **58**, 263–275. [15.6]

Kung, J. P. S. (1986a). *A source book in matroid theory*. Birkhäuser, Boston. [1.0, 15.0]

Kung, J. P. S. (1986b). Numerically regular hereditary classes of combinatorial geometries. *Geom. Dedicata* **21** , 85–105. [14.10]

Kung, J. P. S. (1987). Excluding the cycle geometries of the Kuratowski graphs from binary geometries. *Proc. London Math. Soc. (3)* **55**, 209–242. [14.10, 15.3]

Kung, J. P. S. (1988). The long-line graph of a combinatorial geometry. I. Excluding $M(K_4)$ and the $(q + 2)$-point line as minors. *Quart. J. Math. Oxford Ser. (2)* **39**, 223–234. [14.10]

Kung, J. P. S. (1990). Combinatorial geometries representable over GF(3) and GF(q). I. The number of points. *Discrete Comput. Geom.* **5**, 83–95. [14.10]

Kung, J. P. S. (1993). Extremal matroid theory. In *Graph structure theory* (eds. N. Robertson and P. Seymour), Contemp. Math. **147**, pp. 21–61, Amer. Math. Soc, Providence. [14.10, 15.9]

Kung, J. P. S. (1996). Critical problems. In *Matroid theory* (eds. J. Bonin, J. Oxley, and B. Servatius), Contemp. Math. **197**, pp. 1–127, Amer. Math. Soc., Providence. [15.3, 15.9]

Kung, J. P. S. (2000). On the lines-planes inequality for matroids. *J. Combin. Theory Ser. A* **91**, 363–368. [15.2]

Kung, J. P. S. (2008). Binary matroids with no $PG(3, 2)$-minor. *Algebra Universalis* **59**, 111–116. [14.10]

Kung, J. P. S. and Oxley, J. G. (1988). Combinatorial geometries representable over GF(3) and GF(q). II. Dowling geometries. *Graphs Combin.* **4**, 323–332. [14.10, Appendix]

Kuratowski, K. (1930). Sur le problème des courbes gauches en topologie. *Fund. Math.* **15**, 271–283. [2.3]

Lam, C. W. H., Thiel, L., and Swiercz, S. (1989). The non-existence of finite projective planes of order 10. *Canad. J. Math.* **41**, 1117–1123. [6.1]

Lang, S. (1965). *Algebra*. Addison-Wesley, Reading. [6.7]

Las Vergnas, M. (1970). Sur les systèmes de représentants distincts d'une famille d'ensembles. *C. R. Acad. Sci. Paris Sér. A-B* **270**, A501–A503. [10.4, 11.2]

Las Vergnas, M. (1980). Fundamental circuits and a characterization of binary matroids. *Discrete Math.* **31**, 327. [9.1]

Lawler, E. (1976). *Combinatorial optimization: networks and matroids*. Holt, Rinehart and Winston, New York. Reprinted 2001, Dover, Mineola. [11.3, 13.4]

Lax, R. F. (2004). On the character of S_n acting on the subspaces of $\mathbb{F}_q^n$. *Finite Fields Appl.* **10**, 315–322. [15.5]

Lazarson, T. (1958). The representation problem for independence functions. *J. London Math. Soc.* **33**, 21–25. [6.5]

Lehman, A. (1964). A solution of the Shannon switching game. *J. Soc. Indust. Appl. Math.* **12**, 687–725. [4.3, 9.1, 9.4]

Lemos, M. (1988). An extension of Lindström's result about characteristic sets of matroids. *Discrete Math.* **68**, 85–101. [6.7, 15.1]

Lemos, M. (1989). On 3-connected matroids. *Discrete Math.* **73**, 273–283. [8.7]

Lemos, M. (1994). Matroids having the same connectivity function. *Discrete Math.* **131**, 153–161. [8.2]

Lemos, M. (2002). On the connectivity function of a binary matroid. *J. Combin. Theory Ser. B* **86**, 114–132. [8.2]

Lemos, M. (2004). On the number of non-isomorphic matroids. *Adv. in Appl. Math.* **33**, 733–746. [4.3, 7.4, 15.5]

Lemos, M. (2009). A characterization of graphic matroids using non-separating cocircuits. *Adv. in Appl. Math.* **42**, 75–81. [9.4]

Lemos, M. and Oxley, J. (1998). On packing minors into connected matroids. *Discrete Math.* **189**, 283–289. [13.3]

Lemos, M. and Oxley, J. (1999). On size, circumference and circuit removal in 3-connected matroids. *Discrete Math.* **220**, 145–157. [15.4]

Lemos, M. and Oxley, J. (2001). A sharp bound on the size of a connected matroid. *Trans. Amer. Math. Soc.* **353**, 4039–4056. [4.3]

Lemos, M. and Oxley, J. (2003). On the minor-minimal 3-connected matroids having a fixed minor. *European J. Combin.* **24**, 1097–1123. [13.3]

Li, W. (1983). On matroids of the greatest W-connectivity. *J. Combin. Theory Ser. B* **35**, 20–27. [8.6]

Lindström, B. (1983). The non-Pappus matroid is algebraic. *Ars Combinatoria* **16B**, 95–96. [6.7]

Lindström, B. (1984a). A simple non-algebraic matroid of rank three. *Utilitas Math.* **25**, 95–97. [6.7]

Lindström, B. (1984b). On binary identically self-dual matroids. *European J. Combin.* **5**, 55–58. [10.4]

Lindström, B. (1985a). A Desarguesian theorem for algebraic combinatorial geometries. *Combinatorica* **5**, 237–239. [6.7, Appendix]

Lindström, B. (1985b). *On the algebraic representations of dual matroids.* Department of Math., Univ. of Stockholm, Reports, No. 5. [6.7, Appendix]

Lindström, B. (1985c). On the algebraic characteristic set for a class of matroids. *Proc. Amer. Math. Soc.* **95**, 147–151. [6.7, 15.1, Appendix]

Lindström, B. (1985d). *More on algebraic representations of matroids.* Department of Math., Univ. of Stockholm, Reports, No. 10. [6.7]

Lindström, B. (1986a). The non-Pappus matroid is algebraic over any finite field. *Utilitas Math.* **30**, 53–55. [6.7, Appendix]

Lindström, B. (1986b). A non-linear algebraic matroid with infinite characteristic set. *Discrete Math.* **59**, 319–320. [6.7]

Lindström, B. (1987a). An elementary proof in matroid theory using Tutte's coordinatization theorem. *Utilitas Math.* **31**, 189–190. [6.5]

Lindström, B. (1987b). A reduction of algebraic representations of matroids. *Proc. Amer. Math. Soc.* **100**, 388–389. [6.7]

Lindström, B. (1987c). A class of non-algebraic matroids of rank three. *Geom. Dedicata* **23**, 255–258. [6.7]

Lindström, B. (1988a). A generalization of the Ingleton–Main lemma and a class of non-algebraic matroids. *Combinatorica* **8**, 87–90. [6.7]

Lindström, B. (1988b). Matroids, algebraic and non-algebraic. In *Algebraic, extremal and metric combinatorics* 1986 (eds. M.-M. Deza, P. Frankl, and I. G. Rosenberg). London Math. Soc. Lecture Notes **131**, pp. 166–174. Cambridge University Press, Cambridge. [6.7, 15.1]

Lindström, B. (1989). Matroids algebraic over $F(t)$ are algebraic over F. *Combinatorica* **9**, 107–109. [6.7]

Lindström, B. (1993). On algebraic matroids. *Discrete Math.* **111**, 357–359. [15.1]

Lovász, L. (1977). Matroids and geometric graphs. In *Combinatorial surveys: Proc. Sixth British Combinatorial Conference* (ed. P. J. Cameron), pp. 45–86. Academic Press, London. [11.1]

Lovász, L. and Plummer, M. D. (1986). *Matching theory*. Ann. Discrete Math. **29**. North-Holland, Amsterdam. [11.1]

Lovász, L. and Recski, A. (1973). On the sum of matroids. *Acta Math. Acad. Sci. Hungar.* **24**, 329–333. [11.3]

Lucas, D. (1975). Weak maps of combinatorial geometries. *Trans. Amer. Math. Soc.* **206**, 247–279. [7.3]

Mac Lane, S. (1936). Some interpretations of abstract linear dependence in terms of projective geometry. *Amer. J. Math.* **58**, 236–240. [1.0]

Mac Lane, S. (1938). A lattice formulation for transcendence degrees and p-bases. *Duke Math. J.* **4**, 455–468. [1.0, 6.7]

Mader, W. (1967). Homomorphieeigenschaften und mittlere Kantendichte von Graphen. *Math. Ann.* **174**, 265–268. [14.10]

Mader, W. (1974). Kreuzungfreie a, b-Wege in endlichen Graphe. *Abh. Math. Sem. Univ. Hamburg* **42**, 187–204. [15.4]

Mader, W. (1979). Connectivity and edge-connectivity in finite graphs. In *Surveys in combinatorics* (ed. B. Bollobás). London Math. Soc. Lecture Notes **38**, pp. 66–95. Cambridge University Press, Cambridge. [15.4]

Mahoney, C. (1985). On the unimodality of the independent set numbers of a class of matroids. *J. Combin. Theory Ser. B* **39**, 77–85. [15.2]

Mason, J. H. (1971). Geometrical realization of combinatorial geometries. *Proc. Amer. Math. Soc.* **30**, 15–21. [1.5]

Mason, J. H. (1972a). On a class of matroids arising from paths in graphs. *Proc. London Math. Soc. (3)* **25**, 55–74. [2.4, 3.2, 15.6]

Mason, J. H. (1972b). Matroids: unimodal conjectures and Motzkin's theorem. In *Combinatorics* (eds. D. J. A. Welsh and D. R. Woodall), pp. 207–220. Institute of Math. and its Applications, Southend-on-Sea. [15.2]

Mason, J. H. (1973). Maximal families of pairwise disjoint maximal proper chains in a geometric lattice. *J. London Math. Soc. (2)* **6**, 539–542. [6.9]

Mason, J. H. (1977). Matroids as the study of geometrical configurations. In *Higher combinatorics* (ed. M. Aigner), pp. 133–176. Reidel, Dordrecht. [7.0, 7.2, 11.2]

Mason, J. H. and Oxley, J. G. (1980). A circuit covering result for matroids. *Math. Proc. Cambridge Philos. Soc.* **87**, 25–27. [4.3]

Matiyasevic, Y. V. (1993). *Hilbert's tenth problem.* Translated from the 1993 Russian original by the author. MIT Press, Cambridge. [6.8]

Matthews, L. R. (1977). Bicircular matroids. *Quart. J. Math. Oxford Ser. (2)* **28**, 213–228. [10.4, 11.1, 11.2]

Matthews, L. R. (1978). Matroids from directed graphs. *Discrete Math.* **24**, 47–61. [6.10]

Mayhew, D. (2008). Matroid complexity and nonsuccinct descriptions. *SIAM J. Discrete Math.* **22**, 455–466. [9.4]

Mayhew, D., Newman, M., Welsh, D., and Whittle, G. (2009). On the asymptotic proportion of connected matroids. *European J. Combin.,* to appear. [15.5]

Mayhew, D., Newman, M., and Whittle, G. (2009). On excluded minors for real-representability. *J. Combin. Theory Ser. B* **99**, 685–689. [14.3, 14.4]

Mayhew, D., Oporowski, B., Oxley, J., and Whittle, G. (2009). The excluded minors for the matroids that are either binary or ternary. *European J. Combin.,* to appear. [14.5]

Mayhew, D. and Royle, G. (2008). Matroids with nine elements. *J. Combin. Theory Ser. B.* **98**, 415–431. [15.5]

Mayhew, D., Royle, G., and Whittle, G. (2009a). The internally 4-connected binary matroids with no $M(K_{3,3})$-minor. *Mem. Amer. Math. Soc.,* to appear. [9.4, 12.2, 14.10]

Mayhew, D., Royle, G., and Whittle, G. (2009b). Excluding Kuratowski graphs and their duals from binary matroids (submitted). [14.10]

Mayhew, D., Whittle, G., and van Zwam, S. (2009). An obstacle to a decomposition theorem for near-regular matroids (submitted). [15.1]

Mayhew, D., Whittle, G., and van Zwam, S. (2010). Stability, fragility, and Rota's Conjecture (submitted). [14.8]

Mazoit, F. and Thomassé, S. (2007). Branchwidth of graphic matroids. In *Surveys in combinatorics 2007.* London Math. Soc. Lecture Notes **346**, pp. 275–286. Cambridge University Press, Cambridge. [14.2]

McDiarmid, C. J. H. (1973). Independence structures and submodular functions. *Bull. London Math. Soc.* **5**, 18–20. [11.1]

McDiarmid, C. J. H. (1975a). Extensions of Menger's theorem. *Quart. J. Math. Oxford Ser. (2)* **26**, 141–157. [11.2]

McDiarmid, C. J. H. (1975b). An exchange theorem for independence structures. *Proc. Amer. Math. Soc.* **47**, 513–514. [11.3]

McDiarmid, C. J. H. (1975c). Rado's theorem for polymatroids. *Math. Proc. Cambridge Philos. Soc.* **78**, 263–281. [11.1]

McGuinness, S. (2005). Circuits through cocircuits in a graph with extensions to matroids. *Combinatorica* **25**, 451–463. [4.3, 15.4]

Mendelsohn, N. S. and Dulmage, A. L. (1958). Some generalisations of the problem of distinct representatives. *Canad. J. Math.* **10**, 230–242. [11.2]

Menger, K. (1927). Zur allgemeinen Kurventheorie. *Fund. Math.* **10**, 96–115. [8.5, 12.3]

Mighton, J. (2008). A new characterization of graphic matroids. *J. Combin. Theory Ser. B* **98**, 1253–1258. [9.4]

Miller, W. P. (1997). Techniques in matroid reconstruction. *Discrete Math.* **170**, 173–183. [15.4]

Miller, W. P. (2003). Unique representability and matroid reconstruction. *Adv. in Appl. Math.* **30**, 591–606. [15.4]

Minty, G. J. (1966). On the axiomatic foundations of the theories of directed linear graphs, electrical networks and network programming. *J. Math. Mech.* **15**, 485–520. [9.1, 10.4, 12.3]

Mirsky, L. (1955). *Introduction to linear algebra.* Oxford University Press, London. [6.5]

Mirsky, L. (1971). *Transversal theory.* Academic Press, London. [11.2, 11.3]

Motzkin, T. (1951). The lines and planes connecting the points of a finite set. *Trans. Amer. Math. Soc.* **70**, 451–464. [6.9]

Murty, U. S. R. (1974). Extremal critically connected matroids. *Discrete Math.* **8**, 49–58. [4.3, 15.4]

Murty, U. S. R. (1976). Extremal matroids with forbidden restrictions and minors. In *Proc. Seventh Southeastern Conf. on Combinatorics, Graph Theory and Computing.* Congressus Numerantium **17**, pp. 463–468. Utilitas Mathematica, Winnipeg. [14.10]

Nakasawa, T. (1935). Zur Axiomatik der linearen Abhängigkeit. I. *Sci. Rep. Tokyo Bunrika Daigaku Section A* **2**, 235–255. [1.0, 1.1]

Nakasawa, T. (1936a). Zur Axiomatik der linearen Abhängigkeit. II. *Sci. Rep. Tokyo Bunrika Daigaku Section A* **3**, 45–69. [1.0]

Nakasawa, T. (1936b). Zur Axiomatik der linearen Abhängigkeit. III. *Sci. Rep. Tokyo Bunrika Daigaku Section A* **3**, 123–136. [1.0]

Nash-Williams, C. St. J. A. (1961). Edge-disjoint spanning trees of finite graphs. *J. London Math. Soc.* **36**, 445–450. [11.3]

Nash-Williams, C. St. J. A. (1966). An application of matroids to graph theory. In *Theory of graphs* (Internat. Sympos., Rome), pp. 263–265. Dunod, Paris. [11.2, 11.3]

Negami, S. (1982). A characterization of 3-connected graphs containing a given graph. *J. Combin. Theory Ser. B* **32**, 69–74. [12.1]

Nešetřil, J., Poljak, S., and Turzík, D. (1981). Amalgamation of matroids and its applications. *J. Combin. Theory Ser. B* **31**, 9–22. [11.4]

Nešetřil, J., Poljak, S., and Turzík, D. (1985). Special amalgams and Ramsey matroids. In *Matroid theory* (eds. L. Lovász and A. Recski). Colloq. Math. Soc. János Bolyai **40**, pp. 267–298. North-Holland, Amsterdam. [11.4]

Neumann-Lara, V., Rivera-Campo, E., and Urrutia, J. (1999). A note on covering the edges of a graph with bonds. *Discrete Math.* **197/198**, 633–636. [4.3, 15.4]

Oporowski, B. (2002). Partitioning matroids with only small cocircuits. *Combin. Probab. Comput.* **11**, 191–197. [14.2]

Oporowski, B., Oxley, J, and Thomas, R. (1993). Typical subgraphs of 3- and 4-connected graphs. *J. Combin. Theory Ser. B* **57**, 239–257. [14.9]

Ore, O. (1955). Graphs and matching theorems. *Duke Math. J.* **22**, 625–639. [11.2]

Ore, O. (1967). *The four-color problem.* Academic Press, New York. [5.2, 5.3]

Oum, S. and Seymour, P. (2007). Testing branch-width. *J. Combin. Theory Ser. B* **97**, 385–393. [14.4]

Oxley, J. G. (1978a). Colouring, packing and the critical problem. *Quart. J. Math. Oxford Ser. (2)* **29**, 11–22. [6.2]

Oxley, J. G. (1978b). Cocircuit coverings and packings for binary matroids. *Math. Proc. Cambridge Philos. Soc.* **83**, 347–351. [9.4]

Oxley, J. G. (1979). On cographic regular matroids. *Discrete Math.* **25**, 89–90. [10.3]

Oxley, J. G. (1981a). On connectivity in matroids and graphs. *Trans. Amer. Math. Soc.* **265**, 47–58. [4.3, 8.7]

Oxley, J. G. (1981b). On matroid connectivity. *Quart. J. Math. Oxford Ser. (2)* **32**, 193–208. [8.1, 8.7]

Oxley, J. G. (1981c). On a matroid generalization of graph connectivity. *Math. Proc. Cambridge Philos. Soc.* **90**, 207–214. [8.6]

Oxley, J. G. (1983). On the numbers of bases and circuits in simple binary matroids. *European J. Combin.* **4**, 169–178. [9.4]

Oxley, J. G. (1984a). On the intersections of circuits and cocircuits in matroids. *Combinatorica* **4**, 187–195. [5.4, 9.1, 12.2, 12.3]

Oxley, J. G. (1984b). On singleton 1-rounded sets of matroids. *J. Combin. Theory Ser. B* **37**, 189–197. [7.2, 12.3, Appendix]

Oxley, J. G. (1986a). On the matroids representable over GF(4). *J. Combin. Theory Ser. B* **41**, 250–252. [6.4]

Oxley, J. G. (1986b). On ternary transversal matroids. *Discrete Math.* **62**, 71–83. [10.4, 15.6]

Oxley, J. G. (1987a). On nonbinary 3-connected matroids. *Trans. Amer. Math. Soc.* **300**, 663–679. [12.2, 12.3, 15.9, Appendix]

Oxley, J. G. (1987b). The binary matroids with no 4-wheel minor. *Trans. Amer. Math. Soc.* **301**, 63–75. [12.2]

Oxley, J. G. (1987c). A characterization of the ternary matroids with no $M(K_4)$-minor. *J. Combin. Theory Ser. B* **42**, 212–249. [10.4, 12.1, 12.2, 14.10, Appendix]

Oxley, J. G. (1988). On circuit exchange properties for matroids. *European J. Combin.* **9**, 331–336. [9.1, 12.2]

Oxley, J. G. (1989a). The regular matroids with no 5-wheel minor. *J. Combin. Theory Ser. B* **46**, 292–305. [12.1, 12.2, 14.10]

Oxley, J. G. (1989b). A characterization of certain excluded-minor classes of matroids. *European J. Combin.* **10**, 275–279. [12.2, 12.3]

Oxley, J. G. (1990a). On an excluded-minor class of matroids. *Discrete Math.* **82**, 35–52. [12.1, 14.10, Appendix]

Oxley, J. G. (1990b). A characterization of a class of non-binary matroids. *J. Combin. Theory Ser. B* **49**, 181–189. [15.9]

Oxley, J. G. (1991). Ternary paving matroids. *Discrete Math.* **91**, 77–86. [12.2, Appendix]

Oxley, J. G. (1992). Infinite matroids. In *Matroid applications* (ed. N. White), pp. 73–90. Cambridge University Press, Cambridge. [15.0]

Oxley, J. (1996). Structure theory and connectivity for matroids. In *Matroid theory* (eds. J. Bonin, J. Oxley, and B. Servatius), Contemp. Math. **197**, pp. 129–170, Amer. Math. Soc., Providence. [12.0]

Oxley, J. (1997). A matroid generalization of a result of Dirac. *Combinatorica* **17**, 267–273. [8.7]

Oxley, J. (2001). On the interplay between graphs and matroids. In *Surveys in combinatorics, 2001 (Sussex)*. London Math. Soc. Lecture Notes **288**, pp. 199–239. Cambridge University Press, Cambridge. [15.4]

Oxley, J. (2003). What is a matroid? *Cubo Mat. Educ.* **5**, 179–218. [Preface]

Oxley, J., Prendergast, K., and Row, D. (1982). Matroids whose ground sets are domains of functions. *J. Austral. Math. Soc. Ser. A* **32**, 380–387. [11.1, 11.2]

Oxley, J. G. and Reid, T. J. (1990). The smallest rounded sets of binary matroids. *European J. Combin.* **11**, 47–56. [12.3, Appendix]

Oxley, J. and Row, D. (1989). On fixing elements in matroid minors. *Combinatorica* **9**, 69–74. [12.3]

Oxley, J., Semple, C., and Vertigan, D. (2000). Generalized Δ-Y exchange and k-regular matroids. *J. Combin. Theory Ser. B* **79**, 1–65. [6.5, 11.5, 12.2]

Oxley, J., Semple, C., Vertigan, D., and Whittle, G. (2002). Infinite antichains of matroids with characteristic set $\{p\}$. *Discrete Math.* **242**, 175–185. [14.1]

Oxley, J., Semple, C., and Whittle, G. (2004). The structure of the 3-separations of 3-connected matroids. *J. Combin. Theory Ser. B* **92**, 257–293. [8.2, 8.3]

Oxley, J., Semple, C., and Whittle, G. (2007). The structure of the 3-separations of 3-connected matroids. II. *European J. Combin.* **28**, 1239–1261. [8.3]

Oxley, J., Vertigan, D., and Whittle, G. (1996a). On inequivalent representations of matroids over finite fields. *J. Combin. Theory Ser. B* **67**, 325–343. [14.6, 14.8]

Oxley, J., Vertigan, D., and Whittle, G. (1996b). Conjectures on matroids representable over $GF(3)$ and other fields. In *Matroid theory* (eds. J. Bonin, J. Oxley, and B. Servatius). Contemp. Math. **197**, pp. 411–412, Amer. Math. Soc., Providence. [14.7]

Oxley, J., Vertigan, D., and Whittle, G. (1998). On maximum-sized near-regular and $\sqrt[6]{1}$-matroids. *Graphs and Combinatorics* **14**, 163–179. [14.10]

Oxley, J. and Whittle, G. (1991). A note on the non-spanning circuits of a matroid. *European J. Combin.* **12**, 259–261. [7.3, 9.1]

Oxley, J. and Whittle, G. (1998). On weak maps of ternary matroids. *European J. Combin.* **19**, 377–389. [10.2]

Oxley, J. and Wu, H. (1995). On matroid connectivity. *Discrete Math.* **146**, 321–324. [8.6]

Oxley, J. and Wu, H. (2000a). On the structure of 3-connected matroids and graphs. *European J. Combin.* **21**, 667–688. [8.8]

Oxley, J. and Wu, H. (2000b). Matroids and graphs with few non-essential elements. *Graphs Combin.* **16**, 199–229. [8.8]

Oxley, J. and Wu, H. (2004). The 3-connected graphs with exactly three non-essential edges. *Graphs Combin.* **20**, 233–246. [8.8]

Parsons, T. D. (1971). On planar graphs. *Amer. Math. Monthly* **78**, 176–178. [5.2]

Pemantle, R. (2000). Towards a theory of negative dependence. Probabilistic techniques in equilibrium and nonequilibrium statistical physics. *J. Math. Phys.* **41**, 1371–1390. [15.8]

Pendavingh, R. A. and van Zwam, S. H. M. (2010a). Lifts of matroid representations over partial fields. *J. Combin. Theory Ser. B* **100**, 36–67. [14.7]

Pendavingh, R. A. and van Zwam, S. H. M. (2010b). Confinement of matroid representations to subsets of partial fields *J. Combin. Theory Ser. B*, to appear. [14.7]

Perfect, H. (1968). Applications of Menger's graph theorem. *J. Math. Anal. Appl.* **22**, 96–111. [11.3]

Perfect, H. (1969). Independence spaces and combinatorial problems. *Proc. London Math. Soc. (3)* **19**, 17–30. [11.2]

Piff, M. J. (1969). *The representability of matroids.* Dissertation for Diploma in Advanced Math., University of Oxford. [6.7]

Piff, M. J. (1972). *Some problems in combinatorial theory.* D.Phil. thesis, University of Oxford. [3.2, 6.7, 10.4]

Piff, M. J. (1973). An upper bound for the number of matroids. *J. Combin. Theory Ser. B* **14**, 241–245. [1.1]

Piff, M. J. and Welsh, D. J. A. (1970). On the vector representation of matroids. *J. London Math. Soc. (2)* **2**, 284–288. [6.8, 11.2, 11.3]

Plummer, M. D. (1968). On minimal blocks. *Trans. Amer. Math. Soc.* **134**, 85–94. [4.3]

Poincaré, H. (1900). Second complément à l'analysis situs. *Proc. London Math. Soc.* **32**, 277–308. [5.1]

Poljak, S. and Turzík, D. (1982). A note on sticky matroids. *Discrete Math.* **42**, 119–123. [11.4, 15.9]

Poljak, S. and Turzík, D. (1984). Amalgamation over uniform matroids. *Czech. Math. J.* **34**, 239–246. [11.4]

Purdy, G. (1986). Two results about points, lines and planes. *Discrete Math.* **60**, 215–218. [15.2]

Pym, J. S. and Perfect, H. (1970). Submodular functions and independence structures. *J. Math. Anal. Appl.* **30**, 1–31. [11.3]

Rado, R. (1942). A theorem on independence relations. *Quart. J. Math. Oxford* **13**, 83–89. [11.2]

Rado, R. (1957). Note on independence functions. *Proc. London Math. Soc. (3)* **7**, 300–320. [6.7, 9.1]

Rado, R. (1966). Abstract linear dependence. *Colloq. Math.* **14**, 257–264. [11.3]

Rado, R. (1967). On the number of systems of distinct representatives of sets. *J. London Math. Soc.* **42**, 107–109. [11.2]

Read, R. C. (1968). An introduction to chromatic polynomials. *J. Combin. Theory* **4**, 52–71. [15.2]

Recski, A. (1985). Some open problems of matroid theory, suggested by its applications. In *Matroid theory* (eds. L. Lovász and A. Recski). Colloq. Math. Soc. János Bolyai **40**, pp. 311–325. North-Holland, Amsterdam. [11.3]

Recski, A. (1989). *Matroid theory and its applications in electrical network theory and in statics.* Springer-Verlag, Berlin. [13.4, 15.0]

Reid, T. J. (1988). *On roundedness in matroid theory.* Ph.D. thesis, Louisiana State University. [12.3]

Reid, T. J. (1991). Triangles in 3-connected matroids. *Discrete Math.* **90**, 281–296. [12.3]

Reid, T. J., Wu, H., and Zhou, X. (2008). On minimally k-connected matroids. *J. Combin. Theory Ser. B.* **98**, 1311–1324. [15.4]

Richardson, W. R. H. (1973). Decomposition of chain-groups and binary matroids. In *Proc. Fourth Southeastern Conf. on Combinatorics, Graph Theory and Computing*, pp. 463–476. Utilitas Mathematica, Winnipeg. [8.6]

Robertson, N. and Seymour, P. D. (1984). Generalizing Kuratowski's Theorem. *Congressus Numerantium* **45**, 129–138. [7.1, 9.3, 11.2, 12.4]

Robertson, N. and Seymour, P. D. (1985). Graph minors – a survey. In *Surveys in combinatorics 1985,* London Math. Soc. Lecture Notes **103**, pp. 153–171. Cambridge University Press, Cambridge. [14.1]

Robertson, N. and Seymour, P. D. (1986). Graph minors. V. Excluding a planar graph. *J. Combin. Theory Ser. B* **41**, 92–114. [14.3]

Robertson, N. and Seymour, P. D. (1990). Graph minors. IV. Tree-width and well-quasi-ordering. *J. Combin. Theory Ser. B* **48**, 227–254. [14.2]

Robertson, N. and Seymour, P. D. (1991). Graph minors. X. Obstructions to tree-decomposition. *J. Combin. Theory Ser. B* **52**, 153–190. [14.2, 14.3]

Robertson, N. and Seymour, P. D. (1995). Graph minors. XIII. The disjoint paths problem. *J. Combin. Theory Ser. B* **63**, 65–110. [14.4]

Robertson, N. and Seymour, P. D. (2003). Graph minors. XVI. Excluding a non-planar graph. *J. Combin. Theory Ser. B* **89**, 43–76. [14.1]

Robertson, N. and Seymour, P. D. (2004). Graph minors. XX. Wagner's conjecture. *J. Combin. Theory Ser. B* **92**, 325–357. [14.0, 14.1]

Robertson, N. and Seymour, P. D. (2009). Graph minors. XXI. Graphs with unique linkages. *J. Combin. Theory Ser. B* **99**, 583–616. [14.1]

Robertson, N., Seymour, P., and Thomas, R. (1993). Hadwiger's conjecture for K_6-free graphs. *Combinatorica* **13**, 279–361. [15.3]

Robinson, G. C. and Welsh, D. J. A. (1980). The computational complexity of matroid properties. *Math. Proc. Cambridge Philos. Soc.* **87**, 29–45. [9.4]

Rockafellar, R. T. (1969). The elementary vectors of a subspace of R^N. In *Combinatorial mathematics and its applications* (eds. R. C. Bose and T. A. Dowling), pp. 104–127. University of North Carolina Press, Chapel Hill. [10.4]

Rónyai, L., Babai, L., and Ganapathy, K. M. (2001). On the number of zero-patterns of a sequence of polynomials. *J. Amer. Math. Soc.* **14**, 717–735. [15.5]

Rota, G.-C. (1971). Combinatorial theory, old and new. In *Proc. Internat. Cong. Math.* (Nice, 1970), pp. 229–233. Gauthier-Villars, Paris. [6.0, 6.5, 14.0, 15.2]

Schrijver, A. (1986). *Theory of linear and integer programming*. Wiley, Chichester. [10.1, 13.4]

Schrijver, A. (2003). *Combinatorial optimization*. Springer, Berlin. [1.0, 11.1, 15.0]

Schrijver, A. and Seymour, P. D. (1990). Spanning trees of different weights. In *Polyhedral combinatorics*. DIMACS Ser. Discrete Math. Theoret. Comput. Sci. **1**, pp. 281–288, Amer. Math. Soc., Providence. [15.9]

Segre, B. (1955). Curve razionali normali e k-archi negli spazi finiti. *Ann. Mat. Pura Appl. (4)* **39**, 357–379. [6.5]

Semple, C. and Welsh, D. (2008). Negative correlation in graphs and matroids. *Combin. Probab. Comput.* **17**, 423–435. [15.8]

Semple, C. and Whittle, G. (1996) Partial fields and matroid representation. *Adv. in Appl. Math.* **17**, 184–208. [14.7]

Seymour, P. D. (1975). *Matroids, hypergraphs and the max-flow min-cut theorem*. D.Phil. thesis, University of Oxford. [15.2]

Seymour, P. D. (1976). The forbidden minors of binary clutters. *J. London Math. Soc. (2)* **12**, 356–360. [9.1]

Seymour, P. D. (1977a). A note on the production of matroid minors. *J. Combin. Theory Ser. B* **22**, 289–295. [4.3, 12.3]

Seymour, P. D. (1977b). The matroids with the max-flow min-cut property. *J. Combin. Theory Ser. B* **23**, 189–222. [8.5, 12.3]

Seymour, P. D. (1978). Some applications of matroid decomposition. In *Algebraic methods in graph theory* (eds. L. Lovász and V. T. Sós), Colloq. Math. Soc. János Bolyai **25**, pp. 713–726. North-Holland, Amsterdam. [12.3]

Seymour, P. D. (1979). Matroid representation over GF(3). *J. Combin. Theory Ser. B* **26**, 159–173. [4.3, 6.5, 10.2]

Seymour, P. D. (1980a). Packing and covering with matroid circuits. *J. Combin. Theory Ser. B* **28**, 237–242. [1.4]

Seymour, P. D. (1980b). Decomposition of regular matroids. *J. Combin. Theory Ser. B* **28**, 305–359. [5.2, 6.6, 8.3, 8.6, 8.8, 9.3, 10.1, 10.3, 12.1, 12.2, 13.0, 13.1, 13.3, Appendix]

Seymour, P. D. (1980c). On Tutte's characterization of graphic matroids. In *Combinatorics* 79. *Part I* (eds. M. Deza and I. G. Rosenberg). Ann. Discrete Math. **8**, pp. 83–90. North-Holland, Amsterdam. [10.3 13.4]

Seymour, P. D. (1981a). Recognizing graphic matroids. *Combinatorica* **1**, 75–78. [6.4, 9.4, 15.1]

Seymour, P. D. (1981b). On minors of non-binary matroids. *Combinatorica* **1**, 387–394. [8.3, 9.1, 12.3, 15.9]

Seymour, P. D. (1981c). Matroids and multicommodity flows. *European J. Combin.* **2**, 257–290. [12.2, 12.3, 13.4]

Seymour, P. D. (1981d). On Tutte's extension of the four-colour problem. *J. Combin. Theory Ser. B* **31**, 82–94. [13.4, 15.3]

Seymour, P. D. (1982). On the points–lines–planes conjecture. *J. Combin. Theory Ser. B* **33**, 17–26. [15.2]

Seymour, P. D. (1985a). Minors of 3-connected matroids. *European J. Combin.* **6**, 375–382. [12.2, 12.3]

Seymour, P. D. (1985b). Applications of the regular matroid decomposition. In *Matroid theory* (eds. L. Lovász and A. Recski). Colloq. Math. Soc. János Bolyai **40**, pp. 345–357. North-Holland, Amsterdam. [13.4]

Seymour, P. D. (1986a). Triples in matroid circuits. *European J. Combin.* **7**, 177–185. [9.4, 15.9]

Seymour, P. D. (1986b). Adjacency in binary matroids. *European J. Combin.* **7**, 171–176. [13.4]

Seymour, P. D. (1988). On the connectivity function of a matroid. *J. Combin. Theory Ser. B* **45**, 25–30. [8.2]

Seymour, P. D. (1995). Matroid minors. In *Handbook of combinatorics* (eds. R. Graham, M. Grötschel, L. Lovász), pp. 527–550. Elsevier, Amsterdam; MIT Press, Cambridge. [10.4, 12.0, 12.2, 12.3]

Seymour, P. D. and Walton, P. N. (1981). Detecting matroid minors. *J. London Math. Soc. (2)* **23**, 193–203. [9.4]

Seymour, P. D. and Welsh, D. J. A. (1975). Combinatorial applications of an inequality from statistical mechanics *Math. Proc. Cambridge Philos. Soc.* **77**, 485–495. [15.8]

Shameeva, O. V. (1985). Algebraic representability of matroids. *Moscow Univ. Math. Bull.* **40**, 43–48. [6.7]

Sims, J. A. (1977). An extension of Dilworth's theorem. *J. London Math. Soc. (2)* **16**, 393–396. [11.2]

Sims, J. A. (1980). *Some problems in matroid theory*. D.Phil. thesis, University of Oxford. [3.3]

Stanley, R. P. (1977). Some combinatorial aspects of the Schubert calculus. In *Combinatoire et représentation du groupe symétrique*. Lecture Notes in Math. Vol. 579, pp. 217–251. Springer-Verlag, Berlin. [11.2]

Stonesifer, J. R. (1975). Logarithmic concavity for edge lattices of graphs. *J. Combin. Theory Ser. A* **18**, 36–46. [15.2]

Sturmfels, B. (1987). On the decidability of diophantine problems in combinatorial geometry. *Bull. Amer. Math. Soc.* **17**, 121–124. [6.8, 15.1]

Tan, J. J.-M. (1981). *Matroid 3-connectivity.* Ph.D. thesis, Carleton University. [12.1, 12.3]

Tarski, A. (1951). *A decision method for elementary algebra and geometry.* Second edition. University of California Press, Berkeley. [6.8]

Thas, J. A. (1968). Normal rational curves and k-arcs in Galois spaces. *Rend. Mat. (6)* **1**, 331–334. [6.5]

Thomas, R. (1988). A counter-example to 'Wagner's conjecture' for infinite graphs. *Math. Proc. Cambridge Philos. Soc.* **103** (1988), 55–57. [14.1]

Thomas, R. (1990). A Menger-like property of tree-width: the finite case. *J. Combin. Theory Ser. B* **48**, 67–76. [14.2]

Thomas, R. (1999). Recent excluded minor theorems for graphs. In *Surveys in combinatorics, 1999* (eds. J. D. Lamb and D. A. Preece), pp. 201–222. Cambridge University Press, Cambridge. [15.3]

Truemper, K. (1980). On Whitney's 2-isomorphism theorem for graphs. *J. Graph Theory* **4**, 43–49. [5.3]

Truemper, K. (1982a). Alpha-balanced graphs and matrices and GF(3)-representability of matroids. *J. Combin. Theory Ser. B* **32**, 112-139. [10.2]

Truemper, K. (1982b). On the efficiency of representability tests for matroids. *European J. Combin.* **3**, 275–291. [9.4, 15.1]

Truemper, K. (1984). Partial matroid representations. *European J. Combin.* **5**, 377–394. [6.4, 12.3]

Truemper, K. (1985). A decomposition theory for matroids. I. General results. *J. Combin. Theory Ser. B* **39**, 43–76. [7.2]

Truemper, K. (1986). A decomposition theory for matroids. III. Decomposition conditions. *J. Combin. Theory Ser. B* **41**, 275–305. [12.3]

Truemper, K. (1989). On the delta-wye reduction for planar graphs. *J. Graph Theory* **13**, 141–148. [11.5]

Truemper, K. (1992a). A decomposition theory for matroids. VI. Almost regular matroids. *J. Combin. Theory Ser. B* **55**, 253–301. [14.5]

Truemper, K. (1992b). *Matroid decomposition.* Academic Press, Boston. [10.2, 15.0]

Tseng, F. T. and Truemper, K. (1986). A decomposition of the matroids with the max-flow min-cut property. *Discrete Appl. Math.* **15**, 329–364. [12.3]

Tutte, W. T. (1954). A contribution to the theory of chromatic polynomials. *Canad. J. Math.* **6**, 80–91. [15.3]

Tutte, W. T. (1958). A homotopy theorem for matroids, I, II. *Trans. Amer. Math. Soc.* **88**, 144–174. [3.1, 6.5, 6.6, 10.0, 10.1]

Tutte, W. T. (1959). Matroids and graphs. *Trans. Amer. Math. Soc.* **90**, 527–552. [6.6, 10.0, 10.3]

Tutte, W. T. (1960). An algorithm for determining whether a given binary matroid is graphic. *Proc. Amer. Math. Soc.* **11**, 905–917. [9.4]

Tutte, W. T. (1961a). A theory of 3-connected graphs. *Nederl. Akad. Wetensch. Proc. Ser. A* **64**, 441–455. [8.8, 12.2]

Tutte, W. T. (1961b). On the problem of decomposing a graph into *n* connected factors. *J. London Math. Soc.* **36**, 221–230. [11.3]

Tutte, W. T. (1965a). Lectures on matroids. *J. Res. Nat. Bur. Standards Sect. B* **69B**, 1–47. [6.6, 9.1, 9.2, 10.3, 10.4]

Tutte, W. T. (1965b). Menger's Theorem for matroids. *J. Res. Nat. Bur. Standards Sect. B* **69B**, 49–53. [8.5]

Tutte, W. T. (1966a). *Connectivity in graphs*. University of Toronto Press, Toronto. [5.3]

Tutte, W. T. (1966b). Connectivity in matroids. *Canad. J. Math.* **18**, 1301–1324. [4.3, 8.1, 8.2, 8.4, 8.7, 8.8, 12.1]

Tutte, W. T. (1966c). On the algebraic theory of graph colorings. *J. Combin. Theory* **1**, 15–50. [15.3]

Tutte, W. T. (1984). *Graph theory*. Cambridge University Press, Cambridge. [12.2]

Vámos, P. (1968). On the representation of independence structures (unpublished manuscript). [2.1, 2.2]

Vámos, P. (1971a). A necessary and sufficient condition for a matroid to be linear. In *Möbius algebras* (Proc. Conf. Univ. Waterloo, 1971), pp. 166–173. University of Waterloo, Waterloo. [6.5, 6.8]

Vámos, P. (1971b). Linearity of matroids over division rings (notes by G. Roulet). In *Möbius algebras* (Proc. Conf. Univ. Waterloo, 1971), pp. 174–178. University of Waterloo, Waterloo. [6.8]

van den Heuvel, J. and Thomassé, S. (2009). Cyclic orderings and cyclic arboricity of matroids (submitted). http://arxiv.org/abs/arXiv:0912.2929 [15.9]

van der Waerden, B. L. (1937). *Moderne Algebra Vol. 1*. Second edition. Springer-Verlag, Berlin. [1.0, 6.7]

Vertigan, D. L. (1998). On the intertwining conjecture for matroids. Conference talk, *ACCOTA Workshop*, Oaxaca, Mexico. [14.5]

Wagner, D. G. (2004/06). Matroid inequalities from electrical network theory. *Electron. J. Combin.* **11**, no. 2, Article 1, 17 pp. (electronic). [15.8]

Wagner, D. G. (2005). Rank-three matroids are Rayleigh. *Electron. J. Combin.* **12**, Note 8, 11 pp. (electronic). [15.8]

Wagner, D. G. (2008). Negatively correlated random variables and Mason's Conjecture for independent sets in matroids. *Ann. Comb.* **12**, 211–239. [15.8]

Wagner, D. G. and Wei, Y. (2009). A criterion for the half-plane property. *Discrete Math.* **309**, 1385–1390. [15.8]

Wagner, D. K. (1985). On theorems of Whitney and Tutte. *Discrete Math.* **57**, 147–154. [5.3, 10.3]

Wagner, D. K. (1988). Equivalent factor matroids of graphs. *Combinatorica* **8**, 373–377. [6.10]

Wagner, D. K. (2010). On Mighton's characterization of graphic matroids. *J. Combin. Theory Ser. B*, to appear. [9.4]

Wagner, K. (1937a). Über eine Eigenschaft der ebenen Komplexe. *Math. Ann.* **114**, 570–590. [5.2, 5.4, 10.3, 15.3]

Wagner, K. (1937b). Über eine Erweiterung eines Satzes von Kuratowski. *Deut. Math.* **2**, 280–285. [5.2]

Wagner, K. (1960). Bemerkungen zu Hadwigers Vermutung. *Math. Ann.* **141**, 433–451. [12.1, 12.2]

Walton, P. (1981). *Some topics in combinatorial theory.* D.Phil. thesis, University of Oxford. [12.2]

Walton, P. N. and Welsh, D. J. A. (1980). On the chromatic number of binary matroids. *Mathematika* **27**, 1–9. [12.2, 15.3]

Wanner, T. and Ziegler, G. M. (1991). Supersolvable and modularly complemented matroid extensions. *European J. Combin.* **12**, 341–360. [15.9]

Watkins, M. E. and Mesner, D. M. (1967). Cycles and connectivity in graphs. *Canad. J. Math.* **19**, 1319–1328. [15.9]

Welsh, D. J. A. (1969a). A bound for the number of matroids. *J. Combin. Theory* **6**, 313–316. [1.6, 7.4, 15.5]

Welsh, D. J. A. (1969b). On the hyperplanes of a matroid. *Proc. Cambridge Philos. Soc.* **65**, 11–18. [10.3]

Welsh, D. J. A. (1969c). Euler and bipartite matroids. *J. Combin. Theory* **6**, 313–316. [9.4]

Welsh, D. J. A. (1971a). Generalized versions of Hall's theorem. *J. Combin. Theory Ser. B* **10**, 95–101. [11.2]

Welsh, D. J. A. (1971b). Combinatorial problems in matroid theory. In *Combinatorial mathematics and its applications* (ed. D. J. A. Welsh), pp. 291–306. Academic Press, London. [11.3, 15.2, 15.5, 15.6, 15.9]

Welsh, D. J. A. (1976). *Matroid theory.* Academic Press, London. Reprinted 2010, Dover, Mineola. [Preface, 1.3, 2.1, 5.2, 6.7, 9.4, 11.0, 11.1, 11.2, 11.3, 15.0, 15.2, 15.4]

Welsh, D. J. A. (1980). Colourings, flows and projective geometry. *Nieuw Arch. Wisk. (3)* **28**, 159–176. [15.3]

Welsh, D. J. A. (1982). Matroids and combinatorial optimisation. In *Matroid theory and its applications* (ed. A. Barlotti), pp. 323–416. Liguori editore, Naples. [12.1, 13.4]

White, N. (1971). *The bracket ring and combinatorial geometry.* Ph.D. thesis, Harvard University. [6.9, 9.1]

White, N. (1980a). The transcendence degree of a coordinatization of a combinatorial geometry. *J. Combin. Theory Ser. B* **29**, 168–175. [6.5]

White, N. (1980b). A unique exchange property for bases. *Linear Algebra Appl.* **31**, 81–91. [14.8]

White, N., ed. (1986). *Theory of matroids.* Cambridge University Press, Cambridge. [2.2, 7.3, 15.0]

White, N., ed. (1987a). *Combinatorial geometries.* Cambridge University Press, Cambridge. [1.2, 15.0]

White, N. (1987b). Coordinatizations. In *Combinatorial geometries* (ed. N. White), pp. 1–27. Cambridge University Press, Cambridge. [6.5, 6.6, 6.8]

White, N. (1987c). Unimodular matroids. In *Combinatorial geometries* (ed. N. White), pp. 40–52. Cambridge University Press, Cambridge. [10.4]

White, N., ed. (1992). *Matroid applications.* Cambridge University Press, Cambridge. [15.0]

Whitney, H. (1932a). Non-separable and planar graphs. *Trans. Amer. Math. Soc.* **34**, 339–362. [5.2]

Whitney, H. (1932b). Congruent graphs and the connectivity of graphs. *Amer. J. Math.* **54**, 150–168. [5.3, 8.6]

Whitney, H. (1933). 2-isomorphic graphs. *Amer. J. Math.* **55**, 245–254. [5.3, 13.3]

Whitney, H. (1935). On the abstract properties of linear dependence. *Amer. J. Math.* **57**, 509–533. [1.0, 1.1, 1.8, 2.0, 4.0, 6.0, 6.7, 9.1, 14.0]

Whittle, G. (1987). Modularity in tangential k-blocks. *J. Combin. Theory Ser. B* **42**, 24–35. [15.3]

Whittle, G. (1988). Quotients of tangential k-blocks. *Proc. Amer. Math. Soc.* **102**, 1088–1098. [15.3]

Whittle, G. (1989a). q-lifts of tangential k-blocks. *J. London Math. Soc. (2)* **39**, 9–15. [15.3]

Whittle, G. (1989b). Dowling group geometries and the critical problem. *J. Combin. Theory Ser. B* **47**, 80–92. [6.10, 15.3]

Whittle, G. (1995). A characterisation of the matroids representable over $GF(3)$ and the rationals. *J. Combin. Theory Ser. B* **65**, 222–261. [14.7]

Whittle, G. (1996). Inequivalent representations of ternary matroids. *Discrete Math.* **149**, 233–238. [14.8]

Whittle, G. (1997). On matroids representable over $GF(3)$ and other fields. *Trans. Amer. Math. Soc.* **349**, 579–603. [14.7]

Whittle, G. (1999). Stabilizers of classes of representable matroids. *J. Combin. Theory Ser. B* **77**, 39–72. [12.3, 14.6, 14.8]

Whittle, G. (2005). Recent work in matroid representation theory. *Discrete Math.* **302**, 285–296. [14.5, 14.7, 15.1]

Wiedemann, D. (2006) Cyclic base orders of matroids. Retrieved February 11, 2010 from http://www.plumbyte.com/cyclic_base_orders_1984.pdf [15.9]

Wild, M. (2000). The asymptotic number of inequivalent binary codes and nonisomorphic binary matroids. *Finite Fields Appl.* **6**, 192–202. [15.5]

Wild, M. (2005). The asymptotic number of binary codes and binary matroids. *SIAM J. Discrete Math.* **19**, 691–699. [15.5]

Wilson, R. J. (1973). An introduction to matroid theory. *Amer. Math. Monthly* **80**, 500–525. [Preface]

Witt, E. (1940). Über Steinersche Systeme. *Abh. Math. Sem. Univ. Hamburg* **12**, 265–275. [12.2]

Wong, P.-K. (1978). On certain n-connected matroids. *J. Reine Angew. Math.* **299/300**, 1–6. [8.7, 15.4]

Woodall, D. R. (1974). An exchange theorem for bases of matroids. *J. Combin. Theory Ser. B* **16**, 227–229. [1.4]

Wu, H. (1998a). On contractible and vertically contractible elements in 3-connected matroids and graphs. *Discrete Math.* **179**, 185–203. [10.3]

Wu, H. (1998b). On vertex-triads in 3-connected binary matroids. *Combin. Probab. Comput.* **7**, 485–497. [8.8, 9.4]

Wu, P.-L. (1997). An upper bound on the number of edges of a 2-connected graph. *Combin. Probab. Comput.* **6**, 107–113. [14.9]

Wu, P.-L. (2000). Extremal graphs with prescribed circumference and cocircumference. *Discrete Math.* **223**, 299–308. [4.3]

Wu, P.-L. (2001). On large circuits in matroids. *Graphs Combin.* **17**, 365–388. [14.9]

Wu, Z. (2003). On the number of spikes over finite fields. *Discrete Math.* **265**, 261–296. [6.5, Appendix]

Yu, Y. (2004). Forbidden minors for wye-delta-wye reducibility. *J. Graph Theory* **47**, 317–321. [11.5]

Yu, Y. (2006). More forbidden minors for Wye-Delta-Wye reducibility. *Electron. J. Combin.* **13**, Research Paper 7, 15 pp. (electronic). [11.5]

Zaslavsky, T. (1982a). Signed graphs. *Discrete Appl. Math.* **4** 47–74; Erratum **5** (1983), 248. [6.10]

Zaslavsky, T. (1982b). Voltage-graphic geometries. In *Matroid theory and its applications* (ed. A. Barlotti), pp. 417–423. Liguori editore, Naples. [6.10]

Zaslavsky, T. (1987a). The Möbius function and the characteristic polynomial. In *Combinatorial geometries* (ed. N. White), pp. 114–138. Cambridge University Press, Cambridge. [15.2]

Zaslavsky, T. (1987b). The biased graphs whose matroids are binary. *J. Combin. Theory Ser. B* **42**, 337–347. [6.10]

Zaslavsky, T. (1989). Biased graphs. I. Bias, balance, and gains. *J. Combin. Theory Ser. B* **47**, 32–52. [6.10]

Zaslavsky, T. (1990). Biased graphs whose matroids are special binary matroids. *Graphs Combin.* **6**, 77–93. [6.10]

Zaslavsky, T. (1991). Biased graphs. II. The three matroids. *J. Combin. Theory Ser. B* **51**, 46–72. [6.10]

Zaslavsky, T. (1994). Frame matroids and biased graphs. *European J. Combin.* **15**, 303–307. [6.10]

Zaslavsky, T. (1998a). A mathematical bibliography of signed and gain graphs and allied areas. Manuscript prepared with Marge Pratt. *Electron. J. Combin.* **5** Dynamic Surveys 8, 124 pp. (electronic). [6.10]

Zaslavsky, T. (1998b). Glossary of signed and gain graphs and allied areas. *Electron. J. Combin.* **5** Dynamic Surveys 9, 41 pp. (electronic). [6.10]

Zhao, C. K. (1985). A conjecture on matroids (Chinese). *Neimenggu Daxue Xuebao* **16**, 321–326. [15.2]

APPENDIX

SOME INTERESTING MATROIDS

This appendix is intended to serve as a quick reference for some of the matroids that appear throughout this book. It contains numerous redundancies to facilitate its use. All the matroids listed here are 3-connected except where otherwise noted. Most of the matroids are specified via geometric representations. The list begins with one special 4-element matroid and follows this with certain 5-, 6-, $\cdots$, 13-element matroids. It concludes with several infinite families of matroids.

Recall that a matroid M is regular if and only if M is representable over every field; and M is near-regular if and only if M is representable over every field with at least three elements. Evidently a matroid and its dual have the same automorphism group. Moreover, a matroid with an odd number of elements cannot be self-dual. The automorphism group G of a matroid having ground set E is *doubly transitive* if G is transitive on ordered pairs of distinct elements of E. Similarly, G is *3-transitive* or *5-transitive* if it is transitive on ordered triples or on ordered 5-tuples of distinct elements of E.

$U_{2,4}$

- The 4-point line; isomorphic to $\mathcal{W}^2$, the rank-2 whirl.
- The unique excluded minor for the class of binary matroids (6.5.4, 9.1.3).
- $\mathbb{F}$-representable if and only if $|\mathbb{F}| \geq 3$ (6.5.2); near-regular.
- An excluded minor for the classes of graphic (6.6.7, 10.3.1), cographic (6.6.7), and regular matroids (6.6.6, 10.1.1).
- Transversal, a strict gammoid, a gammoid.
- Identically self-dual.
- Automorphism group is the symmetric group.
- The unique 3-connected non-binary matroid M having an element e such that both $M \backslash e$ and M/e are binary (Oxley 1987a).
- The unique matroid M with at least four elements such that $\{M\}$ is 2-rounded (Oxley 1984b).
- Algebraic over all fields.

$U_{2,5}$, $U_{3,5}$

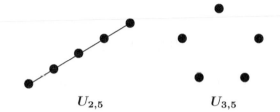

$U_{2,5}$ $U_{3,5}$

- $U_{2,5}$ is the 5-point line; $U_{3,5}$ is five points freely placed in the plane.
- Each is $\mathbb{F}$-representable if and only if $|\mathbb{F}| \geq 4$ (6.5.1, 6.5.2).
- The excluded minors for the class of ternary matroids are $U_{2,5}$, $U_{3,5}$, F_7, and F_7^* (6.5.7, 10.2.1).
- Each is not graphic, not cographic, not regular, and not near-regular.
- Both are transversal, strict gammoids, and gammoids.
- A pair of dual matroids.
- Each has the symmetric group as its automorphism group.
- $U_{3,5}$ is obtained from $U_{2,5}$ by a Δ-Y exchange; $U_{2,5}$ is obtained from $U_{3,5}$ by a Y-Δ exchange
- Each is algebraic over all fields.

$M(K_4)$

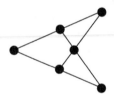

- Isomorphic to $M(\mathcal{W}_3)$, the rank-3 wheel, and to the tipless binary 3-spike.
- Regular.
- Graphic, cographic.
- A minor-minimal non-transversal matroid (Section 11.2).
- A minor-minimal matroid that is not a strict gammoid.
- A minor-minimal non-gammoid.
- Self-dual but not identically self-dual.
- Transitive automorphism group.
- A minor of every 3-connected binary matroid with at least four elements (12.2.13).
- Not base-orderable, not strongly base-orderable.
- A connected non-empty binary matroid is a series–parallel network if and only if it has no $M(K_4)$-minor (12.2.14).
- For all elements e, neither $M(K_4)\backslash e$ nor $M(K_4)/e$ is 3-connected (8.8.4, 12.2.2).
- Algebraic over all fields.

$\mathcal{W}^3$

- The rank-3 whirl.
- $\mathbb{F}$-representable if and only if $|\mathbb{F}| \geq 3$; near-regular.
- Not graphic, not cographic, not regular.
- Transversal, a strict gammoid, a gammoid.
- Self-dual but not identically self-dual.
- Non-transitive automorphism group.
- For all elements e, neither $\mathcal{W}^3 \backslash e$ nor $\mathcal{W}^3 / e$ is 3-connected (8.8.4, 12.2.2).
- The unique relaxation of $M(K_4)$.
- Algebraic over all fields.

Q_6

- $\mathbb{F}$-representable if and only if $|\mathbb{F}| \geq 4$.
- Not graphic, not cographic, not regular, not near-regular.
- Transversal, a strict gammoid, a gammoid.
- Self-dual but not identically self-dual.
- Non-transitive automorphism group.
- The unique relaxation of $\mathcal{W}^3$.
- If N is a matroid having at least four elements, then $\{N\}$ is 1-rounded if and only if $N \cong U_{2,4}$, $M(\mathcal{W}_2)$, or Q_6 (Oxley 1984b).
- Algebraic over all fields.

P_6

- The excluded minors for the class of quaternary matroids are $U_{2,6}$, $U_{4,6}$, P_6, F_7^-, $(F_7^-)^*$, P_8, and $P_8^=$ (6.5.9).
- $\mathbb{F}$-representable if and only if $|\mathbb{F}| \geq 5$ (6.5.8).
- Not graphic, not cographic, not regular, not near-regular.
- Transversal, a strict gammoid, a gammoid.
- Self-dual but not identically self-dual.

- Non-transitive automorphism group.
- The unique relaxation of Q_6.
- Obtained from $U_{2,6}$ by a Δ-Y exchange, and from $U_{4,6}$ by a Y-Δ exchange.
- Algebraic over all fields.

$U_{3,6}$

- Six points freely placed in the plane; the tipless free 3-spike.
- $\mathbb{F}$-representable if and only if $|\mathbb{F}| \geq 4$ (Section 6.5).
- Not graphic, not cographic, not regular, not near-regular.
- Transversal, a strict gammoid, a gammoid.
- Identically self-dual.
- Automorphism group is the symmetric group.
- The unique relaxation of P_6.
- Every 3-connected matroid of rank and corank exceeding two has a minor isomorphic to one of $M(K_4)$, $\mathcal{W}^3$, Q_6, P_6, and $U_{3,6}$ (12.2.19).
- Every 3-connected non-binary matroid of rank and corank exceeding two has a minor isomorphic to one of $\mathcal{W}^3$, P_6, Q_6, and $U_{3,6}$ (12.2.18).
- A 3-connected $GF(q)$-representable matroid with no $U_{3,6}$-minor has at most $(q-2)!$ inequivalent $GF(q)$-representations (14.6.9).
- Algebraic over all fields.

R_6

- Isomorphic to $U_{2,4} \oplus_2 U_{2,4}$.
- Connected but not 3-connected.
- $\mathbb{F}$-representable if and only if $|\mathbb{F}| \geq 3$; near-regular.
- Not graphic, not cographic, not regular.
- Transversal, a strict gammoid, a gammoid.
- Identically self-dual.
- Transitive automorphism group.
- Has P_6 as a relaxation.
- A connected $GF(4)$-representable matroid is uniquely $GF(4)$-representable if and only if it has no 2-separation that is induced by the unique 2-separation of some R_6-minor (Kahn 1988).
- Algebraic over all fields.

F_7, F_7^*

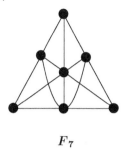

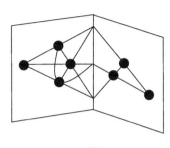

F_7 F_7^*

- The Fano matroid and its dual.
- F_7 is the smallest projective plane, $PG(2,2)$, (Section 6.1) and is isomorphic to the unique $S(2,3,7)$ (Section 12.2) and to the unique binary 3-spike (12.2.20).
- Each is $\mathbb{F}$-representable if and only if $\mathbb{F}$ has characteristic two (6.4.8).
- Each is an excluded minor for $\mathbb{F}$-representability if and only if $\mathbb{F}$ has characteristic other than two (6.5.5).
- Both are excluded minors for the classes of graphic matroids (6.6.7, 10.3.1), cographic matroids (6.6.7), regular matroids (6.6.6, 10.1.1), near-regular matroids (14.7.10), dyadic matroids (14.7.11), and $\sqrt[6]{1}$-matroids (14.7.9).
- Each is not transversal, not a strict gammoid, not a gammoid.
- Their common automorphism group is doubly transitive.
- F_7^* is obtained from F_7 by a Δ-Y exchange, and F_7 is obtained from F_7^* by a Y-Δ exchange.
- Every single-element deletion of F_7 is isomorphic to $M(K_4)$ and every single-element deletion of F_7^* is isomorphic to $M(K_{2,3})$.
- F_7 is a splitter for the class of binary matroids with no F_7^*-minor (12.2.3).
- F_7 and F_7^* are the graft matroids of, respectively, K_4 with every vertex coloured, and $K_{2,3}$ with every vertex coloured except one of degree three.
- A matroid M in $\{F_7, F_7^*\}$ is algebraic over $\mathbb{F}$ if and only if $\mathbb{F}$ has characteristic two (Lindström 1985b,c).

F_7^-, $(F_7^-)^*$

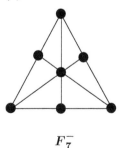

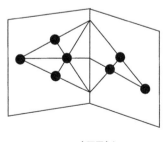

F_7^- $(F_7^-)^*$

- The non-Fano matroid (1.5.13) and its dual; F_7^- is a ternary 3-spike.

- Each is $\mathbb{F}$-representable if and only if $\mathbb{F}$ has characteristic other than two (6.4.8).
- Each is an excluded minor for $\mathbb{F}$-representability if and only if $\mathbb{F}$ has characteristic two and $\mathbb{F} \neq GF(2)$ (6.5.6).
- Each is not graphic, not cographic, and not regular.
- Each is dyadic but is an excluded minor for the classes of near-regular (14.7.10) and $\sqrt[6]{1}$-matroids (14.7.9).
- Each is not transversal, not a strict gammoid, not a gammoid.
- Their common automorphism group is non-transitive, having two orbits.
- The complement of F_7^- in $PG(2,3)$ is $M(K_4)$.
- $(F_7^-)^*$ is obtained from F_7^- by a Δ-Y exchange, and F_7^- is obtained from $(F_7^-)^-$ by a Y-Δ exchange.
- Every single-element deletion of F_7^- and every single-element contraction of $(F_7^-)^*$ is isomorphic to $M(K_4)$ or $\mathcal{W}^3$.
- F_7^- is the unique relaxation of F_7, and $(F_7^-)^*$ is the unique relaxation of F_7^*.
- Each is algebraic over all fields.

O_7

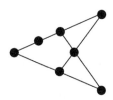

- $\mathbb{F}$-representable if and only if $|\mathbb{F}| \geq 3$.
- Not graphic, not cographic, not regular.
- Near-regular, dyadic, and sixth-root-of-unity.
- Not transversal, not a strict gammoid, not a gammoid.
- The complement in $PG(2,3)$ of $\mathcal{W}^3$.
- Not base-orderable, not strongly base-orderable.
- Algebraic over all fields.

P_7

- The only other ternary 3-spike apart from F_7^-.
- $\mathbb{F}$-representable if and only if $|\mathbb{F}| \geq 3$.
- Not graphic, not cographic, not regular.
- Near-regular, dyadic, and sixth-root-of-unity.

- Not transversal, not a strict gammoid, not a gammoid.
- Has no $M(K_4)$-minor.
- The complement in $PG(2,3)$ of $P(U_{2,4}, U_{2,3})$.
- Base-orderable, strongly base-orderable.
- The excluded minors for the class of ternary gammoids are $U_{2,5}$, $U_{3,5}$, $M(K_4)$, P_7, and P_7^* (Oxley 1987c).
- Algebraic over all fields.

$AG(3,2)$

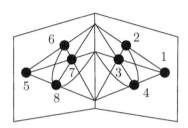 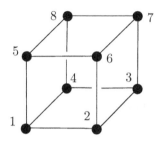

In the figure on the right, the 4-point planes are the six faces of the cube, the six diagonal planes such as $\{1, 2, 7, 8\}$, and two twisted planes: $\{1, 8, 3, 6\}$ and $\{2, 7, 4, 5\}$.

- The binary affine cube; isomorphic to the unique $S(3,4,8)$.
- Isomorphic to the unique tipless binary 4-spike, and to $F_7 \oplus_3 F_7$.
- $\mathbb{F}$-representable if and only if $\mathbb{F}$ has characteristic two.
- Represented over $GF(2)$ by $[I_4 | J_4 - I_4]$.
- Not graphic, not cographic, not regular.
- Not transversal, not a strict gammoid, not a gammoid.
- Identically self-dual.
- Automorphism group is 3-transitive on points and transitive on hyperplanes.
- The sets of circuits and hyperplanes coincide; each circuit has exactly four elements.
- Every single-element deletion is isomorphic to F_7^* and every single-element contraction is isomorphic to F_7.
- This matroid and S_8 are the only 8-element, 3-connected, binary non-regular matroids (12.2.4).
- The unique splitter for the class of binary paving matroids (Acketa 1988).
- The complement of F_7 in $PG(3,2)$.
- Algebraic over $\mathbb{F}$ if and only if $\mathbb{F}$ has characteristic two.

$AG(3,2)'$

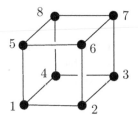

4-point planes are the six faces of the cube, the six diagonal planes such as $\{1,2,7,8\}$, and one twisted plane, $\{1,8,3,6\}$.

- The unique relaxation of $AG(3,2)$; a tipless 4-spike.
- A smallest non-representable matroid.
- Not a (minimal) excluded minor for $\mathbb{F}$-representability for any field $\mathbb{F}$.
- Not graphic, not cographic, not regular, not near-regular.
- Not transversal, not a strict gammoid, not a gammoid.
- Self-dual but not identically self-dual.
- Non-transitive automorphism group.
- Every single-element deletion is isomorphic to F_7^* or $(F_7^-)^*$ and every single-element contraction is isomorphic to F_7 or F_7^-.
- Non-algebraic.

R_8

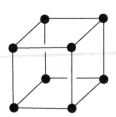

4-point planes are the six faces of the cube and the six diagonal planes.

- The real affine cube (1.5.8); it is represented over all fields of characteristic other than two by $[I_4|J_4 - 2I_4]$; a tipless ternary 4-spike.
- $\mathbb{F}$-representable if and only if the characteristic of $\mathbb{F}$ is not two.
- Not graphic, not cographic, not regular.
- Near-regular, dyadic, and sixth-root-of-unity.
- Not transversal, not a strict gammoid, not a gammoid.
- Identically self-dual.
- Transitive automorphism group.
- Every single-element deletion is isomorphic to $(F_7^-)^*$ and every single-element contraction is isomorphic to F_7^-.
- A relaxation of $AG(3,2)'$.
- The unique matroid that can be obtained from $AG(3,2)$ by relaxing two disjoint circuit–hyperplanes.
- Algebraic over all fields.

F_8

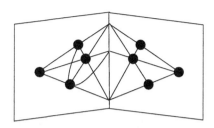

- $P_N(F_7, F_7^-)\setminus E(N)$ where $N \cong U_{2,3}$; a tipless 4-spike.
- A smallest non-representable matroid.
- Not a (minimal) excluded minor for $\mathbb{F}$-representability for any field $\mathbb{F}$.
- Not graphic, not cographic, not regular, not near-regular.
- Not transversal, not a strict gammoid, not a gammoid.
- Self-dual but not identically self-dual.
- Non-transitive automorphism group.
- The only two relaxations of $AG(3,2)'$ are F_8 and R_8.
- Has exactly three non-isomorphic single-element deletions including F_7^* and $(F_7^-)^*$, and has exactly three non-isomorphic single-element contractions including F_7 and F_7^-.
- Non-algebraic.

Q_8

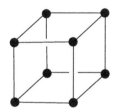

4-point planes are the six faces of the cube and exactly five of the six diagonal planes.

- The unique relaxation of R_8 (6.5.10); a tipless 4-spike.
- A smallest non-representable matroid.
- An excluded minor for $\mathbb{F}$-representability if and only if $|\mathbb{F}| \geq 5$ and $\mathbb{F}$ does not have characteristic two.
- Not graphic, not cographic, not regular, not near-regular.
- Not transversal, not a strict gammoid, not a gammoid.
- Self-dual but not identically self-dual.
- Non-transitive automorphism group.
- Non-algebraic.

L_8

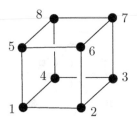

4-point planes are the six faces of the cube plus the two twisted planes, $\{1, 8, 3, 6\}$ and $\{2, 7, 4, 5\}$.

- $\mathbb{F}$-representable if and only if $|\mathbb{F}| \geq 5$.
- Not a (minimal) excluded minor for $\mathbb{F}$-representability for any $\mathbb{F}$.
- Not graphic, not cographic, not regular, not near-regular.
- Not transversal, not a strict gammoid, not a gammoid.
- Identically self-dual.
- Transitive automorphism group.
- Not a tipless spike; every single-element contraction is isomorphic to the free extension of $M(K_4)$.
- A splitter for $EX(\mathcal{W}^3, P_6)$ (Oxley 1990a).
- Algebraic over all fields.

S_8

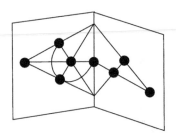

$$\begin{bmatrix} & & & & 0 & 1 & 1 & 1 \\ & & I_4 & & 1 & 0 & 1 & 1 \\ & & & & 1 & 1 & 0 & 1 \\ & & & & 1 & 1 & 1 & 1 \end{bmatrix}$$

This matrix represents S_8 over $GF(2)$.

- The unique deletion of a non-tip element from the binary 4-spike.
- $\mathbb{F}$-representable if and only if $\mathbb{F}$ has characteristic two.
- Not graphic, not cographic, not regular, not near-regular.
- Not transversal, not a strict gammoid, not a gammoid.
- Self-dual but not identically self-dual.
- Non-transitive automorphism group.
- Has a unique element x such that $S_8 \backslash x \cong F_7^*$ and a unique element y such that $S_8 / y \cong F_7$. Moreover, $y \neq x$.
- F_7 has two 3-connected binary single-element coextensions: S_8 and $AG(3, 2)$.
- $\{U_{2,4}, F_7, F_7^*, S_8\}$ is 2-rounded (Section 12.3).
- The complement of $M(K_4) \oplus U_{1,1}$ in $PG(3, 2)$.
- Algebraic over $\mathbb{F}$ if and only if $\mathbb{F}$ has characteristic two.

V_8

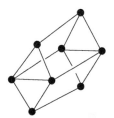

- The Vámos matroid (or Vámos cube) (2.1.25).
- A smallest non-representable matroid (2.2.26, 6.4.10).
- An excluded minor for $\mathbb{F}$-representability if and only if $|\mathbb{F}| \geq 5$.
- Not graphic, not cographic, not regular, not near-regular.
- Not transversal, not a strict gammoid, not a gammoid.
- Self-dual but not identically self-dual.
- Non-transitive automorphism group.
- Not a tipless spike; every single-element contraction is isomorphic to a free 3-spike or to a relaxation of that spike.
- Non-algebraic (Section 6.7).

T_8

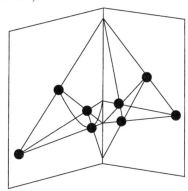

- A ternary tipless 4-spike; the only other such 4-spike is R_8 (Wu 2003).
- Represented over $GF(3)$ by $[I_4|J_4 - I_4]$ (Section 2.2, Exercise 9).
- $\mathbb{F}$-representable if and only if $\mathbb{F}$ has characteristic three.
- An excluded minor for $\mathbb{F}$-representability for all $\mathbb{F}$ whose characteristic is not two or three.
- Not graphic, not cographic, not regular, not near-regular.
- Not transversal, not a strict gammoid, not a gammoid.
- Self-dual but not identically self-dual.
- Non-transitive automorphism group.
- Every single-element contraction is isomorphic to P_7 or F_7^-.
- T_8, R_8, $PG(2,3)$, and $S(5,6,12)$ are the only splitters for the class of ternary paving matroids (Oxley 1991).
- Algebraic over $\mathbb{F}$ if and only if $\mathbb{F}$ has characteristic three (Lindström 1985b,c).

J

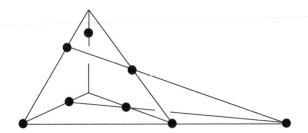

- $\mathbb{F}$-representable if and only if $|\mathbb{F}| \geq 3$.
- Not graphic, not cographic, not regular.
- Near-regular, dyadic, and sixth-root-of-unity.
- Not transversal, not a strict gammoid, not a gammoid.
- Self-dual but not identically self-dual.
- Non-transitive automorphism group.
- J and the relaxation of S_8 are the only two matroids that can be obtained by deleting a non-tip element from a ternary 4-spike.
- J and $S(5,6,12)$ are the only 3-connected splitters for the class of ternary matroids with no $M(K_4)$-minor (12.2.22).
- Not base-orderable, not strongly base-orderable.
- The excluded minors for the class of ternary base-orderable matroids are $U_{2,5}$, $U_{3,5}$, $M(K_4)$, and J (Oxley 1987c).
- The excluded minors for the class of ternary strongly base-orderable matroids are $U_{2,5}$, $U_{3,5}$, $M(K_4)$, J, and P_8 (Oxley 1987c).
- Algebraic over all fields.

P_8

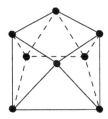

$$\left[\begin{array}{c|cccc} & 0 & 1 & 1 & -1 \\ I_4 & 1 & 0 & 1 & 1 \\ & 1 & 1 & 0 & 1 \\ & -1 & 1 & 1 & 0 \end{array} \right]$$

This matrix represents
P_8 over $GF(3)$ (2.2.25).

- The geometric representation of P_8 over $\mathbb{R}$ is obtained from a 3-dimensional cube having its points at the vertices by rotating a face in its plane through $45°$ about the centre of the face (see also Figure 6.15).
- $\mathbb{F}$-representable if and only if the characteristic of $\mathbb{F}$ is not two (6.4.15).
- An excluded minor for $\mathbb{F}$-representability if and only if $\mathbb{F}$ has characteristic two and $\mathbb{F} \neq GF(2)$ (6.5.6).
- Dyadic; an excluded minor for the classes of near-regular and $\sqrt[6]{1}$-matroids (14.7.9, 14.7.10).

- Not graphic, not cographic, not regular.
- Not transversal, not a strict gammoid, not a gammoid.
- Self-dual but not identically self-dual.
- Automorphism group is transitive on the points but not on the hyperplanes.
- Not a tipless 4-spike; every single-element contraction is isomorphic to P_7 and every single-element deletion is isomorphic to P_7^*.
- A matroid is a $\sqrt[6]{1}$-matroid if and only if it has no minor isomorphic to $U_{2,5}, U_{3,5}, F_7, F_7^*, F_7^-, (F_7^-)^*$, or P_8 (14.7.9).
- Base-orderable but not strongly base-orderable.
- Algebraic over all fields.

$P_8^=$

$$
\left[\begin{array}{c|cccc}
 & 1 & 1 & 1 & 1 \\
I_4 & 1 & 1 & c^{-1} & b \\
 & 1 & b & 0 & b \\
 & 1 & c & 1 & 0
\end{array} \right]
$$

**This matrix represents $P_8^=$ over $\mathbb{F}$ where b and c
are elements of $\mathbb{F} - \{0, 1\}$ such that $b \notin \{c, c^{-1}\}$.**

- Obtained from P_8 by relaxing its only pair of disjoint circuit–hyperplanes.
- $\mathbb{F}$-representable if and only if $|\mathbb{F}| \geq 5$ (6.4.15).
- An excluded minor for $GF(4)$-representability (6.4.12).
- Not graphic, not cographic, not regular, not near-regular.
- Not transversal, not a strict gammoid.
- Self-dual but not identically self-dual.
- Automorphism group is transitive on the points and on the hyperplanes.
- Not a tipless 4-spike; every single-element contraction is isomorphic to the matroid obtained from P_7 by relaxing one of the 3-point lines that meets the other four 3-point lines.
- Algebraic over all fields.

$M(\mathcal{W}_4)$

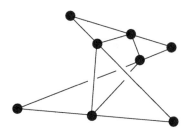

- The rank-4 wheel.
- Regular.

- Graphic, cographic.
- Not transversal, not a strict gammoid, not a gammoid.
- Self-dual but not identically self-dual.
- Non-transitive automorphism group.
- Every 3-connected regular matroid with at least seven elements has an $M(\mathcal{W}_4)$-minor (12.2.21).
- A 3-connected binary matroid M has no $M(\mathcal{W}_4)$-minor if and only if M is isomorphic to a minor of the binary r-spike Z_r for some $r \geq 3$ (12.2.21).
- For M binary having at least four elements, $\{U_{2,4}, M\}$ is 2-rounded if and only if $M \cong M(\mathcal{W}_3)$ or $M(\mathcal{W}_4)$ (Oxley and Reid 1990).
- Algebraic over all fields.

$\mathcal{W}^4$

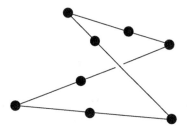

- The rank-4 whirl.
- $\mathbb{F}$-representable if and only if $|\mathbb{F}| \geq 3$; near-regular.
- Not graphic, not cographic, not regular.
- Transversal, a strict gammoid, a gammoid.
- Self-dual but not identically self-dual.
- Non-transitive automorphism group.
- The unique relaxation of $M(\mathcal{W}_4)$.
- Algebraic over all fields.

$M^*(K_{3,3})$, $M(K_{3,3})$

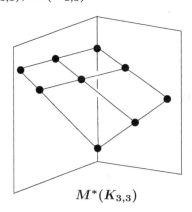

$M^*(K_{3,3})$

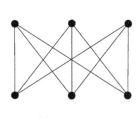

The graph $K_{3,3}$.

- Both are regular.

- $M^*(K_{3,3})$ is an excluded minor for the class of graphic matroids (6.6.7, 10.3.1) and $M(K_{3,3})$ is an excluded minor for the class of cographic matroids (6.6.7).
- Each is not transversal, not a strict gammoid, and not a gammoid.
- Their common automorphism group is transitive.
- $M^*(K_{3,3})$ has no 3-connected regular single-element extensions.
- The excluded minors for the class of cycle matroids of planar graphs are $U_{2,4}$, F_7, F_7^*, $M(K_{3,3})$, $M^*(K_{3,3})$, $M(K_5)$, and $M^*(K_5)$ (6.6.8).
- $M^*(K_{3,3})$ is the complement of $U_{2,3} \oplus U_{2,3}$ in $PG(3,2)$.
- $M^*(K_{3,3})$ is the graft matroid of $\mathcal{W}_4$ with every vertex being coloured except the one of degree four.
- Each is algebraic over all fields.

$AG(2,3)$

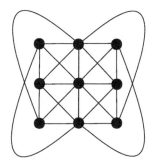

- The ternary affine plane; isomorphic to the unique $S(2,3,9)$.
- $\mathbb{F}$-representable if and only if $\mathbb{F}$ contains a root of the equation $x^2 - x + 1 = 0$. In particular, $\mathbb{C}$-representable but not $\mathbb{R}$-representable.
- $GF(q)$-representable if and only if $q \not\equiv 2 \pmod 3$.
- Not graphic, not cographic, not regular.
- Sixth-root-of-unity, not dyadic, not near-regular.
- Not transversal, not a strict gammoid, not a gammoid.
- Doubly transitive automorphism group.
- Base-orderable and strongly base-orderable.
- The unique maximal simple 3-connected rank-3 ternary matroid having no $M(K_4)$-minor.
- The complement of $U_{2,4}$ in $PG(2,3)$.
- Every single-element deletion is isomorphic to the same matroid, $AG(2,3) \backslash e$.
- $AG(2,3) \backslash e$ is transitive on points and on non-trivial lines, so there is a unique matroid $\Delta_3(AG(2,3) \backslash e)$ that is obtained by performing a Δ-Y exchange on $AG(2,3) \backslash e$.
- A matroid is near-regular if and only if it has no minor isomorphic to $U_{2,5}, U_{3,5}$, $F_7, F_7^*, F_7^-, (F_7^-)^*, AG(2,3) \backslash e, (AG(2,3) \backslash e)^*, \Delta_3(AG(2,3) \backslash e)$, or P_8 (14.7.10).
- Algebraic over all fields (Gordon 1988).

$Q_3(GF(3)^\times)$

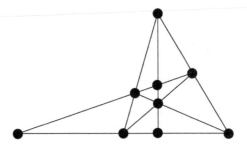

- The rank-3 ternary Dowling geometry.
- $\mathbb{F}$-representable if and only if the characteristic of $\mathbb{F}$ is not two.
- Not graphic, not cographic, not regular.
- Dyadic, not sixth-root-of-unity, not near-regular.
- Not transversal, not a strict gammoid, not a gammoid.
- Non-transitive automorphism group.
- The complement of $U_{3,4}$ in $PG(2,3)$.
- This matroid and $AG(2,3)$ are the only ternary rank-3 simple matroids with more than eight elements that are representable over some field of characteristic other than three (Kung and Oxley 1988).
- Algebraic over all fields.

R_9

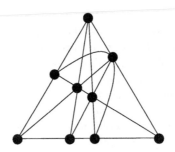

- The ternary Reid geometry.
- $\mathbb{F}$-representable if and only if $\mathbb{F}$ has characteristic three.
- Not graphic, not cographic, not regular.
- Not dyadic, not sixth-root-of-unity, not near-regular.
- Not transversal, not a strict gammoid, not a gammoid.
- Non-transitive automorphism group.
- The complement of $U_{2,3} \oplus U_{1,1}$ in $PG(2,3)$.
- The only 9-element rank-3 simple ternary matroids are $R_9, Q_3(GF(3)^\times)$, and $AG(2,3)$.
- Algebraic over $\mathbb{F}$ if and only if $\mathbb{F}$ has characteristic three (Gordon 1988).

Pappus

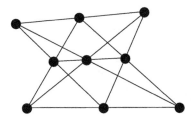

- The Pappus matroid (1.5.15).
- $\mathbb{F}$-representable if and only if $|\mathbb{F}| = 4$ or $|\mathbb{F}| \geq 7$.
- An excluded minor for $GF(5)$-representability.
- Not graphic, not cographic, not regular, not near-regular.
- Not transversal, not a strict gammoid, not a gammoid.
- Transitive automorphism group.
- Can be obtained from $AG(2,3)$ by relaxing three disjoint 3-point lines.
- Algebraic over all fields.

non-Pappus

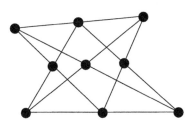

- The non-Pappus matroid (1.5.15).
- Non-representable (6.1.10).
- An excluded minor for $\mathbb{F}$-representability if and only if $|\mathbb{F}| \geq 5$.
- Not graphic, not cographic, not regular, not near-regular.
- Not transversal, not a strict gammoid, not a gammoid.
- Non-transitive automorphism group.
- The unique relaxation of the Pappus matroid.
- Algebraic over a field $\mathbb{F}$ if and only if $\mathbb{F}$ has non-zero characteristic (Ingleton 1971a, Lindström 1986a).

$M(K_5)$, $M^*(K_5)$

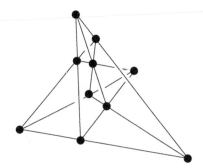

- $M(K_5)$ is the 3-dimensional Desargues configuration shown above.
- Both are regular.
- $M^*(K_5)$ is an excluded minor for the class of graphic matroids (6.6.7, 10.1.1), and $M(K_5)$ is an excluded minor for the class of cographic matroids (6.6.7).
- Each is not transversal, not a strict gammoid, and not a gammoid.
- Their common automorphism group is transitive.
- $M(K_5)$ is the unique rank-4 simple regular matroid with the largest number of elements among such matroids (14.10.3).
- $M(K_5)$ is a tangential 2-block over $GF(2)$ (Section 15.3).
- $M(K_5)$ is a splitter for the class of regular matroids with no minor isomorphic to $M(K_{3,3})$ (Exercise 17 of Section 12.2).
- $M(K_5)$ is the complement in $PG(3,2)$ of $U_{4,5}$, a 5-element circuit.
- Each is algebraic over all fields.

R_{10}

$$
\begin{bmatrix}
 & & -1 & 1 & 0 & 0 & 1 \\
 & & 1 & -1 & 1 & 0 & 0 \\
I_5 & & 0 & 1 & -1 & 1 & 0 \\
 & & 0 & 0 & 1 & -1 & 1 \\
 & & 1 & 0 & 0 & 1 & -1
\end{bmatrix}
$$

This matrix represents R_{10} over all fields.
(In characteristic two, $-1 = 1$.)

- Among regular matroids that are neither graphic nor cographic, the only one with ten elements and the only simple one of rank at most five (Bixby 1977).
- The unique splitter for the class of regular matroids (Section 13.1).
- Not transversal, not a strict gammoid, not a gammoid.
- Self-dual but not identically self-dual.
- Doubly transitive automorphism group (Seymour 1980b).
- Every single-element deletion is isomorphic to $M(K_{3,3})$, and every single-element contraction is isomorphic to $M^*(K_{3,3})$ (6.6.9).
- Every circuit has four or six elements.

- The graft matroid of $K_{3,3}$ in which every vertex is coloured.
- Represented over $GF(2)$ by the ten 5-tuples that have exactly three ones each.
- Algebraic over all fields.

non-Desargues

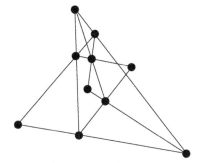

- The non-Desargues matroid; it has rank three.
- Not representable over any division ring (Section 6.1).
- An excluded minor for $\mathbb{F}$-representability if and only if $|\mathbb{F}| \geq 7$.
- Not graphic, not cographic, not regular.
- Not transversal, not a strict gammoid, not a gammoid.
- Non-transitive automorphism group.
- Non-algebraic (Lindström 1985a).

R_{12}

$$\begin{bmatrix} & & 1 & 1 & 1 & 0 & 0 & 0 \\ & & 1 & 1 & 0 & 1 & 0 & 0 \\ & & 1 & 0 & 0 & 0 & 1 & 0 \\ & I_6 & 0 & 1 & 0 & 0 & 0 & 1 \\ & & 0 & 0 & 1 & 0 & 1 & 1 \\ & & 0 & 0 & 0 & 1 & 1 & 1 \end{bmatrix}$$

This matrix represents R_{12} over $GF(2)$.

- Regular.
- Not graphic, not cographic; every proper minor is graphic or cographic (13.1.6).
- Not transversal, not a strict gammoid, not a gammoid.
- The 3-sum of $M^*(K_{3,3})$ and $M(K_5\backslash e)$ across the triangle of $K_5\backslash e$ that is vertex-disjoint from e.
- Self-dual but not identically self-dual.
- Non-transitive automorphism group; its two orbits consist of the union of its two triangles and the union of its two triads (13.1.5).
- A 3-connected regular matroid is either graphic or cographic, or has a minor isomorphic to R_{10} or R_{12} (13.1.2).
- Algebraic over all fields.

$S(5,6,12)$

$$\left[\begin{array}{c|cccccc} & 0 & 1 & 1 & 1 & 1 & 1 \\ & 1 & 0 & 1 & -1 & -1 & 1 \\ I_6 & 1 & 1 & 0 & 1 & -1 & -1 \\ & 1 & -1 & 1 & 0 & 1 & -1 \\ & 1 & -1 & -1 & 1 & 0 & 1 \\ & 1 & 1 & -1 & -1 & 1 & 0 \end{array} \right]$$

This matrix represents $S(5,6,12)$ over $GF(3)$.

- The unique Steiner system with these parameters (Section 12.2).
- The given matrix is a generator matrix for the extended ternary Golay code.
- $\mathbb{F}$-representable if and only if $\mathbb{F}$ has characteristic three.
- Not graphic, not cographic, not regular, not near-regular.
- Not transversal, not a strict gammoid, not a gammoid.
- Identically self-dual.
- Automorphism group is the 5-transitive Mathieu group M_{12}.
- A splitter for $EX(U_{2,5}, U_{3,5}, M(K_4))$ and $EX(M(K_4), P_6, Q_6)$ (Section 12.2).
- Every contraction of three elements is isomorphic to $AG(2,3)$; every contraction of two elements and deletion of two elements is isomorphic to P_8.
- Base-orderable but not strongly base-orderable.
- Tutte connectivity is 5.
- Its circuits and hyperplanes coincide; all have exactly six elements.

T_{12}

$$\left[\begin{array}{c|cccccc} & 1 & 1 & 0 & 0 & 0 & 1 \\ & 1 & 0 & 0 & 0 & 1 & 1 \\ I_6 & 0 & 0 & 0 & 1 & 1 & 1 \\ & 0 & 0 & 1 & 1 & 1 & 0 \\ & 0 & 1 & 1 & 1 & 0 & 0 \\ & 1 & 1 & 1 & 0 & 0 & 0 \end{array} \right]$$

This matrix represents T_{12} over $GF(2)$.

- T_{12} is also represented over $GF(2)$ by the 15×12 matrix whose rows are indexed by the edges of the Petersen graph P_{10} (see Figure 3.5) and whose columns are the incidence vectors of the 5-cycles of P_{10}.
- The edges of P_{10} can be labelled by the 4-circuits of T_{12} so that two edges are adjacent if and only if the corresponding 4-circuits meet in exactly two elements.
- $\mathbb{F}$-representable if and only if $\mathbb{F}$ has characteristic two.
- Not graphic, not cographic, not regular, not near-regular.
- Not transversal, not a strict gammoid, not a gammoid.
- Self-dual but not identically self-dual.
- Transitive but not doubly transitive automorphism group.
- A 3-connected binary matroid with an $M(K_5)$-minor has an $M(K_{3,3})$- or an $M^*(K_{3,3})$-minor, or is isomorphic to $M(K_5)$, T_{12}, or T_{12}/e (Kingan 1997).

- A splitter for the class $EX(U_{2,4}, M(K_{3,3}), M^*(K_{3,3}))$ (Kingan 1997).
- Has exactly two circuit–hyperplanes and these are disjoint.
- The excluded minors for the class of matroids that are binary or ternary are $U_{2,4} \oplus F_7$, $U_{2,4} \oplus F_7^*$, $U_{2,4} \oplus_2 F_7$, $U_{2,4} \oplus_2 F_7^*$, $U_{2,5}$, $U_{3,5}$, and the unique matroids obtained by relaxing a circuit–hyperplane in each of $AG(3,2)$ and T_{12} (14.5.11).
- Algebraic over $\mathbb{F}$ if and only if $\mathbb{F}$ has characteristic two.

$PG(2,3)$

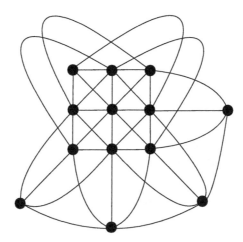

- The second smallest projective plane; isomorphic to the unique $S(2,3,13)$.
- $\mathbb{F}$-representable if and only if $\mathbb{F}$ has characteristic three.
- Not graphic, not cographic, not regular, not near-regular.
- Not transversal, not a strict gammoid, not a gammoid.
- Doubly transitive automorphism group.
- Has every simple ternary matroid of rank at most three as a restriction.
- Algebraic over $\mathbb{F}$ if and only if $\mathbb{F}$ has characteristic three (Gordon 1988).

$M(\mathcal{W}_r), \mathcal{W}^r$

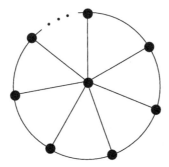

The graph $\mathcal{W}_r$, the r-spoked wheel.

- The rank-r wheel and rank-r whirl ($r \geq 2$).
- The whirl is the unique relaxation of the wheel.

- $M(\mathcal{W}_r)$ is graphic, cographic, and regular.
- $\mathcal{W}^r$ is not graphic, not cographic, and not regular; it is $\mathbb{F}$-representable if and only if $|\mathbb{F}| \geq 3$, so it is near-regular.
- For $r \geq 3$, $M(\mathcal{W}_r)$ is not transversal, not a strict gammoid, and not a gammoid.
- $\mathcal{W}^r$ is transversal, a strict gammoid, and a gammoid.
- Both are self-dual but not identically self-dual.
- Except for $\mathcal{W}^2$, which is isomorphic to $U_{2,4}$, these matroids have non-transitive automorphism groups.
- The only 3-connected matroids for which no single-element deletion and no single-element contraction is 3-connected are the matroids $M(\mathcal{W}_r)$ and $\mathcal{W}^r$ for $r \geq 3$. The rank-2 whirl $\mathcal{W}^2$ is 3-connected, but $M(\mathcal{W}_2)$ is not.
- Both are algebraic over all fields.

$U_{r,n}$

$$E = \{1, 2, \ldots, n\}.$$
$$\mathcal{B}(U_{r,n}) = \{X \subseteq E : |X| = r\}.$$
$$\mathcal{C}(U_{r,n}) = \{X \subseteq E : |X| = r + 1\}.$$

- The rank-r uniform matroid on an n-element set.
- $U_{2,q+2}$ and $U_{q,q+2}$ are excluded minors for $GF(q)$-representability. For general r and n, the $\mathbb{F}$-representability of $U_{r,n}$ has not been determined (Section 6.5).
- Not graphic, cographic, or regular unless r or $n - r$ is 0 or 1, in which case $U_{r,n}$ is graphic, cographic, and regular.
- Transversal, a strict gammoid, a gammoid.
- Identically self-dual if and only if $n = 2r$; otherwise not self-dual, the dual being $U_{n-r,n}$.
- Automorphism group is the symmetric group.
- Every minor is also uniform.
- 3-connected unless $n > 0$ and $0 \in \{r, n - r\}$, or $n > 3$ and $1 \in \{r, n - r\}$.
- Algebraic over all fields.

$PG(r - 1, q)$

- The rank-r projective geometry over $GF(q)$ (Section 6.1); has $\frac{q^r - 1}{q - 1}$ elements.
- The simple matroid associated with $V(r, q)$, the r-dimensional vector space over $GF(q)$.
- $PG(-1, q) \cong U_{0,0}$, $PG(0, q) \cong U_{1,1}$, and $PG(1, q) \cong U_{2,q+1}$.
- For $r \geq 3$, $\mathbb{F}$-representable if and only if $\mathbb{F}$ has $GF(q)$ as a subfield.
- For $r \geq 3$, not graphic, not cographic, not regular.
- For $r \geq 3$, not transversal, not a strict gammoid, not a gammoid.
- For $r \geq 3$, never self-dual.
- For $r \leq 2$, automorphism group is the symmetric group; for $r \geq 3$, automorphism group is doubly transitive.
- Every simple $GF(q)$-representable matroid of rank at most r is a restriction of $PG(r - 1, q)$.

- The restriction to every flat is isomorphic to $PG(k-1, q)$ for some $k \leq r$.
- The simple matroid associated with every proper contraction is isomorphic to $PG(k-1, q)$ for some $k < r$.
- For $r \geq 3$, algebraic over $\mathbb{F}$ if and only if $\mathbb{F}$ and $GF(q)$ have the same characteristic.

$AG(r-1, q)$

- The rank-r affine geometry over $GF(q)$; has q^{r-1} elements when $r \geq 1$.
- Obtained by deleting a hyperplane from $PG(r-1, q)$.
- $AG(-1, q) \cong U_{0,0}$, $AG(0, q) \cong U_{1,1}$,   and $AG(1, q) \cong U_{2,q}$.
- $AG(2, 2) \cong U_{3,4}$; for $AG(2, 3)$ see above. For $r = 3$ and $q \geq 4$, and for $r \geq 4$, $AG(r-1, q)$ is $\mathbb{F}$-representable if and only if $\mathbb{F}$ has $GF(q)$ as a subfield.
- For $r, q \geq 3$, not graphic, not cographic, not regular.
- For $r, q \geq 3$, not transversal, not a strict gammoid, not a gammoid.
- For $r \geq 3$, identically self-dual if and only if $(r-1, q) = (3, 2)$; otherwise never self-dual.
- For $r \leq 2$ or $(r-1, q) = (2, 2)$, automorphism group is the symmetric group. In all other cases, automorphism group is doubly transitive.
- The restriction to every flat is isomorphic to $AG(k-1, q)$ for some $k \leq r$.
- The simple matroid associated with every proper contraction is isomorphic to $PG(k-1, q)$ for some $k < r$.
- 3-connected unless $(r-1, q)$ is $(1, 2)$ or $(2, 2)$.
- For $r = 3$ and $q \geq 5$, and for $r \geq 4$, the matroid $AG(r-1, q)$ is algebraic over $\mathbb{F}$ if and only if $\mathbb{F}$ has the same characteristic as $GF(q)$.

Z_r

$$
\begin{array}{ccccccccc|c}
x_1 & x_2 & \cdots & x_r & y_1 & y_2 & y_3 & \cdots & y_r & t \\
\end{array}
$$

$$
\left[
\begin{array}{cccc|cccccc|c}
 & & & & 0 & 1 & 1 & \cdots & 1 & 1 \\
 & & & & 1 & 0 & 1 & \cdots & 1 & 1 \\
 & I_r & & & 1 & 1 & 0 & \cdots & 1 & 1 \\
 & & & & \vdots & \vdots & \vdots & \ddots & \vdots & \vdots \\
 & & & & 1 & 1 & 1 & \cdots & 0 & 1 \\
\end{array}
\right]
$$

This matrix represents Z_r over $GF(2)$.

- Defined for all $r \geq 3$; the unique rank-r binary spike with tip t (12.2.20).
- $Z_3 \backslash e \cong M(K_4)$ for all e; $Z_3 \cong F_7$; $Z_4 \backslash t \cong AG(3, 2)$; $Z_4 \backslash e \cong S_8$ for all $e \neq t$.
- $Z_r \backslash t$, the tipless binary spike, is self-dual; it is identically self-dual if and only if r is even.
- $\mathbb{F}$-representable if and only if $\mathbb{F}$ has characteristic two.
- Not graphic, not cographic, not regular, not near-regular.
- Not transversal, not a strict gammoid, not a gammoid.
- For all e and f in $E(Z_r) - t$, there is an automorphism of Z_r that fixes t and interchanges e and f, so $Z_r \backslash t$ has a transitive automorphism group.

• For all $e \neq t$, the matroid $Z_r \backslash e$ is self-dual but not identically self-dual.
• For $r \geq 4$, every single-element contraction of $Z_r \backslash t$ is isomorphic to Z_{r-1}.
• A 3-connected binary matroid of rank at least three has no $M(\mathcal{W}_4)$-minor if and only if it is isomorphic to $Z_r, Z_r^*, Z_r \backslash t$, or $Z_r \backslash y_r$ for some $r \geq 3$ (12.2.21).
• Algebraic over $\mathbb{F}$ if and only if $\mathbb{F}$ has characteristic two.

r-spike

$E = \{t, x_1, x_2, \ldots, x_r, y_1, y_2, \ldots, y_r\}$ and $r(E) = r$.
$L_i = \{t, x_i, y_i\}$ is a circuit for all i.
For $1 \leq k \leq r - 1$, the union of any k of $L_1, L_2, \ldots, L_r$ has rank $k + 1$.

• Defined for all $r \geq 3$; a rank-r spike with tip t and legs $L_1, L_2, \ldots, L_r$. Deleting t gives a tipless rank-r spike.
• The non-spanning circuits are $L_1, L_2, \ldots, L_r$, all sets of the form $(L_i \cup L_j) - t$ for $1 \leq i < j \leq r$, and some (possibly empty) collection $\mathcal{C}_3$ of sets of the form $\{z_1, z_2, \ldots, z_r\}$ where $z_i \in \{x_i, y_i\}$ for all i and no two members of $\mathcal{C}_3$ have more than $r - 2$ common elements.
• For $r \geq 4$, the members of $\mathcal{C}_3$ are the circuit–hyperplanes of the spike.
• When $\mathcal{C}_3$ is empty, the spike is free.
• Each of F_7, F_7^-, and P_7 is a 3-spike.
• The collection of spikes includes infinite antichains that are non-representable (14.1.3) and that are representable over every infinite field (14.3.15).
• The free r-spike is representable over all sufficiently large fields and over $GF(p^k)$ for all primes p and all $k \geq 2$. For such p and k, when $p^k \geq 8$, the free r-spike has at least 2^{r-1} inequivalent $GF(p^k)$-representations (14.6.8).
• The free r-spike is obtained by truncating the parallel connection of r copies of $U_{2,3}$ across a common basepoint.
• Not graphic, not cographic, not regular.
• Not transversal, not a strict gammoid, not a gammoid.
• Deleting any element gives a self-dual matroid. The tipless free spike is identically self-dual.
• Every spike is 3-connected and has branch-width three (14.3.14).
• A rank-r matroid M on a $2r$-element set E is a tipless spike if and only if there is a partition of E into r pairs of elements such that the union of every two such pairs is both a circuit and a cocircuit of M (2.1.28)
• For $r \geq 4$, deleting the tip and contracting any other element gives an $(r - 1)$-spike.
• Among r-spikes, exactly one is binary, exactly two are ternary, and exactly $\lfloor (r^2 + 6r + 24)/12 \rfloor$ are quaternary (Wu 2003).
• The free r-spike is algebraic over all fields.

$Q_r(A)$

- The rank-r Dowling geometry associated with the finite group A (Section 6.10).
- Has $r + |A|\binom{r}{2}$ elements including a distinguished basis whose members are called joints; the matroid obtained by deleting these joints is $Q'_r(A)$, the jointless Dowling geometry, which has $|A|\binom{r}{2}$ elements.
- $Q'_r(A)$ has rank r unless $|A| = 1$ and $r \geq 1$ in which case its rank is $r - 1$.
- $Q_0(A) \cong U_{0,0}$; $Q_1(A) \cong U_{1,1}$; and $Q_2(A) \cong U_{2,|A|+2}$.
- $Q_r(GF(2)^\times) \cong M(K_{r+1})$ and $Q'_r(GF(2)^\times) \cong M(K_r)$.
- $Q_r(GF(q)^\times)$ is represented over $GF(q)$ by a matrix whose columns consist of all vectors with one or two non-zero entries, the first of which is a one.
- $Q'_3(GF(3)^\times) \cong M(K_4)$ and $Q'_3(GF(4)^\times) \cong AG(2,3)$.
- For $r \geq 3$, the matroid $Q_r(A)$ is $\mathbb{F}$-representable if and only if A is isomorphic to a subgroup of $\mathbb{F}^\times$ (6.10.10).
- For $|A| = 1$, graphic and regular; cographic if and only if $r \leq 3$. For $|A| \geq 2$ and $r \geq 3$, not graphic, not cographic, not regular.
- For $|A| \geq 2$ and $r \geq 3$, not sixth-root-of-unity, not near-regular; dyadic if and only if $|A| = 2$.
- For $r \geq 3$, not transversal, not a strict gammoid, not a gammoid.
- Self-dual if and only if $(r, |A|) \in \{(0, |A|), (2, 2), (3, 1)\}$; identically self-dual if and only if $(r, |A|) \in \{(0, |A|), (2, 2)\}$.
- For $r \geq 3$, if $Q_r(A) \cong Q_r(A')$, then $A \cong A'$.
- For all $r \geq 1$, the simple matroid associated with every single-element contraction is isomorphic to $Q_{r-1}(A)$.
- The restriction of $Q_r(A)$ to a flat is isomorphic to the direct sum of the cycle matroids of a family of complete graphs and $Q_k(A)$ for some $k \leq r$.
- For $r \leq 2$, automorphism group is the symmetric group; for $|A| = 1$, automorphism group is transitive; otherwise, automorphism group is not transitive although if e and f are both joints or are both non-joints, there is an automorphism that interchanges them.
- Has exactly $((|A| + 1)^r - 1)/|A|$ hyperplanes.

Θ_n

$E = X \cup Y$ where $X = \{x_1, x_2, \ldots, x_n\}$ and $Y = \{y_1, y_2, \ldots, y_n\}$.
$\mathcal{B}(\Theta_n) = \{B \subseteq E : |B| = n;\ |B \cap X| \leq 2;\ B \neq (Y - y_i) \cup x_i \text{ for all } i\}$.

- Defined for all $n \geq 2$; the circuits consist of all 3-element subsets of X, all sets of the form $(Y - y_i) \cup x_i$, and all sets of the form $(Y - y_u) \cup \{x_s, x_t\}$ where s, t, and u are distinct.
- $\Theta_2 \cong U_{1,2} \oplus U_{1,2}$ and $\Theta_3 \cong M(K_4)$.
- $\mathbb{F}$-representable if and only if $|\mathbb{F}| \geq n - 1$.
- For $n \leq 3$, graphic, cographic, and regular; for $n = 4$, not graphic, not cographic, not regular, near-regular; for $n \geq 5$, not graphic, not cographic, not regular, not near-regular.

- For $n \geq 3$, not transversal, not a strict gammoid, not a gammoid.
- Self-dual; identically self-dual if and only if $n = 2$.
- Has X as a modular flat.
- For a matroid M with $E(M) \cap (X \cup Y) = X$ and X a coindependent set in M with $M|X \cong U_{2,n}$, a segment–cosegment exchange on X is $P_X(\Theta_n, M) \backslash X$.
- The set of excluded minors for $\mathbb{F}$-representability is closed under segment–cosegment exchange and its dual operation, cosegment–segment exchange (11.5.15).
- For $n \geq 3$, a 3-connected matroid.
- $\Theta_n / y_i \backslash x_i \cong \Theta_{n-1}$ for all $n \geq 3$ and for all i.
- For $n \leq 3$, automorphism group is transitive; for $n \geq 4$, automorphism group has X and Y as its orbits.
- Algebraic over all fields.

Ψ_r

$E = \{a_1, b_1, a_2, b_2, \ldots, a_r, b_r\}$ and $r(E) = r$. **Non-spanning circuits are all the sets** $\{a_i, b_i, z_{i+1}, z_{i+2}, \ldots, z_{i+k-1}, a_{i+k}, b_{i+k}\}$ **for** $1 \leq k \leq r-3$, **where each** z_j **is in** $\{a_j, b_j\}$, **and all subscripts are calculated mod** r.

- The rank-r free swirl; defined for all $r \geq 3$.
- $\Psi_3 \cong U_{3,6}$.
- The matroid obtained from the rank-r whirl by freely adding an element on each 3-point line is the rank-r jointed free swirl, Ψ_r^+. Deleting all the spokes of the whirl from Ψ_r^+ gives Ψ_r.
- Not binary, not ternary.
- Ψ_3 is $\mathbb{F}$-representable if and only if $|\mathbb{F}| \geq 4$, while Ψ_3^+ is $\mathbb{F}$-representable if and only if $|\mathbb{F}| \geq 5$.
- When $r \geq 4$, the matroids Ψ_r and Ψ_r^+ are representable over exactly the same fields; Ψ_r is representable over all infinite fields and over all $GF(q)$ with $q \geq 4$ unless $q = 2^p$ where $2^p - 1$ is prime. In the exceptional case, Ψ_r is not $GF(q)$-representable for all sufficiently large r (14.6.7).
- When $q > 5$ and q is not of the form 2^p where $2^p - 1$ is prime, Ψ_r^+ has at least 2^r inequivalent $GF(q)$-representations.
- Not graphic, not cographic, not regular, not near-regular.
- Transversal, a strict gammoid, a gammoid.
- Identically self-dual.
- Transitive automorphism group.
- 3-connected but, for all $r \geq 4$, not 4-connected.
- The elements a_i and b_i are clones for all i.
- Algebraic over all fields.

TABLES

NOTATION

INDEX

abstract dual, 140, 150, 330
adjacent, 3
affine
  binary affine cube, 37, 111, 188, 295, 645
  geometry, 170, 661
  matroid, 32, 170, 360
  real affine cube, 37, 86, 111, 295, 646
  ternary affine plane, 558, 653
affinely
  dependent, 32
  independent, 32
algebraic
  closure, 223
  element, 212
  extension, 212, 223
  matroid, 215, 585
  representation, 215
algebraically
  dependent set, 212
  independent set, 212
  representable, 215
algorithm
  Euclidean, 225
  finite, 224
  greedy, 59, 77
  Kruskal's, 58
  k-separation, 523
  matroid intersection, 432, 524
  matroid partitioning, 430
  polynomial-time, 365, 368, 523, 541
almost
  $\mathbb{F}$-representable, 537
  regular, 549
amalgam, 436, 604
  free, 436
  proper, 438
antichain, 525
  infinite, 526, 538
arc
  of a digraph, 6
  n-arc in a projective geometry, 201
asymptotic, 1
  behaviour of matroid property, 595
atom of a poset, 52
augmentation property
  independence, 7

automorphism
  Frobenius, 177
  of a field, 77, 177
  of a matroid, 189
  of a projective geometry, 178
automorphism group, 177, 189, 210, 475,
    492, 548, 558, 639
  doubly transitive, 639
  5-transitive, 475, 639
  3-transitive, 639
  transitive, 189, 210, 548, 558
avoidance graph, 367
axioms
  basis, 15
  circuit, 8, 29
  closure, 26
  cyclic flat, 117
  flat, 31
  girth, 331
  hyperplane, 70
  independent set, 7
  rank, 21, 30
  spanning set, 31

balanced
  cycle, 235, 237
  f-balanced set, 410
  matroid, 600
base, 15
base-orderable matroid, 435
  strongly, 435
basepoint
  of a parallel connection, 253
  of a series connection, 253
  of a 2-sum, 260
basis
  axioms, 15
  bijective basis exchange axiom, 426
  disjoint bases, 430
  exchange axiom, 16, 19, 32, 64
  generating polynomial, 601
  Gröbner, 225
  of a matroid, 15
  of a set, 21
  oracle, 365
  standard basis vector, 2, 81, 108